HANDBOOK OF NEUROCHEMISTRY

VOLUME II

STRUCTURAL NEUROCHEMISTRY

HANDBOOK OF NEUROCHEMISTRY

Edited by Abel Lajtha

HANDBOOK OF NEUROCHEMISTRY

Edited by Abel Lajtha

New York State Research Institute
for Neurochemistry and Drug Addiction
Ward's Island
New York, New York

VOLUME II

STRUCTURAL NEUROCHEMISTRY

SPRINGER SCIENCE+BUSINESS MEDIA, LLC 1969

ISBN 978-1-4899-7301-6   ISBN 978-1-4899-7321-4 (eBook)
DOI 10.1007/978-1-4899-7321-4

Library of Congress Catalog Card Number 68-28097

© 1969 Springer Science+Business Media New York
Originally published by Plenum Press, New York in 1969
Softcover reprint of the hardcover 1st edition 1969

Contributors to this volume:

Leo G. Abood — Center for Brain Research and Department of Biochemistry, University of Rochester, Rochester, New York (page 303)

C. W. M. Adams — Sir William Dunn Professor of Pathology, Guy's Hospital Medical School, London University, and Honorary Consultant Morbid Anatomist, Guy's Hospital, London, Great Britain (page 525)

Joseph Altman — Laboratory of Developmental Neurobiology, Department of Biological Sciences, Purdue University, Lafayette, Indiana (page 137)

Samuel H. Barondes — Departments of Psychiatry and Molecular Biology, Albert Einstein College of Medicine, Bronx, New York (page 435)

S. Berl — Department of Neurology, College of Physicians and Surgeons, Columbia University, New York, New York (page 447)

D. D. Clarke — Chemistry Department, Fordham University, New York, New York (page 447)

T. Z. Csáky — Department of Pharmacology, University of Kentucky College of Medicine, Lexington, Kentucky (page 49)

H. F. Daginawala — Department of Pediatrics, The University of Texas Medical Branch, Galveston, Texas (page 241)

Hugh Davson — Department of Physiology, University College, London, England (page 23)

Eduardo De Robertis — Instituto de Anatomia General y Embriologia, Facultad de Medicina Universidad de Buenos Aires, Buenos Aires, Argentina (page 365)

B. Droz — Departement de Biologie, Commissariat a l'Energie Atomique, Saclay, France (page 505)

K. A. C. Elliott   Department of Biochemistry and the Montreal Neurological Institute, McGill University, Montreal, Canada (page 103)

Ezio Giacobini   Department of Pharmacology, Karolinska Institutet, Stockholm, Sweden (page 195)

John A. Harvey   Department of Psychology, University of Iowa, Iowa City, Iowa (page 115)

Robert Katzman   The Saul R. Korey Department of Neurology, Albert Einstein College of Medicine, Bronx, New York (page 11)

G. A. Kerkut   Department of Physiology and Biochemistry, Southampton University, Southampton, England (page 539)

Edward Koenig   Department of Physiology, State University of New York at Buffalo, Buffalo, New York (page 423)

Harold Koenig   Neurology Service, V. A. Research Hospital, and Department of Neurology and Psychiatry, Northwestern University Medical School, Chicago, Illinois (page 255)

Giulio Levi   Center of Neurobiology, Department of Biochemistry, Instituto Superiore di Sanita, Rome, Italy (page 71)

Richard N. Lolley   Neurochemistry Laboratories, Veterans Administration Hospital, Sepulveda, California, and Department of Anatomy, UCLA School of Medicine, Los Angeles, California (page 473)

P. Maxcy, Jr.   Department of Pediatrics, The University of Texas Medical Branch, Galveston, Texas (page 241)

Henry McIlwain   Department of Biochemistry, Institute of Psychiatry, British Postgraduate Medical Federation, University of London, Maudsley Hospital, London (page 115)

Hanna M. Pappius The Donner Laboratory of Experimental Neuro-chemistry, Montreal Neurological Institute, and Department of Neurology and Neurosurgery, McGill University, Montreal, Canada (page 1)

Giuseppe Porcellati Department of Biological Chemistry, University of Pavia, Pavia, Italy (page 393)

D. A. Rappoport Department of Pediatrics, The University of Texas Medical Branch, Galveston, Texas (page 241)

Georgina Rodríquez de Lores Arnaiz Instituto de Anatomia General y Embriologia, Facultad de Medicina Universidad de Buenos Aires, Buenos Aires, Argentina (page 365)

Steven P. R. Rose Medical Research Council Metabolic Reactions Research Unit, Department of Biochemistry, Imperial College, London, England (page 183)

Herbert Schimmel The Saul R. Korey Department of Neurology, Albert Einstein College of Medicine, Bronx, New York (page 11)

V. P. Whittaker Department of Biochemistry, University of Cambridge, Cambridge, England, and New York State Institute for Basic Research in Mental Retardation, New York, New York (page 327)

PREFACE

That chemicals (although not always called by this name) affect the brain and its functions, such as behavior, has been known for thousands of years. It is therefore surprising that the concept that chemical mechanisms are at least partially responsible for the complex functions of the brain is so recent. Investigation of the closely interlinked biophysical and biochemical properties of the nervous system has achieved many notable successes in recent years and is the most exciting development in 20th-century science.

Although all the morphology, the activity, and the alteration of the brain, whether bioelectric, biochemical, pathological, or structural, constitute an organic and indivisible whole, the ambition of the *Handbook* is to look at only a few aspects of this whole and to focus the discussions on the experiments that the neurochemists have performed.

Neurochemical study of the nervous system has, perhaps of necessity, gone through several phases: the first phase was more analytical and involved study of the composition of the tissue; the second, more recent phase clarified many of the metabolic sequences that occur in this tissue. Clearly, both were essential, but they showed that additional approaches are necessary. The present phase seems to be the study of control processes; present interest focuses on what determines, in a qualitative and quantitative fashion, the processes occurring in the nervous system. Perhaps the next phase will be the study of function, the study of the final stage of integration.

With the great speed of advance of knowledge in our field, it is difficult not to be an optimist. We now talk about things even as complex (and not so long ago seemingly unapproachable) as memory, learning, sleep, emotion, discrimination, and pain, and we not only talk about them but also devise meaningful experiments to study these extremely complex phenomena.

Not only can we study such complex phenomena as outside observers, but also we can manipulate them in a number of ways in order to understand them. It is hardly necessary to emphasize here that it may be too ambitious even at our present stage of knowledge to hope that this book will help in understanding the complexities of the nervous system; what we can hope is that it will help to formulate further research.

The divisions of the various volumes of the *Handbook* are not very sharp; still, the present volume emphasizes structural aspects of neurochemistry. Structural aspects are hardly more important or more complex in any other organ than in the nervous system. This seems to be not only one of the most structurally complex organs, but also one whose structure obviously has great significance for its function and whose mechanisms clearly cannot be understood without understanding the various structures where these mechanisms proceed. Life in all its aspects has great unity, and the facts we learn not only are clearly important for brain research in particular, or for the function of the nervous system, but also have a broader meaning for our concepts of living organisms as a whole. Therefore, from structural as well as other aspects, the results of brain research are applicable to, and contribute to, other disciplines.

It was pointed out in the first volume of this *Handbook*—and, unfortunately, it can be pointed out in any of the future volumes as well—that because of the practical necessity to limit the size of the volumes a great many important contributions could be only very briefly discussed, some not at all. The decision of what to include or exclude was often somewhat arbitrary. For this, mostly the editorial scissors (or short-sightedness) can be blamed, but perhaps it will serve as an excuse that the book is planned as a starting point of, rather than a substitute for, the future search for the wealth of information that has been written in the original publications or in the many reviews.

Among the purposes of the *Handbook* is to attract new talent to the field of brain research by giving to the reader a brief look at what the workers in this area have found out and what questions have been asked, even in a tentative way. Hopefully, many will carry this work further, at times even proving that in the past the wrong questions were asked.

The task, of course, is formidable. The structure proves to be more complex the more we investigate it: the more we find out about the brain, the more we realize what there is to be found out. But what is most significant, the task ahead, which not so long ago seemed impossible, seems less forbidding. We still know very little, but all the facts that seem to be established seem to form a reasonable complex, where all the parts fit and where what we have found so far seems to make sense. There is very little, if any, evidence that would indicate that it is impossible to understand brain function in even its most intricate details. Obviously, the neurochemists who wrote the chapters of these volumes hope that their work will contribute to this understanding. The editor has no doubt that it will and that the fuller understanding of the brain will be one of mankind's greatest achievements.

Abel Lajtha

New York, New York
February 1969

CONTENTS

Chapter 3
The Cerebrospinal Fluid 23
by Hugh Davson

Chapter 4
Choroid Plexus.................................. 49
by T. Z. Csáky

Chapter 5
Spinal Cord.   71
by Giulio Levi

Chapter 13
Brain Mitochondria 303
by Leo G. Abood

Chapter 16
Peripheral Nerve 393
by Giuseppe Porcellati

Chapter 17
Nucleic Acid and Protein Metabolism of the Axon 423
by Edward Koenig

Chapter 18
Axoplasmic Transport 435
by Samuel H. Barondes

Chapter 19
Compartmentation of Amino Acid Metabolism 447
by S. Berl and D. D. Clarke

CONTENTS OF VOLUME I: Chemical Architecture of the Nervous System

CONTENTS OF VOLUME III: Metabolic Reactions in the Nervous System

FURTHER VOLUMES

Chapter 1

WATER SPACES

Hanna M. Pappius

The Donner Laboratory of Experimental Neurochemistry
Montreal Neurological Institute
and Department of Neurology and Neurosurgery
McGill University
Montreal, Canada

I. INTRODUCTION

The interpretation of many physiological and pathological processes requires precise information about the relative sizes of the extracellular and the intracellular compartments of tissues and their dynamic relationships.

The quantitative estimation of the size of fluid compartments in brain has proved to be technically difficult, and much controversy has resulted. The early estimates of an extracellular space (ECS) of considerable size in cerebral tissues was based on the assumption that sodium and chloride were mostly extracellular.[1,4] This was challenged by electron microscopists who initially found a very small, almost nonexistent, ECS in cerebral cortex tissue.[3] The conflict of views on the size of ECS in brain stimulated renewed interest in this field and led to careful reassessment of the assumptions that had been made in the methods used. New techniques were applied to overcome many of the errors inherent in earlier investigations.

II. DEFINITION OF SPACES

Chemically determined ECS of a tissue can be defined as the volume of the tissue water that is available under steady-state conditions for the distribution of the particular extracellular marker used. This requires that the marker be excluded from the intracellular compartment and implies equilibration of the extracellular fluid of the tissue with the plasma *in vivo* or with the fluid in which the tissue is incubated *in vitro*. The concentration of the marker in the reference fluid must be accurately known and remain unchanged during the equilibration period.

1

The space can be calculated by the following equation:

$$\text{Marker space} (\%) = \frac{\text{Concentration of the marker in tissue}}{\text{Concentration of the marker in reference fluid}} \times 100$$

To be meaningful the chemical estimation of fluid spaces eventually must be correlated with the morphological compartments of the tissue, particularly the ECS. The hypothesis that glia serve as a pathway for the movement of ions and small molecules in the central nervous system[4] also suggested the possibility of chemical delineation of the glial and the neuronal spaces.[5,6]

III. PROBLEMS OF METHODOLOGY

A. General

With each of the markers used for the estimation of ECS there are difficulties in achieving true steady-state conditions. Sodium, chloride, and thiocyanate are now known to be intracellular to some extent.[7,8] Thiocyanate is also bound by plasma proteins,[8] and the extent of this binding must be determined to know the concentration of the free equilibrating ion. Intravenously injected sulfate, sucrose, and inulin are rapidly excreted, so either constant infusion of the marker or functional nephrectomy must be used to maintain a constant level of these substances in blood.[8] Sulfate is not completely inert metabolically,[7] and this must be corrected for in experiments of long duration. In studies of cerebral tissue spaces the most serious problems are encountered in establishing real equilibrium conditions and in deciding what is the appropriate reference fluid.[7,8]

B. Experiments *in Vivo*

The penetration of all ECS markers from blood into brain is slow. Thus *in vivo* steady concentration of the markers is approached only after hours of equilibration whereas in muscle it is reached in minutes. Even when steady-state levels are attained it is uncertain whether a true equilibrium between the blood and the ECS has occurred. In other tissues extracellular fluid is thought to be an ultrafiltrate of the plasma. Evidence is accumulating that in the brain the interstitial fluid is more closely related to the CSF. Since even under steady-state conditions the concentrations of ECS markers in plasma ultrafiltrates and in the CSF are very different, the values obtained for spaces vary according to which fluid is taken as a reference. The sodium and chloride contents of the CSF are maintained at levels higher than those in the plasma ultrafiltrate by active secretory processes involved in the formation of the CSF. Markers like inulin, sucrose, and sulfate injected intravenously do not reach concentration levels in the CSF as high as those in the plasma because they are removed by the bulk flow of the CSF faster than they can penetrate into the CSF space from the blood either via the cerebral tissue or the choroid plexus. Thiocyanate and iodide in addition are actively removed from the CSF by a transport system in the choroid plexus. In both these instances a concentration gradient between the blood and the CSF is maintained. This has been referred to as the "sink action" of the CSF. Concentration gradients develop in the reverse direction when the extracellular marker introduced only into the CSF is slowly lost into the blood.

To approach the true equilibrium conditions necessary for estimation of the volume of distribution *in vivo* of added extracellular markers, the same concentration of these substances must be maintained in the blood and in the CSF. This may be achieved by ventriculo-cisternal perfusion of the marker together with intravenous infusion.

The various aspects of the physiology of the blood–brain barrier and of the CSF touched upon very briefly in the foregoing paragraphs have been exhaustively discussed by Davson[7] and Van Harreveld.[8]

C. Experiments *in Vitro*

To avoid the complications due to the blood–brain barrier, *in vitro* preparations have been used by numerous investigators.[5,7,8] The use of tissue slices, however, introduces artefacts due to damage of the tissue resulting from slicing. To minimize the effects of damage, isolated intact frog brain[9] and rabbit retina[10] have been used *in vitro* in studies on cerebral tissue spaces.

IV. CHEMICALLY DETERMINED SPACES

A. Spaces in Vivo

1. *Sodium and Chloride Spaces*

These spaces have been assumed to represent the extracellular compartment of a number of tissues.[7,8] In brain some sodium and chloride is in the intracellular compartment[10,11] and, as already mentioned, the CSF is thought to be more closely representative of the brain extracellular fluid than the ultrafiltrate of plasma.[7,8] In rabbit brain, the chloride space calculated from the normal chloride content of cerebral tissues is 31 or 27%, depending on whether plasma dialysate or CSF is used as reference. The latter figure corrected for intracellular chloride gives an ECS of 21 to 24%.[8,10] Similar calculations based on sodium distribution give an ECS of 24%.[8] A slightly smaller chloride space was reported in rat brain and a somewhat larger one in dog brain.[2,12]

Chloride spaces ranging from 27 to 39% depending on species used were found after intracisternal injection of Cl^{36} but under conditions that undoubtedly resulted in overestimation of the chloride space[13] (see below).

2. Sulfate Space

Isotopically labeled sulfate has been used in a number of studies to estimate the extracellular space of brain. When the marker was injected intravenously and sufficient time was allowed for steady-state conditions to be reached, a sulfate space of less than 5% was found in rat, cat, and dog using plasma as reference fluid.[2,12,14] Sulfate levels in the CSF in normal animals and when steady-state conditions have been reached after injection of labeled sulfate (8–24 hr), were one-sixth to one-fourth of the sulfate concentration of the plasma.[15] Recalculating a sulfate space of 4% on the assumption that the extracellular fluid of brain is similar to CSF, a space of 16 to 24% is obtained.

With ventriculo-cisternal perfusion and intravenous infusion of labeled sulfate to eliminate the concentration gradient between CSF and plasma, a space of 8 to 12% was obtained in cat.[16] This figure may be an underestimation of the space, since only 2–6 hr were allowed for equilibration in these experiments.

3. Inulin Space

When ^{14}C-labeled inulin was injected intravenously into rats and dogs[2,17] and plasma levels were used as reference, a space of 2 to 5% was obtained. In these experiments the inulin concentration in the CSF was only one-fiftieth to one-twentieth of that in the plasma, and calculation of results in terms of the CSF gives meaningless values. Inulin perfused through the ventricles of dogs was shown to diffuse slowly into brain tissue.[18] This system obviously did not reach equilibrium but, by extrapolation, the results indicated an inulin space of about 10 to 12%. When inulin was injected intracisternally and brain tissue and subarachnoid CSF were analyzed simultaneously after a suitable interval of time, inulin spaces ranging from 19 to 39% were found, depending on the species used.[13] Equilibrium conditions were not achieved in these experiments, and therefore the values for inulin space are not reliable. Ventriculo-cisternal perfusion of inulin-containing artificial CSF in nephrectomized rats previously injected with the marker gave an inulin space of 14% after 6 hr equilibration.[19]

4. Sucrose Space

[14]C-labeled sucrose injected intravenously has very limited distribution in cerebral tissues. In the rabbit, the sucrose space was less than 5% using plasma as reference [7] and 8% after subarachnoid perfusion.[20] Three hours of combined ventriculo-cisternal perfusion and intravenous infusion of ^{14}C-sucrose resulted in a space of 10%.[7] After cisternal injection, sucrose space of 22.5% was reported in this species.[13] It is doubtful that true equilibration was attained in any of the studies quoted since kinetic studies have shown that sucrose distribution in brain occurs at quite different rates

in at least two apparent compartments with only the fast component approaching equilibrium in 3 hr.[21]

5. Thiocyanate Space

Thiocyanate penetrates more rapidly from blood into the brain than does sulfate, inulin, or sucrose. The estimation of the volume of distribution of this marker is complicated, however, by its active removal from the CSF.[22] When at any given plasma thiocyanate level steady-state conditions are reached, a true equilibrium between cerebral tissue compartments is not attained. Rather, a concentration gradient for thiocyanate is established between the plasma and the CSF.[23] This gradient decreases with increasing plasma thiocyanate concentration as the transport system for the ion in the choroid plexus becomes saturated. This is the reason that the thiocyanate space of central nervous tissue of rat computed using plasma as reference fluid was found to vary from 4 to 17%, depending on the thiocyanate level in the plasma.[24] Even at relatively high levels of thiocyanate in the plasma (10 mmole) and 10 hr after its infusion the CSF to plasma thiocyanate ratio was found to be only 0.75.[25] Under these conditions the volume of thiocyanate distribution in cat cerebral cortex was estimated to be between 29 and 38%, depending on which body fluid was used as reference, but it was probably closer to the former figure.[25] Thiocyanate is known to distribute in the intracellular compartment to the same extent as chloride.[11] Correcting the above figures for intracellular thiocyanate, an ECS of about 25% is obtained. In cats, after 6 hr of ventriculo-cisternal perfusion of artificial CSF containing 10 mmole thiocyanate and intravenous injection to obtain comparable blood levels, an ECS of 21% was found (corrected for intracellular thiocyanate).[26] A thiocyanate space of 20% was computed for rabbit brain in experiments in which transport of thiocyanate from the CSF was partly inhibited by iodide (competitive inhibition) or by dinitrophenol (inhibition of oxidative metabolism).[27] Thiocyanate spaces similar to ^{36}Cl spaces were found in a number of species after intracisternal injection of ^{14}C-labeled thiocyanate.[13]

6. Iodide Space

Iodide is actively removed from the CSF. The steady-state concentrations of iodide in the CSF and in the extracellular fluid of the brain are in consequence much lower than in the blood.[28] The suggestion that the iodide content of the CSF is a better approximation of brain extracellular fluid iodide than the plasma content was made in some of the earliest publications on this subject.[29] When active transport of iodide was inhibited by perchlorate, the iodide space in rabbit brain was 5% after intravenous infusion of the marker, 15% after intracisternal perfusion, and 20% after 3 hr of combined perfusion and infusion.[7] In rat brain, under conditions approaching even closer true equilibrium between the various fluid compartments, iodide space of 30% was reported.[30]

7. *Space Size in Relation to Brain Size*

In experiments in which chloride, thiocyanate, inulin, and sucrose spaces were measured in nine mammalian species, the size of the space was found to vary as a function of the logarithm of the average brain weight of the species.[13] Further, chloride and thiocyanate spaces, on one hand, and inulin and sucrose spaces, on the other hand, appeared to be of the same magnitude, the latter slightly but significantly smaller than the former. Since the markers were injected intracisternally and subarachnoid fluid and subadjacent cerebral cortex were sampled simultaneously after an interval of time during which CSF levels of the marker decreased steadily, it is unlikely that true equilibrium between the tissue and the CSF was attained. Thus absolute values obtained for the spaces in this study are not reliable. However, the species differences relative to the size of the brain may be valid.

8. *Spaces in Immature Brain Tissue*

Changes in the spaces available for distribution of chloride, thiocyanate, sulfate, and inulin during development have been observed in chicks, rats, and cats.[6,14,31] The apparent volume of distribution of all three markers in terms of their concentration in the plasma decreased with maturation of the central nervous system, but at the same time a fall in the rate of uptake of the markers was noted. Thus it is not clear at present to what extent increase in the blood–brain permeability barrier accounted for the apparent decrease of the spaces in adult brain.

B. **Spaces** *in Vitro*

Cerebral cortex slices incubated *in vitro* swell by up to 30% of their original volume in 1 hr even under optimal aerobic conditions.[5,25] Using inulin, sucrose, and thiocyanate as markers, three distinct fluid compartments have been demonstrated in incubated cerebral tissue: (1) the non-thiocyanate, nonsucrose space, (2) the inulin space, and (3) the compartment that is penetrated by sucrose and thiocyanate but not by inulin. There is strong evidence that damaged tissue elements take up the bulk of the swelling fluid that contains all extracellular markers that may be present in the incubation medium. Thus the inulin space of an incubated slice is not a measure of a true extracellular space of the original tissue. It also includes damaged, swollen, originally intracellular elements and perhaps fluid that has penetrated between tissue structures.[32] On the other hand, the non-inulin space of incubated slices has been regarded as the true intracellular space of undamaged cells.[5,33,34] The equilibration of such markers as sodium, chloride, thiocyanate, and sucrose* with a large fraction of the water in the noninulin space[5,35] has been interpreted as their penetration into an

---

* Pappius and co-workers[5] found sucrose distribution in incubated slices to be of the same order of magnitude as thiocyanate distribution but significantly greater than inulin distribution. Tower and co-workers,[35] on the basis of their results, equate sucrose and inulin spaces.

intracellular, possibly glial, compartment.[5] However, Van Harreveld and his colleagues[8] have evidence that shifts of extracellular ions and water into both the dendritic and glial compartments occur rapidly in cerebral tissues *in vivo* as a consequence of asphyxia. The possibility exists that such a process also results in shifts of fluid containing ions and small molecules into more than one intracellular compartment of the tissue under *in vitro* conditions. This would not invalidate the use of noninulin space as a measure of undamaged intracellular compartments in metabolic studies *in vitro* but would preclude extrapolation of *in vitro* results for quantitative estimation of spaces *in vivo*.

In intact rabbit retina incubated *in vitro* as a prototype of mammalian gray matter, sodium and chloride spaces corresponded closely to the inulin and mannitol spaces and amounted to 31 % of the volume of the tissue.[10] In incubated isolated frog brain, the extracellular space computed from ^{24}Na washout curves was 24 %.[9]

C. Summary

The variability of the size of chemically determined spaces in brain is probably due to critical differences in details of experimental procedures used. When only the results obtained *in vivo* under conditions most closely approaching equilibrium are considered, the variability is diminished and the marker spaces fall into two groups. Sulfate, inulin, and sucrose spaces vary from 14 to 24 %, and sodium, chloride, thiocyanate, and iodide spaces vary from 21 to 30 %. The discrepancy between the two groups is not necessarily due to real differences in the distribution of the markers. Rather, it most likely reflects a slower rate of penetration of sulfate, inulin, and sucrose into brain tissue and, hence, slower and less complete equilibration.

V. EXTRACELLULAR SPACE DETERMINED BY ELECTRICAL IMPEDANCE MEASUREMENTS

A different approach to the problem of measuring the extracellular space in the central nervous system has been developed in recent years when the volume of this fluid compartment was deduced from specific impedance measurements. This method is based on the assumption that the participation of the intracellular ions in current transport is hampered by cell membranes and, therefore, at low frequencies the alternating currents used for impedance measurements in tissue are carried mainly by the extracellular electrolytes. In this formulation the relative volumes of the intracellular and the extracellular compartments are among the determining parameters of the specific impedance of the tissue. This would not be true if many cells in the tissue constituted a low-resistance pathway.[36] Theoretical and practical considerations of this approach to estimating the volume of extracellular fluid have been discussed in detail by Van Harreveld.[8]

In rabbit, rat, and cat, estimates of central nervous tissue extracellular spaces of 18–25% have been derived from measurements of cortical impedance.[8,36,37]

Considerable increases in cortical impedance have been observed after a few minutes of asphyxia, during spreading depression and during perfusion fixation *in vivo* for electron microscopy.[8,36,37] These changes have been interpreted to arise from a shift of extracellular water and electrolytes into intracellular compartments and are evidence that the distribution of the water and electrolytes in cerebral tissues *in vivo* is labile.

VI. ELECTRON MICROSCOPY AND EXTRACELLULAR SPACES IN CEREBRAL TISSUES

Early electron micrographs of cerebral tissues showed close packing of all cellular elements and only small amounts of extracellular space.[3] From such studies a mean extracellular space of about 5% was computed.[7] However, a careful study of electrical impedance and volume changes in brain during preparation for electron microscopy has shown that significant changes in the size of tissue compartments occur during fixation and dehydration.[37] Thus electron micrographs of central nervous tissue prepared with currently conventional fixation methods cannot be accepted as a quantitative representation of tissue compartments *in vivo*.

Cerebral extracellular spaces can be seen in electron micrographs of brain tissues that have been rapidly frozen and then subjected to substitution–fixation with osmium tetroxide in acetone at low temperatures.[8] In contrast, tissues asphyxiated for 8 min showed little extracellular space even with this technique. Glial cells and dendrites were swollen, and a shift of chloride into these structures was demonstrated histochemically in asphyxiated tissue.[8] This further indicates that electrolyte and water distribution in cerebral tissues is very labile and changes drastically under unfavorable physiological conditions.

At present, the morphological estimates of cerebral extracellular space are subject to at least as much artefact as the chemical estimations.

VII. ABNORMAL SPACES

A practical application of extracellular markers has been their use to elucidate basic differences in two types of cerebral swelling. In cerebral edema associated with trauma, the increase in water content of white matter was paralleled by increases of sulfate and thiocyanate spaces,[38] a finding compatible with the accumulation of the extraneous fluid in the extracellular compartment of the tissue. In edema induced by triethyltin poisoning, sulfate and thiocyanate spaces remained unchanged in the presence of increased water and sodium content of white matter,[38] indicating accumulation of an exudate not in equilibrium with the extracellular fluid. This conclusion was confirmed with the electron microscope when intramyelenic vacuoles were demonstrated in edematous brains of triethyltin-treated animals.[38]

VIII. CONCLUSION

After a period of controversy about the size, in fact the very existence, of the ECS of brain, considerable progress has been made toward obtaining a reasonable estimate of this fluid compartment. All methods used to date in the elucidation of this problem depend critically on assumptions that are not easily tested, and for this reason the results must be considered as approximate.

Measurable extracellular space exists in cerebral tissues and represents between 15 and 25% of the total tissue volume, but this space is readily altered by abnormal, and perhaps by physiological, conditions.

IX. REFERENCES

1. J. F. Manery and L. F. Haege, The extent to which radioactive chloride penetrates tissues, and its significance, *Am. J. Physiol.* **134**:83–93 (1941).
2. D. M. Woodbury, P. S. Timiras, A. Koch, and A. Ballard, Distribution of radiochloride, radiosulfate, and inulin in brain of rats, *Federation Proc.* **15**:501–502 (1956).
3. R. L. Schultz, E. A. Maynard, and D. C. Pease, Electron microscopy of neurons and neuroglia of cerebral cortex and corpus callosum, *Am. J. Anat.* **100**:369–407 (1957).
4. E. DeRobertis and H. M. Gerschenfeld, Submicroscopic morphology and function of glial cells, *Intern. Rev. Neurobiol.* **3**:1–65 (1961).
5. H. M. Pappius, The distribution of water in brain tissues swollen *in vitro* and *in vivo*, *Progr. Brain Res.* **15**:135–154 (1965).
6. A. Vernadakis and D. M. Woodbury, Cellular and extracellular spaces in developing rat brain, *Arch. Neurol.* **12**:284–293 (1965).
7. H. Davson, *Physiology of Cerebrospinal Fluid*, J. and A. Churchill, Ltd., London (1967).
8. A. Van Harreveld, *Brain Tissue Electrolytes*, Butterworths, Washington (1966).
9. J. A. Zadunaisky and P. F. Curran, Sodium fluxes in isolated frog brain, *Am. J. Physiol.* **205**:949–956 (1963).
10. A. Ames, III, and F. B. Nesbett, Intracellular and extracellular compartments of mammalian central nervous tissue, *J. Physiol.* (*London*) **184**:216–238 (1966).
11. J. S. Coombs, J. C. Eccles, and P. Fatt, The specific ionic conductances and the ionic movements across the motoneuronal membrane that produce the inhibitory post-synaptic potential, *J. Physiol.* (*London*) **130**:326–373 (1955).
12. R. F. Kibler, R. P. O'Neill, and E. D. Robin, Intracellular acid–base relations of dog brain with reference to the brain extracellular volume, *J. Clin. Invest.* **43**:431–443 (1964).
13. R. S. Bourke, E. S. Greenberg, and D. B. Tower, Variations of cerebral cortex fluid spaces *in vivo* as a function of species brain size, *Am. J. Physiol.* **208**:682–692 (1965).
14. C. F. Barlow, N. S. Domek, M. A. Goldberg, and L. J. Roth, Extracellular brain space measured by S[35] sulfate, *Arch. Neurol.* **5**:102–110 (1961).

15. A. Van Harreveld, N. Ahmed, and D. J. Tanner, Sulfate concentrations in cerebrospinal fluid and serum of rabbits and cats, *Am. J. Physiol.* **210**:777–780 (1966).
16. R. W. P. Cutler, C. F. Barlow, and A. V. Lorenzo, The effect of brain-cerebrospinal fluid diffusion gradients on the determination of extracellular space in cat brain, *J. Neuropathol. Exptl. Neurol.* **26**:167 (1967).
17. A. B. Morrison, The distribution of intravenously-injected inulin in the fluids of the nervous system of the dog and rat, *J. Clin. Invest.* **38**:1769–1777 (1959).
18. D. P. Rall, W. W. Oppelt, and C. S. Patlak, Extracellular space of brain as determined by diffusion of inulin from the ventricular system, *Life Sci.* **2**:43–48 (1962).
19. D. L. Woodward, D. J. Reed, and D. M. Woodbury, Extracellular space of rat cerebral cortex, *Am. J. Physiol.* **212**:367–370 (1967).
20. J. D. Fenstermacher and M. O. Bartlett, Sucrose space measurements in the rabbit central nervous system, *Am. J. Physiol.* **212**:1268–1272 (1967).
21. D. J. Reed and D. M. Woodbury, Kinetics of movement of iodide, sucrose, inulin and radio-iodinated albumin in central nervous system and cerebrospinal fluid of rat, *J. Physiol. (London)* **169**:816–850 (1963).
22. M. Pollay and H. Davson, The passage of certain substances out of the cerebrospinal fluid, *Brain* **86**:137–150 (1963).
23. E. Streicher, D. P. Rall, and J. R. Gaskins, Distribution of thiocyanate between plasma and cerebrospinal fluid, *Am. J. Physiol.* **206**:251–254 (1961).
24. E. Streicher, Thiocyanate space of rat brain, *Am. J. Physiol.* **201**:334–336 (1961).
25. H. M. Pappius, Spaces in brain tissue *in vitro* and *in vivo*, *Progr. Brain Res.* **29**:455–460 (1968).
26. H. M. Pappius, unpublished observation.
27. M. Pollay, Cerebrospinal fluid transport and the thiocyanate space of the brain, *Am. J. Physiol.* **210**:275–279 (1966).
28. L. Z. Bito, M. W. B. Bradbury, and H. Davson, Factors affecting the distribution of iodide and bromide in the central nervous system, *J. Physiol. (London)* **185**:323–354 (1966).
29. G. B. Wallace and B. B. Brodie, The distribution of iodide, thiocyanate, bromide and chloride in the central nervous system and spinal fluid, *J. Pharmacol. Exptl. Therap.* **65**:220–226 (1939).
30. D. M. Woodbury, Distribution of various substances in the brain as affected by alterations in active transport of cations and anions across the choroid plexus, *Progr. Brain Res.* **29**:297–313 (1968).
31. A. Lajtha, The development of the blood–brain barrier, *J. Neurochem.* **1**:216–227 (1957).
32. K. A. C. Elliott, Brain swelling and fluid and electrolyte distribution, *in The Chemical Pathology of Brain* (J. Folchi-Pi, ed.), pp. 277–293, Pergamon Press, New York (1961).
33. S. Varon and H. McIlwain, Fluid content and compartments in isolated cerebral tissues, *J. Neurochem.* **8**:262–275 (1961).
34. G. Levi, A. Cherayil, and A. Lajtha, Cerebral amino acid transport *in vitro*—III, *J. Neurochem.* **12**:757–770 (1965).
35. R. S. Bourke and D. B. Tower, Fluid compartmentation and electrolytes of cat cerebral cortex *in vitro*—I, *J. Neurochem.* **13**:1071–1097 (1966).
36. J. B. Ranck, Analysis of specific impedance of rabbit cerebral cortex, *Exptl. Neurol.* **7**:153–174 (1963).
37. A. H. Nevis and G. H. Collins, Electrical impedance and volume changes in brain during preparation for electron microscopy, *Brain Res.* **5**:57–85 (1967).
38. L. Bakay and J. C. Lee, *Cerebral Edema,* Charles C. Thomas, Springfield, Illinois (1965).

Chapter 2

WATER MOVEMENT

Robert Katzman and Herbert Schimmel

The Saul R. Korey Department of Neurology
Albert Einstein College of Medicine
Bronx, New York

Water is the most abundant compound in the brain. In gray matter, water constitutes 80–88 % of the total fresh weight; in white matter, 65–72 %. This difference correlates with the amount of myelin present.

Whereas special transport mechanisms representing a portion of the "blood–brain barrier" are probably involved in the movement of most cations and anions, water appears to move freely between blood and brain in response to osmotic changes in either compartment. This fact is clinically important in water intoxication and in dehydration and is now widely used in the osmotic therapy of intracranial hypertension.

I. ANALOGY WITH LIPID BILAYERS

The fact that water moves freely across membrane systems that are barriers to most substances not readily soluble in lipids is not surprising in view of present work on reconstituted lipid membranes.[1,2] This work has shown that water can cross a lipid bilayer membrane set up between two saline solutions even though the membrane is impervious to the small charged ions such as Na^+ or K^+. The rate of movement across such protein-free membranes is similar to that across plasma membranes.

The water movement across the lipid bilayer membrane is not related to the electrical resistance of these membranes, being essentially unchanged as their electrical resistance is reduced from the initial values of 10^8 to 10^3 Ω/cm^2 by the addition of certain protein factors. Hence, the water movement appears to be across the lipid bilayer itself, and there is no need to postulate special "pores" for the water flow in this system. The molecular mechanism of water movement across these high-resistance lipid barriers is unknown.

One further result obtained from the lipid bilayer studies deserves emphasis. In many biological systems, the water permeability of membranes is greater under an osmotic gradient than if diffusion of water is measured by the introduction of the tracers ^2H_2O (D_2O) or ^3H_2O (T_2O). This has led to the postulation of long pores through which the hydrodynamic flux is greater than diffusional flux.[3] In the initial work with the lipid bilayer membranes, a difference between the permeability to osmotic and diffusional flow also was noted. This, however, has been shown to be due to the presence of unstirred layers that impede diffusion more than osmotic flow. In many cellular systems, such unstirred layers in effect are built in if the tissue is structurally intact. However, it is not certain whether all the differences noted in biological systems between these two forms of water flux can be explained on this basis.

II. DIFFUSION OF LABELED WATER BETWEEN BLOOD AND BRAIN

Bering[4] measured the rate of appearance of D_2O in brain tissue after parenteral administration; half-exchange occurred within seconds. His results are shown in Table I. Presumably, the small differences between different regions of the brain reflect the density of and average distance between capillaries in the tissue. In the CSF, the half-times of exchange are much greater due to the very great distance between capillaries and the large volume in which the D_2O must be distributed.

III. FLUX OF WATER BETWEEN BLOOD AND BRAIN UNDER HYDROSTATIC OR OSMOTIC PRESSURE GRADIENTS

An interesting experiment was reported by Coulter[5] in which water flux was measured either by introducing fluid into the CSF or removing it, thereby altering intracranial pressure and determining the time required for restoration of normal pressure. However, this experiment was subsequently criticized by Edstrom[6] and Fenstermacher and Johnson[7] on the grounds that the increased intracranial pressure would open the arachnoid valves[8] between CSF and blood, permitting the more rapid removal of CSF and giving spurious values (by a factor of 70) for water flux.

The movement of water between blood and brain under osmotic loads was measured by Fenstermacher and Johnson.[7] These investigators measured changes in intracranial volume in rabbits given intravenous osmotic loads. Intracranial volume changes were followed by watching the meniscus of heparinized Ringer's solution in a horizontally placed pipette connected to the surface of the brain in plastic tubing. Serum osmolality was raised abruptly by about 50 mM following intravenous injection of either raffinose, sucrose, glucose, urea, or formamide, and the serum osmolality was maintained constant for 10 min. Maximum water movement was produced by

TABLE I

Time for Various Portions of Central Nervous System and Cerebrospinal Fluid to Reach $\frac{1}{2}$ Serum Concentration of D_2O After Intravenous Injection of D_2O as 0.8 % NaCl (Dogs)[a]

Tissue	Half-time D_2O appearance in dogs
Cerebral gray matter	12 sec
Cerebral white matter	20 sec
Cerebellum	12 sec
Spinal cord	25 sec
Ventricular cerebrospinal fluid	8 min
Cisternal cerebrospinal fluid	3 min
Lumbar cerebrospinal fluid	7 min

[a] Table 3 from Bering, Jr.[4]

sucrose and raffinose, the other molecules probably crossing capillary membranes to an appreciable extent. In fact, it is known from other experiments that sucrose does not appreciably cross the blood–brain barrier, whereas urea does, so that these results are reasonable. If one expresses these results in terms of reflection coefficients σ,[9] assuming σ for raffinose to be 1.00, then the values obtained experimentally by Fenstermacher and Johnson[7] were

Average reflection coefficients ($\pm$ standard deviation)

Raffinose	1.00 ± 0.022
Sucrose	0.98 ± 0.014
Glucose	0.89 ± 0.008
Urea	0.44 ± 0.054
Formamide	0.37 ± 0.025

From these experiments, it was found that the average net water loss from the 12-g rabbit brain produced by a 1 mM osmotic gradient of raffinose was 14.5×10^{-7} liters/min. If the rabbit brain contains 75 % water, then the 12-g brain will contain 9 g or 9×10^{-3} liters of water. Hence, water loss will be $(14.5 \times 10^{-7})/(9 \times 10^{-3}) = 1.6 \times 10^{-4}$ liters of H_2O/liter of brain H_2O or 0.016 %/min/mM osmotic gradient.

IV. COMPARISON OF OSMOTIC AND DIFFUSIONAL FLOW

Using Fenstermacher and Johnson's results gives a 1 % change in brain water per minute for a 62 mM osmotic gradient. At first glance, this value may appear to be radically different from the results obtained by Bering,[4] where the half-time of water exchange was 12 sec for cerebral gray matter and 20 sec for white. However, if one were to use the concentration gradient of 55 M for water in the above computation, then there would be a 50% change in water content in about 3 sec.

A more refined calculation can be made that would permit comparison of Fenstermacher and Johnson's permeability coefficients obtained from osmotic experiments with the half-time values obtained by Bering for diffusion of water. The calculation follows:

In the diffusion experiment, c_{se} is the steady concentration of D_2O introduced into the capillaries at $t = 0$ (time zero); c_{br} is the concentration of D_2O in the brain, a function of t; A is the capillary membrane area per unit of volume; and P_d is the permeability coefficient per unit area of membrane. Then

$$\frac{dc_{br}}{dt} = AP_d(c_{se} - c_{br}) \tag{1}$$

This equation states that the rate of change of concentration, dc_{br}/dt, is equal to the flow of material into a unit volume under a concentration gradient $(c_{se} - c_{br})$ across a membrane area A with a permeability coefficient P_d. This is an application of Fick's first law assuming perfect mixing in the brain compartment. Equation (1) may be readily integrated into the following form:

$$\frac{c_{se} - c_{br}}{c_{se}} = e^{-AP_d t} \tag{2}$$

If $c_{br} = c_{se}/2$ at the half-time τ, then $e^{-AP\tau} = 1/2 = e^{-0.7}$ and

$$\tau = 0.7/AP_d \tag{3}$$

From Fenstermacher and Johnson's experiment, a value of AP_f, where P_f is the permeability coefficient to osmotic flow, can be obtained. Fenstermacher and Johnson's experiments gave a flux of 14.5×10^{-7} liter min^{-1} mM^{-1} into 9 ml rabbit brain water. Expressing this as a flux of mM of water per second per mM gradient, we get $AP_f = 14.5 \times 10^{-7} \times 10^6/20 \times 1/60 \times 0.009 = 0.14$ sec^{-1}. If we use this value of AP_f for AP_d in Eq. (3), a τ of 5 sec is obtained. This is one-half to one-fourth the values obtained in Bering's experiment. Another way of expressing this is in terms of the permeability coefficients. Since half-times and permeability coefficients are inversely related [see Eq. (3)], the permeability coefficient derived from Fenstermacher and Johnson's osmotic flow experiments is two to four times greater than that derived from Bering's diffusion experiment. This result is similar to that found in experiments where permeabilities measured by diffusion and osmotic flow are compared, e.g., red blood cells and lipid membranes.

It is interesting to compute the value of P_f so that comparison can be made with the lipid membrane model. Coulter[5] and Crane[10] give values for A of 52 and 240 cm^2/g, respectively. Allowing for the fact that there is only 0.7 ml of water per gram and assuming $AP_f = 0.14$ sec^{-1}[7], one obtains values of 3.5×10^{-3} and 0.7×10^{-3} cm sec^{-1}, respectively, as compared to 1.0×10^{-3} cm sec^{-1} for the lipid membrane model. The above computation shows that the estimate of A is critical to computing a value of P_f. In fact, the density of capillaries and their area will vary markedly from one region of the brain to another, and the value of P_f can be given only when the value of A has been estimated for the same brain in which water movement has been studied.

V. EFFECT OF VOLUME DIFFUSION ON HALF-TIMES

The assignment of the resistance to flow primarily to the capillary wall and the assumption of mixing in the brain compartment are extreme simplifications of the actual situation. Except for the extracellular space that appears to account for less than 20 % of the brain volume and less than 25 % of the water content, the flow of water whether in or out must take place through cell walls that can be expected to impede flow. Furthermore, the diffusion in the medium, whether to the cell or within the cell, will take a finite time.

Let us first consider the effect of volume diffusion. In a very simplified model we may assume that each capillary of diameter a will feed a volume contained within a cylinder of diameter b. The ratio $(a/b)^2$ will represent the proportion of brain volume that is occupied by the capillaries if $a \ll b$.[11] Figure 1 gives the results of a computation of half-times based on a capillary

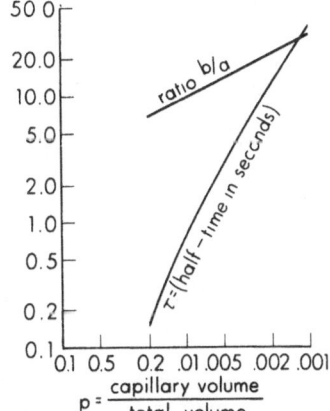

Fig. 1. Diffusion of water in water from cylinder of diameter a into concentric cylinder of diameter b. Half-time computation corresponds to the value $a = 10\,\mu$. For other values of a multiply τ in graph by $a^2/100$ where a is measured in μ. Note $p \simeq a^2/b^2$. See text.

diameter of 10μ and cylinder diameters varying from 70 to 300μ corresponding to capillary volumes ranging from 2 % down to 0.1 %. The range of half-times is from 0.15 to 32 sec. For a capillary volume of 1 %, the value is 0.8 sec.

In this calculation, we have assumed that the capillary volume is uniform throughout the brain. In fact, it is well known that this is not so, the gray matter being far more vascular than the white. The effect of this inhomogeneity would be to increase the half-time as compared to a computation made from the whole brain. Suppose that the capillaries in gray matter occupy 2.3 % of the gray matter volume and that the capillaries in white matter occupy 0.3 % of the white matter volume[12] but that the gray matter was only one-third of the total brain volume. Here, the capillary volume for the brain as a whole would be 1 %. However, the half-time computed in this instance would be 3 sec rather than the 0.8 sec obtained for the case of a 1 % capillary volume in a homogeneous brain.[12]

In our earlier calculations, we assumed that the half-times were inversely proportional to the membrane permeabilities and were a measure of the capillary membrane resistance. From our calculations of bulk diffusion without mixing, one gets an estimate of the order of magnitude of the impedance due to this factor. It can be seen that this factor appears small if calculated for homogeneous capillary densities but begins to become significant as regional inhomogeneities are considered. Moreover, in our calculations we have assumed that the diffusion of water in the tissue is similar to that of diffusion of water in water. Ling,[13] however, has found in measurements of water diffusion into frog eggs that the diffusion profile can be explained best by assuming no membrane impedance with all impedances due to the bulk diffusion of water. However, the actual diffusion coefficients required for this assumption are one-third to one-half those of water in water. He attributes the coefficients to the interaction of water and protein in the tissue. If such lower diffusion coefficients were applicable to our system, then it would appear that a large part of the impedance to water flow in the brain could be attributed to bulk diffusion in cytoplasm. The actual behavior in brain tissue is extremely complex and is the result of impedances attributed separately to the capillary wall, cell membrane, and bulk diffusion. The calculation of P_f and P_d that has been made, assuming arbitrary values for the areas of capillary walls, must be treated as a useful convention rather than as an appropriate simplified model. It not only assumes that the impedance due to bulk diffusion is negligible, which may or may not be the case, but also assumes that the impedance due to cell membranes is negligible, which is unlikely.

VI. PREDICTION OF EFFECTS OF OSMOTIC TREATMENTS

Fenstermacher and Johnson[7] developed an equation [their (2a)] that together with their experimental results may be used to predict the approximate shrinkage of water volume in brain tissue following an increase of osmolality of the plasma.

The value for water movement obtained by Fenstermacher and Johnson was based upon CNS volume changes in rabbits occurring within 12 min after introduction of the osmotic load. It is interesting to inquire as to whether the value obtained for water movement under these experimental conditions can be applied more generally in order to predict the kinetics of brain shrinkage with different degrees of alteration of plasma osmolality and for different time periods.

It is useful to review briefly the assumptions underlying the equation, since these will be only partially met in the treatment and experimental results to be discussed. It is assumed that a test or treatment substance of concentration C_{ps} is introduced into the plasma at time zero ($t = 0$) and that this elevated concentration is kept constant. An osmotic pressure will be developed that will cause a flow of water to the plasma and a reduction of brain tissue water volume from its initial V_0 to a later value V with $V < V_0$. We shall denote the ratio $(V_0-V)/V$ by S, and it will represent the relative shrinkage. It also is assumed that the brain tissues contain an initial concentration C_{pi} of impermeant molecules whose concentration will increase as water leaves, offsetting the effect of the treatment substance until maximum shrinkage [which is found to be $S_\infty = C_{ps}/(C_{ps} + C_{pi})$], if the test or treatment substance cannot pass through the capillary walls to the brain compartment. If the test or treatment substance does permeate to the brain tissue, then the effective osmotic pressure will correspond to a concentration $\sigma\, C_{ps}$ of impermeant test substance, and $S_\infty = \sigma\, C_{ps}/(\sigma\, C_{ps} + C_{pi})$, where $1 \geqslant \sigma \geqslant 0$, and σ is called the reflection coefficient.[9]

Fenstermacher and Johnson's[7] Eq. 2a, based on these assumptions, can be transposed into the form

$$t \text{ (in minutes)} = 20\left[S(1 - S_\infty) + (1 - S_\infty)^2 \log \frac{S_\infty}{S_\infty - S}\right]$$

$$(4)$$

$$S = (V_0 - V)/V_0 \qquad S_\infty = \frac{\sigma C_{ps}}{\sigma C_{ps} + C_{pi}}$$

The coefficient 20 corresponds to the reciprocal of the product of the filtration coefficient per unit of volume ($14.5 \times 10^{-7}/0.009$) by the concentration of impermeants (310 mM). This coefficient should be adjusted if other values are assumed for these parameters. It should be noted that in the computation of S_∞ and in the derivation of Eq. (4) the countervailing effect of any test substance that has passed into the brain tissue is ignored (see the discussion of urea below).

Figure 2 is based on Eq. (4) and gives plots of S as a function of various values of S_∞ that in turn corresponds to the various values of $\sigma\, C_{ps}$.

Katzman and Schimmel

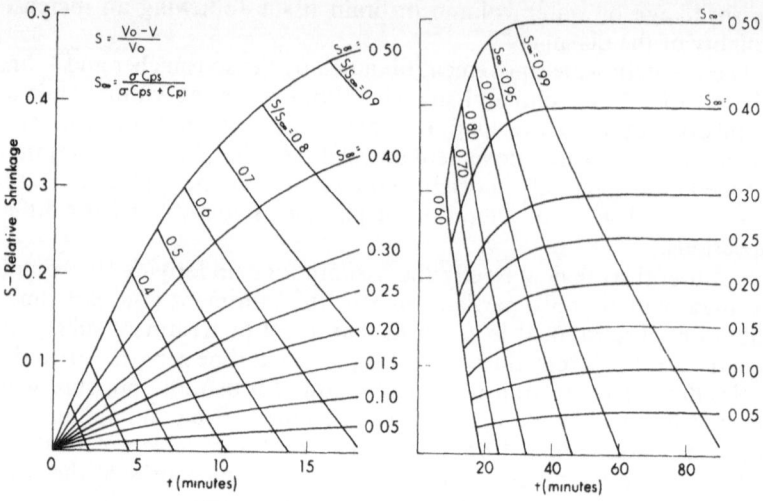

Fig. 2. Theoretical shrinkage of tissue subjected to osmotic treatment. Lines emerging from coordinate origin correspond to different concentrations of test molecules. Diagonal lines running downward from left to right correspond to time required to achieve a fixed proportion of the ultimate shrinkage at $t \to \infty$. See text for definitions and assumptions.

For a fixed value of C_{pi} (310 mM), the values of S_∞ and σC_{ps} are related as follows:

S_∞	σC_{ps} mM	S_∞	σC_{ps} mM
0.05	17	0.25	103
0.10	34	0.30	133
0.15	55	0.40	207
0.20	78	0.50	310

For treatment purposes, values of σC_{ps} are usually below 34 mM while even for experimental purposes they are usually below 100 mM. The higher values of S_∞ have been included in Fig. 2 to bring out more sharply the effects of increased concentration of test substances. The plots in Fig. 2 show that as concentration of treatment substance is increased not only does the shrinkage S_∞ for a given time increase in an absolute sense, but also in a relative sense. This is brought out by the diagonal lines moving upward from right to left, which correspond to times required for achieving a fixed portion of the maximum shrinkage. Thus, the diagonal labeled 0.5 corresponds to an effective half-time for shrinkage. At lowest concentrations, it is equal to 14 min. Where maximum shrinkage is one-half the initial volume, the corresponding half-time is 6 min. The corresponding times for achieving 0.95 maximum shrinkage are 60 and 20 min.

The range of practical interest in Fig. 2 is for the cases up to $S_\infty = 0.3$, as this corresponds to $\sigma C_{ps} = 135$ mM, if $C_{pi} = 310$ mM. Most hypertonic treatments on human beings whether with urea, mannitol, or glycerol would correspond to an S_∞ of 0.1 or smaller.

Let us compare these predicted rates of shrinkage with the data obtained by Reed and Woodbury with urea.[14] In evaluating the data, one notes that urea permeates the brain so that shrinkage cannot be properly described by Eq. (4). However, the uptake of urea by the brain is substantially slower than the shrinkage process. Thus, 20–30 min are required for 25 % of the ultimate uptake and about 1 hr for 50 % of the uptake.[7,14] The corresponding 25 and 50 % shrinkage times ($S_\infty = 0.1$) are 5 and 12 min (see Fig. 2a). Figure 3 of Reed and Woodbury[14] reproduced here (Fig. 3) shows the type of experimental animal data that confirms the shrinkage effects. The actual shrinkages measured seem to be small, but then the hypertonicity is not clearly defined. The urea dosage is 2 g/kg injected intraperitoneally in rats. At 20 min, there is a relative shrinkage (S) without and with diamox, respectively, of 0.014 (0.9/65) and 0.019 (1.25/65). Maximum shrinkages are only 20 % greater and occur at 60 min.

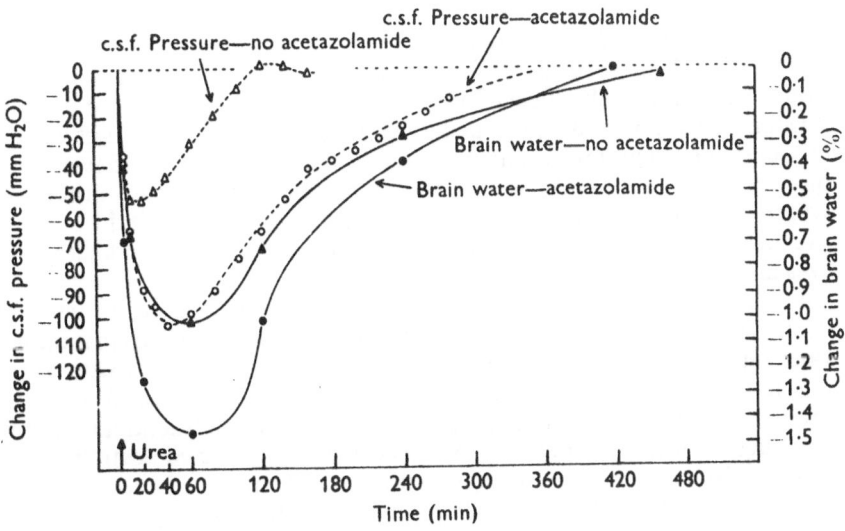

Fig. 3. The comparison of the changes in brain water and CSF pressure in rats treated with urea and acetazolamide (urea given intraperitoneally in dose of 1 or 2 g/kg in 5% glucose solution; acetazolamide given intraperitoneally in dose of 5 mg/kg). From Reed and Woodbury.[22]

VII. RELATIONSHIP OF CHANGES IN BRAIN WATER TO CHANGES IN INTRACRANIAL PRESSURE—CLINICAL APPLICATIONS

One effect of the rapid change in brain water content is to produce an alteration in intracranial pressure. Advantage has been taken of this fact clinically in the use of osmo-therapy for reduction in intracranial pressure. Interest in such therapy dates back to the mid-1930's when sucrose was used as a dehydrating agent.[15,16] In subsequent years, the injection of 50% sucrose in patients suffering from increased intracranlal pressure was used from time to time, but the technique did not gain wide popularity. In 1956, Javid and Settlage[17] introduced the use of urea in invert sugar. They found that with intravenous infusion of this preparation the intracranial pressure could be reduced for a number of hours. Other workers noted that although this was an exceedingly effective therapeutic technique, sometimes a rebound with an increase in intracranial pressure occurred after several hours. This rebound was attributed to the slow diffusion of urea into brain tissue. After several hours, the concentration of urea in brain tissue became as great or greater than that of the plasma; water could then flow from plasma to brain, increasing intracranial pressure. This "rebound" is usually not clinically significant, but occasionally it may be. Javid and Settlage[17] have questioned whether this phenomenon really occurs or whether the occasional instances of high intracranial pressure several hours after urea infusion simply reflect chance variations in the elevated CSF pressure once the urea effect has been dissipated.

Mannitol is often preferred over urea as the agent in osmo-therapy. Since mannitol enters the tissue somewhat more slowly than urea, it does not produce a rebound. The effect of mannitol may last 6–12 hr after intravenous infusion. Recently, oral glycerol has come into vogue. This compound, although somewhat unpleasant in its ingestion even when masked by orange juice, produces a slower rise in serum osmolality and a more prolonged hypertonic state. Moreover, glycerol has the advantage that if it does diffuse into brain tissue, it is metabolized by the cell to CO_2 and H_2O and, therefore, loses any osmotic effect.

The alteration in water content by osmotic agents may not be uniform throughout the brain. It has been found on electron micrographs that during water intoxication, clear glia cells (astrocytes) appear to show greater swelling than do other tissue elements.[18,19] In experimental cerebral edema, it is found that the effect of hypertonic solutions is to lower intracranial pressure by reduction of the water content of the normal or unaffected portions of the brain.[20] In experimental edema in which the edema is secondary to local irritation or trauma such as that produced by the freezing lesion, there apparently is sufficient alteration in the blood–brain barrier so that the capillary walls are no longer acting as semipermeable membranes. However, in the diffuse edema produced by triethyltin, oral glycerol will reduce the water content of the brain rather uniformly.[21]

In many of the clinical and experimental studies of the effects of urea, mannitol, and glycerol the time course of change of intracranial pressure has been reported. Such changes in intracranial pressure represent the net result of a complex series of events including the shrinkage of the brain due to rapid water movement across capillaries. By comparison, the CSF water probably changes slowly due to the much greater volume of the CSF compartments, the greater distance between capillaries on either side of the ventricles, cisterna magna, etc. Pressure changes lead to alteration in CSF flow from the arachnoid villi,[8] but they produce little change in rate of formation of CSF, so that CSF volume will change. In addition, one must consider the properties of the cranial cavity, which is a rigid container with some elastic elements in it. In such a complex situation, there might be no necessary relationship between changes in brain water and that of CSF pressure. This problem was studied by Reed and Woodbury,[13,22] who indeed found a discrepancy between the two. They found that after the administration of hypertonic urea solution, CSF pressure fell within minutes, whereas brain water content was not minimal until 90 min and required 8 hr to return to the control level (Fig. 3). Reed and Woodbury postulated that this difference was due to the time required for CSF volume to be adjusted by formation of new CSF, since acetazolamide markedly affected their results. This discrepancy between change in brain water and CSF pressure with the more rapid adjustment of the CSF pressure is consistent with the discrepancy with the values of water flow found by Coulter,[5] who altered hydrostatic pressure, and Fenstermacher and Johnson,[7] who altered osmotic pressure as we have already discussed.

VIII. REFERENCES

1. T. Hanai and D. A. Haydon, The permeability to water of bimolecular lipid membranes, *J. Theoret. Biol.* **11**:370–382 (1966).
2. A. Cass and A. Finkelstein, Water permeability of thin lipid membranes, *J. Gen. Physiol.* **50**:1765–1784 (1967).
3. A. Mauro, Nature of solvent transfer in osmosis, *Science* **126**:252–253 (1957).
4. E. A. Bering, Jr., Water exchange of central nervous system and cerebrospinal fluid, *J. Neurosurg.* **9**:275–287 (1952).
5. N. A. Coulter, Filtration coefficient of the capillaries of the brain, *Am. J. Physiol.* **195**:459–464 (1958).
6. R. Edstrom, in *International Review of Neurobiology* (C. C. Pfeiffer and J. R. Smythres, eds.), p. 153, Academic Press, New York (1964).
7. J. D. Fenstermacher and J. A. Johnson, Filtration and reflection coefficients of the rabbit blood–brain barrier, *Am. J. Physiol.* **211**:341–346 (1966).
8. K. Welch and V. Friedman, The cerebrospinal fluid values, *Brain* **83**:454–469 (1960).
9. A. Katchalsky and P. F. Curran, *Nonequilibrium Thermodynamics in Biophysics,* Cambridge University Press, Cambridge (1965).
10. C. Crone, The permeability of capillaries in various organs as determined by use of the indicator diffusion method, *Acta Physiol. Scand.* **58**:292–305 (1963).

11. H. Schimmel and R. Siagus, tutorial in preparation.

12. R. Katzman, H. Schimmel, and C. E. Wilson, Diffusion of inulin as a measure of extra-cellular fluid space in brain, Leo M. Davidoff Festschrift, *Proc. Rudolf Virchow Med. Soc. City N.Y. Suppl.* **26**:254 (1968).

13. G. N. Ling, M. M. Ochsenfeld, and G. Karreman, Is the cell membrane a universal rate-limiting barrier to the movement of water between the living cell and its surrounding medium? *J. Gen. Physiol.* **50**:1807–1820 (1967).

14. D. J. Reed and D. M. Woodbury, Effect of hypertonic urea on cerebrospinal fluid pressure of brain volume, *J. Physiol.* **164**:252–264 (1962).

15. L. T. Bullock, M. I. Gregersen, and R. Kinney, The use of hypertonic sucrose solution intravenously to reduce cerebrospinal fluid pressure without a secondary rise, *Am. J. Physiol.* **112**:82–86 (1935).

16. J. H. Masserman, Effects of the intravenous administration of hypertonic solutions of sucrose, *Johns Hopkins Hosp. Bull.* **57**:12–21 (1935).

17. M. Javid and P. Settlage, Effect of urea on CSF pressure in human subjects, *J. Am. Med. Assoc.* **160**:943–949 (1956).

18. J. A. Zadunaisky, F. Wald, and E. D. P. DeRobertis, Osmotic behavior and ultrastructural modifications in isolated frog brain, *Exptl. Neurol.* **8**:290–309 (1963).

19. S. A. Luse and B. Harris, Brain ultrastructure in hydration and dehydration, *Arch. Neurol.* **4**:139–153 (1961).

20. R. A. Clasen, P. M. Cooke, S. Pandolfi, G. Carnecki, and G. Bryor, Hypertonic urea in experimental cerebral edema, *Arch. Neurol.* **12**:424–434 (1965).

21. S. Mandell, J. M. Taylor, D. G. Kotsilimbas, and L. C. Scheinberg, The effect of glycerol on cerebral edema induced by tri-ethyltin sulphate in rabbits, *J. Neurosurg.* **24**:984–986 (1966).

22. D. J. Reed and D. M. Woodbury, Effect of urea and acetazolamide on brain volume and cerebrospinal fluid pressure, *J. Physiol.* **164**:265–273 (1962).

Chapter 3

THE CEREBROSPINAL FLUID

Hugh Davson

Department of Physiology
University College
London, England

I. LOCATION

The cerebrospinal fluid is contained within, and surrounds, the brain and spinal cord (Fig. 1). Thus, the internal fluid is contained within the ventricles —a set of intercommunicating cavities—while the external fluid occupies the subarachnoid spaces, i.e., the spaces on the surface of the brain and cord contained by the pia, internally, and the arachnoid membrane, externally. The internal fluid connects with the external fluid by way of one or more foramina. Thus, in man the principal connection is by way of the *foramen of Magendie*, which is essentially a gap in the roof of the IVth ventricle that permits flow out of this into the large adjacent subarachnoid space, the *cisterna magna*. In lower animals this foramen is not present, so that connection is made through the two *foramina of Luschka*, which are holes in the lateral recesses of the IVth ventricle opening into the subarachnoid spaces at the base of the brain. Each foramen is situated in the angle between the pons and medulla and opens into the *cisterna pontis*.

II. ORIGIN AND FATE

The main source of the fluid is undoubtedly the choroid plexuses within the ventricles. These are highly vascular outpouchings of the pia into the cavities, retaining the ependymal covering, which becomes the *choroidal epithelium*. It is considered that this epithelium transports solutes and water from the stroma of the plexus into the ventricular cavity. Blockage of the pathway between ventricles and subarachnoid spaces, e.g., by plugging the aqueduct of Sylvius, causes a rise in ventricular pressure which, if maintained, leads to an internal hydrocephalus, i.e., dilatation of the ventricles at the expense of the surrounding nervous tissue. There is some evidence that fluid from the nervous tissue—extracellular fluid—may contribute to the total

23

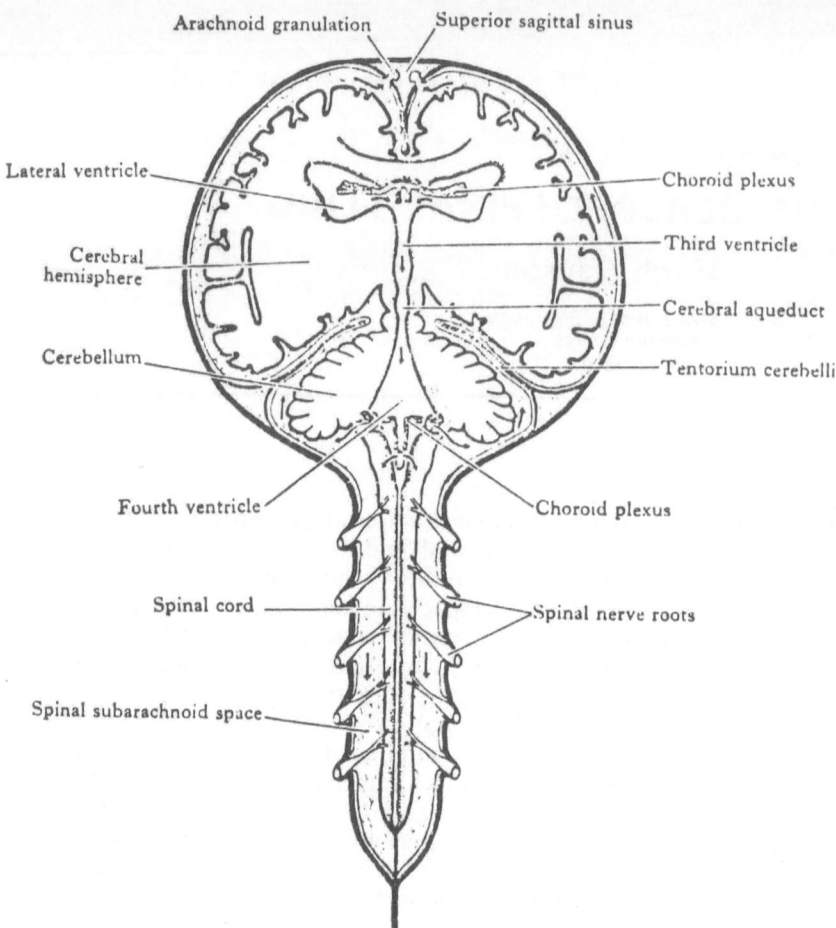

Fig. 1. Illustrating the subarachnoid spaces and circulation of the cerebrospinal fluid. From Millen and Woollam.[110]

amount in the system.[1,2] The cerebrospinal fluid is formed continuously and so must be drained away at the same rate; the site of drainage is on the surface of the brain through the so-called *arachnoid villi*, which are essentially perforations through the walls of the large dural sinuses that establish connection between the subarachnoid space and the lumen of these vessels. The connection is valvular, so that reflux of blood from the dural sinus into the subarachnoid space is inhibited, while flow in the opposite direction, namely, of cerebrospinal fluid into the blood, is permitted.[3] The force governing the flow is the generally greater pressure of the cerebrospinal fluid than the dural sinus pressure. The rate of production of fluid—and therefore the rate of drainage when the system is in a steady state—varies with the size of the animal but is a fairly constant fraction of the total volume of the fluid. Thus, the mean rates in rabbit, dog, and man are 10.1, 50, and 520 μl/min, respectively, but expressed as a percentage of the total volume they are 0.63, 0.40, and 0.37 %/min, respectively. The tendency for the fractional turnover rate to decrease with increasing size of the brain, suggested by the above figures, is probably valid. Thus, the rat's rate of secretion is 2.2 μl/min, which corresponds to some 2.5 %/min.[4,5]

III. FUNCTIONAL INSULATION FROM BLOOD

As we shall see, the cerebrospinal fluid has a composition that is characteristically different from that of the blood plasma, or an ultrafiltrate of blood plasma; therefore, it is not a simple tissue fluid like lymph or pericardial fluid. It comes into relation with the vascular system, directly in the regions of the choroid plexuses and less directly with the capillary systems of the central nervous tissue. Thus, the blood constituents should be able to diffuse passively from choroid plexus to cerebrospinal fluid in the ventricles and from the capillaries of the nervous tissue through the intercellular spaces and thence across the pial and ependymal linings into the subarachnoid and ventricular fluids, respectively. Finally, there could be passive diffusion from blood capillaries of the dura through this tissue and into the subarachnoid spaces. Studies of the passage of a variety of solutes from blood to cerebrospinal fluid have shown, however, that such passive diffusional processes are limited in rate, especially when the substances concerned are not lipid soluble. It is customary to speak, then, of a *blood–CSF barrier*, indicating the hindrance to passage of solutes from blood to the fluid. Because of the indirect pathway from blood to cerebrospinal fluid, namely, from blood to central nervous extracellular space and thence to cerebrospinal fluid, we must expect the passage from blood to brain to be similarly restricted—otherwise the effects of restraint to passage by the direct route across the choroid plexuses would be frustrated by this indirect route. In fact, when measurements of passage from blood to brain tissue are made, it is found that a similar selectivity is shown, lipid-insoluble molecules usually passing only very slowly from blood to nervous tissue. These findings are the basis of the concept of a blood–brain barrier.

We may ask what is the basis of this partial insulation of the cerebrospinal fluid from the blood, an insulation that permits it to retain a characteristic chemical composition different from that of blood plasma and an insulation that restrains the rates of exchange of many solutes between blood and cerebrospinal fluid? The epithelial layer covering the choroid plexuses is presumably the basis for the blood–CSF barrier in the ventricles; epithelia exhibiting secretory activity, such as the kidney tubular epithelium and frog skin, exhibit low passive permeability to lipid-insoluble substances. This is presumably because of the tight junctions, or zonulae occludentes that seal the spaces between adjacent epithelial cells.[6] A further barrier might be the ependymal lining of the ventricles, which could act as an additional restraint on passage from blood in the nervous tissue to the cerebrospinal fluid. In fact, however, there is a great deal of evidence against such a function of the ependyma; substances that fail to pass out of the blood into the brain, e.g., inulin, when injected into the ventricles will pass into the brain, and this most probably occurs across the ependymal linings of these cavities.[7]

The low selectivity of the ependyma well may be due to the absence of well-defined zonulae occludentes between adjacent cells.[8] It is possible, however, that there are regional variations in the character of the ependyma, so that it is conceivable that it shows selectivity in some areas and not in others.[9]

The cerebrospinal fluid in the subarachnoid spaces is separated from the brain by the pia and from the dura by the arachnoid; both of these membranes are composed of characteristic epithelial cells, which are in marked contrast to the mesothelial lining of the dura.[10,11] It seems likely, then, that the insulation of the subarachnoid fluid from the dura is achieved by the arachnoid. Whether the pial layer restrains exchange of solutes between subarachnoid space and the underlying nervous tissue is an open question. It is very frequently assumed that exchanges across this layer are relatively unrestricted, similar to the transependymal exchanges. This would account for the effectiveness of drugs when applied directly to the surface of the brain and for the rapidity with which subarachnoidally injected substances pass into the central nervous tissue. As with the ependyma, however, there may be local variations in selectivity, especially when cord and brain are compared. In general, then, it would seem that the cerebrospinal fluid is insulated from rapid exchanges with blood by virtue of the selectively permeable choroidal epithelium and the arachnoid membrane. Rapid exchanges between blood in the nervous tissue and the cerebrospinal fluid are prevented by the blood–brain barrier certainly, and possibly additional factors are the ependymal and pial layers on the brain surfaces, but this is an unnecessary hypothesis since the blood–brain barrier is a highly efficient restraining mechanism. The anatomical basis of the blood–brain barrier remains an open question, but it seems likely that the capillaries of the brain and spinal cord have special features that bring about a highly restricted permeability.[12,13,13a,13b]

IV. CHEMICAL COMPOSITION

The cerebrospinal fluid, taken for analysis from humans, is almost invariably taken from the lumbar subarachnoid space, while that taken from animals is taken from the cisterna magna. In general, the lumbar fluid may be regarded as older than the cisternal fluid because the lumbar subarachnoid space is a backwater from the mainstream of fluid, which passes out of the IVth ventrical into the cisterna magna and basal cisterns of the brain. Insofar as changes in composition take place during the fluid's stay in the cerebrospinal system, we may expect the lumbar fluid to differ in composition from the cisternal fluid. Although some changes do take place, and these will be discussed later, the magnitude of these is not so great as to vitiate comparisons of cisternal fluid from, say, the rabbit, with lumbar fluid from man.

A. Protein

The concentration of protein in cerebrospinal fluid is very low by comparison with that in plasma. Thus, in human lumbar fluid it is given as 31.3 mg/100 ml by Hill *et al.*[14] and 40 mg/100 ml by Plum and Fog.[15] These proteins presumably are derived principally from the plasma since the separate fractions, with one or two exceptions, are identical to similar fractions in the plasma. The proportions in which they occur in the two fluids are not the same, however; and in pathological conditions there may be an exceptional preponderance of one or other type, e.g., in neurosyphilis and multiple sclerosis the concentration of γ-globulin is unusually high.

B. Crystalloids

The cerebrospinal fluid is, by comparison with blood plasma, a virtually protein-free fluid, so that if we are to compare the concentrations of crystalloidal substances in the cerebrospinal fluid with a reference fluid we should choose a filtrate or dialysate of plasma for the comparison rather than plasma itself. The dialysate represents the fluid that is in passive equilibrium with plasma and is the fluid that would be formed were purely mechanical forces of filtration employed to produce it, by contrast with the active transport that is involved during the elaboration of specific secretions. Because the proteins of plasma are ionized, the concentrations of ions in the plasma are not the same as those in its dialysate, the operation of the Gibbs–Donnan equilibrium demanding, e.g., that the concentration of a univalent positive ion, such as Na^+, be some 4% higher in the plasma than in the dialysate. The inverse pertains for a univalent negative ion such as Cl^-. Thus, if we represent the ratio :concentration in dialysate/concentration in plasma as R_{dial}, we will find experimentally that R_{dial} for Na^+ and K^+ is in the region of 0.96 and that for Cl^- and HCO_3^- is 1.04. With divalent ions the theoretical ratios will be squared, namely, 0.92 and 1.08, respectively, but it must be emphasized that these ratios are theoretically deduced on the basis of the known base-binding characteristics of the plasma proteins, whereas complicating factors, such as specific "binding" of ions to the proteins may lead to quite different values of R_{dial}. This is specially manifest where Ca^{2+} and Mg^{2+} are concerned. In Table I the concentrations of a number of solutes in plasma and cerebrospinal fluid of the rabbit are shown. The ratio of the concentrations :concentration in CSF/concentration in plasma, indicated by R_{CSF}, is tabulated and compared with the experimentally determined values of R_{dial}. The fluids were withdrawn approximately simultaneously from the animal, and the concentrations and ratios may be taken as reasonably close to the steady-state values for the animal concerned.

TABLE I

Concentrations of Various Solutes (meq/kg H₂O) in Cerebrospinal Fluid and Plasma of the Rabbit, and Distribution Ratios. R_{CSF} = Conc. in CSF/Conc. in Plasma, and R_{dial} = Conc. in Dialysate/Conc. in Plasma[a]

Substance	Plasma	CSF	R_{CSF}	R_{dial}
Na	148	149	1.005	0.945
K	4.3	2.9	0.675	0.96
Mg[b]	2.02	1.74	0.92	0.80
Ca[b]	5.60	2.47	0.45	0.65
Cl	106	130	1.23	1.04
HCO₃	25	22	0.92	1.04
Glucose	8.3	5.35	0.64	0.97
Urea	8.35	6.5	0.78	1.00
Osmolality	298.5	305.2	1.02	0.995
pH	7.46	7.27	—	—

[a] From Davson.[16]
[b] From Bradbury (unpublished).

The results show that the cerebrospinal fluid is different in many respects from a simple protein-free dialysate or filtrate from blood plasma. Thus the concentration of K^+ is only 2.9 meq/kg H₂O compared with 4.3 for plasma and 4.1 for a plasma dialysate. The concentration of Mg^{2+} in cerebrospinal fluid is 1.74 meq/kg H₂O compared with 2.02 in plasma. This suggests a deficiency of Mg^{2+} in the cerebrospinal fluid but, because of the specific binding of Mg^{2+} to plasma in a nondiffusible form, the value of R_{dial} is 0.80 whereas the value of R_{CSF} is actually higher, namely, 0.92, indicating an *excess* of Mg^{2+} above that in a plasma dialysate. This excess is very much more marked in man (Table II). In general, a careful examination of Table I reveals that the cerebrospinal fluid of the rabbit contains excesses of Na^+, Cl^-, and Mg^{2+} over the concentrations in a plasma dialysate and deficiencies of K^+, Ca^{2+}, and HCO_3^-, in glucose and urea. The pH is more acid than the plasma while the total osmolality is significantly higher. Table II shows some figures for human lumbar fluid and plasma. In general the same tendencies are present, although the excess of Mg^{2+} in the cerebrospinal fluid is much greater, while the concentration of Ca^{2+} is approximately equal to that in a dialysate of plasma, i.e., there is no deficiency as in the rabbit.

1. *CSF as a Secretion*

The results indicate unequivocally that the cerebrospinal fluid is not a simple filtrate or dialysate from plasma, so that it is correct to describe it as a specific secretion. By analogy with other transport processes we may postulate a primary active transport of certain solutes, e.g., Na^+, followed by passive diffusion of water to produce an approximately isosmolal fluid. As to which solutes are actively transported and which are passively distributed, a decision requires careful studies of the effects of changed plasma concentration on the steady-state distribution, while the influence of any electrical potential gradients across the choroid plexuses cannot be ignored. Thus, when an electrode is placed in the ventricle or on the surface of the brain and another is placed in the bloodstream, a difference of potential may be recorded, probably identical with the so-called d.c. potentials of the electrophysiologists. In the rabbit the cerebrospinal fluid is some 12 mV positive in respect to blood,[17] and in the cat it was 5.3 mV.[18] In the unanesthetized goat, Held, Fencl, and Pappenheimer[19] found a value of 5 mV on the average. If ions were passively distributed across the choroid plexuses in accordance with this potential difference, there would be a tendency for the cations such as Na^+ and K^+ to be in deficit and the anions, such as Cl^- and HCO_3^- to be in excess in the cerebrospinal fluid, the excesses and deficiencies being considerably greater than demanded of a simple Gibbs–Donnan distribution between plasma and its dialysate. While there is every reason to believe that the potential across the choroid plexuses will affect the movements of ions across the choroidal epithelium, and thus influence the steady-

TABLE II

Concentrations of Various Solutes (meq/kg H_2O) in Plasma and Lumbar Cerebrospinal Fluid of Human Subjects[a]

Substance[b]	Plasma	CSF	R_{CSF}
Na	150	147	0.98
K	4.63	2.86	0.615
Mg	1.61	2.23	1.39
Ca	4.70	2.28	0.49
Cl	99	113	1.14
HCO_3	26.8	23.3	0.87
Br	2.45	0.90	0.37
Inorganic P (mg/100 ml)	4.70	3.40	0.725
Osmolality	289	289	1.0
pH	7.397	7.307	—
P_{CO_2} (mm Hg)	41.1	50.5	

[a] From Davson.[16]
[b] Cl^- from Fremont-Smith[104]; Na^+ and K^+ from Salminen and Luomanmaki[105]; Mg^{2+} and Ca^{2+} from Hunter and Smith[26]; Br^- from Hunter, Smith, and Taylor[106]; HCO_3, pH, and P_{CO_2} from Bradley and Semple[107]; inorganic phosphate from Ragazzini[108]; osmolality from Hendry.[109]

state distribution between plasma and cerebrospinal fluid, experimentally it may be shown that this potential is not a serious factor so far as the observed distributions are concerned, and it will certainly not explain the distributions of Na^+, K^+, Cl^-, and HCO_3^- normally observed. Thus the potential is very easily modified by changes in CO_2 tension of the blood,[20] whereas the steady-state concentrations of K^+, Mg^{2+}, etc., in the cerebrospinal fluid are remarkably invariable.

2. Homeostasis of Chemical Composition

As just indicated, the concentrations of many of the constituents of cerebrospinal fluid exhibit a remarkable independence of the concentration in the plasma. This is remarkable because there is no doubt that these constituents can cross the blood–CSF barrier. Thus, when isotopic K^+, e.g., ^{42}K, is injected into the bloodstream, the isotope appears in the cerebrospinal fluid. If enough time is allowed, the steady-state distribution of the isotope between plasma and fluid is the same as the steady-state distribution of the inactive K^+. Thus, exchanges of the K^+ ion may take place, yet when the concentration of K^+ in the plasma is raised or lowered chronically, the concentration in the cerebrospinal fluid is barely affected.[21–24] An essentially similar independence of fluid concentration from that in plasma has been demonstrated for Mg^{2+}.[25–28]

The simplest explanation for this homeostasis is to assume a carrier-mediated transport of the ion into the ventricle, the transport depending on the attachment of the ion to a hypothetical carrier. On the blood side of the system the ion tends to attach to the carrier, and at the cerebrospinal fluid side it tends to be detached. If the spaces on the carrier are all occupied over a large range of plasma concentrations, i.e., if the carrier sites are said to be "saturated," then transport becomes independent of plasma concentration. Such a hypothesis seems to account for many of the facts regarding K^{+} [4,24,29–34] and Ca^{2+}.[35]

V. RELATION OF CEREBROSPINAL FLUID TO BRAIN AND CORD

The virtual independence of the cerebrospinal fluid from the blood, so far as its ionic composition is concerned, raises an interesting problem regarding the relation of the cerebrospinal fluid to the extracellular fluid of the brain and cord. If this extracellular fluid were similar to extracellular fluid of other parts of the body, such as skeletal muscle, its composition would be markedly different from that of cerebrospinal fluid; its concentration of K^+ would be about 30 % higher, its concentration of Cl^- some 10 % lower, etc. There is no doubt that exchanges between the cerebrospinal fluid and the extracellular fluid of brain are relatively unrestricted. Thus, if the concentration of K^+ in the cerebrospinal fluid is raised there is an immediate passage of K into the

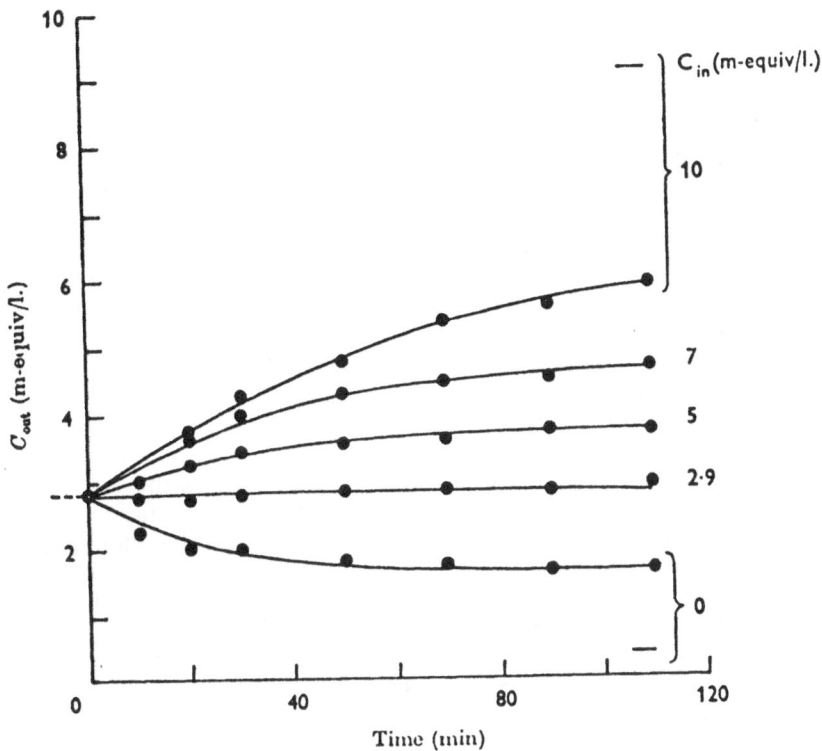

Fig. 2. The concentration of K^+ in the outflowing perfusion fluid C_{out} at various times after the start of perfusion of fluids of different K^+ content from lateral ventricle to cisterna magna of the rabbit. The dashed line on the ordinate represents the normal concentration of K^+ in rabbit CSF. The numbers at the right-hand side of the figure indicate the values of C_{In}, the concentration of K^+ in the inflowing fluid. The short horizontal marks bracketed to the curves marked 10 and 0 are the values of C_{out} that would have been observed, with these values of C_{In}, had there been no losses of K^+ to the tissue. From Bradbury and Davson.[34]

brain; if the concentration is lowered there is an immediate passage out of the brain into the fluid. This point is well demonstrated by Fig. 2, which describes the effects of perfusion of the ventricles of the rabbit with artificial cerebrospinal fluids containing different concentrations of K^+. The concentration of K^+ in the emerging perfusate, C_{out}, is plotted as a function of time from the beginning of the perfusion. The figures to the right of the curves indicate the inflowing concentration of K^+. If the concentration of the inflowing fluid is 2.9 meq/liter, the concentration in the outflowing fluid remains the same, i.e., there has been neither gain nor loss of K^+. This concentration represents, in fact, the concentration in the normal rabbit's cerebrospinal fluid. If the inflowing concentration is higher, e.g., 7 meq/liter, then the concentration in the outflowing fluid builds up with time to a steady-state value of about 4 meq/liter, which is considerably less than that of the inflowing fluid and indicates a continuous loss to the brain. If the concentration in the inflowing fluid is zero, then the concentration in the outflowing fluid reaches a steady state considerably above zero, namely, about 1.8 meq/liter, and this represents a continuous outflux from the brain into the K^+-free perfusion fluid. (Initially the outflowing concentration is higher. This is because the fluid first emerging is mainly the original cerebrospinal fluid. As this is replaced by perfusion fluid the concentration falls, but not to zero because of the influx from the brain. We may note that the constant secretion of newly formed cerebrospinal fluid may be expected to raise the concentration in the K^+-free perfusion fluid, independently of any gains from the brain tissue. This effect is relatively small and is indicated by the short horizontal line at the base of the bracket. In a similar way, the secretion of fluid will tend to lower the concentration in the high-K^+ perfusion fluids independently of losses to the nervous tissue. This effect is small, as indicated by the short horizontal line included in the top bracket, which shows the reduction in concentration due to this effect when the highest K^+ concentration, namely, 10 meq/liter, was employed.) The curves of Fig. 2 indicate that the concentration of K^+ in the extracellular fluid of the brain is the same as that in the cerebrospinal fluid, namely, some 2.9 meq/liter. A similar type of experiment with magnesium led to the same conclusion,[27] and the respiratory studies of Fencl, Miller, and Pappenheimer[36] likewise demanded that the acid–base characteristics of the cerebrospinal fluid reflected those of the extracellular fluid of brain.

A. Regional Variations in Composition

If cerebrospinal fluid is withdrawn from different parts of the cerebro-spinal system, e.g., lumbar subarachnoid space, the cisterna magna, and the lateral ventricle, we may analyze the fluids so obtained and determine whether the composition is uniform throughout the system or whether changes take place during the passage from lateral ventricles through cisterna magna and the other subarachnoid spaces. If the composition of the freshly secreted fluid were identical with that of the extracellular fluid of the brain and cord, we should not expect any significant differences in composition as we passed from lateral ventricle to lumbar subarachnoid space. If, on the other hand, there were differences, then these differences should be more pronounced when the freshly formed fluid in the ventricles was compared with the subarachnoid fluid taken either from the lumbar sac or from the surface of the cerebral cortex, since fluid in these two last regions is definitely "older" than that in the ventricles or cisterna magna. When ventricular fluid was compared with the cortical subarachnoid fluid, Bito and Davson[37] found a general uniformity in concentrations of many solutes, but K^+ and urea were marked exceptions. The concentration of urea in the mixed fluids is normally some 80% of that in plasma, as we have already seen (Table I).

When cisternal and cortical subarachnoid fluids were compared in dogs and cats, it was found that the older (cortical subarachnoid) fluid had the higher concentration. This suggested that the freshly secreted fluid had originally a relatively low concentration of urea while that in the extracellular fluid was much higher and perhaps close to that in a plasma filtrate. As a result of passive exchanges between cerebrospinal fluid and brain, the concentrations tended to equalize. When K^+ was studied, a reverse situation was encountered; here ventricular, cisternal, and cortical subarachnoid fluids were studied, and the following figures were obtained in the dog:

Ventricle	Cisterna magna	Cortical subarachnoid
2.93 ± 0.08	2.98 ± 0.06	2.65 ± 0.10

comparing with a plasma concentration of 4.57 meq/liter. It appears from these results that all fluids are greatly different from plasma but that there is a tendency for the concentrations to move away from the plasma concentration with passage of time as the fluid passes from ventricle to cisterna magna to cortical subarachnoid space. This suggests that the extracellular fluid of brain has a rather smaller concentration of K^+ than that in the newly formed cerebrospinal fluid.

B. Freshly Secreted Cerebrospinal Fluid

By exposing the choroid plexus of a lateral ventricle and covering it with oil, the formation of fluid may be observed under the microscope and collected for analysis. According to the studies from Ames' laboratory[38–40] the fluid so collected does differ materially in composition from that collected in bulk from the cisterna magna. As Table III shows, there is a tendency for the composition of the cerebrospinal fluid to move away from the composition of a filtrate from plasma, so far as Cl^-, K^+, and Ca^{2+} are concerned, although Mg^{2+} behaves in the opposite fashion. These results, too, indicate that the extracellular fluid of brain is not identical with cerebrospinal fluid but that it is nevertheless very similar, and certainly both fluids are greatly different from a simple plasma dialysate.

C. Secretion of Extracellular Fluid

The conclusion to be drawn from these studies on regional similarities and variations in composition is that extracellular fluid is sufficiently similar to cerebrospinal fluid to justify the assumption that it is formed as a secretion, presumably by the capillary endothelial cells or by those astrocytes that make an intimate contact with the central nervous capillaries. This means that the fluid balance of the brain and cord is maintained in a fundamentally different manner from that in which it is maintained in the subcutaneous tissue of a limb. For example, in the latter situation the tendency for fluid to accumulate or not will depend on the mean capillary pressure and the difference of colloid–osmotic pressure between extracellular fluid and plasma, in accordance

TABLE III

Concentration of Ions (meq/kg H$_2$O) in Plasma, Plasma-Ultrafiltrate, Choroid-Plexus Fluid, and Cisterna Magna Fluid of the Cat[a]

	Cl	Na	K	Ca	Mg
Plasma	132	163	4.4[b]	2.62	1.35
Plasma ultrafiltrate	136	151	3.3	1.83	0.95
Choroid plexus fluid	138	158	3.28	1.67	1.47
Cisterna magna fluid	144	158	2.69	1.50	1.33

[a] From Ames, Sakanone, and Endo.[39]
[b] Value for K$^+$ in plasma considered too high because of white cells, etc., being present.

with the Starling hypothesis. Excess of fluid, together with plasma proteins that have escaped into the fluid, will be carried away by the lymphatic system. In the brain, on the other hand, the formation of an extracellular fluid will depend on active transport processes across a relatively impermeable system represented by the blood–brain barrier. By analogy with other secretory systems, the formation of fluid will be governed by metabolic activities of secreting cells and will be virtually independent of capillary pressure. Because of the relative impermeability of the blood–brain barrier, i.e., of the secreting system, escape of proteins into the extracellular spaces will be much more restricted than in a limb. In fact, under normal conditions it may never occur at all, so that a highly efficient lymphatic system, in close association with the vascular system, becomes unnecessary. Thus, in the central nervous system, significant fluctuations in the fluid content of the tissue may be ruled out by virtue of the much stricter control over the formation of the extracellular fluid that the secretory process provides, by comparison with the state of affairs in a limb where large fluctuations may occur normally as a result of altered capillary permeability, venous stasis, etc.

Finally, the provision of a secreted extracellular fluid, produced by active transport processes that are presumably similar to those that lead to the formation of the cerebrospinal fluid, permits the maintenance of an ionic environment for the central neurons that is not only different from that of plasma but also is far more invariant. Thus fluctuations in concentration of K$^+$ in plasma of more than 100 % can occur physiologically in many experimental animals, and pathologically much larger fluctuations may occur in man, yet the homeostatic mechanisms described above ensure that the cerebrospinal fluid is almost completely immune from these fluctuations, and hence we may assume that the extracellular fluid is likewise immune. The importance of this immunity from change is shown by injecting small quantities of K^+ into the ventricles of the unanesthetized animal; these result in respiratory and cardiovascular excitement[41] and may induce electro-encephalically recorded changes reminiscent of epilepsy.[42]

D. Lymphatic Role of the Cerebrospinal Fluid

We have indicated that the central nervous tissue has no lymphatic system. This raises the question as to how proteins that escape from the capillaries, or appear as breakdown products of cells, are able to return to the vascular circulation or, what is more important, how they are cleared from the nervous tissue. It is true that the highly selective blood–brain barrier reduces escape of colloidal material from blood to a minimum. Nevertheless, some escape may be expected both normally and pathologically. It is here that the lymphatic role of the cerebrospinal fluid comes into play. It is an interesting fact that in man the shape of the ventricles and subarachnoid spaces is such that no part of the central nervous parenchyma is more than about 2 cm distant from the cerebrospinal fluid.[43] Diffusion of solutes through the parenchyma seems to be no more restricted than diffusion through an aqueous medium[44] in spite of the tortuosity of the intracellular paths available. The ventricular linings apparently exert little restraint on the migration of solutes, including large molecular-weight substances like inulin, from brain into the cerebrospinal fluid. It is likely, but not absolutely proved, that the pial lining likewise offers little restraint, so that both crystalloidal and colloidal material, appearing in the extracellular fluid, should be able to diffuse into the cerebrospinal system.

By virtue of its circulation—i.e., by its continual reformation and drainage back into the blood—and because of the complete absence of restraint on even large particles as they pass in the cerebrospinal fluid through the arachnoïd villi into the dural sinuses, the cerebrospinal fluid acts as a permanent "sink" or "drain" for the central nervous parenchyma. We here see the virtue of a circulating, as opposed to a stagnant, cerebrospinal fluid, a virtue that emphasizes the more than mechanical role the fluid plays in the economy of the brain.

1. Sink Action

This quasi lymphatic role of the cerebrospinal fluid has been described as a sink action[43,45,46] and has been adduced to explain some of the features of the blood–brain barrier. Thus the volume of distribution of a large water-soluble molecule like sucrose in the brain is of the order of 10–12%,[44] and this may correspond with the extracellular space of brain. This volume of distribution may only be achieved, however, if the sucrose is presented by way of the cerebrospinal fluid or if the excised tissue is employed. If a steady level of sucrose is maintained in the blood for long periods of time, the volume of distribution never rises to this value but achieves a steady level in the region of 2%.[44,47] The reason for this is the sink action of the cerebrospinal fluid. Because of the blood–CSF barrier, and because of the continuous formation and drainage of the fluid, the concentration in the cerebrospinal fluid never comes close to that in the blood plasma. Instead, a steady state is achieved after a few hours in which the concentration is only some 2 % of the plasma level. Thus, sucrose passes out of the capillaries of the brain tissue and, although this happens slowly because of the blood–brain barrier, in the absence of other factors the concentration would eventually build up to that in the plasma and the volume of distribution would be equivalent to the extracellular space, namely, some 10–12 %. In fact, however, the continuous diffusion from this extracellular space into the adjacent cerebrospinal fluid, with its very low concentration, ensures that the concentration in the extra-cellular fluid does not build up to that in the plasma. A steady state is achieved with a much lower concentration and thus an apparent volume of distribution much smaller than the true one that would only be reached in the absence of a circulating cerebrospinal fluid. Figure 3 illustrates a theoretical picture of the situation presented to explain, on this basis, the low apparent sulphate space of brain,[45] while Fig. 4 shows the results of an experimental investigation. It was pointed out by Bito, Bradbury, and Davson[48] that the true volume of distribution of solutes penetrating slowly across the blood-brain barrier would be given neither by presenting the solute through the blood—because of the sink action of the cerebrospinal fluid—nor yet by pre-senting the solute through the cerebrospinal fluid. In this latter situation a much closer approximation would be achieved, because of the absence of significant restraint to diffusion from cerebrospinal fluid into the adjacent tissues, but the steady state achieved would not represent the true volume of distribution because, under these conditions, the concentration of the solute in the blood would be virtually zero and consequently the blood would be removing the solute from the extracellular space as it diffused in from the cerebrospinal fluid. This removal would be slow, because of the brain-blood barrier, but it would be significant, and the true volume of distribution would only be given when this loss was prevented, i.e., by maintaining the same

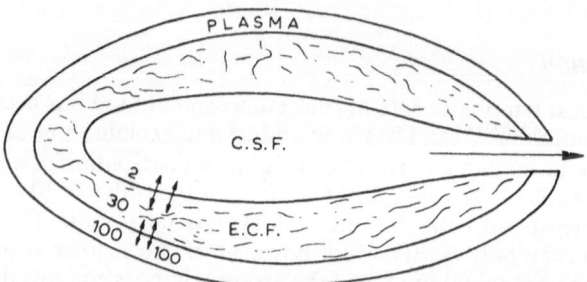

Fig. 3. Illustrating the manner in which the cerebrospinal fluid may impose a low concentration on an extracellular tag. From Davson and Bradbury.[45]

Fig. 4. The apparent brain sucrose space after 1-, 2-, and 3-hr perfusion. Curve C represents sucrose perfusion of brain via blood with nonradioactive ventriculocisternal perfusion running. The encircled λ is the 3-hr value using blood sucrose perfusion but with no ventricular perfusion. Curve B is the apparent sucrose space when a sucrose level is maintained only in the ventriculocisternal perfusion. Curve A represents both blood and ventriculocisternal perfusion. From Oldendorf and Davson.[43]

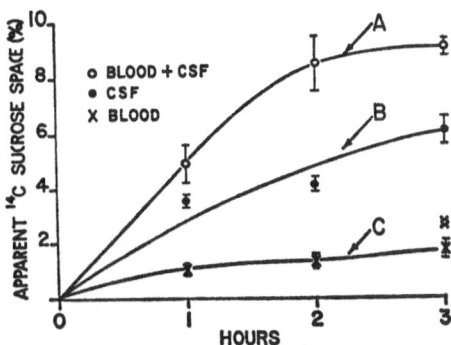

concentration of the solute in the plasma as in the cerebrospinal fluid. Experiments in which this experimental arrangement were employed demonstrated the correctness of the hypothesis with respect to ^{131}I,[48,49] and Fig. 4 illustrates the situation with respect to sucrose. Here Curve C shows the penetration of sucrose into the brain when the ventricular system is perfused with a mock cerebrospinal fluid and a steady level of sucrose is maintained in the blood plasma. The volume of distribution, or apparent sucrose space, is small because of the sink action of the cerebrospinal fluid perfusing the ventricular system. Curve B shows the penetration when the sucrose was in the mock cerebrospinal fluid, the concentration in the blood being effectively zero. The penetration is greater to achieve a higher apparent space. When sucrose was presented by both routes the Curve A was obtained, giving the true volume of distribution of approximately 9%, and this is, presumably, a measure of the extracellular space of the rabbit brain.

2. *Active Transport Out of the Cerebrospinal Fluid*

When steady levels of certain substances are maintained in the blood plasma, the penetration into the brain and cerebrospinal fluid is remarkably low. This is true of the anions I^- and CNS^{-}[50,51] and also of certain organic anions such as para-aminohippurate.[44] The remarkable feature of this barrier to penetration is the relative ease with which the same ions escape from the ventricles and subarachnoid space when injected there. The escape is presumably into the adjacent tissue and thence into the blood.[52] The explanation for these two phenomena—namely, slow penetration into the central nervous system and relatively rapid elimination—was provided by Pappenheimer, Heisey, and Jordon,[53] when they showed that the choroid plexuses were capable of actively transporting Diodrast and phenolsulphonephthaleine out of the ventricles into the blood. Subsequent studies by a number of authors have extended the number of substances that exhibit the phenomenon and the species in which it occurs[54-61]; not only can the phenomenon be demonstrated in the intact animal but also in the isolated choroid plexus[55-57]; furthermore, the process is not confined to anions but is also manifest in respect to organic amines.[58-61] From the point of view of this discussion, the findings are of particular interest. They indicate how the efficiency of the sink action of the cerebrospinal fluid may be increased by an active process that ensures that the concentration of certain substances is maintained at negligibly low levels in the cerebrospinal fluid, thus permitting the highly efficient removal of these substances from the brain, should they enter from the blood or appear as a result of activity within the central nervous system. Thus thiocyanate is a central nervous poison but will only exhibit toxic effects on intravenous administration when the level in the blood has been maintained at such a high level as to suppress the active transport process in the choroid plexuses. It seems likely that the active process that removes amines from the cerebrospinal fluid has some physiological significance. Such amines are presumably transmitters at synapses, and without an active process it is possible that too high a concentration might build up in the cerebrospinal fluid. The importance of the sink action of the cerebrospinal fluid in respect to such ions as I^- and CNS^- is manifest when the active process in the choroid plexuses is poisoned. In this event the concentration builds up in both cerebrospinal fluid and brain when a steady level is maintained in the blood.[48,62] As to whether the nervous tissue itself likewise has an active mechanism for keeping the concentration of these ions below levels at which they are toxic, or physiologically a nuisance, cannot be stated with any confidence. Attempts at proving the point, one way or the other, have not been completely conclusive but have pointed, on balance, away from this conclusion.[48]

3. *Enzymes*

Enzymes are proteins of the globulin type, so that their concentrations in the cerebrospinal fluid are low. As with some of the plasma globulins, however, the enzymes may be derived from the nervous tissue and, under certain pathological conditions, if not normally, may be present in higher proportions in the cerebrospinal fluid than in the plasma. The clinician is interested in those enzymes that may give a clue to abnormal activity in the central nervous system, e.g., necrosis following infarction, demyelinating disease, or the rapid multiplication of cells in a tumor. Of the enzymes that have been studied, glutamic-oxalacetic transaminase (GOT) and lactic dehydrogenase (LDH) have proved to be of real diagnostic value, since in necrotic lesions following infarction their concentration in the cerebrospinal fluid rises.[15,63-67] From the point of view of this discussion, these clinical findings are of interest since they once again reveal the sink action of the cerebrospinal fluid, manifest this time in respect to products of the nervous tissue itself.

VI. BLOOD–BRAIN BARRIER

This account of the cerebrospinal fluid is designed to clarify, where possible, the significance of the fluid in relation to the physiology and biochemistry of the central nervous system as a whole, thereby providing a theoretical basis for the interpretation of the large and growing amount of information that we have regarding exchanges of metabolites within the system. This account has emphasized what the author considers to be a prime function of the fluid, namely, its contribution to the maintenance of a suitable environment for the central neurons in regard to its chemical composition and to the relative invariance of this composition in the face of alterations in the blood concentrations of various constituents and in the face of metabolic activity within the central nervous system itself. This function is achieved by virtue of the secretory activity of the choroidal epithelial cells that produces a fluid of relatively invariant composition continuously and an activity that removes certain substances, such as thiocyanate and organic amines, that may enter the fluid adventitiously. This secretory activity, however, would most probably be quite inadequate if it were not assisted by powerful restraints on the passage of many solutes from the blood into the nervous tissue, i.e., by the presence of what has been termed the blood–brain barrier. If exchanges of solutes between plasma and the extracellular space of brain were as free and unrestricted as exchanges between plasma and the extracellular space of muscle, then the slow formation and drainage of cerebrospinal fluid (its half-life is of the order of 3 hr) would be quite inadequate to fulfill this function of maintaining a special environment of the central neurons. We must emphasize once again, therefore, the fundamental importance of the blood–brain barrier as a restraint on the free exchange of solutes between blood and nervous tissue. Since there has been a tendency in recent years to discount the importance of the barrier, in fact to deny its existence,[68–70] it is important that the evidence on which the concept is based should be kept clearly in mind.

A. Goldmann's Experiments

Essentially, then, the evidence derives from Goldmann's[71,72] two types of experiment. In his first he showed that trypan blue, when injected into the bloodstream, passed into all tissues of the body except the brain and spinal cord; in his second experiment he showed that trypan blue, injected into the cerebrospinal fluid, passed into the nervous tissue. Thus the tissue could take up the dye, but not when this was presented through the blood. Subsequent studies of a more quantitative nature, too numerous to be referred to individually here (see, for example, Bakay[73] and Davson[74] for reviews), have amply substantiated the concept by showing that, whereas many solutes such as sulphate, sucrose, inulin, and iodide pass slowly into the brain from the blood and achieve a low volume of distribution at a steady state, the same substances, when presented by way of the cerebrospinal fluid or by placing excised tissue in a Ringer's solution containing the solutes, will pass rapidly into the tissue and achieve steady-state volumes of distribution that are high and, in many instances, may be equated to the extracellular space of the tissue.

The arguments against the existence of a blood–brain barrier have been naive and have been based on a failure to appreciate the fundamental experimental facts on which the concept was based. Thus, because in electron microscope preparations the brain appeared to have a low interstitial space,[68,75,76] it was argued that failure of sucrose, say, to penetrate to any extent into the brain, was due to the small volume of the extracellular space rather than to any barrier to penetration from the blood. The fallacy in this argument is immediately evident and has been repeatedly emphasized.[16,44–46,77] Thus, presentation of sucrose by way of the cerebrospinal fluid reveals quite a large volume of distribution and consequently quite a large extracellular space, so the fact that the volume of distribution is so much smaller when the presentation is by way of the blood means that there is a barrier to passage from the blood—the blood–brain barrier—and this barrier, in conjunction with the circulation of the cerebrospinal fluid, serves to keep the apparent volume of distribution low and thus to indicate an erroneous value of the extracellular space. Thus, even if the electron microscopist's picture of the structure of the brain tissue were correct—and there is reason to believe from Van Harreveld's studies[78,79] that it is largely artefactual—this would in no way invalidate the concept of the blood–brain barrier.

B. Sugars and Amino Acids

The concentration of glucose in the cerebrospinal fluid is considerably lower than in plasma (Table I). This usually has been attributed to consumption within the central nervous system. Thus, if we assumed that, when secreted, the cerebrospinal fluid had the same concentration as in the plasma, the fluid during its slow passage through the ventricles and subarachnoid spaces might lose glucose by consumption of the cells lining these cavities and by diffusion into the extracellular fluid, if the concentration in this fluid were less than in plasma. Analysis of ventricular, cisternal, and lumbar fluids shows a successively diminishing concentration.[80] This supports the concept of losses to the brain tissue, but by no means proves that the freshly secreted fluid has the same concentration as in plasma. Glucose, being highly lipid insoluble, would not be expected to cross the blood–CSF barrier or blood–brain barrier easily. In fact, however, passage is fairly rapid[51,81–84] and exhibits the phenomenon of facilitated transfer, in the sense that the transport has some of the characteristics of an enzymatically controlled process, as though it depended on attachment to certain highly specific carrier sites.[81,82,85,86] We may regard this as an adaptation to permit the rapid supply of glucose to nervous tissue that must be necessary to maintain its high metabolic requirements.

The behavior of amino acids in the economy of the central nervous system is a highly complex subject that will be the theme of later chapters. Here we need only emphasize that the concentrations in the cerebrospinal fluid of most amino acids are lower than in plasma withdrawn from the animal at the same time (Table IV). An exception is glutamic acid, and this is undoubtedly related to its importance physiologically and its synthesis, along with some other amino acids such as GABA, by the nervous tissue from glucose. Exchanges of amino acids across the blood–CSF and blood–brain barriers certainly take place, the rapidity of such exchanges varying greatly from one amino acid to another. As with sugars, the passage across the barriers exhibits some of the features of a carrier-mediated transport process.[88–93]

C. Cerebral Edema

We have already indicated that the fluid economy of the brain is governed by fundamentally different principles from those exhibited in subcutaneous tissue. In this latter instance the volume of extracellular fluid is determined by capillary pressure, colloid osmotic pressures of plasma and extracellular fluid, and the efficacy of the lymphatic drainage mechanism. In the central nervous system there is no lymphatic system, and its role as a scavenger that clears the tissue of unwanted products is taken over by the

TABLE IV

Concentrations of Amino Acids (μmole/g H_2O) in Plasma and Cerebrospinal Fluid of Dogs[a]

Amino acid	Plasma	Cerebrospinal fluid
Alanine	0.29	0.13
Arginine	0.11	0.04
Aspartic acid	0.02	n.m.[b]
Cystine	0.18	0.02
Glutamic acid	0.13	0.16
Glutamine	n.m.	0.03
Glycine	0.29	0.23
Histidine	0.07	0.09
Isoleucine	0.10	n.m.
Leucine	0.12	0.06
Lysine	0.23	0.18
Methionine	0.03	n.m.
Phenylalanine	0.04	0.07
Serine	0.27	0.05
Threonine	0.13	0.02
Tyrosine	0.05	0.05
Valine	0.04	0.01
Total	2.11	0.80

[a] From Bito et al.[87]
[b] Denotes "not measured."

cerebrospinal fluid. As to what extent the cerebrospinal system acts to carry away extracellular fluid is a matter on which there is very little evidence. We have argued that the composition of this extracellular fluid must be controlled by secretory activity, presumably that of the astrocytes that make such intimate contact with the blood capillaries, but this does not mean that there is a continuous secretion of the fluid similar to the continuous secretion of cerebrospinal fluid. Bering and Sato's[1] studies on dogs, in which the normal drainage channels were obstructed, indicated that some 40 % of the cerebrospinal fluid formed under these conditions was derived from nonventricular sources, i.e. it might be formed by the nervous tissue. A recent study of Pollay and Curl,[2] in which only the aqueduct of Sylvius was perfused, indicated the formation of fluid that could not have been derived from the choroid plexuses and presumably came from the adjacent nervous tissue. Whatever the final picture that emerges in the light of later work, it is quite clear that the water economy of the nervous tissue is strictly controlled and that the cerebrospinal fluid may well act as a recipient for overflow and as a source for extracellular fluid when this, through some failure in the economy, becomes deficient. Edema in a limb results from simple physical causes, predictable on the basis of the Starling hypothesis. In the brain, on the other hand, the same causes need not produce the same effect, and in fact they probably do not. Edema, nevertheless, does occur, and the elucidation of the factors determining it is of fundamental clinical interest. The phenomenon has been the subject of a symposium[94] and a book[95] and can only be briefly touched on here. A fundamental feature of the edema that manifests itself in areas adjacent to damaged tissue is the accumulation of a protein-containing fluid, just as in other parts of the body. This is presumably due to a breakdown of the blood–brain barrier that allows the escape of protein. However, it is by no means certain that the edema in the area adjacent to an injury is due to breakdown of the barrier there. The protein-containing fluid might well have migrated from the damaged area,[96] whereas the edematous condition, revealed by an increased water content of the tissue adjacent to the injury, could be shown to precede the migration of this protein-containing fluid. Thus, the early stages of edema, at any rate, reveal an increased water content that might well be the result of expansion of the extracellular space, which is not due to simple Starling factors of increased capillary permeability to proteins and inadequate lymphatic drainage. There is difference of opinion as to whether this extra fluid content of the tissue is extracellular or not. Those who have examined the edematous brain in electron microscope preparations have been struck by the tremendous engorgement of the astrocytes, by contrast with the normal or shrunken size of the neurons, and they have suggested that the edema is mainly intracellular.[75,97] It appears as though the astrocytes, which may be concerned in the normal secretion of extracellular fluid, were deranged in some way so that they tended to take up fluid, presumably from the blood capillaries, but failed to transport it into the extracellular spaces and thus became engorged.

Electron microscope pictures of fixed tissue can give such illusory impressions of events actually taking place that it would be inadvisable to base a theory of pathological processes entirely on them. At any rate studies on the distribution of extracellular markers suggest that there is a considerable expansion of extracellular space in edema caused by cold injury.[98]

Thus it appears that there has been a disturbance in the mechanism that controls the relations between blood and extracellular fluid, i.e., a disturbance in the blood–brain barrier. This, indeed, may be a simple generalized increase in passive permeability of the barrier to all solutes, so that the active transport processes, that normally maintain an accurate control over the net amounts of solutes and fluid that enter the fluid in unit time, are unable to cope with the new conditions. The most striking evidence on this point is the failure of hyperosmolal solutions, when injected intravenously, to reverse the edematous condition (see, for example, Pappius and Dayes[99]). Thus, because of the blood–brain barrier, intravenous injection of hyperosmolal urea causes a shrinkage of the brain[100–103] that is due to the more rapid passage of water from brain extracellular space into the blood capillaries than passage of the urea into the extracellular space. The shrinkage thus depends on the integrity of the blood–brain barrier. The failure of hyperosmolal urea to cause edematous tissue to shrink is unquestionable evidence for a dysfunction of the barrier under these conditions.

VII. GENERAL CONCLUSIONS

The significance of the cerebrospinal fluid is now emerging. Originally considered by the physiologist as little more than a fluid cushion that supported the brain and protected it from stress, and by the pathologist as a useful fluid whose closeness to the central nervous parenchyma allowed it to reflect some of the diseases that afflicted it, the cerebrospinal fluid has now been revealed as a specific secretion that serves to maintain the stable and peculiar environment that is required for the adequate functioning of the central neurons. Of itself it would be unable either to establish or to maintain this environment, but in conjunction with a similar fluid, perhaps formed by similar active processes, namely, the extracellular fluid of the brain and cord, such an environment is established and maintained. It thus is impossible to discuss the cerebrospinal fluid without at the same time discussing the extracellular fluid of the central nervous system, and this, of course, necessitates an appreciation of the various factors governing the exchanges between blood and both cerebrospinal fluid and extracellular fluid.

VIII. REFERENCES

1. E. A. Bering and O. Sato, Hydrocephalus: changes in formation and absorption of cerebrospinal fluid within the cerebral ventricles, *J. Neurosurg.* **20**:1050–1063 (1963).

2. M. Pollay and F. Curl, Secretion of cerebrospinal fluid by the ventricular ependyma of the rabbit, *Am. J. Physiol.* **213**:1031–1038 (1967).

3. K. Welch and V. Friedman, The cerebrospinal fluid valves, *Brain* **83**:454–469 (1960).

4. H. Cserr, Potassium exchange between cerebrospinal fluid, plasma, and brain, *Am. J. Physiol.* **209**:1219–1226 (1965).

5. M. Segal and H. Davson (in preparation).

6. M. G. Farquhar and G. E. Palade, Junctional complexes in various epithelia, *J. Cell Biol.* **17**:375–412 (1963).

7. D. P. Rall, W. W. Oppelt, and C. S. Patlak, Extracellular space of brain as determined by diffusion of inulin from the ventricular system, *Life Sci.* **2**:43–48 (1962).

8. M. W. Brightman, The distribution within the brain of ferritin injected into cerebrospinal fluid compartments—I, *J. Cell Biol.* **26**:99–123 (1965).

9. K. Fleischhauer, Regional differences in the structure of the ependyma and subependymal layers of the cerebral ventricles of the cat, *in Regional Neurochemistry* (S. S. Kety and J. Elkes, eds.), pp. 279 to 283, Pergamon Press, London (1961).

10. E. Nelson, K. Blinzinger, and H. Hager, Electron microscopic observations on subarachnoid and perivascular spaces of the Syrian hamster brain, *Neurology* **11**:285–295 (1961).

11. J. D. Waggener and J. Beggs, The membranous coverings of neural tissues: an electron microscopy study, *J. Neuropathol.* **26**:412–426 (1967).

12. L. A. Rodriguez, Experiments on the histologic locus of the hemato-encephalic barrier, *J. Comp. Neurol.* **102**:27–45 (1955).

13. M. Shakib and J. C. Cunha-Vaz, Studies on the permeability of the blood–retinal-barrier. IV. Junctional complexes of the retinal vessels and their role in the permeability of the blood–retinal-barrier, *Exptl. Eye Res.* **5**:229–234 (1966).

13a. T. S. Reese and M. J. Karnovsky, Fine structural localization of a blood–brain barrier to exogenous peroxidase, *J. Cell Biol.* **34**:207–217 (1967).

13b. T. S. Bodenheimer and M. W. Brightman, A blood–brain barrier to peroxidase in capillaries surrounded by perivascular spaces, *Am. J. Anat.* **122**: 249–267 (1968).

14. N. C. Hill, B. F. McKenzie, W. F. McGuckin, N. P. Goldstein, and H. J. Svien, Proteins, glycoproteins and lipoproteins in the serum and cerebrospinal fluid of healthy subjects, *Proc. Mayo Clinic* **33**:686–698 (1958).

15. C. M. Plum and T. Fog, Studies in multiple sclerosis, *Acta Psychiat. Neurol. Scand.* **34**, Suppl. 128 (1959).

16. H. Davson, *Physiology of Cerebrospinal Fluid*, Churchill, London (1967).

17. H. H. Lieschke, Uber Bestandspotentiale im Gebiete der Medulla oblongata, *Pfluegers Arch. Ges. Physiol.* **262**:517–531 (1956).

18. H. J. Mottschall and H. H. Loeschke, Messungen des transmeningealen Potentials der Katze bei Anderung des CO_2- Drucks und der H-Ionen-Konzentration im Blut, *Pflugers Arch. Ges. Physiol.* **277**:662–670 (1963).

19. D. Held, V. Fencl, and J. R. Pappenheimer, Electrical potential of cerebrospinal fluid, *J. Neurophysiol.* **27**:775–781 (1964).

20. H. H. Loeschke, Uber den Einfluss von CO_2 auf die Nestandspotentiale der Hirnhaute, *Pfluegers Arch. Ges. Physiol.* **262**:532–536 (1956).

21. J. Bekaert and G. Demeester, The influence of glucose and insulin upon the potassium concentration of serum and cerebrospinal fluid, *Arch. Intern. Physiol.* **59**:262–264 (1951).

22. J. Bekaert and G. Demeester, The influence of the infusion of potassium chloride on the cerebrospinal fluid concentration of potassium, *Arch. Intern. Physiol.* **59**:393–394 (1951).

23. E. S. Cooper, E. Lechner, and S. Bellet, Relations between serum and cerebrospinal fluid electrolytes under normal and abnormal conditions, *Am. J. Med.* **18**:613–621 (1955).

24. M. W. B. Bradbury and C. R. Kleeman, Stability of the potassium content of cerebrospinal fluid and brain, *Am. J. Physiol.* **213**:519–528 (1967).
25. H. Cohen, The magnesium content of the cerebrospinal and other body fluids, *Quart. J. Med.* **20**:173–186 (1927).
26. G. Hunter and H. V. Smith, Calcium and magnesium in human cerebrospinal fluid, *Nature* **186**:161–162 (1960).
27. M. W. B. Bradbury, Magnesium and calcium in cerebrospinal fluid and in the extracellular fluid of brain, *J. Physiol.* **179**:67–68 (1965).
28. F. K. Herbert, The total and diffusible calcium of serum and the calcium of cerebrospinal fluid in human cases of hypocalcaemia and hypercalcaemia, *Biochem. J.* **27**:1978–1991 (1933).
29. P. H. Leiderman and R. Katzman, Effect of adrenalectomy, desoxycorticosterone and cortisone on brain potassium exchange, *Am. J. Physiol.* **175**:271–275 (1953).
30. R. Katzman and P. H. Leiderman, Brain potassium exchange in normal adult and immature rats, *Am. J. Physiol.* **175**:263–270 (1953).
31. R. Katzman, L. Graziani, R. Kaplan, and A. Escriva, Exchange of cerebrospinal fluid potassium with blood and brain, *Arch. Neurol.* **13**:513–524 (1963).
32. F. R. Domer, Transport of ^{42}K from blood to cerebrospinal fluid in cats, *J. Physiol.* **158**:366–373 (1961).
33. F. R. Domer and M. Whitcomb, Studies of ^{42}K movement between the blood and the cerebrospinal fluid of cats, *J. Pharmacol.* **145**:52–57 (1964).
34. M. W. B. Bradbury and H. Davson, The transport of potassium between blood, cerebrospinal fluid and brain, *J. Physiol.* **181**:151–174 (1965).
35. L. Graziani, A. Escriva, and R. Katzman, Exchange of calcium between blood, brain, and cerebrospinal fluid, *Am. J. Physiol.* **208**:1058–1064 (1965).
36. V. Fencl, T. B. Miller, and J. R. Pappenheimer, Studies on the respiratory response to disturbances of acid–base balance, with deductions concerning the ionic composition of cerebral interstitial fluid, *Am. J. Physiol.* **210**:459–472 (1966).
37. L. Z. Bito and H. Davson, Local variations in cerebrospinal fluid composition and its relationship to the composition of the extracellular fluid of the cortex, *Exptl. Neurol.* **14**:264–280 (1966).
38. J. De Rougemont, A. Ames, F. B. Nesbitt, and H. F. Hofmann, Fluid formed by choroid plexus, *J. Neurophysiol.* **23**:485–495 (1960).
39. A. Ames, M. Sakanoue, and S. Endo, Na, K, Ca, Mg and Cl concentrations in choroid plexus fluid and cisternal fluid compared with plasma ultrafiltrate, *J. Neurophysiol.* **27**:672–681 (1964).
40. A. Ames, K. Higashi, and F. B. Nesbett, Relation of potassium concentration in choroid plexus fluid to that in plasma, *J. Physiol.* **181**:506–515 (1965).
41. J. R. Pappenheimer, S. R. Heisey, E. F. Jordan, and J. De C. Downer, Perfusion of the cerebral ventricular system in unanesthetized goats, *Am. J. Physiol.* **203**:763–774 (1962).
42. E. C. Zuckermann and G. H. Glaser, Hippocampal epileptic activity induced by localized ventricular perfusion with high-potassium cerebrospinal fluid, *Exptl. Neurol.* **20**:87–110 (1968).
43. W. H. Olendorf and H. Davson, Brain extracellular space and the sink action of cerebrospinal fluid, *Arch. Neurol.* **17**:196–205 (1967).
44. H. Davson and E. Spaziani, The blood–brain barrier, *J. Physiol.* **149**:135–143 (1959).
45. H. Davson and M. Bradbury, The extracellular space of the brain, *Progr. Brain Res.* **15**:124–134 (1965).
46. H. Davson, The cerebrospinal fluid, *Ergeb. Physiol.* **52**:21–73 (1963).

47. D. J. Reed and D. M. Woodbury, Kinetics of movement of iodine, sucrose, inulin and radio-iodinated serum albumin in the central nervous system and cerebrospinal fluid of the rat, *J. Physiol.* **169**:816–850 (1963).
48. L. Z. Bito, M. W. B. Bradbury, and H. Davson, Factors affecting the distribution of iodide and bromide in the central nervous system, *J. Physiol.* **185**:323–354 (1966).
49. D. L. Woodward, D. J. Reed, and D. M. Woodbury, Extracellular space of rat cerebral cortex, *Am. J. Physiol.* **212**:367–370 (1967).
50. G. B. Wallace and B. B. Brodie, The distribution of iodide, thiocyanate, bromide and chloride in the central nervous system and spinal fluid, *J. Pharmacol.* **65**:220–226 (1939).
51. H. Davson, A comparative study of the aqueous humour and cerebrospinal fluid in the rabbit, *J. Physiol.* **129**:111–133 (1955).
52. H. Davson, The rates of disappearance of substances injected into the subarachnoid space of rabbits, *J. Physiol.* **128**:52–53 (1955).
53. J. R. Pappenheimer, S. R. Heisey, and E. F. Jordan, Active transport of Diodrast and phenolsulfonphthalein from cerebrospinal fluid to blood, *Am. J. Physiol.* **200**:1–10 (1961).
54. M. Pollay and H. Davson, The passage of certain substances out of the cerebrospinal fluid, *Brain* **86**:137–150 (1963).
55. K. Welch, Concentration of thiocyanate by the choroid plexus of the rabbit *in vitro, Proc. Soc. Exptl. Biol. N.Y.* **109**:953–954 (1962).
56. K. Welch, Active transport of iodide by choroid plexus of the rabbit *in vitro, Am. J. Physiol.* **202**:757–760 (1962).
57. B. Becker, Cerebrospinal fluid iodide, *Am. J. Physiol.* **201**:1149–1151 (1961).
58. L. S. Schanker, L. D. Prockop, J. Schou, and P. Sisodia, Rapid efflux of some quaternary ammonium compounds from cerebrospinal fluid, *Life Sci.* **10**:515–521 (1962).
59. Y. Tochino and L. S. Schanker, Active transport of quaternary ammonium compounds by the choroid plexus *in vitro, Am. J. Physiol.* **208**:666–673 (1965).
60. Y. Tochino and L. S. Schanker, Transport of serotonin and norepinephrine by the rabbit choroid plexus *in vitro, Biochem. Pharmacol.* **14**:1557–1566 (1965).
61. Y. Tochino and L. S. Schanker, Serum and tissue factors that inhibit amine transport by the choroid plexus *in vitro, Am. J. Physiol.* **210**:1229–1233 (1966).
62. M. Pollay, Cerebrospinal fluid transport and the thiocyanate space of the brain, *Am. J. Physiol.* **210**:275–279 (1966).
63. K. G. Wakim and G. A. Fleisher, The effect of experimental cerebral infarction on transaminase activity in serum, cerebrospinal fluid and infarcted tissue, *Proc. Mayo Clinic* **31**:391–399 (1956).
64. J. B. Green, H. A. Oldewurtel, D. S. O'Doherty, F. M. Forster, and L. P. Sanchez-Longo, Cerebrospinal fluid glutamic oxalacetic transaminase activity in neurologic disease, *Neurology* **7**:313–322 (1957).
65. J. Lieberman, O. Daiber, S. I. Dulkin, O. E. Lobstell, and M. R. Kaplan, Glutamic oxalacetic transminase in serum and cerebrospinal fluid of patients with cerebrovascular accidents, *New Engl. J. Med.* **257**:1201–1207 (1957).
66. R. Katzman, R. A. Fishman, and E. S. Goldensohn, Glutamic-oxalacetic transaminase activity in spiral fluid, *Neurology* **7**:833–855 (1957).
67. S. M. Aronson, A. Saifer, G. Perle, and B. W. Volk, Cerebrospinal fluid enzymes in central nervous system lipidoses, *Proc. Soc. Exptl. Biol. N.Y.* **97**:331–334 (1958).
68. E. A. Maynard, R. L. Schultz, and D. C. Pease, Electron microscopy of the vascular bed of the rat cerebral cortex, *Am. J. Anat.* **100**:409–433 (1957).

69. C. E. Lumsden, The cytology and cell physiology of the neuroglia and of the connective tissue in the brain with reference to the blood–brain barrier, *Excerpta Med. VIII* **8**:832–834 (1955).

70. R. Edstrom, An explanation of the blood–brain barrier phenomenon, *Acta Psychiat. Kbhandl.* **33**:403–416 (1958).

71. E. E. Goldmann, Die aussere und innere Sekretion des gesunden und kranken Organismus im Lichte der "vitalen Farbung," *Beitr. Klin. Chirurg.* **64**:192–265 (1909).

72. E. E. Goldmann, Vitalfarbung am Zentralnervensystem, *Abhandl. Preuss. Akad. Wiss., Phys: Math. Kl*(1), 1–60 (1913).

73. L. Bakay, *The Blood–Brain Barrier with Special Regard to the Use of Isotopes*, Charles C. Thomas, Springfield, Illinois (1956).

74. H. Davson, *Physiology of the Ocular and Cerebrospinal Fluids*, Churchill, London (1956).

75. H. M. Gerschenfeld, F. Wald, J. A. Zadunaisky, and E. D. P. De Robertis, Function of astroglia in the water-ion metabolism of the central nervous system. An electron microscope study, *Neurology* **9**:412–425 (1959).

76. E. Horstmann and H. Meves, Die Feinstruktur des molekularen Rindengraues und ihre physiologische Bedeutung, *Z. Zellforsch.* **49**:569–604 (1959).

77. H. Davson, C. R. Kleeman, and E. Levin, The blood–brain barrier, in *Drugs and Membranes* (C. A. M. Hogben and P. Lindgren, eds.), pp. 71–94, Pergamon Press, Oxford (1963).

78. A. Van Harreveld, J. Crowell, and S. K. Malhotra, A study of extracellular space in central nervous tissue by freeze-substitution, *J. Cell Biol.* **25**:117–137 (1965).

79. A. Van Harreveld and S. K. Malhotra, Extracellular space in the cerebral cortex of the mouse, *J. Anat.* **101**(2):197–207 (1967).

80. K. Chevassut, Glycolysis in cerebrospinal fluid and its clinical significance, *Quart. J. Med.* **21**:91–106 (1927).

81. C. Crone, *Om diffusionen af Nogle Organiske Nonelektrolyter fra Bold til Hjernevaev*, Ejnar Munksgaard, Kobenhavn (1961).

82. C. Crone, Facilitated transfer of glucose from blood into brain tissue, *J. Physiol.* **181**:103–113 (1965).

83. G. G. Myers and M. G. Netsky, Relation of blood and cerebrospinal fluid glucose, *Arch. Neurol.* **6**:18–26 (1962).

84. J. E. Sifontes, R. D. B. Williams, E. M. Lincoln, and H. Clemons, Observations on the effect of induced hyperglycaemia on the glucose content of the cerebrospinal fluid in patients with tuberculous meningitis, *Am. Rev. Tuberc.* **67**:732–754 (1953).

85. R. A. Fishman, Carrier transport of glucose between blood and cerebrospinal fluid, *Am. J. Physiol.* **206**:836–844 (1964).

86. E. Eidelberg, J. Fishman, and M. L. Hams, Penetration of sugars across the blood–brain barrier, *J. Physiol.* **191**:47–57 (1967).

87. L. Bito, H. Davson, E. Levin, M. Murray, and N. Snider, The concentrations of amino acids and other electrolytes in cerebrospinal fluid, *in vivo* dialysate of brain and blood plasma of the dog, *J. Neurochem.* **13**:1057–1067 (1966).

88. P. Wiechert, Uber den Einfluss von Aminosauren auf die Permeabilitat der Blut-Liquor-Schranke, *Acta Biol. Med. Ger.* **11**:68–76 (1963).

89. A. Lajtha, and J. Toth, Uptake and transport of amino acids by the brain, *J. Neurochem.* **8**:216–225 (1961).

90. A. Lajtha and J. Toth, The efflux of intracerebrally administered amino acids from the brain, *J. Neurochem.* **9**:199–212 (1962).

91. A. Lajtha and J. Toth, The brain barrier system. V. Stereospecificity of amino acid uptake, exchange and efflux, *J. Neurochem.* **10**:909–920 (1963).

92. E. Levin, G. J. Nogueira, and C. A. G. Argiz, Ventriculo-cisternal perfusion of amino acids in cat brain. I. Rates of disappearance from the perfusate, *J. Neurochem.* **13**:761–767 (1966).

93. G. J. Noguiera, C. A. G. Argiz, and E. Levin, Disappearance of different substances in contact with the external surface of the brain, *Scientia* **21**:734–735 (1965).

94. I. Klatzo and F. Seitelberger (eds.), *Brain Edema. Proceedings of the Symposium, September 11–13, 1965, Vienna*, Springer-Verlag, Vienna (1967).

95. L. Bakay and J. C. Lee, Ultrastructural changes in the edematous central nervous system. III. Edema in shark brain, *Arch. Neurol.* **14**:644–660 (1966).

96. L. Bakay and I. U. Haque, Morphological and chemical studies in cerebral edema. I. Cold induced edema, *J. Neuropathol.* **23**:393–418 (1964).

97. A. Hirano, H. M. Zimmerman, and S. Levine, Fine structure of cerebral fluid accumulation. IV, *Arch. Neurol.* **12**:189–196 (1965).

98. E. Streicher, P. J. Ferris, J. D. Prokop, and I. Klatzo, Brain volume and thiocyanate space in local cold injury, *Arch. Neurol.* **11**:444–448 (1964).

99. H. M. Pappius and L. A. Dayes, Hypertonic urea, *Arch. Neurol.* **13**:395–402 (1965).

100. M. W. B. Bradbury and R. V. Coxon, The penetration of urea into the central nervous system at high blood levels, *J. Physiol.* **163**:423–435 (1962).

101. C. R. Kleeman, H. Davson, and E. Levin, Urea transport in the central nervous system, *Am. J. Physiol.* **203**:739–747 (1962).

102. M. Javid and P. Settlage, Effect of urea on cerebrospinal fluid pressure in human subjects, *J. Am. Med. Assoc.* **160**:943–949 (1956).

103. D. J. Reed and D. M. Woodbury, Effect of hypertonic urea on cerebrospinal fluid pressure and brain volume, *J. Physiol.* **164**:252–264 (1962).

104. F. Fremont-Smith, Pathogenesis of the changes in the cerebrospinal fluid in meningitis, *Arch. Neurol. Psychiat.* **25**:206–208 (1932).

105. S. Salminen and K. Luomanmaki, Distribution of sodium and potassium in serum, cerebrospinal fluid, and serum ultrafiltrate in some diseases, *Scand. J. Clin. Lab. Invest.* **14**:425–429 (1962).

106. G. Hunter, H. V. Smith, and L. M. Taylor, On the bromide list of permeability of the barrier between blood and cerebrospinal fluid—an assessment, *Biochem. J.* **56**:588–597 (1954).

107. R. D. Bradley and S. J. G. Semple, A comparison of certain acid–base characteristics of arterial blood, jugular venous blood and cerebrospinal fluid in man, and the effect on them of some acute and chronic acid–base disturbances, *J. Physiol.* **160**:381–391 (1962).

108. F. Ragazzini, Variazioni dei livelli emato-liquorali del fosforo inorganico e del potassio in corso di meningite tubercolare, *Riv. Clin. Pediat.* **50**:381–388 (1952).

109. E. B. Hendry, The osmotic pressure and chemical composition of human body fluids, *Clin. Chem.* **8**:246–265 (1962).

110. J. W. Millen and D. H. M. Woollam, *The Anatomy of the Cerebrospinal Fluid*, Oxford University Press, London (1962).

Chapter 4

CHOROID PLEXUS*

T. Z. Csáky

Department of Pharmacology
University of Kentucky College of Medicine
Lexington, Kentucky

I. INTRODUCTION

A. Morphology

The choroid plexuses are specialized highly vascular anatomical structures which protrude into the cerebral ventricles. A choroid plexus is found in each lateral ventricle, as well as in the third ventricle and fourth ventricle. The surface of the choroid plexus consists of small villi each covered with a single layer of large cuboidal epithelial cells. Underneath the epithelial layer a central core is found consisting of loose connective tissues in which a central capillary is embedded (Fig. 1).

The epithelial covering of the choroid plexus consists of modified ependymal cells displaying characteristics similar to those of the epithelium of typical transporting organs such as the kidney and intestine. The ventricular surface of these cells is the "brush border," which under the electron-microscope appears as a system of infoldings of the cell membrane called "microvilli." Occasionally some epithelial cells have cilia, more in young and less in old animals, which may facilitate the constant motion of the CSF.[1]

The cytoplasm of the cell contains a large number of mitochondria. The basal membrane shows numerous infoldings. This is also the case with basal membranes of other epithelial cells which are involved in vigorous trans-porting functions, such as the kidney tubule, ciliary, and submaxillary gland (Fig. 2). For this reason it has been suggested that the infoldings have a functional significance in the transport of water.[2]

The choroid plexuses display a rich blood supply; accordingly, circulation in these is quite rapid: about 3 ml/min/g tissue.[3,4] The perfusion rate of the choroid plexus is thus about five times greater than that of the whole brain (0.54 ml/min/g)[5] and about twice that of the kidney.[6]

* The author is indebted to the U. S. Public Health Service for grants that enabled him to conduct the original research, part of which is reported in this chapter.

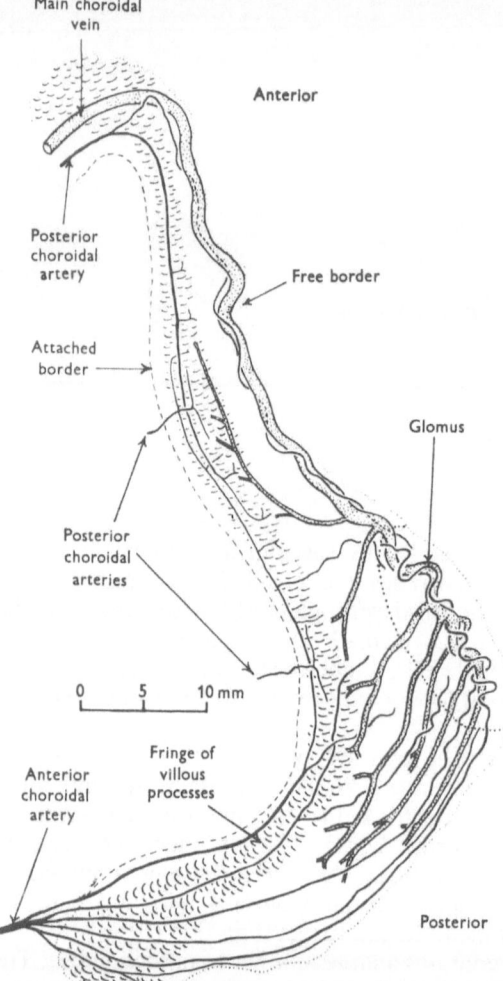

Fig. 1. Principal features of the human choroid plexus. Lateral ventricle. From Millen and Woollam.[18]

Because of its great vascularity, the volume of the choroid plexus significantly increases with every arterial filling and decreases with venous emptying. In this way a wave of pressure is produced which extends throughout the ventricle and contributes heavily to the mixing and circulation of the CSF.[7,8] This suggestion has gained further support by the experiments of Wilson and Bertan[9] who, by ligating the anterior choroidal artery in dogs, obtained a significant reduction of intraventricular pressure in artificial hydrocephalus.

Using various histological techniques, numerous granulations have been observed in the choroid epithelium. Many of these were thought to be

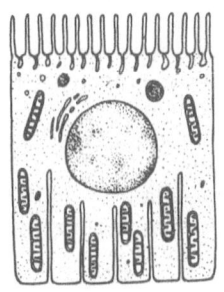 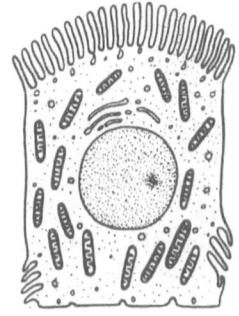

PROXIMAL CONVOLUTED TUBULE   ADULT CHOROID PLEXUS

Fig. 2. Diagram illustrating the comparative structural features
of an epithelial cell from the kidney and choroid plexus. Note the
brush border and the infoldings of the basal membrane. From
Tennyson and Pappas.[66]

involved in the secretory function of the choroid plexus. Recent studies,
however, have failed to reveal the existence of true secretory granules.[10]
Earlier claims to this effect may have been due to artifacts or to mistaking
such structures as unusual forms of mitochondria, Golgi apparatus, or
endoplasmic reticulum for secretory granules.

B. Metabolism and Enzyme Content

An outstanding feature of the choroid plexus is a vigorous aerobic
metabolism. Krebs and Rosenhagen[11] measured a Q_{O_2} of -19.7 as compared with the gray matter of the brain cortex: -9.0. Friedenwald et al.[12]
estimated a Q_{O_2} of 9–12 in the isolated choroid plexus of the pig.

Corresponding with the high rate of metabolism, various respiratory
enzymes are found in high concentration both by histochemical[13] and by
direct chemical analyses.[12,14,15] Acid phosphatase, "indophenol oxidase,"
cytochrome oxidases, succinodehydrogenase, and fumarase were found in
high concentration in the epithelial cells, but not in the stroma. Therefore it
is logical to assume that the epithelium and not the stroma is responsible for
the very high rate of aerobic metabolism.

Of those enzymes not primarily involved in aerobic metabolism, the
presence of two should be particularly mentioned, as these are likely to be
connected with active transport processes: Carbonic anhydrase, which was
found in unusually high concentration in the choroid plexus,[16] and sodium-
potassium-stimulated ATPase.[17]

C. Function

The morphological structure of the epithelium, the unusually high rates
of blood flow and metabolism, and the high concentrations of enzymes

usually connected with vigorous active transport processes indicate that the primary function of the choroid epithelium is that of transport. Indeed, it is clear that the plexuses are involved in the maintenance of homeostasis of the CSF and that in this function they transport both from the blood to the ventricles and from the ventricles into the blood.

Therefore in discussing the possible functions of the choroid plexus only the transport will be included in this chapter.

Recent monographs by Millen and Woollam[18] and Davson[19] could be consulted for more detailed information concerning the structure and function of choroid plexus.

II. EXPERIMENTAL TECHNIQUES

The transport function of the choroid plexus can be studied experimentally either by indirect or direct approaches.

In the indirect approach, the net transport of various substances into and out of an artificially perfused cerebral ventricle is examined. By assuming that the choroidal epithelium is the main site where substances are exchanged between blood and CSF, indirect conclusions can be drawn regarding the transport properties of this tissue.* An elegant quantitative method for perfusion of the ventricle of the unanesthetized goat was worked out by Pappenheimer et al.[20] The method enables one to assess both volume and concentration changes and to calculate in-and-out fluxes accurately. Yet it is clear that the results apply to the entire surface of the wall of the ventricle and not exclusively to the choroid plexus.

The direct approach can be employed either *in vitro* or *in vivo*.

A. *In Vitro* Technique

In vitro there are two useful preparations available, which are described below.

1. The freshly excised plexus can be incubated in an appropriate nutrient medium at optimum temperature with proper oxygenation in the presence of the solute, the transport of which is to be studied. After a certain time the solute concentration is determined, both in the tissue water and in the medium water. A tissue-to-medium concentration (T/M) of more than unity indicates accumulation. If accumulation is inhibited by low temperature, lack of oxygen, metabolic inhibitors, or by lack of sodium in the medium, it is assumed that the accumulation is due to active transport. Such a technique is useful as an orienting experiment. Its main disadvantage is that no indication is given as to the direction of the active transport. Moreover, substances may

---

* Great caution should be exercised, however, in making this assumption, e.g., it was shown that a great portion of the potassium outflux from the CSF was regulated not by the choroid plexus but by the uptake of the cation by brain cells across the ependyma.[71]

occasionally be transferred transcellularly by an active process without an appreciable intracellular accumulation.

2. In large animals one of the main choroidal arteries can be cannulated and a nutrient medium perfused through it. The effluent is collected from the choroidal vein. At the same time, the ventricular side of the plexus is bathed in an appropriate medium. Such a preparation allows the study of directional fluxes. Csáky and Rigor[21] used such a method in the study of active sugar transport in the choroid plexus of the horse.

B. *In Vivo* Technique

1. The simplest approach is to expose the lateral ventricle of an anesthetized animal and visually observe the "sweating" of the choroid plexus. Cushing[22] concluded from such observations that the choroid plexus secretes the CSF. Because of the relatively low rate of secretion and rapid evaporation, the analysis of fluid so collected cannot yield reliable data. Rougemont et al.[23] succeeded in overcoming this obstacle by keeping the plexus under oil. By inserting a capillary tube under the oil onto the choroid plexus they could collect for analysis small samples of authentic, freshly secreted choroid plexus fluid.

2. The plexus can be bathed *in vivo* in a nutrient medium and the transport of substances studied by taking samples from the choroidal vein by micropuncture.[4]

III. BLOOD-TO-CSF TRANSPORT

That the choroid plexuses are involved in the secretion of CSF had been suggested some hundred years ago.[24,25] The first clear-cut experimental evidence to this effect was, however, provided by Dandy[26] who experimentally produced unilateral hydrocephalus in the lateral ventricle of the dog by obstructing the corresponding foramen of Monro. However, no hydrocephalus developed following the obstruction if the choroid plexus was previously removed from the ipsilateral ventricle. More direct evidence of the secretory function of the choroid plexus was provided recently by Welch,[4] who measured the blood flow and the simultaneous volume loss in the lateral choroid plexus of the rabbit. From his measurements he concluded that 1 g of choroid plexus tissue produced 0.37 ml/min of CSF.* Table I lists the rate of CSF secretion in various species.

---

* It is interesting to note that the dogmatic statement of Dandy[26] that the choroid plexus is the *sole* source of CSF was challenged by many, and perhaps most vigorously by Bering.[27,28] Direct proof that Dandy's postulate was incorrect has been provided by Pollay and Curl.[29] These authors found fluid secretion in the perfused aqueduct of Sylvius and in the anterior fourth ventricle. These regions do not have choroid plexuses. Assuming that the ependymal secretion rate is uniform throughout the ventricular system, these investigators concluded that about two-thirds of the CSF is produced by the choroid plexus and one-third by the non-choroidal ependyma.

TABLE I

Rate of Formation of CSF in Various Species

Species	Total rate (μl/min)	Choroid plexus secretion (μl/min/mg)	Reference
Rabbit	12.67	0.37	4, 29
	10.1	0.43	30
Cat	20	0.50	31
	22	0.55	32
Dog	50	0.625	33
	50	0.77[a]	28
Goat	160	0.36	3
Man	330	0.17–0.23	3

[a] Calculated assuming that the total secretion was performed by the choroid plexus.

IV. MECHANISM OF CSF FORMATION BY THE CHOROID PLEXUS

The production of cerebrospinal fluid by the choroid plexus is a secreting process, not a simple filtration. The fact that some solutes are present in freshly secreted CSF in higher concentration than in the plasma ultrafiltrate is perhaps the most convincing evidence that the choroid epithelium performs an uphill, active, transport. The data shown in Table II indicate that the concentrations of sodium, chloride, and magnesium are higher in the fresh choroid plexus secretion than in the plasma ultrafiltrate. Considering that there exists an electrical potential difference across the choroidal epithelium (see below), the luminal surface being electropositive to the blood, clearly, the transport of Na and Mg has to be performed against both concentration and electrical gradients.

Even if this were not the case, a simple calculation can reveal that filtration cannot be the sole force in producing the CSF. Davson[19] calculated that a pressure of 54 mm Hg would be needed to maintain the concentration difference of potassium by filtration between the blood and the CSF. If we add to this about 30 mm Hg pressure, which is needed to overcome the colloid osmotic pressure of the blood proteins, a total of about 84 mm Hg would be needed to produce CSF by filtration. Such high pressures do not occur in the capillary system.

The transport function of the choroid plexus is not restricted to the process of secretion. Reabsorption also occurs by active processes. Thus the choroidal epithelium—not unlike the epithelium of the kidney tubule—can

TABLE II

Concentration of Various Solutes in the Plasma (PL), Plasma Ultrafiltrate (PUF), Mixed Cerebrospinal Fluid from the Lateral Ventricle (CSF), and in the Fluid Freshly Secreted by the Choroid Plexus (CPF) of the Cat

	Na^a	K^a	Cl^a	Ca^a	Mg^a	$Urea^b$	$Glucose^c$
			(mM/kg H_2O)				
PL	163	4.40	132	2.62	1.35	11.6	6.2
PUF	151	3.30	136	1.83	0.95		
CSF	158	2.69	144	2.69	1.33	6.7	3.68
CPF	158	3.28	138	1.67	1.47		
$R = \dfrac{CPF}{PUF}$	1.05	0.99	1.01	0.92	1.55		0.60

[a] Ames, Sallanous, and Endo.[34]
[b] Davson[19] (CPF/PL).
[c] Sadler and Welch.[35]

transport in both directions either by passive, carrier-mediated equilibrating, or by active processes.

V. PERMEABILITY TO WATER

Labeled water exchanges rapidly between blood and cerebral spinal fluid.[36,37] Heisey, Held, and Pappenheimer[3] found that in the perfused ventriculo-cisternal system of the goat, the coefficient of osmotic flow (produced by hypertonic NaCl) was 2×10^{-11} cm^3/dyne-sec. The diffusion coefficient for TOH was about 100 times less. A figure for osmotic flow produced by sucrose in the choroid plexus of the rabbit was found by Welch, Sadler, and Gold[38] to be 0.28×10^{-11} cm^3/dyne-sec. From the value of Heisey and co-workers[3] an equivalent pore radius (r) of 36 Å was calculated[80] for the entire ependymal wall of the ventricle (cf. with $r = 40$ Å for the toad bladder under the influence of vasopressin). Similar calculations, using the values for osmotic flow in the rabbit choroid plexus reported by Welch, Sadler, and Gold[38] indicated $r = 14$ Å. This is closer to the $r = 8.2$ Å which Heisey and co-workers[3] calculated from the theory of restricted diffusion for urea and creatinine. In every case the values for r indicate a relatively porous membrane of the choroid plexus. Whether the difference between the observations of Heisey and co-workers, and Welch and co-workers, is due to the fact that one group examined the entire wall of the ventricle and the other only the choroid plexus, or due to the solute used to produce the osmotic difference is not quite clear, although Welch, Sadler, and Gold concluded from their findings that the reflection coefficient of NaCl for the choroid epithelial membrane, like that of sucrose, is close to unity.

VI. APPARENT UPHILL MOVEMENT OF WATER

Pappenheimer and co-workers[3] in the perfused ventricular system of the goat, and Welch, Sadler, and Gold[38] in the separately perfused choroid plexus, found a secretion of fluid into a hypotonic ventricular solution, indicating that water moved against its activity gradient. It is not necessary, however, to assume that a true active transport of water took place in these cases. Apparent uphill movement of water molecules can be explained with a coupled solute and water flow leading essentially to secretion of an isosmotic fluid. This occurs commonly in epithelial cells.[39,40] The solute transport is the primary event in these cases and an uphill movement of water can occur as a consequence of an active solute transport without true active transport of the water molecules themselves. Consequently, in examining the apparent uphill transport of water, one has to examine carefully the solute transport. The most likely candidate for such active solute transport is usually the sodium ion.

VII. TRANSPORT OF ELECTROLYTES

A. Sodium

There are no accurate measurements available regarding the flux of sodium across the choroidal epithelium. It is, however, clear that transport of sodium from the blood into the CSF is definitely not a simple diffusion process, and there is good evidence that sodium is actively secreted into the CSF. The Na content of the CSF is higher than that of the arterial plasma ultrafiltrate, and it is unlikely that the concentration difference is the result of filtration and reabsorption processes whereby sodium is left behind because the ventricular ependyma is rather permeable to sodium.[36,41] The choroid plexus epithelium is a contributing factor to the high Na in the CSF, as the fluid freshly secreted by the lateral choroid plexus and collected under oil is richer in Na than the plasma ultrafiltrate (see Table II).[23,34,42]

The rate of net sodium secretion depends upon the Na concentration in the plasma. If the latter were to be reduced by infusion of isosmotic glucose, the Na and Cl in the fresh choroid plexus secretion would be reduced but its osmolality would be made up by glucose.[42]

An electrical potential difference exists between the blood and the cerebral spinal fluid, the latter being positive. Direct measurements, placing one electrode into the cerebral spinal fluid and another into the blood, revealed a potential difference of some 5–10 mV.[43,44] Welch and Sadler[45] proved that the potential difference between the blood and the CSF most probably originates in the choroid plexus. They inserted a capillary electrode into individual epithelial cells of the choroid plexus and measured the potential between this electrode and the electrode put in an artificial CSF bathing the epithelial cells from the ventricular site. They measured an average potential between the cells and the ventricular fluid of 64 mV, the inside of the cells being negative; then they placed the outside electrode into the connective tissue of the plexus and measured a potential of 50 mV, the inside of the cell again being negative. From the difference they calculated that the potential across the epithelium must have been approximately 14 mV, the ventricular site being positive. This is in close agreement with the measurements quoted above. Consequently, it is likely that the choroid plexus sodium pump is electrogenic.

The potential difference may produce an uneven distribution of ions. The data illustrated in Table III indicate that, in the case of Na and Mg, transport definitely occurs against a considerable electrochemical gradient, while the distribution of other ions (with the possible exception of Ca) does not require an uphill transport.

B. Potassium

The potassium concentration in fresh choroid plexus fluid is approximately the same as in the plasma:[42] thus K, unlike Na, does not appear to be secreted by the active function of the choroid plexus. However, the transport of K from the blood into the CSF appears to be more a carrier-mediated process than a simple diffusion. Data presented in Fig. 3 indicate[42] that as the plasma concentration of potassium was artificially increased, the potassium concentration of the freshly secreted CSF increased to a certain

TABLE III

Theoretical Passive Distribution (R) of Ions Between Blood and CSF, Based on a Potential Difference of 5 mV[44] as Compared with the Actual Values Between Plasma Ultrafiltrate and Freshly Secreted Choroid Plexus Fluid[43]

Ions	Theoretical	Actual
Na	0.83	1.05
K	0.83	0.91
Ca	0.69	0.90
Mg	0.69	1.56
Cl	1.21	1.01

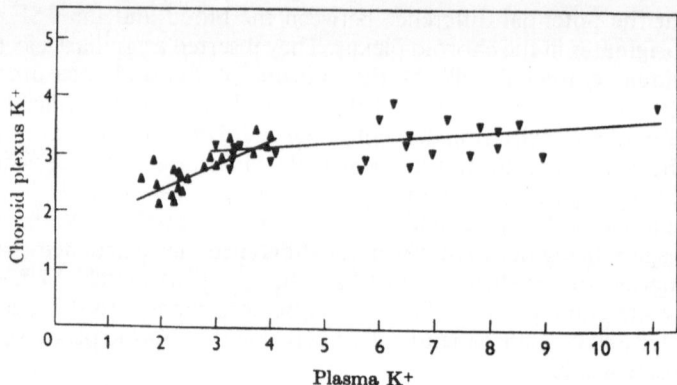

Fig. 3. Correlation between the concentration of the K^+ in the plasma and in the freshly secreted cerebrospinal fluid. From Ames, Higashi, and Nesbett.[42]

level, leveling off at higher plasma concentrations, indicating a saturation-type transport kinetics, usually indicative of carrier mediation.

C. Calcium and Magnesium

The concentration of calcium in the cerebral spinal fluid is much less than in the plasma but not too different from that in the plasma ultrafiltrate. The difference is most likely due to the high degree of binding of Ca to plasma protein. The experiments of Paupe[46] are interesting. He compared the total and ionized calcium of serum with that in the CSF, assaying the ionized calcium by the action of the fluid upon the isolated heart. He found little or no correlation between the total calcium concentrations in the serum and CSF, but a very good correlation between the concentration of the ionized calcium in the two fluids.

The magnesium content of freshly secreted CSF is much higher than that in the blood, indicating an uphill secretion. The significance of this choroidal magnesium pump is obscure.

D. Chloride

The chloride content of the CSF of most species is higher than that of the blood plasma, although the difference between blood chloride content and the freshly secreted choroid plexus content is less striking.

It is not necessary to assume that a chloride pump is involved in the maintenance of an uneven chloride distribution. The potential difference between plasma and CSF, the latter being positive, could be solely responsible for the uneven distribution of the chloride (see Table III). The apparent involvement of carbonic anhydrase is sometimes quoted as evidence for an active chloride secretion in the choroid epithelium,[47] because inhibitors of this enzyme (e.g., acetazolamide) almost abolish the chloride concentration

difference between the plasma and the CSF[48] and because carbonic anhydrase may be directly involved in chloride transport.[49] This evidence,
however, should be treated with some caution, as acetazolamide also reduces
the transepithelial potential,[45] which in turn may affect the distribution.

E. Iodine, Bromides, and Thiocyanate

These ions are actively transported from the CSF into the blood (see
below). Consequently, their concentration in the former is remarkably low.

VIII. TRANSPORT OF NONELECTROLYTES

Nonelectrolyte transport can occur by (a) simple diffusion, (b) carrier-
mediated diffusion, (c) active transport, (d) bulk flow either through pores or
by pinocytosis.

Welch and Sadler[50] studied the transport of urea, creatinine, mannitol,
sucrose, and inulin from the ventricular surface of the lateral choroid plexus
of the rabbit, irrigated with an artificial solution, in the choroidal venous
blood. They distinguished between transfer by unrestricted diffusion and
another type of transfer with a single permeability coefficient for all compounds studied. This latter flow was not particularly altered by the artificial
increase or decrease of the choroidal secretion rate. Although Welch and
Sadler[50] suggest that this bulk solute transfer may be due to pinocytosis,
they did not completely eliminate the possibility of flow through pores by
solvent drag, particularly in view of the fact that the choroid epithelium is,
in this respect, quite leaky (see above).

Welch and Sadler's[50] approach is the only direct one, but some deduction regarding choroid plexus permeability can be drawn from experiments
using the ventriculo-cisternal perfusion technique of Pappenheimer, Heisey,
and Jordan,[51] because a large portion of the ventricular surface is occupied
by the choroid plexus (according to Heisey, Held, and Pappenheimer,[3]
50%; according to Bering and Sato,[27] 30%). In addition, the choroid
plexus has a much more vigorous rate of blood perfusion than any other part
of the ventricular wall.

By adding various solutes to the perfusate and measuring their rate of
disappearance against the bulk absorption of fluids (established by inulin
clearance), conclusions can be drawn about the nature of the transfer. Using
this technique, Bradbury and Davson[30] found that urea and creatinine
exchange by simple diffusion across the ventricular wall, while a saturable
carrier mediation is involved in the exchange of D-glucose, D-xylose, and
D-fructose. These sugars most likely share the same carrier, as they mutually
inhibit each other's passage.

The transport of glucose across the choroidal epithelium is a particularly
interesting phenomenon. The glucose content of the CSF is about half of
that in the blood plasma. Moreover, it roughly follows changes in blood

concentration, going up in hyperglycemia and decreasing in hypoglycemia. Even in freshly collected choroid plexus fluid, the ratio of glucose in the CSF/plasma is approximately 0.6 and remained the same over a wide range of blood plasma glucose levels.[35] Should the inflow of glucose across the choroidal epithelium be a passive process, it would be determined by the concentration in the plasma and the reflection coefficient of glucose, according to the following equation :[52]

$$J_s = \bar{c}_s \,(1 - \delta)\, J_v$$

where J_s is the flow of glucose, $\bar{c}_s$ is the average concentration of glucose in the membrane, δ is the reflection coefficient of glucose, and J_v is the volume flow. This simple explanation of the observed glucose distribution between blood and CSF was upset by the finding that glucose and galactose are definitely concentrated in the isolated choroid plexus.[53] The concentrating mechanism displays many characteristics of a typical sugar pump, being inhibited by excess glucose or galactose, by metabolic poisons, digitalis, lack of oxygen, or sodium. If the active process is involved primarily in the secretion of glucose from the blood into the cerebral spinal fluid, it would be difficult to arrive at a fixed ratio of 0.6 over a wide range of plasma-glucose concentrations. An alternate mechanism, involving a passive (or mediated) blood-to-CSF outflow coupled with an active reabsorption, was suggested. According to the theory, the plasma–CSF concentration is determined by a low capacity CSF-to-blood glucose pump.[21]

IX. ENERGY-DEPENDENT ACCUMULATION OF SUBSTANCES IN THE CHOROID PLEXUS

If substrate accumulates within a tissue and the accumulation is inhibited by the depression of the energy production in the tissue, it is reasonable to assume that the tissue has a pumping mechanism for that particular solute unless the substance is bound to some tissue constituent. Accumulation within plexus tissue was shown for a number of substances: inorganic ions such as iodide,[54,55] and quaternary amines such as hexamethonium, decamethonium, and N-methylnicotinamide.[56] Apparently the same pumping mechanism can cause the accumulation of norepinephrine and serotonin.[57] Morphine,[58] dihydromorphine, levorphanol,[59] and other narcotic analgesics[60] are actively taken up by the plexus tissue. Sugars such as glucose and galactose are also accumulated by the plexus in vitro.[53]

Iodide accumulation is strongly inhibited by perchlorate, fluoroborate, thiocyanate, nitrate, bromide, and even by such complex organic ions as penicillin.[54] Accumulation of serotonin and norepinephrine is a saturable process and is inhibited by an anaerobiosis, by dinitrophenol, and by ouabain.

X. TRANSPORT FROM THE CSF INTO THE BLOOD

A number of substances are transported via the choroid plexus epithelium by active process from the CSF into the bloodstream. Pappenheimer, Heisey, and Jordan[51] showed that Diodrast and phenolsulfonphthalein, both weak organic acids, are rapidly absorbed from the CSF against a higher concentration from the perfused ventricles into the bloodstream. Their absorption kinetics was carrier mediated as it could be saturated by high concentration of substrate and inhibited competitively by *p*-aminohippurate. Similar active transport is found in the isolated dogfish choroid plexus for phenol red and chlorophenol red.[61] Among the other organic acids which are actively reabsorbed, penicillin should be mentioned in particular. This substance is rapidly absorbed from the dog's cisternal fluid against an apparent higher concentration in the blood. The rate of absorption, however, was slowed by simultaneous administration of high concentration of phenolsulfonphthalein, aminohippurate, or probenecid.[62]

Quaternary ammonium bases, such as hexamethonium, decamethonium, and *N*-methylnicotinamide are absorbed from the CSF with unexpected rapidity. Moreover, they mutually inhibit each other's absorption.[63] In view of the finding that the choroid plexus *in vitro* accumulates the same quaternary base (see above), it is reasonable to assume that these substances are actively transported by the choroid epithelium from the CSF into the blood.

Morphine and related compounds also leave the CSF rapidly after intraventricular injection[64] and, because they are accumulated in the choroid epithelium, it is assumed that an absorptive pump is present in the cells for these substances.

Similarly, it is likely that an active absorptive pump in the plexus is responsible for the rapid disappearance of the labeled norepinephrine from the ventricles of the rat. Not more than half of the injected quantity can be detected in the brain, even in an analysis performed very shortly after intrathecal administration of the amine.[65]

Simple sugars are transported by the choroidal epithelium from the ventricle into the blood. Thus in the isolated plexus of the horse, labeled galactose was found in higher concentration in the choroidal vein than in the artificial cerebrospinal fluid bathing the ventricular surface.[21]

The active absorbing function of the choroid plexus epithelium is thus very similar to that of the kidney tubule except that the direction is reversed (with the possible exception of glucose). The result of this function is the maintenance of a steady, relatively low, occasionally even very low concentration of a variety of substances in the CSF. This function can be considered to be a result of a "pump-and-leak" ratio, meaning that a given solute can passively diffuse through the choroid epithelium from the blood into the CSF and then be actively transported by the same cells from the CSF into the blood. The steady-state transport rate ($\bar{V}$) will then be the function of the rate of passive leak ($V_{out} = d(-s)$) and the active reabsorption (V_{in}).

$$\bar{V} = V_{out} - V_{in}$$

Exact assessment of the two parallel functions is missing and calculations in this regard are sorely needed.

XI. PINOCYTOSIS

Pinocytosis is a process whereby the plasma membrane forms invaginations, engulfing the content of the extracellular space. These invaginations then separate from the plasma membrane, creating vesicles. Eventually, the membrane of the vesicle disappears, discharging its content into the intracellular space.

Membrane-bound vesicles can be seen in the epithelium of the choroid plexus.[66] Moreover, if a thorium oxide suspension were injected into the CSF, the particles would attach to the brush border and later become engulfed into the pinocytotic vesicle. In the case of thorium oxide, a transcellular transport could not be demonstrated, but Brightman[67] injected ferritin in the CSF of the rat and found that it was absorbed through the choroid epithelium by pinocytosis.

It is not clear whether a considerable solute absorption can take place via pinocytosis. Welch and Sadler[50] speculated about such a large "bulk" nondiffusional solute flow which would require that each epithelial cell transfer its own volume in about 15 min by pinocytosis. Taking into account that not all vesicles seen with the electron microscope at any given time in a fixed preparation originate at the brush border side,[66] this would require an enormous number of pinocytotic vesicles. Moreover, there is no way to assess how fast the pinocytotic vesicles travel from one border of the epithelium to the other. In the case of thorium oxide, this passage apparently occurs rather slowly; only after 15–20 min after the administration of the suspension can one detect a number of intracellular vesicles loaded with thorium oxide.

Unfortunately, there is no method available for a specific inhibition of pinocytosis. This, and the lack of a method for quantitation, make it impossible to assess at this time the contribution of pinocytosis to the actual transport processes which take place in the choroid plexus.

XII. EFFECTS OF DRUGS ON CHOROIDAL TRANSPORT

Active transport processes in the choroidal epithelium require the expenditure of energy, which the cell generates through vigorous oxidative metabolism. Consequently, any inhibition of metabolism is usually followed by a depression of the transport, secretion and reabsorption alike. Special enzymes may be involved in active transport: sodium-potassium-activated ATPase ("pump" ATPase), and carbonic anhydrase. Inhibition of these enzymes may also decrease the rate of choroidal transport. In the following, a few examples are given to illustrate how a metabolic inhibitor (dinitrophenol), a pump ATPase inhibitor (digitalis), and the inhibition of carbonic anhydrase affect the choroidal transport.

A. 2, 4-Dinitrophenol (DNP)

This poison inhibits the mitochondrial oxidative phosphorylation, and because of this it depresses the production of ATP which is essential for active cation transport. Thus, ultimately, the function of the sodium pump is inhibited. As active sodium transport is involved in the secretion of CSF by the choroid plexus, it can be expected that DNP diminishes the rate of secretion. Pollay and Davson[68] found a 45% reduction in secretion after adding 5×10^{-5} M DNP to the ventriculo-cisternal perfusate. Welch,[4] on the other hand, did not find any change even with DNP concentrations as high as 9×10^{-3} M. It is interesting, though, that the same preparation of Welch[4] was not inhibited by cyanide, which is a potent inhibitor of respiration by poisoning the cytochrome system. One wonders whether enough energy could be derived in this case from a nonoxidative, fermentation process. If this was the case it is not surprising that DNP did not inhibit.

On the absorptive side of transport, 2×10^{-3} M DNP produced a 50% inhibition in the *in vitro* accumulation of iodide, while the same effect was achieved with 7×10^{-3} M cyanide.[54] 10^{-3} M DNP depressed the *in vitro* concentrative uptake of serotonin and norepinephrine by 38% and 55% respectively,[57] and that of glucose and galactose by 96% and 94% respectively[53] (see Table IV).

B. Digitalis

Cardioactive steroids are specific inhibitors of the sodium-potassium-stimulated ATPase which is involved in the active transport of sodium in all tissues.[69] All transport mechanisms of the choroid plexus are inhibited by digitalis.

Ouabain in concentration of 10^{-4} to 10^{-5} M decreases the rate of secretion measured by ventriculo-cisternal perfusion,[70,71] or by arteriovenal fluid loss in the choroid plexus,[4] or by direct collection of the secretions of

TABLE IV

Inhibition of the *in Vitro* Accumulation of Various Substrates by Drugs in the Choroid Plexus

Species	Substrate	Ouabain		2,4-Dinitrophenol		Acetazolamide		Reference
		Conc.	% Inh.	Conc.	% Inh.	Conc.	% Inh.	
Rabbit	Iodide	10^{-7}M	50	2×10^{-3}M	50	10^{-1}M	50	54
Dog	Glucose	10^{-4}M	30	10^{-3}M	96			53
Dog	Galactose	10^{-4}M	40	10^{-3}M	94			53
Rabbit	Serotonin	10^{-4}M	23	10^{-3}M	38			57
Rabbit	Norepinephrine	10^{-4}M	30	10^{-3}M	55			57

the plexus.[72] Vates, Bonting, and Oppelt[17] found good correlation between the *in vivo* inhibition of the CSF secretion and the *in vitro* inhibition of the sodium-potassium-stimulated ATPase in the choroid plexus (see Fig. 4).

Direct effect of ouabain upon sodium fluxes across the choroidal epithelium has not been measured. However, the drug, in concentration of 10^{-5} M, reduced by about 50% the transepithelial resting potential in the choroid plexus.[45] Since the sodium pump is involved in the generation of the potential, this finding is strong indirect evidence that ouabain inhibits active choroidal sodium transport.

Ouabain also depresses the ventricle-to-blood transport of iodide.[70]

Table IV illustrates that ouabain inhibits the active *in vitro* accumulation of various substances in the choroid plexus.

C. Acetazolamide (Diamox)

This drug is a powerful inhibitor of the enzyme carbonic anhydrase, which is present in large amounts in the choroidal epithelium. As Diamox is known to produce diuresis by inhibiting the reabsorption of electrolytes and water in the kidney, it was speculated that it may also inhibit the secretion of CSF. This was found to be true in experiments, but at the same time great variation was observed in the quantitative action of the drug.[4,71,73–79] The latter could be attributed to differences in the species used, dosage, and the site of administration (intravenous vs. intraventricular), but perhaps the most significant factor was the acid-base state of the experimental animal.

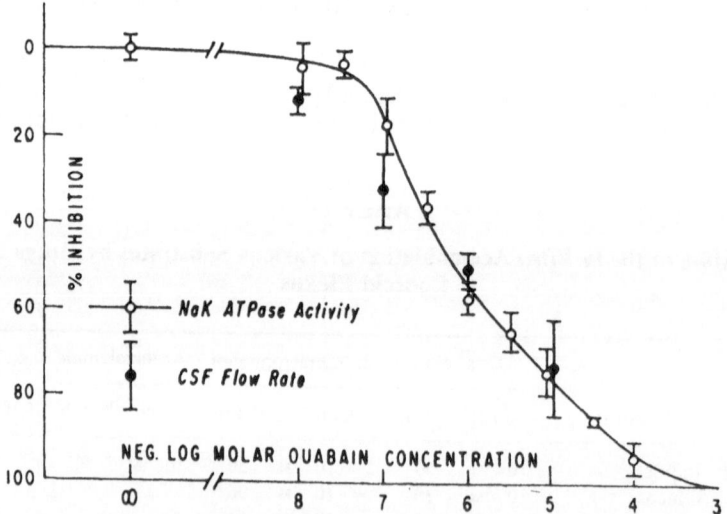

Fig. 4. Comparison between the inhibition of the Na-K-stimulated ATPase activity *in vitro* and of the CSF flow *in vivo* by intraventricular ouabain perfusion. From Vates, Bonting, and Oppelt.[17]

While acidosis does not influence the effect of acetazolamide, metabolic and even more so, respiratory alkalosis (which by itself inhibits the secretion of CSF) potentiates the inhibitory effect of the drug on the CSF production.[33]

The mode of action of Diamox in the CSF is puzzling. This drug is a potent inhibitor of the enzyme carbonic anhydrase which catalyzes the following reaction:

$$CO_2 + H_2O \rightleftharpoons H^+ + HCO_3^-$$

This reaction proceeds spontaneously but its rate is markedly increased by the enzyme. The proton (H^+) thus generated can exchange with other univalent cations, (e.g., Na^+) while the anion (HCO_3^-) is able to exchange with other anions, such as Cl^-. Whether such an exchange mechanism is involved in the ion transport connected with the secretion of CSF is not certain.

One striking effect of carbonic anhydrase inhibition is the decrease of the chloride concentration in the CSF so that it approaches the chloride concentration of the blood.[48,75] However, Ames, Higashi, and Nesbett,[72] suggest that the reduction of the chloride in the CSF by Diamox is not due to the direct effect of the drug upon the choroid plexus, as it did not change the chloride content of the secretion collected under oil from the surface of the plexus, although the rate of formation was markedly diminished. Welch, Sadler, and Gold[38] found that under the influence of acetazolamide the rate of volume flow decreased, but at the same time the secreted fluid was hypotonic to the blood, as if Diamox had caused an uncoupling of the flow of water and solute.

In view of the alleged causative involvement of choroid plexus secretion in the development of hydrocephalus, it is not surprising that great efforts are being made to find a suitable drug that could, to some extent selectively, reduce the fluid secretion by the plexus. At present, unfortunately, such a selective remedy is not available.

XIII. REFERENCES

1. W. C. Worthington, Jr., and R. S. Cathcart, III, Ependymal cilia: distribution and activity in the adult human brain, *Science* **139**:221–222 (1963).
2. D. C. Pease, Infolded basal plasma membranes found in epithelia noted for their water transport, *J. Biophys. Biochem. Cytol.* **2**, Suppl. 4, 203–208 (1956).
3. S. R. Heisey, D. H. Held, and J. R. Pappenheimer, Bulk flow and diffusion in the cerebrospinal fluid system of the goat, *Am. J. Physiol.* **203**:775–781 (1962).
4. K. Welch, Secretion of cerebrospinal fluid by choroid plexus of the rabbit, *Am. J. Physiol.* **205**:617–624 (1963).
5. S. S. Kety and C. F. Schmidt, The nitrous oxide method for the quantitative determination of cerebral blood flow in man: theory, procedure and normal values, *J. Clin. Invest.* **27**:476–483 (1948).
6. H. W. Smith, *The Kidney: Structure and Function in Health and Disease*, Oxford University Press, New York (1951).

7. E. A. Bering, Jr., Choroid plexus and arterial pulsation of cerebrospinal fluid. Demonstration of the choroid plexuses as a cerebrospinal fluid pump, *Arch. Neurol. Psychiat.* **73**:165–172 (1955).

8. E. A. Bering, Jr., Circulation of the cerebrospinal fluid: demonstration of the choroid plexus as the generator of the force for flow of fluid and ventricular enlargement, *J. Neurosurg.* **19**:405–413 (1962).

9. C. B. Wilson and V. Bertan, Interruption of the anterior choroidal artery in experimental hydrocephalus, *Arch. Neurol.* **17**:614–619 (1967).

10. O. Järvi, U. d. Einwirkung von Pilokarpin und Atropin auf das mikroskopische Bild des Adergeflechtes, *Acta. Soc. Med. "Duodecim" Ser. A* **23**:84–96 (1940).

11. H. A. Krebs and H. Rosenhagen, Über den Stoffwechsel des plexus chorioideus, *Z. Ges. Neurol. Psychiat.* **134**:643 (1931).

12. J. S. Friedenwald, H. Herrmann, and R. Buka, The distribution of certain oxidative enzymes in the choroid plexus, *Bull. Johns Hopkins Hosp.* **70**:1–18 (1942).

13. E. H. Leduc and G. B. Wislocki, The histochemical localization of acid and alkaline phosphates, non-specific esterase and succinic dehydrogenase in the structures comprising the hematoencephalic barrier of the rat, *J. Comp. Neurol.* **97**:241–279 (1952).

14. R. D. Stiehler and L. B. Flexner, A mechanism of secretion in the choroid plexus. The conversion of oxidation–reduction energy into work, *J. Biol. Chem.* **126**:603–617 (1938).

15. R. G. Fisher and J. H. Copenhaver, Jr., The metabolic activity of the choroid plexus, *J. Neurosurg.* **16**:167–176 (1959).

16. T. H. Maren, Carbonic anhydrase: Chemistry, physiology and inhibition, *Physiol. Rev.* **47**:595–781 (1967).

17. T. Z. Vates, Jr., S. L. Bonting, and W. W. Oppelt, Na-K activated adenosine triphosphatase and formation of cerebrospinal fluid in the cat, *Am. J. Physiol.* **206**:1165–1172 (1964).

18. J. W. Millen and D. H. M. Woollam, *The Anatomy of the Cerebrospinal Fluid,* Oxford University Press, London (1962).

19. H. Davson, *Physiology of the Cerebrospinal Fluid,* Little, Brown, Boston (1967).

20. J. R. Pappenheimer, S. R. Heisey, E. F. Jordan, and J. DeC. Downer, Perfusion of the cerebral ventricular system in unanesthetized goats, *Am. J. Physiol.* **200**:1–10 (1961).

21. T. Z. Csáky and B. M. Rigor, The choroid plexus as a glucose barrier, *in Brain Barrier Systems, Progr. Brain Res.* **29**:147–158 (1968).

22. H. Cushing, *Studies in Intracranial Physiology and Surgery,* Oxford Medical Publications, London (1926).

23. H. De Rougemont, A. Ames, III, F. D. Nesbett, and H. F. Hoffmann, Fluid formed by choroid plexus, *J. Neurophysiol.* **23**:485–495 (1960).

24. J. Faivre, Recherches sur la structure du coronarium et des plexus choroides chez l'homme et les animaux, *Gazette Med. Paris* **9**:555 (1854).

25. H. von Luschka, *Die Adergeflechte des menschlichen Gehirnes,* George Reimer, Berlin (1855).

26. W. E. Dandy, Experimental hydrocephalus, *Ann. Surg.* **70**:129–142 (1919).

27. E. A. Bering, Jr., and O. Sato, Hydrocephalus: Changes in formation and absorption of cerebrospinal fluid within the cerebral ventricles, *J. Neurosurg.* **20**:1050–1063 (1963).

28. E. A. Bering, Jr., Cerebrospinal fluid production and its relationship to cerebral metabolism and cerebral blood flow, *Am. J. Physiol.* **197**:825–828 (1959).

29. M. Pollay and F. Curl, Secretion of cerebrospinal fluid by the ventricular ependyma of the rabbit, *Am. J. Physiol.* **213**:1031–1038 (1967).

30. M. W. B. Bradbury and H. Davson, The transport of urea, creatinine and certain monosaccharides between blood and fluid perfusing the cerebral ventricular system of rabbits, *J. Physiol.* **170**:195–211 (1964).

31. H. Davson, C. R. Kleeman, and E. Levin, Quantitative studies of the passage of different substances out of the cerebrospinal fluid, *J. Physiol.* **161**:126–142 (1962).

32. L. Graziani, A. Escriva, and R. Katzman, Exchange of calcium between blood, brain, and cerebrospinal fluid, *Am. J. Physiol.* **208**:1058–1064 (1965).

33. W. W. Oppelt, T. H. Maren, E. S. Owens, and D. P. Rall, Effects of acid–base alteration on cerebrospinal fluid production, *Proc. Soc. Exptl. Biol. N.Y.* **114**:86–89 (1963).

34. A. Ames, III, M. Sallanous, and S. Endo, Na, K, Ca, Mg and Cl concentrations in choroid plexus fluid and cisternal fluid compared with plasma ultrafiltrate, *J. Neurophysiol.* **27**:672–681 (1964).

35. K. Sadler and K. Welch, Concentration of glucose in new choroidal cerebrospinal fluid of the rabbit, *Nature* **215**:884 (1967).

36. E. A. Bering, Jr., Studies on the role of the choroid plexus in tracer exchanges between blood and cerebrospinal fluid, *J. Neurosurg.* **12**:385–392 (1955).

37. W. H. Sweet, B. Selverstone, S. Solloway, and D. Stetten, Jr., Studies of formation, flow and absorption of cerebrospinal fluid. Studies with heavy water in the normal man, *American College of Surgeons Surgical Forum*, pp. 376–381, Saunders, Philadelphia (1950).

38. K. Welch, K. Sadler, and G. Gold, Volume flow across choroidal ependyma of the rabbit, *Am. J. Physiol.* **210**:232–236 (1966).

39. J. M. Diamond, The mechanism of isotonic water transport, *J. Gen. Physiol.* **48**:15–42 (1964).

40. J. M. Diamond and W. H. Bossert, Standing-gradient osmotic flow. A mechanism for coupling of water and solute transport in epithelia, *J. Gen. Physiol.* **50**:2061–2083 (1967).

41. W. H. Sweet, G. L. Brownell, J. A. Scholl, D. R. Bowsher, P. Benda, and E. E. Stickley, The formation flow and absorption of cerebrospinal fluid; newer concepts based on studies with isotopes, *Neurol. Psych. Childhood Res. Publ. Assoc. Nerv. Ment. Dis.* **34**:101–159 (1954).

42. A. Ames, III, K. Higashi, and F. B. Nesbett, Relation of potassium concentration in choroid plexus fluid to that in plasma, *J. Physiol.* **181**:506–515 (1965).

43. H. Loeschke, Über Bestandspotentiale im Gebiete der Medulla oblongata, *Pflügers Arch. Ges. Physiol.* **262**:517–531 (1956).

44. D. Held, V. Fencl, and J. R. Pappenheimer, Electrical potential of cerebrospinal fluid, *J. Neurophysiol.* **27**:942–959 (1964).

45. K. Welch and K. Sadler, Electrical potential of choroid plexus of the rabbit, *J. Neurosurg.* **22**:344–349 (1965).

46. J. Paupe, Comparison entre fractions calciques de liquides céphalo-rachidiens et de sérums normaux chez l'homme, *Compt. Rend. Soc. Biol.* **151**:318–320 (1957)

47. C. A. M. Hogben, P. Wistrand, and T. H. Maren, Role of active transport of chloride in formation of dogfish cerebrospinal fluid, *Am. J. Physiol.* **199**:124–126 (1960).

48. L. Birzis, C. H. Carter, and T. H. Maren, Effect of acetazolamide on CSF pressure and electrolytes in hydrocephalus, *Neurology* **8**:522–528 (1958).

49. B. Becker, Carbonic anydrase and the formation of aqueous humour, *Am. J. Ophthalmol.* **47**:342 (1959).

50. K. Welch and K. Sadler, Permeability of the choroid plexus of the rabbit to several solutes, *Am. J. Physiol.* **210**:652–660 (1966).

51. J. R. Pappenheimer, S. R. Heisey, and E. F. Jordan, Active transport of Diodrast and phenolsulfonphthalein from cerebrospinal fluid to blood, *Am. J. Physiol.* **200**:1–10 (1961).

52. A. Katchalsky and P. F. Curran, *Nonequilibrium Thermodynamics in Biophysics*, Harvard University Press, Cambridge (1965).

53. T. Z. Csáky and B. M. Rigor, A concentrative mechanism for sugars in the choroid plexus, *Life Sci.* **3**:931–936 (1964).

54. B. Becker, Cerebrospinal fluid iodide, *Am. J. Physiol.* **301**:1149–1151 (1961).

55. K. Welch, Concentration of thiocyanate by the choroid plexus of the rabbit, *Am. J. Physiol.* **205**:617–624 (1963).

56. Y. Tochino and L. S. Schanker, Active transport of quaternary ammonium compounds by the choroid plexus *in vitro, Am. J. Physiol.* **208**:666–673 (1965).
57. Y. Tochino and L. S. Schanker, Transport of serotinin and norepinephrine by the rabbit choroid plexus *in vitro, Biochem. Pharmacol.* **14**:1557–1566 (1965).
58. A. E. Takemori and M. W. Stenwick, Studies on the uptake of morphine by the choroid plexus *in vitro, J. Pharmacol. Exptl. Therap.* **154**:586–594 (1966).
59. C. C. Hug and M. M. Ziegler, Transport of narcotic analgesics by choroid plexus and renal tissue *in vitro, Federation Proc.* **25**:415 (1966).
60. C. C. Hug, Transport of narcotic analgesics by choroid plexus and kidney tissue *in vitro, Biochem. Pharmacol.* **16**:345 (1967).
61. D. P. Rall and W. Sheldon, Transport of organic acid dyes by the isolated choroid plexus of the spiny dogfish *S. acanthias, Biochem. Pharmacol.* **11**:169 (1961).
62. R. A. Fishman, Active transport and the blood–brain barrier to penicillin and related organic acids, *Trans. Am. Neurol. Assoc.* **89**:51–55 (1964).
63. L. S. Schanker, L. D. Prockop, J. Schou, and P. Sisodia, Rapid efflux of some quaternary ammonium compounds from cerebrospinal fluid, *Life Sci.* **10**:515–521 (1962).
64. T. K. Adler, CNS localization, biological disposition, and analgetic action of morphine and codeine after intraventricular injection of mice. *Federation Proc.* **23**:283 (1964).
65. J. Glowinski, I. J. Kopin, and J. Axelrod, Metabolism of noradrenaline-^{3}H in the rat brain, *J. Neurochem.* **12**:25–30 (1965).
66. V. M. Tennyson and G. D. Pappas, Electronmicroscope studies of the developing telencephalic choroid plexus in normal and hydrocephalic rabbits, *in Disorders of the Developing Nervous System* (W. S. Fields and M. M. Desmond, eds.), pp. 267–318. Charles C. Thomas, Springfield (1961).
67. M. M. Brightman, The distribution within the brain of ferritin injected into cerebrospinal fluid compartments, *J. Cell Biol.* **26**:99–123 (1965).
68. M. Pollay and H. Davson, The passage of certain substances out of the cerebrospinal fluid, *Brain* **86**:137–150 (1963).
69. I. M. Glynn, The action of cardiac glycosides on ion movements, *Pharmacol. Rev.* **16**:381–407 (1964).
70. H. Davson and M. Pollay, Influence of various drugs on the transport of ^{131}I and PAH across the cerebrospinal fluid–blood barrier, *J. Physiol.* **167**:239–246 (1963).
71. H. Cserr, Potassium exchange between cerebrospinal fluid, plasma, and brain, *Am. J. Physiol.* **209**:1219–1226 (1965).
72. A. Ames, III, K. Higashi, and F. B. Nesbett, Effects of P_{CO_2}, acetazolamide, and ouabain on volume and composition of choroid plexus fluid, *J. Physiol.* **181**:516–524 (1965).
73. R. D. Tschirgi, R. W. Frost, and J. L. Taylor, Inhibition of cerebrospinal fluid formation by a carbonic anhydrase inhibitor, 2-acetylamino-1,3,4-thiadiazole-5-sulfonamide (Diamox), *Proc. Soc. Exptl. Biol. Med. N.Y.* **87**:373–376 (1954).
74. S. Kister, Carbonic anhydrase inhibition. The effect of acetazolamide on cerebrospinal fluid flow, *J. Pharmacol. Exptl. Therap.* **117**:402–405 (1956).
75. H. Davson and C. P. Luck, The effect of acetazolamide on the chemical composition of the aqueous humour and cerebrospinal fluid of some mammalian species and on the rate of turnover of ^{24}Na in these fluids, *J. Physiol.* **137**:279–293 (1957).
76. R. A. Fishman, Factors influencing the exchange of sodium between plasma and cerebrospinal fluid, *J. Clin. Invest.* **38**:1698–1708 (1959).
77. C. A. Van Wart, J. R. Dupont, and L. Kraintz, Effect of acetazolamide on passage of protein from cerebrospinal fluid to plasma, *Proc. Soc. Exptl. Biol. N.Y.* **106**:113–114 (1961).
78. H. Davson and M. Pollay, The turnover of ^{24}Na in the cerebrospinal fluid and its bearing on the blood–brain barrier, *J. Physiol.* **167**:247–255 (1963).

79. W. W. Oppelt, C. S. Patlak, and D. P. Rall, Effect of certain drugs on cerebrospinal fluid production in the dog, *Am. J. Physiol.* **206**:247–250 (1964).
80. W. G. Stein, *The Movement of Molecules Across Cell Membranes,* Academic Press, New York (1967).

Chapter 5

SPINAL CORD

Giulio Levi

Center of Neurobiology
Department of Biochemistry
Istituto Superiore di Sanita'
Rome, Italy

I. INTRODUCTION

The biochemistry of the spinal cord (SC) has generally received less attention than that of other areas of the CNS, perhaps because of the relative inaccessibility of this tissue or because it is considered a less important region of the CNS. No recent review on the biochemistry of SC has appeared, to the knowledge of the author. This chapter summarizes the available data on the structural biochemistry of this organ. Only the data referring to normal tissue are considered; some data relative to the biochemical development of the SC also are included. Results obtained with histochemical techniques are mentioned briefly, only to give some insight, when possible, on the cellular localization and distribution of compounds that are generally measured in large samples of tissue. Restriction of space does not allow more than limited comparisons between the biochemical composition of SC and that of other regions of the CNS.

Since the data available were often fragmentary, it was thought preferable to organize the chapter in sections, each covering completely a given subject (e.g., acetylcholine and enzymes involved in its synthesis and breakdown). It will be noted that some topics are treated more thoroughly than others and that some are not treated at all; it is hoped that the gaps in information and the obsolescence of some of the data will stimulate further, more systematic research on the biochemistry of SC.

II. INORGANIC CONSTITUENTS

Table I shows the amount of water and inorganic constituents of SC. The rather low water content of adult SC (about 70% of fresh tissue weight) is due to the large amount of white matter present in this tissue. In fact the

TABLE I

Inorganic Constituents of Spinal Cord (mg/g wet wt.)

Species	Age	Water	Na	K	Cl	Ca	Mg	Fe	Reference
Man	0–1 yr	818	1.44	0.85		0.08			1
	Adult	739	1.37	0.75		0.10			1
		720	2.07	2.11					2
		644	2.01	3.61	1.52	0.18	0.38	0.05	3
Ox	Adult	631	1.34	2.61	1.30	0.32	0.48	0.05	3
		680[a]	1.77[a]	2.77[a]					4
		780[b]	2.23[b]	3.51[b]					4
Dog	Adult	670	1.36	2.79	1.27				5
Rabbit	Adult	685	1.43		1.47				6
Guinea pig	Fetus	910	3.08	1.99		0.24			7
	Adult	740	1.79	2.77		0.24			7
Rat	1 day	841							8
	70 days	680							8

[a] White matter only.
[b] Gray matter only.

water content is significantly higher in gray matter than in whole SC or in SC white matter (Table I) and is about 90% of the fat-free tissue weight;[5] moreover, the drop in water content normally occurring from fetal to adult life is higher in SC than in most other CNS areas. An inverse relationship between body size and water content has been demonstrated in the dog SC; in fact, total lipids increase with increasing body size.[9]

There is some variation in the sodium and potassium concentrations reported by the various authors. The level of both these electrolytes is apparently higher in gray than in white matter (Table I) but is practically the same if it is referred to the tissue water instead of the tissue wet weight.[4]

Sodium and potassium concentrations are similar in different animal species and in different areas of the CNS.[2] In the SC of guinea pig fetuses sodium is much higher than in newborn and adult animals. The drop appears as a primary phenomenon, since it is only partially due to the decrease of tissue water and is independent of myelination.[7]

Iron is present in low concentration[3,10] and the true level of inorganic iron is probably even lower since postmortem dissociation of iron bound to organic substances may occur.[10] In other CNS areas the iron level is up to five times higher.

Small amounts of copper[11] (about 25 $\mu g/g$ in human SC) and of iodine[12] (0.25–2.8 $\mu g/g$ in the SC of different animal species) also have been demonstrated. Iodine level seems inversely related to body size.[12]

III. CARBOHYDRATES

A. Carbohydrate Levels and Transport

Table II shows the concentration of glucose and other monosaccharides, of three glycolytic metabolites, and of glycogen in SC. The presence of sorbitol and related enzymes suggested the existence of a "sorbitol pathway" for the transformation of glucose into fructose.[13] The concentration of glycogen is about half of the average concentration in the cerebrum,[15] yet glycogen is more stable in SC than in other CNS areas in hypoglycemic animals.[16] During postnatal development glycogen decreases four to five times in SC, whereas it increases three to four times in cerebral cortex. The change occurring in SC may be partly explained by the corresponding increase of white matter.[15]

A uniquely high concentration of glycogen is present in a particular structure of lumbo-sacral SC of all birds, called "glycogen body" and first described by Terni.[17]

The derivation of "glycogen body" cells and the function of the stored glycogen are still open to discussion. Glycogen constitutes as much as 60–80 % of the lipid-free dry weight of this structure in the chick.[18] During the development of the chick embryo the level of glycogen contained in the "glycogen body" increases, and at hatching time it accounts for 10% of the total body glycogen.[18] However, the metabolism of this glycogen is very slow, probably in relation to the very low level of glycogenolitic enzymes.[19]

The rate of transport through the cell membranes seems to be an important factor regulating the level and the metabolic utilization of glucose in

TABLE II

Carbohydrate Levels in Rat Spinal Cord

Compound	μmole/g wet wt.
Glucose	0.41[13]a
Glucose-1-P	0.50[14]
Glucose-6-P	1.64[14]
Fructose	0.15[13]
Fructose-6-P	0.10[14]
Fructose-1-6-P	0.04[14]
Sorbitol	1.18[13]
Phosphoglycerate	0.80[14]
Lactate	2.00[14]
Pyruvate	0.40[14]
Glycogen: cat (mg/g)	0.25[15]
dog (mg/g)	0.29[15]

a References are given in parentheses.

TABLE III

Activity of Enzymes of Glucose and Glycogen Metabolism in Spinal Cord

	mmoles of substrate converted/g lipid-free dry wt./hr		
	Rabbit (white matter)[a]	Rabbit (ventral horn)	Man (whole cord)[c]
Hexokinase (2.7.1.1)	1.5[23] 1.3[24]	6.3[b(28)]	
Glucose phosphate isomerase (5.3.1.9)	64[23,24]	30[b(28)]	7.0[27]
Phosphofructokinase (2.7.1.11)	9.8[23] 7.7[24]		
Fructosediphosphate aldolase (4.1.2.13)	2.8[23] 3.7[26]		
Lactate dehydrogenase (1.1.1.27)	5.3[23] 5.5[24]	32[b(28)]	3.6[27]
Glucose-6-P dehydrogenase (1.1.1.49)	0.67[24]	0.98[b(28)]	3.9[27]
Phosphogluconate dehydrogenase (1.1.1.43)		0.99[b(28)]	0.91[27]
Phosphoglucomutase (2.7.5.1)	7.9[23] 14[24] 19[25]	16[25]	0.91[27]
UDP-glucose pyrophosphorylase (2.7.7.9)	1.8[25]	2.2[25]	
Glycogen-UDP glucosyl-transferase (2.4.1.11)	0.40[25]	0.39[25]	
Glycogen phosphorylase (2.4.1.1)	5.1[23] 6.6[25]	10[25]	1.8[27]

[a] References are given in parentheses. The values reported for different fiber tracts in References 23, 24, and 25 were averaged.

[b] Values obtained in isolated ventral horn cell bodies.

[c] Values reported for thoracic and lumbar spinal cord were averaged and converted from a wet weight basis to a lipid-free dry weight basis assuming a correction factor of 10.

the nervous tissue. Rafaelsen[20] reported a glucose uptake of 3.2 mg/g wet wt./hr by rat SC slices incubated in the presence of 3 mg/ml glucose. Comparable values were reported by others.[21,22] Rafaelsen found that insulin (10^{-1} i.u./ml) can increase up to 80% the uptake of glucose and to a much lesser extent that of fructose and galactose but has no effect on the uptake of arabinose, xylose, and ramnose. Since no stimulatory effect was observed on several pathways of glucose utilization, it was suggested that insulin might increase cell permeability to glucose in SC.[20] Conclusive evidence is lacking, however. Other authors[22] were not able to reproduce these results unless they used the same type of anesthesia used by Rafaelsen.

B. Enzymes of Glucose and Glycogen Metabolism*

Table III shows that enzymes involved in glucose metabolism are present in both gray and white matter of SC. However, some differences in the distribution can be noted—e.g., hexokinase activity is higher in ventral horn cells than in white matter, suggesting a higher rate of glycolysis in gray matter; the activity of glucose-6-phosphate dehydrogenase, which is generally taken as a measure of the activity of the hexose monophosphate shunt (pentose cycle), shows the opposite pattern. In general, the activity of the pentose cycle enzymes is higher in SC than in other CNS areas and also is higher in white than in gray matter, as shown for transketolase (2.2.1.1),[29] glucose-6-phosphate dehydrogenase,[24,27] and phosphogluconate dehydrogenase[27]; it is likely that these enzymes are related to the oligodendroglia and to the metabolism of myelin.[29] Pyruvate decarboxylase (4.1.1.1) is less active in the cord than in other CNS areas and its association with neurons was suggested.[30] A high activity of the pentose cycle was correlated with early phases of cellular differentiation and proliferation in the chick embryo SC.[31]

Lactate dehydrogenase consists of several isoenzymes whose distribution pattern varies in different regions of the CNS[32,33]; the pattern of SC is similar to that of medulla, pons, mesencephalon, and cerebellum.[32]

Table III shows that enzymes related to the synthesis and breakdown of glycogen are also present in both gray and white matter of SC. It is worthwhile noting that glycogen phosphorylase activity is lower in white than in gray matter.[25]

C. Enzymes of Krebs Cycle and Respiratory Enzymes— O₂ Uptake

Quantitative studies on these groups of enzymes are scanty and are mentioned only briefly here.

The activity of succinate dehydrogenase (1.3.99.1), malate dehydrogenase (1.1.1.37), cytochrome oxidase (1.9.3.1) and reduced NAD

---

* The names used for the enzymes are the trivial names recommended by the International Union of Biochemistry. The systematic number of each enzyme is reported in parentheses at its first appearance in the text or in the tables.

dehydrogenase (1.6.99.3) was shown to be about half in SC white matter than in cerebral cortex.[14] In SC white matter, fumarate hydratase (4.2.1.2) activity (9.3 mmole of substrate converted per gram of lipid-free dry weight per hour) is about double that of isocitrate dehydrogenase (1.1.1.41).[34]

Some oxidative enzymes were more widely studied with histochemical methods. They have generally lower activity in SC than in other CNS areas and have low or no activity in white matter.[35-37] The distribution in gray matter is not homogeneous; large and medium multipolar ventral horn neurons[36] and neurons of the Clark Stilling column[38] show higher activity than the rest of neuropil. A high level of succinate dehydrogenase and cytochrome oxidase has been described in synaptic regions, probably in relation to the large number of mitochondria present.[37]

In agreement with the low activity of oxidative enzymes, the respiration of SC is rather low, about half or less than half of cerebral hemispheres. The rate of respiration *in vitro* (Table IV) tends to decrease with time and depends on the substrate present in the medium, being highest in presence of glucose. From birth to adult age, the oxygen consumption decreases in SC, whereas it increases in cerebral cortex[41]; this may be only partially explained by the larger deposition of myelin in SC. Interestingly enough, the same developmental pattern was observed in glycogen levels. Respiration rates severalfold lower than those reported in Table IV are sometimes found in the literature,[22,42] even when the experimental conditions were apparently the same. One factor determining this variation may be the different thickness of the SC slices used.

The presence of high potassium concentration in the incubation medium, which is known to stimulate the respiration rate of brain slices, has no effect[42] or has only a moderate effect[40] on the O_2 uptake of SC.

IV. LIPIDS

A. Lipid Composition of Spinal Cord

The most systematic studies on SC lipids are still those of Branté in 1949[8] and of McColl and Rossiter in 1952[43]; from that time new classes of lipids have been characterized and new methods of isolation have been developed, which makes it sometimes difficult to compare older with newer data. In many cases we could not avoid using the older terms of lecithins and cephalins because these phospholipids had not yet been further characterized in some of the most relevant publications quoted.

Table V shows the lipid composition of the SC of ten animal species; values for eight other species were reported.[43] For clarity all the values are given as percent of dry weight, which sometimes involves recalculation of the original data. In a few cases this may have brought some slight distortion.

Total lipids account for about two thirds of the SC dry weight (Table V) and increase with the body size of the animal.[9]

TABLE IV

Factors Affecting the Respiration of Spinal Cord *In Vitro*

Incubation time[39]a (rat)	Q_{O_2}	Substrate[40] (rat)	Q_{O_2}	Age[41] (dog)	$Q_{O_2}{}^b$
0–30 min	6.3	None	2.5	First week	2.7
30–60 min	5.2	Glucose 10 mM	6.1	Second week	2.7
60–90 min	4.8	Fructose 10 mM	5.0	Third week	3.1
		Glutamate 10 mM	3.5	Adult	1.7
		α-Ketoglutarate 10 mM	3.5		

[a] References are given in parentheses.
[b] Recalculated assuming 30% dry wt.

In almost all the species studied the amount of cephalins appears to exceed that of lecithins by a factor of 1.5–3.[43] However, the values reported in Table V for cephalins and lecithins include their plasmalogens, and in later studies[45] it was shown that about 90% of cephalins and only a negligible fraction of lecithins are present in the plasmalogen form. Therefore, if plasmalogens are excluded, the ratio between cephalins and lecithins is inverted. The largest fraction of cephalins consists of ethanolamine phosphatide.[45]

The concentration of lecithins in SC is lower than in brain in birds and mammals, is similar in amphibians and reptiles, and is higher only in fish. Cephalin concentration is in most cases similar in brain and SC.[50]

Cholesterol, sphingomyelin, and cerebrosides (currently, though inexactly called "myelin lipids" for their high concentration in myelin) represent about two thirds of the total lipids; they tend to be somewhat lower in submammalian than in mammalian species.[43] The SC concentration of "myelin lipids" consistently exceeds that of brain in all the species except menhaden and turtle, where cerebrosides are higher in brain.[43,50]

An inverse correlation seems to exist between the concentrations of sphingomyelin and cerebrosides in the various animal species; in general relatively higher sphingomyelin levels correspond to relatively lower levels of cerebrosides and vice versa.[50]

In adult animals cholesterol occurs almost exclusively in free form,[51] but in earlier stages of life esterified cholesterol may be almost 50% of the total; esterified cholesterol is still present in the SC of 5-yr-old children.[49] In men the free form increases after birth for few months and then remains at a rather constant level; in the hen there is a progressive increase from the fourteenth day of incubation (chick embryo) to the thirtieth to fortieth day of life.[49] The level of cholesterol is much higher in white matter, but the cholesterol of gray matter has a higher turnover.[52]

TABLE V
Lipids of the Spinal Cord (% dry wt.)[a]

	Man	Cow[b]	Cat	Rabbit	Guinea pig	Rat	Hen	Pigeon	Turtle	Frog
Total lipids	60[44]	76[8] 81[47]		69[b][8]		57[8]	78[48]			
Cholesterol	17[b][8]	16[8] 16[47]	18[43]	16[8] 18[43]	16[43]	12[8]	21[48] 13[49]	13[43]	15[43]	11[43]
Total phospholipids	26[45]	34[8] 36[47]	29[43]	33[43]	27[43]	26[8] 36[45]	36[45] 32[48]	19[43]	28[43]	21[43]
Lecithins	6.3[b][8]	6.3[8]	4.7[43]	5.4[b][8] 5.6[43]	5.8[43]	6.9[8]		3.8[43]	7.4[43]	8.7[43]
Cephalins[c]	17[b][8]	17[8]	15[43]	16[b][8] 17[43]	14[43]	13[8]		8.6[43]	13[43]	7.7[43]
Sphingomyelin	11[b,d][8]	10[d][8]	9.6[43]	9.7[b,d][8] 12[43]	7.1[43]	6.1[d][8]	9.1[48]	6.1[43]	7.4[43]	5.1[43]
Total plasmalogen	8.8[45]					17[45]	18[45]			
Cephalin plasmalogen	8.8[45]					15[45]	14[45]			
Cerebrosides[e]		20[8]	19[43]	17[b][8] 18[43]	13[43]	14[8]	22[48]	14[43]	7.8[43]	20[43]
Diphosphoinositide		2.1[8]		2.1[8]						
Proteolipids[f]	4.8[b][46]	4.0[47]								

[a] References are given in parentheses. Some of the values presented were recalculated assuming a dry weight of 30% for the whole spinal cord and of 35% for spinal cord white matter.
[b] White matter only.
[c] Includes cephalin plasmalogens.
[d] Includes some plasmalogen and long-chain ether phospholipid.
[e] Includes glycolipids.
[f] Measured as protein.

Sphingomyelin and cerebrosides contain, in addition to sphingosine and dihydrosphingosine, small quantities of C_{20} sphingosine and of another unidentified base and even smaller quantities of C_{16} dihydrosphingosine. Sphingomyelin contains 22.7% and cerebrosides 24.7% dihydrosphingosine in the rabbit SC, a value that is much higher than in the brain.[53]

The amount of plasmalogen found in human SC is about half of that in the rat and in the hen.[45] In the latter two species it constitutes about half of the total phospholipids.

A caudo-cranial decrease of proteolipids in human antero-lateral columns, but not in dorsal columns, has been described.[46] Proteolipids are higher in white than in gray matter. However, their concentration is not proportional to that of "myelin lipids" but rather to the total axon circumference of the white matter area considered. Consequently they are relatively higher in areas in which the fiber diameter is small (such as corpus callosum) than in areas (such as SC white columns) where large fibers prevail.[47]

B. Developmental Changes in Spinal Cord Lipids

Maturation of gray matter and especially myelination determine a large increase of total lipids. The deposition of newly formed lipids does not proceed simultaneously in gray and white matter,[8] and in white matter different fiber tracts undergo myelination at different developmental stages; myelination occurs earlier in phylogenetically older systems such as SC.[54,55]

Table VI shows the changes occurring in various lipid classes in phases of active myelination. Although the range of age covered is different in the three animals, some common pattern can be outlined. Generally "myelin lipids" increase more than do other lipids; in contrast, plasmalogen and phosphatidylserine seem to decrease in the developing lamb SC white matter. A decrease of plasmalogen with age also has been described in human SC, where it tends to increase again in very advanced age.[44] In the chick embryo SC the largest increase of most lipid classes occurs between the eighteenth and twenty-first day, which is also the period of most active myelination as judged by histochemical methods.[57] In the chick embryo only minor changes were demonstrated for sphingomyelin, which is the compound increasing the most in rat and in lamb SC (Table VI).

In humans a steady level of total SC lipids is reached after the tenth year of life.[44] In the rat SC lipid deposition is completed around the fortieth day of life (Table VI).

In the rat a great increase of glycerol-phosphate-dehydrogenase (1.1.1.8) occurs between the twentieth and the fortieth day of life, which is a period of active myelination.[58]

C. Phosphorylated Compounds Related to Phospholipids

Porcellati and co-workers[59–61] described the presence in the nervous tissue of a series of free phosphorylated compounds that are classified in this section because they seem to be related to the metabolism of phospholipids.

TABLE VI

Developmental Changes in Spinal Cord Lipids[a]

Age, days	Lamb[b,56]				Rat[8]						Chick embryo[57]			
	7	14	28	182	1	11	17	24	41	70	14	16	18	21
Total lipids	70	63	73	100	42	64	77	90	99	100	63	71	81	100
Cholesterol	80	77	86	100	26	47	44	71	90	100	53	58	76	100
Total phospholipids	71	83	98	100	46	70	85	93	101	100	74	83	84	100
Phosphatidylcholine	102	104	104	100			93	126	107	100	71	74	73	100
Ethanolamine phospholipid	175	148	345	100	58[c]	79[c]	82[d]	88[d]	92[d]	100[d]	72	82	80	100
Phosphatidylserine	36	43	55	100	5[e]	38[e]	82[e]	67[e]	116[e]	100[e]	55[f]	66[f]	79[f]	100[f]
Sphingomyelin	140	144	146	100							85	98	90	100
Total plasmalogen	64	62	73	100										
Cerebrosides							39	73	115	100	45	51	58	100

[a] References are given in parentheses. All the figures reported were recalculated from the original, taking as 100 the values given for the latest stage of development studied.
[b] White matter only.
[c] In the original publication the values were comprehensive of lecithins and cephalins.
[d] In the original publication the values were given for "cephalins."
[e] Includes some plasmalogen and long-chain ether phospholipid.
[f] Includes phosphatidylinositol.

TABLE VII

Concentration of Some Free Phosphoric Esters in the Spinal Cord (μmole/g wet wt.)

Animal species	PS[59]a	PE[59]	PC[59]	GPE[59]	GPC[59]	L-SEP[60]	L-TEP[60]
Scyliorhinus canicula	None					None	2.60
Cyprinus carpio	None			0.45		0.78	2.21
Ameiurus nebulosus		0.24		0.06	0.06	0.40	4.20
Rana esculenta		0.19	0.12	1.07	1.11	2.05	None
Testudo hermanni		0.98		2.02			
Hemidactilus turcicus	None	0.18	0.16	0.39	0.36		
Gallus domesticus	0.06	0.49	0.32	0.59	0.14		
Rattus rattus (albus)	Present	0.29	0.31	0.29	0.31		
Felis catus		1.13		2.84	Present		
Bos taurus		1.12	0.87	1.61	1.20		
*Machacus rhesus*b		0.67		1.62	1.49		
*Homo*c		0.21		1.98	Present		

a References are given in parentheses. Abbreviations: PS, phosphorylserine; PE, phosphorylethanolamine; PC, phosphorylcholine; GPE, glyceryl-phosphorylethanolamine; GPC, glycerylphosphorylcholine; L-SEP, serine ethanolamine phosphate; L-TEP, threonine ethanolamine phosphate.
b From lyophilized preparations.
c Extracted 20 hr after death.

Table VII shows the distribution of these compounds in the SC of 12 animal species representative of 6 classes of vertebrates. More species were studied in the original publications,[59,60] where brain levels also were reported.

The distribution of serine ethanolamine phosphate and of threonine ethanolamine phosphate is interesting because in some species only the former, and in others only the latter, compound is present. The same type of distribution occurs in the brain, and it seems to be related to the evolutionary development of the species studied[60]; microsomal fractions or extracts of fish brain contain enzymes capable of hydrolyzing these two phospho-diesters.[61]

It is possible that the described substances are utilized for the synthesis of lecithins and cephalins; alternatively they might be products of the catabolism of these phospholipids.[61]

D. Prostaglandins

Among the compounds released spontaneously and after electrical stimulation from the isolated frog SC a group of unsaturated hydroxy-carboxylic acids called prostaglandins has been isolated.[62]

Prostaglandins are released also by cerebral cortex and by diaphragm after nerve stimulation, and they may be related to nervous activity.[62]

V. PROTEINS

A. Amino Acids

1. Free Amino Acid Pools

Many recent studies have appeared on the composition of free amino acid pools in the brain and in different brain areas. In striking contrast, to the author's knowledge no such study has been done on SC, and the most complete analysis of free amino acids in this organ appears in a publication dating back to 1957.[63] Table VIII shows the levels of free amino acids in the SC of three animal species. Generally, as in the brain, the concentration of Glu and related amino acids is high as compared to that of most other amino acids. In the hen, the level of GABA and, to a lesser extent, that of other amino acids (Asp, Glu, Gln, Ser, Gly, Thr, and Tau) is lower in SC than in brain.[63] The level of Gly in cat SC (Table VIII) is higher than its cerebral level in many species (cf. Table IIA in Tallan).[68] Asp, GABA, Glu, Gln, and Gly are 1.5–3 times more concentrated in gray than in white matter. Glu and GABA are higher in dorsal than in ventral gray matter; Asp and Gly show the opposite pattern. Moreover, Glu is higher in dorsal portions of white matter and Gly in ventral[64] (Table VIII). The possible role of Gly and Asp as inhibitory or excitatory transmitters, respectively, is discussed in Volume III of this series.

TABLE VIII

Concentration of Free Amino Acids in Spinal Cord (μmole/g wet wt.)

Amino acid	Cat[64]a Gray matter. Ventral	Gray matter. Dorsal	White matter Ventral	White matter Dorsal	Rat	Hen[63]
Alanine						1.0
Aspartate	3.1	2.1	1.3	1.1	1.1[59] 2.2[65] 3.2	2.0[67]
GABA	1.0	2.2	0.44	0.43	1.6[65]	0.86
Glutamate	5.3	6.5	3.9	4.8	2.8[65]	6.9
Glutamine	5.4	5.3	3.8	3.6		1.4
Glycine	7.1	5.7	4.4	3.0		1.2
Leucine					0.08[66]	0.18
Lysine					0.64[66]	
Serine						0.81
Taurine					1.3[65] 1.1[59]	0.84 0.54[67]
Threonine					0.42[59]	0.50 0.26[67]
Tyrosine						0.14
Valine						0.23

a References are given in parentheses.

In the rat, Lys tends to increase from cranial to caudal areas of the CNS and is more concentrated in SC than in the other six areas analyzed.[69]

Differences among various species in the SC and brain levels of Asp, Thr, Tau, and urea are outlined in a recent comparative study.[59]

Acetylated and phosphorylated amino acids have been described in SC. Most of free Asp seems to occur in the form of acetylaspartate.[68] Concentrations of 4.9,[65] of 3.2,[70] of 0.8 (lumbar cord), and of 1.5 (thoracic cord) μmole/g[71] were reported. Similar values were shown for N-acetyl-aspartylglutamate.[71] Some phosphorylated amino acids have been treated in the lipid section.

2. Amino Acid Transport

SC slices are able to take up amino acids from the incubation medium against a concentration gradient. The uptake rate and the level of steady-state accumulation are lower than the average of other CNS areas, which may be partly due to the large amount of myelin present.[69]

In vivo a "barrier system," similar to that of brain, seems to be operating in SC. The passage of amino acids from the blood is much restricted, and a net uptake of Leu, Lys, and Tau could be obtained only by increasing many-fold their concentration in the blood.[66] A net passage of Tau did not occur in other regions of the CNS than SC and pons medulla, which would point to the existence of regional differences in the permeability of the "barrier system."[66]

3. Enzymes Related to the Metabolism of Amino Acids

The enzymes studied most are those related to the metabolism of Glu and GABA.

Glutamate decarboxylase (4.1.1.15) activity is lower in SC than in brain and is lower in monkey than in rabbit SC (1.4 and 2.3 μmole of GABA formed per gram wet weight per hour, respectively); the enzyme activity seems to be inversely related to body size.[72] No activity could be detected in the rhesus monkey SC white matter[72,73]; in gray matter an increasing ventro-dorsal gradient of enzyme activity has been demonstrated with an almost fourfold difference between most ventral and most dorsal areas (6.8 to 25.6 μmole of product formed per gram wet weight per hour).[73] Also, similar if not as steep is the ventro-dorsal gradient of activity shown for GABA aminotransferase (2.6.1.19; 197 to 315 μmole of product formed per gram dry weight per hour).[74] The ratio between aminotransferase and decarboxylase (which is always higher than 1 in the CNS) decreases from a value of 29 to 12 from most ventral to most dorsal areas of SC gray matter.[73] Interestingly enough, also Glu and GABA show a ventro-dorsal increase of concentration (Table VIII).

Another enzyme of the "GABA cycle," succinate semialdehyde dehydrogenase (1.2.1.16), is also higher in dorsal than in ventral gray matter of monkey SC (987 and 647 μmole of product formed per gram dry weight per hour, respectively); the activity of white matter is about ten times lower.[75]

Aspartate aminotransferase (2.6.1.1) and glutamate dehydrogenase (1.4.1.2) activities were found to be 32 and 3.6 mmole/g lipid-free dry weight per hour, respectively, in isolated rabbit ventral horn cells.[28]

The presence of D-amino acid oxidase (1.4.3.3) in the CNS has been described[76,77]; in mammals the activity is highest in cerebellum, lowest in hemispheres, and intermediate in SC, where it decreases in the following order: rat > sheep > dogfish > frog > mouse > duck > flounder, with a twentyfold difference between rat and flounder.[76]

B. Peptides

1. Phosphopeptides

A phosphopeptide-like substance structurally related to two phospho-diesters, serine ethanolamine phosphate and threonine ethanolamine phosphate, has been demonstrated in the SC of birds and of some amphibians.[78] Its concentration is very low in the brain, and it is absent in the NS of mammals.[79] This substance would yield upon hydrolysis Ser, ethanolamine, Glu, Val, Thr (?), Pi, and probably other nitrogen-containing components. It was shown that some vertebrate species contain only one or both the phosphodiesters and not the phosphopeptide, and it is likely that the distribution of these compounds among the various species is related to evolutionary processes.[78] This phosphopeptide is probably involved in the metabolism of phospholipids and phosphatidylpeptides in highly myelinated tissues.[78]

2. Substance P

Substance P is a polypeptide, not yet isolated in pure form, present in the nervous system of all vertebrates; the topic has been recently reviewed[80] and is treated extensively in another section of this book.

In SC the concentration of Substance P is high in dorsal roots and funiculi gracilis and cuneatus and low in lateral and ventral white matter and in ventral roots, a distribution that is opposite to that of choline acetyltransferase.[81] The concentration is generally higher in gray than in white matter.

3. Encephalitogenic Peptides

Many investigators used SC (mostly bovine) as a major source of encephalitogenic material. There is no general agreement among the various authors as to the physical and chemical characteristics of this material, but any detailed description is beyond the scope of this presentation. A recent article summarizes the present state of research in the field.[82]

High yield of encephalitogenic substance is obtained from basic protein fractions of SC. Basic proteins are approximately 1% of SC proteins.[82] The biologically active component seems to be one or more relatively small, basic polypeptide rather than larger proteins; the molecular weight claimed ranges from approximately 3000[83] to approximately 16,000.[84] This peptide may be bound to other protein or lipid molecules in the living animal[82] or may represent a degradation product of larger molecules.[84] Its amino acid composition is typical for the very high content of Gly and rather high content of Ser, Lys, and Arg.[83,85]

C. Proteins

1. Protein Content

The protein content of adult rat SC is about 10% of the wet weight, with insignificant differences among cervical, thoracic, and lumbar segments.[86]

From the tenth day before birth to about a month after birth the amount of protein increases tenfold if it is related to wet weight and about three- to fourfold if it is related to dry weight.[86] In the CNS development, proteins increase and reach a steady level earlier in phylogenetically older regions, such as SC.[87] The amino acid composition also changes during development. In guinea pigs the most significant changes occur between late fetal and early postnatal stages and consist of a two- to threefold increase of Glu, Phe, and

TABLE IX

Phosphorus, Labile Phosphates, and Nucleic Acids in Spinal Cord (mg/g wet wt.)

	Pi	Total P	Lipid P	Acid-soluble P	Phosphocreatine	ATP	DNA P	RNA P
Guinea pig[98]a		5.0	3.4	1.2			0.16(DNA + RNA)	
Rat[14]	0.17	4.7[99]	3.4[b]	1.2	2.7	0.52	0.12[b]	0.20[b]

a References are given in parentheses.

b White matter only. Other data reported for DNA and RNA content in spinal cord are as follows: (1) in men (mg/g dry wt.)[100]—DNA, 4.3; RNA, 7.3; (2) in cats (mg/g ethanol ether treated dried tissue)[101]—DNA, 2.7; RNA, 14; (3) in rabbit isolated, Carnoy fixed, ventral horn neurons (mg/cm³)[102]—RNA, 25.

Leu and in a similar decrease of Val and Ser. Also, His decreases more than three times, but the change occurs more slowly and continues in adult life.[88] Most likely these changes reflect changes in the relative amount of different proteins or the appearance of new proteins. Before myelination the amino acid composition of white and gray matter proteins is similar.[88]

The author is not aware of studies on individual proteins in SC except for the identification of the S-100 protein[89] in amount comparable to that found in other areas (B. W. Moore, personal communication). Proteolipids have been briefly described in the lipid section.

2. Protein Turnover

The half-life time of SC proteins (10–19 days in the monkey) is 20–40% higher than in other areas of the CNS, indicating a lower turnover rate.[90] The incorporation of methionine S^{35} into proteins is higher in SC than in cerebral cortex and hypothalamus in young animals but lower in adult animals.[91] Autoradiographic studies with methionine S^{35} showed that the incorporation of label in SC is less than in other CNS areas and is more than in peripheral nerve.[92] In SC the highest incorporation of label from methionine S^{35}[92] and leucine H^3[93] was found in ventral horn motoneurons, where it was greater than in several other types of neurons, including pyramidal cells; very little incorporation was found in SC white matter.[93]

With biochemical methods it was confirmed that the uptake of C^{14} or H^3 from injected glutamine or lysine, respectively, is lower in SC as a whole and in SC gray matter than in other CNS areas but is higher in ventral horn motoneurons than in any other area of the CNS on a whole-tissue basis.[94,95]

The metabolic breakdown of proteins received much less attention than protein synthesis. The presence of two groups of proteinases, similar to those described in the brain, one active at acid and one at neutral pH, has been described in the monkey SC.[96] The enzyme activity is somewhat lower than in other areas of the CNS, which is in agreement with the higher half-life time of SC proteins. Also in the hen proteinase activity is lower in SC than in brain.[97]

VI. PHOSPHORUS, LABILE PHOSPHATES, AND NUCLEIC ACIDS

The few data available on the concentration of labile phosphates and nucleic acids in SC have been summarized in Table IX. The data on DNA and RNA are not very consistent and were obtained with different methods by the various authors. The level of RNA is lower in SC than in other CNS areas, while DNA seems more evenly distributed[101]; however, a detailed regional analysis also revealed differences in DNA concentrations.[103] In experiments performed in isolated motoneurons the amount of RNA per cell was related to the cell surface and not to the cell volume.[102]

Studies on the incorporation of labeled thymidine into DNA excluded any renewal of spinal neurons in adult mice; in fact, only a small number of glial cells became labeled.[104]

VII. PHARMACOLOGICALLY ACTIVE SUBSTANCES

A. Acetylcholine and Related Enzymes

The concentration of acetylcholine (ACh) is much higher in SC gray than white matter, and within white matter, it is higher in areas with efferent axons and almost absent in areas with only afferent axons.[105] Average values reported for dog lumbar SC in micrograms per gram are: whole cord, 1.6; gray matter, 3.0; and white matter, 0.02–1.0.[105] These values refer to anesthetized and eserinized animals; therefore, the ACh concentration in the living SC may be much lower; in the SC of nonanesthetized cats (with praetrigeminal mediopontine section), a level of 0.38 μg/g was found[106]; a similar value had been reported for human SC (0.45 μg/g).[107] The concentration of ACh in SC is in the same range as in other regions of the CNS and lower than in some peripheral nerves and ganglia.[105]

Since the discovery of ACh and of its role in the nervous system many studies have dealt with the enzymes responsible for its synthesis and degradation, and often the enzymatic activities were taken as indicative of the presence of cholinergic transmission. However, quantitative biochemical data are rather scanty and often inconsistent.

Choline acetyltransferase (2.3.1.6) is lower in white than in gray matter and shows an increasing dorso-ventral gradient of activity[108–110]; maximum activity was found in ventral roots (73 μmole/g acetone powder per hour) and almost none in dorsal roots.[108] In ventral horn, dorsal horn and white matter activities of 31, 9, and 2 μmole/g dry wt./hr, respectively, were measured.[109] Much lower values were reported in earlier studies,[110] because the experimental conditions used did not allow maximum rates of reaction.

The activity of acetylcholinesterase (3.1.1.7) (AChE) is severalfold higher than that of choline acetyltransferase, but the distribution pattern of the two enzymes in the CNS is similar.[111] The activity is higher in gray than in white matter[111] and is not homogeneously distributed throughout gray matter. AChE was measured in isolated ventral horn neurons: within one group of motoneurons two types of cells were differentiated on the basis of AChE activity. The difference between the two groups of cells is more relevant if the enzyme activity is referred to single cells instead of units of volume.[112] In isolated motoneurons AChE activity is about ten times greater in the cytoplasm, dendrites, and axons than in the nucleoplasm, and no activity is present in the nucleolus.[113]

In adult animals AChE activity is lower in SC than in brain. For example, caudate nucleus activity, which is the highest in the brain, is six to seven times greater than that of SC gray matter.[111] However, the opposite relation exists in earlier stages of development, as shown in the sheep[114] and in the rat[86] CNS. In both animals AChE reaches its peak of activity in SC when the brain level is still very low; then it decreases slowly in SC and progressively increases in brain. In the rat SC the maximum activity is reached in the second day of postnatal life (Table X) and in the sheep during the eleventh to twelfth week of gestation. In the chick embryo SC a fivefold increase in AChE activity occurs at an early developmental stage (fifth to eleventh day).[115] The increase in AChE activity seems to correspond to the onset of nervous function.[86,114,115]

Earlier studies did not discriminate between AChE and cholinesterase; this distinction is especially important in regions with low AChE activity. In SC gray matter cholinesterase accounts for 20% of total ACh hydrolysis; in white matter, for 33–43% depending on the fiber tracts considered.[111] Cholinesterase is more evenly distributed in the CNS than AChE.[111]

With histochemical methods AChE was generally demonstrated in neuronal structures: perykarya, surface membranes, axons, dendrites, and synaptic regions.[116–119] The activity is higher in motoneurons than in sensory neurons.[116,118] Cholinesterase is localized in nonneuronal structures: either in blood vessels[117,119] or in blood vessels and glial cells.[118,120] Species differences are rather frequent; e.g., in the CNS of amphibians and fish AChE also can be found in blood vessels[120]; cholinesterase is extremely low or absent in the SC and brain of amphibians and reptiles.[116,120]

B. Monoamines and Related Enzymes

Table XI shows the levels of catecholamines and 5-HT in the SC of various animal species. Dihydroxyphenylalanine (DOPA) has been also included in this table, since it is the direct precursor of hydroxytyramine (dopamine).

TABLE X

Developmental Changes of AChE Activity in Rat Spinal Cord[86]

	AChE activity	
Age, days	μmole/g wet wt.	μmole/100 mg protein
Before birth		
11	0.49	4.8
15	3.0	18
19	9.7	39
After birth		
2	14	46
9	12	16
44	9.8	9.5

TABLE XI

Concentration of Monoamines and of Dihydroxyphenylalanine (DOPA) in Spinal Cord

Species	ng/g wet weight			
	DOPA	Dopamine	Noradrenaline	5-HT
Man	350[121]a	320[121]	130[121]	
Ox	180[121]	200[121]	100[121]	
Hog	240[121]	180[121]	110[121]	
Cat	400[121] 50[122]	450[121] 50[122]	190[121] 50[122]	
Rabbit	330[121] 50[122]	840[121] 20[122]	160[121] 30[122]	330[124]
Guinea pig	50[122]	60[122]	170[122]	
Rat	670[121] 50[122]	1000[121] 20[122]	150[121] 110[122]	
Hen		60[123]	157[123]	440[123]
Turtle	50[122]	80[122]	80[122]	
Frog				2530[125]

a References are given in parentheses.

The lower values reported in Table XI for DOPA and dopamine were obtained with more specific methods[122]; other authors found that dopamine is barely measurable in normal SC.[126]

In general no large differences among the various species seem to be present except for the very high level of 5-HT in the frog SC; 5-HT is very high also in frog brain, particularly in mesencephalon[125]; a much lower concentration of 5-HT was reported in isolated frog SC.[127]

The caudo-cranial increase in noradrenaline (NA) and 5-HT in SC (Table XII)[128,129] may be related to a corresponding increase of gray matter[129]; NA concentration is in fact very low in white matter.[128,129] In the SC of hen, in contrast with mammals, the segmental distribution of NA, dopamine, and 5-HT is very even, and the concentration of 5-HT is significantly lower than in brain.[123]

The concentration of NA seems to be higher in lateral horns than in ventral and dorsal horns; significant differences were demonstrated in the cow[128] but not in the dog.[130] Other authors found much higher values of dopamine and NA in dorsal than in ventral horns but did not report separate values for lateral horns.[121]

Adrenaline level in SC is very low as compared to that of NA (approximately 20 ng/g wet wt.).[121,122]

Rather high levels of tyramine have been described in the rabbit and dog SC (3.3 and 3.4 μg/g, respectively).[131] Tyramine is almost absent in tissues other than the CNS, and its role might be related to nervous function.[131]

The segmental distribution of DOPA decarboxylase (4.1.1.26), the enzyme considered responsible for the formation of catecholamines and 5-HT, follows rather closely that of NA and 5-HT (Table XII). In the hen, in agreement with the even distribution of NA and 5-HT, no segmental difference in the activity of DOPA decarboxylase was found.[123]

It is beyond the scope of this review to discuss the role of NA and 5-HT as neurotransmitter substances. However, some experiments performed on SC that are related to this problem will be briefly mentioned.

Isolated SC spontaneously releases NA[132] and 5-HT[127]; the release is greatly enhanced by electrical stimulation of the descending spinal tracts, and is evident only in animals treated with monoamine-oxidase inhibitors.

In *in vivo* experiments, a depletion of NA and 5-HT in SC can be obtained after prolonged electrical stimulation.[133] An almost complete disappearance of NA[126] and 5-HT[134] occurs in the caudal segment of transected SC. DOPA decarboxylase shows the same pattern, suggesting that NA, 5-HT, and the enzyme responsible for their formation are localized in the descending tracts of SC.[135]

With histochemical methods based on the different fluorescence of catecholamines and 5-HT, a high concentration of amines was demonstrated in synaptic terminals.[133] NA and 5-HT terminals have their highest concentration in lateral horns, where they surround most of the nervous cells; many are present also in ventral horns and much less in dorsal horns.[136] Moderate fluorescence is seen also in axons.[133]

NA and 5-HT terminals derive from nerve tracts descending in the anterior and lateral column of SC, probably having their origin in the medulla oblongata. The neuron systems of NA and 5-HT are differently distributed.[133]

TABLE XII

Segmental Distribution of Monoamines and of DOPA-Decarboxylase in Spinal Cord[128]

Area	Noradrenaline[a]		5-HT[a]		DOPA-decarboxylase[b]	
	Rabbit	Cat	Rabbit	Cat	Rabbit	Cat
C_1–C_3	90	80	250	180	40	26
Cervical enlargement	90	120	340	370	40	24
Th_3–L_1	70	110	220	250	38	27
Lumbar enlargement	140	230	340	740	41	43
Conus medullaris	260	280	590	1410	54	77

[a] ng/g wet wt.
[b] μg of dopamine formed per gram wet weight 45 min after the addition of DOPA. Other authors[129] reported levels of noradrenaline ranging from 160 ng/g (C_2–C_3) to 400 ng/g (S_1–S_3) in the cat spinal cord.

C. Polyamines

The presence in the CNS of two polyamines, spermidine and spermine, has been described.[137] The concentration of spermidine is higher in white matter and that of spermine, in gray matter. Spermine is more concentrated in cervical than in lumbar SC (20 and 9 μg/g wet wt., respectively), whereas spermidine is evenly distributed (average, 134 μg/g). The physiological role of these compounds is unknown.

VIII. HYDROLASES AND CARBONIC ANHYDRASE

A. Phosphohydrolases

1. *Alkaline Phosphatase (3.1.3.1)*

Enzyme activities (in milligrams Pi liberated per gram wet weight per hour) of about 1 and 6.6 were reported in guinea pig SC[138] and in rat SC white matter, respectively.[14] In guinea pigs the activity was somewhat higher at cervical and especially at lumbar enlargements than in the thoracic tract.[138]

The histochemical localization varies according to the method used for demonstration,[119,139] and the heterogeneous results obtained suggest the existence of more than one alkaline phosphatase[139] and its involvement in more than one biological process.[140] Alkaline phosphatase activity is high in early stages of differentiation of the mouse CNS and decreases earlier in SC than in other areas, paralleling the earlier nervous differentiation of this tissue.[141]

2. *Acid Phosphatase (3.1.3.2)*

The activity of acid phosphatase in guinea pig SC is about 40% of that of alkaline phosphatase and has the same segmental distribution.[138] A discrete activity is generally found in white matter[140] (glial cells and large axons) and a high activity in the cytoplasm and processes of motoneurons.[119]

3. *5-Nucleotidase (3.1.3.5)*

A biochemical study of 5-nucleotidase showed an activity of approximately 1.8 mg of Pi liberated per gram wet weight per hour in guinea pig SC, the activity being somewhat higher at the level of lumbar enlargement.[138] Histochemically, the concentration was shown to be higher in gray than in white matter, and within gray matter, higher in dorsal than in ventral neurons.[142]

4. *ATPase (3.6.1.3)*

ATPase was measured in rat SC in a recent study.[143] The enzyme could be activated about 100% by Mg^{2+} or by Ca^{2+}; a further doubling of activity could be obtained by adding Na^+ and K^+ to the Mg^{2+} but not to the Ca^{2+} activated enzyme, and this further activation was ouabain-sensitive.[143]

No significant segmental difference in the activity of Ca^{2+}-activated ATPase was found in guinea pig SC.[138]

Histochemically only the Mg^{2+}-activated moiety can be demonstrated.[143] Its activity is higher in gray than in white matter and is highest around the central canal and in the substantia gelatinosa; high activity is present also in septa, blood vessels, and meningeal linings.[143] The presence of ATPase in motoneurons is uncertain.[119]

B. Nonspecific Esterases

The physiological substrates of this group of enzymes are still unknown. In a recent biochemical approach it was shown that their activity is two to three times lower in SC than in brain.[144] Nonspecific esterases have been demonstrated histochemically in motoneurons.[119]

C. Carbonic Anhydrase

Carbonic anhydrase (4.2.1.1) is probably involved in the secretion of cerebrospinal fluid and in active ion transport.[145] In the adult CNS a caudo-cranial increase of enzyme activity was demonstrated in a number of species, with the lowest levels in the most caudal segments of SC.[146] In humans an increasing gradient of carbonic anhydrase is evident only from the cord through the pons, the activity of hemispheres being the same as that of the pons.[146]

A correlation seems to exist between the appearance of enzyme activity and the beginning of nervous function.[147]

IX. VITAMINS

The level of free thiamine (Vitamin B_1) appears very homogeneous in the SC of various mammalian species (average, 0.5 µg/g wet wt.)[148] whereas large variations were found in the amount of thiamine esters: in calves and dogs bound thiamine was not detectable; in rabbits levels of 1.7 and 0.4 µg/g were measured in spring and in winter, respectively.[148] No seasonal variations could be detected in the rat. In this animal a concentration of 2.8 µg/g was reported.[149]

In the SC of young men the concentration of Vitamin C was found to be 180 µg/g wet wt.; somewhat lower values were measured in other mammals. Vitamin C tends to decrease with age.[150] Both Vitamins B_1 and C are lower in SC than in other CNS areas.

Riboflavin level in SC (2.7 µg/g wet wt.) is about the same as in cerebral cortex and is slightly higher in cervical and lumbar than in thoracic regions.[151]

X. CONCLUDING REMARKS

In the present chapter the author has attempted to unify and summarize the literature concerning the biochemistry of the SC. In concluding, a brief comment is made on what seems to be some of the limitations of part of this work and on what might represent fruitful approaches to the problem.

The work presented may be considered as a vast platform of data that will be of great utility in future investigations. However, the picture of the structural biochemistry of SC is still far from being complete. One of the reasons may be found in the fact that in many investigations SC was utilized as a suitable object for the study of definite chemical properties of the CNS (such as the composition of white matter), but only rarely the aim of elucidating the chemical basis of nerve structures specific of SC, or of correlating biochemical structural and functional aspects of SC, was in the fore. Furthermore, the interest and the value of part of the investigations performed so far is somewhat limited by the fact that SC was treated as a homogeneous tissue or, at best, as a cylinder of white matter with a core of gray matter, which is indeed a very rough approximation. In fact, biochemical heterogeneity was found not only, as was to be expected, within parts of gray matter, but also in different zones of white matter at a given segmental level of SC. Heterogeneity could be demonstrated even within groups of histochemically undistinguishable motoneurons.

It would seem profitable to have future investigations concentrate more on this most valuable segment of the CNS, which presents considerable interest and advantages from a biochemical, functional, and phylogenetical point of view. In fact, SC in spite of its remarkable complexity has a much simpler organization than even small areas of higher brain centers. Moreover, this phylogenetically ancient region, which has undergone much less structural changes from lower to higher vertebrates as compared to newer regions of the CNS, lends itself far better than the brain to comparative studies, which will certainly shed light on the practically unexplored field of neurochemistry dealing with biochemical evolutionary changes in homologous nerve structures.

Microtechniques devised for the analysis of small, anatomically well-defined areas of tissue represent a step forward in the study of neurochemical problems, and it would be desirable that in the future a larger use is made of methods that permit carrying the investigation to the cell unit or even to the subcellular level.

XI. REFERENCES

1. K. Seige and V. Thierbach, Zur Biomorphose und Biochemie des menschlichen Rukenmarks. III. Asche- und Mineralgehaltinder Trockensubstanz-Stickstoff, Phosphor und Schwefel im lipoid und wasswefreien Rukstand, Z. Alternsforsch. 15:46–61 (1961).
2. A. Löwenthal, Déterminations de la teneur du système nerveux central en matière sèche,

potassium et sodium, *in Chemical Pathology of the Nervous System, Proc. 3rd Intern. Neurochem. Symp., Strasbourg, 1958,* pp. 299–306, Pergamon Press, Oxford (1961).

3. A. Weil, Vergleichende Studien über den Gehalt verschiedenartiger Nervensubstanz an Aschenbestandteilen, *Hoppe–Seyler's Z. Physiol. Chem.* **89**:349–359 (1914).

4. F. Davies, R. E. Davies, E. T. B. Francis, and R. Whittam, The sodium and potassium content of cardiac and other tissues of the ox, *J. Physiol. (London)* **118**:276–281 (1952).

5. N. Tupikova and R. W. Gerard, Salt content of neural structures, *Am. J. Physiol.* **119**:414–415 (1937).

6. J. F. Manery and A. B. Hastings, The distribution of electrolytes in mammalian tissues, *J. Biol. Chem.* **127**:657–676 (1939).

7. M. Wender and M. Hierowski, The concentration of electrolytes in the developing nervous system with special reference to the period of myelination, *J. Neurochem.* **5**:105–108 (1960).

8. G. Branté, Studies on lipids in the nervous system. With special reference to quantitative chemical determination and topical distribution, *Acta Physiol. Scand.* **18** *Suppl.* 63 (1949).

9. J. Kreiner, The quantitative myelination of brains and spinal cords on dogs of various sizes, *Acta Anat.* **33**:50–64 (1958).

10. M. Wollemann, A photometrical method for testing the presence of iron in the central nervous system, *Acta Morphol. Acad. Sci. Hung.* **1**:127–132 (1951).

11. P. J. Warren, C. J. Earl, and R. H. S. Thompson, The distribution of copper in human brain, *Brain* **83**:709–717 (1960).

12. A. Schittenhelm and B. Eisler, Über die Verteilung des Jodes im Zentralnervensystem bei Mensch und Tier, *Z. Ges. Exptl. Med.* **86**:290–293 (1933).

13. K. H. Gabbay, L. O. Merola, and R. A. Field, Sorbitol pathway: presence in nerve and cord with substrate accumulation in diabetes, *Science* **151**:209–210 (1966).

14. L. G. Abood, R. W. Gerard, J. Banks, and R. D. Tschirgi, Substrate and enzyme distribution in cells and cell fractions of the nervous system, *Am. J. Physiol.* **168**:728–738 (1952).

15. A. Chesler and H. E. Himwich, The glycogen content of various parts of the central nervous system of dogs and cats at different ages, *Arch. Biochem.* **2**:175–181 (1943).

16. A. Chesler and H. E. Himwich, Effect of insulin hypoglycemia on glycogen content of parts of the central nervous system of the dog, *Arch. Neurol. Psychiat.* **52**:114–116 (1944).

17. T. Terni, Ricerche sulla cosidetta sostanza gelatinosa (corpo glicogenico) del midollo lombo-sacrale degli uccelli, *Arch. Ital. Anat. Embriol.* **21**:55–86 (1924).

18. W. L. Doyle and R. L. Watterson, The accumulation of glycogen in the "glycogen body" of the nerve cord of the developing chick, *J. Morphol.* **85**:391–403 (1949).

19. A. M. Lervold and J. Szepsenwol, Glycogenolysis in aliquots of glycogen bodies of the chick, *Nature* **200**:81 (1963).

20. O. J. Rafaelsen, Action of insulin on carbohydrate uptake of isolated rat spinal cord, *J. Neurochem.* **7**:33–44 (1961).

21. M. E. Smith, Glucose metabolism of central nervous tissues in rats with experimental allergic encephalomyelitis, *Nature* **209**:1031–1032 (1966).

22. J. L. R. Candela and D. Martin-Hernandez, Action of insulin *in vitro* on the glucose uptake of the spinal cord of the cat, *Experientia* **15**:439–440 (1959).

23. M. V. Buell, O. H. Lowry, N. H. Roberts, M. L. W. Chang, and J. I. Kapphahn, The quantitative histochemistry of the brain. V. Enzymes of glucose metabolism, *J. Biol. Chem.* **232**:979–993 (1958).

24. D. B. McDougal, D. W. Schulz, J. V. Passonneau, J. R. Clark, M. A. Reynolds, and O. H. Lowry, Quantitative studies of white matter. I. Enzymes involved in glucose-6-phosphate metabolism, *J. Gen. Physiol.* **44**:487–498 (1961).

25. B. M. Breckenridge and E. J. Crawford, The quantitative histochemistry of the brain. Enzymes of glycogen metabolism, *J. Neurochem.* **7**:234–240 (1961).

26. D. B. McDougal, R. T. Schimke, E. M. Jones, and E. Touchill, Quantitative studies of white matter. II. Enzymes involved in triose phosphate metabolism, *J. Gen. Physiol.* **47**:419–433 (1964).

27. N. Robinson and B. M. Phillips, Glycolytic enzymes in human brain, *Biochem. J.* **92**:254–259 (1964).

28. O. H. Lowry, Enzyme concentrations in individual nerve cell bodies, in *Metabolism of the Nervous System* (D. Richter, ed.), pp. 323–328, Pergamon Press, Oxford (1957).

29. P. M. Dreyfus, The regional distribution of transketolase in the normal and the thiamine deficient nervous system, *J. Neuropathol. Exptl. Neurol.* **24**:119–129 (1965).

30. P. M. Dreyfus and G. Hauser, The effect of thiamine deficiency on the pyruvate decarboxylase system of the central nervous system, *Biochim. Biophys. Acta* **104**:78–84 (1965).

31. A. M. Burt, Glucose metabolism and chick neurogenesis. I. 6-phosphate dehydrogenase activity in the embryonic brachial cord, *Develop. Biol.* **12**:213–232 (1965).

32. V. Bonavita and R. Guarnieri, Lactate-dehydrogenase isoenzymes in nervous tissue. III. Regional distribution in ox brain, *J. Neurochem.* **10**:755–764 (1963).

33. A. Löwenthal, D. Karcher, and M. Van Sande, Electrophoretic patterns of lactate dehydrogenase isoenzymes in nervous tissues, *J. Neurochem.* **11**:247–250 (1954).

34. D. B. McDougal, Quantitative histochemistry of selected central tracts, *Neurology* **8**:58–59 (1958).

35. R. L. Friede, L. M. Fleming, and M. Knoller, A comparative mapping of enzymes involved in hexosemonophosphate shunt and citric acid cycle in the brain, *J. Neurochem.* **10**:263–277 (1963).

36. S. L. Manocha and G. H. Bourne, Histochemical mapping of succinic dehydrogenase and cytochrome oxidase in the spinal cord, medulla oblongata and cerebellum of squirrel monkey (*Saimiri sciureus*), *Exptl. Brain Res.* **2**:216–229 (1966).

37. K. Nandy and G. H. Bourne, A histochemical study of the localization of succinic dehydrogenase, cytochrome oxidase and diphosphopyridine nucleotide-diaphorase in the synaptic regions of the spinal cord in the rat, *J. Histochem. Cytochem.* **12**:188–193 (1964).

38. K. Nandy and G. H. Bourne, A histochemical study of the localization of the oxidative enzymes in the neurones of the spinal cord in rats, *J. Anat. (London)* **98**:647–653 (1964).

39. A. J. Hudson, J. H. Quastel, and P. G. Scholefield, The effect of heated snake venom on the phosphate metabolism of the rat spinal cord, *J. Neurochem.* **5**:177–184 (1960).

40. A. J. Hudson and M. M. Kini, The preparation and some metabolic properties of rat spinal cord slices, *Can. J. Biochem.* **38**:965–968 (1960).

41. H. E. Himwich and J. F. Fazekas, Comparative studies of the metabolism of the brain of infant and adult dog, *Am. J. Physiol.* **132**:454–459 (1941).

42. L. Hertz and T. Clausen, Effects of potassium and sodium on respiration: Their specificity to slices from certain brain regions, *Biochem. J.* **89**:526–533 (1963).

43. J. D. McColl and R. J. Rossiter, A comparative study of the lipids of the vertebrate central nervous system, II. Spinal cord, *J. Exptl. Biol.* **29**:203–210 (1952).

44. K. Seige, Zur Biomorphose und Biochemie des menschlichen Rükenmarks. II. Untersuchungen der Alternswandlungen des Vorkommens der Gesamtlipoide und Untersuchungen verschiedener Fettfraktionen, *Z. Alternsforsch.* **14**:126–147 (1960).

45. G. R. Webster, Studies on the plasmalogens of nervous tissue, *Biochim. Biophys. Acta* **44**:109–116 (1960).

46. L. Amaducci, The distribution of proteolipids in the human nervous system, *J. Neurochem.* **9**:153–160 (1962).

47. L. Amaducci, A. Pazzagli, and G. Pessina, The relation of proteolipids and phosphatidopeptides to tissue elements in the bovine nervous system, *J. Neurochem.* **9**:509–518 (1962).

48. C. D. Joel, H. W. Moser, G. Majno, and M. L. Karnovsky, Effect of bis-(monoiso-propylamino)-fluoro-phosphine oxide (Mipafox) and of starvation on the lipids in the nervous system of the hen, *J. Neurochem.* **14**:479–488 (1967).
49. C. W. M. Adams and A. N. Davison, The occurrence of esterified cholesterol in the developing nervous system, *J. Neurochem.* **4**:282–289 (1959).
50. W. C. McMurray, J. D. McColl, and R. J. Rossiter, A comparative study of the lipids of the invertebrate and vertebrate nervous system, in *Comparative Neurochemistry* (D. Richter, ed.), pp. 101–107, Pergamon Press, Oxford (1964).
51. C. H. Williams, H. J. Johnson, and J. L. Casterline, Cholesterol content of spinal cord and sciatic nerve of hens after organophosphate and carbonate administration, *J. Neurochem.* **13**:471–474 (1966).
52. F. Chevallier and L. Petit, Incorporation of cholesterol into the central nervous system and its autoradiographic localization, *Exptl. Neurol.* **16**:250–254 (1966).
53. H. P. Schwarz, I. Kostyk, A. Marmolejo, and C. Sarappa, Long-chain bases of brain and spinal cord of rabbits, *J. Neurochem.* **14**:91–97 (1967).
54. P. Flechsig, *Die Leitungsbahnen im Gehirn und Rükenmark des Menschen*, Engelemann, Leipzig (1876).
55. M. F. Lucas Keene and E. E. Hewer, Some observations on myelination in the human central nervous system, *J. Anat. (London)* **66**:1–13 (1931).
56. A. N. Davison and J. M. Oxberry, A comparison of the composition of white matter lipids in swayback and border disease of lambs, *Res. Vet. Sci.* **7**:67–71 (1966).
57. H. I. El-Eishi, Biochemical and histochemical studies of myelination in the chick embryo spinal cord, *J. Neurochem.* **14**:405–412 (1967).
58. R. H. Laatsch, Glycerol phosphate dehydrogenase activity of developing rat central nervous system, *J. Neurochem.* **9**:487–492 (1962).
59. G. Porcellati, On the occurrence and distribution of the free phospholipid phosphoric esters and some amino compounds on the nervous tissues of some animal species, *Riv. Biol.* **56**:209–226 (1963).
60. G. Porcellati, A. Floridi, and A. Ciammarughi, The distribution and the biological significance of L-serine ethanolamine and L-threonine ethanolamine phosphates, *Comp. Biochem. Physiol.* **14**:413–418 (1965).
61. G. Porcellati, Distribution and metabolism of threonine ethanolamine and serine ethanolamine phosphodiesters in nervous tissue, *Biochim. Biophys. Acta* **90**:183–186 (1964).
62. P. W. Ramwell, J. E. Shaw, and R. Jessup, Spontaneous and evoked release of prostaglandins from frog spinal cord, *Am. J. Physiol.* **211**:998–1004 (1966).
63. G. Porcellati and R. H. S. Thompson, The effect of nerve section on the level of free amino acids of nerve tissue, *J. Neurochem.* **1**:340–347 (1957).
64. L. T. Graham, Jr., R. P. Shank, R. Werman, and M. H. Aprison, Distribution of some synaptic transmitter suspects in cat spinal cord: Glutamic acid, aspartic acid, γ-amino-butyric acid, glycine and glutamine, *J. Neurochem.* **14**:465–472 (1967).
65. Y. Nagata, Y. Yokoi, and Y. Tsukada, Studies on free amino acid metabolism in excised cervical sympathetic ganglia from the rat, *J. Neurochem.* **13**:1421–1431 (1966).
66. J. Kandera, G. Levi, and A. Lajtha, Control of cerebral metabolite levels. II. Amino acid uptake and levels in various areas of the rat brain, *Arch. Biochem. Biophys.* **126**:249–260 (1968).
67. R. W. A. Baker and G. Porcellati, The separation of nitrogen-containing phosphate esters from brain and spinal cord by ion-exchange chromatography, *Biochem. J.* **73**:561–566 (1959).
68. H. H. Tallan, A survey of the amino acids and related compounds in the nervous tissue, in *Amino Acid Pools* (J. T. Holden, ed.), pp. 471–485, Elsevier, Amsterdam (1962).

69. G. Levi and A. Lajtha, Cerebral amino acid transport *in vitro*. II. Regional differences in amino acid uptake by slices from the central nervous system of the rat, *J. Neurochem.* **12**:639–648 (1965).

70. H. H. Tallan, Studies on the distribution of *N*-acetyl-L-aspartic acid in brain, *J. Biol. Chem.* **224**:41–45 (1957).

71. A. Curatolo, P. D'Arcangelo, A. Lino, and A. Brancati, Distribution of *N*-acetyl-aspartic and *N*-acetyl-aspartyl-glutamic acids in nervous tissue, *J. Neurochem.* **12**:339–342 (1965).

72. I. P. Lowe, E. Robins, and G. S. Eyerman, The fluorimetric measurement of glutamic decarboxylase and its distribution in brain, *J. Neurochem.* **3**:8–18 (1958).

73. R. W. Albers and R. O. Brady, The distribution of glutamic decarboxylase in the nervous system of the rhesus monkey, *J. Biol. Chem.* **234**:926–928 (1959).

74. R. A. Salvador and R. W. Albers, The distribution of glutamic-γ-aminobutyric transaminase in the nervous system of the rhesus monkey, *J. Biol. Chem.* **234**:922–925 (1959).

75. F. N. Pitts, Jr., and C. Quick, Brain succinate semialdehyde dehydrogenase. I. Assay and distribution, *J. Neurochem.* **12**:893–900 (1965).

76. D. B. Goldstein, D-Amino acid oxidase in brain: Distribution in several species and inhibition by pentobarbitone, *J. Neurochem.* **13**:1011–1016 (1966).

77. A. H. Neims, W. D. Zieverink, and J. D. Smilack, Distribution of D-amino acid oxidase in bovine and human nervous tissues, *J. Neurochem.* **13**:163–168 (1966).

78. B. Curti and G. Porcellati, Some properties of a phosphopeptide isolated from nervous tissues. VI. The relationship with some phosphorus-containing compounds of biological interest, Communication No. 231/bis, *Proc. 8th Natl. Congr. Ital. Biochem. Soc., Padoa,* October 3–6 (1962).

79. G. Porcellati and I. Montanini, Su di alcune caratteristiche di un fosfopeptide isolato dal tessuto nervoso. VII. Distribuzione ed attivita' metabolica in alcuni tessuti di pollo, *Boll. Soc. Ital. Biol. Sper.* **39**:115–118 (1963).

80. F. Lembeck and G. Zetler, Substance *P*: A polypeptide of possible physiological significance, especially within the nervous system, *Intern. Rev. Neurobiol.* **4**:159–213 (1962).

81. A. H. Amin, T. B. B. Crawford, and J. H. Gaddum, The distribution of substance *P* and 5-hydroxytryptamine in the central nervous system of the dog, *J. Physiol. (London)* **126**:596–618 (1954).

82. C. E. Lumsden, D. M. Robertson, and R. Blight, Chemical studies on experimental allergic encephalomyelitis. Peptide as the common denominator in all encephalitogenic "antigens," *J. Neurochem.* **13**:127–162 (1966).

83. P. R. Carnegie and C. E. Lumsden, Encephalitogenic peptides of spinal cord, *Nature* **209**:1354–1355 (1966).

84. A. Nakao, W. J. Davis, and E. Roboz-Einstein, Basic proteins from the acidic extract of bovine spinal cord. II. Encephalitogenic, immunologic and structural interrelationships, *Biochim. Biophys. Acta* **130**:171–179 (1966).

85. M. W. Kies, E. B. Thompson, and E. C. Alvord, Jr., The relationship of myelin proteins to experimental allergic encephalomyelitis, *Ann. N.Y. Acad. Sci.* **122**:148–160 (1965).

86. G. J. Maletta, A. Vernadakis, and P. S. Timiras, Pre- and postnatal development of the spinal cord: Increased acetylcholinesterase activity, *Proc. Soc. Exptl. Biol. Med.* **121**:1210–1211 (1966).

87. B. Kelley, Age and nitrogen content of rabbit brain parts, *Am. J. Physiol.* **185**:299–301 (1956).

88. M. Wender and Z. Waligora, The content of amino acids in the proteins of the developing nervous system of the guinea pig. III. Spinal cord, *J. Neurochem.* **11**:243–247 (1964).

89. B. W. Moore, A soluble protein characteristic of the nervous system, *Biochem. Biophys. Res. Commun.* **19**:739–744 (1965).

90. A. Lajtha, Protein metabolism of the nervous system, *Intern. Rev. Neurobiol.* **6**:1–98 (1964).

91. L. F. Pantchenko, Protein turnover at various levels of the central nervous system and in the liver in growing and full grown animals, *J. Physiol. URSS* **44**:243–248 (1958).

92. J. Fisher, J. Kolousek, and Z. Lodin, Incorporation of methionine (sulphur-35) into the central nervous system, *Nature* **178**:1122–1123 (1956).

93. J. Altman, Regional utilization of leucine-H^3 by normal rat brain: Microdensitometric evaluation of autoradiograms, *J. Histochem. Cytochem.* **11**:741–750 (1963).

94. D. H. Ford and R. Rhines, Uptake of C^{14} into the brain and other tissues of normal and dysthyroidal male rats after injection of C^{14}-L-glutamine, *Acta Neurol. Scand.* **43**:33–47 (1967).

95. D. H. Ford, E. Pascoe, and R. Rhines, Effect of high pressure oxygen on the uptake of DL-lysine-H^3 by brain and other tissues of the rat. *Acta Neurol. Scand.* **43**:129–148 (1967).

96. A. Lajtha, Observations on protein catabolism in brain, in *Regional Neurochemistry* (S. S. Kety and J. Elkes, eds.), pp. 25–36, Pergamon Press, Oxford (1961).

97. G. Porcellati, A. Millo, and I. Manocchio, Proteinase activity of nervous tissues in organophosphorus compound poisoning, *J. Neurochem.* **7**:317–320 (1961).

98. A. J. Samuels, L. L. Boyarsky, R. W. Gerard, B. Libet, and M. Brust, Distribution, exchange and migration of phosphate compounds in the nervous system, *Am. J. Physiol.* **164**:1–15 (1951).

99. H. H. Donaldson (ed.), *The Rat. Data and Reference Tables*, p. 321, Memoirs of the Wistar Institute of Anatomy and Biology, No. 6, Philadelphia (1924).

100. K. Seige and V. Thierbach, Zur Biomorphose und Biochemie des menschlichen Rükenmarks. IV. Untersuchungen über den Nukleinsäurengehalt in den verschiedenen Altersstufen, *Z. Alternsforsch.* **16**:211–218 (1961).

101. Lj. Mihailović, D. B. Janković, M. Petković, and D. Mančić, Distribution of DNA and RNA in different regions of cat's brain, *Experientia* **14**:9–10 (1958).

102. J. E. Edström, The content and concentration of ribonucleic acid in motor anterior horn cells from the rabbit, *J. Neurochem.* **1**:159–165 (1956).

103. R. Landolt, H. H. Hess, and C. Thalheimer, Regional distribution of some chemical structural components of the human nervous system. I. DNA, RNA and ganglioside sialic acid, *J. Neurochem.* **13**:1441–1452 (1966).

104. E. K. Adrian, Jr., and B. E. Walker, Incorporation of thymidine-H^3 by cells in normal and injured mouse spinal cord, *J. Neuropath. Exptl. Neurol.* **21**:597–609 (1962).

105. F. C. MacIntosh, The distribution of acetylcholine in the peripheral and the central nervous system, *J. Physiol. (London)* **99**:436–442 (1941).

106. G. C. Pepeu, Le amine biogene del sistema nervoso centrale, *La Settimana Medica* **54**:57–74 (1966).

107. G. S. Barsoum, The acetylcholine equivalent of nervous tissues, *J. Physiol. (London)* **84**:259 262 (1935).

108. C. O. Hebb and A. Silver, Choline acetylase in the central nervous system of man and some other mammals, *J. Physiol. (London)* **134**:718–728 (1956).

109. R. E. McCaman and J. M. Hunt, Microdetermination of choline acetylase in nervous tissue, *J. Neurochem.* **12**:253–259 (1965).

110. W. Feldberg and M. Vogt, Acetylcholine synthesis in different regions of the central nervous system, *J. Physiol. (London)* **107**:372–381 (1948).

111. A. S. V. Burgen and L. M. Chipman, Cholinesterase and succinic dehydrogenase in the central nervous system of the dog, *J. Physiol. (London)* **114**:296–305 (1951).

112. E. Giacobini and B. Holmsted, Cholinesterase content of certain regions of the spinal cord as judged by histochemical and Cartesian diver technique, *Acta Physiol. Scand.* **42**:12–27 (1958).

113. E. Giacobini, Intracellular distribution of cholinesterase in the anterior horn cells of rat, *Arch. Ital. Biol.* **99**:163–177 (1961).

114. M. D. Nachmansohn, Choline esterase in brain and spinal cord of sheep embryos, *J. Neurophysiol.* **3**:396–402 (1940).

115. B. S. Wenger, Cholinesterase activity in different spinal cord levels of the chick embryo, *Federation Proc.* **10**:268–269 (1961).

116. L. W. Chacko and J. A. Cerf, Histochemical localization of cholinesterase in the amphibian spinal cord and alterations following ventral root section, *J. Anat. (London)* **94**:74–81 (1960).

117. K. Nandy and G. H. Bourne, The effects of D-lysergic acid diethylamide tartrate (LSD-25) on the cholinesterases and monoamine oxidase in the spinal cord: a possible factor in the mechanism of hallucination, *J. Neurol. Neurosurg. Psychiat.* **27**:259–267 (1964).

118. G. B. Koelle, The histochemical localization of cholinesterases in the central nervous system of the rat, *J. Comp. Neurol.* **100**:211–228 (1954).

119. Söderholm, Histochemical localization of esterases, phosphatases and tetrazolium reductases in the motor neurons of the spinal cord of the rat and the effect of nerve division, *Acta Physiol. Scand.* **65**: *Suppl.* 256, 3–60 (1965).

120. M. W. Brightman and R. W. Albers, Species differences in the distribution of extraneuronal cholinesterases within the vertebrate central nervous system. *J. Neurochem.* **4**:244–250 (1959).

121. E. G. McGreer and P. L. McGeer, Catecholamine content of spinal cord, *Can. J. Biochem.* **40**:1141–1152 (1962).

122. A. H. Anton and D. F. Sayre, The distribution of dopamine and dopa in various animals and a method for their determination in diverse biological material. *J. Pharmacol. Exptl. Therap.* **145**:326 (1964).

123. G. R. Pscheidt and B. Haber, Regional distribution of dihydroxyphenylalanine and 5-hydroxytryptophan decarboxylase and of biogenic amines in the chicken central nervous system, *J. Neurochem.* **12**:613–618 (1965).

124. N. E. Andén, T. Magnusson, B. E. Roos, and B. W. Werdinius, 5-Hydroxyindolacetic acid of rabbit spinal cord normally and after transection, *Acta Physiol. Scand.* **64**:193–196 (1965).

125. D. Davila, M. Rabadjija, D. J. Palaić, and Z. Supek, Content and distribution of 5-hydroxytryptamine in the central nervous system of the frog, *J. Neurochem.* **12**:59–60 (1965).

126. T. Magnusson and E. Rosengren, Catecholamines of the spinal cord normally and after transection, *Experientia* **19**:229–230 (1963).

127. N. E. Andén, A. Carlsson, N. A. Hillarp, and T. Magnusson, 5-hydroxytryptamine release by nerve stimulation of the spinal cord, *Life Sci.* **3**:473–478 (1964).

128. N. E. Andén, Distribution of monoamines and dihydroxyphenylalanine decarboxylase activity in the spinal cord, *Acta Physiol. Scand.* **64**:197–203 (1965).

129. E. G. Anderson and L. O. Holgerson, The distribution of 5-hydroxytryptamine and norepinephrine in cat spinal cord, *J. Neurochem.* **13**:479–485 (1966).

130. M. Vogt, The concentration of sympathin in different parts of the central nervous system under normal conditions and after the administration of drugs, *J. Physiol. (London)* **123**:451–481 (1954).

131. S. Spector, K. Melmon, W. Lovenberg, and A. Sjoerdsma, The presence and distribution of tyramine in mammalian tissues, *J. Pharmacol Exptl. Therap.* **140**:229–235 (1963).

132. N. E. Andén, A. Carlsson, N. A. Hillarp, and T. Magnusson, Noradrenaline release by nerve stimulation of the spinal cord, *Life Sci.* **4**:129–132 (1965).

133. A. Dahlstrom and K. Fuxe, Evidence for the existence of monoamine-containing neurons in the central nervous system. II. Experimentally induced changes in the intraneuronal

amine level of bulbospinal neuron systems, *Acta Physiol. Scand.* **64**:*Suppl.* 247, 1–36 (1965).

134. A. Carlsson, T. Magnusson, and E. Rosengren, 5-Hydroxytryptamine of the spinal cord normally and after transection, *Experientia* **19**:359 (1963).

135. N. E. Andén, T. Magnusson, and E. Rosengren, On the presence of dihydroxyphenylalanine decarboxylase in nerves, *Experientia* **20**:328–329 (1964).

136. A. Carlsson, B. Falck, K. Fuxe, and N. A. Hillarp, Cellular localization of monoamines in the spinal cord, *Acta Physiol. Scand.* **60**:112–119 (1964).

137. H. Shimizu, Y. Kakimoto, and I. Samu, The determination and distribution of polyamines in mammalian nervous system, *J. Pharmacol. Exptl. Therap.* **143**:199–204 (1964).

138. C. Fieschi and S. Soriani, Enzymic activities in the spinal cord after sciatic section. Alkaline and acid phosphatases, 5-nucleotidase and ATPase, *J. Neurochem.* **4**:71–77 (1959).

139. K. Nandy and G. H. Bourne, Alkaline phosphatases in brain and spinal cord, *Nature* **200**:1216–1217 (1963).

140. R. L. Friede, Alkaline and acid phosphatases and non specific esterases, in *Topographic Brain Chemistry*, pp. 178–225, Academic Press, New York (1966).

141. A. D. Chiquoine, Distribution of alkaline phosphomonoesterases in the central nervous system of the mouse embryo, *J. Comp. Neurol.* **100**:415–439 (1954).

142. K. Nandy and G. H. Bourne, Adenosine triphosphatase and 5-nucleotidase in spinal cord, *Arch. Neurol.* **11**:547–554 (1964).

143. R. Fried, Sodium-potassium activated ATPase from spinal cord of normal and injured rats, *J. Neurochem.* **12**:815–832 (1965).

144. E. Poulsen and W. N. Aldridge, Studies on esterases in the chicken central nervous system, *Biochem. J.* **90**:182–189 (1964).

145. R. D. Tschirgi, The blood–brain barrier, in *Biology of Neuroglia* (W. F. Windle, ed.), pp. 130–138, C. C. Thomas, Springfield, Illinois (1958).

146. W. Ashby, On the quantitative incidence of carbonic anhydrase in the central nervous system, *J. Biol. Chem.* **155**:671–679 (1944).

147. W. Ashby and E. M. Schuster, Carbonic anhydrase in the brain of newborn in relation to functional maturity, *J. Biol. Chem.* **184**:109–116 (1950).

148. G. de Muralt, Aneurine libre et totale dans les nerfs périphériques et le système nerveux central de quelques mammifères, *Intern. Z. Vitaminforsch.* **19**:74–101 (1947).

149. P. M. Dreyfus, The quantitative histochemical distribution of thiamine in normal rat brain, *J. Neurochem.* **4**:183–190 (1959).

150. F. Plaut and M. Bülow, Über Unterschiede im C-Vitamingehalt verschiedener Teile des Nervensystems, *Z. Ges. Neurol. Psychiat.* **153**:182–192 (1935).

151. H. Leeman and E. Pichler, Über den Lactoflavingehalt des Zentralnervensystems und Seine Bedentung, *Arch. Phychiat. Nervenkrankh.* **114**:265–289 (1942).

Chapter 6

THE USE OF BRAIN SLICES

K. A. C. Elliott

Department of Biochemistry and
the Montreal Neurological Institute
McGill University
Montreal, Canada

I. INTRODUCTION

Otto Warburg[1] introduced the tissue slice technique as a method for studying, by simple, easily controlled *in vitro* methods, the metabolism of tissues the structure of which had not been extensively disrupted. When we use this technique we assume, or hope, that results obtained may represent actual physiological activities or potentialities of the tissue. The technique has been described and discussed previously in commonly available volumes,[2,3] and I shall not repeat all the material written, nor the references cited, in the previous articles.

II. PERMISSIBLE THICKNESS

In the preparation and manipulation of tissue slices, damage to the tissue, at least to the superficial layers of cells, is inevitable. Of the total cells in a slice the proportion that are damaged in preparation of the slice will be lower the thicker the slice. From this point of view, therefore, thick slices are preferable. The maximum permissible thickness for the study of aerobic metabolism is the thickness that will allow oxygen, diffusing in from the suspending medium, to satisfy the oxygen demands of all intermediate layers and to reach the innermost layer of the slice at a concentration great enough to permit maximum oxygen uptake rate. Warburg calculated this thickness assuming the applicability of a diffusion constant for oxygen that was arrived at by Krogh and assuming that maximum oxygen consumption by cells will occur even at very low concentrations. According to this calculation, for a rate of oxygen consumption of about 3 ml/g of fresh tissue per hour, the maximum permissible thickness is about 0.5 mm when the partial pressure of oxygen is about 1 atm. Warburg's calculations seemed to be confirmed experimentally with slices from rat liver,[4] and in my own experience with rat cerebral cortex the oxygen uptake rate, about 3 ml/g-hr, is not increased by using thinner slices. Various experiments have validated the assumption that cells will respire at their maximum rate at very low oxygen concentrations. The assumption about the applicable diffusion constant, however, appears questionable. Longmuir and Bourke[5,6] report that, with slices of liver, heart, or kidney, limitation of oxygen consumption is much less dependent on slice thickness than the calculation indicates. It appears that there is a carrier mechanism for oxygen transport into liver, kidney, and heart slices but not into brain. It seems likely that the use of thicker slices may be permissible, and this should be tested for each tissue to be studied. (For a recent discussion and determination of the diffusion constant of oxygen in liver see MacDougall and McCabe.[7])

III. PREPARATION AND WEIGHING OF SLICES

Various devices have been described and used for cutting tissue slices. The Stadie–Riggs[8] microtome is the most satisfactory instrument for preparing slices of known uniform thickness, and the modification of this instrument, illustrated in Fig. 1, is the simplest and best that I have used. It can be made with precision equipment by any skilled workshop technician. With this instrument it is essential that the hole in the middle piece have a considerably larger radius than the pedestal on which the tissue is placed so that the slicer can be tilted with respect to the pedestal and tissue. The distance from the cutting region to the screw should be amply long enough to allow the razor blade to slide, between the upper and lower pieces, completely beyond the depression in the upper piece so that the slice is left adhering to the top of the depression from which it can be lifted off with fine or flat coverslip forceps. It is convenient to have two or three upper pieces with depressions of say 0.010, 0.015, and 0.020 in. deep. With these, and using the standard 0.01-in. thick blade, slices approximately 0.4, 0.5, or 0.6 mm thick can be prepared. The preparation of brain tissue for *in vitro* studies by means of an instrument that strikes the tissue with a blade has been described by McIlwain and Buddle.[9] By this "chopper" technique large amounts of slices can be produced rapidly; the slices consist of cross sections of the organ and thus contain mixed gray and white matter from various regions.

I consider it essential that slices, at least of brain tissue, should be cut in a humid chamber. Figure 2 shows the type of humid chamber used in my laboratory. Its humidity is maintained by the large area (the entire base) covered with several layers of water-saturated filter paper and by faint positive pressure maintained by a stream of air bubbling in through water. It is very difficult to cut satisfactory slices of any tissue in a dry atmosphere without moistening the tissue. With brain tissue I find it impossible. Slices of brain can be cut in a dry atmosphere if the tissue and blade are moistened with a saline solution, but brain slices rapidly absorb considerable amounts

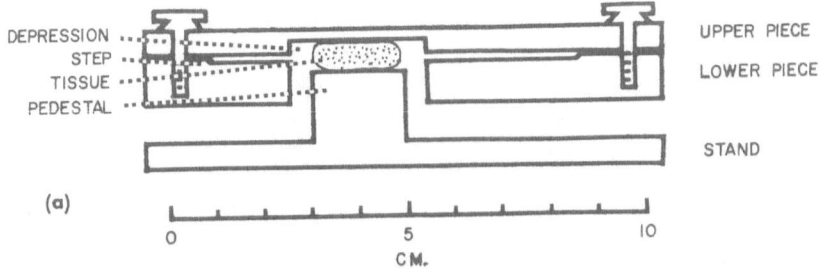

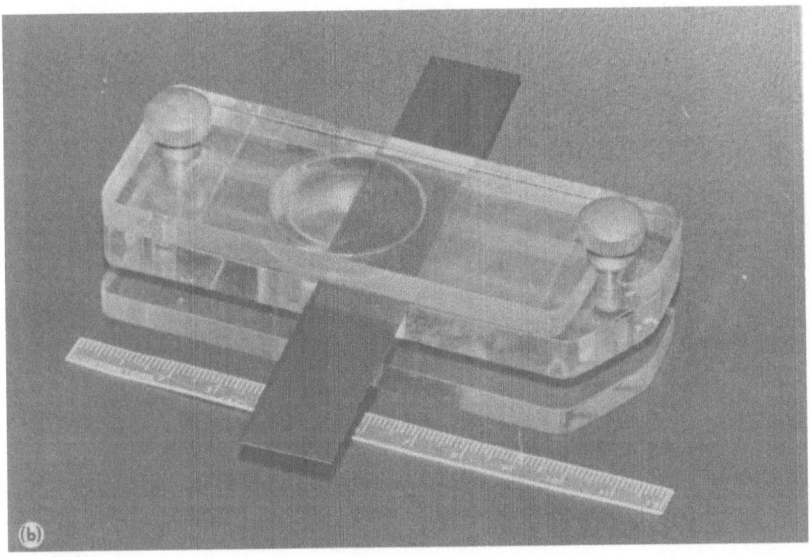

Fig. 1. Tissue slicer modified from Stadie and Riggs.[8] (Designed and made at the Medical Research Council Laboratories, Carshalton, Surrey, England, by J. Cremer and R. C. Emery.) The "step" and the "depression" are exaggerated in part (a).

of any aqueous solution with which they come in contact. Therefore it is not possible to determine the true fresh weight of slices that have been moistened. It becomes necessary then to determine the dry weight, the nitrogen content, or some other parameter of the slices after incubating. This is inaccurate because of loss of solid matter by leaching into the medium during the experiment and also sometimes loss of fragments of the slice. Drying or digesting the tissue also makes it impossible to carry out further procedures with the slices.

The slices, as they are cut, are collected on watch glasses covered by petri dishes in the humid chamber. They are loaded, in the chamber, onto a light (aluminum foil) pan and immediately weighed on a torsion balance. The true fresh weight of tissue to be used is thus obtained. The slices then may

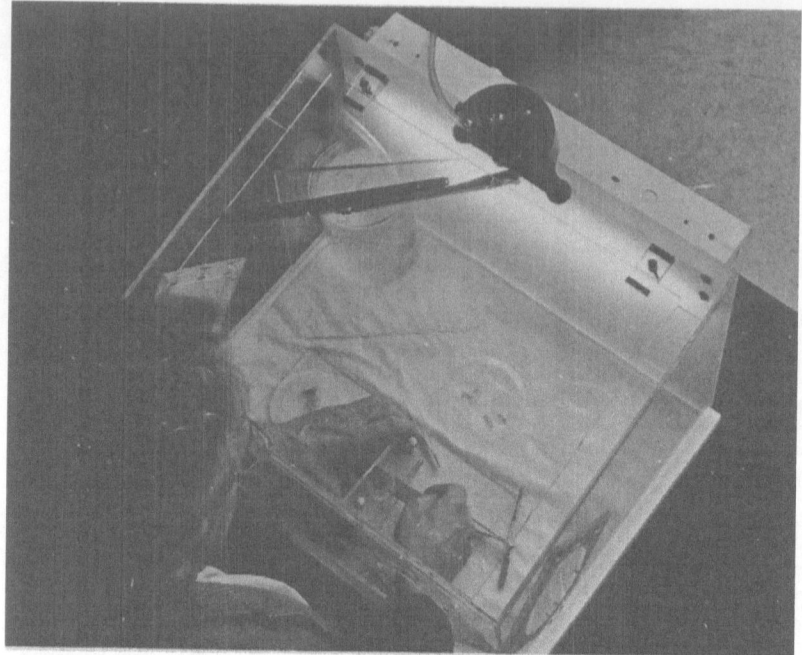

Fig. 2. Humid chamber.

be moistened so that they will slide easily down the side of the experimental vessel.

I much prefer to express results in milliliters or micromoles per gram of fresh tissue, which is possible when slices are cut without moistening and weighed fresh. Since the terms Q_{O_2} and Q_{CO_2} were defined by Warburg as cubic millimeters of gas evolved per milligram dry weight of tissue (determined at the end of the incubation) per hour, this term should not be used for other units. It is best to use terms such as "micromoles per gram fresh weight per hour" that require no separate definition. The term "fresh weight" should be used since "wet weight" suggests that the tissue actually has been wetted.

It is often necessary to reweigh the slices at the end of the experimental procedure. For this purpose the contents of the experimental vessel are poured through a small glazed perforated disc in a small funnel held in a Büchner flask. Very light suction is applied. The slices remain on the disc, and adherent moisture is removed with edges of torn filter paper. They are then weighed. (The suspending medium may be caught in a test-tube under the funnel in the flask.)

The weight of the drained tissue, determined at the end of an incubation, allows one to determine the approximate amount of swelling fluid that has been absorbed by the slices. In aerobic experiments this fluid has about the same composition as that of the suspending medium and occupies mainly extracellular spaces at the cut edges of slices, presumably spaces that have become available as a result of the damage caused by the slicing.[10] When the uptake of a substance from the medium, or loss to the medium, is being studied it is proper to calculate[11] a correction for the amount of this substance that is in the swelling fluid and not truly in the tissue. This correction can never be perfectly accurate. Though most of the swelling fluid is in the extraneous spaces a certain amount, depending on conditions, is intracellular.

IV. SUSPENDING MEDIA

Since the point of using slices is to maintain, as far as possible, normal membrane structure and behavior in the tissue, the suspending medium should be isosmotic with the normal environment of the cells and, presumably, should contain electrolytes in the concentrations normally present in extracellular fluids unless the particular study involves deliberate variations in electrolytes. The most appropriate medium is bicarbonate-buffered. I use as "normal" medium a solution containing the following, in millimoles per liter: NaCl, 122; KCl, 3.1; $MgSO_4$, 1.2; $CaCl_2$, 1.3; KH_2PO_4, 0.4; $NaHCO_3$, 25; and glucose, 10. This solution approaches the composition of cerebrospinal fluid. It has a lower concentration of calcium than that in the medium of Krebs and Henseleit[12] but approximately the same as that of the ionized calcium in plasma. The bicarbonate-buffered solution must be kept in equilibrium with a gas containing about 5% carbon dioxide to maintain a pH of 7.4 at 37°C.

Special apparatus, procedures, and assumptions are involved in the measurement, in the presence of bicarbonate-buffered solution, of oxygen consumption by tissue that produces carbon dioxide and acid changes. To avoid these complications phosphate-buffered solutions are commonly used with absorption of carbon dioxide by alkali-soaked filter paper. Phosphate-buffered solutions, however, are unphysiological in their high phosphate content. Phosphate tends to precipitate calcium, and this tendency is variable. This limits the concentration of phosphate that can be present, and hence the buffering capacity, unless calcium is completely omitted. It seems safe to replace the bicarbonate in the above solution with 11 mM sodium phosphate. It is best to add the calcium to the cold mixture immediately before the experiment to avoid precipitation. If one adjusts the initial pH of the phosphate-buffered medium to 7.6, the pH after equilibration with tissue is commonly about 7.4 and at the end of an incubation about 7.2. Trishydroxymethyl aminomethane (tris) is totally foreign to tissues but seems to be nontoxic and does not precipitate nor complex calcium ion. It can be used as buffer, but its pK is 8.1, which is considerably above the desirable pH. Sundry other buffers have been used. Certain doubts and speculations about the appropriate composition of the suspending medium are mentioned below.

Apparently valid measurements of some aspects of metabolism can be studied over a limited period with "dry" slices, i.e., with slices not suspended in aqueous medium.[13,14] The effects of drugs that have been administered *in vivo*, for instance, may be studied *in vitro* by this method without the concentration of the drug in the tissue becoming diluted by a suspending solution.

V. METHODS OF STUDY

Slices were originally used in the Warburg manometric apparatus for studies of oxygen consumption and carbon dioxide production. Modifications of the apparatus have allowed measurements, on a single sample of tissue in bicarbonate-buffered medium, of oxygen consumption, respiratory carbon dioxide production, and evolution or absorption of carbon dioxide due to acid–base changes.[15–19]

Many other kinds of study have been made with slices. These usually involve separation of slices from the suspending medium and chemical analyses on and/or determination of radioactivity in the suspending medium and in the slices. If manometric measurements of gas exchanges are not required, simple Erlenmeyer flasks filled with the appropriate gas and shaken in an ordinary water bath may be used. (The temperature control then need not be as precise as is necessary for gasometric measurements.)

VI. CRITICISM

The physiological meaningfulness of results obtained with tissue slices, particularly of brain, is subject to a number of doubts. In cutting slices of brain tissue it seems inevitable that parts of very many cells must be damaged though we like to assume that the damage affects mainly distal parts of cell processes and that these cut ends perhaps seal themselves. The damage to total cell surfaces actually may be relatively small.[20]

It would scarcely be practicable to provide a suspending medium that contained all the substances that are normally present and possibly significant in extracellular fluids. Even the electrolyte composition of the saline solutions that are used because they presumably resemble extracellular fluids may not be the most appropriate. We should perhaps consider that many mitochondrial and other enzyme systems may, as a result of cell damage, come in contact with the suspending medium instead of their normal intracellular environment. *In vivo*, when neurons of the brain are rather constantly active with consequent extrusion of potassium ions into narrow extracellular spaces, the fluid with which cells are in immediate contact may have a higher potassium concentration than that in cerebrospinal fluid. The high potassium media in which brain metabolism is increased,[21] may, for these two reasons, actually be more physiological than the "normal" media. (In the case of liver slices it was shown that a high potassium–high magnesium medium resembling *intracellular* fluid gave "better" results in studies on glycogen formation.[22])

Slices that have been incubated in solutions containing electrolytes in the concentrations present in cerebrospinal fluid or plasma always contain considerably less potassium and more sodium and more water than are present in untreated tissue. The unphysiological mechanical and metabolic conditions involved in the preparation of brain slices and setting up of experiments with them allows release of potassium, presumably from intracellular spaces, and this potassium is leached out and largely replaced with sodium when the slices are immersed in an ordinary high sodium–low potassium medium. If subsequent metabolic conditions are good, as when oxygen and glucose are provided or when active anaerobic glucolysis can occur, some of the sodium in the slice is replaced by potassium, but the potassium concentration in the slice never reaches that in native tissue. During an hour's incubation brain slices absorb up to 40%* of their weight of water from ordinary saline media in the presence of oxygen and even more in the absence of oxygen or in the presence of glutamate or high potassium.[23] The following is an example of recorded average figures.[11] Fresh rat cerebral cortex contained, in milliequivalents per liter of tissue water, K 136 and Na 56. After 1 hr of incubation in bicarbonate-buffered saline medium the tissue weight had increased by 40%. Thus the water content per 100 mg of fresh tissue had increased from about 80 to 120 mg. The electrolyte levels after incubation were K 44 and Na 123 per liter of total water in the slices or, if the water and electrolyte content of the absorbed medium were deducted, the "true" tissue water contained K 63 and Na 112 meq/liter.

Since a slice has no capillary circulation, all exchanges between the tissue and its environment must occur by diffusion over distances up to half the thickness of the slice. To maintain an adequate oxygen supply to the innermost layers, an unphysiologically high and actually toxic[24,25] oxygen tension has to be maintained in contact with the outer layers The concentration of any other rapidly utilized substrate provided in the medium also has to be higher than what might be the physiological level. Any substance that is produced by the tissue will be present at a higher concentration within most of the slice than in the medium. Carbon dioxide, lactic acid, and hydrogen ion concentrations are known to be higher within a metabolizing brain slice than in the surrounding medium.[26]

By using slices of cerebral cortex we at least deal with "pure" gray matter. It is difficult to prepare slices of pure gray or pure white matter from any other region of the brain of a small animal. Further, cerebral cortex itself is not uniform; areas of cerebral cortex vary appreciably in metabolic activity.[27] Due to the layered structure of cerebral cortex the cellular composition of tissue studied will vary with the level (surface or lower) and thickness of each slice. In no case can we consider that we are studying a single

---

* Later work[10] showed only about 30% swelling if the tissue was kept cold during the relatively anaerobic conditions that prevailed during setting up of experiments.

type of cell. We have, on the contrary, always a variety of different kinds of neurons and glial cells and different proportions of cell bodies and processes.

Until fairly recently we have had no indication of the physiological status of brain slices, the extent to which damage due to slicing has altered physiological potentialities and the effects on these potentialities of the various conditions that we have imposed on the tissue deliberately or unavoidably.

VII. TYPES OF INFORMATION GAINED WITH SLICES

In spite of the above criticism I know of no instance in which information obtained by the slice technique has been found to be misleading after more physiological experimentation. On the contrary a great deal of valuable information has been obtained, some of which has later been confirmed by methods, particularly studies *in vivo* or with whole organs, that apparently are more valid.

The original studies of Warburg[1] with slices showed the range of oxygen consumption rates among various tissues, the high potentiality of growing tissue for anaerobic energy-producing metabolism, glucolytic lactic acid production, and the tendency of cancer tissues to continue active metabolism by the anaerobic process while at the same time carrying on active oxidative metabolism in the presence of oxygen. (This latter pattern is now known not to be entirely specific to cancer tissue nor characteristic of all cancer tissues.) Other fundamental biochemical processes, such as the cycles of reactions concerned in urea formation[12] and in the tricarboxylic acid cycle,[28] have been deduced from studies on slices.

Rather detailed information (summarized in Elliott and Wolfe[29]) has been recorded concerning oxygen consumption, glucose breakdown, and the metabolism of other substrates by brain and the effects on these of electrolytes, inhibitors, and other factors.

Brain slices, if provided with glucose and other appropriate conditions including a little calcium in the suspending medium, will take up oxygen at a constant rate for hours. This rate and its constancy depend upon the medium used and other experimental details, on the particular part of the brain studied, and on the species. It varies somewhat even between areas of the cerebral cortex and decreases with increasing size of the species. Glucose (or lactate or pyruvate, which are derived from glucose) is the preferred substrate. Glutamic acid or noncarbohydrate endogenous material can be consumed, but this consumption is largely suppressed in the presence of glucose. The oxygen uptake rate of unstimulated cerebral cortex slices is not higher than that of several other tissues. However, brain tissue respiration can be stimulated up to 100% by a high potassium concentration in the medium[21] or by oscillating electric potentials. (See Chapter 7 by Harvey and McIlwain in this volume.)

It is difficult to compare the rate of respiration measured *in vitro* with that occurring *in vivo* because the metabolism of slices of a particular part of the brain cannot represent the metabolism of the whole brain while measurements *in vivo* refer to the whole brain. However, I have made calculations based on experiments with cerebral cortex slices and with suspensions of cortex and whole brain[29] that indicate that the average rates for the whole brains of cat, monkey, and man would be respectively about 1.6, 1.2, and 0.95 ml of oxygen per gram of fresh tissue per hour. These rates correspond to rates found *in vivo* for brains in which the activity was depressed by anesthetic, hemorrhage, or diabetic or insulin coma. Brains of normal or conscious animals respire about twice as actively, and the rate found with slices stimulated by high potassium or electrically may well correspond to the rate for normally active brain. Whereas glucose is virtually the sole source of carbon to the brain, studies on slices and on brain perfused *in situ,* on ganglia, and on nerve, have shown that exogenous glucose is not all directly oxidized to CO_2, via the tricarboxylic acid pathway, but much of the glucose carbon enters pools of amino acids and other substances. During physiological activity or hyperactivity, direct glucolysis is active but oxidative metabolism of exogenous glucose is largely suppressed and endogenous materials are oxidized. With slices this change in metabolic pattern has been shown to occur on stimulation by potassium (or dinitrophenol).[29a] The increased metabolism of brain slices caused by high potassium or electric stimulation is sensitive to inhibition by low concentrations of narcotic and other agents.[30,32]

Comparative studies of the metabolism of slices of cerebral cortex, cerebellar cortex, white matter, and brain tumors, with DNA determinations, have given preliminary estimates of the range of metabolic activity of neurons and neuroglia.[31] It appears that individual oligodendrocytes may metabolize more actively than the small neurons in the cerebellum but less actively than the average neuron of the cerebral cortex.

Studies on slices indicate that brain tissue has a high potential for anaerobic activity, producing lactic acid from glucose. This "anaerobic glycolysis" is normally largely, though not instantaneously, suppressed under aerobic conditions, but it can be stimulated by high potassium or by electrical impulses. The latter effects are in accord with the observation of high lactic acid production in brain and other nervous tissues during convulsions or hyperactivity. The function of the lactic acid production is probably largely the generation of high energy phosphate during periods of excessive chemical energy demand, but it is also likely to be involved in regulating local circulation by the effect of acid in dilating blood vessels and perhaps in regulating nervous activity by a depressive action of lowered tissue pH. The glycolytic activity of brain tissue is extremely sensitive to changes in conditions.

Slices have been used for more detailed studies of energy metabolism, its regulation in relation to electrolytes, and high energy phosphate compounds and the effects of drugs.[32,33] The technique has been used in the study of transport of electrolytes[11] and amino acids and other substances.[34] Work on many aspects of brain metabolism is still being pursued.[35] An adaptation of the slice technique has contributed much to understanding the interrelations of the activities of the hypothalamus, the pituitary, and the adrenal cortex.[36]

With the slice technique J. H. Quastel and his collaborators first showed the biosynthesis, and the existence in "free" and "bound" forms, of a neuro-transmitter substance, acetylcholine.[37,38] The slice technique has proved the metabolism of carbon atoms of glucose via the neurophysiologically active amino acids related to the Krebs cycle, particularly glutamic, aspartic, and γ-aminobutyric acids,[39] and has shown their uptake and binding, which is especially extensive in the case of γ-aminobutyric acid,[40] by brain tissue. Recently the release of catecholamines[41] and serotonin[42] from bound forms in brain has been studied with slices. Studies on the water distribution in and the swelling of brain slices and their extensive uptake of fluid into damaged regions[23,10] arose out of studies on cerebral edema and has led to further understanding of this subject. Suggestive findings have been made with slices on the biphasic effects of drugs on energy metabolism and acetylcholine synthesis.[43,44]

There is no doubt about the usefulness of the tissue slice technique for discovery of potentialities of tissues and preliminary elucidation of processes.

VIII. FURTHER DEVELOPMENTS

In studies on nervous tissue we are interested in the biochemical aspects of functional activity—action potentials, conduction, synaptic transmission, excitation and inhibition, and various very rapid events. Perhaps the most significant advances in tissue slice technique are being made by McIlwain and his colleagues, who have developed methods for rapid handling of excised tissue and methods for electrical stimulation and recording of resting and action potentials in excised tissue. These procedures are described by Harvey and McIlwain in Chapter 7 in this volume.

IX. REFERENCES

1. O. Warburg, *The Metabolism of Tumours* (F. Dickens, trans.), Richard R. Smith, New York (1931).
2. J. Field, II, *in Methods in Medical Research* (V. R. Potter, ed.), Vol. 1, pp. 289–307, Yearbook Publications, Chicago (1948).
3. K. A. C. Elliott *in Methods in Enzymology* (S. P. Colowick and N. O. Kaplan, eds.), Vol. 1, pp. 3–9, Academic Press, New York (1955).
4. F. A. Fuhrman and J. Field, II, Factors determining the metabolic rate of excised liver tissue, *Arch. Biochem.* **6**:337–349 (1945).
5. I. S. Longmuir and A. Bourke, Application of Warburg's equation to tissue slices, *Nature* **184**:634–635 (1959).
6. I. S. Longmuir, Tissue oxygen transport, *in Third International Conference on Hyperbaric Medicine*, pp. 46–51, National Academy of Sciences, Washington, D.C. (1966).

7. J. D. B. MacDougall and M. McCabe, Diffusion coefficient of oxygen through tissues, *Nature* 215:1173–1174 (1967).

8. W. C. Stadie and B. C. Riggs, Microtome for the preparation of tissue slices for metabolic studies of surviving tissues *in vitro, J. Biol. Chem.* 154:687–690 (1944).

9. H. McIlwain and H. L. Buddle, Techniques in tissue metabolism. 1. A mechanical chopper, *Biochem. J.* 53:412–420 (1953); H. McIlwain, Techniques in tissue metabolism. 5. Chopping and slicing tissue samples, *Biochem. J.* 78:213–218 (1961).

10. H. M. Pappius, I. Klatzo, and K. A. C. Elliott, Further studies on swelling of brain slices, *Can. J. Biochem. Physiol.* 40:885–898 (1962).

11. H. M. Pappius and K. A. C. Elliott, Factors affecting the potassium content of incubated brain slices, *Can. J. Biochem. Physiol.* 34:1053–1067 (1956).

12. H. A. Krebs and K. Henseleit, Untersuchungen uber die Harnstoffbildung im Tierkörper, *Z. Physiol. Chem.* 210:33–66 (1932).

13. R. Rodnight and H. McIlwain, Techniques in tissue metabolism. 3. Study of tissue fragments with little or no added aqueous phase, and in oils, *Biochem. J.* 57:649–661 (1954).

14. E. A. Hosein, M. Emblem, S. Rochon, and S. E. Morch, Tissue slice respiration in the absence of suspension medium, *Arch. Biochem. Biophys.* 99:414–417 (1962).

15. M. Dixon and D. Keilin, An improved method for the measurement of tissue respiration, *Biochem. J.* 27:86–95 (1933).

16. K. A. C. Elliott and E. F. Schroeder, The metabolism of lactic and pyruvic acids in normal and tumour tissue. 1. Methods and results with kidney cortex, *Biochem. J.* 28:1920–1939 (1934).

17. W. H. Summerson, A combination simple manometer and constant volume differential manometer for studies in metabolism, *J. Biol. Chem.* 131:579–595 (1939).

18. W. W. Umbreit, R. H. Burris, and J. F. Staufer, *Manometric Techniques,* Burgess, Minneapolis (rev. 1957).

19. M. Dixon, *Manometric Methods,* 3rd ed., University Press, Cambridge, England (1951).

20. H. McIlwain *in Symposium on Barriers in the Nervous System,* Elsevier, Amsterdam (1967).

21. F. Dickens and G. D. Greville, The metabolism of normal and tumour tissue. XIII. Neutral salt effects, *Biochem. J.* 29:1468–1483 (1935).

22. G. F. Cahill, Jr., J. A. Ashmore, S. Zottu, and A. B. Hastings, Studies on carbohydrate metabolism in rat liver slices. IX. Ionic and hormonal effects on phosphorylase and glycogen, *J. Biol. Chem.* 224:237–250 (1957).

23. H. M. Pappius and K. A. C. Elliott, Water distribution in incubated slices of brain and other tissues, *Can. J. Biochem. Physiol.* 34:1007–1022 (1956).

24. K. A. C. Elliott and B. Libet, Studies on the metabolism of brain suspensions. 1. Oxygen uptake, *J. Biol. Chem.* 143:227–246 (1942).

25. F. Dickens, *in Neurochemistry* (K. A. C. Elliott, I. H. Page, and J. H. Quastel, eds.), 2nd ed., pp. 851–869, Charles C. Thomas, Springfield, Illinois (1962).

26. K. A. C. Elliott and M. K. Birmingham, The effect of pH on the respiration of brain tissue; the pH of tissue slices, *J. Biol. Chem.* 177:51–58 (1949).

27. K. A. C. Elliott and N. Henderson, Metabolism of brain tissue slices and suspensions from various mammals, *J. Neurophysiol.* 11:485–490 (1948).

28. H. A. Krebs and W. A. Johnson, The role of citric acid in intermediate metabolism in animal tissues, *Enzymologia* 4:148–156 (1937).

29. K. A. C. Elliott and L. S. Wolfe, *in Neurochemistry* (K. A. C. Elliott, I. H. Page, and J. H. Quastel, eds.), 2nd ed., pp. 177–211, Charles C. Thomas, Springfield, Illinois (1962).

29a. H. Gainer, C. L. Allweis, and I. L. Chaikoff, Precursors of CO_2 produced by rat brain slices stimulated electrically and by 2,4-dinitrophenol, *J. Neurochem.* 9:432–442 (1962).

30. J. H. Quastel, *in Neurochemistry* (K. A. C. Elliott, I. H. Page, and J. H. Quastel, eds.), 2nd ed., pp. 790–812, Charles C. Thomas, Springfield, Illinois (1962).

31. I. H. Heller and K. A. C. Elliott, The metabolism of normal brain and human gliomas in relation to cell type and density, *Can. J. Biochem. Physiol.* **33**:395–403 (1955).

32. F. Bilodeau and K. A. C. Elliott, The influence of drugs and potassium on respiration and potassium accumulation by brain tissue, *Can. J. Biochem. Physiol.* **41**:779–792 (1963).

33. H. M. Pappius, D. M. Johnson, and K. A. C. Elliott, Acid labile phosphate content of incubated brain slices, *Can. J. Biochem. Physiol.* **37**:999–1010 (1959).

34. R. M. Johnstone and P. G. Scholefield, *in Neurochemistry* (K. A. C. Elliott, I. H. Page, and J. H. Quastel, eds.), 2nd ed., pp. 376–396, Charles C. Thomas, Springfield, Illinois (1962).

35. J. E. Cremer, Studies on brain cortex slices, Differences in the oxidation of ^{14}C-labelled glucose and pyruvate revealed by the action of triethylin and other toxic agents, *Biochem. J.* **104**:212–222 (1967); J. E. Cremer, The influence of various inhibitors on the retention of potassium ions and amino acids with glucose or pyruvate as substrate, *Biochem. J.* **104**:223–228 (1967).

36. M. Saffran, A. V. Schally, and B. G. Benfey, Stimulation of the release of corticotropin from the adenohypophysis by a neurohypophysial factor, *Endocrinology* **57**:439–444 (1955).

37. J. H. Quastel, M. Tennenbaum, and A. H. M. Wheatley, Choline ester formation in, and choline esterase activities of, tissues *in vitro, Biochem. J.* **30**:1668–1681 (1936).

38. P. J. G. Mann, M. Tennenbaum, and J. H. Quastel, Acetylcholine metabolism in the central nervous system. The effects of potassium and other cations on acetylcholine liberation, *Biochem. J.* **33**:822–835 (1939).

39. A. Beloff-Chain, R. Catazaro, E. B. Chain, I. Masi, and F. Pocchiari, Fate of uniformly labelled ^{14}C-glucose in brain slices, *Proc. Roy. Soc. (London) B,* **144**:22–28 (1955–1956).

40. K. A. C. Elliott and N. M. van Gelder, Occlusion and metabolism of γ-aminobutyric acid by brain tissue, *J. Neurochem.* **3**:28–40 (1958).

41. R. J. Baldensarini and I. J. Kopin, The effect of drugs on the release of nor-epinephrine-H^3 from central nervous system tissues by electrical stimulation *in vitro, J. Pharmacol. Exptl. Therap.* **156**:31–38 (1967).

42. T. N. Chase, G. R. Breese, and I. J. Kopin, Serotonin release from brain slices by electrical stimulation: Regional differences and effect of LSD, *Science* **157**:1461–1463 (1967).

43. J. L. Webb and K. A. C. Elliott, Effects of narcotics and convulsants on tissue glycolysis and respiration, *J. Pharmacol. Exptl. Therap.* **103**:24–34 (1951).

44. H. McLennan and K. A. C. Elliott, Effects of convulsant and narcotic drugs on acetylcholine synthesis, *J. Pharmacol. Exptl. Therap.* **103**:35–43 (1951).

Chapter 7

ELECTRICAL PHENOMENA AND ISOLATED TISSUES FROM THE BRAIN

John A. Harvey

Department of Psychology
University of Iowa
Iowa City, Iowa

and

Henry McIlwain

Department of Biochemistry
Institute of Psychiatry
British Postgraduate Medical Federation, University of London
Maudsley Hospital, London

I. INTRODUCTION

Under appropriate conditions of preparation and incubation tissue specimens from the mammalian brain have been shown to possess electrophysiological responses, ionic content, and metabolic properties *in vitro* that approximate those observed *in situ*. Information therefore has become available concerning the linkage between ionic movement and metabolism during the maintenance of the resting membrane potential in mammalian tissue and the events occurring during excitation and subsequent recovery.[1,2] The present chapter describes some of the methods employed as well as the data obtained in such investigations; for fuller accounts of several of the topics discussed, see McIlwain,[1,2] and McIlwain and Rodnight.[4]

II. TECHNIQUES

A. Tissue Preparation

1. *Source of Specimens*

Specimens obtained from guinea pig or rat have been examined most frequently; in these species the brain can be rapidly exposed and removed from the skull. The cortex of the cerebral hemispheres is a common source of tissue samples obtained as thin sheets, approximately $350\,\mu$ thick, by cutting parallel to the outer cortical surface. Slices thus obtained from the dorsal surface of the hemispheres constitute neocortical tissue that is characterized by six more or less distinct layers of perikarya and by a thickness in the adult animal of some 1.2 mm in the guinea pig and 1.9 mm in the rat. From each hemisphere three or more successive slices thus can be obtained differing from each other in metabolic properties, chemical composition, and neuronal elements. In addition, the outer slices possess the natural surface of the brain that may have a specific functional relationship with the cerebrospinal fluids that normally bathe it. Rapid entry of some compounds into the brain *in vivo* has been shown to occur across this surface.[3]

In addition to neocortex, the cerebral hemispheres also can provide specimens of the piriform cortex that occupies the anterior base of the hemispheres and parts of the lateral surface where it is separated from neocortex by the rhinal fissure. This phylogenetically older cortex, paleocortex, receives olfactory impulses via the lateral olfactory tract, which is located on its surface and easily visible especially in the guinea pig. Samples also have been prepared, as slices, from a variety of regions including subcortical white matter, hippocampus, hypothalamus, and medulla.

2. Cutting and Weighing

The procedures described below have been found to produce specimens of brain that most closely approximate *in situ* tissues with respect to ionic content, electrophysiological responses, and associated metabolic events. In preparing such specimens one should (1) avoid all unnecessary manipulations of the brain at the time of removal from the skull as well as during procedures employed to obtain a slice, especially those that would tend to crush, smear, or lead to contamination of the tissue; (2) avoid any contact of brain or sample with fluids prior to placing the slice in its complete medium, fully oxygenated at 38°C and this requires the use of a dry-cutting method, i.e., without added fluid, employing a bow cutter and recessed glass guide; and (3) keep the time between cessation of blood flow to the brain and placing of slice in its complete medium to a minimum, e.g., of 1–4 min.

Dry cutting the cerebrum *in situ* comes closest to meeting these requirements but restricts samples available and requires two persons to obtain the tissue sample. This procedure, however, allows one to bring the time between exsanguination, or opening of the skull in anesthetized animals, and placing the slice in complete medium to 0.5–2 min. Slices prepared in such short intervals may possess characteristics that are lost if longer intervals are employed (see below). Removal of brain prior to dry slicing is most commonly employed and is now described. For greater detail and more extensive treatment see McIlwain and Rodnight.[4]

The animal, rat or guinea pig, is stunned with a light blow to the cervical region and exsanguinated by cutting across the carotids with a pair of scissors. The skin overlying the skull is removed with scissors, and a cut is made at the posterior end of the skull through the muscle and spinal cord. The scissors are then employed to make a cut through the occipital bone along the saggital suture and above the cerebellum. With the scissors still inserted and slightly open, a twisting motion will extend the cut as a crack along the rostral extent of the saggital suture. The parietal and frontal bones are now removed with care being taken to avoid contact with or pressure to the brain. If the crack does not extend into the frontal bones, these still can be prized up with the ends of the scissors extended to either side of the bone. The dura is often removed by these procedures; if not, a spatula can be inserted between cortex and dura from the most caudal portion of the cerebrum and a sideways motion will cut the dura that then can be lifted to either side. A cut with the spatula now can be made between the olfactory bulbs and cerebrum anteriorly and between the cerebrum and cerebellum posteriorly. Inserting the spatula beneath the base of the brain from the posterior end of the cerebrum and moving it sideways will cut the remaining connections, and the brain then can be lifted out of the skull cavity and placed immediately in a Petri dish that already contains a moistened filter paper on the inside cover.

Cutting is done best on a stable platform that is elevated to a convenient height from the top of a laboratory bench. A slightly moistened piece of filter paper is placed on the platform to receive the portion of brain from which the slice will be obtained. The brain is cut longitudinally with a scalpel or sharp spatula to yield the two hemispheres. To obtain the prepiriform area, an incision is made with a scalpel along the rhinal fissure and just dorsal to it. This cut is angled in a manner that allows the prepiriform area and lateral olfactory tract to be in an approximately horizontal position and not tilted to the left or right when this portion of the brain is placed, outer surface upward, on the cutting platform. A depression will exist between the juncture of the prepiriform area and piriform lobe, and this can be reduced by placing four narrow strips of filter paper on top of each other to build up a mound under this depression.

With the anterior end of this tissue facing the investigator, the dry, recessed glass guide is placed on the piriform lobes and brought down with minimum pressure to lie flat on the surface of the area. The dry bow cutter now is used to begin a slice at the anterior end and terminated upon reaching the piriform lobes. This somewhat pear-shaped slice, with its base at the caudal level, will contain the lateral olfactory tract on its surface. The weight of this slice is approximately 35–40 mg with a few mg contributed by the lateral olfactory tract. The slice is then picked up on a bent wire rider, rapidly weighed to the nearest milligram, and immediately placed in its complete, oxygenated medium at 38°C. This can be the experimental vessel itself or an evaporating dish containing 50 ml of medium for subsequent transfer to various electrode assemblies.

Slices of neocortex can be obtained by taking the remaining portion of the cerebral hemisphere, removing the brainstem with a spatula, and cutting slices of approximately 50–60 mg as above.

B. Incubation Conditions

1. Media

A bicarbonate buffered medium(pH, 7.4) has been found to be adequate for most purposes involving measurement of ion movement, electrophysiological responses, and tissue phosphates. The composition of this medium is (millimolar): NaCl, 124; KCl, 1–5; KH_2SO_4, 1.2; $NaHCO_3$, 26; $CaCl_2$, 0.75; and glucose, 10. This medium is kept in equilibrium with a gas phase of $O_2 + CO_2$ (95:5). For measurement of fluid spaces, inulin (1 %, wt. vol.) can be added.

The concentrations of various constituents are based on their levels in plasma and cerebrospinal fluid; however, departures from this composition should be noted. (1) The use of oxygen in place of air permits the use of thicker slices, and damage to structural elements is minimized. No deleterious effects are known, although oxygen tensions that are still higher have been reported to depress respiration. Under normal conditions slices of 350 μ thickness can be adequately maintained for several hours, and diffusion of oxygen and metabolites is not rate-limiting.[5, 6] (2) For most experiments it is not necessary to employ media with high molecular weight substances to replace the osmotic effects of proteins. Plasma proteins can be replaced by bicarbonate as above; alternatively, phosphate, glycylglycine, or quaternary ammonium salts have been used and permit manometric measurement of respiration. All of these buffers can provide a media iso-osmotic with respect to the bloodstream. (3) Calcium in plasma or cerebrospinal fluid is normally between 2 to 3 mM, and this has led to the use of 2.8 mM calcium in most media. However, approximately 50 % of the calcium in body fluids is bound, indicating that a lower concentration of calcium should be used in artificial media. The brain *in situ* contains approximately 2 μeq Ca/g fresh weight of tissue. When tissue slices are incubated in media containing 0.75 mM Ca, the calcium content of the slice is found to be 2.5 μeq/g, and correcting this value for additional fluid uptake during incubation (see below) gives a value of 2 μeq/g, close to that found *in situ*. Incubation of slices in 2.8 mM Ca on the other hand gives a tissue content of 4.5 μeq Ca/g or more than double that found *in situ*.[7] Tissue incubated in 0.75 mM Ca also exhibits a sodium, potassium, and chloride content differing from that *in situ* by only a small, though significant extent, and exhibiting less intracellular swelling on incubation.[8]

The fluid/tissue ratio employed *in vitro* may be deliberately chosen and modified, and the following considerations should be noted. The normal flow of blood through the brain averages about 50 ml/100 g of tissue per minute, and thus 3 ml would have passed through 100 mg of tissue in 1 hr. Three to five ml of incubation media are normally employed with 50–100 mg of tissue for experiments of 1–3 hr. The presence of these volumes of media, 30–100 times the volume of a slice, represents one possible arrangement: It can favor both the accumulation and depletion of tissue metabolites. Alternative arrangements are discussed by McIlwain and Rodnight.[4]

2. *Electrodes for Electrical Stimulation and Quick Transfer of Slices*

Rapid transfer electrodes are employed to stimulate tissue electrically or when changes in ions or metabolites are being determined over short intervals. Two grid electrodes, each having 4–5 windings of enameled silver wire, are mounted as jaws of a clamp. The slice, floating in an evaporating dish containing oxygenated medium at 38°C, is picked up between the jaws, which are held together by a spring and transferred to a 20-ml beaker containing 5 ml of medium in equilibrium with its gas phase in a 38°C thermostat. A sponge rubber seal at the top of the electrode assembly keeps the fluid surface exposed to O_2-rich gas, reduces evaporation of fluids, and holds in place a polyethylene tube that delivers gas bubbles through the medium. The wires of both jaws are bared at the point at which they are to serve as electrodes, and by them the entire tissue sample can be brought within a voltage gradient. The impedance of the electrodes to alternating condenser pulses of 10 V peak potential and 0.4 msec time constant at 100 pulses/sec has been found to be equivalent to approximately 20 Ω. The wires of the grid electrodes if desired can be bared only at specific points, thus allowing localized stimulation of tissue. This procedure is especially useful for stimulation of the lateral olfactory tract and allows one to observe effects produced by transmission of excitation from the localized point of stimulation. When tissue is to be analyzed for ions or metabolites the slice can be rapidly released from the electrodes into its own medium for subsequent manipulations. Similar electrode assemblies have been developed for use with manometric vessels.[4]

3. *Slice Chamber*

Tissues also can be placed in a slice chamber for measurement of resting membrane potentials, presynaptic axon potentials, and postsynaptic potentials. The chamber allows the following manipulations:

1. An initial incubation of tissue immersed in well-oxygenated media, under conditions that approximate those of metabolic experiments. Such preincubation has been found necessary for reestablishment of tissue content of labile metabolites, ions, and membrane potentials (see below).
2. After preincubation, the chamber allows the tissue to be supported at the surface of the incubation medium, so that electrical observations can be made without excess aqueous fluids and with the tissue in a fixed position in relation to electrodes.
3. An extensive outer ducting of the chamber allows the tissue to be adequately oxygenated while remaining accessible to electrodes from above, and the electrodes are allowed considerable movement.[4,9,10]

The tissue can be stimulated at localized points with a pair of ball-tipped silver wires having a tip diameter of 0.5 mm and a 1 mm tip distance. Such electrodes have been employed for stimulation of lateral olfactory tract where transmitted potentials can be recorded. For neocortex the entire tissue can be brought between a voltage gradient by means of grid electrodes described above. For recording extracellular potentials a ball-tipped silver wire identical to the stimulating electrodes can be employed or fine stainless steel needles. Glass micropipettes of less than 0.5-μ tip diameter and containing 2.7 M KCl are employed for intracellular recording. Similar micropipettes with tip diameters of 2–8 μ also can be employed for localized extracellular recording.[4]

4. Radioisotopes and Ionic Fluxes

Radioisotopes of the ions under investigation can be used conveniently for determination of ionic influx and efflux under normal conditions and during stimulation and subsequent recovery. Choice of isotopes is dependent on the nature of the experiment. The following isotopes, with their respective half-life, have been employed: ^{24}Na, 14.9 hr; ^{22}Na, 2.6 yr; ^{42}K, 12.4 hr; ^{40}K, 1.4×10^9 yr; ^{36}Cl, 4×10^5 yr; ^{47}Ca, 5.8 days; and ^{45}Ca, 152 days. The short-lived isotopes are convenient when subsequent measures of stable constituents of tissue extracts are required. After determination of radioactivity in extract and media the samples can be placed in a refrigerator until radioactivity is essentially background level. Their drawback is the rapidity at which experiments must be conducted once the isotope is received, the requirement for immediate measurement of radioactivity to ensure sufficient counts above background, and the need for calculating the radioactive decay occurring between initial addition of isotope and final measurement of radioactivity. The long-lived isotopes reduce these problems but require scrupulous attention to possible contamination of working area and glassware.

In typical experiments for measurement of ion influx, slices 35–60 mg fresh weight are incubated for 40 min or longer in order to achieve stable levels of ions and phosphates and complete equilibrium of inulin with the tissue. The sodium or chloride salts of the above-listed isotopes then can be added in a volume of 0.1 ml or less by means of a syringe to give a final concentration of approximately 1.5 μC/ml of medium. Control slices receive an equivalent concentration of the nonradioactive salt. Alternatively the radioisotope can be incorporated into glucose bicarbonate media so that addition does not alter the osmotic or ionic composition of the medium. Slices are then removed at intervals of 0.5–60 min after addition of isotope, handled as described below, and a sample of the tissue extract and media in which it was incubated is measured for radioactivity by means of a liquid scintillation counter. For the calculations described below, determination of ion and inulin content of tissue and media also are required.

The effects of electrical stimulation or of depolarizing agents on ion influx can be determined in more than one way. An initial response to stimulation is obtained by applying electrical pulses or chemical agents before or at the same time as addition of isotope. It should be noted that tissue content of ions will be changing rapidly during this time. Alternatively one can determine influx during the steady state of stimulation. Maximum changes in Na and K content of tissue following electrical stimulation occur within 6 min. Addition of isotopes after 10 min electrical stimulation allows one therefore to measure influx in the absence of net changes in ion content.

For determination of efflux a quantity of isotope, e.g., three times that used for influx, is added to the medium after 30 min incubation and a further interval of 30 min or more is employed to allow sufficient exchange of the isotope with the tissue. The tissue is then transferred to a beaker containing nonradioactive medium. If a tissue holder, such as the rapid transfer electrode, is employed the holder and slice may have to be rinsed in nonradioactive medium to remove radioactivity from the holder, prior to transferring to a nonradioactive beaker for subsequent measurements of efflux. Samples of media then can be taken over successive intervals and then the radioactive content of an aliquot of tissue extract determined. This allows one to calculate the total radioactivity present in the slice at time zero and the amount released at various intervals. It sometimes may be advantageous to transfer the slice repeatedly from one beaker to another at various intervals in order to collect the radioactivity. Extraneous materials may be leached from the tissue under these conditions and appropriate control experiments are needed.

Electrical stimulation or effects of added agents can be compared with nonstimulated efflux in several ways. Stimulation can be applied at zero time and continued during the course of sampling from media, or stimulation can be applied over a specified interval of time in order to see the changes in efflux that it produces.

5. Draining and Reweighing of Tissue

It is important to know the final weight of the tissue so that the amount of fluid uptake can be calculated. The slice is picked up from its beaker with a bent wire rider, care being taken to avoid contact with the sides of the beaker where condensates having a different composition of ions from that of the media will have collected. The slice folded over the rider is brought into contact with a clean glass surface so that first one and then the other side of the slice touch the glass repeatedly during 1–2 sec. The slice is now weighed to the nearest milligram and then homogenized in, e.g., 4 ml of 6% (wt. vol.) trichloroacetic acid for analysis of tissue constituents. At the same time aliquots of media are diluted to required amounts. Methods for subsequent determination of ions, inulin, phosphates, and other constituents have been described.[4]

C. Calculations

1. Ionic Content

The initial weight of the slice, prior to any contact with experimental fluids, is its fresh weight. The final weight of the slice after draining represents the wet weight. The difference between these two weights is regarded as representing the amount of additional fluid taken up by the slice during incubation. The ion content of the tissue (T) is expressed as microequivalents per gram of fresh weight. The inulin of the tissue after reaching a stable level is assumed to be, in the tissue fluids to which it has access, at the same concentration as that of the medium. Then the inulin space (i.s.) of the tissue, expressed as microliters per gram of fresh weight, can be obtained by

$$\text{i.s.} = \frac{\text{mg inulin/g tissue}}{\text{mg inulin}/\mu\text{l media}}$$

If the concentration of ion in the inulin space is assumed to be the same as that of the media then the ion content of the inulin space (I) expressed as microequivalents per gram is obtained by $I = \text{i.s.} \times M$, where M is the ion content of the media (microequivalents per microliter). The content of ion in the noninulin space expressed as microequivalents per gram is therefore $T - I$.

The noninulin space (n.i.s.) is calculated from data on the density of tissue (e.g., $1000 \ \mu\text{l/g}$) and its water content (e.g., $800 \ \mu\text{l/g}$). The difference between the initial fresh weight of tissue and its final wet weight gives the amount of additional fluid (a.f.) expressed as microliters per gram, which has become associated with the slice during incubation. Some of this fluid is extracellular and therefore taken into account by the calculation of inulin space. The remaining portion is intracellular, in the sense that it is not accessible to inulin (see below) and therefore must be included in the calculation. Therefore the volume of fluid in the noninulin space (microliters per gram) is calculated as

$$\text{n.i.s.} = 800 + \text{a.f.} - \text{i.s.}$$

The concentration of ions in the noninulin space is then obtained as

$$\text{Ions in n.i.s. (mM)} = \frac{\text{Content ion in n.i.s. } (\mu\text{eq/g})}{\text{n.i.s. } (\mu\text{l/g})}$$

Alternatively the noninulin space can be expressed by its weight (grams per gram tissue), and its ion content can be given as microequivalents per gram noninulin space. Chloride, sucrose, and other spaces can be calculated in the same manner.

2. Ionic Fluxes

Assuming that the radioisotope of a particular ion is diffusing at the same rate as the common isotope, the following calculations can be made to determine influx. The radioactivity of the tissue is expressed as counts per minute per gram tissue fresh weight and of the medium as counts per minute per microliter. An ion space (S) for the tissue (microliters per gram) can then be calculated as

$$S = \frac{\text{Counts/min/g tissue}}{\text{Counts/min}/\mu\text{l medium}}$$

The total amount of ion (E) exchanging with the slice (microequivalents per gram) is obtained by

$$E = S \times M$$

When the radioisotopically estimated ion space is greater than the inulin space a portion of the ion can be assumed to have entered the noninulin space. Assuming that the specific activity of the medium at this time is the same as that of the inulin space, then the ion exchanging with the noninulin space (microequivalents per gram) is

$$(E - I)$$

The quantity of ion in the noninulin space that has not exchanged during this time is $(T - E)$.

The inulin space of course does not reach the same specific activity as the medium before any radioisotope begins entering the noninulin space. Both processes occur at the same time but at different rates. Exchange of ^{22}Na with a slice as expressed by the microequivalents exchanging per gram fresh weight of tissue consists of two fairly distinct phases. Approximately 90 μeq/g were found to have exchanged within 30 sec while only an additional 3 μeq/g had exchanged during the next 30 sec.[11] It is assumed that the high rate of the first component represents equilibration of isotope with the inulin space, whereas the second component represents primarily equilibration between inulin and noninulin space. The equilibration of the radioisotope with the inulin space, as determined by the time required for the radioisotope space to equal the inulin space, occurs in 90 sec.[11,12] At this time the amount of isotope exchanging with the slice is highly correlated with the inulin space of the slice. At longer intervals of 120 sec this correlation no longer exists.[12] It appears therefore that these calculations of ion exchange with noninulin space represent reasonably valid estimates of influx.

For calculations of efflux one first determines the total radioactivity that was in the slice at time zero by adding the total number of counts that have come out of the tissue to the counts still remaining at the end of the experiment. Efflux then is simply expressed as the fraction of radioactivity that has left the tissue at specified intervals. Since the ion content of the tissue is known, the fraction of radioactivity leaving the tissue is equal to the fraction of ions leaving the tissue. For potassium, which is primarily intracellular, such a procedure can give one a reasonable estimate of efflux from noninulin space. However, for sodium or other ions, which are primarily extracellular, one can only determine efflux from the total slice.

III. ION EXCHANGE

A. Maintenance of Normal Ionic Gradients

The process of tissue sampling produces an artificial cut surface with consequent damage to cell elements. Neocortex contains a neuronal surface of approximately 10^4 cm^2/g tissue.[13] A slice 350-μ thick and weighing 100 mg would have a surface area of 30 cm^2/g, and thus the artificial cut would at most affect 0.3 % of the total cell surface. Slices of cerebral cortex lose a smaller proportion of their dry weight than slices of other tissue such as liver, and this is most likely due to their possession of long and very narrow axons and dendritic processes from which leakage does not so readily occur. In addition, sealing of such cells can occur by reforming lipoprotein membranes, which may be aided by contact with physiological saline.[14]

In most experiments differences can be seen between outer and inner slices (see Section II.A.1.), and this is attributable to the possession by the latter of an extra artificial cut that produces a greater extracellular space. Differences in inulin space of inner and outer slices indicate that each artificial cut adds 63 μl/g tissue to the inulin space, and thus 6.3 % of formerly intracellular tissue volume is now exposed to external media.[8] For a 100-mg slice of 350 μ thickness this represents a depth of 22 μ being affected, assuming a localized change. Such factors tend to produce greater content of sodium and chloride and lesser content of potassium in tissue slices as compared with their content *in situ*.

In addition to changes in fluid content resulting from the artificial cut, other changes in fluid volume, both extracellular and intracellular, have been noted during initial contact of specimens with medium and on subsequent incubation. These have been divided into three types.[15]

1. Adhering Medium

After 10 sec exposure to medium a redrained slice shows an approximately 10 % increase in weight. Thus adhering medium adds some 100 μl of inulin space per gram tissue.

2. Swelling During Preparation

If samples are placed briefly in an adequately oxygenated medium at room temperature a further, irreversible, increase in water uptake occurs that is intracellular in the sense that it is not associated with the uptake of inulin. This water uptake can occur at the rate of 25 μl/g tissue/min. The same situation is occurring in tissue between the time of cessation of blood flow and placing of the slice in a fully oxygenated medium even if dry-cutting methods are employed. The intracellular swelling noted above was measured when the medium contained 2.8 mM Ca. Use of 0.75 mM Ca in the medium reduces such swelling considerably so that tissues cut *in situ* and prepared within 4 min showed little or no intracellular swelling.[8]

3. Swelling on Incubation

There is a slow increase in the weight of the tissue on incubation under adequate metabolic conditions, which corresponds to the uptake of 1 μl water/g tissue/min. This swelling is paralleled by a corresponding increase in the inulin space and therefore can be assumed to be extracellular.

Such changes in fluid content of the slice apply to neocortex, and differences can exist with other samples. Thus the prepiriform cortex exhibits less swelling under conditions identical with those employed for neocortex.[11] Certain substances added to the medium also can alter the fluid uptake into inulin or noninulin spaces of the tissue.[15]

Regardless of the method of preparation the sample usually, upon first being placed in its complete medium, displays an increase in tissue content of sodium and chloride and a decrease in potassium,[16] a membrane potential near zero,[17] and a large rise in inorganic phosphate and a decrease in high-energy phosphates.[1,2,4] Within 20–30 min of incubation the increase in sodium and chloride of the tissue reaches a stable level, potassium is re-assimilated into the tissue to increase its content,[16] the membrane potentials indicate a repolarization of the cells,[17] and there is a decrease in inorganic phosphate and an increase in high-energy phosphates.[2] If the tissue is prepared rapidly *in situ,* an active extrusion of sodium occurs and a greater reassimilation of potassium[16] so that the tissue content of these ions approximates more closely the content *in situ,* if 0.75 mM Ca is employed in the media.[8]

B. Ion Content and Fluxes Under Normal Conditions

The content of ions *in situ* and *in vitro* for neocortex is given in Table I. The closeness of the sodium, potassium, chloride, and calcium contents of slices *in vitro* to values obtained *in situ* are in many ways remarkable when one considers that the permeability of neural tissue to these ions is inherently variable.

It may be noted that potassium efflux is somewhat higher than the corresponding values for influx, under conditions where there is no net change in potassium content of the slice. This implies that a proportion of the tissue potassium exchanges at a still slower rate than that quoted in the table, and this would correspond to approximately 20 % of the potassium being relatively nonexchangeable in the adult animal. Efflux data for sodium have not been obtained in the form allowing for calculation of rate.[12] However, extrusion of sodium during preincubation of tissue has been noted to occur at approximately 200 μeq/g tissue.[16] Calculation of the non-exchanging fraction of noninulin sodium during influx under normal conditions has indicated the existence of at least two pools of sodium; the component exchanging more slowly is not altered by stimulation.[12] In contrast to sodium and potassium, calcium exchange is quite slow.[7]

C. Effects of Electrical Stimulation on Ion Movement— Recovery

Electrical stimulation produces a net entry of sodium and an equivalent net loss of potassium in tissue slices. The net changes in ion content occur

TABLE I

Ion Composition of Cerebral Tissue and Body Fluids Under Normal Conditions *in Situ* and *in Vitro* and Associated Ionic Fluxes[a]

Measurement	Na	K	Cl	Ca[b]
Content *in situ* (μeq/g)	45	96	41	2
Content *in vitro* (μeq/g)[c]	74	82	58	2
Noninulin space content (μeq/g)	28	68	16	1.7
Concentration in noninulin space (mM)	40	114	23	3
Influx into noninulin space (μeq/g/hr)	200–400	330	—	1.2
Efflux from noninulin space (μeq/g/hr)	200–400[d]	400	—	1.2
Ratio of concentration in noninulin space to concentration in media[e]	0.25	18.9	0.11	4
E_m (mV)[f]	+37	−79	−59	−19
Extracellular concentration *in situ* (mM)[g]	151	3	126	0.86
Intracellular concentration *in situ* calculated for −60 mV potential (mM)[h]	35	135	13	2.4

[a] Data taken from Lolley,[7] Keesey *et al.*,[8] Gibson and McIlwain,[9] Harvey and McIlwain,[11] Keesey and Wallgren,[12] Bachelard,[16] and Cummins and McIlwain.[18]

[b] Of these values for tissue about 40 % exists in the unbound form.

[c] Content of outer slices corrected for adhering fluid.

[d] Estimated from influx during steady state.

[e] Ratio is based on content of bicarbonate buffered medium described in text.

[f] Calculated as described in text.

[g] Calculated from values given for guinea pig plasma[7,16] and corrected by the distribution ratios for these ions between plasma and cerebrospinal fluid.[34]

[h] Based on chloride space calculated from the assumption that it is distributed in electrochemical equilibrium with membrane potentials of −60 mV obtained *in situ*.

during the first 6 min of maximal stimulation and then remain constant.[8,16] Entry of sodium into the noninulin space during the initial phase of stimulation proceeds at rates of 460 μeq/g/hr and results in a doubling of the noninulin content of sodium.[8] Influx of sodium into the noninulin space during this time or during the time that there is no further net change in sodium content of the noninulin space proceeds at the rate of 1060 μeq/g/hr.[8] Thus stimulation has produced an approximately fivefold increase in influx. To account for these measures of influx and changes in net content, efflux must have increased to approximately 600 μeq/g/hr (1060 minus 460) during the first few minutes of stimulation and then to 1060 μeq/g/hr after 6 min when no further change in net content was noted, and therefore influx must be expected to be balanced by an equal efflux of sodium.

Net loss of potassium proceeds at 480 μeq/g/hr when tissue is stimulated as above. Thus movements of sodium and potassium balance each other. Potassium influx increases approximately twofold on stimulation to 645 μeq/g/hr.[18] Note that this rate of influx would balance the estimated rate of sodium efflux 600 μeq/g/hr during this period. The measures of potassium efflux are based on efflux from total slice and not from the noninulin space and give values of 600–750 μeq/g/hr.

If the capacity of the outer membrane of neocortical cells[19] is taken as 10^{-5} F/cm^2 and the normal resting potential of neurons (see below) as -60×10^{-3} V, then the amount of charge required to produce complete depolarization would be 6×10^{-7} coul. This is equivalent to 6×10^{-12} mole of monovalent cation. Taking the observed net entry of sodium at 13×10^{-8} eq/g/sec and assuming 10^4 cm^2 of neuronal membrane per gram of tissue gives a rate of sodium entry of 13×10^{-12} eq/cm^2/sec, or double the amount required to completely depolarize the tissue. This undoubtedly is due to the fact that some of the sodium entering the cell exchanges with potassium, particularly in the early phases when permeability to both ions has been increased. Similar calculations with squid axon indicate a sodium entry and potassium loss of some $3–4 \times 10^{-12}$ eq/cm^2 impulse, a value three times greater than the 1.2×10^{-12} eq/cm^2/sec calculated to be required for depolarization of the axon.[20]

In contrast to the rapid movements of sodium and potassium, calcium influx and efflux is increased only by approximately 0.07 μeq/g on stimulation and there is no change in net content.[7] Chloride also shows little change on stimulation, with only a slight increase being noted in the noninulin space. Its changes seem rather to depend on the total noninulin content of sodium and potassium,[8] but the subject has not been studied extensively.

Upon cessation of stimulation there is a rapid recovery of normal noninulin content of sodium and potassium, which is complete or nearly complete within 10 min. The initial rates of net change in the noninulin content of these ions have been found to be 160 μeq/g/hr for sodium and 230 μeq/g/hr for potassium.[8] Recovery of potassium content in the tissue is initially more rapid than for sodium, and just after termination of stimulation it may be as high as 500 μeq/g/hr.[18] The recovery of normal sodium and potassium content in the slice is less in media 2.8 mM in calcium, although the initial rates of extrusion of sodium and entry of potassium are the same.[16] Increased rates of influx and efflux have been noted to persist after termination of electrical stimulation.

D. Associated Phenomena

1. Substrates

In the absence of glucose the tissue content of sodium is increased and that of potassium decreased. Concomitantly, there is an increased intracellular accumulation of water. Although the potassium content of the tissue is lowered by omission of glucose, exchange with ^{42}K of media proceeds at approximately the same rate as in its presence, and again stimulation increases this exchange.[18] This indicates that the permeability alteration caused by electrical stimulation is largely independent of the major energy metabolism of the tissue which, as shall be seen, is primarily required for restoration of normal ion content. Similar conclusions have been reached with squid axon.[21]

The substrates that can replace glucose without appreciably altering the sodium or potassium content of the noninulin space or the degree of tissue swelling have been examined.[22] Succinate and fumarate lead to both a decrease in potassium of some 40 μeq/g and an increase in sodium of 60 μeq/g but do not affect swelling of the tissue.

Absence of added oxidizable substrate led to a 50 μeq/g fall in potassium, 34 μeq/g rise in sodium, and swelling of 44%. In spite of this, electrical stimulation of the tissue for 10 min led to a noticeable decrease in potassium and to an equivalent increase in sodium. However, only partial recovery of sodium and potassium occurs following stimulation with substrates other than glucose, and there is no recovery in the absence of oxidizable substrates.[22]

In the absence of oxygen, changes in ion content are more rapid. With no added substrate for 3 hr or under anoxia the concentrations of sodium approximated that of the medium, but potassium concentration in noninulin space remained two to three times above that of the medium. Rapid movement of water is also seen to occur, which is mainly intracellular and amounts to about 50% of the initial tissue weight. With metabolic inhibitors, such as NaF or 2,4-dinitrophenol, the sodium gradient also is abolished but the potassium concentration in the noninulin space of the tissue remains 10–13 times higher than that of the medium.[22]

2. Excitatory and Inhibitor Agents

Several amino acids that are excitatory *in vivo*[25] have cognate effects on slices of cortex incubated *in vitro*, as reflected in their ability to produce depolarization (see below). They also promote a rapid entry of sodium and loss of potassium from the slices. These substances include L-glutamate, L-aspartate, L-cysteate, DL-homocysteate, and L-α-aminoadipate.[11,17,23–25] Rates of sodium and potassium entry induced by L-glutamates are impressively large—600 and 660 μeq/g/hr, respectively—and thus are comparable to that produced by electric stimulation of tissue.[11,24] Sodium influx following addition of 5 mM L-glutamate was found to occur at an initial rate of 1230 μeq/g/hr into noninulin space, suggesting that sodium efflux had also been increased to approximately 630 μeq/g/hr (1230 minus 600). In addition, N-methyl-DL-aspartate produced similar changes in ion content and sodium influx but at a lower concentration (0.1 mM), which is in agreement with its reported greater potency *in vivo*.[11]

Tetrodotoxin has been demonstrated to prevent the ionic and metabolic changes normally produced by electrical stimulation of tissue slices. Although there is no detectable effect of tetrodotoxin on ion content or respiration of tissue under normal conditions, it prevents the entry of sodium into the noninulin space during excitation produced by either electric stimulation or by glutamate. The effects of maximal electric stimulation of tissue slices are completely blocked at concentrations of 0.05 μM tetrodotoxin in the medium.[26,27,28] Tetrodotoxin at 0.3 μM also blocks the increased influx of sodium ions normally produced by 5 mM glutamate.[28] It has been suggested that tetrodotoxin exerts its effect by a highly specific blockage of sodium channels normally activated during excitation.[27]

IV. CONSEQUENT METABOLIC CHANGES

It has been seen that the movement of sodium and consequent loss of potassium on depolarization can occur independently of energy-requiring processes but that these are necessary for both the normal maintenance of ionic gradients and for subsequent recovery from the effects of depolarization. The stabilization of sodium and potassium in tissue *in vitro* occurs with the re-establishment of energy-rich phosphates in the tissue and the increased activity of the sodium pump presumably via activation of the Na-K-Mg ATPase.[1] The sequence of phosphate changes on excitation consists of a transitory fall in adenosine triphosphate, followed by its recovery as phospho-creatine falls. These changes as measured in the intact tissue involve loss of 1200–1500 μeq $\sim$P/g tissue per hour during the 5–10 sec concerned. Within the first 10 sec of pulses cerebral glycolysis increases in rate, and the sustained increase in glycolysis and respiration that ensue can provide an additional energy-rich phosphate at about 500 μeq $\sim$ P/g/hr,[29] though for a limited period only. The increase in respiratory rate that also results from stimulation is sustained, however, and is capable of yielding $\sim$P at similar rate. The reason for the fall in energy-rich phosphate diminishing after 5–10 sec while the level of adenosine triphosphate is maintained is thus the augmented rate of formation of the ATP. These rates of change in phosphates are well within the known range of enzymic activity in brain. Further, it has been estimated that only 26% of the total energy available from respiration is required to move sodium and potassium in the steady state of unstimulated tissue and that some 35% of the additional energy available during stimulation would be required for the additional ion movement.[12]

V. ASSOCIATED ELECTRICAL PHENOMENA

A. Resting Membrane Potential

1. Under Normal Conditions

Membrane potentials of cells in the neocortex have been recorded employing the slice chamber and micropipette electrodes of less than 0.5-μ tip diameter, described in Section II.B.3. On penetration of the tissue by the micropipette, negative potentials are obtained that can be held for periods of at least 30 sec[9,10,17,30] and often for 20 min. A typical resting membrane potential of -60 mV has been obtained in neocortical tissue by averaging the maximum potential obtained during each penetration.[17] This value corresponds well with potentials recorded *in situ*.[30] Injury discharges have been noted on penetration in a small proportion of cases.[10,30]

These potentials are not present when the slice is first placed in the medium but develop within 20–30 min of incubation,[17] during which time the tissue also is acquiring its normal content of ions and high-energy phosphates.[2,16] The potentials are lost again in the absence of oxidizable substrate or under anoxic conditions but can be restored by returning the slice to its complete oxygenated medium. Under optimum conditions potentials can be recorded for periods of 2–3 hr and are not affected by repeated washes. Omission of sodium or potassium from the medium prevents the development of potentials while the omission of calcium or magnesium delays the development of potentials.[17] Added agents can affect these resting potentials in neocortex. Thus, clupein lowered and chlorpromazine raised the average resting potential.[10]

2. Following Stimulation and Recovery

Placing a specimen of neocortex between grid electrodes and applying electric pulses results in loss of the membrane potential. Recordings are often only obtainable 30–60 sec after termination of pulses. Readings at this time revealed that the membrane potential was still reduced by 50 % or more from resting averages and returned to normal within 3–4 min.[10] Acidic amino acids such as L-glutamate also produce depolarization, as does an increased concentration of potassium in the media.[9,17,24] The effects of electric stimulation on the resting membrane potential can be altered by added agents, which can prevent depolarization (phenobarbital), increase the rate of recovery of normal membrane potential (chlorpromazine), or decrease the rate of recovery (clupein).[10]

3. Relation Between Membrane Potential and Ionic Gradients

Applying the Nernst equation in the form

$$E_m = \frac{RT}{nF} \ln \frac{[C]_e}{[C]_i}$$

allows one to calculate the equilibrium potential (E_m) for the various ions listed in Table I. R, T, and F have their usual significance, n is the valence number of the ion, while $[C]_e$ and $[C]_i$ indicate the extra- and intracellular activities, respectively, for the ion. Taking concentration proportional to activity, calculated values for E_m appear in Table I. For an alternative calculation of the equilibrium potential that takes into account the effects of sodium, potassium, and chloride see Gibson and McIlwain.[9]

It can be seen that the equilibrium potential for potassium is markedly more negative than the observed maximum potentials in neocortex of -60 mV. Presumably the opposing sodium gradient, whose equilibrium potential is $+37$ mV, contributes to the lower observed potential. The positive value for the sodium potential is in the direction of the observed positive potentials recorded at the peak of the spike discharge *in situ* and *in vitro*.[30] Chloride, on the other hand, is distributed in electrochemical equilibrium with the observed membrane potential.

The calculations for calcium depend on whether one includes the bound portion in the tissue. In either case, its equilibrium potential deviates markedly from the resting membrane potential, possibly indicating that it may also contribute to the final value of the membrane potential although, as has been noted, its distribution is not markedly altered during excitation.

Hodgkin[20] has applied the Nernst equation in the form

$$E_m = \frac{RT}{F} \ln \frac{[K]_e + b[Na]_e}{[K]_i + b[Na]_i}$$

where b gives the apparent permeability of sodium relative to that of potassium. The greater permeability of the membrane to potassium as compared with sodium is reflected in the calculated value for b of 0.05 at rest.[9] During stimulation the concentrations of sodium and potassium give a value for b of 0.8, indicating a 16-fold increase in sodium permeability relative to that of potassium. These changes in the relative permeability of sodium are best understood in squid axon, where b at rest is 0.01–0.08, increasing on excitation to values as high as 30.[20,31]

B. Electrical Responses to Stimulation

The resting membrane potentials demonstrable by intracellular electrodes, and the responses to excitation shown metabolically, gave expectation that excitation of the isolated tissues was accompanied by active membrane phenomena that could be demonstrated by electrical recording. With appropriate electrode systems, tissues maintained in chemically defined media indeed have shown a rich variety of such activities. Specimens from several parts of the brain of experimental animals, and also from the human brain, have proved susceptible to study in this fashion. Specimens of mammalian neocortex, piriform cortex, lateral olfactory tract, hippocampus, corpus callosum, optic tract, and superior colliculus have afforded electrical responses *in vitro*. The responses have included conduction in fiber tracts; discrete postsynaptic responses following single stimuli; repetitive discharges triggered by a single stimulus, and complex patterns of response to repetitive stimuli.

1. Conducted Responses

When a specimen has been chosen that includes a fiber tract that *in situ* would conduct a sensory input to the brain, or conduct impulses between different parts of the brain, stimulation *in vitro* also has initiated impulse conduction. This was first shown in the lateral olfactory tract of the guinea pig. Here and in the rat (Fig. 1) the tract runs for about a centimeter, superficially

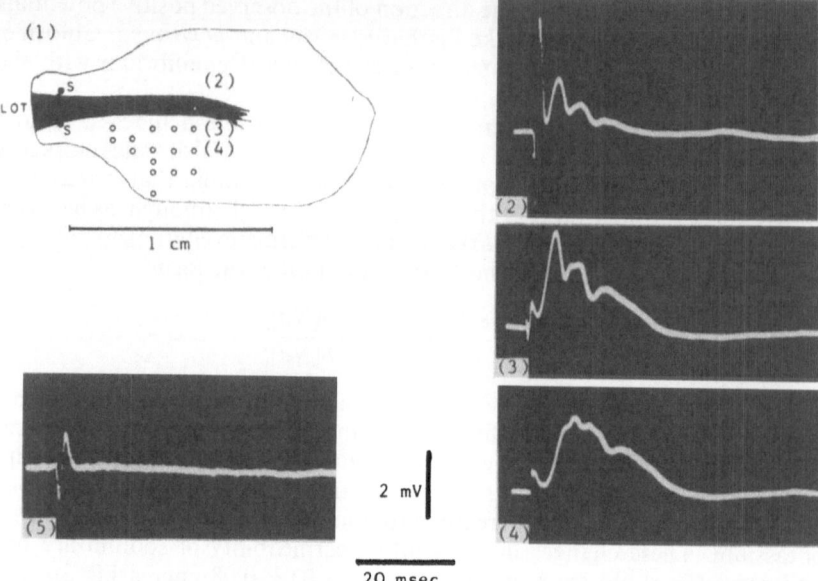

Fig. 1. Electrical responses to stimulation of isolates from the piriform cortex of the rat.[37] (1) The specimen used, which was 0.35 mm in thickness, in the plane perpendicular to the diagram. L.O.T. denotes lateral olfactory tract. The filled points S give the positions of stimulating electrodes, and the open circles give the positions examined with recording electrodes. From positions (2), (3), and (4) were recorded the signals in the right-hand diagrams (2), (3), and (4), respectively. Recording (5) was obtained in a different experiment from a position approximating (3), but while the tissue was exposed to diethylether. The slow waves previously evident were abolished but reappeared on removing ether.[37]

on the piriform lobe. A slice 0.35-mm thick was taken from the lobe, as described in Section II.A.2, maintained with oxygenated glucose saline, and stimulated with silver wire tips at a point close to the position previously occupied by the olfactory bulb. A similar wire tip, at different distances from the stimulating electrodes, acted as a recording electrode and demonstrated an impulse progressing at 10–12 m/sec along the tract.[32] This response was at greatest amplitude on the lateral olfactory tract itself, as shown in Fig. 1, and differed from other responses that are described below in its relative invariance when the tissue was exposed to some drugs and to a variety of stimulus patterns. Impulse-conduction has been observed also in the optic tract and corpus callosum.

2. Postsynaptic Responses

Such responses, in the piriform cortex of the rat, also are shown in Fig. 1, following stimulation via the lateral olfactory tract. They were characterized as follows. (a) When the recording electrode was moved from position (2) on the tract through position (3) to (4), the initial (fastest, conducted) response diminished in amplitude to about one sixth of its former value. A slower group of responses developed, which occupied some 35 msec in place of the 2 msec or so occupied by the conducted response. The slower response was preponderantly negative in potential when measured at the tissue surface, and this data favored an interpretation in terms of stimulation of new units in the cortex rather than in terms of slowly conducting fibers from the tract. (b) Moreover, on repetitive stimulation at intervals of 10–20 msec, the group of preponderantly negative responses increased in amplitude and became more complex, when recorded from a single point a few millimeters from the tract. (c) The negative response also was more susceptible to a number of pharmacological agents than was the conducted response. Position (5) in Fig. 1 shows it to have been completely eliminated by diethyl-ether; the conducted response, though diminished, remains. Phenobarbitone could greatly diminish the slow negative response without affecting the conducted impulse.[37]

The postsynaptic responses elicited by stimulation of the lateral olfactory tract also were studied, using micropipette electrodes[33] at defined depths in the cortical samples. Using surface electrodes, discharges also have been recorded in other cerebral preparations, including the superior colliculus, following stimulation from the optic tract.[38]

3. Dendritic and Other Responses from the Neocortex

Examining neocortical samples with electrode arrangements similar to those of Fig. 1 displayed a variety of types of response (Fig. 2).[35,38] In quite thin tissues successive negative responses could be elicited at the parts of the specimen derived from the outer cortical surface. These responses spread for a few millimeters in the guinea pig, but spread was slow: An initial wave lasted 10 msec and a subsequent one, 150 msec. They were concluded to derive from the apical dendrites and to correspond to the direct cortical response observable *in vivo*.

When thicker neocortical specimens were penetrated with capillary micropipette electrodes used extracellularly, stimulation a few millimeters away led to spike potentials that could be recorded 150–180 μ below the surface. These potentials were some 1 msec in duration and were preponderantly negative but with positive phases. They varied considerably when elicited by successive, otherwise identical, stimuli; this was suggested to be due to varying degrees of synaptic activation by the cell elements that had first responded. A number of stimuli delivered at 1 sec intervals could induce repetitive firing, which continued for up to 1 sec at some 30–60 discharges/sec. The repetitive discharge could be favored by certain changes in the composition of the medium bathing the tissue, in particular by replacement of its chloride content by other ions. Comparable phenomena also were observed in the isolated hippocampus of the guinea pig.

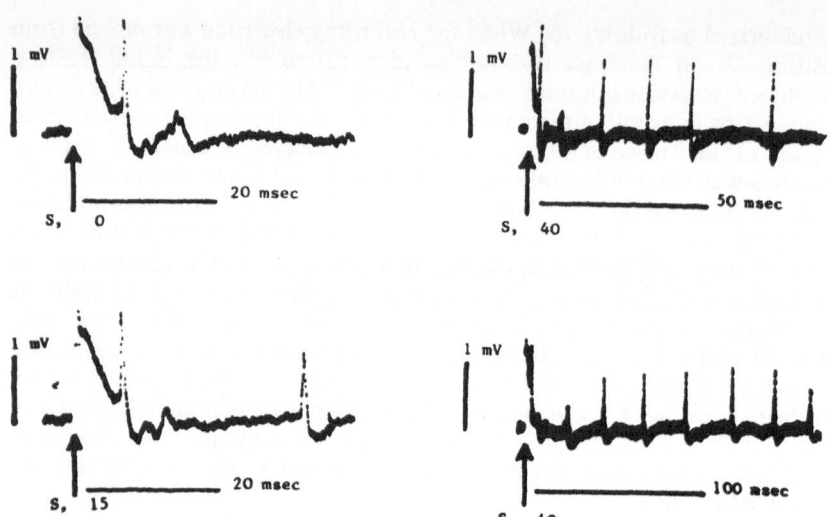

Fig. 2. Repetitive firing induced by surface stimulation of a guinea pig neocortical specimen approximately 6 × 8 mm and carrying the outer 0.35 mm of the cortex; the chloride of a bicarbonate-glucose medium partly replaced by ethylsulfonate. Stimuli S at 1 per second, shown by the arrows followed by a number indicating the number of stimuli previously delivered to the specimen recording extracellularly by micropipette.

VI. REFERENCES

1. H. McIlwain, *Chemical Exploration of the Brain. A Study of Cerebral Excitability and Ion Movement*, Elsevier, Amsterdam (1963).
2. H. McIlwain, *Biochemistry and the Central Nervous System*, Churchill, Ltd., London (1966).
3. L. J. Roth and C. F. Barlow, Drugs in the brain, *Science* **134**:22–31 (1961).
4. H. McIlwain and R. Rodnight, *Practical Neurochemistry*, Churchill, Ltd., London (1962).
5. O. Warburg, *The Metabolism of Tumors* (F. Dickens, trans.), Constable, London (1930).
6. J. Field, Respiration of tissue slices, *Methods Med. Res.* **1**:289–307 (1948).
7. R. N. Lolley, The calcium content of isolated cerebral tissues and their steady-state exchange of calcium, *J. Neurochem.* **10**:665–676 (1963).
8. J. C. Keesey, H. Wallgren, and H. McIlwain, The sodium, potassium and chloride of cerebral tissues: maintenance, change on stimulation and subsequent recovery, *Biochem. J.* **95**:289–300 (1965).
9. I. M. Gibson and H. McIlwain, Continuous recording of changes in membrane potential in mammalian cerebral tissues *in vitro*, recovery after depolarization by added substances, *J. Physiol. (London)* **176**:261–283 (1965).
10. H. H. Hillman, W. J. Campbell, and H. McIlwain, Membrane potentials in isolated and electrically stimulated mammalian cerebral cortex, *J. Neurochem.* **10**:325–339 (1963).
11. J. A. Harvey and H. McIlwain, Excitatory acidic amino acids and the cation content and sodium flux of isolated tissues from the brain, *Biochem. J.* **108**:269–274 (1968).
12. J. C. Keesey and H. Wallgren, Movements of radioactive sodium in cerebral-cortex slices in response to electrical stimulation, *Biochem. J.* **95**:301–310 (1965).

13. J. P. Schade and C. F. Baxter, *in Inhibition in the Nervous System and Gaba* (E. Roberts and C. F. Baxter, eds.), pp. 207–213, Pergamon Press, New York (1960).
14. L. V. Heilbrunn, *Outline of General Physiology*, Saunders, London (1952).
15. S. Varon and H. McIlwain, Fluid content and compartments in isolated cerebral tissues, *J. Neurochem.* **8**:262–275 (1961).
16. H. S. Bachelard, W. J. Campbell, and H. McIlwain, The sodium and other ions of mammalian cerebral tissues, maintained and electrically stimulated *in vitro, Biochem. J.* **84**:225–232 (1962).
17. H. H. Hillman and H. McIlwain, Membrane potentials in mammalian cerebral tissues *in vitro:* dependence on ionic environment, *J. Physiol.* (*London*) **157**:263–278 (1961).
18. J. T. Cummins and H. McIlwain, Electrical pulses and the potassium and other ions of isolated cerebral tissues, *Biochem. J.* **79**:330–341 (1961).
19. J. C. Eccles, *The Physiology of Nerve Cells*, Johns Hopkins Press, Baltimore (1957).
20. A. L. Hodgkin, Ionic movements and electrical activity in giant nerve fibers, *Proc. Roy. Soc.* (*London*) *Ser. B.* **148**:1–37 (1958).
21. A. L. Hogdkin and R. D. Keynes, Active transport of cations in giant axons from *Sepia* and *Loligo, J. Physiol.* (*London*) **128**:28–60 (1955).
22. P. Joanny and H. H. Hillman, Substrates and the potassium and sodium levels of guinea pig: cerebral cortex slices *in vitro:* effects of application of electrical pulses, of inhibitors and of anoxia, *J. Neurochem.* **10**:655–664 (1963).
23. R. J. Woodman and H. McIlwain, Glutamic acid, other amino acids and related compounds as substrates for cerebral tissues: their effects on tissue phosphates, *Biochem. J.* **81**:83–93 (1961).
24. H. F. Bradford and H. McIlwain, Ionic basis for the depolarization of cerebral tissues by excitatory acidic amino acids, *J. Neurochem.* **13**:1163–1177 (1966).
25. D. R. Curtis and J. C. Watkins, The excitation and depression of spinal neurones by structurally related amino acids, *J. Neurochem.* **6**:117–141 (1960).
26. H. McIlwain, Tetrodotoxin and the excitability and cation movements of isolated mammalian cerebral tissues, *J. Physiol.* (*London*) **190**:39–40 (1967).
27. H. McIlwain, Tetrodotoxin and the cation content, excitability and metabolism of isolated mammalian cerebral tissues, *Biochem. Pharmacol.* **16**:1389–1396 (1967).
28. H. McIlwain, J. A. Harvey, and G. Rodriguez, Tetrodotoxin on the sodium and other constituents of cerebral tissues, excited electrically and with glutamate, unpublished data.
29. H. McIlwain and M. Tresize, The glucose, glycogen and aerobic glycolysis of isolated cerebral tissues, *Biochem. J.* **63**:250–257 (1956).
30. C. L. Ki and H. McIlwain, Maintenance of resting membrane potentials in slices of mammalian cerebral cortex and other tissues *in vitro, J. Physiol.* (*London*) **139**:178–190 (1957).
31. P. F. Baker, A. L. Hodgkin, and T. I. Shaw, The effects of changes in internal ionic concentrations on the electrical properties of perfused giant axons, *J. Physiol.* (*London*) **164**:355–374 (1962).
32. C. Yamamoto and H. McIlwain, Electrical activities in thin sections from the mammalian brain maintained in chemically-defined media *in vitro, J. Neurochem.* **13**:1333–1343 (1966).
33. C. R. Richards and R. Sercombe, Post-synaptic potentials observed in mammalian prepiriform cortex *in vitro, Biochem. J.* **102**:30–31 (1966).
34. H. Davson, *in Handbook of Physiology* (J. Field, ed.), Sec. 1, Vol. III, p. 1761, American Physiological Society, Washington, D.C. (1960).
35. C. D. Richards and H. McIlwain, Electrical responses in brain samples, *Nature* (*London*) **215**:704–707 (1967).
36. C. Yamamoto and N. Kawai, Seizure discharges evoked *in vitro* in thin section from guinea pig hippocampus, *Science* **155**:341 (1967).

37. W. J. Campbell, H. McIlwain, C. D. Richards, and A. R. Somerville, Responses *in vitro* from the piriform cortex of the rat, and their susceptibility to centrally-acting drugs, *J. Neurochem.* **14**:937 (1967).

38. C. Yamamoto and N. Kawai, Origin of the direct cortical response as studied *in vitro* in thin cortical sections, *Experientia* **23**:821 (1967).

Chapter 8

DNA METABOLISM AND CELL PROLIFERATION*

Joseph Altman

Laboratory of Developmental Neurobiology
Department of Biological Sciences
Purdue University
Lafayette, Indiana

I. INTRODUCTION

In considering the problem of DNA metabolism and cell proliferation in the central nervous system we shall review in this chapter results obtained with three major techniques: (1) histological and cytological studies of cell division, which in some investigations were coupled with the auxiliary procedure of blocking mitosis with a mitotic poison, such as colchicine; (2) gross and microchemical studies of differences in DNA concentration in the brain in different species, at different phases of development or in different regions of the brain; and (3) studies in which a radioactively labeled specific precursor of DNA was administered to animals to tag newly forming cells and these were rendered visible and their fate traced *in situ* with autoradiography. Because in the history of this topic the histological approach antedated the other two, we shall introduce the subject by describing briefly the results of early histological studies. These early studies generated concepts that had a lasting influence on investigations in this area, but these have been supplemented, modified and, to some extent, superseded by results obtained with more modern techniques.

In the substantive part of this survey we shall deal (1) with DNA metabolism and cell proliferation during embryonic development of the nervous system, (2) with its continuation (in mammals) in the course of the postnatal maturation of the brain, and (3) with DNA metabolism and cell proliferation in the mature brain under normal and abnormal conditions.

* This study was supported by the U.S. Atomic Energy Commission and the National Institute of Mental Health.

A. Classical and Modern Views of Cell Proliferation in the Brain

1. *Embryonic Cell Proliferation in the Brain*

Intensive cell proliferation of specializing ectodermal cells signals the beginning of the embryonic development of the nervous system. In vertebrates, under the inductive influence of the underlying archenteron roof, specializing neuroepithelial cells form the neural plate mid-dorsally, along the longitudinal axis of the embryo. The rapid proliferation of these cells leads to an increase in the width of the plate, which becomes a several-cell thick neuroepithelium. The cell population of the neuroepithelium increases considerably when the medial depression of the plate deepens and the neural groove stage is reached. The early phase in the development of the primitive neuroepithelium is completed when the thickened marginal neural folds fuse mid-dorsally and the closed neural tube is formed. It is assumed generally that all the ectodermal elements of the nervous system originate directly or indirectly from cells of the primitive neuroepithelium (Fig. 1). (The peripheral nervous system derives directly from the neural crest and placodes.) The

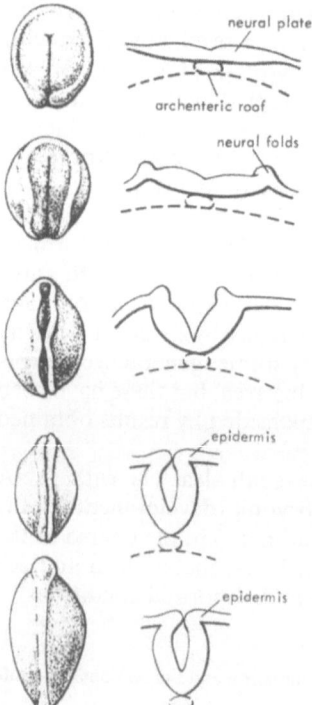

Fig. 1. Diagrammatic illustration of early stages in the embryonic development of the nervous system. Left: dorsal view; right: in transverse section. From Källén.[1]

structural differentiation of the caudal and rostral parts of the nervous system (the spinal cord and brain) is initiated after closure of the neural tube when some of the cells formed leave the proliferative neuroepithelium and give rise to differentiating and maturing regions, such as the mantle and marginal layers of the spinal cord. It is beyond the scope of this chapter to provide a description of the complex migratory movements and the mode of specialization of cells in different regions of the developing embryonic nervous system. For such descriptions the reader is referred to recent surveys in this area.[1–4] Some aspects of the problem are discussed in the fourth section of this chapter.

2. Postnatal Cell Proliferation in the Brain

In a recent textbook of histology the general topic of cell division is introduced with the following statement: "Many organs of the mature mammal show relatively few cells in division, and the central nervous system shows none."[5] In the paragraph from which this quotation is taken the following statement is also made: "Since, in mammalian cells, the small size and large number of chromosomes make detailed observation difficult, most studies of cell division have been made on cells of lower forms." The commonly held view that cell multiplication is absent in the brains of mammals may be due partly to the difficulty of recognizing multiplying cells in the brain with traditional histological techniques and to an even greater extent to a difficulty of tracing the fate of those cells that were for some time known to multiply at some specific sites in the brain. The presence in the brains of infant mice, rats (Fig. 2), cats, and dogs of many mitotic cells (Fig. 3) around the ependymal wall of the ventricles has been familiar to most neuroanatomists, and it was reported by several observers that mitotic cells persist into adulthood in the subependymal layer of the anterior portion of the lateral ventricle in rats,[6,7] mice,[8,9] and man.[10–12] The fate of these mitotic cells could not be determined. Some investigators postulated that these mitotic cells gave rise to glia cells; others assumed that cell proliferation at this site represents merely an abortive prolongation of embryonic activity.

Another well-known site of intensive mitotic activity in the brains of infant mice, rats, kittens, dogs, and man is the subpial germinal zone of the cerebellar cortex, the external granular layer.[13–18] In this instance many observers ventured to voice the inescapable conclusion that these cells are a source of neurons, namely, the increasing number of granule cells of the internal granular layer in the rapidly growing cerebellum, a brain structure that increases in bulk thirtyfold from birth to maturity in the rat.[19] Nevertheless, because differentiated neurons do not multiply, the conclusion that neurogenesis ceases before birth was not seriously shaken.

The use of colchicine, a poison that temporarily blocks mitosis in metaphase and leads to an increase in the number of recognizable mitotic cells in any multiplying cell population (Fig. 4), reinforced the conclusion that proliferative sites persist around brain ventricles after birth.[7] However,

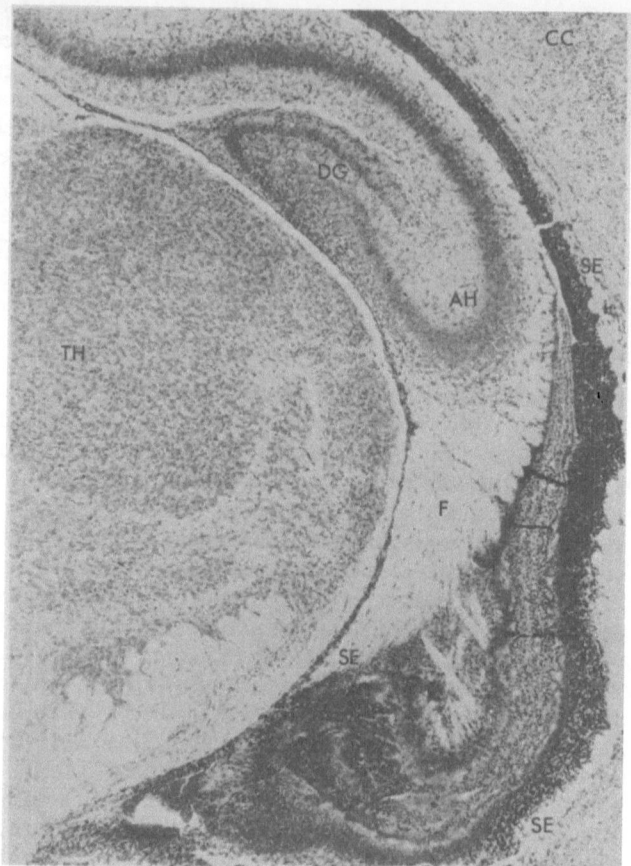

Fig. 2. Low-power photomicrograph of a brain section from the region of the dorsal hippocampus in a newborn rat. Note the considerable area occupied by the subependymal layer of the lateral ventricle (SE). AH, Ammon's horn of hippocampus; CC, cerebral cortex; DG, dentate gyrus of hippocampus; F, fimbria of hippocampus; TH, thalamus. From Altman and Das.[87]

histological and cytological techniques alone could not provide evidence about the significance of postnatal cell proliferation until other techniques were introduced to study this problem. These new results indicate that neurogenesis does not cease at birth and that in many brain regions gliogenesis is essentially a postnatal phenomenon.

3. Theory of Sequential Cell Proliferation in the Brain

In order to provide a framework for this survey of cell proliferation in the developing and maturing brain, the attempt will be made to formulate,

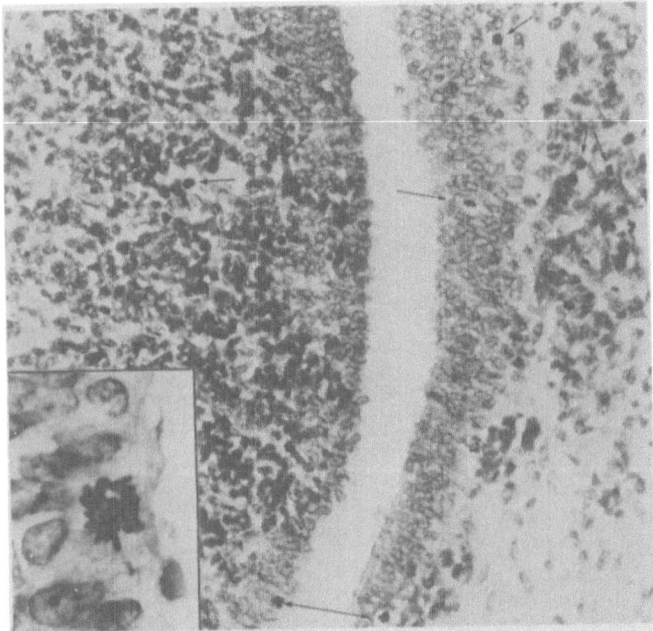

Fig. 3. Ependymal and subependymal layers surrounding the lumen of the lateral ventricle in a 4-day-old rat injected with colchicine and killed 4 hr after injection. Arrows point to some of the mitotic cells. Insert shows a mitotic cell at high magnification. From Altman and Das.[124]

on the basis of new results that are described later in detail, a tentative, generalized picture of the sequential nature of cell proliferation in the brain. We shall pay particular attention in this brief introductory account to the differentiation of the more complex, hierarchically organized cephalic components of the brain rather than to the differentiation of the spinal cord.

It will be assumed in this discussion that brain tissue proper (as opposed to meninges and blood vessels) is composed of four cellular elements: macroneurons, microneurons, neuroglia, and microglia. Macroneurons are the long-axoned nerve cells constituting the afferent, relay, commissural, and efferent elements of the nervous system, or its input–output core. Microneurons are short-axoned (or anaxonic) interneurons that, interposed among the macroneurons of a given brain region, are presumed to have restricted local integrating and modulatory functions. Neuroglia, composed of astrocytes and oligodendrocytes, are the supporting elements of the brain. They are not directly involved in the gathering and processing of sensory information or the control of muscular action but are responsible for such ancillary functions as nourishing nerve cells, providing axons with insulating myelin, and the like. In contrast to these three classes of cells, all of which are

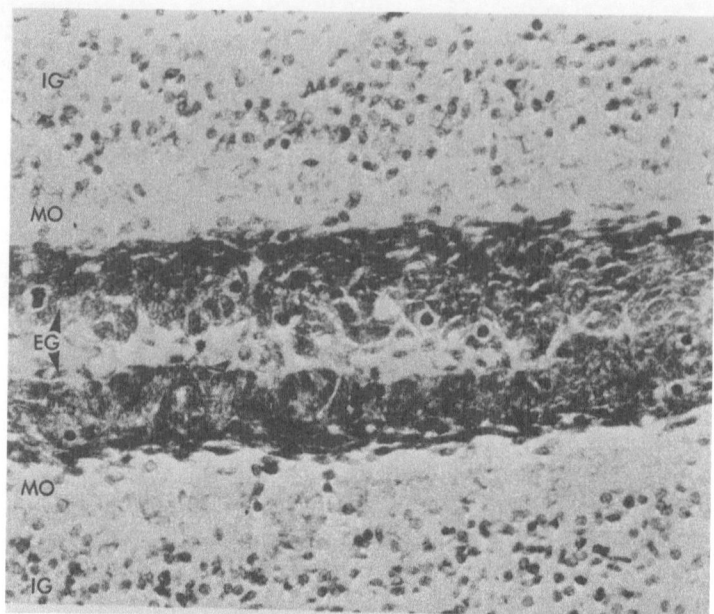

Fig. 4. Cells with arrested mitosis in external granular layer (EG) of the cerebellum of a rat injected with colchicine at 2 days of age and killed 4 hr later. IG, internal granular layer; MO, undeveloped molecular layer. From Altman and Das.[124]

derived from the primitive neuroepithelium, microglia are of mesenchymal origin.[20] Microglia play a major role as scavenger cells under pathological conditions.

It will be further assumed that the three neuroepithelial elements are hierarchically related to one another and are formed successively during the differentiation of a brain region in the following manner (Fig. 5). After closure of the neural tube the proliferating primitive neuroepithelium gives rise to both undifferentiated and differentiating cells. The bulk of the differentiating cells, which leave the neuroepithelium and lose their proliferative capacity, become macroneurons. The undifferentiated cells continue proliferating in the neuroepithelium, and their progeny provide more macroneurons to the differentiating brain. When the production of cells that are to form the macroneuronal core of that brain region is nearing completion, the cells near the lumen of the ventricle specialize as ependymal cells, thereby changing the character of the ventricular wall. Other cells retain their proliferative capacity and form a new type of germinal pool, the secondary germinal matrix. Examples of this matrix, which we will consider below, are the subependymal layer of the forebrain ventricles and the external granular layer of the cerebellum.

This secondary pool again gives rise to undifferentiated and differentiating cells. The differentiating cells leave the germinal matrix and,

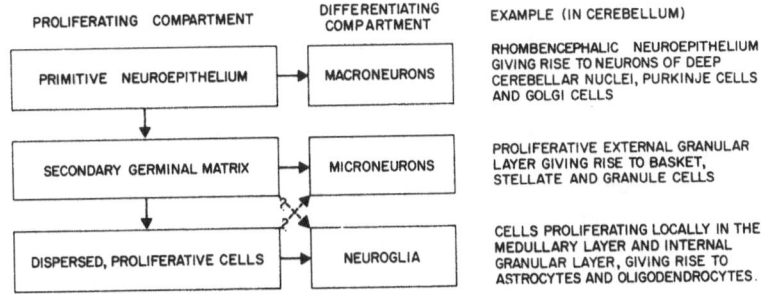

Fig. 5. Hierarchical organization of the three proliferative pools of the developing nervous system and the type of differentiating cells arising from them.

moving through brain regions composed of macroneurons, establish synaptic contacts with them and become differentiated into microneurons. In this manner the fine wiring of the brain commences. Some of the cells of this secondary matrix may become differentiated into neuroglia. The undifferentiated cells continue proliferating within the germinal matrix to provide more microneurons or to give rise to a third source of proliferative cells. This third class of stem cells leaves the pool and is dispersed throughout the brain. These dispersed stem cells can be identified as the locally proliferating cells that are the primary source of differentiating neuroglia. With the insertion of these supporting and insulating elements the development of the given brain region is completed. Evidence that has recently become available indicates that in the forebrain of altricial mammals the proliferation of the precursors of microneurons and neuroglia and their differentiation are essentially postnatal phenomena.

II. BIOCHEMICAL STUDIES OF DNA METABOLISM

The demonstration that DNA concentration per diploid nuclei tends to be constant in cells of the body[21–23] and the brain[24] opened the way to making estimates, by determining DNA concentration, of the number of cells in the brain and of changes in cell concentration, at different stages of development, in different regions, and under certain pathological conditions. The earlier, less specific techniques[25,26] were gradually improved to give more reliable and consistent results.[27–30] In the rat DNA content was estimated as 7×10^{-12} g/nucleus.[24]

A. Estimates of Cell Concentration in the Brain

1. Embryonic Development of the Brain

Flexner and Flexner[31] reported that the amount of DNA per unit weight of cortex is seven times higher in fetal guinea pigs (at 25 days of gestation) than at term or in adults, the decline becoming asymptotic at 45 days of gestational age. This was in agreement with histological estimates of changes in the number of nerve cells per nuclei per unit volume in the cortex of developing guinea pigs.[32] These findings were interpreted to reflect the rapid increase in spacing among cell nuclei during development, due to the growth of the cytoplasm during neuronal differentiation. Comparable results were obtained in developing chick embryos. The concentration of DNA per dry weight, in the embryo as a whole, increases between the seventh and eleventh to twelfth day of incubation, this period representing the stage of rapid cell proliferation. Then, during the later phases of incubation up to the period of hatching, DNA concentration per unit weight falls rapidly.[33, 34] Within the central nervous system of the chick the decrease was reported to occur between 8 and 21 days of incubation, although this was relatively small, from 13.0 mg/100 g nerve tissue to 10.0 mg.[35]

2. Postnatal Development of the Brain

Albrecht[36] reported a decrease in DNA concentration from birth to adulthood in the brain of mice, from 380 mg/100 g wet weight at 1–2 days to 240 mg/100 g at 3–6 months of age. Dahl and Samson[37] reported that in the brains of rats the concentration of DNA declines from birth to the fifth day, then rises to a peak at about 15 days, and declines again thereafter. A more complex series of changes was reported in the brains of mice by Uzman and Rumley[38]: considerable rise was observed from 7 to 14 days, then a drop occurred, followed by a small increase from the twentieth day to adulthood. Mandel and his collaborators[39,40] (Fig. 6) determined the amount of DNA during postnatal development in several mammalian species. They found that the amount of DNA per brain was at the adult level at birth in the precocious guinea pig, though not in another precocial species, the chick, in which total DNA was doubled between hatching and adulthood. Adult levels of DNA were reached in the brains of rats at 15 days of age, at 3 months in the rabbit, at 5 weeks in the dog, and after 1 year of age in man. Mandel and his collaborators also studied DNA per unit weight in the brains of (adult?) fish, amphibians, reptiles, birds, and mammals. Generally speaking, the concentration was higher in fish than in the other vertebrate orders studied, but a clear phylogenetic trend did not emerge.

3. Regional Studies

Several investigators have measured the concentration of DNA per unit weight in different areas of the brain, directed at establishing regional differences in cell number and concentration.[41,42] Mihailovic et al.[43] reported that DNA per unit weight of dried brain tissue was comparable in most regions of the cat brain, with no differences between cerebral gray and white matter, except that it was considerably higher in the cerebellum and hypothalamus. May and Grenell[44] found no differences in DNA concen-

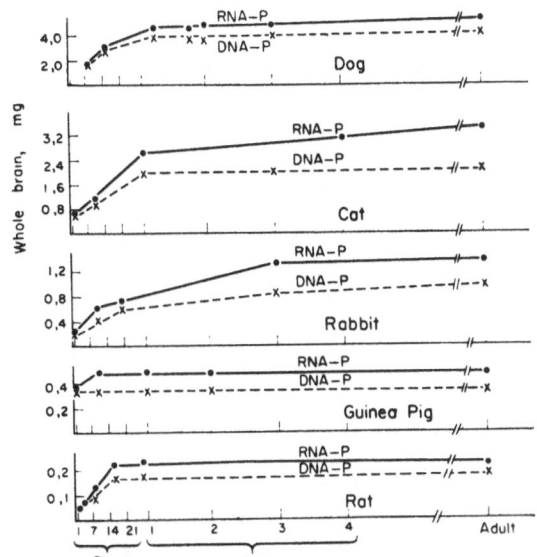

Fig. 6. Increase in DNA (and RNA) of the whole brain in different mammalian species during postnatal development. From Mandel *et al.*[40]

tration in cerebral gray and white matter in the rat in the hypothalamus, thalamus, or medulla, but in the cerebellum DNA content was four times higher than in these brain structures. The high DNA content of the cerebellum was reported by several other investigators in a variety of mammalian species.[30,45–47] High DNA levels in the cerebellum are clearly related to the lárge concentration of densely packed small cells, in particular the granule cells, in the internal granular layer of the cerebellar cortex. Mandel *et al.*[47] also reported high DNA values for the olfactory bulb, another brain region with high concentration of granule cells.

Attempts also were made to use the technique of quantitative DNA determination for estimating the number of cells in different subdivisions or layers of a particular brain region. Kissane and Robins[30] studied differences in DNA content in the molecular, granular, and medullary layers of the cerebellar cortex. They found that cell concentration was 10–20 times higher in the granular layer than in the other two layers of the cerebellum. Similar laminar studies were carried out in the cerebral cortex of the monkey[48] and the rat.[49] In a comparison of different cortical areas, Robins *et al.*[48] estimated that cell concentration in the visual cortex was twice as high as in the motor cortex. Comparable results were obtained by Landolt *et al.*[50] These latter results can be attributed to differences in the granular visual cortex and agranular motor cortex both in cell-packing density and cell size. Lowest DNA concentration per unit tissue weight was obtained in the large-celled red nucleus.[50]

4. Pathological Studies

When a peripheral nerve is cut, the distal portion of the nerve (consisting of axons that lose their continuity with the cell bodies) will degenerate. This degeneration involves the disintegration of the axon and its myelin sheath, and it is followed by regenerative processes, which include the proliferation of Schwann cells and the regrowth of the axons. Logan et al.[42] found that after sectioning or crushing of the sciatic nerve there was an increase in DNA in the severed nerve that reached a peak of about threefold of normal levels 16–32 days after sectioning. These results were comparable to histological findings of increase in cell number in the sciatic nerve of rabbits.[51]

B. Some Shortcomings of the Technique

The rationale of studies involving DNA determination is that by establishing absolute DNA concentration in the brain during development, in a particular species or given brain region, estimates can be made of the number or concentration of cells. The initial attempts, in which nonspecific techniques were used, suffered from providing inaccurate and unreliable results, but these technical problems were subsequently largely overcome by improved and more specific procedures.[27] All estimates of cell number are predicated on the assumption that the DNA concentration of all diploid cells, irrespective of cell type or size, is constant. However, evidence has been accumulating that not all cells in the brain are diploid. The studies of Lapham and his collaborators[52–54] indicate that though the majority of glia cells and granule cells are diploid in the cerebellar cortex of man, Purkinje cells are tetraploid (containing twice the normal amount of DNA in their nucleus). Astrocytes may change from the diploid to the tetraploid state under pathological conditions, and the DNA concentration of tumor cells in the brain may also differ from normal.[24] Therefore, under certain conditions and in certain brain regions it may be hazardous to make direct estimates of cell number from data on DNA content. Another shortcoming of the technique is that it does not permit the separation of nuclei from different cell types, neuron from glia, or one type of neuron and glia from another, making the interpretation of the result difficult without corollary histological investigation. Moreover, these techniques are not suited for the determination of DNA turnover where there is no net gain in DNA concentration. Finally, the technique is insensitive in the sense that where no gain in DNA content has been indicated, for instance, in the guinea pig after birth,[31,32,39,40] the more sensitive autoradiographic technique reveals continuing, though modest, cell proliferation.[55]

III. THYMIDINE-³H AUTORADIOGRAPHY

The application of fine-resolution thymidine-³H autoradiography to the study of DNA metabolism and cellular proliferation in the central nervous system has revolutionized this area of investigation. With its aid, controversial problems raised by histological and chemical studies have been resolved, and questions that could not be handled by other techniques are being studied extensively and profitably. These include such problems as estimating rates of cell proliferation; the site and time of origin of cells; steps in the cell generation cycle; the migratory paths and the mode of differentiation of newly formed cells. It is safe to say that we have only begun to reap the benefits of this technique and that it will lend itself to the investigation of many other problems as well.

A. Technique of Thymidine-³H Autoradiography

In cytological and histological autoradiography the incorporation of radioactively tagged precursors of a substance by cell organelles, cells, or tissues is rendered visible *in situ* by bringing the fixed and processed biological material in contact with a nuclear emulsion that is optimally sensitive to the type of radiation emitted by the radioactive isotope. After a suitable exposure period, the emulsion is developed and the blackened grains reveal the site and amount of radioactive emissions. In most autoradiographic investigations the dipping technique is employed.[56-62] It consists of immersing the tissue slides into melted nuclear bulk emulsion, forming a radiosensitive film that is in intimate contact with and adhering to the tissue. In this manner the blackened silver grains can be directly related to their source in the tissue. In most autoradiographic investigations with thymidine, tritium is the preferred radioactive isotope, not only because it is economical and has an advantageous half-life (12.5 years), but more importantly because it emits weak β-rays, most of which (about 90%) will not penetrate more than 1μ into the emulsion.[63,64] The limited path of the emitted electrons allows the accurate localization of the source of radiation.

To label cells that are preparing to multiply or to label their progeny, tritiated thymidine is injected systemically (intravenously, intraperitoneally, or subcutaneously) into the animal. This exogenous labeled nucleoside is readily and quickly incorporated into all cell nuclei that are synthesizing DNA in preparation for cell division.[65-68] Since DNA is metabolically stable,[69] the tagged cell, unless it undergoes further divisions, will retain its radioactivity indefinitely. Thus, the vicissitudes of a class of tagged cells, their migrations, mode of differentiation, and life span can be traced by killing a group of identically treated animals at different periods after injection. Systemically injected thymidine-³H is catabolized by the liver[70]; the unincorporated nucleoside is cleared from the blood in about 30 min and may be detected as catabolized products.[70,71] Therefore, a single injection of thymidine-³H represents a flash-labeling procedure; only cells that are synthesizing DNA in the 30-min span following injection incorporate the precursor. If the labeled cells undergo further mitotic divisions, then with each division the amount of labeled DNA (the number of blackened grains) will be halved (Fig. 7). Therefore, the dilution of label over nuclei and the

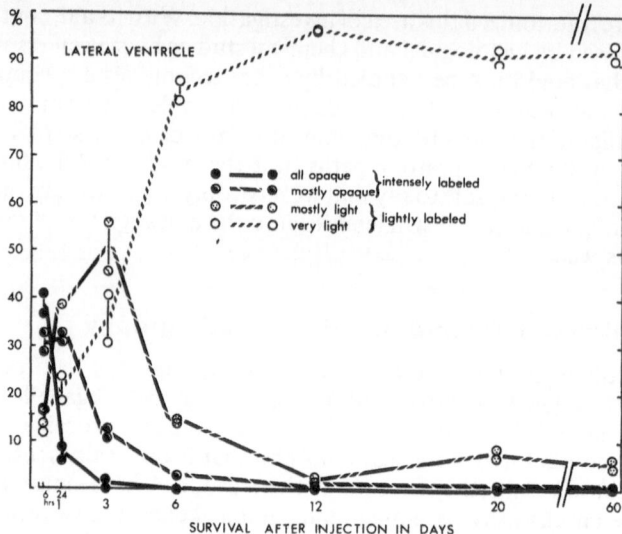

Fig. 7. Percentage distribution of intensely and lightly labeled cells in the subependymal layer of lateral ventricle in rats injected with thymidine-^{3}H at 13 days of age and killed at different intervals afterward. From Altman.[125] See also Fig. 11.

increase in the number of labeled cells after longer survival can be reliably used to study the kinetics of cell proliferation.[72–74] For purposes of quantitative study, it is of paramount importance to hold constant the amount of thymidine-^{3}H injected per body weight, to use a radiochemical with the same specific activity, to use an identical type of nuclear emulsion, and to expose the autoradiograms for the same period after developing.

1. *Hazards of Thymidine-^{3}H Autoradiography*

When thymidine-^{3}H is administered to an organism the procedure involves the introduction of an exogenous precursor of DNA which, in addition, may be a source of radiation damage. Administration of unlabeled thymidine was reported to stimulate mitotic activity in the duodenal epithelium of the mouse,[75] to increase the duration of metaphase,[76] and to suppress colony growth in HeLa cells.[77] Prolonged exposure of cultured cells to thymidine-^{3}H produces radiation damage,[78,79] and injection of thymidine-^{3}H into chick eggs, which also leads to prolonged exposure, was also shown to have similar effects.[80] Chronic administration of thymidine-^{3}H was shown to adversely affect the reproductive capacity of mice[81] and even with flash-labeling thymidine-^{3}H administered in doses above 1 μC/g body weight causes radiation damage to mouse spermatogonia[82] and rat liver.[83,84] Long-term studies indicate that flash labeling with 1–5 μC/g does not shorten the life span of mice,[85] and there is general agreement that with a single low-dose injection no clearly discernible radiation effects are produced in most organs of the body.[77,79,80,86] In our studies we have observed that labeled neurons of the hippocampus have the appearance of normal cells 8 months after labeling.[87] The possibility that subtle harmful effects are produced even with single injections cannot be ruled out, but thymidine-^{3}H can be used to study DNA metabolism and cellular proliferation provided that parallel control studies are carried out on material from noninjected animals to eliminate the possibility of artificially induced effects.

B. Evaluation of Autoradiograms

1. Kinetics of Cell Proliferation

The available evidence indicates that all interphase cells that are synthesizing DNA at the time of injection with thymidine-^{3}H become labeled. This period, during which thymidine-^{3}H is incorporated into DNA, is called the synthetic or S phase, which is said to be followed by a period, the postsynthetic or G_2 phase, during which thymidine is no longer incorporated. Subsequently, the cell enters the mitotic or M phase, followed by a postmitotic, presynthetic period, called the G_1 phase.[88]

In studying the kinetics of cell proliferation, animals are sacrificed at different intervals after injection with a single, standardized dose of thymidine-^{3}H (usually in the range of 1–10 μC/g body weight). Counting of labeled mitoses in tissues taken from animals killed at different intervals after injection is the most direct method for determining the duration of the cell generation cycle.[89,90] While cells in interphase become labeled within a few minutes after injection, labeled mitoses are not seen in the primitive neuroepithelium of the 11-day-old mouse embryo until 45 min after injection.[91] According to Atlas and Bond,[91] the duration of the G_2 phase, the period taken before neuroepithelial cells incorporating thymidine-^{3}H reach the mitotic phase, is about 1 hr. The minimum mitotic phase was also estimated at about 1 hr. For the synthetic or S phase a value of about 5.5 hr was obtained, and the total generation cycle (the time from the start of the first wave of labeled mitoses to the beginning of the second wave) was about 11 hr. The presynthetic or G_1 period was estimated (by subtracting the duration of G_2, S, and M phases) to be about 3.5 hr (Fig. 8). These calculations suggest that during this phase of development the cells of the neuroepithelium pass through two generation cycles per day. Comparable though not identical results were obtained by Langman and his associates[92,93] by counting labeled metaphases in the neuroepithelium of the 14–15-day-old mouse embryo (Fig. 9).

When labeled mitoses are low in number or are not easy to identify, other methods may be used for estimating the cell generation cycle. One method is counting the number of blackened grains over nuclei. Since each time that a cell divides its radioactivity is diluted by one-half, the halving time in grain counts in an adequate sample can be considered to equal the cell generation

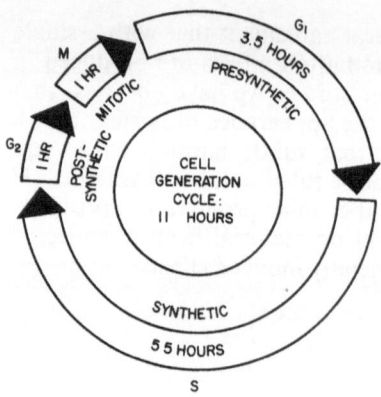

Fig. 8. Diagrammatic illustration of the duration of phases of the cell generation cycle in the neuroepithelium of the 11-day-old mouse embryo. Data based on Atlas and Bond.[91]

cycle.[94,95] If cells do not leave the population (due to migration or to cell death), a comparable method is the determination of the period taken to double the number of labeled cells. Among other methods, we may mention the technique of repeated injections[96,97] in which the animals are injected at intervals to label all the multiplying cells of a proliferating population. The time at which all the cells become labeled, which would include cells that entered the G_2 phase at the time of the first injection, represents the duration of the cell generation cycle.

2. Tracing the Fate of Proliferating Cells

The production of new cells may lead to an increase of the cell population or to renewal of the population.[98] An increase in cell population as a consequence of rapid cell multiplication is commonly seen in growing organs during development. In tissues or organ systems in which there is continuous cell loss, cell multiplication may serve the function of replacing the lost cells with new ones to maintain the tissue in steady state. The time taken to

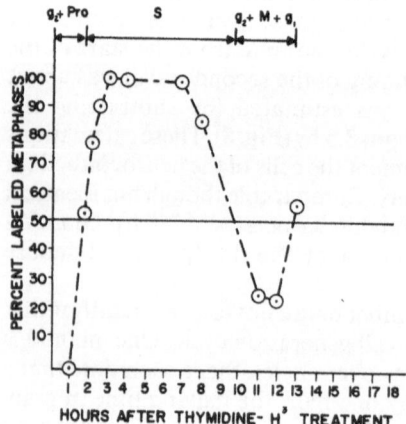

Fig. 9. Percentage of labeled metaphases in the neuroepithelium of the 15-day-old mouse embryo. From Langman and Welch.[93]

replace the entire number of cells in the population is called the turnover time. If in such a renewing system the rate of cell production falls below the rate of cell loss, the population decreases, leading to a transient or permanent decline in tissue steady state.

The fate of newly produced cells differs in different tissue compartments and at different stages of development in the same compartment.[86,99,100] In a proliferative compartment (such as the primitive neuroepithelium surrounding the lumen of the neural tube) those cells that retain their proliferative capacity after mitosis are called stem cells. If they remain at or near the site of their origin they may be specifically referred to as stationary stem cells. If both cells after mitosis retain this capacity, the population of the proliferative compartment increases. The rapid growth in size of the primitive neuroepithelium before and for some time after closure of the neural tube is an example. If one of the dividing cells leaves the proliferative compartment, the compartment remains in steady state. If in a large proportion of cases both cells leave the compartment, the size of the germinal matrix gradually will decrease.

Of those cells that leave the proliferative compartment some may retain their proliferative capacity while others, those that begin to differentiate, tend to lose it. An example of differentiating cells are the neuroblasts that migrate into the mantle layer of the spinal cord. These cells appear to entirely lose their proliferative capacity.[103] The emigrating cells that do not differentiate can form a secondary proliferative compartment as do, for instance, some of the stem cells that leave the neuroepithelium of the lateral recess of the fourth ventricle and form the external granular layer of the cerebellar cortex.[101,102] Otherwise, these undifferentiated cells, the dispersed stem cells, may become scattered among the differentiated cells of the tissue, retaining their proliferative capacity for a long period.

In the developing nervous system, proliferative compartments are geographically separated from the differentiating compartments (Fig. 20). The time taken for cells originating in a proliferative compartment to reach the differentiating compartment is called the transit time.[99] For example, the transit time from the external granular layer, the proliferative compartment of the maturing cerebellar cortex, to one of its differentiating compartments, the internal granular layer, was established[104,105] (Fig. 10). In some brain regions the cells migrating from one compartment to the other may be forming a discrete migratory zone, such as the subcallosal band that can be traced over the caudate nucleus from the anterior lateral ventricle to the olfactory bulb.[9,106,107,182] Partially differentiated cells en route to their final destination also may form distinct zones, such as the basal zone of the granular layer of the hippocampal dentate gyrus in the infant rat, guinea pig, and cat[108] (Fig. 20).

3. Dating Cell Proliferation and Cell Differentiation

A study of cell proliferation in the nervous system indicates that it is restricted to undifferentiated cells; such differentiated cells as neurons, and

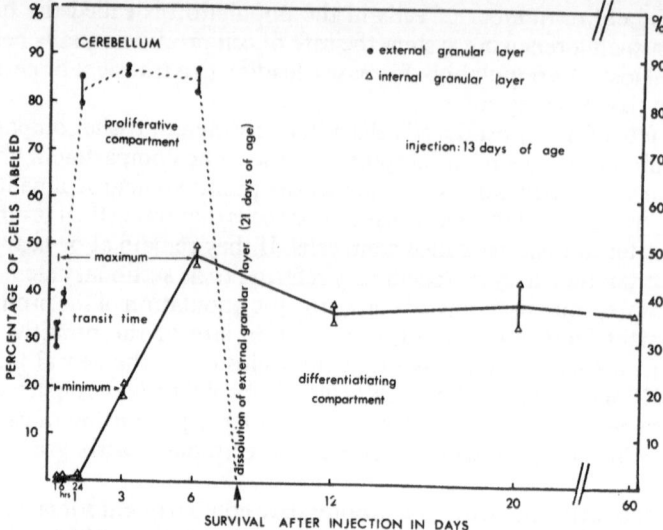

Fig. 10. Percentage of labeled cells in the external and internal granular layers of the cerebellar cortex in animals injected at 13 days of age and killed at different intervals afterward. Since local proliferation at this age is minimal in the differentiating compartment of the internal granular layer, the considerable growth in the number of labeled cells between 3 and 6 days after injection must be due to migration from the proliferative compartment. This allows estimation of the transit time through the molecular and Purkinje cell layers. Modified after Altman.[104]

most probably differentiated neuroglia, do not multiply. Therefore, if brains are available from mature animals that were injected serially either pre- or postnatally, it is relatively easy to establish when cells of a brain region were formed and when they commenced to differentiate (Fig. 11).

The preponderance of lightly labeled cells in a brain region indicates that they are descendants of precursor cells that have undergone many subdivisions, i.e., they have not differentiated for some time after injection. The preponderance of intensely labeled cells (combined with unlabeled cells) in the same brain region in animals injected at some later date indicates that the precursor cells were undergoing their last divisions (or no longer divided) at that date. This latter date signals the time of termination of cell division and the commencement of differentiation. With these two dates available a curve can be constructed of rates of cell proliferation and differentiation in a selected brain region by examining the brains of animals injected at intervening periods and plotting the changes in the proportion of cells with different degrees of nuclear labeling. In this context it should be pointed out that the preponderance of unlabeled cells at a brain site in a mature animal injected pre- or postnatally may be due to two opposite reasons: the cells

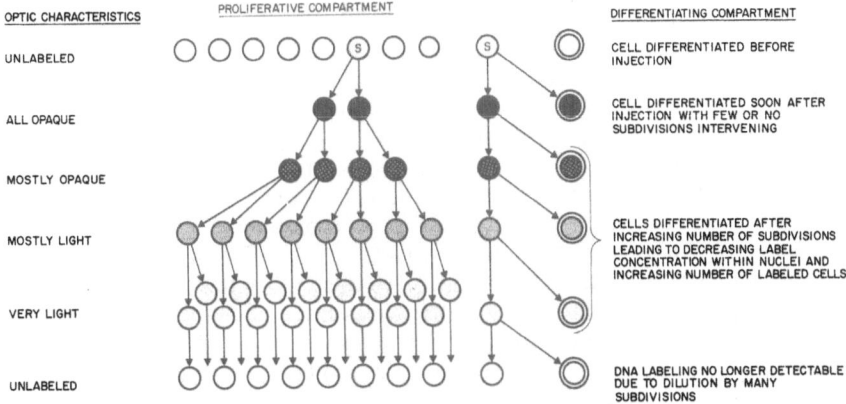

Fig. 11. Rough categories of nuclear labeling (left column) based on visual judgment. The categories are employed for convenience and do not represent a true interval scale of cell generations (middle column). In animals injected at specified ages, neuron differentiation in various brain regions can be dated on the basis of predominant labeling characteristics (right column). See text.

were formed before injection or the cells have undergone several divisions after injection that led to label dilution beyond detectability. For instance, if the cerebellum of mature animals injected soon after birth with thymidine-^{3}H is examined, all Purkinje cells and the majority of granule cells appear unlabeled. In the cerebellum of mature animals injected at about 2 wk of age, all Purkinje cells are unlabeled but a large proportion of the granule cells are intensely or lightly labeled.[104] This indicates that the absence of label in granule cells in animals injected soon after birth is due to excessive label dilution because the precursors of granule cells have undergone many divisions before differentiation. Examining brains from a series of postnatally injected animals in this manner reveals that, in contrast, the absence of labeling in Purkinje cells is due to the fact that they are formed prenatally. With this procedure it is possible to date not only the formation and differentiation of cells in different brain regions but also of different cell types in one brain region.[183] This was accomplished in a preliminary manner for the different cells of the cerebellar cortex (Fig. 16). Interpretation of results may be more difficult in systems in which cell proliferation is not continuous but intermittent, with long rest periods between peaks of mitotic activity.

IV. CELL PROLIFERATION IN THE EMBRYONIC NERVOUS SYSTEM

A. Cell Proliferation in the Primitive Neuroepithelium

According to the classical description of His,[109] the early neural tube of vertebrates is composed of three layers: (1) the ependymal layer on the border of the lumen, (2) a surrounding cellular zone, called the mantle layer, and (3) a superficial cell-free or cell-sparse zone, the marginal layer. The "ependymal" layer (correctly, the neuroepithelium) was divided into two zones: the innermost zone composed of round cells, with frequent mitotic activity, called germinal cells, and an outer zone composed of slender cells, referred to as spongioblasts. His suggested that the germinal cells give rise to neuroblasts that migrate into the mantle layer and differentiate there into neurons, whereas the spongioblasts are precursors of glia cells. Thus, neurons and glia cells were conceived to arise from different precursors. The mantle layer was considered the rudiment of the gray matter and the marginal layer, the precursor of the white matter. In contrast, Schaper[110] suggested that all the cells of the neuroepithelium represented one class of cells, those in the inner zone representing the mitotic elements and those in the outer zone, the nondividing elements. Differentiation into neuroblasts and spongioblasts was assumed to occur after these "indifferent cells" migrated into the mantle layer.

Most neuroembryologists accepted the view of His until Sauer,[111] on the basis of a detailed cytological study, came to the conclusion that the "germinal cells" are the mitotic and the "spongioblasts" are the interkinetic phases of the same radially arranged columnar cells of the neuroepithelium. He claimed that the cells that are ready to undergo mitosis change from a columnar to a round form, but because they are anchored by rigid terminal bars to one another near the lumen, the expanding nuclei must move toward the lumen. After mitosis these nuclei resume their normal shape and are squeezed away from the lumen by other nuclei rounding up in preparation for division (Fig. 12).

Recent studies with thymidine-^{3}H autoradiography[91,103,112–115] confirmed Sauer's conclusion of the interkinetic migration of the nuclei of cells of the pseudostratified neuroepithelium. In embryos killed 1 hr after administration of thymidine-^{3}H, only nuclei in the outer zone of the neuroepithelium are labeled; nuclei in the zone surrounding the wall of the neural groove or neural tube, whether mitotic or nonmitotic, are not labeled. In embryos killed 2 hr after injection many mitotic cells in the inner zone are labeled; 4–5 hr after injection all mitotic cells are labeled. However, 8–10 hr after injection few cells in the inner zone are labeled; the majority of labeled cells are seen again in the outer zone. These results show clearly that nuclei in the S phase (the period of DNA synthesis) are located in the outer zone and that they move toward the lumen for mitosis and then return to the outer zone.

Whereas in the neural groove stage the labeled nuclei remain in the growing neuroepithelium, after closure of the neural tube a few of these cells leave the neuroepithelium to gradually form the mantle layer.[103] If embryos of this age are examined a few hours after injection, the cells of the mantle layer in the spinal cord are not labeled (even when multiple injection is used), indicating that they are not multiplying. By the twelfth hour after injection, however, a few labeled cells appear in the mantle layer, together with a much larger proportion of labeled cells that begin to accumulate again around the

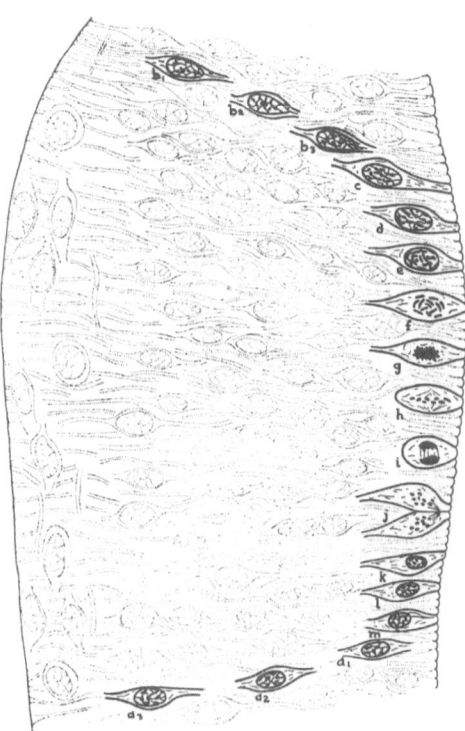

Fig. 12. Neuroepithelium in the alar plate of a 10-mm pig embryo, illustrating the interkinetic migration of cell nuclei preparing for mitosis. Mitotic nuclei are displaced toward the lumen (right border). From Sauer.[111]

lumen. This suggests that most of the progeny of labeled cells remain at this stage of development in the primitive neuroepithelium but that a few migrate outward to form the mantle layer where they lose their proliferative capacity and presumably begin to differentiate. At later stages of development (7–9 days of incubation in the chick) an increasing proportion of labeled cells may be seen entering the mantle layer, with the result that the neuroepithelium gradually decreases and the gray matter increases in size. The character of the neural tube changes and the cells that do not leave differentiate into cuboidal, ciliated ependymal cells of the spinal canal.

B. Cell Proliferation in the Secondary Germinal Matrix

In the rostral part of the neuraxis, cortical appendages are formed around outpouchings and recesses of the original columnar lumen. Among these are the lateral ventricles, around which the cerebral hemispheres are formed; the olfactory ventricles (the rostral extension of the lateral ventricles), around which the olfactory ventricles are formed; the lateral recesses of the fourth ventricle, where the cerebellar cortex arises; and a few others. The embryogeny of these appendicular cortical structures differs in many respects from the embryogeny of the axial spinal cord. As in the latter, the lumen of the ventricle is initially surrounded by a pseudostratified columnar neuro-epithelium whose cell nuclei display interkinetic movements.[93,101] However, structural organization and cell movements in these regions around the neuroepithelium differ from that seen in the spinal cord.

1. Embryogeny of the Cerebral Cortex

Around the wall of the lateral ventricle, the primitive neuroepithelium is surrounded by a cell-rich region composed of migratory cells and not by the mantle layer as in the spinal cord. These cells, which are presumed by most investigators to be neuroblasts, migrate radially toward the surface and gradually form the laminated cortex.[93,116–122] Those cells that arrive in the cortex in the early phases of its development form the deeper layers of the cortex; those that are produced later migrate through the layers of the cortex composed of older cells and give rise to the more superficial layers of the cortex. Gradually the primitive neuroepithelium becomes reduced in size. At that period, some cells around the lumen begin to differentiate and become ciliated ependymal cells; others retain their proliferative capacity and form a secondary germinal matrix, the subependymal layer.[9,11] In the mouse, this subependymal layer appears at 14 days of gestation.[93] This layer is a truly stratified structure that does not show the interkinetic nuclear movements seen in the primitive neuroepithelium.

The subependymal layer, which is pronounced in the lateral and rostral wall of the lateral ventricle, first increases in size and then gives rise to a second wave of cells that migrate over the white matter of the cerebrum and through the layers of the cortex. The proliferation and migration of these cells reach their height in the mouse, rat, and cat after birth.[104,108,123,124,182] These cells give rise to microneurons and, directly or indirectly, to neuro-glia.[125] If the assumption is correct that the primitive neuroepithelium is the source of the first contingent of neurons, it may be hypothesized that they form the long-axoned macroneurons. The second contingent of neurons, the interneurons, arises from the secondary germinal matrix, the subependymal layer (Fig. 13).

2. Embryogeny of the Cerebellar Cortex

Another example of a secondary germinal matrix is the external granular layer of the cerebellum. This proliferative zone, unlike the subependymal layer, is not situated in the vicinity of the ventricular lumen but is found in a subpial position over the surface of the cerebellar cortex. The origin of the initial contingent of cells forming the external granular layer was traced to the rhombic lip surrounding the lateral recess of the fourth ventricle. In the early phases of embryogenesis, the cells situated around the lateral recess of the fourth ventricle give rise to various rhombencephalic structures[102,126] and to the large neurons (those of the deep cerebellar nuclei and Purkinje cells) of the cerebellum.[127–130] Another group of cells, which do not

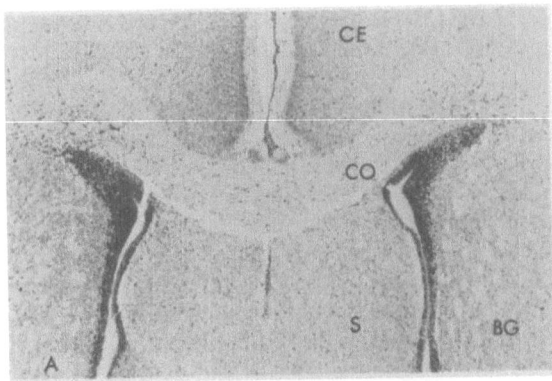

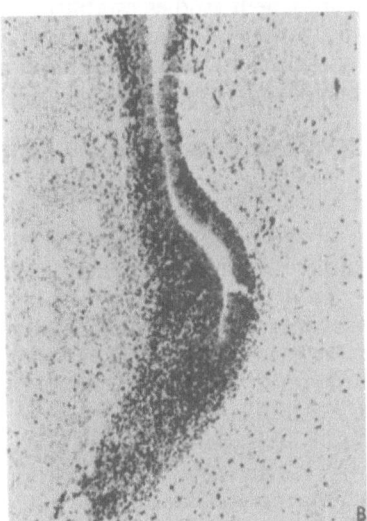

Fig. 13. (A) Autoradiogram showing labeled cells in the subependymal layer of the lateral ventricle in a coronal brain section from the level of the septum (S). The rat was injected with thymidine-³H on the fifth, sixth, seventh, and eighth postnatal day and killed on the ninth day. BG, basal ganglia; CO, corpus callosum; CE, cerebral cortex. (B) Lower portion of lateral ventricular wall at higher magnification. From Altman.[108]

differentiate, appear to move through the body of the rudimentary cerebellum and form the superficial external granular layer. The external granular layer makes its appearance in the chick on the sixth day of incubation[101] and in the mouse on the thirteenth day of gestation.[102] In the chick, the cells of the external granular layer do not display DNA synthesis on the sixth day of incubation. This begins on the seventh day.[101] In the mouse, rat, and cat intensive cell proliferation in the external granular layer does not begin until the time of birth. Because the proliferation of cells derived from the secondary germinal matrix has been studied in greater detail in animals injected with thymidine-³H postnatally, a summary of these results is given in the next section.

V. REGIONAL STUDIES OF POSTNATAL CELL PROLIFERATION IN THE BRAIN

Pilot studies with thymidine-^{3}H autoradiography[8,9,107] confirmed earlier observations that the subependymal layer of the lateral ventricle[6,7,10-12] is a site of cell proliferation in the brains of adult mammals. Evidence was also obtained with this technique of the neogenesis of neuroglia and microglia cells in the normal[106,107] and abnormal nervous system[131-134] of adult rats and cats. Initial studies indicated that the nuclei of neurons were not labeled,[72,98,133-137] in agreement with the accepted belief that neurogenesis is a prenatal phenomenon. The suggestive evidence that new neurons may be formed in the brains of adult mammals[131] was shown to hold for the granule cells of the dentate gyrus of the hippocampus.[87,107] Subsequent studies established that a large proportion of the microneurons in a variety of brain regions (such as the cochlear nucleus, olfactory bulb, and in agreement with earlier findings, in the cerebellar cortex) are formed after birth in various altricial mammals.[104,108,123,124,182,183]

A. Postnatal Neurogenesis

1. Postnatal Neurogenesis in the Cerebellum

The presence of the mitotically active external granular layer over the surface of the cerebellar cortex in infant mice, rats, cats, dogs, man and other mammalian and submammallian species is well known. This germinal layer disappears in mice and rats toward the end of the third week of life, after 2 months in cats, and past 1 year of age in man. The controversy among early anatomists[13-16] as to whether these cells give rise to neuroglia or neurons of the cerebellar cortex, or to cells of the molecular or granular layer, was largely resolved by autoradiographic studies[96,97,104,105,123,124,127,128] (Fig. 14).

In the rat,[104,124] cell proliferation in the external granular layer commences about the time of birth when it is a relatively thin tissue, covering the poorly folded surface of the undeveloped cerebellar cortex. In the ensuing 3 weeks the surface of the cerebellum is profusely folded and the weight and area increase considerably.[138] Evidence was obtained that soon after birth the rate of cell proliferation increases. This leads initially to a supply of proliferating cells to the expanding external granular layer covering the surface of the growing cerebellar cortex and to an increase in the cell thickness of this germinal layer.[183] Two separate zones can be distinguished in the external granular layer: a subpial, outer zone, which is the proliferative compartment, and an inner zone, which is a migratory compartment. The cells of the outer zone tend to be of round shape and many of them show mitotic activity; the cells of the inner zone are typically spindle-shaped, with no mitotic cells. If animals are killed 1-6 hr after injection, the great majority of the cells in the outer zone become intensely labeled, whereas few if any cells in the inner zone will be tagged (Fig. 15). In animals that are killed 24 hr after injection, the bulk of the cells in the inner zone also will be labeled,

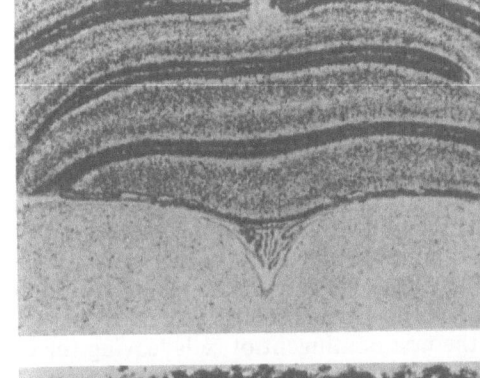

Fig. 14. (A) Autoradiogram of the cerebellum from a rat injected with thymidine-^{3}H, as described in Fig. 13. Dark bands are apposed surfaces of the external granular layer, with practically all cells labeled, as shown at higher magnification in part 'B'. Many cells in the internal granular layer are labeled but not the Purkinje cells situated at the base of the molecular layer (MO). EG, external granular layer; IG, internal granular layer. (B) High magnification of autoradiogram of cerebellum from a rat, as described in Fig. 14(A). From Altman.[108]

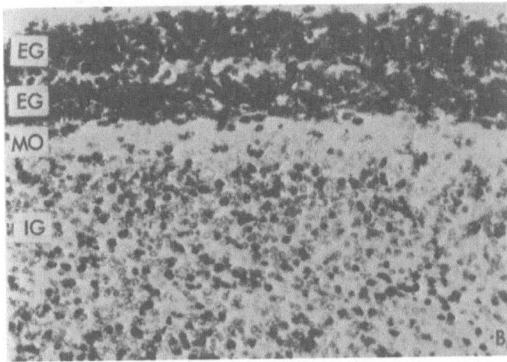

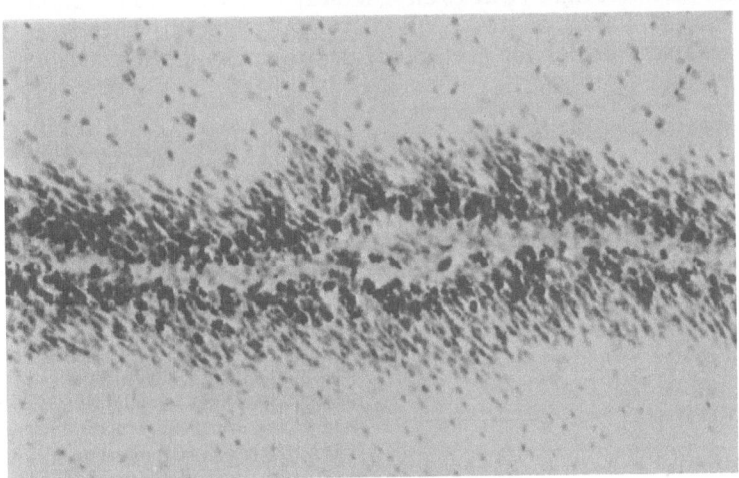

Fig. 15. Autoradiogram of the apposed surfaces of the external granular layer of the cerebellum in an animal injected at 13 days of age with thymidine-^{3}H and killed 6 hr later. Note that the cells in the subpial, proliferative zone are labeled but not the spindle-shaped cells in the lower migratory zone.

accompanied by considerable label dilution within nuclei in the cells of the proliferative zone. In animals with longer survival periods after injection, the migration of cells may be traced through the molecular layer, past the Purkinje cells, and into the internal granular layer. In addition to cell proliferation in the external granular layer, considerable proliferation is indicated by autoradiography in the medullary layer. These cells are presumably the precursors of medullary glia cells.

The fate of the cells formed in the external granular layer can be traced by studying nuclear labeling in different cell types of the cerebellar cortex in mature animals that were injected at different periods after birth.[183] An examination of the cerebellum in animals injected at 6 hr, 2, 6, and 13 days of age, which survived for 60, 120, and 180 days after injection (Fig. 16), indicates that the first contingent of cells leaving the external granular layer became differentiated as basket cells of the molecular layer. Well-labeled basket cells were seen in animals injected at 2 and 6 days of age; only lightly labeled basket cells were seen in animals injected soon after birth, and they were generally unlabeled in the animals injected at 13 days of age. This indicates that basket cells differentiate between 2 and 6 days of age. Stellate cells in the upper part of the molecular layer were not seen intensely labeled except in the animals injected at 13 days, indicating that these cells are formed much later than the basket cells. The formation of granule cells spans a longer period. Intensely labeled cells were seen in growing number in animals injected at 6 days but their number greatly increased in the animals injected at 13 days. Application of another technique, counting of all labeled cells and establishing the proportion of labeled to unlabeled cells in mature animals that were injected at 13 days, indicated that 45 % of the granule cells were labeled. Many of these cells were lightly labeled. Therefore, if one assumes that some of the progeny of the originally labeled cells lost their

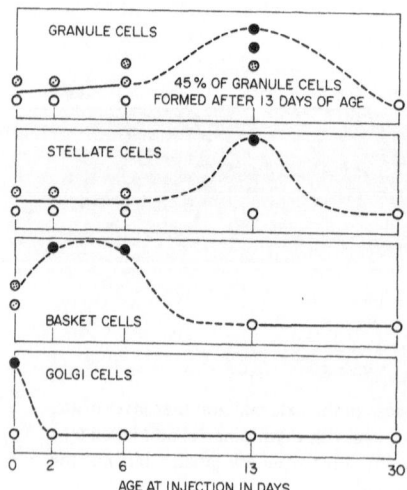

Fig. 16. Approximate dating of neuron differentiation in the cerebellum on the basis of categorization of nuclear labeling (see Fig. 11). Only a few of the Golgi cells are formed after birth; the basket cells in the lower half of the molecular layer are formed before the stellate cells in the upper half; and about half of the granule cells in the granular layer are formed during the third week. For a quantitative analysis of these data, see Altman.[183]

labeling as a result of dilution, this finding would indicate that at least 45 % of the granule cells are formed in the third (and last) week of cerebellar neurogenesis. Apparently the granule and basket cells are the last differentiating neuronal elements of the cerebellar cortex. The speed of migration from the external granular into the internal granular layer was estimated to be about 60–70 μ/day; the minimal transit time from the proliferative to differentiating compartment was estimated to be about 3 days (Fig. 10). A hypothesis of possible interactions among these short-axoned neurons and the prenatally formed Purkinje cells is presented elsewhere.[140]

2. Postnatal Neurogenesis in the Hippocampus

Prenatal neurogenesis in the hippocampal region and early postnatal neurogenesis in the dentate gyrus were studied in detail by Angevine.[141] Angevine's studies showed that in the vicinity of the hilus of the dentate gyrus (CA 3), intensely labeled pyramidal cells of Ammon's horn are seen in greatest concentration on embryonic days 12–16 and intensely labeled polymorph cells on embryonic days 12–14. His study indicated that granule cells in the dentate gyrus may be formed up to 20 days after birth. Our observation[107] that a few labeled granule cells are always present in the dentate gyrus in rats injected with thymidine-^{3}H in adulthood was followed up by a systematic study of postnatal hippocampal neurogenesis in the rat with injections ranging from birth to adulthood.[87,104,123,124] It was established that, except for an outer 1–2-cell-thick zone in the granular layer of the dentate gyrus, the bulk of granule cells in this structure are formed postnatally. Cells multiplying in the subependymal layer of the lateral ventricle were traced migrating by way of the fimbria, past the pyramidal cells of Ammon's horn to the polymorph cell layer (Fig. 17). The examination of brains from animals sacrificed 1 or 6 hr after injection indicated continuing cell proliferation in the polymorph cell layer. From here the cells could be traced migrating to the basal border of the granular layer, where they were identified as the dark, small, spindle-shaped cells that abound here in large numbers and may form a distinct zone in infant and adolescent rats, guinea pigs,[142] and cats.[108] In the course of 2–3 wk, the majority of these small cells became differentiated into large, typical granule cells (Fig. 18). The steady migration of neuroblasts from the polymorph layer to the basal border of the granular layer leads to an increase in the extent and width of the granular layer. The arrangement of cells within the layer tends to follow chronologically their date of differentiation: cells that are formed before birth form the outer zone of the granular layer (near the border of the molecular layer); cells that arrive subsequently are added from inside, with the latest forming cells occupying a position at the base of the granular layer, near the polymorph layer (Figs. 19 and 20). The granule cells that are formed in adulthood are typically in this basal position.[87] Unlike postnatal cerebellar neurogenesis, postnatal hippocampal neurogenesis is a prolonged process. Postnatal neurogenesis of granule cells in the dentate gyrus of the

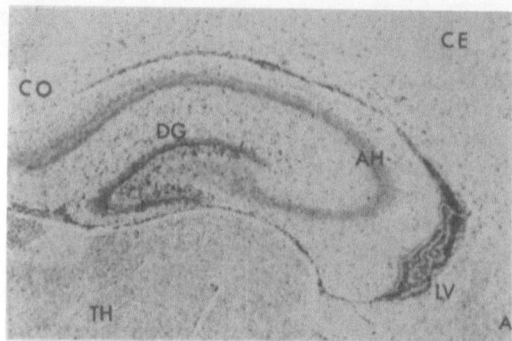

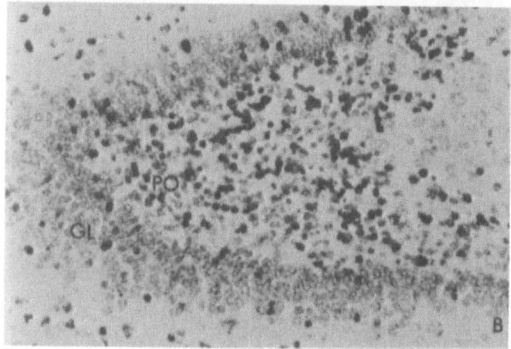

Fig. 17. (A) Autoradiogram of the hippocampal region from a rat injected with thymidine-³H, as described in Fig. 13. AH, Ammon's horn; CE, cerebral cortex; CO, corpus callosum; DG, dentate gyrus; LV, lateral ventricle; TH, thalamus. (B) Portion of dentate gyrus at higher magnification. Note the accumulation of labeled cells in the maturational zone of the polymorph layer (PO). Few labeled cells migrated into the granular layer (GL). From Altman.[108] Compare with Fig. 20.

hippocampus is characteristic, to different degrees, of all mammalian species, including the precocial guinea pig.[142]

In addition to granule cells, in all probability many of the stellate cells in the molecular layer of the dentate gyrus are formed after birth (unpublished observations).

3. Postnatal Neurogenesis in Other Brain Regions

Many of the granule cells of the cochlear nucleus[102,124] and olfactory bulb[104,124] (Fig. 21) are formed postnatally in the mouse and rat, an observation which raised the possibility that short-axoned neurons in all brain regions may be formed after birth.[108,123] Because microneurons, such as stellate or Golgi-Type II cells, are difficult to distinguish from glia cells in Nissl-stained material (a difficulty aggravated by the accumulation óf blackened silver grains over them), the postnatal origin of these cell types is easier to demonstrate where they form a discrete, homogeneous layer, as in the cerebellum, hippocampus, cochlear nucleus, and olfactory bulb.

In other brain regions, as the neocortex in lower mammals, where microneurons tend to be scattered among other kinds of cells, the quantitative evaluation of the proportion of microneurons formed after birth is very

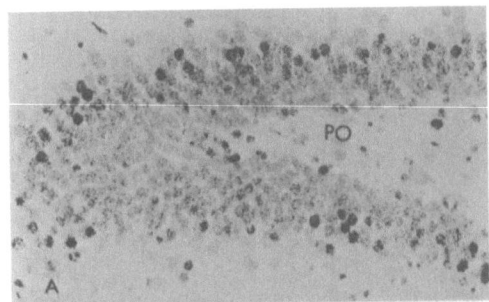

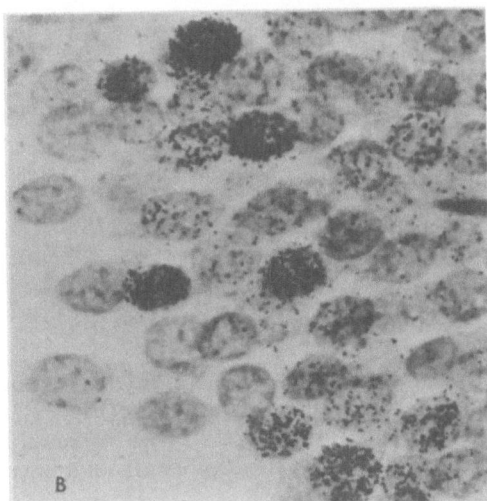

Fig. 18. (A) Autoradiogram of the granular layer of the dentate gyrus in a rat injected repeatedly with thymidine-³H during the first week of life and killed several months afterward. Note that granule cells in the outer zone are not labeled; these are formed prenatally. Many of the granule cells at the base of the granular layer are also unlabeled, due to label dilution. PO, polymorph layer. (B) Labeled granule cells at higher magnification.

difficult. There is moderate regional cell proliferation in the cortex after birth[104,124] and the migration of labeled cells from the subependymal matrix of the lateral ventricle, by way of the cerebral white matter, to the cortex was demonstrated.[125] In a recent unpublished investigation in which rats were given multiple injections of thymidine-³H during their first week of life and killed 3 months later, we established that 36% of the cells in the dorsal cortex were tagged. Some of these were intensely labeled, many more lightly, indicating that at least this proportion of cells was of postnatal origin. The majority of these labeled cells in the cortex were undoubtedly glia cells but some of them, found in greater concentration in the upper layers of the cortex, were typical small neurons.

Of particular interest in this context is the fact that cell proliferation is maintained for the longest period after birth in the anterior part of the lateral ventricle, where a mitotic subependymal layer is present in the adolescent or adult mouse or rat,[7-12] characterized by a high rate of cell production.[107]

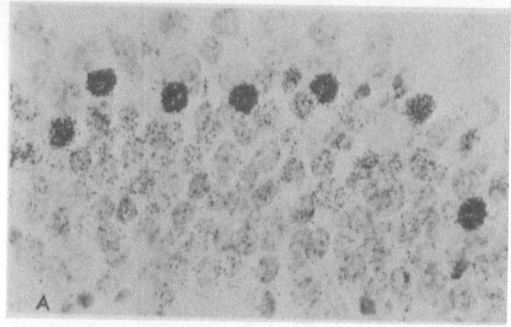

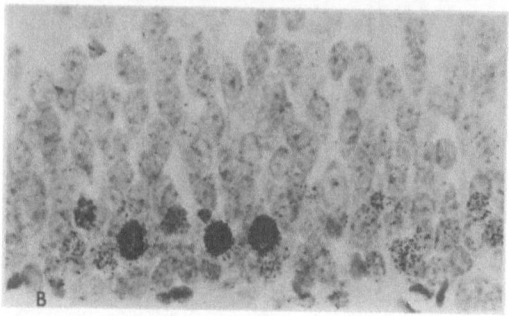

Fig. 19. (A) Autoradiogram of the granular layer of dentate gyrus in a rat injected at 2 days of age and killed 2 months later. Note intensely labeled cells in the upper region of the layer below a zone of unlabeled cells. These intensely labeled cells presumably began differentiating soon after injection. Below these are lightly labeled cells that have undergone many subdivisions before their differentiation. (B) Distribution of intensely and lightly labeled cells near the bottom of the granular layer in an animal injected at 13 days and killed 2 months later. From Altman.[108]

The size of the subcallosal migratory stream that can be traced from this region in the adult rat is impressive (Fig. 22). As a corollary of the continuation of cell proliferation in this region into adulthood, a recent study showed[138] that the growth of the cerebrum in adolescent and young adult rats, though

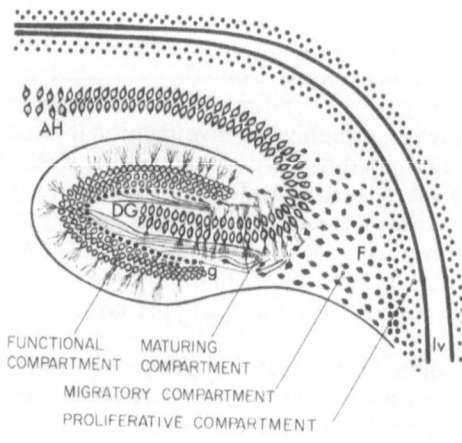

Fig. 20. Schematic illustration of neurogenesis in the hippocampal dentate gyrus (DG). Cells proliferating in the subependymal layer of the lateral ventricle (LV) migrate by way of fimbria (F) through the dendritic fields of the pyramidal cells of Ammon's horn (AH) and the polymorph cells. This is presumably the maturing compartment where axonal synapses are established by the differentiating granule cells.[141] These maturing neuroblasts first accumulate in the basal border of the granular layer, then migrate into this layer, and with the development of their dendrites become functionally mature neurons (g). Cell multiplication can take place in all regions except the granular layer.

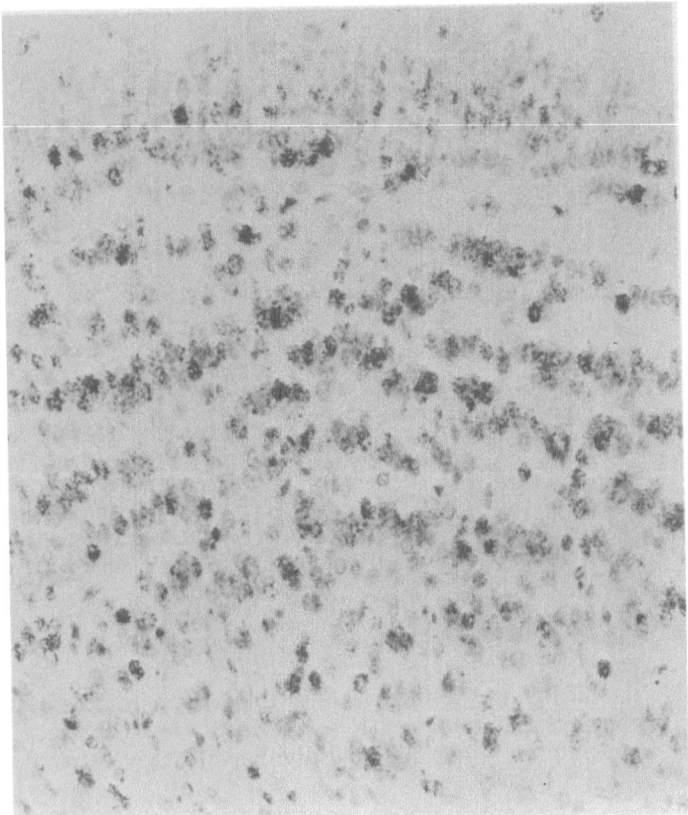

Fig. 21. Autoradiogram of the olfactory bulb with a large proportion of the granule cells labeled from a rat injected at birth and killed 12 days later.

limited in the lateral direction, is quite pronounced longitudinally, an observation reported previously by Sugita.[19] A possible interpretation of this is the continuing growth of the anterior parts of the forebrain. An attempt was made to trace the fate of the cells produced in the subependymal matrix of the lateral ventricle in the anterior section of the brain, and it was found (unpublished) that in animals injected soon after birth many large neurons in the caudate–putamen (basal ganglia complex) as well as other cell types were labeled. These are the largest neurons that are formed after birth that we have identified so far (Fig. 23). It remains to be determined whether these cells are long-axoned or larger type "short-axoned" neurons (with axons terminating within the basal ganglia).

Though the position of these basal ganglia neurons strongly suggests that they may be derived from the proliferative matrix of the anterior lateral ventricle, examination of serial coronal sections in the anterior portion of the

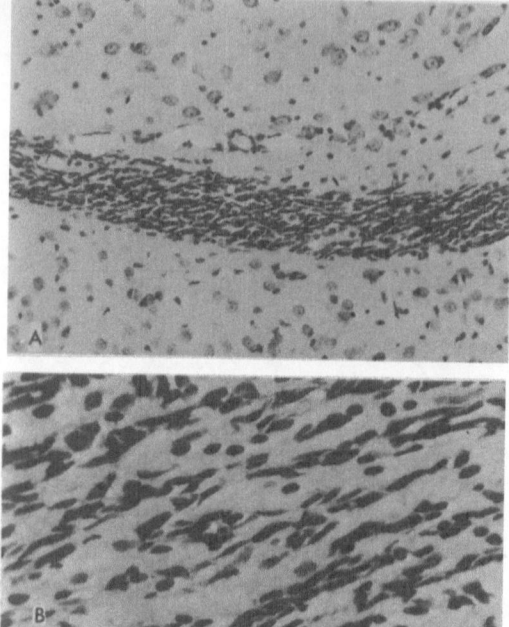

Fig. 22. (A) Low-power photomicrograph of the stream of undifferentiated cells migrating from the anterior lateral ventricle over the caudate nucleus. From an adult rat (3 months). (B) Cells at higher magnification. These cells migrate to the olfactory bulb.[182]

brain makes it clear that the destination of the majority of these cells must be the olfactory cortex and olfactory lobes. Evidence was obtained of the continued acquisition of a considerable number of newly formed granule

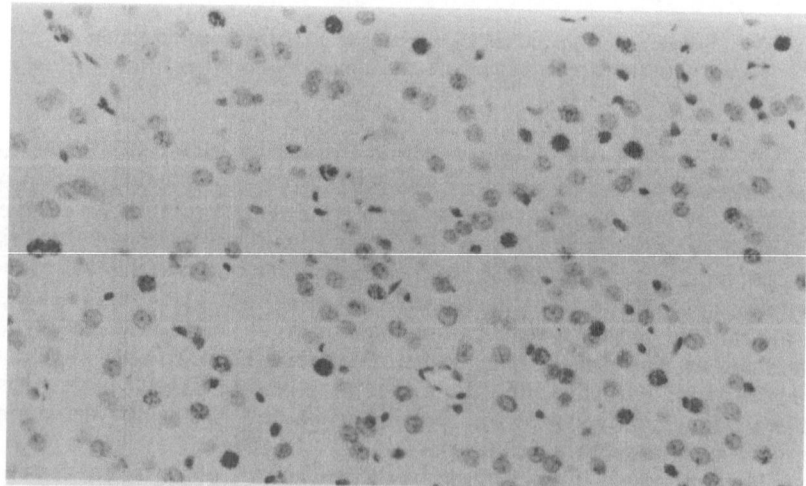

Fig. 23. Autoradiogram of caudate nucleus neurons from a rat injected repeatedly during the first week after birth and killed in adulthood.

cells in young adult rats.[182] Considering the large number of immigrating cells and the relatively small size of the receiving area, it is conceivable that the addition of these cells does not have a growth but a renewal function (replacement of dying cells in the olfactory bulb?).

B. Postnatal Gliogenesis

The occurrence of labeled cell nuclei in the white matter of the spinal cord, cerebellum, and cerebrum in infant, adolescent, and adult mice, rats, and cats was observed by several investigators.[8,9,73,106,107,131–137,143] Some considered these cells undifferentiated spongioblasts, while others considered them multiplying glial elements. The proliferation of precursors of glia cells can be reconciled with the observation with conventional histological procedures that myelination of fiber tracts in the developing brain is preceded by an increase in the number of glia cells,[144–150] in particular oligodendrocytes, a phenomenon referred to as "myelination gliosis."[147]

Our studies[104,107,125,131,132,152] indicated that there is considerable local cell proliferation in fiber tracts in all regions of the spinal cord and brain in infant and adolescent rats and a reduced rate of proliferation in adults. However, it also was shown that the labeled cells found in these regions after short survival following injection are not necessarily stationary precursors of perifascicular glia; rather, the bulk of these cells are undifferentiated neuroblasts and spongioblasts that use these fiber tracts as migratory paths (Fig. 24). In animals that were injected with thymidine-³H during the first 2 wk of life and survived for a short period after injection, intensely labeled cells were seen in great number in such areas as the fimbria of the hippocampus and the white matter of the cerebral hemispheres. However, in animals that lived for several months after injection only an occasional intensely labeled cell was seen in this region, with some lightly labeled ones and the majority of cells unlabeled. This indicates not only that the labeled cells originally seen here were transitory elements (they were shown to add cells to the hippocampus and cortex) but also that perifascicular glia cells in these fiber tracts do not begin differentiating until after the second week of life. Indeed, intensely labeled cells were seen in appreciable number in these regions in animals injected at 30 days of age. These observations can be related to the finding that the myelination of fibers in the cerebral white matter begins slowly during the third week after birth in rats.[146,153,154] The continuing cell proliferation in the cerebral white and gray matter in the adult can be related, in part, to protracted myelination,[155,156] and it may account for the observed increase in the number of glia cells and glia/neuron ratio in adults.[157,158]

That a large proportion of the postnatally multiplying cells eventually differentiate into neuroglia can be demonstrated by studying the disposition and mode of specialization of labeled cells in animals that survived for 2, 4, and 8 months after injection (unpublished). For instance, after repeated injections with thymidine-³H during the first week of life, 47% of the cells of the cervical spinal cord were labeled in adulthood. The majority of these

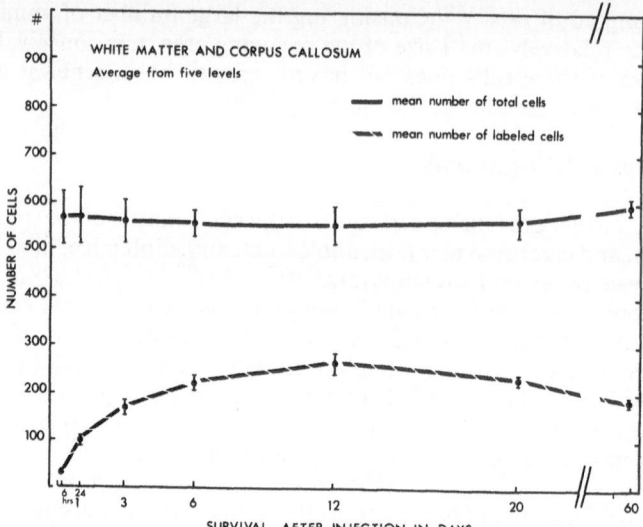

Fig. 24. Mean number of total cells and mean number of labeled cells in sampled areas (182 μ^2) in the white matter and corpus callosum in rats that were injected at 13 days of age and survived for different periods. Note that the addition of the large number of new cells by the twelfth day has not increased the total cell population. From Altman.[125]

were identified as astrocytes and oligodendrocytes. Considering that most of these cells were lightly labeled, this indicates that at least one-half of the glia cells in the spinal cord are of postnatal origin. In the cortex of the same animal virtually all the glia cells were lightly labeled, indicating that cortical gliogenesis is altogether a postnatal phenomenon.

C. Meninges, Capillaries, and Choroid Plexus

In animals that were injected with thymidine-^{3}H at birth or 2 days of age and that lived for several months afterward, virtually all the cell nuclei of the pia and arachnoid membranes were lightly labeled. In animals injected at 6 days of age these cells tended to show intermediate labeling, and in animals injected at 13 days there were many intensely labeled cells in the meninges but more than half of them were unlabeled. In the animals injected at 30 days, with brief or prolonged survival afterward, only an occasional cell was labeled, usually intensely. These observations (unpublished) indicate that most of the cells of the pia and arachnoid of the adult brain are formed postnatally but before the end of the first month, with only moderate turnover afterward (except under pathological conditions).

Likewise, the majority of endothelial cells of the larger capillaries of the forebrain were lightly labeled in adult animals that were injected at birth or 2 days of age, although the endothelial cells of the fine capillaries often were unlabeled. More such cells were labeled in the animals injected at 6 days of age. In animals injected at 13 days, intensely labeled endothelial cells were quite common, with the majority of them lightly labeled. In animals injected at 30 days the vast majority of endothelial cells were unlabeled, with occasional ones intensely labeled. These data show that vascularization of the forebrain begins after birth and nears completion at the end of the first month. This is reconcilable with the earlier quantitative studies of Craigie[159] and Horstmann[160] of progressive vascularization in the cortex of rats.

The choroid plexus is a prominent feature in the ventricles of the developing brain. It is a highly convoluted tissue in which fine capillaries are encased in connective tissue elements which, in turn, are surrounded by a monocellular layer of columnar epithelial cells. The latter are specialized cells derived from the ependyma of the ventricles. A recent autoradiographic study[161] indicates that the choroid plexus grows for some period after birth in the rat, and the following sequence of events was established for its postnatal growth: ependymal cells multiply during the first week of life and at a declining rate in the second week, near the lips of the opening of the ventricles, which leads to their invagination into the ventricular space. Due to continuing multiplication, the infolding chains of epithelial cells become progressively longer, the "older" epithelial cells being moved farther and farther from the opening. Then the newly forming epithelial tubules are invaded by capillaries of the tela choroidea and by leptomeningeal membranes of the brain. Subsequently, from the vascular beds free mesenchymal cells or macrophages migrate into the choroid tubules. It is of some interest that the growth of the choroid plexus is most pronounced at those ventricular sites where there is intense subependymal cell proliferation (as the lateral ventricles, the roof of the third ventricle, and the lateral recess of the fourth ventricle).

VI. CONDITIONS ALTERING CELL PROLIFERATION IN THE BRAIN

Cell proliferation in the developing and adult brain is radically altered by pathological processes and, to a lesser extent, by variable functional demands on the brain under "normal" conditions by differential rearing conditions and other behavioral influences. Among the pathological conditions affecting cell proliferation are mechanical injuries, viral and bacterial infections and vascular diseases and such harmful influences as ionizing radiation, endocrine imbalance, malnutrition, and the like. Among the behavioral effects are different modes of rearing the developing organism, as raising animals in impoverished or enriched environments and subjecting them to prolonged training procedures.

A. Pathological Effects

1. Lesions and Injury

Alterations in the concentration of neuroglia and microglia cells are among the most common consequences of brain injury, and recent autoradiographic studies confirm the deduction that this must be due to induced cell proliferation. An increased number of labeled glia cells in injured neural tissue was recognized early.[133,134] In one study,[132] electrolytic lesions were produced unilaterally in the lateral geniculate body in rats, and thymidine-^{3}H was injected into the lesion site. Labeled microglia and neuroglia cells were found concentrated in large numbers around the lesion by the first day after operation. The increase in the number of labeled cells, combined with decrease in label concentration within cells 1 month after the operation, indicated prolonged cell proliferation induced by the lesion. While labeled microglia cells and astrocytes predominated in the directly affected region, labeled oligodendrocytes were seen in increased numbers in brain areas structurally and functionally associated with the lateral geniculate body (Fig. 25). In a subsequent study,[107] in which thymidine-^{3}H was injected systematically, it was shown that the majority of the supernumerary glia cells at sites of "spontaneous" gliosis in the brains of "normal" rats were labeled.

Increase in glial proliferation in the facial[162] and hypoglossal nucleus[163] were reported after severing the respective nerves. In the facial nucleus a

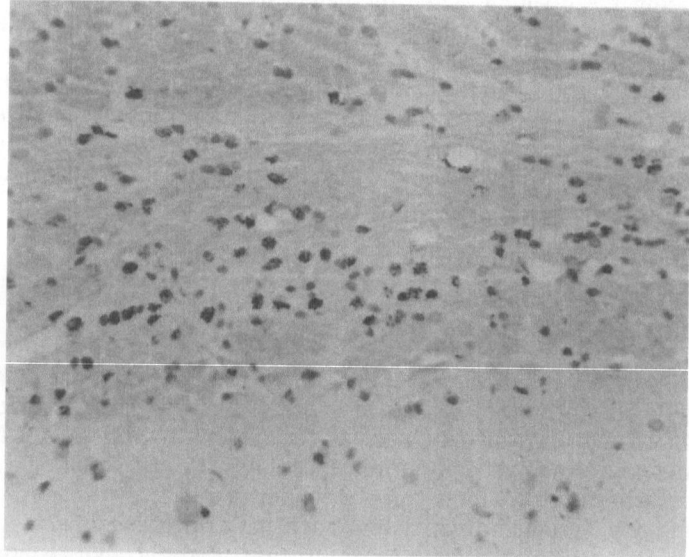

Fig. 25. Autoradiogram of the region of the visual radiation in a rat killed 1 month after combined lateral geniculate body lesion and intracranial injection of thymidine-[3]H. From Altman.[132]

thirtyfold increase in labeled glia cells was seen by the fourth day after operation and a tenfold increase in the hypoglossal nucleus. Most of these labeled cells were identified as microglia. More recently,[164] extensive proliferation of neuroglia, mainly oligodendrocytes, was reported in the white matter of the cat cerebral cortex after local compression designed to produce edema. The identity of proliferating cells after brain injury deserves further investigation. In addition to glial proliferation, most observers reported the proliferation of endothelial cells under such conditions, leading to increased vascularization of the affected areas.

2. Viruses with Affinity for Dividing Cells

We shall not deal here with the proliferation of microglia and neuroglia in the brain in response to viral, bacterial, or parasitic infection but have singled out a specific class of agents, feline ataxia virus and rat virus, which attack directly the multiplying cells of the brain and may entirely destroy an active proliferative matrix.[165,166] If infant, nonimmunized rats, hamsters, and ferrets are injected with rat virus, or kittens with feline ataxia virus, the external granular layer of the cerebellum may be totally destroyed. If the animals are permitted to survive, they develop a hypoplasic cerebellum, characterized by the scarcity or subtotal elimination of the postnatally formed microneurons of the cerebellar cortex. Evidence was presented that these viruses attack actively proliferating cells of the body and brain. The drastic effect in the cerebellum may be due to the circumstance that practically all the cells of the external granular layer are multiplying at the time of inoculation and none can escape destruction before immune reactions set in. The effect is irreversible and results in lasting locomotor ataxia. Other proliferative sites in the brain, such as the subependymal layer of the cerebral ventricles, may also be affected, although less severely.

3. Ionizing Radiation

Another agent that specifically destroys newly forming cells is ionizing radiation. It is well known that undifferentiated, multiplying cells, unlike mature cells, are extremely sensitive to ionizing radiation. The studies of Hicks[116–118] showed that as a result of whole-body irradiation of pregnant rats with 200 r, a large proportion of the multiplying cells of the developing embryonic nervous system is destroyed and that migratory cells may be destroyed with even lower doses. The demonstration that microneurons and neuroglia are formed after birth prompted us to study the effect of irradiation of the brain, and of the cerebellum alone, with repeated doses of 200-r x-ray from the day of birth upon the development of the cerebellar cortex.[138,167,168] The results showed that the granule cell population of the cerebellum may be reduced in proportion to the number of daily irradiations with 200 r and that after irradiation with 8–10 × 200 r, the cerebellum develops essentially devoid of basket, stellate, and granule cells but with the normal complement of Purkinje cells (Fig. 26). Irradiation with a few daily doses of 200 r produces

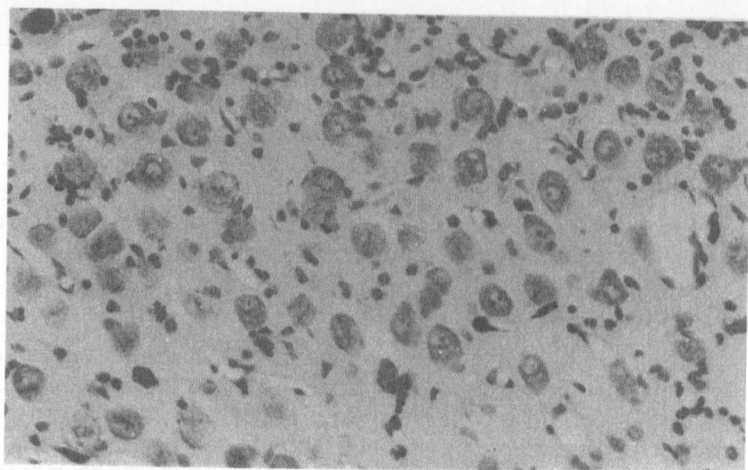

Fig. 26. Typical appearance of the cerebellar cortex in an adult rat whose cerebellum was irradiated daily with 200r x-ray from days 1–10 after birth. The internal granular layer is obliterated, and only a few scattered small cells are seen among the Purkinje cells.

transient ataxia and lasting subtler locomotor deficits; 8–10 × 200 r leads to permanent ataxia.

4. *Vitamins and Hormones*

Excess vitamin A administered to pregnant rats at a time antedating the closure of the neural groove produces severe malformations in the developing nervous system.[169] This effect was shown to be due partly to prolongation of the mitotic cycle in cells of the primitive neuroepithelium[93] (Fig. 27). When pregnant mice were given excess vitamin A on different days at later stages of

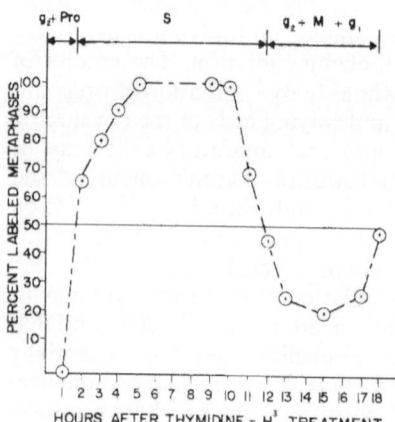

Fig. 27. Percentage of labeled metaphases in the neuroepithelium of a 15-day-old mouse embryo treated with excess vitamin A on days 14 and 15 of gestation. Note lengthening cell generation cycle when compared with a normal embryo of the same age, as shown in Fig. 9. From Langman and Welch.[93]

gestation, the formation of neurons in different layers of the cortex was affected differentially in accordance with the timetable of the origin of cells in these layers.[93]

Administration of bovine growth hormone to pregnant rats from the seventh to twentieth day of gestation was reported[170] to result in offspring with unchanged body weight and a significant increase in brain weight associated with increased DNA content of the brain, increased cortical cell density, and neuron/glia ratio. These effects were not observed by another group of investigators,[171] who reported merely a richer arborization of the dendrites of cortical neurons after treatment with pituitary growth hormone.

5. Malnutrition

Malnutrition during infancy permanently retards brain growth in rats, as gauged by reduced wet or dry brain weight in adults,[171,172] reduction in myelin,[173] and decrease in other brain constituents.[174] Of particular interest is a recent biochemical study by Winick and Noble[175] that indicates that the primary effect of malnutrition during infancy is a reduction in the number of cells in the brain rather than a reduction in cell size. The brains (and other organs) of adult rats from large or small litters (18 pups as opposed to the normal 9–12 to a nursing mother) were compared, on the assumption that the abnormally large litter size led to a reduction in caloric intake. They found a proportional reduction in protein, RNA, and DNA content in the experimental group, indicating that cell number was decreased by caloric intake in the brain but cell size was unaffected. Malnutrition after weaning, to which a young-adult and an adult group were subjected, led to reduction in brain weight and protein and RNA content but not DNA, indicating that during these periods only cell size but not cell number were affected. In the latter group the effects could be reversed by feeding, but the cell losses sustained during infancy were irreversible. In a complementary study,[176] it was shown that by increasing caloric intake over normal levels (by raising 3–6 pups with single mothers), the DNA content of the brain was increased with proportional increases in protein and RNA. This indicated that during this early period of development the total number of cells in the brain could be increased above normal but cell size was not affected. If these results can be confirmed, the conclusion of the authors is justified that the cell population in the brain is determined not only by genetic but also by extrinsic factors, such as nutrition. This would indicate that brain development may be enhanced by favorable environmental conditions.

B. Behavioral Effects

Recent studies suggest that differential manipulation of infant rats may affect cell proliferation in the brain. Whether these are due to direct influences or indirect ones by way of altered endocrine balance, changes in nutrition, and the like, remain to be determined. Handling of infant rats before weaning was shown to affect their behavior in adulthood.[177] In a pilot study[178] we

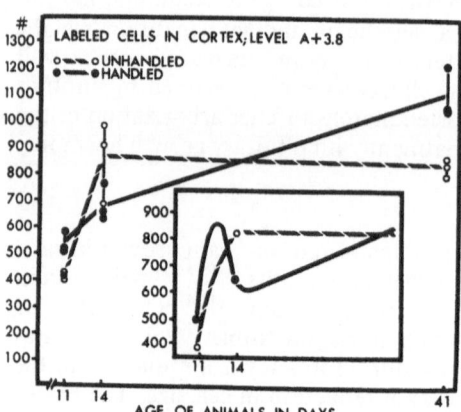

Fig. 28. Number of labeled cells at a coronal level of neocortex in triplets of handled and unhandled rats at different ages after cessation of handling and injection on the eleventh day. The number of labeled cells was higher immediately after injection and 30 days later in the handled animals. The labeled cells at 11 days are the locally multiplying cells; those seen at 41 days presumably also include those cells that have migrated into the cortex. The insert offers an explanation of the reversed relationship at 14 days of age (before the arrival of the migratory cells), attributing it to decline in the numbers of labeled cells because of label dilution in the locally produced cells due to high rate of multiplication. From Altman, Das, and Anderson.[178]

found that brain growth is retarded by handling during the first 10 days of life, as determined within a few days after cessation of the handling procedure. However, the effect is not a lasting one, and this recovery was attributed to a facilitation of cell proliferation in various brain regions after handling, as determined by thymidine-^{3}H autoradiography (Fig. 28). Cell proliferation also may be affected by behavioral treatments applied to young-adult or adult rats by raising them in an enriched or impoverished environment. Increase in brain weight and thickness of cortex was reported in animals raised in an enriched environment when compared with animals raised in a restricted environment.[179] This was associated with an increase in the number of glia cells[180,181] and a facilitation of cell proliferation.[152] Studies in progress in our laboratory indicate an increase in the length of the cerebrum, not only in enriched rats, but also in adult rats undergoing prolonged training on a complex visual discrimination task. Though the brain constituents contributing to this increase have not been identified, it is worth noting that the facilitated growth was observed along the axis that shows continuing growth in adult rats,[138] a growth attributed to a continued high rate of cell proliferation in the anterior part of the forebrain late into adulthood. The possible effect of use and disuse on cell proliferation requires further investigation.

VII. REFERENCES

1. B. Källén, *in Organogenesis* (R. L. DeHaan and H. Ursprung, eds.), pp. 107–128, Holt, Rinehart and Winston, New York (1965).

2. R. L. Watterson, *in Organogenesis* (R. L. DeHaan and H. Ursprung, eds.), pp. 129–159, Holt, Rinehart and Winston, New York (1965).

3. R. Levi-Montalcini, Events in the developing nervous system, *Progr. Brain Res.* **4**:1–29 (1964).

4. P. Weiss (ed.), *Genetic Neurology*, University of Chicago Press, Chicago (1950).

5. W. Bloom and D. W. Fawcett, *A Textbook of Histology*, 8th ed., p. 33, Saunders, Philadelphia (1962).

6. E. Allen, Cessation of mitosis in the central nervous system of the albino rat, *J. Comp. Neurol.* **22**:547–568 (1912).

7. W. A. Bryans, Mitotic activity in the brain of the adult rat, *Anat. Rec.* **133**:65–71 (1959).

8. B. Messier, C. P. Leblond, and I. Smart, Presence of DNA synthesis and mitosis in the brain of young adult mice, *Exptl. Cell. Res.* **14**:224–226 (1958).

9. I. Smart, The subependymal layer of the mouse brain and its cell production as shown by radioautography after thymidine-H^3 injection, *J. Comp. Neurol.* **116**:325–347 (1961).

10. A. Opalski, Ueber lokale Unterschiede in Bau der Ventrikelwände beim Menschen, *Z. Neurol. Psychiat.* **149**:221–254 (1933).

11. J. Kershman, The medulloblast and the medulloblastoma, *Arch. Neurol. Psychiat.* (*Chicago*) **40**:937–967 (1938).

12. J. H. Globus and H. Kuhlenbeck, Subependymal cell plate (matrix) and its relationship to brain tumors of the ependymal type, *J. Neuropathol.* **3**:1–35 (1944).

13. H. Obersteiner, Der feinere Bau der Kleinhirnrinde beim Menschen und bei Tieren, *Biol. Centralbl.* **3**:145–155 (1883).

14. A. Schaper, Die morphologische und histologische Entwicklung des Kleinhirns, *Morphol. Jahrb.* **21**:625–708 (1894).

15. S. R. Cajal, *Histologie du Système Nerveux*, Vol. 2, pp. 80–119, Maloire, Paris (1911).

16. W. H. F. Addison, The development of the Purkinje cells and of the cortical layers in the cerebellum of the albino rat, *J. Comp. Neurol.* **21**:459–488 (1911).

17. D. P. Purpura, R. J. Shofer, E. M. Housepian, and C. R. Noback, Comparative ontogenesis of structure-function relations in cerebral and cerebellar cortex, *Progr. Brain Res.* **4**:187–221 (1944).

18. J. Raaf and J. W. Kernohan, A study of the external granular layer in the cerebellum. The disappearance of the external granular layer and the growth of the molecular and internal granular layers in the cerebellum, *Am. J. Anat.* **75**:151–172 (1944).

19. N. Sugita, Comparative studies on the growth of the cerebral cortex. I. On the changes in the size and shape of the cerebrum during the postnatal growth of the brain, *J. Comp. Neurol.* **28**:495–510 (1917).

20. P. del Rio-Hortega, *in Cytology and Cellular Pathology of the Nervous System* (W. Penfield, ed.), Vol. 2, pp. 483–534, Hafner, New York (1932, reprinted 1965).

21. A. Boivin, R. Vendrely, and C. Vendrely, L'acide désoxyribonucléique du noyau cellulaire dépositaire des caractères héréditaires; arguments d'ordre analytique, *Compt. Rend. Acad. Sci.* **22**:1061–1063 (1948).

22. R. Vendrely and C. Vendrely, La teneur du noyau cellulaire en acide désoxyribonucléique à travers les organes, les individus et les espèces animales, *Experientia* **5**:327–329 (1949).

23. A. E. Mirsky and H. Ris, Variable and constant components of chromosomes, *Nature* **163**:666–667 (1949).

24. I. H. Heller and K. A. C. Elliott, Desoxyribonucleic acid content and cell density in brain and human brain tumors, *Can. J. Biochem. Physiol.* **32**:584–592 (1954).

25. G. Schmidt and S. J. Thannhauser, A method for the determination of desoxyribonucleic acid, ribonucleic acid and phosphoproteins in animal tissues, *J. Biol. Chem.* **161**:83–89 (1945).

26. W. C. Schneider, Phosphorus compounds in animal tissues. I. Extraction and estimation of desoxypentose nucleic acid and pentose nucleic acid, *J. Biol. Chem.* **161**:293–303 (1945).

27. J. E. Logan, W. A. Mannell, and R. J. Rossiter, Estimation of nucleic acids in tissue from the nervous system, *Biochem. J.* **51**:470–479 (1952).

28. H. A. DeLuca, R. J. Rossiter, and K. P. Strickland, Incorporation of radioactive phosphate into the nucleic acids of brain slices, *Biochem. J.* **55**:193–200 (1953).

29. P. J. Heald, *Phosphorus Metabolism of the Brain,* Pergamon Press, Oxford (1960).

30. J. M. Kissane and E. Robins, The fluorometric measurement of deoxyribonucleic acid in animal tissues, with special reference to the central nervous system, *J. Biol. Chem.* **233**:184–188 (1958).

31. J. B. Flexner and L. B. Flexner, Biochemical and physiological differentiation during morphogenesis. XIV. The nucleic acids of the developing cerebral cortex and liver of the fetal guinea pig, *J. Cell. Comp. Physiol.* **38**:1–16 (1951).

32. V. B. Peters and L. B. Flexner, Biochemical and physiological differentiation during morphogenesis. VIII. Quantitative morphologic studies on the developing cerebral cortex of the fetal guinea pig, *Am. J. Anat.* **86**:133–161 (1950).

33. D. V. N. Reddy, M. E. Lombardo, and L. R. Cerecedo, Nucleic acid changes during the development of the chick embryo, *J. Biol. Chem.* **198**:267–270 (1952).

34. W. Tung-Yue, H. Kuang-Mei, and H. T. Blumenthal, The effect of growth hormone on the nucleic acid content of the developing chick embryo, *Endocrinology* **53**:520–526 (1953).

35. J. Szepsenwol, J. Mason, and M. E. Shontz, Phospholipids and nucleic acids in embryonic tissues of the chick, *Am. J. Physiol.* **180**:525–529 (1955).

36. W. Albrecht, Änderung des Gehalts der verschiedenen Phosphatfraktionen in Gehirn und Muskel der Maus während des Wachstums, *Z. Naturforsch.* **11B**:248–252 (1956).

37. D. R. Dahl and F. E. Samson, Metabolism of rat brain mitochondria during postnatal development, *Am. J. Physiol.* **196**:470–472 (1959).

38. L. L. Uzman and N. K. Rumley, Changes in the composition of the developing mouse brain during early myelination, *J. Neurochem.* **3**:170–184 (1958).

39. P. Mandel, R. Bieth, and R. Stoll, Developpement biochimique du cervau de l'embryon de poulet. I. Acides nucléiques, *Bull. Soc. Chim. Biol.* **31**:1335–1340 (1949).

40. P. Mandel, H. Rein, S. Harth-Edel, and R. Mardell, *in Comparative Neurochemistry* (D. Richter, ed.), pp. 149–163, Pergamon Press, Oxford (1964).

41. D. Bodian and D. Dziewiatkowski, The disposition of radioactive phosphorus in normal, as compared with regenerating and degenerating nervous tissue, *J. Cell. Comp. Physiol.* **35**:155–177 (1950).

42. J. E. Logan, W. A. Mannell, and R. J. Rossiter, Chemical studies of peripheral nerve during Wallerian degeneration. 3. Nucleic acids and their protein-bound phosphorus compounds, *Biochem. J,* **51**:482–487 (1952).

43. L. Mihailovic, B. D. Jankovic, M. Petkovic, and D. Mančić, Distribution of DNA and RNA in different regions of cat's brain, *Experientia* **14**:9–10 (1958).

44. L. May and R. G. Grenell, Nucleic acid content of various areas of the rat brain, *Proc. Soc. Exptl. Biol. Med.* **102**:235–239 (1959).

45. A. V. Palladin, *in Biochemistry of the Developing Nervous System* (H. Waelsch, ed.), pp. 177–184, Academic Press, New York (1955).

46. K. A. C. Elliott and I. H. Heller, *in Metabolism of the Nervous System* (D. Richter, ed.), pp. 286–290, Pergamon Press, London (1957).

47. P. Mandel, S. Harth, and T. Borkowski, *in Regional Neurochemistry* (S. S. Kety and J. Elkes, eds.), pp. 160–174, Pergamon Press, London (1961).

48. E. Robins, D. E. Smith, and K. M. Eydt, The quantitative histochemistry of the cerebral cortex. I. Architectonic distribution of ten chemical constituents in the motor and visual cortices, *J. Neurochem.* **1**:54–57 (1956).

49. M. N. Baranov and L. A. Pevzner, Microchemical and microspectrophotometric studies on the intralaminar distribution of nucleic acids in the brain cortex under various experimental conditions, *J. Neurochem.* **10**:279–283 (1963).

50. R. Landolt, H. H. Hess, and C. Thalheimer, Regional distribution of some chemical structural components of the human nervous system. I. DNA, RNA and ganglioside sialic acid, *J. Neurochem.* **13**:1441–1452 (1966).

51. M. Abercrombie and M. L. Johnson, Quantitative histology of Wallerian degeneration. I. Nuclear population in rabbit sciatic nerve, *J. Anat.* **80**:37–50 (1946).

52. L. W. Lapham and M. A. Johnstone, Cytologic and cytochemical studies of neuroglia. II. The occurrence of two DNA classes among glial nuclei in the Purkinje cell layer of normal adult human cerebellar cortex, *Arch. Neurol.* **9**:194–202 (1963).

53. L. W. Lapham, The tetraploid DNA content of normal human Purkinje cells, *J. Neuropathol. Exptl. Neurol.* **25**:131–132 (1965).

54. L. W. Lapham, Tetraploid DNA content of Purkinje neurons of human cerebellar cortex, *Science* **159**:310–312 (1968).

55. J. Altman and G. D. Das, Postnatal neurogenesis in the guinea pig, *Nature* **214**:1098–1101 (1967).

56. L. F. Belanger and C. P. Leblond, A method for locating radioactive elements in tissues by covering histological sections with a photographic emulsion, *Endocrinology* **39**:8–13 (1946).

57. B. Messier and C. P. Leblond, Preparation of coated radioautographs by dipping sections in fluid emulsion, *Proc. Soc. Exptl. Biol. Med.* **96**:7–10 (1957).

58. B. M. Kopriwa and C. P. Leblond, Improvements in the coating technique of radioautography, *J. Histochem. Cytochem.* **10**:269–284 (1962).

59. J. Altman, *in Response of the Nervous System to Ionizing Radiation* (T. J. Haley and R. S. Slater, eds.), pp. 336–359, Little, Brown, Boston (1964).

60. L. G. Caro, *in Methods in Cell Physiology* (D. M. Prescott, ed.), Vol. 1, pp. 327–363, Academic Press, New York (1964).

61. M. M. Salpeter, *in Methods in Cell Physiology* (D. M. Prescott, ed.), Vol. 2, pp. 229–253, Academic Press, New York (1966).

62. A. R. Stevens, *in Methods in Cell Physiology* (D. M. Prescott, ed.), Vol. 2, pp. 255–310, Academic Press, New York (1966).

63. P. J. Fitzgerald, M. L. Eidinoff, J. E. Knoll, and E. B. Simmel, Tritium in radioautography, *Science* **114**:494–498 (1951).

64. R. P. Perry, *in Methods in Cell Physiology* (D. M. Prescott, ed.), Vol. 1, pp. 305–326, Academic Press, New York (1964).

65. P. Reichard and B. Estborn, Utilization of desoxyribosides in the synthesis of polynucleotides, *J. Biol. Chem.* **188**:839–846 (1951).

66. J. H. Taylor, P. S. Woods, and W. L. Hughes, The organization and duplication of chromosomes as revealed by autoradiographic studies using tritium-labeled thymidine, *Proc. Natl. Acad. Sci. (U.S.)* **43**:122–128 (1957).

67. W. L. Hughes, V. P. Bond, G. Brecher, E. P. Cronkite, R. B. Painter, H. Quastler, and F. G. Sherman, Cell proliferation and migration as revealed by autoradiography after injection of thymidine-H[3] into male rats and mice, *Am. J. Anat.* **106**:247–285 (1958).

68. M. Amano, B. Messier, and C. P. Leblond, Specificity of labeled thymidine as a deoxyribonucleic acid precursor in radioautography, *J. Histochem. Cytochem.* **7**:153–155 (1959).

69. W. L. Hughes, *in The Kinetics of Cellular Proliferation* (F. Stohlman, ed.), pp. 83–96, Grune & Stratton, New York (1959).

70. V. R. Potter, *in The Kinetics of Cellular Proliferation* (F. Stohlman, ed.), pp. 104–117, Grune & Stratton, New York (1959).

71. J. R. Rubini, E. P. Cronkite, V. P. Bond, and T. M. Fliedner, The metabolic fate of tritiated thymidine in man, *J. Clin. Invest.* **39**:909–919 (1960).

72. C. P. Leblond, B. Messier, and B. Kopriwa, Thymidine-H^3 as a tool for the investigation of renewal of cell populations, *Lab. Invest.* **8**:296–308 (1959).

73. B. E. Walker and C. P. Leblond, Sites of nucleic acid synthesis in the mouse visualized by radioautography after administration of C^{14}-labeled adenine and thymidine, *Exptl. Cell. Res.* **14**:510–531 (1958).

74. R. B. Painter, R. M. Drew, and W. L. Hughes, Inhibition of HeLa growth by intranuclear tritium, *Science* **127**:1244–1245 (1958).

75. R. C. Greulich, I. L. Cameron, and J. D. Thrasher, Stimulation of mitosis in adult mice by administration of thymidine, *Proc. Natl. Acad. Sci. (U.S.)* **47**:743–748 (1961).

76. H. J. Barr, An effect of exogenous thymidine on the mitotic cycle, *J. Cell. Comp. Physiol.* **61**:119–127 (1963).

77. R. B. Painter, R. M. Drew, and R. E. Rasmussen, Limitations in the use of carbon-labeled and tritium-labeled thymidine in cell culture studies, *Radiation Res.* **21**:355–366 (1964).

78. R. M. Drew and R. B. Painter, Action of tritiated thymidine on the clonal growth of mammalian cells, *Radiation Res.* **11**:535–544 (1959).

79. G. Marin and D. M. Prescott, The frequency of sister chromatid exchanges following exposure to varying doses of H^3-thymidine or x-rays, *J. Cell. Biol.* **21**:159–167 (1964).

80. M. E. Sauer and B. E. Walker, Radiation injury resulting from nuclear labeling with tritiated thymidine in the chick embryo, *Radiation Res.* **14**:633–642 (1961).

81. R. C. Greulich, Deleterious influence of orally administered tritiated thymidine on reproductive capacity of mice, *Radiation Res.* **14**:83–95 (1961).

82. H. A. Johnson and E. P. Cronkite, The effect of tritiated thymidine on mouse spermatogonia, *Radiation Res.* **11**:825–831 (1959).

83. J. W. Grisham, Inhibitory effect of tritiated thymidine on regeneration of the liver in the young rat, *Proc. Soc. Exptl. Biol. Med.* **105**:555–558 (1960).

84. J. Post and J. Hoffman, Some effects of tritiated thymidine as a deoxyribonucleic acid label in rat liver, *Radiation Res.* **14**:713–720 (1961).

85. H. A. Johnson and E. P. Cronkite, The effect of tritiated thymidine on mortality and tumor incidence in mice, *Radiation Res.* **30**:488–496 (1967).

86. J. D. Thrasher, *in Methods in Cell Physiology* (D. M. Prescott, ed.), Vol. 2, pp. 323–357, Academic Press, New York (1966).

87. J. Altman and G. D. Das, Autoradiographic and histological evidence of postnatal hippocampal neurogenesis in rats, *J. Comp. Neurol.* **124**:319–335 (1965).

88. A. Howard and S. R. Pelc, Synthesis of deoxyribonucleic acid in normal and irradiated cells and its relation to chromosome breakage, *Heredity Suppl.* **6**:261 (1953).

89. H. Quastler, *in Kinetics of Cellular Proliferation* (F. Stohlman, ed.), pp. 431–439, Grune & Stratton, New York (1959).

90. D. E. Wimber, Duration of the nuclear cycle in *Tradescantia paludosa* root tips as measured with H^3-thymidine, *Am. J. Bot.* **47**:828 (1960).

91. M. Atlas and V. P. Bond, The cell generation cycle of the eleven-day mouse embryo, *J. Cell. Biol.* **26**:19–24 (1965).

92. J. Langman, R. L. Guerrant, and B. G. Freeman, Behavior of neuroepithelial cells during closure of the neural tube, *J. Comp. Neurol.* **127**:399–412 (1966).

93. J. Langman and G. W. Welch, Excess vitamin A and development of the cerebral cortex, *J. Comp. Neurol.* **131**:15–26 (1967).

94. E. Koburg, *in Cell Proliferation* (L. F. Lamerton and R. J. M. Fry, eds.), pp. 62–76, Blackwell, Oxford (1963).

95. G. G. Steel, *in Cell Proliferation* (L. F. Lamerton and R. J. M. Fry, eds.), pp. 37–57, Blackwell, Oxford (1963).

96. S. Fujita, M. Shimada, and T. Nakamura, H³-thymidine autoradiographic studies on the cell proliferation and differentiation in the external and the internal granular layers of the mouse cerebellum, *J. Comp. Neurol.* **128**:191–208 (1966).

97. M. Shimada, Cytokinetics and histogenesis of early postnatal mouse brain as studied by ³H-thymidine autoradiography, *Arch. Histol. Japan.* **26**:413–437 (1966).

98. C. P. Leblond and B. E. Walker, Renewal of cell populations, *Physiol. Rev.* **36**:255–276 (1956).

99. H. Quastler, in *Cell Proliferation* (L. F. Lamerton and R. J. M. Fry, eds.), pp. 18–34, Blackwell, Oxford (1963).

100. D. E. Wimber, in *Cell Proliferation* (L. F. Lamerton and R. J. M. Fry, eds.), pp. 1–17, Blackwell, Oxford (1963).

101. J. Hanaway, Formation and differentiation of the external granular layer of the chick cerebellum, *J. Comp. Neurol.* **131**:1–14 (1967).

102. E. T. Pierce, Histogenesis of the dorsal and ventral cochlear nuclei in the mouse. An autoradiographic study, *J. Comp. Neurol.* **131**:27–53 (1967).

103. A. Martin and J. Langman, The development of the spinal cord examined by auto-radiography, *J. Embryol. Exptl. Morphol.* **14**:25–35 (1965).

104. J. Altman, Autoradiographic and histological studies of postnatal neurogenesis. II. A longitudinal investigation of the kinetics, migration and transformation of cells incorporating tritiated thymidine in infant rats, with special reference to neurogenesis in some brain regions, *J. Comp. Neurol.* **128**:431–473 (1966).

105. S. Fujita, Quantitative analysis of cell proliferation and differentiation in the cortex of the postnatal mouse cerebellum, *J. Cell. Biol.* **32**:277–288 (1967).

106. I. Smart and C. P. Leblond, Evidence for the division and transformations of neuroglia cells in the mouse brain as derived from radioautography after injection of thymidine-H³, *J. Comp. Neurol.* **116**:349–367 (1961).

107. J. Altman, Autoradiographic investigation of cell proliferation in the brains of rats and cats, *Anat. Rec.* **145**:573–591 (1963).

108. J. Altman, in *The Neurosciences, A Study Program* (G. C. Quarton, T. Melnechuk, and F. O. Schmitt, eds.), pp. 723–743, Rockefeller University Press, New York (1967).

109. W. His, Die Neuroblasten und deren Entstehung in embryonalen Mark, *Sächs. Wiss.* **15**:313–372 (1889).

110. A. Schaper, Die frühesten Differenzierungsvorgänge in Centralnervensystem, *Arch. Entwicklungsmech. Organ.* **5**:81–132 (1897).

111. F. C. Sauer, Mitosis in the neural tube, *J. Comp. Neurol.* **62**:377–405 (1935).

112. M. B. Sauer and B. E. Walker, Radioautographic studies of interkinetic nuclear migration in the neural tube, *Proc. Soc. Exptl. Biol. Med.* **101**:557–560 (1959).

113. R. L. Sidman, I. L. Miale, and N. Feder, Cell proliferation and migration in the primitive ependymal zone; an autoradiographic study of histogenesis in the nervous system, *Exptl. Neurol.* **1**:322–333 (1959).

114. S. Fujita, Kinetics of cellular proliferation, *Exptl. Cell. Res.* **28**:52–60 (1962).

115. B. Källén and K. Valmin, DNA synthesis in the embryonic chick central nervous system, *Z. Zellforsch.* **60**:491–496 (1963).

116. S. P. Hicks, C. J. D'Amato, and M. J. Lowe, Development of the mammalian nervous system. I. Malformations of the brain, especially in the cerebral cortex, induced in rats by radiation. II. Some mechanisms of the malformations of the cortex, *J. Comp. Neurol.* **113**:435–470 (1959).

117. S. P. Hicks, C. J. D'Amato, M. A. Coy, E. D. O'Brien, J. M. Thurston, and D. L. Joftes, Migrating cells in the developing nervous system studied by their radiosensitivity and tritiated thymidine uptake, *Brookhaven Symp. Biol.* **14**:246–261 (1961).

118. S. P. Hicks and C. J. D'Amato, *in Advances in Teratology* (D. H. M. Woollam, ed.), pp. 195–250, Logos Press, London (1966).

119. J. D. Angevine and R. L. Sidman, Autoradiographic study of cell migration during histogenesis of cerebral cortex in the mouse, *Nature* **192**:766–768 (1961).

120. M. Berry and J. T. Eayrs, Histogenesis of the cerebral cortex, *Nature* **197**:984–985 (1963).

121. M. Berry and A. W. Rogers, The migration of neuroblasts in the developing cerebral cortex, *J. Anat. (London)* **99**:691–709 (1965).

122. M. Berry and J. T. Eayrs, The effects of x-irradiation on the development of the cerebral cortex, *J. Anat. (London)* **100**:707–722 (1966).

123. J. Altman and G. D. Das, Post-natal origin of microneurones in the rat brain, *Nature* **207**:953–956 (1965).

124. J. Altman and G. D. Das, Autoradiographic and histological studies of postnatal neuro-genesis. I. A longitudinal investigation of the kinetics, migration and transformation of cells incorporating tritiated thymidine in neonate rats, with special reference to postnatal neurogenesis in some brain regions, *J. Comp. Neurol.* **126**:337–390 (1966).

125. J. Altman, Proliferation and migration of undifferentiated precursor cells in the rat during postnatal gliogenesis, *Exptl. Neurol.* **16**:263–278 (1966).

126. E. Taber Pierce, Histogenesis of the nuclei *griseum pontis, corporis pontobularis* and *reticularis tegmenti ponti* (Bechterew) in the mouse. An autoradiographic study, *J. Comp. Neurol.* **126**:219–239 (1966).

127. L. L. Uzman, The histogenesis of the mouse cerebellum as studied by its tritiated thymidine uptake, *J. Comp. Neurol.* **114**:137–159 (1960).

128. I. L. Miale and R. L. Sidman, An autoradiographic analysis of histogenesis in the mouse cerebellum, *Exptl. Neurol.* **4**:277–296 (1961).

129. S. Fujita, Analysis of neuron differentiation in the central nervous system by tritiated thymidine autoradiography, *J. Comp. Neurol.* **122**:311–328 (1964).

130. E. Taber Pierce, Histogenesis of deep cerebellar nuclei studied autoradiographically with thymidine-H³ in the mouse, *Anat. Rec.* **157**:301 (1967).

131. J. Altman, Are new neurons formed in the brains of adult mammals?, *Science* **135**:1127–1128 (1962).

132. J. Altman, Autoradiographic study of degenerative and regenerative proliferation of neuroglia cells with tritiated thymidine, *Exptl. Neurol.* **5**:302–318 (1962).

133. E. K. Adrian and B. E. Walker, Incorporation of thymidine-H³ by cells in normal and injured mouse spinal cord, *J. Neuropath. Exptl. Neurol.* **21**:597–609 (1962).

134. H. Koenig, M. B. Bunge, and R. P. Bunge, Nucleic acid and protein metabolism in white matter, *Arch. Neurol.* **6**:177–193 (1962).

135. B. Messier and C. P. Leblond, Cell proliferation and migration as revealed by radio-autography after injection of thymidine-H³ into male rats and mice, *Am. J. Anat.* **106**:247–288 (1960).

136. H. Koenig, Autoradiographic studies of deoxyribonucleic acid (DNA) turnover in the feline neuraxis, *J. Histochem. Cytochem.* **8**:337 (1960).

137. B. Schultze and W. Oehlert, Autoradiographic investigation of incorporation of H³-thymidine into cells of the rat and mouse, *Science* **131**:737–738 (1960).

138. J. Altman, W. J. Anderson, and K. A. Wright, Gross morphological consequences of irradiation of the cerebellum in infant rats with repeated doses of low-level x-ray, *Exptl. Neurol.* **21**:69–91 (1968).

139. J. Altman, Autoradiographic and histological studies of postnatal neurogenesis. V. A longitudinal investigation of the kinetics, migration and transformation of cells incorporating tritiated thymidine in adolescent and adult rats, with special reference to postnatal neurogenesis in some brain regions, in preparation.

140. J. Altman, Postnatal neurogenesis and the problem of neural plasticity, *in Developmental Neurobiology* (W. A. Himwich, ed.), Charles C. Thomas, Springfield, Ill., in press.

141. J. B. Angevine, Time of neuron origin in the hippocampal region. An autoradiographic study in the mouse, *Exptl. Neurol. Suppl.* **2**:1–70 (1965).

142. J. Altman and G. D. Das, Postnatal neurogenesis in the guinea pig, *Nature* **214**:1098–1101 (1967).

143. O. R. Hommes and C. P. Leblond, Mitotic division of neuroglia in the normal adult rat, *J. Comp. Neurol.* **129**:269–278 (1967).

144. E. A. Linell and M. I. Tom, The postnatal development of the oligodendroglia cell in the brain of the white rat and the possible role of this cell in myelogenesis, *Anat. Rec.* **48**: 27 (1931).

145. L. R. Morrison, Role of oligodendroglia in myelogenesis, *Arch. Neurol. Psychiat.* (*Chicago*) **28**:204–205 (1932).

146. F. Tilney, Behavior in its relation to the development of the brain. II. Correlation between the development of the brain and behavior in the albino rat from embryonic states to maturity, *Bull. Neurol. Inst. N.Y.* **3**:252–358 (1933).

147. H. N. Roback and H. J. Scherer, Über die feinere Morphologie des frühkindlichen Hirnes unter besonderer Berücksichtigung der Gliaentwicklung, *Virchows Arch. Pathol. Anat.* **294**:365–413 (1935).

148. J. P. M. Bensted, J. Dobbing, R. S. Morgan, R. T. W. Reid, and G. P. Wright, Neuroglia development and myelination in the spinal cord of the chick embryo, *J. Embryol. Exptl. Morphol.* **5**:428–431 (1957).

149. A. Dekaban, Oligodendroglia and axis cylinders in rabbits before, during, and after myelination, *Anat. Rec.* **126**:111–122 (1956).

150. R. L. Friede, A histochemical study of DPN-diaphorase in human white matter, with some notes on myelination, *J. Neurochem.* **8**:17–30 (1961).

151. A. N. Davison and J. Dobbing, Myelination as a vulnerable period in brain development, *Brit. Med. Bull.* **22**:40–44 (1966).

152. J. Altman and G. D. Das, Autoradiographic examination of the effects of enriched environment on the rate of glial multiplication in the adult rat brain, *Nature* **204**:1161–1163 (1964).

153. J. B. Watson, Animal education: An experimental study on the physical development of the white rat, correlated with the growth of its nervous system, *Contrib. Phil.* **4**:5–122 (1903).

154. S. Jacobson, Sequence of myelinization in the brain of the albino rat. A. Cerebral cortex, thalamus and related structures, *J. Comp. Neurol.* **121**:5–29 (1963).

155. P. Flechsig, *Anatomie des menschlichen Gehirns und Rückenmarks auf myelogenetischer Grundlage,* Thieme, Leipzig (1920).

156. T. Kaes, *Die Grosshirnrinde des Menschen in ihren Massen und in ihren Fasergehalt,* Fischer, Jena (1907).

157. K. R. Brizzee, J. Vogt, and X. Kharetchko, Postnatal changes in glia/neuron index with a comparison of methods of cell enumeration in the white rat, *Progr. Brain Res.* **4**:136–149 (1964).

158. K. R. Brizzee, N. Sherwood, and P. S. Timiras, Comparison of cell populations at various depth levels in cerebral cortex of young adult and aged Long-Evans rats, *J. Gerontol.,* in press.

159. E. H. Craigie, Postnatal changes in vascularity in the cerebral cortex of the male albino rat, *J. Comp. Neurol.* **39**:301–324 (1925).

160. E. Horstmann, *in Structure and Function of the Cerebral Cortex* (D. B. Tower and J. P. Schade, eds.), pp. 59–63, Elsevier, Amsterdam (1960).

161. J. Altman, Postnatal development of the choroid plexus and the migration of mesenchymal elements to the brain through leptomeningeal spaces. An autoradiographic study in rats with tritiated thymidine, unpublished.

162. G. W. Kreutzberg, Autoradiographische Untersuchung über die Beteiligung von Gliazellen an der axonalen Reaktion im Facialiskern der Ratte, *Acta Neuropathol.* **7**:149–161 (1966).

163. J. Sjöstrand, Glial cells in the hypoglossal nucleus of the rabbit during nerve regeneration, *Acta. Physiol. Scand.* **67**: *Suppl. 270*, 1–17 (1966).

164. B. Schultze and P. Kleihues, DNS-Synthese der Neuroglia in experimentellen Hirnödem, *Experientia* **23**:941 (1967).

165. L. Kilham and G. Margolis, Viral etiology of spontaneous ataxia of cats, *Am. J. Pathol.* **48**:991–1004 (1966).

166. G. Margolis and L. Kilham, Rat virus, an agent with affinity for the dividing cell, *Natl. Inst. Neurol. Dis. Blind. Monogr.* **2**:361–367 (1966).

167. J. Altman, W. J. Anderson, and K. A. Wright, Selective destruction of precursors of microneurons of the cerebellar cortex with fractionated low-dose x-rays, *Exptl. Neurol.* **4**:481–497 (1967).

168. J. Altman, W. Anderson, and K. A. Wright, Quantitative histological evaluation of the effects of irradiation of the cerebellum in infant rats with repeated doses of low-level x-ray, in preparation.

169. J. Langman and G. W. Welch, Effect of vitamin A on development of the central nervous system, *J. Comp. Neurol.* **128**:1–15 (1966).

170. S. Zamenhof, J. Mosley, and E. Schuller, Stimulation of the proliferation of cortical neurons by prenatal treatment with growth hormone, *Science* **152**:1396–1397 (1966).

171. S. Hatai, The effect of partial starvation on the brain of the white rat, *Am. J. Physiol.* **12**:116–127 (1904).

172. H. H. Donaldson, The effect of underfeeding on the percentage of water, on the ether-alcohol extract, and on medullation in the central nervous system of the albino rat, *J. Comp. Neurol.* **21**:139–145 (1911).

173. J. Dobbing, The influence of early nutrition on the development and myelination of the brain, *Proc. Roy. Soc. (London) Ser. B* **159**:503–509 (1964).

174. J. W. Benton, H. W. Moser, P. R. Dodge, and S. Carr, Modification of the schedule of myelination in the rat by early nutritional deprivation, *Pediatrics* **38**:801–807 (1966).

175. M. Winick and A. Noble, Cellular response in rats during malnutrition at various ages, *J. Nutrition* **89**:300–306 (1966).

176. M. Winick and A. Noble, Cellular response with increased feeding in neonatal rats, *J. Nutrition* **91**:179–182 (1967).

177. V. H. Denenberg, in *The Behavior of Domestic Animals* (E. S. E. Hafez, ed.), pp. 109–138, Bailliere, Tindall, Cox, London (1962).

178. J. Altman, G. Das, and W. J. Anderson, Effects of infantile handling on morphological development of the rat brain: An exploratory study, *Devel. Psychobiol.* **1**:10–20 (1968).

179. E. L. Bennett, M. C. Diamond, D. Krech, and M. R. Rosenzweig, Chemical and anatomical plasticity of brain, *Science* **146**:610–619 (1964).

180. M. C. Diamond, D. Krech, and M. R. Rosenzweig, The effects of an enriched environment on the histology of the rat cerebral cortex, *J. Comp. Neurol.* **123**:11–120 (1964).

181. M. C. Diamond, F. Law, H. Rhodes, M. R. Rosenzweig, D. Krech, and E. L. Bennett, Increases in cortical depth and glia numbers in rats subjected to enriched environment, *J. Comp. Neurol.* **128**:117–126 (1966).

182. J. Altman, Autoradiographic and histological studies of postnatal neurogenesis. IV. Cell proliferation and migration in the anterior forebrain, with special reference to persisting neurogenesis in the olfactory bulb, *J. Comp. Neurol.*, in press.

183. J. Altman, Autoradiographic and histological studies of postnatal neurogenesis. III. Dating the time of production and onset of differentiation of cerebellar microneurons in rats, *J. Comp. Neurol.*, in press.

Chapter 9

NEURONS AND GLIA: SEPARATION TECHNIQUES AND BIOCHEMICAL INTERRELATIONSHIPS

Steven P. R. Rose

Medical Research Council Metabolic Reactions Research Unit
Department of Biochemistry
Imperial College
London, England

I. INTRODUCTION

In recent years it has become evident that, because of the individuation of specific cells within the central nervous system (CNS) at the physiological, functional, and structural level, their biochemistry is likely to differ in certain critical aspects, too. Thus, it has become necessary to devise techniques to study biochemical variables within individual cells or cell classes. While both neurons and glia certainly differ among themselves biochemically as well as functionally, and the physiological role of different glial subclasses, oligo-dendroglia, astrocytes, and microglia, has been extensively studied,[1] so far technical sophistication has scarcely proceeded beyond the point of separating the two major classes, neurons and glia, and examining some of their interrelationships.

Available methods include histochemistry, the use of specialized brain regions and of abnormal tissue enriched in particular cell species, physical separation of cells on a macro- or a microscale, and the use of nonmammalian systems that offer special ease of access to particular cell types. The first of these methods is discussed by Lehrer[2] and the last in Chapter 10 of this volume by Giacobini,[3] while regional variations form the theme of several of the preceding chapters (e.g., Vol. I, Chapter 17, Vol. II, Chapter 5). Here we consider techniques and data derived from the physical separation of mammalian neurons and glia and their implication for neuronal–glial interrelationships.

II. METHODOLOGY

Separation of cells into individuals or classes presents two problems: that of disaggregating the tissue so as to separate the maximum number of cells but disrupt the minimum number and that of isolating the separated cells according to appropriate criteria, e.g., neurons from glia. In the microdissection techniques discussed by Lehrer,[2] both problems are approached simultaneously, but in macroseparations the two are sequential.

A. Separation

Separation procedures can depend on mechanical disaggregation or the use of chemicals or enzymes. The disaggregating agent tetraphenylboron (TPB) has been used by Rappaport.[4] Satake and Abe[5] suspend chopped tissue in ice-cold 1/1/1 acetone–glycerol–water for 30 min before centrifuging and disaggregating the pellet with a loose-fitting Teflon homogenizer in 3/1 glycerol–0.25 M sucrose. Enzymatic techniques seem to produce damaged cells.[6]

Mechanical techniques were used by Korey in preparing a glial fraction from lamb white matter; the tissue was dispersed in sucrose with a modified laboratory blendor and filtered on a silk grid.[7] Roots and Johnston[8] passed material from ox thalamus or lateral vestibular nuclei through nylon meshes under pressure. The author's own procedure[6] involves passing cleaned rat cortex through a 110-μ nylon mesh into a medium consisting of 10% Ficoll buffered with KCl and phosphate, followed by filtration under suction on a 40-μ stainless steel grid. A similar procedure has been used by Bocci.[9]

All these procedures result in mixtures containing greater or lesser proportions of (1) undisrupted tissue, (2) separated neurons and glia, (3) red blood cells, and (4) cell debris.

B. Isolation

While some authors have carried their preparations no further than the disaggregated cell stage,[4,10] the major purpose of separation must be to isolate the cell types. Feasible methods include electrophoresis, phase separation, gel filtration, or density gradient or differential centrifugation; in fact, only the centrifugation methods have been applied. Techniques that provide neurons alone, or along with a rather heterogeneous nonneuronal fraction, are those of Satake and Abe[5] and of Bocci.[9] Korey's method[7] is for glia. The author's own method[6] provides both neuronal and glial fractions in a one-stage centrifugation step on a discontinuous gradient. The cell suspension, in 10% Ficoll, is layered over 1.45 M sucrose and 30% Ficoll buffered with KCl and phosphate. After equilibrium centrifugation, unbroken tissue and myelin floats to the surface and red cells and broken nuclei are pelleted to the bottom of the tubes. The glial fraction is at the 10–30% Ficoll interface and the neuronal at the 30% Ficoll–1.45 M sucrose interface; the fractions can be further washed and resuspended. Two to ten percent of the total neurons originally present are collected in the neuronal fraction, or 3×10^7 cells from 20 rat brains, using the biggest available centrifuge buckets. Variations on this method also have been reported.[11]

C. Nature of the Isolated Cells: Comparison with Microdissection Techniques

Studies based on the use of isolated cells, whether prepared by micro-dissection or macromethods, are subject to two major criticisms. The first relates to the validity of the characterization of the fractions isolated as "neuronal" or "glial." The tissue disaggregation techniques inevitably involve the breakage of axons and dendrites. The "neuron" obtained is, in reality, a neuronal perikaryon, possessed of but a fraction of its original processes, although it is probable that the shorn ends, where processes have been torn off, rapidly reseal so as to prevent further loss of cellular material. The numbers of glial processes removed during separation are presumably much smaller.

As only the larger neurons can be collected by microdissection, they represent a highly selected, albeit pure, neuronal population. The macro-methods produce a more "normal" neuronal population but include as contamination glia, capillary endothelial cells, and some more or less extensively damaged cells. Glial contamination in the author's procedure, with carbonic anhydrase as a glial marker,[12] is less than 11 %.[6]

Both micro- and macromethods produce glial fractions that appear morphologically very similar under the light microscope. Electron microscopy of the glial fraction prepared by the author's technique[6] shows that while there are no neuronal perikarya present, there is contamination by pinched-off nerve endings, axonal and dendritic material, and other subcellular frag-ments. No doubt the microdissected glial fraction must contain some at least of such contaminants, but no electron micrographic studies of such fractions have been made. However, it may be preferable to regard this fraction as "neuronil" rather than glial.

The second major criticism relates to the integrity of the isolated cells. Do the broken cells reseal? Roots and Johnston[8] published electron micro-graphs that demonstrated that cells isolated by their techniques were without external membranes at all. However, it is possible that this was either an artefact of the fixation process or depended on the nature of the medium in which separation took place, as electron micrographs of cells prepared by the author's method do show many cells with substantial regions of continuous external cell membranes. Other data bearing on the question of the integrity of the external membrane would be the osmotic properties of the cells, their "leakiness" to soluble enzymes and cofactors, their ability to accumu-late ions against a concentration gradient, and their ability to maintain a resting potential. Insofar as these questions have been examined, the data are ambiguous. Thus, the cells do behave like osmometers,[13] and soluble enzymes such as lactate dehydrogenase do not leak out.[6] Both the glial and the neuronal fraction can accumulate potassium and amino acids against a gradient,[13,14] at a lower rate, however, than the intact slice. Hillman and Hydén have observed resting membrane potentials with microdissected neurons under rather special conditions[15]; with the much smaller cortical neurons, however, it has not been possible to find a potential.[14] Finally, cells isolated at least from very young animals,[6] or after TPB dissociation,[4] are capable of being grown in tissue culture, which presumably means that a percentage at least must remain viable during the isolation procedure.

D. Choice of Method

Choice must depend on the type of question asked. The advantage of macrotechniques over microdissection is the possibility of performing a wider range of conventional biochemical assays and metabolic determinations. Against this must be balanced the loss of a precise knowledge of exactly which cells are being studied.

Among the macrotechniques, those of Satake and Abe,[5] of Bocci,[9] and of Rappaport[4] produce cells apparently deficient in cytoplasmic content. Thus, Bocci reports a recovery of only 1.6–2.2% of the original lactate dehydrogenase in his spinal cord neurons and Purkinje cells, while metabolically, cells prepared by the Satake and Abe or Rappaport methods show almost no respiration with glucose, or CO_2 or lactate production, or incorporation of amino acid into protein.[13] Morphologically, though, the cells appear satisfactory, and those of Rappaport are viable under culture conditions.[4]

Cells prepared by the author's technique seem under phase microscopy less well preserved but contain a substantial complement of cytoplasmic enzymes and respire *in vitro* at a rate some 70–80% of slice, generating CO_2 and lactate and amino acids from glucose and synthesizing ATP and protein.[6,13,16–18] In general, the mixture of neurons and glia, prior to isolation, behaves metabolically like a slightly damaged slice, and metabolic conclusions drawn from studies of the cells in isolation may be extrapolated back with caution. For *in vitro* studies, therefore, this method is probably preferable.

III. BIOCHEMICAL DATA

The purpose of the separation techniques is to throw light on the functional relationship of the cell types by studying their properties in isolation. Hence, the significant findings are those of differences in neuronal and glial biochemistry; these may be chemical, enzymatic or metabolic, *in vivo* or *in vitro*. While several techniques have contributed to such studies, we concentrate here primarily on data relating to the mammalian brain and obtained by macroseparation.

A. Chemical Differences

Neurons prepared by the author's technique are relatively enriched in RNA and DNA on a protein basis, compared with glia, although no data exist on base ratios that may be compared with the studies made by Hydén by microdissection.[19] Hydén found that, compared to the neurons, glial RNA was enriched in guanine and low in adenine.[19,20] This makes it more

DNA-like in type and is presumably accounted for by the fact that there is a much greater amount of cytoplasmic ribosomal RNA in the neurons, evident from electron microscopic studies and from Nissl staining. However, because of the lability of such base ratios and their apparent dependence on external conditions,[21–23] it is doubtful what significance can be attached to them. One of the few proteins whose distribution has been studied, the glutamate-enriched S-100 protein, has been shown, in microdissected cells, to be glial rather than neuronal.[24]

With respect to lipid composition, Davison et al.[25] found glial lipid to resemble early myelin; by contrast with the neuron, it was much lower in sphingomyelin but enriched in cerebroside, which was lacking in the neuron. Freysz et al.[11,26] on the other hand, found in cells isolated after acetone/glycerol/water treatment that the glia were enriched both in total lipid and phospholipid; cholesterol was nearly nine times higher in the glia, and phosphatides six times. Some of these differences may relate to the different separation procedures used.

Among low-molecular weight components, the amino acid pools have been examined by the author[13,17,18] and striking differences found. On a protein basis the pool sizes of the amino acids glutamate, glutamine, GABA, aspartate, and alanine are all greater, by a factor of two or more, in the neuronal than in the glial fraction.

B. Enzymatic Differences

Most of the allocation of enzymes to particular cell types has been done on a histochemical basis and has been reviewed by Friede.[27] It is hard to assess the significance of differences in enzyme levels, as most are present in more than saturating amounts, and quite large differences in absolute activity may be metabolically without significance. Attention therefore should be focused on enzymes that are totally lacking in one cell type or catalyze rate-limiting reactions in particular metabolic pathways.

Perhaps the only truly exclusively located enzyme is carbonic anhydrase, which is almost absent from neurons but present in substantial quantities in glia[12]; it may be regarded as a glial marker[6] although its role in the cell is unfortunately obscure. No obvious candidate for a neuronal marker exists. Virtually all the enzymes of glucose and amino acid metabolism have been shown to be present in both cell types, although Hess and Pope[28] concluded that in the rat the respiratory enzymes were mainly neuronal. Hydén and Pigon found cytochrome oxidase nearly threefold and succinoxidase twice as active in rabbit Dieter's nucleus glia as in neurons,[20] but these figures were markedly dependent on the functional state of the cell.[29] Isolation by the author's technique produced cells in which lactate and succinate dehydrogenase, aspartate aminotransferase, and glutamate decarboxylase activity were similar,[6,13] as was triphosphoinositide phosphomonoesterase,[30] while cytochrome oxidase was 80% and glutamate dehydrogenase was four times more active in the neurons than the glia.[6,18]

C. Metabolic Differences

Metabolic calculations are made more complex by the need to establish a reference point. Activity on a cell basis or per unit protein or dry weight or the relative contribution of the two cell classes to the overall metabolism of the tissue may form the basis of comparison. Such calculations are hindered by the discrepancies in cell size, the uncertainty of the relative numbers of neurons and glia in the brain, and the need to assess what proportion of the "glial" metabolism in fact derives from neuronal dendrites embedded inextricably in the glial mass. All such problems are substantially more complex when the additional factors introduced by a comparison of the *in vivo* with the *in vitro* situation is attempted. In general, macromolecular metabolism has been studied *in vivo* and small molecules *in vitro*.

1. In Vivo

In vivo studies generally have involved autoradiography and hence turnover of macromolecules. Attention has been devoted primarily to RNA and DNA turnover, and it is clear that in the adult DNA turnover is almost exclusively glial (but see Altman and Das[31]) while RNA turnover proceeds more rapidly in the neurons.[32] RNA turnover is markedly subject to variation, depending on the environmental status of the animal.[32] Volpe and Giuditta[33] labeled RNA *in vivo* and then separated the neurons and glia on a gradient and found the rate of incorporation into neuronal RNA was more rapid than into glial. After about 14 hr, however, the neuronal RNA specific activity declined below the glial. The rate of protein synthesis, too, is higher in the neurons than the glia.[16]

2. In Vitro

a. Glucose Oxidation. There is some discrepancy in the data on oxygen uptake by isolated cells. With the microdiver technique, estimates vary from 263[34] to 1080[35] μmoles/g fresh wt./hr for neurons and 36[35] to 41[34] for glia. With macroseparated cells, the figures are 70[6] to 144[36] for neurons and 41[36] to 74[6] for glia. Such differences in part are based on differences in origin of the tissue. Attempts to assess the proportion of total oxidative metabolism contributed by neurons have varied from 20[34] to 95%,[37] depending on conflicting estimates of the glial/neuronal index and the proportion to be allowed for dendritic contamination. While Lowry et al.[38] have claimed that dendritic metabolism accounts for a high proportion of the metabolism of gray matter, and it is known that synaptosomes can metabolize glucose, the Swedish school argue that these dendritic elements contribute little to the oxygen uptake of the glial fraction.[20,34,39] However, both cell types clearly actively metabolize glucose.

b. Amino Acid Metabolism. The known compartmentation of amino acid metabolism[40] makes it an appropriate candidate for study in isolated cells. The author has followed the incorporation of radioactivity from U–^{14}C glucose, pyruvate, and glutamate into amino acids in neurons and glia. Both from glucose and from pyruvate, the absolute incorporation into and specific activities of the amino acids are lower in the neurons than the glia by a factor of 3–5. Thus, the rate at which the glia generate amino acids is substantially greater than that of the neurons. On the other hand (see above) the pool size of the amino acids is much greater in the neurons than the glia.[17,18] This observation may help account for the known facts of amino acid compartmentation and suggests speculation about a possible glial–neuronal amino acid transfer mechanism.

   c. Protein Synthesis. Few data exist at present on protein synthesis by the isolated cells, except that, as *in vivo*, incorporation of labeled amino acids into protein is two- to threefold faster in the neurons than in the glia.[16] No details of the types of protein produced, however, are available.

IV. NEURONAL–GLIAL INTERRELATIONSHIPS

   On the basis of this cursory survey, and evidence discussed elsewhere in this book, certain comments can be made on the role of the neuronal/glial interrelationship. The biochemistry of the neuron is characteristic of a rapidly metabolizing cell faced with the problems of transporting essential metabolites over long distances and of fabricating a number of specialized membrane components, proteins, and transmitters. Thus, the problem of the neuronal/glial interrelationship resolves itself inevitably into one of the function of the glia.
   A wide variety of roles has been suggested for these cells[1]: that they act as structural support or insulation; that they form a neuronal extracellular space, transporting metabolites between the neurons and the capillaries; that they are "high sodium" cells, absorbing the potassium released by the neuron during impulse propagation and, as a corollary, that they are "potassium pumps," transferring the ion from one firing neuron to another not synaptically connected to it, hence causing a spread of current; that they fabricate metabolites that the neuron cannot and subsequently transfer them; that they act as synaptic regulators, either fabricating or absorbing transmitter substances; or that they act as the units of memory storage—all this apart, of course, from the known role of the oligodendroglia in myelination.[1] Here we only can summarize these possibilities from the biochemical viewpoint.
   It is necessary to distinguish a separate role for the several different glial subspecies. The small microglia are not of ectodermal origin and are probably analogous to macrophages in other tissues. They are generally assigned a scavenger role in pathological conditions. Other glial types are astrocytes and oligodendroglia. Biochemical preparations that attempt to distinguish them are Hamberger's[39] microdissection into "perineuronal" and "capillary" glia, and certain histochemical techniques.[2] While these do suggest differences in the levels of certain oxidative enzymes, and perhaps lend support to the suggestions that oligodendroglia have a higher oxidative

metabolism than astrocytes,[41] these data neither confirm nor refute a role of the astrocytes in regulating ionic environment.

The rationale for such a role is based on the suggestive relationship of the astrocytes to the capillaries and the apparent problem of the lack of extracellular space in brain. However, it is now generally accepted that the brain does have a nonneuronal, nonglial, extracellular space between 10 and 15%,[42] quite sufficient to account for ionic changes during passage of impulses and to provide a channel for passage of metabolites. Kuffler[43] has shown that the extracellular space of the leech central nervous system does indeed fill this role, although the problem of extrapolating from leech to mammal must be stressed. In the absence of either supporting evidence or manifestly overriding theoretical need one must reject the need for a passage of metabolites through glia, an energy-requiring and rather complex process. While it remains possible that the astrocytes may take up certain substances selectively, there are no relevant biochemical data to this.

Direct evidence for transfer of metabolites between neurons and glia does not exist, but there are numbers of suggestive pieces of evidence. First, there are the series of reciprocal changes that Hydén and his group have observed in cells microdissected from animals under various conditions of behavioral stimulation[19-22] during which he found increases of 5% in neuronal RNA and 12% in neuronal protein while glial RNA decreased by 60%: Glial cytochrome oxidase decreased by 74% and neuronal cytochrome oxidase increased by 60%. Similar reciprocal changes in enzyme levels during the sleep/wakefulness cycle also have been found.[29]

The magnitude of these effects is surprising, and it is equally difficult to be sure of the meaning at least of the enzymatic changes, because it would be surprising if the oxidative systems of both cell species were not already present in substantial excess. Hydén and Lange[29] suggest that the changes in protein and enzyme levels might be mediated by the direct transfer of RNA molecules or nucleotides from glia to neurons, and the advantage of this system would be that it would confer stability on the neuronal and glial unit. However, the changes Hydén observes are between neuronal perikarya and the surrounding glial fraction. This glial preparation also contains a certain amount of dendrites and axonal material, which contain RNA and possess a protein metabolism. Thus, the changes in RNA and protein also could be accounted for on the basis of intracellular transfer of material between different regions of the neuron, rather than an intercellular transfer.[44] Another possible transfer of metabolites may occur with the amino acids, fabricated in the glia but stored in the neurons, as discussed above. An alternative model, in which the glia have a regulating role in amino acid and transmitter metabolism at the synapse, instead of the cell body, has been suggested by de Robertis.[45]

However, such models still are largely inferential in nature and lack adequate biochemical verification or, for that matter, clear logical necessity. At present, it is still not possible either to answer many of the fundamental questions on the role of the glia nor to accept as fully satisfactory any of the several hypotheses of the nature of neuronal/glial interrelationships. Indeed, it is probably true that we do not yet even have available a full set of methodological keys to the problem.

V. NOTE ADDED IN PROOF

Since this chapter was written there have been a number of further developments in separation techniques. Fewster et al.[46] have used a modification of the method described here for isolation of glial cells from rat and bovine white matter, and have determined the lipid content of the isolated cells.[47] Flangas and Bowman[48] have described a modified method, utilizing a stainless steel tissue press, for preparing cell suspensions in bulk for zonal centrifugation. However, Cremer et al.[49] have reported difficulties in replicating the original method and have claimed that their neuronal fraction is contaminated with capillary endothelial cells. However, detailed cell counts suggest[13] that Cremer's estimate of such contamination is excessive. These cells represent a more significant contamination of the neuronal fraction than do the glia, but of all cells visible in the fraction 64% could be identified as neuronal, compared to 20% in the starting material. Some problems, however, can arise in the purification if care is not taken to standardize the Ficoll, as batches of this material vary widely in their properties.

With respect to amino acid metabolism, interesting additional data on compartmentation of glutamate, glutamine, GABA, and aspartate have been provided, by use of ablation and degeneration techniques, by Margolis et al.[50] The results suggestive of differential pool sizes and rates of turnover described above in vitro[18] have now been extended to the in vivo situation as well.[51]

VI. REVIEWS

R. Galambos (ed.), Glial cells, Neurosci. Res. Bull. II(6):1–64 (1964).

H. Hydén, in The Cell (J. Brachet and A. E. Mirsky, eds.), Vol. IV, pp. 215–325, Academic Press, New York (1960).

S. W. Kuffler and J. G. Nicholls, The physiology of neuroglia cells, Ergeb. Physiol. 57:1–90 (1966).

S. P. R. Rose, in Applied Neurochemistry (A. Davison and J. Dobbing, eds.), pp. 332–355, Blackwells, Oxford (1968).

W. F. Windle (ed.), Biology of Neuroglia, Charles C. Thomas, Springfield, Illinois (1958).

VII. REFERENCES

1. R. Galambos (ed.), Glial cells, Neurosci. Res. Bull. II(6):1–64 (1964).
2. G. Lehrer and H. Maker, Isolated cells, in Handbook of Neurochemistry, Vol. V (A. Lajtha, ed.), Chapter 12, Plenum Press, New York, in preparation.
3. E. Giacobini, Chemistry of isolated invertebrate neurons, in Handbook of Neurochemistry (A. Lajtha, ed.), Vol. II, pp. 195–239, Plenum Press, New York (1969).

4. C. Rappaport, Further studies on the disassociation of adult mouse tissues, *Proc. Soc. Exptl. Biol. Med.* **121**:1016–1021 (1966).

5. M. Satake and S. Abe, Preparation and characterisation of nerve cell perikaryon from rat cerebral cortex, *J. Biochem.* (*Tokyo*) **59**:72–75 (1966).

6. S. P. R. Rose, Preparation of enriched fractions from cerebral cortex containing isolated, metabolically active, neuronal and glial cells, *Biochem. J.* **102**:33–43 (1967).

7. S. R. Korey, *in Biology of the Neuroglia* (W. F. Windle, ed.), pp. 203–217, Charles C. Thomas, Springfield, Illinois (1958).

8. B. I. Roots and P. V. Johnston, Neurons of ox brain nuclei: Their isolation and appearance by light and electron microscopy, *J. Ultrastruct. Res.* **10**:350–361 (1964).

9. V. Bocci, Enzyme and metabolic properties of isolated neurones, *Nature* **212**:826–827 (1966).

10. T. C. Johnson and M. W. Luttges, The effects of maturation on *in vitro* protein synthesis by mouse brain cells, *J. Neurochem.* **13**:545–552 (1966).

11. L. Freysz, R. Bieth, C. Judes, M. Sesenbrenner, M. Jacob, and P. Mandel, *J. Neurochem.* Distribution quantitative des divers phospholipides dans les neurones et les cellules gliales isolés du cortex cérébral de rat adulte, **15**:307–315 (1968).

12. E. Giacobini, *in Morphological Correlates of Neural Activity* (M. M. Cohen and R. S. Snyder, eds.), pp. 15–29, Harper & Row, New York (1964).

13. S. P. R. Rose, Some properties of isolated neuronal cell fractions, *J. Neurochem.* (1969), in press.

14. H. F. Bradford and S. P. R. Rose, Ionic accumulation and membrane properties of enriched preparations of neurons and glia from mammalian cerebral cortex, *J. Neurochem.* **14**:373–375 (1967).

15. H. Hillman and H. Hydén, Membrane potentials in isolated neurons *in vitro* from Deiter's nucleus of rabbit, *J. Physiol.* **177**:398–410 (1965).

16. B. Tiplady and S. P. R. Rose, unpublished results.

17. S. P. R. Rose, Amino acid metabolism in isolated neuronal and glial cells, *Biochem. J.* **102**:21P (1967).

18. S. P. R. Rose, Glucose and amino acid metabolism in isolated neuronal and glial cell fractions *in vitro*, *J. Neurochem.* **15**:1415–1429 (1968).

19. H. Hydén, *in The Cell* (J. Brachet and A. E. Mirsky, eds.), Vol. IV, pp. 215–325, Academic Press, New York (1960).

20. H. Hydén and A. Pigon, A cytophysiological study of the functional relationships between oligodendroglial cells and nerve cells of Deiter's nucleus, *J. Neurochem.* **6**:57–72 (1960).

21. H. Hydén and E. Egyházi, Nuclear RNA changes of nerve cells during a learning experiment in rats, *Proc. Natl. Acad. Sci.* **48**:1366–1372 (1962).

22. H. Hydén and E. Egyházi, Glial RNA changes during a learning experiment in rats, *Proc. Natl. Acad. Sci.* **49**:618–624 (1963).

23. J. Jarlstedt, *in Macromolecules and the Function of the Neuron* (Z. Lodin and S. P. R. Rose, eds.), pp. 321–333, Excerpta Medica, Amsterdam (1968).

24. H. Hydén and B. McEwen, A glial protein specific for the nervous system, *Proc. Natl. Acad. Sci.* **55**:354–358 (1966).

25. A. N. Davison, M. L. Cuzner, N. C. Banik, and J. Oxberry, Myelinogenesis in the rat brain, *Nature* **212**:1373–1374 (1966).

26. L. Freysz and M. Sensenbrenner, *Proc. First Meeting Intern. Soc. Neurochem.*, Strasbourg, 1967.

27. R. L. Friede, *Topographic Brain Chemistry*, Academic Press, New York (1966).

28. H. H. Hess and A. Pope, Intralaminar distribution of cytochrome oxidase activity in human frontal isocortex, *J. Neurochem.* **5**:207–217 (1958).

29. H. Hydén and P. W. Lange, Rhythmic enzyme changes in neurons and glia during sleep, *Science* **149**:654–656 (1965).

30. J. G. Salway, M. Kai, and J. N. Hawthorne, Triphosphoinositide phosphomonoesterase in nerve cell bodies, neuroglia and subcellular fractions from whole rat brain, *J. Neurochem.* **14**:1013–1025 (1967).

31. J. Altman and G. D. Das, Autoradiographic and histological studies of postnatal neurogenesis, *J. Comp. Neurol.* **126**:337–390 (1966).

32. W. E. Watson, An autoradiographic study of the incorporation of nucleic acid precursors by neurons and glia during nerve stimulation, *J. Physiol.* **180**:754–765 (1965).

33. P. Volpe and A. Giuditta, Kinetics of RNA labelling in fractions enriched with neuroglia and neurons, *Nature* **216**:154 (1967).

34. L. Hertz, Neuroglial localisation of potassium and sodium effects on respiration in brain, *J. Neurochem.* **13**:1373–1387 (1966).

35. M. H. Epstein and J. S. O'Connor, Respiration of single cortical neurons and of surrounding neuropile, *J. Neurochem.* **12**:389–395 (1965).

36. S. R. Korey and M. Orchen, Relative respiration of neuronal and glial cells, *J. Neurochem.* **3**:277–285 (1959).

37. H. H. Hess, in *Regional Neurochemistry* (S. Kety and L. Elkes, eds.), pp. 200–212, Pergamon Press, Oxford (1961).

38. O. H. Lowry, N. R. Roberts, K. Y. Leiner, M. L. Wu, A. L. Farr, and R. W. Albers, The quantitative histochemistry of brain. III. Ammons horn, *J. Biol. Chem.* **207**:39–49 (1954).

39. A. Hamberger, Differences between isolated neuronal and vascular glia with respect to respiratory activity, *Acta Physiol. Scand.* **58**: *Suppl. 203*, 1–67 (1963).

40. D. Garfinkel, A simulation study of the metabolism and compartmentation in brain of glutamate, aspartate, the Krebs cycle and related metabolites, *J. Biol. Chem.* **241**:3918–3929 (1966).

41. L. G. Abood, R. W. Gerard, J. Banks, and R. D. Tschirgi, Substrate and enzyme distribution in cells and cell fractions of the nervous system, *Am. J. Physiol.* **168**:728–738 (1952).

42. H. Pappius, Water spaces, in *Handbook of Neurochemistry* (A. Lajtha, ed.), Vol. II, pp. 1–10, Plenum Press, New York (1969).

43. S. W. Kuffler and J. G. Nicholls, The physiology of neuroglia cells, *Ergeb. Physiol.* **57**:1–90 (1966).

44. J. E. Edström, in *Macromolecules and the Function of the Neuron* (Z. Lodin and S. P. R. Rose, eds.), p. 331, Excerpta Medica, Amsterdam (1968).

45. E. de Robertis, in *Nerve as Tissue* (K. Rodahl and B. Issekutz, eds.), pp. 88–115, Hoeber, New York (1966).

46. M. E. Fewster, A. B. Scheibel, and J. F. Mead, The preparation of isolated glial cells from rat and bovine white matter, *Brain Res.* **6**:401–408 (1967).

47. M. E. Fewster and J. F. Mead, Fatty acid and fatty aldehyde composition of glial cell lipids isolated from bovine white matter, *J. Neurochem.* **15**:1303–1312 (1968).

48. A. L. Flangas and R. E. Bowman, Neuronal perikarya of rat brain isolated by zonal centrifugation, *Science* **161**:1025–1027 (1968).

49. J. E. Cremer, P. V. Johnston, B. I. Roots and A. J. Trevor, Heterogeneity of brain fractions containing neuronal and glial cells, *J. Neurochem.* **15**:1361–1370 (1968).

50. R. K. Margolis, A. Heller, and R. Y. Moore, Effects of changes in cellular composition following neuronal degeneration on amino acids in the brain, *Brain Res.* **11**:19–31 (1968).

51. S. P. R. Rose, Isolated brain cell fractions and the Waelsch effect, *FEBS Abstracts* (1969), p. 182.

Chapter 10

CHEMISTRY OF ISOLATED INVERTEBRATE NEURONS*

Ezio Giacobini

Department of Pharmacology
Karolinska Institutet
Stockholm, Sweden

I. ADVANTAGES OF INVERTEBRATE OVER VERTEBRATE PREPARATIONS AND DIFFERENT TYPES OF ISOLATED INVERTEBRATE NEURON PREPARATIONS

In neurobiology, perhaps more than in any other field of biology, there are considerable advantages in dealing with "simpler" systems and with "model" systems. Even in the lower invertebrates the degree of organization and differentiation of the nervous system is sufficiently high for the fundamental mechanisms of transmission, conduction, and integration to be studied. However, some scientists still hesitate to compare the properties of neurons of the vertebrate central nervous system with those of the invertebrate nervous system.

Physiological, biochemical, and pharmacological differences do exist between vertebrate and invertebrate neurons, but they consist mainly of differences in pharmacological sensitivity and some chemical characteristics rather than in the fundamental mechanism of function and mode of response of the cells to transmitters and drugs.

The electrophysiological properties of all neurons appear to be almost identical and comparable with those of electrogenetic cells in general. The advantages of using invertebrate neurons are especially pronounced when it comes to the analysis of individual cells. These advantages are summarized in Table I.

The nervous systems of invertebrates offer many examples of cells that easily can be isolated by means of simple tools under a dissection microscope

---

* This chapter is dedicated to the memory of the late Professor K. Linderstrøm-Lang of the Carlsberg Laboratory, Copenhagen, who, by contributing to the development of most precise and sensitive methods of quantitative cytochemistry, made these cellular investigations possible.

TABLE I

Main Advantages of Invertebrate Over Vertebrate Cell Preparations for Neurochemical or Neuropharmacological Studies

1. They are more easily separated from the nervous system by dissection.
2. The cell bodies are generally larger (diameter up to 1 mm). This means more material for chemical analysis.
3. They retain their functional activity after dissection and survive for several hours—even days.
4. The functional activity can be tested under direct inspection by means of either intra- or extracellular recording. The recording can be made simultaneously in different parts of the cell, and the cell can be physiologically stimulated.
5. The environment of the cell can be changed by adding or substituting ions or by adding metabolites, inhibitors, toxins, drugs, etc. in known concentrations. The alterations in the composition of the external medium may be readily controlled.
6. The membrane is readily accessible to different molecules (see point 5 above).
7. There are good possibilities of studying neuron–glia relationships. Easier isolation is possible of large amounts of glial material.
8. Easy identification occurs of the same neuron in different preparations.
9. Smaller number of neurons and simpler organization of different pathways exists.
10. There is a presence of large synapses, some of which can be isolated for direct chemical analysis, and a presence of large axons (diameter up to 1.7 mm).

at a relatively low magnification. The same cells also can be maintained after isolation *in vitro* in a satisfactory functional condition. The cell body of invertebrate neurons is often larger than that of vertebrate neurons (the diameter can be as much as 800 μ, see Table IIa) and is easily visualized after cutting the translucent membrane that covers the ganglion. By means of fine and sharp needles, the single elements may be dissected out of the ganglion. This operation can be performed with fresh preparations, or the ganglion may be frozen and sliced into thin sections (10–20 μ) that are then lyophilized. From these sections single cells can be cut out using micro-instruments of the type described by Lowry.[1]

Fortunately, in some cases nature has provided us with neurons that are "already isolated" in the nervous system of the animal. The best-known example of such a preparation is the stretch receptor of crustacea.

When one is dealing with problems of correlation between chemical or physical aspects and function in the nerve cells, it is of primary importance that the unavoidable mechanical injury, the ionic distortion, and the deficiency in oxygen and natural metabolites, occurring during manipulation, is minimized. In this respect mammalian preparations are rather vulnerable.

Another peculiar property of the invertebrate nervous system is the presence of elements with different electrical characteristics, like slow and fast components, already naturally separated and identifiable. This gives us the opportunity of studying physico-chemically different types of neuronal membrane.

The relatively simple organization of invertebrate material like ganglia makes it generally easier to stimulate or inhibit selectively a few cells or a single cell. Therefore, the study of pathways of inhibition and excitation in small groups of cells is particularly favorable since the single unit of the circuit may be studied in isolation. The arthropod nervous system, e.g., offers the unique possibility of studying a complete circuit of different neurons controlling a series of movements. Cellular changes associated with the process of learning also can be investigated in isolated systems, like the ganglia of the cockroach,[2] or in single cells.

Four main groups of cell preparation can be separated from the nervous systems of invertebrates: (1) single large cell bodies with a short proximal segment of the axon; (2) single giant axons; (3) neuromuscular junctions; and (4) giant synapses.

A fifth group of preparations consists of glial material and will be discussed separately. Tables IIA and IIB give details of the four above-mentioned groups together with the related species, the dimensions of the preparation, and the particular structure studied. The preparations reported in Tables IIA and IIB so far have been used mostly for physiological and pharmacological, rather than chemical, studies.

Isolated cell body preparations can be obtained by free-hand dissection under low magnification (dissection microscope) of fresh material (e.g., the stretch receptor neuron) or from lyophilized slices. They may contain other components besides the cell body and the proximal segment of the axon. Contamination with these components may be difficult to avoid and therefore should be taken into consideration when evaluating the results. The most common contaminants are glial and Schwann cells, axons or fibers belonging to other neurons, and synaptic boutons sticking to the cell body. The presence of other axons or fibers can be avoided by careful dissection, and the synaptic boutons can be eliminated by presynaptic denervation (when this is practicable).

It is more difficult to avoid the contamination with glia since it is practically impossible to obtain neuronal material *completely* free of glia and vice versa. The two constituents, glia and neurons, are so tightly intermingled as to make complete separation virtually impossible. However, the invertebrate nervous system offers much better possibilities from this point of view than the CNS of vertebrates (Table I).

In some cases, in order to preserve the functional integrity of the nerve cell to be studied, the chemical analysis is made (by means of physical methods) inside the living structure. For example, cell preparations from Helix and Aplysia have been used by several authors[3,4] for spectroscopic analysis. These authors were able to follow in the living neuron (inside the ganglion) the activity of respiratory enzymes and compare it with the electrical response.

The stretch receptor neuron is probably the only "isolated" and functional cell preparation to have been extensively investigated. In fact, most of the examples reported in the literature refer to studies made on single cells still connected to the intact ganglion *in vivo* or *in vitro*.

Table IIB lists a series of axon preparations including several types of so-called "giant" axons having diameters up to several hundred microns. Such preparations have been used mainly for electrophysiological and pharmacological studies, but they also should be suitable for chemical analysis. Equally valuable may be the "giant" synapses found in the invertebrate nervous system and the neuromuscular junctions (Table IIB).

The enormous variety of invertebrate cell preparations and the amount of knowledge obtained concerning their physiology is well documented in the comprehensive handbook of Bullock and Horridge.[5]

TABLE IIA

Invertebrate Preparations for Cytochemical Studies—Cell Bodies

Species	Preparation	Cell body diameter (μ)
Crustacea		
Astacus fluviatilis (crayfish)	Stretch receptor (slow-adapting neuron)	50–75
Astacus astacus (crayfish)	Stretch receptor (slow-adapting neuron)	50–85
Homarus americanus (lobster)	Stretch receptor (slow-adapting neuron)	75–120
Insecta		
Periplaneta americana (cockroach)	Thoracic ganglion	50–100
Mollusca		
Aplysia	Visceral ganglion	400–800
Helix aspersa (snail)	Visceral ganglion	40–320
Helix pomatia (snail)	Parietal ganglion	260–400
Tritonia	Cerebral ganglion	500–1000
Archidoris	Cerebral ganglion	100–300
Anisodoris	Cerebral ganglion	100–300
Octopus (squid)	Cerebral ganglion	10–40
Loligo (squid)	Cerebral ganglion	50–120
	Pedal ganglion	150–750
Annelida		
Lumbricus (earthworm)	Ventral nerve cord	30–50
Hirudo (leech)	Subesophageal ganglion Ventral nerve cord	35–50

TABLE IIB

Invertebrate Preparations for Cytochemical Studies; Giant Axons, Giant Synapses, and Neuromuscular Junctions

Species	Preparation	Giant axons diameter (μ)
Crustacea		
Homarus americanus (lobster)	Peripheral nerve—inhibitory and motor axons—ventral nerve cord	40–75 (up to 12 cm long)
Carcinus maenas (crab)	Leg—peripheral nerve	30
Procambarus clarkii (crayfish)	Ventral cord—giant axon—nerve cord	100–250
Insecta		
Periplaneta americana (cockroach)	Giant fiber system—giant axon— nerve cord	20–45
Mollusca		
Aplysia	Peripheral nerves	25–50
Loligo pealii (squid)	Giant nerve fiber system—giant axon	500–700
Loligo forbesi (squid)	Giant nerve fiber system—giant axon	700–900
Sepia officinalis (cuttlefish)	Giant nerve fiber system—giant axon	100–250
Annelida		
Lumbricus (earthworm)	Ventral cord—dorsal giants	75–100
Hirudo medicinalis (leech)	Ventral cord—dorsal giants	60–90
Myxicola	Ventral cord—dorsal giants	100–1700 (largest known nerve fiber)
Cambarus (crayfish)	Septal segmental—axon to axon	150–250
Loligo pealii (squid)	Stellate ganglion—axon to axon	50–1000
Lumbricus (earthworm)	Septal segmental—axon to axon	150–200
Lobster, crayfish, crabs, cockroach, snail	Various locations	250–500

II. CORRELATIONS BETWEEN FUNCTIONAL ACTIVITY AND CHEMICAL CHANGES IN INDIVIDUAL INVERTEBRATE NEURONS

In a critical evaluation of the advantages and disadvantages of single cell preparations, we have discussed elsewhere[6a,6b,7] the basic requirements for cytochemical analysis. We have stressed the fact that either the functional integrity of the neuron should be maintained after dissection or the metabolism of the nerve cell should be stopped abruptly by means of rapid freezing.[1] The first alternative is of greater importance for direct neurophysiological and neuropharmacological correlations, while the second represents a necessary step in the analysis of labile metabolites.

The importance of comparing the effect of physiological or electrical stimulation on individual neurons of comparable size, weight, and function also has been emphasized, and criticism has been made of the experiments performed on neurons from different regions of the vertebrate CNS.[8] The most serious objection is that it is difficult to correlate functional activity and chemical changes in neurons randomly sampled from areas of the brain containing many thousands of different neurons. Kuffler and Nicholls[8] stress the point that "in pools of neurons studied electrophysiologically a stimulus usually activates some cells and suppresses the firing of others," quoting the studies of Hubel and Wiesel.[9] In view of this, it cannot be assumed that an "increased functional activity is equally present in all the cells of a particular area."[8]

The electrophysiological techniques and the sampling methods presently available are still incapable of providing the basis for the selection and isolation of any particular individual cell in a group of neurons of the CNS following a physiological stimulus. This would require a preparation in which (1) the impulse activity and the other electrical parameters are related to the physiological stimulation or to the functional state of the cell monitored before, during, and after the isolation and (2) the cell is isolated without any functional, structural, or chemical changes.

The use of invertebrate neurons, as previously pointed out,[6a,6b,10] eliminates most of the problems mentioned above since the cell can be stimulated and the impulse activity recorded under easily controllable conditions, while the metabolism can be studied in the frozen and lyophilized cell or even during impulse activity.[5,11] In this way the functional state of the cell to be chemically analyzed is checked before or even during the analysis itself.

Another possibility could be direct observation of chemical and physico-chemical changes in isolated neurons, still connected to the nervous system.[12]

III. THE CRUSTACEAN STRETCH RECEPTOR NEURON (SRN) AS A MODEL FOR BIOCHEMICAL AND PHARMACOLOGICAL STUDIES IN SINGLE CELLS

A. General Features

The stretch receptor neuron of the crustacea is probably the most extensively investigated invertebrate nerve cell. This cell exhibits most of the advantages of invertebrate over vertebrate cell preparations.

The SRN represents one of the few examples of a neuron that is completely isolated from the rest of the nervous system. This cell is regarded by neurophysiologists as an excellent model of a neuron, showing all the typical electrical characteristics and properties of nerve cells in general.

The SRN was discovered in 1951 by Alexandrowitz.[13] The receptors occur in pairs on both sides of each abdominal segment and can be easily dissected (Fig. 1). They are capable of producing regular and sustained impulse activity when physiologically stimulated *in vitro*, their activity can

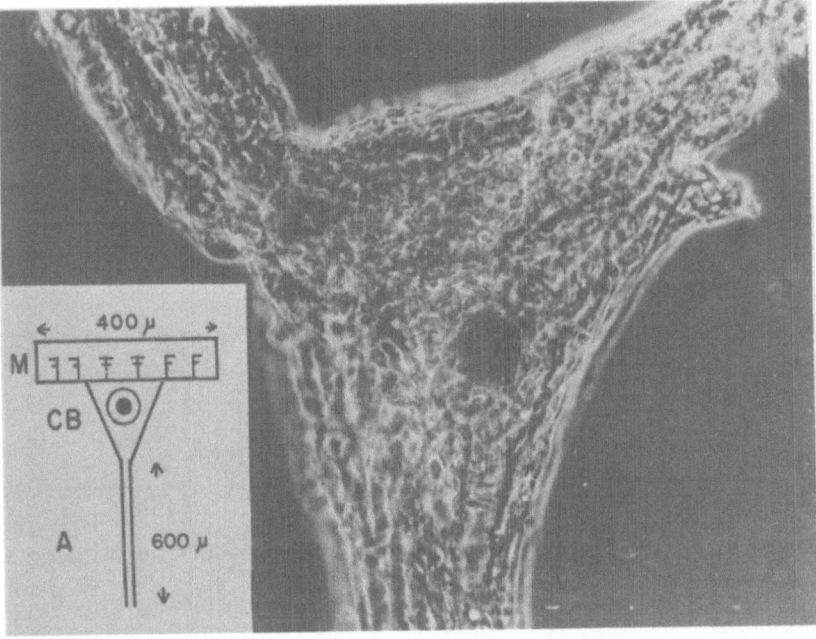

Fig. 1. Photomicrograph of an unstained functional preparation of the slowly adapting stretch receptor cell of the crayfish. The thin muscle fiber (upper part) and the nucleus with the nucleolus in the center of the cell body and the axon (lower right) are visible. The glial tissue surrounding the cell body has not been teased out. Left corner: simplified diagram of the elements of the standard functional preparation with its dimensions. M, muscle fiber (dendritic zone); CB, cell body with nucleus and nucleolus; and A, proximal segment of the axon. The SRN has been dissected by hand and suspended in Van Harreveld solution.

be easily recorded both extra- and intracellularly, and they can survive for a period of at least 24 hr in a completely isolated state. The SRN does not show any substantial morphological difference from other types of nerve cell, except for the large dendrites, which are embedded in a modified muscle fiber and surrounded by connective tissue (Fig. 1). In the dendritic region the muscle fiber does not show any typical muscular structure. The dendrites, when stretched, give rise to a generator potential, and they contain very peculiarly located and conspicuous mitochondria, a characteristic with a particular relevance to both the neurophysiologist and the biochemist, as a potential source of energy.

The standard preparation used in our experiments[10] is shown in Fig. 1. It consists of the cell body of the slowly adapting SRN organ of the crayfish, a segment of axon approximately 600 μ long and a portion of modified thin muscle bundle, about 200 μ long, in which the dendrites are embedded. This has been shown to be the "minimal" preparation compatible with survival and functional activity.[10] The contribution made by the modified muscle fiber to the total metabolism of the preparation can be estimated in separate experiments.

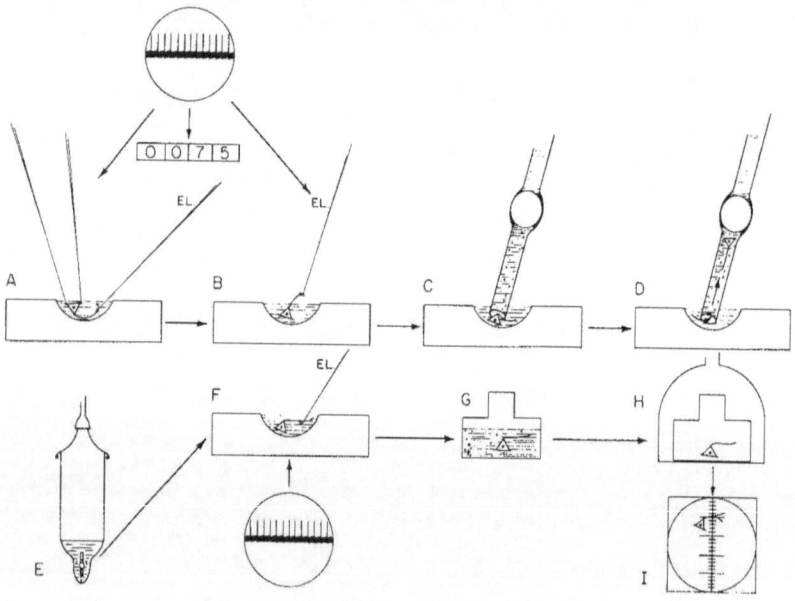

Fig. 2. Simplified diagram of the experimental procedure for measurement of O_2 uptake and enzyme activity in single nerve cell preparations at rest and during impulse activity. A and B: test of the functional activity of the isolated cell. C and D: suction of the cell into the microdiver. E: measurement of O_2 uptake or enzyme activity. F: retesting the functional activity after the chemical analysis. G and H: extraction of lipids or freeze-drying of the cell. I: weighing the cell with the quartz fiber microbalance. From Giacobini *et al.*[10] Copyright 1963 by the American Association for the Advancement of Science.

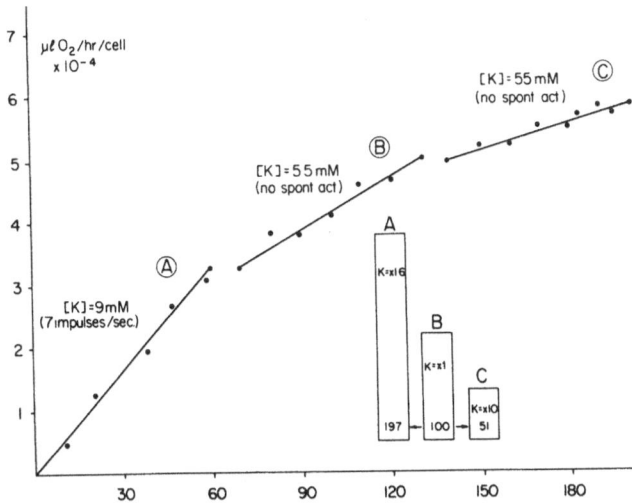

Fig. 3. A three-stage experiment using the Cartesian diver. O_2 uptake in a single crustacean nerve cell preparation in a 9 mM K-solution (1.6 times normal) (A), in a 5.5 mM K-solution (physiological) (B), and in 55 mM K-solution (10 times normal) (C). Bar graphs show the final relationship as percentages. Bar graph B is O_2 uptake with physiological K-concentration taken as 100. The figures 197 and 51 represent the percentage changes in the O_2 uptake. From Giacobini.[6a]

Before each chemical analysis, the receptor is isolated by microdissection and its ability to respond to physiological stimuli (stretch), with repetitive impulse activity, is tested (Fig. 2). The preparation is then reduced to the minimal functional dimensions, and its physiological integrity is checked by recording impulse activity from the axon with a 20 μ wire electrode (Fig. 2). In some cases, although no forceps can be applied to the muscle to produce stretch, after the preparation has been cut to the stated minimal dimensions, stretch can still be applied by raising the axon, due to the surface tension of the solution (Fig. 2). The impulse activity is displayed on an oscilloscope and audiomonitored (Fig. 2). In isolated conditions both in a small dish (containing physiological Van Harreveld solution) or inside the Cartesian diver, the cell can still be stimulated simply by varying the K and Ca concentrations in the medium (Figs. 2 and 3). This method is utilized when the cell metabolism is measured inside the Cartesian diver (Fig. 2), where it would be difficult to apply a mechanical stimulus (Fig. 3). The frequency of the impulse activity then must be recorded before and after the metabolic measurements (Fig. 2). This represents one of the few experiments in which it is possible to apply stimuli and measure metabolism in a living single nerve cell. The fluorescence method of Chance et al.[14,15] allows only measurements of fluorescence changes whereas the Cartesian diver can be used for respiration as well as

TABLE III

The Crayfish Stretch Receptor Neuron—Physical and Metabolic Characteristics[a]

	Total	Cell body	Axon	Dendritic region	Muscle	Glia
Dimensions (diameter in μ)[a]						
Small crayfish		58.6 (23)	8–15 (21)	55 (13)	33 (11)	
Large crayfish		85 (10)	13–20 (12)	87.7 (12)	57 (11)	
Volume ($\mu^3 \times 10^3$)[a]	~500					
Weight (μg)						
Fresh	3.5 (8)					
Dry	0.58 (49)					
Lipid-free dry	0.55 (46)	0.18 (19)	0.085 (20)	0.23 (12)	0.21 (12)	0.055 (10)
Percent of total	100	33	15	42	38	10
Water content						
Percent of fresh wt.	83.4					
O_2 uptake						
μl × 10^{-4}/hr	4.8 ± 1.8 (36)	2.72[b]	0.75 ± 0.05 (28)	1.06 ± 0.09 (14)	0.65 ± 0.03 (18)	0.27 ± 0.08 (9)
Percent of total	100	66[b]	16	22	14	6
μl/mg/hr (dry wt.)	0.9	1.5	0.9	0.5	0.3	0.5
RNA[c]		2233 ± 105 (17)				2935 ± 225 (16)
Concentration percent (wt./vol.)		0.5	0.06			
Cations and electric properties	$[Na^+]_i$ (mM/liter)	$[K^+]_i$ (mM/liter)	$[Na^+]_o$ (mM/liter)	$[K^+]_o$ (mM/liter)	E_m^d (mV)	E_a^d (mV)
	12–17.4[e]	180–265[e]	205.3	5.4	61–80	–16 – 20
Ratio	$[K^+]_i/[Na^+]_i = 15$		$[K^+]_o/[Na^+]_o = 41$			
Anions	$[Cl]_i = 12.7^e$ mM/liter					

[a] In parentheses, number of samples. Species studied: *Orconectes virilis, Astacus astacus,* and *Astacus fluviatilis.* The dimension measurements were taken on the mounted receptor in relaxed conditions. The dendritic region of the muscle shows the maximal thickness. The volume was calculated using the above dimensions.
[b] Computed data.
[c] Ref. 18, picograms, lobster.
[d] Refs. 37, 38; E_m = resting potential; E_a = action potential.
[e] Ref. 24, axoplasm.

enzyme and metabolic measurements.[6a,6b,10] The quantitative methods of biochemical analysis in cellular units used in our laboratory are shown in Fig. 6. A critical review of these microtechniques has been given elsewhere.[7]

Table III lists some physical and metabolic characteristics of the SRN. The dimensions of the cell body and axon of the SRN are quite large compared with other invertebrate cells (Tables IIA and IIB). The SRN of crayfish is somewhat smaller than that of the lobster, but we found it preferable since it contains less glial tissue surrounding the cell body itself. For our studies we selected several types of crayfish, and we found that the slowly adapting neurons of the crayfish Astacus astacus or Astacus fluviatilis, were especially useful. In small crayfish (Table III) the cell body is somewhat smaller than in large crayfish, but it has the advantage of containing less glial tissue. However, the ratio cell body/muscle diameter is higher in small animals (3–4 cm body length) than in larger ones (Table III). In our samples the lipid-free dry weight (Table III), mean value = 0.55 μg, varied between 0.24 and 0.84 μg. The weight of the cell was determined with the quartz fiber balance of Lowry.[1] Up to eight SRN can be dissected from the same animal. Generally, both right and left slowly adapting neurons are dissected, and one of them is used as a control. The modified muscle fiber and glia constitute the main contamination of this preparation. The contribution of these tissues to the whole metabolism of the preparation has to be taken into account when calculating concentrations of metabolites or enzyme activity. In terms of dry weight the muscle part represents about one-third of the total and the glia about one-tenth; however, in terms of metabolism (O_2 uptake) the contributions of the muscle and glia are only 14 and 6% respectively (Table III). The ratio of the metabolism in the nervous to nonnervous part is therefore on the order of 4:1.

Generally, single cell preparations from vertebrates have a weight on the order of 5–20,000 pg compared to more than 500,000 pg in the SRN (for comparison see Giacobini).[7] The percentage water content of the SRN is of the same order as other crustacean nervous tissues (for comparison see Treherne).[16] Arachnida and insecta generally have a lower water content, i.e., 61–83% fresh wt.[16] The volume of the cell body of the unfixed slowly adapting nerve cell of the lobster is about 500,000 μ^3. According to Grampp and Edström,[17,18] it contains an average of 2333–2800 pg of RNA, i.e., 0.5% (wt./vol.). This concentration is somewhat lower than that in nerve cells of mammals.[19] The isolated axon of the same preparation has an RNA concentration of only 0.06%, which is of the same order of magnitude as has been reported for the axon of the Mauthner neuron.[20] Generation and conduction of up to 100,000 spikes during several hours do not alter the total amount of RNA in the nerve cell body but significantly increase the adenine: uracil and purine: pyrimidine ratios.[17,18] Likewise, there is no significant impairment of activity in the presence of puromycin under conditions where inhibition of protein synthesis is complete.[21] Actinomycin D, in concentrations completely inhibiting RNA synthesis, has no effect on the impulse activity during periods up to 12 hr and after a total number of impulses up to

1 million.[21] Neither puromycin nor actinomycin affected the respiratory activity of crayfish SNR[22] (Table IX).

The data on the cation concentrations in the SRN (Table III) refer to studies on the total preparation[6a,23] and to studies with axoplasm.[24] The Cl^- concentrations refer to axoplasm values obtained by Wallin.[24]

B. Oxygen Consumption and Energy Metabolism Studied in Single Invertebrate Neurons

In the SRN the O_2 uptake in microliters per milligram dry weight per hour at rest is of the same order of magnitude as in several vertebrate preparations. Some vertebrate preparations, like the superior cervical ganglion of the rat[25] and the sciatic nerve of rat,[26] show comparatively higher O_2 uptake.

A marked increase in oxygen consumption can be observed and is related to the frequency of impulse activity.[6a,10,27] This effect, which may be as high as 120% at 15–18 impulses/sec, is a function of the impulse frequency (Fig. 4) and is not dependent on the external K^+ (Fig. 3) and Ca^{2+} concentrations. This could be proved by obtaining different frequencies of stimulation in the same cell, in the presence of different concentrations of K^+ and Ca^{2+}.[6a,25] The lowest rate of respiration found in the literature is that of unmyelinated mammalian C fibers[28] and corresponds to about one-fifteenth of the SRN respiration. The oxygen uptake increases after stimulation (3 impulses/sec) by about 65% while in the SRN it increases by up to 120%. The ionic imbalance that occurs after activity is particularly large in C fibers. However, according to Ritchie[28] the oxygen consumption in the sciatic nerves in vitro is fully adequate for complete recovery.

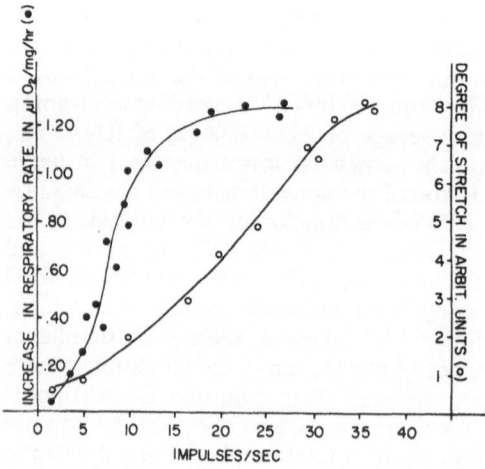

Fig. 4. Relationship between increase in respiratory rate, frequency of impulse activity, and degree of physiological stimulation (stretch) in single SRN preparations. From Giacobini.[41]

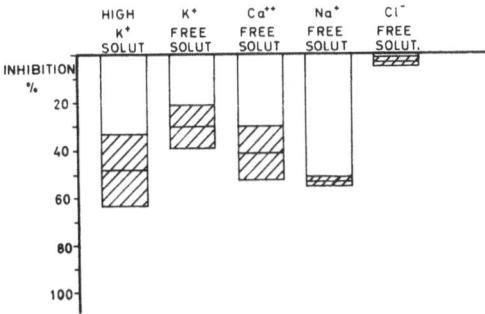

Fig. 5. Percent inhibition of the O_2 uptake of single SRN preparations in various electrolyte solutions. The mean value of at least ten experiments is reported, together with the confidence interval. From Giacobini.[79]

It can be calculated that the increase in O_2 consumption observed in the SRN in relation to impulse activity corresponds to a heat production of about 10 $\mu cal/g$/impulse. Abbott, Hill, and Howarth[29] estimated a heat production of about 9 $\mu cal/g$/impulse in a nonmedullated nerve of the spider crab (maia).

The dissected cervical superior ganglion represents an excellent physiological preparation in which O_2 uptake can be studied in vivo.[25] The values reported[25] for this ganglion may be corrected for the contribution of the fibers. However, it is extremely difficult to calculate the metabolic activity of a cellular component from the data of the whole structure. This is true for both cortex[30,31,32] and ganglia.[25] The fact that the respiration of extensively damaged cell preparations like ganglion cells,[33] Deiter's cells,[34,35] and cells from the cat cortex[36] and cortex slices[31] compares favorably with that of functional preparations like the SRN,[10] intact ganglia,[25] or nerve fibers[26] is not surprising. It is well known that high respiratory activity can be derived from grossly altered tissues such as homogenate fractions obtained by centrifugation.[32] Unfortunately, mammalian isolated nerve cells do not show the known electrophysiological responses to stimulation associated with brain ganglia or invertebrate preparations in vivo and therefore are of very little value for studies correlating metabolism with functional activity in neurons.

Following alterations of the ionic environment (K^+, Na^+, Ca^{2+}, Cl^-) that exert known effects upon the electrical properties of the cell membrane,[37,38] the O_2 uptake is reduced to different levels (except following omission of Cl^-; Fig. 5).[27] This means that in the SRN the energy metabolism (O_2 uptake) is highly dependent on the ionic concentration of the external medium. The decrease in O_2 uptake was not found to be connected with any characteristic changes in the electrical properties of the membrane. However, both hyperpolarization and depolarization decrease this parameter equally.

C. Studies on the Variations of Oxidoreduction Processes

Measurements of oxidized and reduced forms of pyridine nucleotides have been performed using two independent techniques[14,15,29,39] (Fig. 6). The microfluorimetric method of Lowry et al.[39] (Fig. 6) was applied at the single cell level for the first time by Giacobini and Grasso[40] (Fig. 7) and

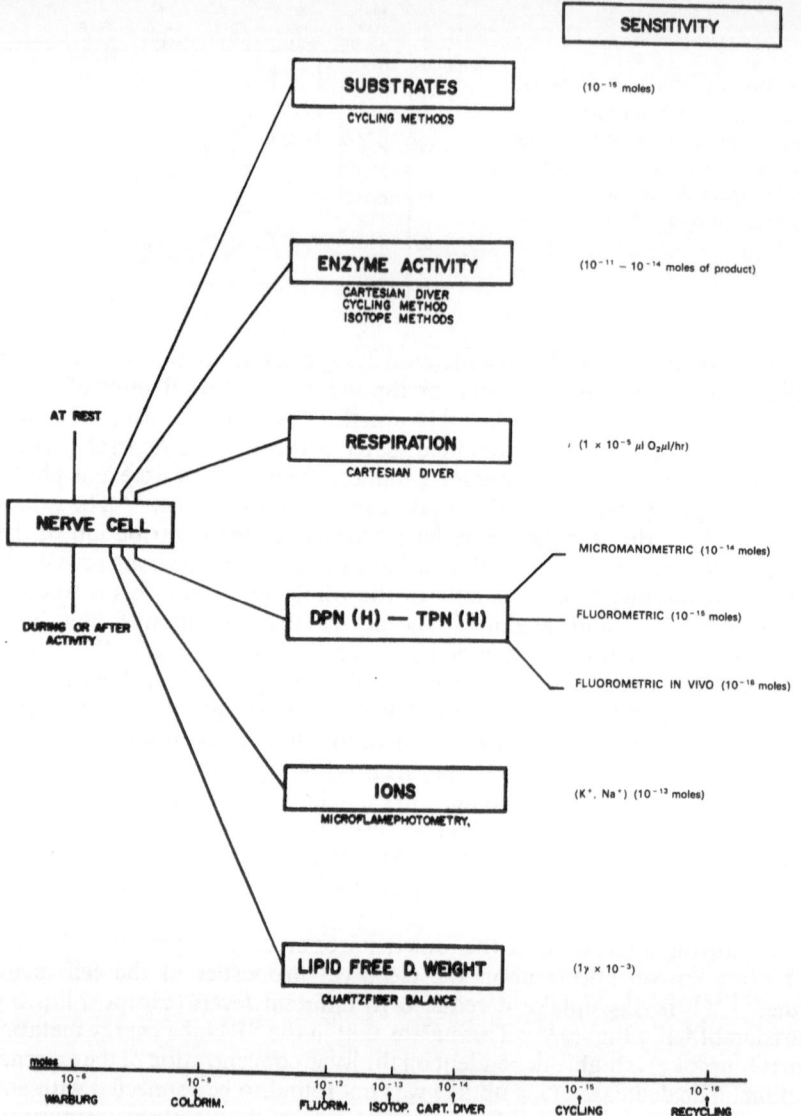

Fig. 6. A schematic survey of methods for biochemical analysis at the cellular level together with their sensitivities (bottom), expressed in moles of product of sample (only the methods used in the author's laboratory are reported). From Giacobini.[41]

later by Giacobini.[41] The results of these studies are summarized in Table IV. The table shows that after physiological stimulation the ratio TPN/TPNH falls from 0.69 to 0.21, i.e., by more than 60%. Similarly, the ratio DPN/DPNH fell from 4.4 to 1.5. Table IV also shows that the sum of the triphosphopyridine nucleotides (oxidized + reduced) and the sum of the diphosphopyridine nucleotides (oxidized + reduced) remain almost constant under the above conditions. Their ratio also remains constant. The ratio of the oxidized nucleotides to the reduced nucleotides decreases, however, from 4.6 to 1.3 after 500,000 impulses. The total PN diminish only slightly in the above conditions. The dynamic variations in the pyridine nucleotides after activity, which also can be seen in Fig. 8, leave no doubt that the observed changes in the level of these metabolites must in some way be involved with the mechanism leading to impulse activity. The only difference between the two groups of cells examined (resting and stimulated cells dissected from the same segment and animal) is the presence of impulse activity.

The microfluorimetric method of Chance et al.[14,15] for measuring reduced pyridine nucleotides offers the possibility of following dynamic changes in oxidation–reduction reactions under different experimental conditions, including stimulation, in a single neuron. The potential of this very elegant method and the possibility of recording the action potential of the cell at the same time as the chemical analysis makes it unique. This technique was applied for the first time to single isolated neurons in 1964 by Terzuolo et al.[42]

No measurable change in fluorescence was observed with the method of Chance et al.[14,15] during prolonged stimulation (up to 70 impulses/sec) over

TABLE IV

The Crayfish Stretch Receptor Neuron—Variations of Pyridine Nucleotides in the Slowly Adapting Nerve Cell at Rest and After Prolonged Physiological Stimulation[a]

μmoles/kg/wet wt.	At rest	After 12×10^4 impulses	After 50×10^4 impulses	Fluorimetry *in vivo* during impulse activity[b]
TPN	10.7 ± 2.3	4.6 ± 3.2	4.2 ± 1.5	
TPNH	15.4 ± 1.4	14.6 ± 0.6	20.1 ± 0.9	
TPN/TPNH	0.69	0.31	0.21	No changes
DPN	255 ± 44	180 ± 36	160 ± 19	
DPNH	58.1 ± 6	90 ± 14.8	105 ± 11	
DPN/DPNH	4.4	2	1.5	No changes
TPN + TPNH	26.1	19.2	22.3	
DPN + DPNH	313.1	270	265	

[a] Giacobini and Grasso[40] and Giacobini.[41]
[b] Rossini et al.,[43] impulse activity 17 impulses/sec over a period of "several hours."

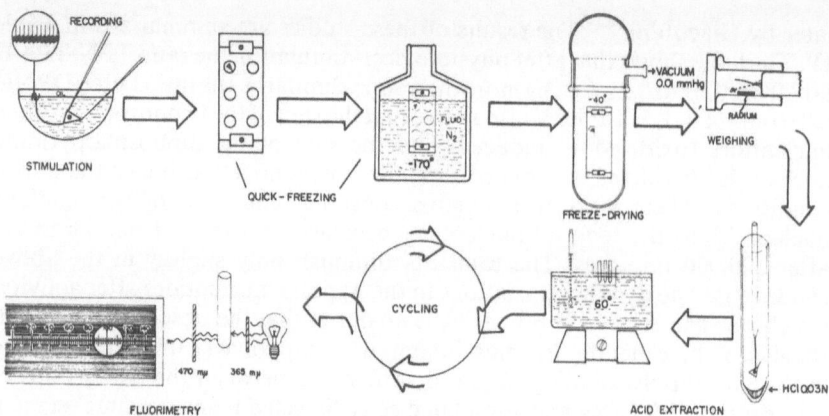

Fig. 7. Schematic description of the analytical procedure used for determination of metabolic intermediates, pyridine nucleotides, and phosphate compounds in single nerve cell preparations. From Giacobini and Grasso.[40]

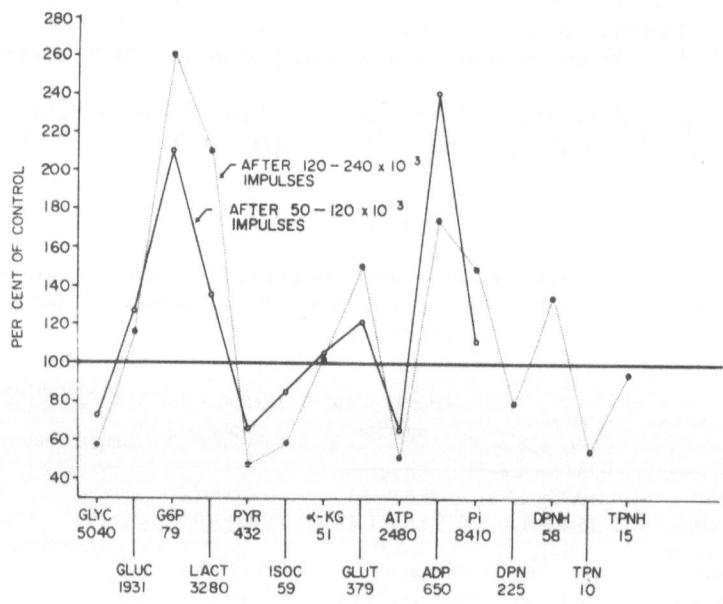

Fig. 8. The percentage variation from the resting state (control) in the level of 15 chemical components after two different periods of physiological stimulation of the slowly adapting SRN.[41]

a period of several hours.[11,43] Under the same conditions, as shown in Table IV,[40,41] the level of oxidized nucleotides is reduced by 30% after 120,000 impulses and by 38% after 500,000 impulses. The level of reduced nucleotides is simultaneously increased by 42 and 69% respectively. This discrepancy may be due to the different methods used, as pointed out in their conclusion by Terzuolo et al.[11] The disadvantage of the direct microfluorimetric method of Chance et al.[14,15] in comparison with the indirect microfluorimetric method[39,40] is that it is unable to give absolute quantitative data for the single nucleotides. The relative changes in the ratio of the nucleotides that may be determined with the method of Chance et al.[14,15] are quite insufficient for kinetic studies of metabolite changes. The effect of metabolites and metabolic inhibitors on pyridine nucleotides is discussed in Section III,H of this chapter.

Oxidoreduction levels of intracellular pigments have been reported by Chalazonitis et al.[44a,44b,45] They applied spectrophotometric methods to the giant neurons of Aplysia (Table IIA). The principle of these methods is based on the measurement of the variation of the intensity of transmitted monochromatic light corresponding to a specific absorption in vivo for each examined pigment. The measurements are made in conditions of functional activity. The absorption spectrum of the cell is registered as a function of time, utilizing the cinespectrometric technique. The advantage of this technique is that it is possible to simultaneously register the transmittance of several enzymes and to measure their instantaneous variations. The same authors[44a,44b,45] have recently described a differential microspectrophotometer utilizing a double-beam spectrometer.

Chalazonitis and Takeuchi[46] demonstrated a relationship between oxygen concentration and membrane potential in the Aplysia nerve cells. An increase in the oxygen concentration in the medium surrounding the cell preparation caused an increase in the resting potential from 52 to 58 mV. A decrease in the oxygen concentration had the effect of depolarizing the neuron. The increased oxygen concentration also increased the threshold of electrical excitability. The Aplysia nerve cells contain a hemoglobin-like pigment, which allowed the authors to follow the intracellular O_2 concentration. In conclusion, the findings of Arvanitaki, Chalazonitis, and their co-workers[44a,44b.45] support a relationship between the metabolic activity of the neuron as determined by oxidation–reduction systems, O_2 concentration, and membrane potential. This is in agreement with the data previously reported by Giacobini et al.[6a,10,27,40]

D. Studies on the Variation of Enzymatic Activity During Functional Activity

Since no previous experiments had been done on the influence of physiological stimulation of the enzyme activities of individual neurons in isolated conditions, three different metabolic pathways and physiological systems were investigated in our laboratory using different types of micro-chemical technique (see Fig. 6). These were the acetylcholine system, the respiratory enzymes, and some enzymes of the pentose shunt. As shown in Table V, acetylcholine's metabolizing enzymes, acetylcholinesterase (AChE) and cholinacetylase (ChAc), are present in the crustacean stretch receptor nerve cells. Acetylcholine increases the receptor discharge at very low concentrations (10^{-6} M and less) and at lower concentrations in the presence of eserine.[47,48] This effect is blocked by atropine, which like ACh increases the impulse frequency of the receptor.[47,48] Both "muscarinic" and "nico-tinic" compounds exert a stimulating effect upon the SRN.[49] Therefore, it is suspected that these neurons possess cholinoceptive receptors. In spite of this evidence it is generally accepted[47] that "it would seem unlikely that ACh is involved in the transmission of excitation from stretched muscle to neuron. It is however quite likely that ACh is normally involved in the transmission of excitation from regulatory nerve endings to the sensory neuron." The results of McLennan and York[49] confirm this view and indicate that the cholinoceptive receptors of crayfish SRN do not conform precisely to the classical pharmacological division into muscarinic and nicotinic types and that the excitation exerted by several compounds is mediated by a single type of receptor.

Both AChE and BuChE (butyrylcholinesterase) could be measured by means of the Cartesian diver[41] (Table V) and were found in rather high concentrations in the SRN. The suggestion of Maynard and Maynard[59] that the fast-adapting neuron should exhibit higher ChE activity than the slowly adapting neuron could not be confirmed. These authors used a histochemical method without applying selective inhibitors and substrates. ChAc (Table V) could be demonstrated for the first time in an isolated invertebrate neuron, in the cell body, as well as the dendritic region and axon of the SRN.[41]

It was of particular physiological interest to determine whether impulse activity could modify the activity of AChE, an enzyme that often has been claimed[51] to be related to conduction of nerve impulses. No changes could be detected in cells stimulated during several hours with up to 500,000 impulses.[41]

On the other hand, both respiratory enzyme activity (cytochrome and succino oxidase) and glucose-6-phosphate dehydrogenase were significantly increased during activity—cytochrome oxidase activity more markedly than succino oxidase activity (Table V). These enzymes were determined with a Cartesian diver technique[52] using several substrates (reducing agents).[41] Ascorbic acid was found to be the most powerful of all the reducing agents used for cytochrome C reduction; in fact, more than twice as powerful as p-phenylenediamine. Furthermore, compared to the other reducing agents it showed a relatively low rate of auto-oxidation. Under the above experimental conditions it was possible to detect significant differences between experiments in the presence or absence of ascorbate.[41] By substituting DPNH for succinate in the incubation solution used for measuring succino oxidase activity, it was possible to demonstrate that both succinate and DPNH are

TABLE V

The Crayfish Stretch Receptor Neuron—Enzyme Activities in the Slowly Adapting Nerve Cell at Rest and After Prolonged Physiological Stimulation

Metabolic pathway	Enzyme	At rest	During activity	Percent increase
Acetylcholine metabolism[a]	AChE	9.2 ± 0.1	10.1 ± 0.03	0
	BuChE	7.1 ± 0.04	8 ± 0.04	0
	ChAc	15 ± 0.8		
Respiratory enzymes[b]	Cytochrome oxidase	29.4 ± 0.8	51.2 ± 0.7	74
	Succino oxidase	54 ± 0.9	68.4 ± 0.6	27
Pentose shunt[c]	G6PDH	303 ± 68	642 ± 82	112
	6PGDH	76.8 ± 22	85.7 ± 10.7	12

[a] Giacobini.[41] AChE (acetylcholinesterase) and BuChE (butyrylcholinesterase) activity in microliters CO_2 per milligram wet weight per hour. ChAc (cholineacetylase) in micromoles ACh $\times 10^{-7}$ per milligram per hour wet weight.
[b] Giacobini.[41] Cytochrome oxidase and succino oxidase activity in microliters O_2 milligrams per hour per wet weight.
[c] Giacobini and Jongkind.[60] G6PDH (glucose-6-phosphatedehydrogenase) and 6PGDH (6-phosphogluconatedehydrogenase) activity in micromoles per mg per hour dry weight.

oxidized—DPNH at a much lower rate than succinate. This finding is in agreement with the changes in DPN levels determined with the micro-fluorimetric technique[40] (Table IV) and indicates that both pathways are possible but succinate is the preferred one.

Burrin and Bechey[53] demonstrated the presence of cytochrome oxidase and cytochromes A and A_3 in phosphorylating mitochondria isolated from Carcinus maenas. Changes in the respiratory enzyme activity in several mammalian nerve cell preparations have been described by Hydén and Pigon,[52] Hydén and Lange,[54] Hamberger and Hydén,[55] and Hamberger and Sjöstrand[56] in Deiter's cells of rabbit. In these experiments the respiratory enzyme activity was measured both in neurons and in the surrounding glial cells. (Criticisms of these experiments were discussed in Section II.) These above authors interpreted their results as showing a clear difference between the respiratory activity of the neuron and that of the surrounding glia. Their interpretation was that the nerve cell reacted to functional stimuli by increasing the capacity of the electron transport system in parallel with an increased consumption of energy. The glia, in contrast, showed no change in this respect. It is doubtful whether such results can be compared with those from studies in isolated functional preparations.

The finding[43] that terminal inhibitors of the cytochrome chain (e.g., CN) do not increase the PNH level suggested the presence of an accessory electron transport system, not poisoned by CN and identified spectrophotometrically as b_5, present in the microsomal portion of crustacean nerve tissue. The increase in reduced pyridine nucleotides found by Rossini et al.[43] with ouabain (see Section III,H) and the absence of a similar increase with CN suggests a functional linkage between the rate of electron transport in the microsomal component of the respiratory chain and the Na^+- and K^+-activated ATPase present in this fraction. Since experimental evidence for such a hypothesis was lacking, a study was carried out in our laboratory on microsomal preparations from rat brain using ESR (electron spin resonance) in order to find out whether free radicals were present and in some way connected with activation of microsomal ATPase. Previous work[57] using frog skin suggested that unpaired electrons participate in active cation transport, but no studies of this kind on brain microsomes have been previously reported. In freshly prepared brain microsomes an ESR signal at $G = 2.0039 \pm 0.002$ and with a peak-to-peak width of 11 ± 1 gauss was found. The signal saturated readily as the microwave power was increased.[58] The ESR signal in brain microsomes decreased in amplitude in the presence of $ATP + Na^+ + K^+ + Mg^{2+}$ but not with ATP or ions alone. Ouabain reversed this effect, but neither ADP or AMP had any effect. Thus, we observed a decrease in the signal under optimal conditions for ATPase activity and a return of the signal to the original level when the inhibitor was present. The experiment of Kometiani[57] suggested a direct, not inverse as here, relationship between the number of unpaired spins and the extent of active cation transport. This work,[57] however, required the lyophilization of samples, a procedure itself known to cause the formation of free radicals. The experiment carried out in our laboratory suggests, therefore, a possible connection between free radicals, microsomal ATPase, and active cation transport.

Chalazonitis and Arvanitaki[3,4,12] applied spectroscopic methods to living Aplysia neurons and were able to follow the activity of succinic dehydrogenase and cytochrome systems within the neuron during functional activity or in granules isolated from these nerve cells. Chalazonitis and Gola[4] found that the most active region of the nerve cells, in samples of Helix, was the pigmented apex that showed both high cytochrome and high succinic dehydrogenase activity. Chalazonitis[3] found that the electrical response of Aplysia neurons differs in respect to CO_2, pH, or O_2 sensitivity, according to the cell type. The electrical response of these cells to visible or infrared light also differs. This fact was related to the different natures of the pigment systems, one possibly involving cytochromes and the other carotenes.[59] This very elegant work, done at the cellular level and in physiological conditions, demonstrates the possibility of following changes in enzyme activity within a $20 \mu^2$ area. Furthermore, it shows the heterogeneity in pigmentation and sensitivity to CO_2, pH, light, etc. The relationship between the metabolic activity (respiration and electron transport) determined by oxidation–reduction systems, O_2 concentration, and membrane potentials found by these authors agrees well with our own findings (see Tables IV, V, and VI and Fig. 4).

It is generally accepted that brain tissue of crustaceans, like that of most invertebrates, derives its energy mostly from respiration and aerobic glycolysis. Enzymes, coenzymes, and metabolites belonging to the hexose

TABLE VI

The Crayfish Stretch Receptor Neuron—Levels of Metabolites in the Slowly Adapting Nerve Cell at Rest and After Prolonged Physiological Stimulation

Metabolic pathway	Metabolite[a]	At rest (μmoles/kg wet wt.)[b]	Percent variation during activity[c]
Glycolysis[d]	Glycogen	5040 ± 944	−57
	Glucose	1931 ± 280	+17
	G6P	79 ± 15	+160
	Pyruvate	432 ± 210	−48
	Lactate	3280 ± 600	+115
Krebs cycle[e]	Isocitrate	59 ± 4	−58
	α-ketoglutarate	51 ± 6	0
	Glutamate	379 ± 22	+50
Pentose shunt[f]	6PG	18 ± 2	+48
Phosphate compounds[g]	ATP	2480 ± 565	−52
	ADP	650 ± 240	+140
	PA	4200 ± 480	−51
	Pi	8410 ± 850	+52

[a] G6P = glucose-6-phosphate, 6PG = 6-phosphogluconate, ATP = adenosine-5′-triphosphate, ADP = adenosine-5′-diphosphate, PA = arginine phosphate, and Pi = inorganic phosphate.
[b] Levels of endogenous substrates at rest.
[c] Levels of endogenous substrates after 160,000 impulses; for PA, after 50,000 impulses.
[d] Giacobini and Grasso.[40]
[e] Giacobini and Marchisio.[65,65b]
[f] Giacobini and Jongkind.[60]
[g] Giacobini and Grasso[40] and Giacobini.[41]

monophosphate pathway (PS, pentose shunt) also have been demonstrated in such tissues. Several authors, however, have concluded that the PS pathway is nonoperative or of very minor importance. Therefore, it was of great interest to test whether the activity of two enzymes involved in the oxidative part of the pentose cycle showed any variation during functional activity.

Giacobini and Jongkind[60] (Table V) found that G6PDH activity was significantly increased already after about 50,000 impulses. 6PGDH activity, on the other hand, did not significantly change under the same conditions. 6-Phosphogluconate (Table VI) also was found to be increased by 48 % after a period of regular impulse activity.

Recently the significance of the pentose shunt has been reevaluated.[61] The rates of synthesis of phosphoribosyl pyrophosphate and its derivatives (nucleotides and nucleic acids) have been estimated in bacterial and mammalian tissues by these authors.[61] In contrast to what was previously believed, the nucleic acid synthesis accounts for only a small fraction of the pentose-5P formed by decarboxylation of glucose-6P.[61]

The fraction of glucose that is converted to RNA pentose, therefore seems to be of a much lower magnitude than previously estimated. The above authors, assuming that the contribution of the pentose cycle to glucose metabolism *in vivo* is of the order of 1%, found that only one-sixth of the pentose-5-P was converted to phosphoribosepyrophosphate (to be used for the synthesis of nucleotides and nucleic acids). The importance of the pentose shunt, from a functional point of view, therefore should be related to the synthesis of lipids, particularly membrane lipids. An increased demand for pentose shunt activity, as shown in our experiments[60] (Table V), would then indicate an increased demand for lipid synthesis (reflected in an enhanced TPN reduction, see Table IV). Such a role is suggested in Fig. 10, summarizing different biochemical changes discussed in this chapter. In this figure the pentose shunt is connected to the synthesis of informational macromolecules (through RNA and protein synthesis) and membranes.

E. Variations in the Levels of Metabolites and Phosphate Compounds During Functional Activity

Although our knowledge of crustacean metabolism is relatively limited, it appears that both acetate and glucose can be oxidized. In the invertebrate nerve cell, glycogen is probably one of the principal, if not *the* principal, form of stored energy, and glucose from this source represents the most important substrate for the production of energy through glycolysis and the citric acid cycle. Other sources of energy are probably available in the neurons and apparently mobilized when energy is needed.

Glucose levels vary both in the blood and the nervous system in relation to physiological states[62] and the ecology[63] of the animal.

Puyear et al.[64] demonstrated that glucose is metabolized in crayfish (intermolt) via three pathways: (1) glycolysis–pyruvate decarboxylation sequence, (2) glucuronate pathway, and (3) pentose phosphate pathway. This view also is supported by studies using metabolic inhibitors. Quantitatively speaking, the pentose phosphate pathway participates less than the glycolytic and glucuronate pathways in glucose catabolism in intermolt crayfish.[64]

A study with metabolic inhibitors[22] indicated glycolysis as a probable source of substrates for oxidative processes linked to impulse activity in the slowly adapting receptor cell of crayfish (see Table VIII). Giacobini and Grasso[40] investigated the levels of several glycolytic intermediates with a newly described biochemical technique (Figs. 6 and 7), allowing the estimation of these substrates at the cellular level.[39,40] Krebs' cycle metabolites,[65a,65b] pentose shunt substrates,[60] and phosphate compounds[40] also were determined in the same cell preparation (SRN), both at rest and after prolonged physiological stimulation.[65a,65b]

The results of this investigation are summarized in Table VI and Fig. 8. It should be emphasized here that the experimental conditions in which the cell was firing were far from anoxic and therefore these results were not obtained in anaerobic conditions. The ratio between the glycogen consumed and the lactate formed was about 0.5 after 120,000 impulses. These variations suggest an increased glycolysis in the firing cell as compared with the control

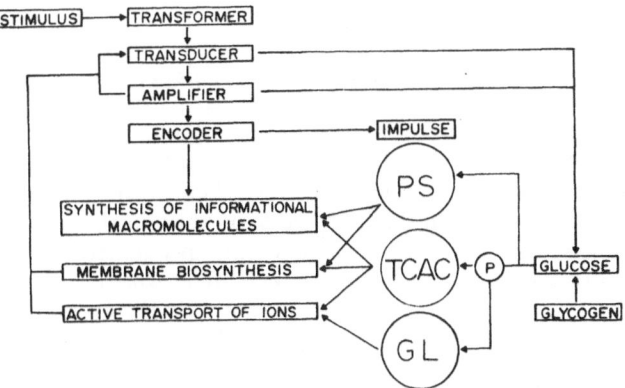

Fig. 9. A general scheme of neurophysiological and metabolic correlations in the slowly adapting cell of the SRN. See text. From Giacobini.[41]

cell at rest. Since the only experimental condition differing between the two groups of cells examined is the presence of impulse activity, it seems reasonable to assume that glycolysis is involved in some way with the mechanism leading to impulse activity (see Fig. 9). If appropriate corrections are made in the substrate levels for the nonnervous components present in the preparation (modified muscle and glia, see Fig. 1) the significance of these results does not change. The increase in DPNH (Table IV) could be attributed to the higher activity of glyceraldehyde-3-phosphatedehydrogenase. The almost steady level of glucose is probably a consequence of glycolytic control at the phosphorylase step. This suggestion is supported by the significant increase of G6P, the highest percentage change of any of the products measured. An activated gluconeogenesis also has to be taken into consideration when interpreting these results.

In a parallel study[65b,66] the physiological significance of different steps of the tricarboxylic acid (TCA) cycle in connection with the impulse activity

Fig. 10. Relationship between energy metabolism and physiological events in the SRN of crustacea. See text.[79]

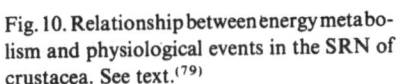
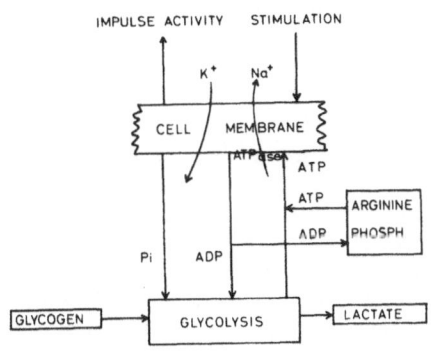

of this neuron was investigated. A significant decrease in pyruvate level was found after 50,000 impulses (Table VI). In terms of resting values these variations represented a 50 % decrease (Fig. 8). This decrease in pyruvate may account for the increase in lactate demonstrated by the study of Giacobini and Grasso[40] after a corresponding number of impulses.

The glutamate content of the SRN was found to be lower than in other crustacean[65a] or mammalian nerve tissues.[67] Our results also suggested a higher concentration of glutamate in the axon than in the cell body of the SRN. Recently Coté et al.[68] found an increase in specific activity of glutamate in stimulated frog nerves, pointing to an increased turnover of the citric acid cycle with a concomitant increased flux of CO_2 into the cycle.

The results of Giacobini and Marchisio[65a,66] indicate that in the SRN, as in frog or mammalian nerve, pyruvate is oxidized to yield energy for the maintenance of excitability and that the TCA cycle is probably also involved in this mechanism. ATP and PA (Table VI) represent the two fundamental sources of energy involved in the impulse mechanism of the SRN (Fig. 9). This appears to be clear when we look at the continuous ATP and PA decrease during prolonged stimulation (Table VI and Fig. 8). The decrease in ATP levels is concomitant with the increase in ADP, while variations of both ATP and PA are consistent with the variations of Pi.

On the basis of investigations in giant axons with metabolic inhibitors such as cyanide, azide, or dinitrophenol, which block both sodium extrusion and potassium uptake in the resting state but have only a slight effect on the sodium movements during passage of the impulse,[69] it was concluded that the permeability system allowing ionic transport across the neuronal membrane is not metabolically dependent, unlike the secretory system operating during the phase of recovery. At the cellular level, such an attempt has been practically impossible because of the lack of sufficiently sensitive techniques for measurement of metabolites. The early finding of Shanes and Brown[70] in crab nerves emphasized the importance of glycogen as a source of energy. This finding now seems in perfect agreement with the results obtained in the isolated SRN. The functional importance of glycogen for the impulse activity of the SRN is further suggested by the fact that its concentration is 40 % higher in the fast adapting cell as compared to that in the slowly adapting cell.[40] Giacobini[7] found similarly that ATP and PA are almost 50 % higher in the fast adapting cell.

The importance of phosphate compounds in relation to the active transport of Na^+ has been emphasized by the work of Caldwell[71] and Caldwell et al.[72] in squid giant axons and of Caldwell and Keynes[73] in the electric organ of Electrophorus.

Concerning the results shown in Table VI, two groups of experiments are relevant. The first includes the optical studies of biochemical events in the electric organ of Electrophorus by Aubert et al.[74] and the studies of Maitra et al.[75] on the glycolytic metabolism following electrical activity in the same preparation. Aubert et al.[74] found that when slices of the electric organ were stimulated electrically, three different effects could be detected optically:

a large decrease in absorption, a decrease in absorption having a definite peak and arising from oxidation of a haemoprotein, and a diphasic change in fluorescence emission, interpreted as arising from oxidation reduction of the phosphopyridine nucleotides. The above observations, taken together with the results of chemical analysis and thermal studies, support the view[74] that the events following stimulation of the electric organ follow this sequence: first changes in internal ionic concentration (resulting from the flow of electric current), second activation of the sodium pump (driven by phosphate-bound energy and drawing initially on the depots of creatine phosphate) and, third, activation of glycogenolysis in order to restore the ATP–ADP ratio. The changes in the DPNH–DPN ratio observed in this preparation as well as in the SRN by us[40] contrast with the results obtained by Terzuolo et al.[11] and Rossini et al.[43] using the same technique as Aubert et al.[74] and implying an involvement of glycogenolysis in transport mechanisms and a generation of impulse activity. The studies of Maitra et al.[75] on glycolytic metabolism following electrical activity in the same organ are in agreement with our own results (Table VI).

The variations of glucose, G-6P, lactate, ATP, ADP, and PA in the crustacean nerve cell after brief stimulation (Table VI) are very similar to the pattern observed in the electric organ.[75] Pi, however, first diminishes after 28 sec of stimulation and then increases following 4–5 sec recovery.[75] Large increases of fructose-1,6-diphosphate, glyceraldehyde-3-phosphate, and 3-phosphoglyceric acid were found to occur at the same time as the maximum fluorescence emission of the slices of the Electrophorus electricus.[75] The fluorescence increase was identified as being due to the DPN reduction at the glyceraldehyde-3-phosphate dehydrogenase step. These authors[75] concluded by suggesting that the activation of glycolysis may be due to a concerted effect of increased ADP concentration upon phosphofructokinase, 1,3-diphosphoglyceratekinase, and glyceraldehyde-3-phosphate dehydrogenase.

Our results (Table VI) are similar to the above and suggest that the impulse activity in the crustacean nerve cell is linked to an energy-requiring system (Fig. 9). This energy is furnished by ATP splitting and glycolysis directly or through the usual step of arginine phosphate supporting the ATP resynthesis.

If glycogen is considered, hypothetically, to be the only source of energy in the SRN, and if its decrease is compared with the oxygen consumption of the neuron in the same conditions of stimulation, a discrepancy is found, revealing that glucose is probably not the only source of energy in this cell.

On the basis of changes of the respiratory quotient, it has been suspected since the early studies of Gerard[76] that the metabolism of nerves may differ qualitatively as well as quantitatively in resting and active conditions. Furthermore, impulse activity also increases substrate uptake.[76]

Dettbarn and Hoskin[77] studied the production of $^{14}CO_2$ by lobster nerves from 1-^{14}C-glucose and 6-^{14}C glucose under conditions of stimulation. They found that less than 10% of the O_2 used by the lobster nerves corresponded to the added glucose oxidized to CO_2. It also appeared that the extra metabolism of added glucose attributable to neuronal activity was not qualitatively different from resting metabolism. This conclusion, however, did not exclude the possibility of the utilization of endogenous substrates other than glucose.

It is relevant to this discussion to draw attention to the data of Rossini et al.[43] and Terzuolo et al.[11] on ATP levels. These authors measured ATP levels in SRN using the luciferin luciferase method. This method has a sensitivity of 10^{-12} M, which is relatively low compared to the fluorimetric technique (cycling) but adequate for measuring this compound in large neurons like the SRN. The average ATP level of SRN was found to be about 1.86 μM/kg dry wt.,[43] and the contribution to the total ATP by the muscle present in the preparation is about 35%.[11]

In contrast to the results reported in Table VI, these authors did not detect any changes in ATP levels after 150,000–250,000 impulses (compare the 52% drop in Table VI). Arginine phosphate in our experiments fell by more than 50% after only 60,000 impulses (Table VI), indicating a possible time relationship of the energy releasing events in the SRN (Fig. 9). According to Terzuolo et al.[11] the ATP level was altered under some particular experimental conditions such as anaerobiosis and after cyanide poisoning or after 1 hr of exposure to K^+-rich and Ca^{2+}-free solutions but not after exposure to Na^+-free solutions or in Ca^{2+}-free media in the presence of ouabain.

F. Variations of Intracellular Cation Concentrations

The impulse activity of the SRN is strongly influenced by the concentrations of inorganic ions within the cell and in the fluid surrounding it. Although the literature contains reports regarding the concentrations of intra- and extracellular ions in invertebrate nervous tissues (Table X), very few reliable figures are available for intracellular ions measured in single cells.

Table X summarizes the available information on the ionic content (Na^+, K^+, Ca^{++}, and Cl^-) of some invertebrate nervous tissue preparations. It can be seen that in all the species studied the internal K^+ concentration exceeds that of Na^+ by a factor of 1.7 to 15. In crustacea, including crayfish, the ratio K_i^+/Na_i^+ is about 15. In insecta and mollusca this ratio is somewhat lower (3.3–8.4). When measurements are carried out in axoplasm this ratio, in some cases, may fall to 1.7 (peripheral nerve of Carcinus maenas) or 3.5 (Aplysia). However, in crayfish the data of Wallin[24] indicate approximately the same ratio in the axoplasm as Giacobini et al.[23] found in the total neuron.

The external K^+ concentration, on the other hand, is much lower than that of Na^+ in all the species reported in Table VIII. The ratio Na_0^+/K_0^+ may be as high as 46. In crustacea it is generally in the order of 40, but in insecta and mollusca this ratio is usually somewhat smaller. Chloride is also more concentrated extracellularly (blood or haemolymph) than intracellularly.

It is important to note that the estimated concentrations of ions within the cells have so far been markedly different from those of the whole nerve: Intracellular Na^+, e.g., often has been found to be lower and intracellular K^+ much higher than in whole nerve. Most of these values have been calculated on the assumption that the extracellular concentrations of the electrolytes were the same as in the external medium available to the experimenter (haemolymph, blood, etc.). Estimation of the intracellular concentrations, with the exception of the direct measurement of extruded axoplasm in the squid, has been carried out on large fragments of tissue that contain both extra- and intracellular fluid. In other cases, the intracellular concentrations of various cations have been estimated from calculation of the kinetics of their exchange with the external medium.

The SRN is a suitable preparation for such measurements and, compared to isolated axons, has the obvious advantage of being a whole neuron in which several functional parameters, including the generation of impulse activity, can be studied in parallel with ionic changes. Furthermore, the variations in exterior ionic concentrations can be readily controlled.

In the course of our study, in order to decide whether some of the results obtained with inhibitors, or metabolites, could be related to active transport processes, it was felt necessary to determine Na^+ and K^+ in cells treated with these inhibitors or in conditions of modified ionic environment.

An integrating microflamephotometer, suitable for analysis of either individual cells weighing $0.5-0.005$ μg or microsamples of biological material in the order of 1 nliter or less, was constructed in our laboratory.[79,80] The

Fig. 11A. Assembly for handling the sample. Note in the middle the sample carrier with the injection motor and sample holder with the platinum wire. Burner and photomultipliers are at the right.

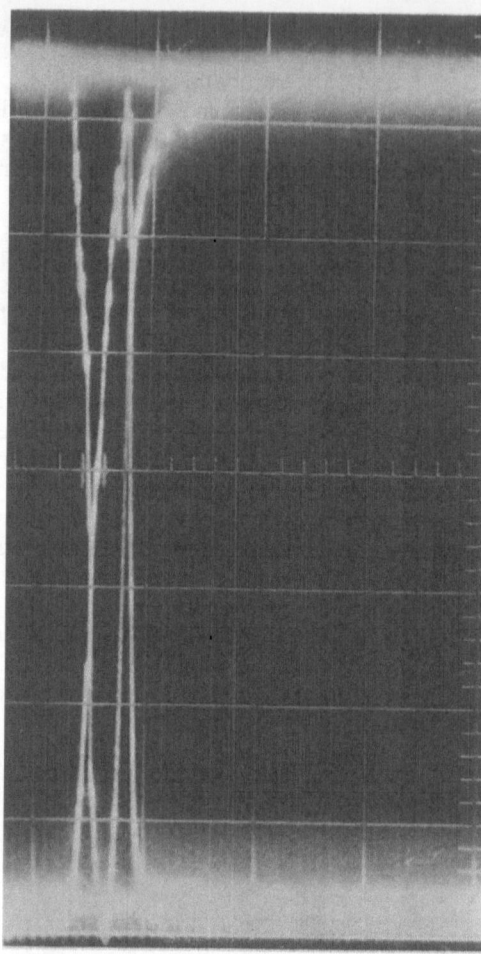

Fig. 11B. Record of voltage output of the photomultipliers. Single SRN from the crayfish. The inverted upper trace is for potassium, the lower for sodium. On Y axis: 0.5 V/cm for potassium; 0.05 V/cm for sodium. On X axis: 0.2 sec (cm). Each square is 1×1 cm. From Carlsson et al.[80]

instrument has a sensitivity of 1×10^{-14} M for Na^+ and K^+. Figure 11A shows the sample carrier with the injection motor and the sample holder with the platinum wire. The material to be examined is placed on the wire under microscopic inspection (Fig. 11A) and introduced into the flame where excitation takes place. The time integral of the intensity of the light emitted at the chosen analytical spectral wavelength of an element is proportional to the amount of the element present. Figure 11B shows the records of voltage output of the photomultipliers during the analysis of a single neuron from the crayfish.

With the help of this technique, it has been possible to follow changes in Na^+ and K^+ content occurring during the initial period following isolation of single invertebrate neurons or during incubation under conditions of reduced movement of Na^+ and K^+.[23,79]

After isolation of the SRN the ionic events can be divided into two periods. During the first (initial) period, a continuous recovery from the initial distortion of the ionic content takes place. The intracellular cationic content is influenced during this period by at least two processes acting simultaneously and counteracting each other; one is dependent on the manipulation of the cell and the other is on active transport. If the active transport mechanism is inhibited (e.g., with ouabain or by omitting Na^+ or K^+) the recovery does not take place. The second period starts about 1 hr later and is characterized by stationary levels of Na^+ and K^+, which—in normal solution—are close to physiological levels. This period may continue for at least 5 hr.[23]

The effect of periods of excitation following physiological stimulation also could be studied in isolated neurons.[23] This technique can be combined with conventional neurophysiological analysis and with metabolic studies using the Cartesian diver technique.

Tetrodotoxin, a selective inhibitor of both inwardly and outwardly directed passive Na^+ movements had no effect on the Na^+ and K^+ levels in our cell.[23] Brief physiological stimulation applied to our preparations also had no effect on the ionic content.

In spite of very careful dissection, some of the glial satellites still may be sticking to the preparation. The sheath cells (Schwann cells) surrounding the nerve fiber of the squid have higher concentrations of Na^+ and lower concentrations of K^+ than the axon (Table X).[81] This suggests that the Schwann cells or the glial Na^+ may help to maintain adequate ionic concentrations for active transport in the plasmalemma- or axolemma-glia cell space during the initial period.

The 40% reduction in the O_2 consumption of the SRN when subjected to ouabain (10^{-5} M) or in Na^+-free or K^+-free media[6b,79] is in good agreement with the intracellular changes in these ions detected with our technique and supports the involvement of an energy-dependent active transport mechanism.

G. Studies on the Effect of Nervous Activity on RNA and Protein Synthesis

The RNA contents of single SRNs of lobster have been examined by Grampp and Edström[17] and Edström and Grampp[18] and are shown in Table III. These data have already been discussed in Section III,A.

The RNA concentration of about 0.5%, found in the SRN, is somewhat lower than that found in the nerve cells of mammals.[19] The finding by Grampp and Edström[17] of low concentrations of RNA in isolated axons of the SRN is also of interest. As is well known, RNA has now been found in several types of axon, including mammalian axons. The adenine/uracil ratio and the purine/pyrimidine ratio were 1.22 and 1.21 respectively.

The finding that the total amount of RNA in the SRN was not affected by stimulation and activity may be due to any of the following reasons:[17] The electrophysiological activity was insufficient to cause measurable changes in the total RNA (it should be noted, however, that such a stimulation should be quite effective both with respect to length and selectivity if compared with the short periods of mild stimulation applied by other authors);[52] the size of the pool of RNA precursors accessible to the cell is ignored, but as the cell is kept for up to 7 hr in a saline medium the absence of organic hexogenous substrates might limit RNA synthesis; changes in the total amount of RNA may be too small to be measured because of the slow turnover at low temperature; or measurable changes in the quantity of RNA do not occur because of the absence of effects on the general metabolism *in vivo*, to which other RNA changes can be related.

However, it should be emphasized that the base composition of the RNA of the SRN was altered during stimulation as reported in the first study of Grampp and Edström.[17] This must be due to changes in the proportion of different types of RNA or RNA fractions. In a later investigation, Edström and Grampp[18] were unable to record changes in RNA content and composition or RNA synthesis during extended periods of stimulation causing intense spike generation and firing in the SRN. The impulse activity present during periods of up to 24 hr was completely unaffected by inhibition of the RNA and protein synthesis, in spite of the fact that about 10–20% of total RNA was lost from the cell bodies during this period.

These results are fully in agreement with those reported by Toschi and Giacobini[21] (Table IX) on the action of puromycin and actinomycin-D. In these experiments the complete inhibition of protein and RNA synthesis did not affect either the respiration or the impulse activity of the cell during stimulation periods of up to 12 hr corresponding to more than 1 million impulses (Table IX). Such findings may be relevant when trying to understand the relationship of RNA and protein synthesis to learning mechanisms. One should not forget that the SRN is a receptor cell and as such has not been demonstrated to possess "learning" capacity. However, it is widely accepted that the process of elaboration of the incoming information starts already at the level of the receptor.

Judging from our results[21] (Table IX) and those of Edström and Grampp[17,18] it seems quite unlikely that impulse activity in the isolated SRN preparation is coupled, at least during periods of time up to 24 hr, to RNA and protein synthesis. If these data are valid for other preparations and neurons, it seems equally unlikely that, if learning processes are dependent on or coupled to an increase in the frequency of the impulse activity, they are probably independent of RNA and protein synthesis. Furthermore, the above experiments with the SRN do not support the results previously obtained by other authors in CNS neurons.

Induction of increased RNA synthesis in neurons of the abdominal chain of the earthworm by electrical activity was reported by D'Yakonova et al.[82] Structural changes in RNA, histochemically stained, in the SRN during stimulation were reported by Kogan and Zakuskin.[83] According to these authors the Nissl bodies of these neurons show different structural patterns in the excited and inhibited states. Similar changes in the Nissl bodies have been·correlated with a number of different functional states in several vertebrate neurons.[84]

The correlation between structure, function, and RNA metabolism in central neurons of insects has been studied by Cohen[85] and is discussed in the book by Wiersma.[85] According to Cohen[85] the structural arrangement of the RNA aggregation is followed by "release" of ribosome-messenger RNA from the endoplasmic reticulum, formation of polyribosomes, and manufacture of protein specified by the new messenger RNA.

It now seems possible to identify individual cells in invertebrate nervous structures, like the cockroach ganglion, involved in a specific behavioral pattern, coordinated locomotor pattern, or learned response. The invertebrate and vertebrate CNS show common properties in the RNA-protein synthesis mechanism, and it seems that these properties may be influenced by their functional state. The differences between invertebrate and vertebrate neurons seem more to be of a temporal and quantitative nature and not to be differences in specificity.

H. Effect of Metabolites, Metabolic Inhibitors, Neurotoxins, and Drugs on Metabolism and Impulse Activity

The effect of naturally occurring substrates added to the immediate environment of nerve cells depends on a "pharmacological" effect at the level of receptors in the nerve cell membrane or on a "metabolic" effect due to a direct interaction of these substances with the cell metabolism. This effect may act generally on energy-providing processes or on energy processes linked to impulse activity. The "pharmacological" interpretation is supported by the results of those authors who investigated the effect of iontophoretically applied substrates on single CNS neurons.

L-glutamic acid, e.g., has been found to be strongly excitatory for spinal neuron activity.[86] Several substances have been tested on the SRN, mainly in connection with the problem of GABA inhibition—among these, succinic and α-ketoglutaric acid.[87]

Giacobini and Marchisio[65a,65b,66] studied the effect of several intermediates of the tricarboxylic acid cycle and of pyruvate and glutamate on the impulse activity and O_2 uptake (Fig. 2) of the SRN using the Cartesian diver. Their results are summarized in Table VII.

Glutamate was the only substrate, out of seven intermediates of the Krebs cycle studied, showing a constant and powerful inhibitory effect on the impulse activity of the SRN. The block occurs instantaneously after extracellular application of the substrate. An excitatory phase precedes the inhibitory effect; this is similar to the effect found by Krnjevic and Phyllis[88] in cortical neurons. GABA shows a constant inhibitory action in concentrations of 0.01–0.06 mM. As with glutamate, an excitatory phase preceded

TABLE VII

The Crayfish Stretch Receptor Neuron—Effect of Metabolites on Oxygen Uptake, Impulse Activity, and Fluorescence in Vivo

Metabolite	Final concentration (mM) [a]	Percent changes of O_2 uptake [a]	Final concentration (mM)	Percent changes of fluorescence [b]	Final concentration (mM)	Changes of impulse frequency [a] times initial frequency
Glucose	0.6	+150	5–80	0	1	2
Gluconate			27–80	−12; −28		0
Pyruvate	35	0	10–80	0	43–75	0
Citrate	35	+600			18–50	37
Isocitrate	25	+120			9–27	13
α-Ketoglutarate	27.5	+300	40–53	+20; +44	5–27	2
Glutamate	18	+20	50	+18; +24	28	Block
GABA	0.04–0.1	0	10	−6; −21	0.04	Block
Succinate	75	+180	1.20–80	0	4–96	15
Fumarate	43	0	40	0	35–57	0
l-Malate	18	+210	53	0	14–43	0
Oxalacetate	7	−25	53	−7; −33	7–16	0

[a] Giacobini and Marchisio.[65a,65b,66]
[b] Rossini et al.[43]

the inhibition. Hydroxylamine, an inhibitor of pyridoxal-kinase, the enzyme responsible for the conversion of vitamin B_6 to the corresponding phosphate, potentiated the inhibitory effects of glutamate and GABA.[66] Glutamate had very little effect and GABA had none on the O_2 consumption of the SRN.

Several possible explanations of the inhibitory action of glutamate were suggested by Giacobini and Marchisio[65a,65b,66]: (1) a direct inhibitory action on some receptor sites producing an overdepolarization with subsequent cathodal block; (2) a direct effect on the energy metabolism; (3) an indirect effect on the GABA-containing inhibitory synapses; and (4) an indirect effect mediated through an excitatory action on the muscle receptors present in the preparation.

The second hypothesis is improbable since intracellular injection of glutamate is without effect.[89] The above suggestions clearly illustrate the difficulty of interpreting pharmacological effects even in a single neuron completely isolated from surrounding cells and in conditions of controlled environment.

The "metabolic" interpretation also suffers from the fact that it is not known whether the substrates employed freely permeate the cell membrane. In the case of the respiratory effect of some of these compounds, we should not, however, discard the possibility of a metabolic action. The capacity of some intermediates (citrate, isocitrate, α-ketoglutarate, etc.) to bind Ca^{2+} also could explain the effect of some of the substrates on the impulse activity.

The fact that nervous tissue respiration may be stimulated by the reduction of extracellular calcium without the presence of spontaneously initiated impulse activity[90] is not supported by our results in the SRN, where Ca-free solutions have been found to markedly diminish O_2 uptake.

The significance of such pharmacological studies is evident, but the relationship of the different steps of the TCA cycle to generation and propagation of the impulse in SRN can be analyzed only by determining the subsequent variations of substrate levels after physiological stimulation, as previously reported in Table VI.

The effects of different substrates added to the external medium on the fluorescence of the SRN have been reported by Rossini et al.[43] The average fluorescence level of the cell is equivalent to that of a 0.7 M solution of DPNH.[43] The fluorescence of the SRN preparation is not equally distributed throughout the whole surface area of the neuron but is much lower in the dendrite and the axon and in the tissue surrounding the neuron.[43] In SRNs in which the neurons have been treated with formaldehyde to demonstrate catecholamines,[91] the so called nonspecific (protein) fluorescence is confined to the cell body and to the dendritic zone. According to the data of Rossini et al.[43] (Table VII), most glycolytic and TCA cycle intermediates did not significantly change the PNH level. Some of them, however, (Table VII) produced an increase of fluorescence (α-ketoglutarate and glutamate), whereas others produced a decrease (gluconate, GABA, and oxalacetate). Phosphate compounds, like ATP, ADP, and phosphocreatine, were inactive.

TABLE VIII

The Crayfish Stretch Receptor Neuron—Effect of Inhibitors of Anaerobic and Oxidative Metabolism on O_2 Uptake, Impulse Activity, and Fluorescence in Vivo

Type of inhibition	Inhibitor	Concentration[a] suppression or activation O_2 uptake (mM)	Percent[a] inhibition O_2 uptake	Concentration[a] suppression impulse activity (mM)	Concentration used in microfluorimetric experiment (mM)	Percent[b] change of fluorescence	Approximate[b] time to suppress impulse activity (min)
Electron transfer	Amytal	10	100	1.3	4–8	+12; +40	20–5
	Na-cyanide	10	100	6.5	0.25–5	0	120–60
	Antimycin A	0.2	100	0.02	0.5–11 (μM)	+55; +200	60–10
	Rotenone			0.03 (μM)			
	Sulfide				1	0	120
Uncoupling of oxidative phosphorylation	2-4-DPN		Activation	0.75			
Glycolysis	Na-arsenate	30	10	21	0.1–10	0	120–20
	Dicumarol	1	Activation	0.01	1	−10	Several hours
	2-deoxy-d-glucose	10	100	2.0			
	Na-iodoacetate	2–10	40	5.0	0.1–10	−13; −58	120–60
	Phloridzin				0.1–1	0	120–30
	2-Methyl-1-4-naphthoquinone				0.005–0.01	0; −16	120
	Meta-arsenite				0.1–1		
Citric acid cycle	Na-malonate			0.02	2–53	0	30–10
	Monofluoro-acetate				1–15	0	300–120
	Oxaloacetate					0	120–80
Oxidative phosphorylation	Na azide	20	25	25	1–53	−25; −33	120
	Oligomycin	3 (μM)	100	2 (μM)	2–10	0	120–10
	Atractylate	5–10	Activation	7			

[a] Giacobini.[22]
[b] Terzuolo et al.[11] and Rossini et al.[43]

The explanation that the membrane permeability may limit the action of these substrates cannot apply to all of them since it is known that at least some permeate readily through the membrane (glucose and succinate), as shown by their pronounced effect on O_2 uptake (Table VII). Phosphoglyceric acids, isocitrate, and citrate did not induce more pronounced changes in PNH level than those occurring in Ca^{2+}-free solutions,[43] i.e., inexcitability and drop of PNH level.[43] It also should be noticed that the effects of gluconate, α-ketoglutarate, and glutamate were obtained only with high concentrations, i.e., more than twice those producing changes in O_2 uptake and impulse frequency (Table VII). For α-ketoglutarate, the increase in PNH level was maximal during the first 20–40 min but it was still present after more than 2 hr.[43]

Succinate, which produced a marked change of O_2 uptake, did not increase the PNH level.[43] The lack of effect of the administration of succinate and glucose on fluorescence of the SRN is not consistent with the results obtained on the PNH level of mitochondria in the presence of the same substrates. The combined administration of several substrates, like glucose or pyruvate, after succinate, in order to achieve the well-known potentiation, did not produce any change in PNH level.[43]

The data emerging from the studies of Giacobini[22,92] on impulse activity and O_2 uptake, and Rossini et al.[43] on the fluorescence, of the SRN are summarized in Tables VIII and IX.

The effects of 20 metabolic and ion transport inhibitors acting at various enzymatic levels within several pathways were investigated by Giacobini.[22,92] A study on the effect of inhibitors on a living cell like SRN has some limitations. First, the action of these inhibitors in vivo (e.g., on mitochondrial preparations) may not be the same as that on isolated cells, where more complex metabolic functional systems are acting. This discrepancy will be seen in some of the compounds tested and reported in Tables VIII and IX. Other circumstances may affect the interpretation of data obtained with inhibitors on individual cell preparations. For example, we observe that neither azide nor cyanide inhibits O_2 uptake as much as might be expected from experiments in vitro.

The effect of the respiratory inhibitors, especially rotenone and antimycin A, suggests a coupling of respiration and impulse activity in the crustacean SRN, as previously shown with direct experiments. This also is consistent with the reported presence and activity of respiratory enzymes and cytochromes (Table V) and the fact that the activity of oxidative enzymes is increased during impulse activity. Generally, the results with the inhibitors suggest that glycolysis, the tricarboxylic acid cycle, and the electron transfer system are the major metabolic pathways for crustacea. These findings agree with those from studies on the metabolism of crustacea. It can be seen from the results summarized[22,92] in Tables VIII and IX that the impulse activity is often more sensitive to inhibitors than is respiration. This simply could be interpreted as indicating the presence of some safety mechanism for respiratory processes, needed for other functions as well as for impulse

TABLE IX

The Crayfish Stretch Receptor Neuron—Effect of Different Inhibitors on O$_2$ Uptake, Impulse Activity, and Fluorescence *in Vivo*

Type of inhibition	Inhibitor	Concentration[a] suppression or activation O$_2$ uptake	Percent[a] inhibition O$_2$ uptake	Concentration[b] suppression impulse activity	Concentration[b] used in micro-fluorescence experiment	Percent[b] change of fluorescence	Approximate time to suppress impulse activity (min)
Sulfhydril groups (mercaptide-forming agents)	p-Hydroxy-mercuri benzoate	2	40	0.25			
	α-Iodoacetamide	0.01	20–70		1–10	0; –38	120–60
	o-Iodosobenzoate	0.01	30–70		0.1–1	0	120–10
Ion transport	Ouabain		No effect	0.12	0.1	+13	20–25
	Digoxin		Activ.	0.09			
	Tetrodotoxin	50–100 (nM)	No effect	50 (nM)			
Alcohols	Ethylalcohol	21–65	Activ.	1.75			
Protein synthesis	Puromycin	125–250 μg/ml	No effect	No effect			
RNA synthesis	Actinomycin D	1–2.5 μg/ul	No effect	No effect			

[a] Giacobini[22] and Giacobini, Hovmark, and Kometiani.[23]
[b] Terzuolo et al.[11] and Rossini et al.[43]

TABLE X

Concentrations (mM/l) of Intra- and Extracellular Ions in Invertebrate Nervous Tissues[e]

Species	Preparation	Na^+ (i)	Na^+ (o)	K^+ (i)	K^+ (o)	K_i^+/Na_i^+	Na_o^+/K_o^+	Ca^{2+} (i)	Ca^{2+} (o)	Cl^- (i)	Cl^- (o)	References
Crustacea												
Homarus americanus	Periph. nerve		231[c]	203[d]	8[c]	10	29[c]	65[d]	14[c]	107[d]	270[c]	(16)
	Ventral nerve cord		465	307	10		46					(16)
Carcinus maenas	Periph. nerve	30–53, 152[a]	460	412–432, 260[a]	10	1.7[a]	46	27, 6.5[a]	13[c]	145[a]	557[c]	(16,96)
Cancer pagurus	Periph. nerve		502[c]	134[d]	12[c]		42[c]		14[c]		509[c]	(16)
Maia squinado	Periph. nerve	65[a]	488[c]	132[d]	12[c]		40[c]		14[c]		554[c]	(16)
Orconectes virilis	Stretch receptor	12	205	180	5.4	15	38					(23,79)
Procambarus alleni	Stretch receptor	12	205	180	5.4	15	38			12.7[a]		(23,79)
Astacus astacus	Stretch receptor	12–17.4[a]	208	180–265[a]	5.2	15[a]	40					(16,23,24,79)
Arachnida												
Limulus polyphemus	Periph. nerve		303[c]	123[d]	9[c]		34		6.3[c]		308[c]	(16)
Insecta												
Periplaneta americana	Nerve cord	67	214	225	3–17	3.3	13–71	15	4.2[c]		144[c]	(16)
Carausius morosus	Nerve cord	86	212	556	124, 18[c]	6.5	1.7–12[c]	30[d]	7.5[c]	133[c]	101[c]	(16)
Mollusca												
Aplysia	Axoplasm	67[a]	587[b]	232[a]	12[b]	3.5[b]	5	0.56[c]	14[b]	46–173[a]	646[b]	(97)
Loligo pealii	Giant axon	44	354	369	17	8.4	20	14[a]		140[a]		(16,96)
Sepiotenthis sepioidea	Giant axon: axoplasm	52	344[a]	335		6.5				135		(81)
	Schwann cell	312		220		0.7				167		(81)

[a] Axoplasm concentration in millimoles per kilogram tissue.
[b] Blood concentration in millimoles per kilogram tissue.
[c] Haemolymph concentration in millimoles per kilogram.
[d] Gross content in millimoles per kilogram tissue.
[e] i, inside concentration; o, outside concentration.

propagation. The only exception to this is the effect of the inhibitors of ion transport, but it should be noted that these compounds not only affect respiration but also are inhibitors of ATPase activity.

The marked effect of the glycolytic inhibitors and of Na^+-malonate[22,92] strongly suggests a link between glycolysis and the Krebs cycle on the one hand and the maintenance of impulse activity on the other. This view is also supported by the fact that glucose restores activity after the blocks caused by several inhibitors, as reported by Giacobini.[22,92] If we are to decide whether the inhibitory action of ouabain on O_2 uptake is related to active transport processes in the SRN, it will be necessary to determine Na^+ and K^+ intracellularly in cells treated with different inhibitors. The results of such investigations were reported in Section III,F.

Rossini et al.[43] and Terzuolo et al.[11] studied the effect of several inhibitors on the fluorescence of the SRN preparation, using the method of Chance et al.,[14,15] and their results are summarized in Tables VIII and IX. With the exception of Na^+ iodoacetate, glycolytic inhibitors had little or no effect on the fluorescence of SRN, and even this inhibitor had to be applied in concentrations 10 times greater than those shown to decrease the fluorescence of other tissues. This concentration produces a reduction of the ATP level in the SRN.[43] Dicoumarol consistently reduced the PNH level and decreased the ATP level, but the concentration used by Rossini et al.[43] did not influence impulse activity for several hours. In our hands 100 times this concentration completely suppressed impulse activity in a few minutes[22] (Table VII).

Arsenite, an inhibitor of the citric acid cycle, did not alter the fluorescence of the SRN although it did suppress electrical excitability.[43] Monofluoroacetate and transaconitate, in concentrations of up to 15 and 30 mM respectively, did not influence the electrical activity, and the former also was found to have no effect on the ATP level of the cell. In contrast to these findings, oxaloacetate consistently induced a decrease of fluorescence (Table VIII). This inhibitor acts on the mitochondrial succinate dehydrogenase but also can change the cytoplasmic redox potential. The increase in PNH produced by antimycin (Table VIII) could not be investigated because the compound itself is highly fluorescent.[43]

The increase in fluorescence obtained with amytal[43] is in agreement with results obtained from studies in vivo, showing that barbiturates inhibit the reduction of DPNH dehydrogenase. The absence of changes in reduced PN level in the presence of cyanide suggested the presence of a cytochrome unaffected by cyanide. Such a cytochrome has already been found by Pappenheimer and Williams[93] in the microsomes of sea urchin and in crustacean muscle by Mattisson.[94] The results, together with the finding that ouabain increases the reduced pyridine nucleotides,[43] may indicate a relationship between electron transport in the microsomal component of the respiratory chain and the Na^+- K^+-activated ATPase as discussed in Section III,C.

In connection with the significance of inhibitor studies on transport mechanism, it should be emphasized that ouabain and digoxin rapidly blocked impulse activity at concentrations of about 10^{-4} M.[22,79] Comparing the effect of ouabain at this concentration and controls in physiological solutions it was found that the ouabain-treated SRN preparations were able to fire, without addition of exogenous substrates, on the average a total of about 11,000 impulses, whereas the latter fired a total of 360,000 impulses. If glucose was added after the block of the impulse activity due to ouabain, the impulse activity recovered and the cell could fire on average 1.3 million impulses. ATP had no restoring effect. In the presence of ouabain (10^{-5} M), the impulse activity of the SRN was still present after 40 min. However, in some cases the cell could fire for almost 5 hr. In spite of this, a few seconds after the addition of this inhibitor we noticed changes in the height of the spike and early changes in the repetitive firing, in agreement with those reported by Lieberman in isolated axons.[95] This suggests that the action of this drug on transport mechanisms may be detected much earlier than previously reported. The injection of ouabain into the SRN (cell body) did not increase its effect.[79] This should be expected if ouabain is acting on the membrane-localized ATPase.

The O_2 uptake was shown to be more sensitive to ouabain than the impulse activity. A concentration of 10^{-4} M inhibited the cell respiration by 30–80%. It also was found that the effect on O_2 uptake was diphasic and concentration dependent[22,79]; low concentrations (below 10^{-4} M) increased respiration.[22,79]

These results clearly indicate that the O_2-dependent fraction of the active transport in SRN is quite significant and in the order of at least 40% of the total cell respiration. The SRN also was found to be more sensitive to ethyl alcohol than are nerve tissue slices, increasing its O_2 consumption two- to fourfold.

I. General Remarks

Summarizing the results reported in this section, it appears that the prolonged physiological stimulation of the stretch receptor of the crayfish initiates a sequence of biochemical events now readily measurable at a cellular level (Figs. 2, 6, 7, and 10).

O_2 consumption (Fig. 2), enzyme activity (Figs. 2, 6, and 7), oxidation–reduction processes, pyridine nucleotides, phosphate compounds (Figs. 6 and 7), and cations (Fig. 11B) have been measured. The results obtained in the author's laboratory indicate the importance of glycogen as a source of energy for impulse activity and in this invertebrate suggest a scheme of metabolic control similar to that previously reported for the electric organ of electrophorus[75] (Fig. 9). From Fig. 9 it has to be postulated that ATPase is present in the membrane of the nerve cells, that arginine phosphate is the energy store, and that glycolysis is the source of energy.

Figure 10 is a schematic presentation of the findings in the author's laboratory and shows the chemical and physiological sequence involved in the transformation of the mechanical stimulus to an electrical signal and finally to the production of a nerve impulse in a receptor cell.

The external stimulus is represented by the mechanical stretch of the dendrites, which act as a transducer. This step probably triggers a chemical mechanism, which will provide the necessary energy for the amplification of the output of the transducer. The amplified output of the transducer is then applied to the encoding element that generates the impulses. In the latter step a chemical mechanism may be involved. It is not known whether the receptor has the capacity to synthesize informational macromolecules like other neurons; however, it is generally accepted that in the nervous system the elaboration of the incoming information begins at the level of the receptor.

The three major metabolic pathways involved in the chemical processes outlined above are shown in Fig. 10. They are the pentose shunt, the tricarboxylic acid cycle, and glycolysis. On the basis of our results it is suggested that these pathways contribute differently to the three series of biochemical events related to the process of excitation in the neuron, i.e., (1) active transport of ions, (2) membrane biosynthesis, and (3) synthesis of informational macromolecules. The quantitative contribution of each of these events and their temporal sequence are not known yet.

IV. ACKNOWLEDGMENT

The investigations made by the author and his co-workers were supported by grants NB 04561-01-02-03-04 and 05 from the U.S. Public Health Service, from the Swedish Medical Research Council, project 12X-246-01-02-03 from the Foundations Therese and Johan Anderssons Minne and Gustaf and Tyra Svenssons Minne, from the Karolinska Institute (Reservationsanslaget), from the Swedish Society for Medical Research, and from the National Research Council of Italy.

V. REFERENCES

1. O. H. Lowry, The quantitative histochemistry of the brain, *J. Histochem. Cytochem.* **1**:420–427 (1953).
2. E. M. Eisenstein and G. H. Krasilovsky, *in Invertebrate Nervous Systems* (C. A. G. Wiersma, ed.), pp. 329–332, The University of Chicago Press, Chicago (1967).
3. N. Chalazonitis, Chémopotentiels des neurones géants fonctionnellement différenciés, *Arch. Sci. Physiol.* **13**(1):1–38 (1959).
4. N. Chalazonitis and M. Gola, Analyses microspectrophotométriques relatives à quelques catalyseurs respiratoires dans le neurone isolé (Helix pomatia), *Arch. Sci. Physiol.* **158**:1908 (1964).
5. Th. H. Bullock and G. A. Horridge, *Structure and Function in the Nervous Systems of Invertebrates,* W. H. Freeman, San Francisco (1965).

6a. E. Giacobini, in *Second International Meeting of Pharmacology* (Prague), *Symposium on Drugs and Enzymes* (B. B. Brodie, ed.), Vol. 1, pp. 55–63, Pergamon Press, New York (1965).

6b. E. Giacobini, Metabolism and function studied in single neurons, *Annal. Inst. Super. Sanità* 1:500–520 (1965).

7. E. Giacobini, in *Neurosciences Research* (S. Ehrenpreis and O. Solnitsky, eds.), Vol. 1, Academic Press, New York (1968).

8. St. W. Kuffler and J. G. Nicholls, *The Physiology of Neuroglial Cells*, Springer Verlag, Berlin (1966).

9. D. H. Hubel and T. N. Wiesel, Receptive fields and functional architecture in two non-striate visual areas (18 and 19) of the cat, *J. Neurophysiol.* 28:229–289 (1965).

10. E. Giacobini, E. Handelman, and C. A. Terzuolo, An isolated neuron preparation for studies of metabolic events at rest and during impulse activity, *Science* 140:74–75 (1963).

11. C. A. Terzuolo, B. Chance, E. Handelman, L. Rossini, and P. Schmelzer, Measurements of reduced pyridine nucleotides in a single neuron, *Biochim. Biophys. Acta* 126:361–372 (1966).

12. N. Chalazonitis and A. Arvanitaki, Chromoprotéides et succinoxydase dans divers grains isolables du protoplasme neuronique, *Arch. Sci. Physiol.* 10:291–319 (1956).

13. J. S. Alexandrowitz, Muscle receptor organs in the abdomen of *Homarus vulgaris* and *Palinurus vulgaris, Quart. J. Microscop. Sci.* 92:163–199 (1951).

14. B. Chance, P. Cohen, F. Jobsis, and B. Schoener, Intracellular oxidation-reduction states in vivo, *Science* 137:499–508 (1962).

15. B. Chance and A. V. Legallais, A spectrofluorometer for recording of intracellular oxidation–reduction states, *IEEE Trans. Biomed. Electron.* 10:40 (1963).

16. J. E. Treherne, *The Neurochemistry of Arthropods*, Cambridge University Press, Cambridge (1966).

17. W. Grampp and J. E. Edström, The effect of nervous activity on ribonucleic acid of the crustacean receptor neuron, *J. Neurochem.* 10:725–731 (1963).

18. J. E. Edström and W. Grampp, Nervous activity and metabolism of ribonucleic acids in the crustacean stretch receptor neuron, *J. Neurochem.* 12:735–741 (1965).

19. H. Hydén, in *The Cell* (J. Brachet and A. Mirsky, eds.), Vol. IV, p. 215, Academic Press, New York (1960).

20. J. E. Edström, D. Eichner, and A. Edström, The ribonucleic acid of axons and myelin sheaths from Mauthner neurons, *Biochem. Biophys. Acta* 61:178–184 (1962).

21. G. Toschi and E. Giacobini, Puromycin and the impulse activity of crayfish stretch receptor neuron, *Life Sci.* 4:1831–1834 (1965).

22. E. Giacobini, The effect of metabolic and ion transport inhibitors on the impulse activity and the oxygen uptake of an isolated crustacean neurone, *Acta Physiol. Scand.* 66:34–48 (1966).

23. E. Giacobini, S. Hovmark, and Z. Kometiani, Intracellular variations of Na^+ and K^+ in isolated nerve cells. *Acta Physiol. Scand.* 71:391–400 (1967).

24. B. G. Wallin, Intracellular ion concentrations in single crayfish axons, *Acta Physiol. Scand.* 70:419–430 (1967).

25. M. G. Larrabee, Oxygen consumption of excised sympathetic ganglia at rest and in activity, *J. Neurochem.* 2:81–101 (1958).

26. P. F. Cranefield, F. Brink, and D. W. Bronk, The oxygen uptake of the peripheral nerve of the rat, *J. Neurochem.* 1:245–249 (1957).

27. E. Giacobini, Neurophysiological and biochemical correlations in isolated nerve cell preparations at rest and during impulse activity, *Abst. 2nd Intern. Meeting Pharmacol.* (Prague), *Suppl. 12*, 107 (1963).

28. J. M. Ritchie, The oxygen consumption of mammalian non-myelinated nerve fibers at rest and during activity, *J. Physiol.* 188:309–329 (1967).

29. B. C. Abbott, A. V. Hill, and J. V. Howarth, The positive and negative heat production associated with a nerve impulse, *Proc. Roy. Soc.* **B148**:149–187 (1958).

30. S. S. Kety and C. F. Schmidt, The nitrous oxide method for the quantitative determination of cerebral blood flow in man: Theory, procedure and normal values, *J. Clin. Invest.* **27**:476–483 (1948).

31. K. A. C. Elliott and I. H. Heller, *in Metabolism of the Nervous System* (D. Richter, ed.), p. 286, Pergamon Press, London (1958).

32. S. R. Korey and M. Orchen, Relative respiration of neuronal and glial cells, *J. Neurochem.* **3**:277–285 (1959).

33. E. Giacobini, The distribution and localization of cholinesterases in nerve cells. Academic dissertation, *Acta Physiol. Scand.* **45**: Suppl. 156 (1959).

34. A. Hamberger, Oxidation of tricarboxylic acid cycle intermediates by nerve cell bodies, *J. Neurochem.* **8**:31–35 (1961).

35. H. Hydén and P. W. Lange, The steady state and endogenous respiration in neuron and glia, *Acta Physiol. Scand.* **64**:6–14 (1965).

36. M. H. Epstein and J. S. O'Connor, Respiration of single cortical neurons and of surrounding neuropile, *J. Neurochem.* **12**:389–395 (1965).

37. Ch. Edwards, C. A. Terzuolo, and Y. Washizu, The effect of changes of the ionic environment upon an isolated crustacean sensory neuron, *J. Neurophysiol.* **26**:948–957 (1963).

38. C. Eyzaguirre and S. W. Kuffler, Processes of excitation in the dendrites and in the soma of single isolated sensory nerve cells of the lobster and crayfish, *J. Gen. Physiol.* **39**:87–119 (1955).

39. O. H. Lowry, J. V. Passonneau, F. X. Hasselberger, and D. W. Schulz, Effect of ischemia on known substrates and cofactors of the glycolytic pathway in brain, *J. Biol. Chem.* **239**:18–30 (1964).

40. E. Giacobini and A. Grasso, Variations of glycolytic intermediates, phosphate compounds and pyridine nucleotides after prolonged stimulation of an isolated crustacean neurone, *Acta Physiol. Scand.* **66**:49–57 (1966).

41. E. Giacobini, *in Neurosciences Research* (S. Ehrenpreis and O. Solnitsky, eds.), Vol. 2, Academic Press, New York (1969).

42. C. A. Terzuolo, G. Bonewell, E. Giacobini, E. Handelman, and S. Lin, Metabolic studies in a single isolated nerve cell, *Federation Proc. Abstr.* **23**:130 (1964).

43. L. Rossini, H. P. Cohen, E. Handelman, S. Lin, and C. A. Terzuolo, Measurements of oxidoreduction processes and ATP levels in an isolated crustacean neuron, *Ann. N.Y. Acad. Sci.* **137**:864–876 (1966).

44a. N. Chalazonitis, M. Gola, and A. Arvanitaki, Oscillations lentes du potentiel de membrane neuronique, fonction de la pO_2 intracellulaire. Neurones autoactifs d'*Aplysia depilans,* *Compt. Rend. Soc. Biol.* **159**:2451 (1965).

44b. N. Chalazonitis, M. Gola, and A. Arvanitaki, Microspectrophotométrie différentielle sur des neurones géants *in vivo* (*Aplysia depilans*). Measure de la diffusibilité de l'oxygène, *Compt. Rend. Soc. Biol.* **159**:2440 (1965).

45. N. Chalazonitis and M. Gola, Enregistrements simultanés de la pO_2 intracellulaire et de l'autoactivité électrique du neurone géant (*Aplysia depilans*). *Compt. Rend. Soc. Biol.* **159**:1770 (1965).

46. N. Chalazonitis and H. Takeuchi, Variations de l'excitabilité directe somatique, en hyperoxie (neurones géants d'*Aplysia fasciata* et *Helix pomatia*), *Compt. Rend. Soc. Biol.* **1588**:2400 (1964).

47. E. Florey, Chemical transmission and adaptation, *J. Gen. Physiol.* **40**:533–545 (1957).

48. C. A. G. Wiersma, E. Furshpan, and E. Florey, Physiological and pharmacological observations on muscle receptor organs of the crayfish, cambarus clarkii girard, *J. Exptl. Biol.* **30**:136–150 (1953).

49. H. McLennan and D. H. York, Cholinoceptive receptors of crayfish stretch receptor neurones, *Comp. Biochem. Physiol.* **17**:327–333 (1966).

50. E. A. Maynard and D. M. Maynard, Cholinesterase in the crustacean muscle receptor organ, *J. Histochem. Cytochem.* **8**:376–379 (1960).

51. D. Nachmansohn, *Chemical and Molecular Basis of Nerve Activity*, Academic Press, New York (1959).

52. H. Hydén and A. Pigon, A cytophysiological study of the functional relationship between oligodendroglial cells and nerve cells of Deiters' nucleus, *J. Neurochem.* **6**:57–72 (1960).

53. D. H. Burrin and R. B. Beechey, Cytochrome oxidase and cytochromes a and a_3 in crab mitochondria, *Biochem. J.* **87**:48–53 (1963).

54. H. Hydén and P. W. Lange, A kinetic study of the neuron–glial relationship, *J. Cell Biol.* **13**:233–237 (1962).

55. A. Hamberger and H. Hydén, Inverse enzymatic changes in neurons and glia during increased function and hypoxia, *J. Cell Biol.* **16**:521–525 (1963).

56. A. Hamberger and J. Sjöstrand, Respiratory enzyme activities in neurons and glial cells of the hypoglossal nucleus during nerve regeneration, *Acta Physiol. Scand.* **67**:76–88 (1966).

57. Z. P. Kometiani, Free radicals and active ion transport, *Biophysics* **10**:389 (1965).

58. Z. Kometiani and R. H. Cagan, An electron spin resonance signal in brain microsomes, *Biochim. Biophys. Acta* **135**:1083–1086 (1967).

59. M. Gola and N. Chalazonitis, Measures spectrophotométriques de la saturation en oxygéne de l'hémoprotéine d'*Aplysia depilans*, *Compt. Rend. Soc. Biol.* **159**:1777 (1965).

60. E. Giacobini and J. F. Jongkind, The physiological significance of the pentose shunt in invertebrate neurons, *Acta Physiol. Scand.* **73**:255–256 (1968).

61. J. Katz and R. Rongstad, The labeling of pentose phosphate from glucose-^{14}C and estimation of the rates of transaldolase, transketolase, the contribution of the pentose cycle, and ribose phosphate synthesis, *Biochemistry* **6**:2227–2247 (1967).

62. M. A. McWhinnie and J. D. O'Connor, Metabolism and low temperature acclimation in the temperate crayfish, *Orconectes virilis*, *Comp. Biochem. Physiol.* **20**:131–145 (1967).

63. J. A. Riegel, Blood glucose in crayfishes in relation to moult and handling, *Nature* **186**:727 (1960).

64. R. L. Puyear, C. H. Wang, and A. W. Pritchard, Catabolic pathways of carbohydrate in the intermolt crayfish, *Pacifastacus leniusculus*, *Comp. Biochem. Physiol.* **14**:145–153 (1965).

65a. E. Giacobini and P. C. Marchisio, Pyruvate, glutamate and tricarboxylic acid intermediates in the crustacean stretch receptor neurone after prolonged impulse activity, *Acta Physiol. Scand.* **66**:248–248 (1966).

65b. E. Giacobini and P. C. Marchisio, The action of tricarboxylic acid cycle intermediates and glutamate on the impulse activity and respiration of the crayfish stretch receptor neurone, *Acta Physiol. Scand.* **66**:58–66 (1966).

66. E. Giacobini and P. C. Marchisio, Glutamate and the slow adapting stretch receptor neuron of the crayfish (SRN): Its effect on impulse activity and respiration, and its level before and after physiological stimulation, *IV Intern. Meet. Neurobiol.* (*Stockholm*) (C. Euler, S. Skoglund, and V. Siderberg, eds.), pp. 395–399, Pergamon Press, New York (1968).

67. H. Waelsch, S. Berl, C. A. Rossi, D. D. Clarke, and D. P. Purpura, Quantitative aspects of CO_2 fixation in mammalian brain *in vivo*, *J. Neurochem.* **11**:717–728 (1964).

68. L. J. Coté, S. C. Cheng, and H. Waelsch, CO_2 fixation in the nervous system, *J. Neurochem.* **13**:721–729 (1966).

69. A. L. Hodgkin and R. D. Keynes, Movements of cations during recovery in nerve. *Symp. Soc. Exptl. Biol.* **8**:423 (1954).

70. A. M. Shanes and D. E. S. Brown, The effect of metabolic inhibitors on the resting potential of frog nerve, *J. Cell. Comp. Physiol.* **19**:1–13 (1942).

71. P. C. Caldwell, The phosphorus metabolism of squid axons and its relationship to the active transport of sodium, *J. Physiol.* **152**:545–560 (1960).

72. P. C. Caldwell, A. L. Hodgkin, R. D. Keynes, and T. I. Shaw, The effects of injecting "energy-rich" phosphate compounds on the active transport of ions in the giant axons of loligo, *J. Physiol.* **152**:561–590 (1960).

73. P. C. Caldwell and R. D. Keynes, Phosphagen break-down and lactic acid formation on stimulation of the electric organ of *Electrophorus*, *J. Physiol.* **169**:37P–38P (1963).

74. X. Aubert, B. Chance, and R. D. Keynes, Optical studies of biochemical events in the electric organ of *Electrophorus*, *Proc. Roy. Soc.* **160**:211–245 (1964).

75. P. K. Maitra, A. Ghosh, B. Schoener, and B. Chance, Transients in glycolytic metabolism following electrical activity in electrophorus, *Biochim. Biophys. Acta* **88**:112–119 (1964).

76. R. W. Gerard, Nerve metabolism, *Physiol. Rev.* **12**:469–592 (1932).

77. W. D. Dettbarn and F. C. G. Hoskin, Changes of glucose metabolism during lobster-nerve activity, *Biochim. Biophys. Acta* **50**:568–570 (1961).

78. B. Carlsson, B. E. Giacobini, and S. Hovmark, A modified microflamephotometric technique for Na^+ and K^+ determinations in individual somatic cells (Abstr. Scand. Physiol. Congr., Åbo 1966), *Acta Physiol. Scand. Suppl.* 277, **68**:32 (1966).

79. E. Giacobini, Energy metabolism and ion transport studied in single neurons (Proc. Symp. Biophys. Physiol. Biol. Transport, Rome), *Protoplasma* **63**:52–55 (1967).

80. B. Carlsson, E. Giacobini, and S. Hovmark, Measurement of intracellular Na^+ and K^+ in single cells by means of a modified microflamephotometric technique, *Acta Physiol. Scand.* **71**:379–390 (1967).

81. J. Villegas, L. Villegas, and R. Villegas, Sodium, potassium and chloride concentrations in the Schwann cell and axon of the squid nerve fiber, *J. Gen. Physiol.* **49**:1–7 (1965).

82. T. L. D'Yakónova, B. N. Veprintsev, A. F. Chapas, and V. Ya. Brodskii, Induction of RNA synthesis in neurons by electrical activity, *Biofizika* **10**:826 (1965).

83. A. B. Kogan and S. L. Zaguskin, Relationship between ribonucleic acid patterns and electrical activity of single stretch neuron of crayfish muscle during excitation and inhibition, *J. Evol. Biol. Physiol.* **1**:59–66 (1965).

84. H. Hydén, Protein metabolism in the nerve cell during growth and function, *Acta Physiol. Scand. Suppl.* 17, **6**:1–136 (1943).

85. M. J. Cohen, in *Invertebrate Nervous Systems* (C. A. G. Wiersma, ed.), pp. 65–78, The University of Chicago Press, Chicago (1967).

86. D. R. Curtis, J. W. Phillis, and J. C. Watkins, The depression of spinal neurones by γ-amino-*n*-butyric acid and -alanine, *J. Physiol.* **146**:185–203 (1959).

87. C. Edwards and S. W. Kuffler, The blocking effect of γ-aminobutyric acid (GABA) and the action of related compounds on single nerve cells, *J. Neurochem.* **4**:19–30 (1959).

88. K. Krnjevic and J. W. Phillis, The action of certain amino acids on cortical neurones, *J. Physiol. (London)* **159**:62–63 (1961).

89. J. S. Coombs, J. C. Eccles, and P. Fatt, The specific ionic conductances and the ionic movement across the motoneuronal membrane that produce the inhibitory postsynaptic potential, *J. Physiol.* **130**:326–373 (1955).

90. F. Brink, D. W. Bronk, and M. G. Larrabee, Chemical excitation of nerve, *Ann. N.Y. Acad. Sci.* **47**:457–485 (1946).

91. B. Falck and Ch. Owman, A detailed methodological description of the fluorescence method for the cellular demonstration of biogenic amines, *Acta Univ. Lund*, Sect. II(7), 1–23 (1965).

92. E. Giacobini, The effect of metabolic inhibitors on the respiration of an isolated neuron preparation (Abstr. XI Scand. Physiol. Congr., Copenhagen), *Acta. Physiol. Scand. Suppl.* 213, **59**:48 (1963).

93. A. M. Pappenheimer and C. M. Williams, Cytochrome b_5 and the dihydrocoenzyme I-oxidase system in the *Cecropia* silkworm, *J. Biol. Chem.* **209**:915–929 (1954).

94. A. G. M. Mattisson, The pattern of cellular respiration and its relation to the ultrastructure of the cell. A comparative study on invertebrate muscles, Thesis, University of Lund (1962).

95. E. M. Lieberman, Structural and functional sites of action of ultraviolet radiations in crab nerve fibers. II. Localization of the sites of action of UV radiation by experiments with Ca^{2+} and ouabain, *Exceptl. Cell Res.* **47**:508–517 (1967).

96. G. A. Kerkut, *in Invertebrate Nervous Systems* (C. A. G. Wiersma, ed.), pp. 5–37, The University of Chicago Press, Chicago (1967).

97. C. A. G. Wiersma, *in Invertebrate Nervous Systems* (C. A. G. Wiersma, ed.), pp. 363–364, The University of Chicago Press, Chicago (1967).

Chapter 11

NUCLEI

D. A. Rappoport, P. Maxcy, Jr., and H. F. Daginawala

Department of Pediatrics
The University of Texas Medical Branch
Galveston, Texas

I. INTRODUCTION

The elucidation of the mechanism of gene action has focused interest on the biochemical processes in the cell nucleus with emphasis on the control of protein synthesis. Nuclei have been isolated from liver, thymus, HeLa cells, and ascites tumors, among others, because these tissues consist mainly of homogeneous cells that are easily disrupted and permit isolation of intact nuclei in good yields. These techniques recently have been reviewed.[1] Attempts to isolate nuclei from the central nervous system by these methods have been only partially successful since neurons and glia yield a mixture of the various cell nuclei usually contaminated with fragments of dendrites, axons, and myelin. Only recently have adequate methods been described for the isolation of uncontaminated brain nuclei.[2–8] Consequently, the bulk of the existing information on the biochemistry of cell nuclei is from tissue other than the CNS, and the present summary is based on the thesis that, until specific and reliable biochemical data on brain nuclei become available, nuclei from all tissues are closely similar.

II. MORPHOLOGY

In general, the cells of the nervous system may be classified into three groups: neuronal, neuroglial and endothelial cells. There are many different types of neuronal cells identified in the CNS on the basis of their shape, location, and function.[9] Neuronal nuclei, while varying considerably in size, have similar morphological characteristics and a single, centrally located, dense, and spherical nucleolus. Stained nuclei exhibit a pale nucleoplasm and a diffuse chromatin network.

Glial cells are divided into three groups: the astrocytes (macroglia), oligodendroglia, and microglia. Astrocytes are the largest glial cells and show an oval nucleus containing two or more paracentral "nucleolar-like"* bodies. The nucleoplasm of astrocytes is relatively more dense than neuronal nuclei but less dense than that of other glial cells. Oligodendroglia have small, ovoid nuclei with dense chromatin and contain two or more paracentral "nucleolar-like" bodies. In some instances, it is extremely difficult to distinguish between astrocytes and oligodendroglial nuclei. Microglial nuclei have an irregular, elongated shape with a very dense nucleoplasm that obscures the nucleolus. In isolated nuclei, it is almost impossible to distinguish oligodendroglial from microglial nuclei; however, neuronal and glial nuclei can be differentiated by their morphological characteristics.[4,6–8]

The interphase nuclei of the CNS, as in other tissues, is delineated from the cytoplasm by a double membrane. The inner membrane envelopes the nucleoplasm while the other appears as an extension of the endoplasmic reticulum. Under electron microscopy, the nuclear double membrane exhibits numerous pores 500 Å in diameter.[11,12]

The most distinctive structure of the nucleus is the nucleolus. Under phase contrast microscopy, the nucleoli show irregular dense and light areas. These appear, under electron microscopy, as a matrix of dense clusters of fibers adjacent to areas with diffuse filaments among which are dispersed ribonucleoprotein (RNP) granules 150 Å in diameter.[13] The nucleolus has no limiting membrane but resembles a fibrilar spongy matrix whose dense portion is referred to as the "nucleolonema" and the light, less dense areas as the "pars amorpha." The nucleolonema consists of entwined threadlike 50–80-Å diameter filaments composed of RNP, and the entire fiber has a diameter of approximately 1000 Å.[10] Interdispersed in the pars amorpha are fibrils of 50-Å diameter, also made up of RNP. Nucleoli are partially surrounded by a chromatin mass from which chromatin fibers penetrate into the nucleolar matrix.

Nucleoplasm, viewed through an electron microscope, exhibits a matrix of dispersed and condensed fibers of chromatin 100–200 Å in diameter[14] among which are dispersed granules of RNP 150-Å diameter. Within the chromatin network, there are filaments of 100-Å diameter of RNP that appear at times to be associated with the RNP granules.[15] The chromatin fibers are deoxyribonucleoprotein filaments with other components (Table I) and constitute the dispersed chromosomes of the interphase nucleus. The highly dispersed fibrils of chromatin have been designated as euchromatin,[16] and some investigators consider this as "active" chromatin, i.e., active as a DNA template for RNA polymerase synthesis of RNA.[17] The dense network of fibers in the nucleoplasm has been referred to as heterochromatin.[16] This form of chromatin presumably has repressed genes.[18,19] It is thought that heterochromatin is converted to euchromatin when genes are derepressed.[20] Euchromatin and heterochromatin have been isolated from

---

* It is not known if these bodies are true nucleoli as described by the 1965 Nucleolus Nomenclature Committee.[10]

TABLE I

Chromatin Composition[a]

Tissue	RNA	Histone	Nonhistone protein	Total protein	Reference
Rat					
brain		1.71	0.71	2.42	68
liver		1.49	0.56	2.05	68
liver	0.043	1.0	0.67	1.67	69
Mouse brain					
Neonate				4.35	70
6 month				1.56	70
15 month				2.52	70
30 month				4.68	70
Chicken brain					
4-day embryo	0.29	0.90		3.0	71
8-day embryo	0.21	0.94		3.0	71
Adult	0.18	0.78		2.9	71

[a] All the data based on unit DNA.

various tissues of the rat[21] and from calf thymus.[22] Harbers and Vogt[21] have shown that 60–70% of the total DNA in rat brain is in the form of heterochromatin as compared to 25–40% for kidney, 20–30% for liver, and 50–60% for lung tissue.

III. NUCLEAR COMPONENTS

Nuclear constituents can be isolated in native form by successive salt extractions in which each step is monitored with electron microscopy in addition to the usual analysis of components.[23,24] This technique demonstrates which of the intranuclear structures are removed at each phase of the extraction process (Table II). Initial extraction of nuclei (from rat liver) with dilute salt solutions (pH 6.0) yielded an extract of "nuclear sap" and a residue. This sap contained nuclear ribosomes, tRNA, some mRNA, and a soluble protein (globulin). Extraction of the residue with the same salt solutions, but at pH 8.0, solubilized a protein and also removed 30 S particles of RNP which contained mRNA. This represented an extract of chromatin in the initial residue. In the final step the concentrated salt solutions removed DNP as DNA and protein and some mRNA from the chromatin residue. The final residue consisted of insoluble RNP from the nuclear matrix and nucleolonema and an insoluble protein from the nuclear membrane.

In order to separate nucleoli, isolated nuclei are ruptured in isotonic sucrose solution containing Ca^{2+} using sonication or tightly fitted homogenizers, and this preparation is subjected to sucrose gradient centrifugation.[25] Analysis of rat and guinea pig liver nucleoli are presented in Table III. There are no analytical reports on brain nucleoli isolations.

TABLE II

Summary of Extracted Rat Liver Nuclear Components[a]

Extract or Residue	Components	Total yield Percent of nuclei dry weight		Nuclear origin
0.14 M Salts, pH 6	Ribosomes, tRNA, mRNA, protein (globulin)	RNA	1.5	Nucleoplasm
		Protein	23	
0.14 M Salts, pH 8	mRNA (30 S particle), protein			Chromatin
2 M salts	DNA	DNA	22	Chromatin
	mRNA	RNA 10	0.5	Nuclear and nucleolar
	Histones	Histones	33	
	Proteins	Proteins		
Residue	RNP	RNA	1.0	Nucleolar and from nuclear membrane
	Protein	Protein	4.0	
	Protein	Protein	5.0	

[a] From Georgiev.[24]

TABLE III

Composition of Nucleoli

Source	Percent dry weight			RNA/DNA	Reference
	RNA	DNA	Protein		
Rat liver	9.4	6.5	84.0	1.4	72
Regenerating rat liver					
6 hr	7.6	4.6	87.8	1.7	72
18 hr	15.5	5.4	79.1	2.9	72

Nucleolar RNA extracted from rat liver and fractionated on sucrose gradients gave a profile of 28, 35, and 45 S RNAs (in the order of decreasing quantities) and some 4 S RNA. Base analysis of these fractions gave a ratio $(A + U/G + C)$ of 0.57, corresponding to the base ratio of 28 S preribosomal RNA. Nucleolar DNA has a base ratio $(A + T/G + C)$ of 1.20, which compares well with that of nuclear DNA ratio of 1.15.[26]

Many investigators have studied the sites of synthesis and inter-relationships of the nuclear and nucleolar nucleic acids. Georgiev,[24] among others, fractionated RNA from isolated rat liver nuclei with phenol at various temperatures. He subjected each fraction to sucrose gradient centrifugation, and the isolated RNAs were tested for DNA complementarity by competitive hybridization (Table IV). The low-temperature extract contained 35–40 S RNA with a base composition similar to rRNA (GC rich). The 40–55°C extract contained a 45 S, GC-rich RNA and a heterogeneous 25–30 S RNA fraction that was AU rich but contained a small amount of 28 S rRNA. The final fraction contained a single component of RNA of the AU type and, apparently, this component was bound to the chromatin; hence, solubilization in phenol occurred only at the higher temperature. Pulse-labeling experiments

TABLE IV

Hot Phenol Extraction of Rat Liver Nuclei[a]

Fraction, °C	S value	$A + U/G + C$	Type of RNA from competitive DNA hybridization
10–40	35–40	0.60	Precursor rRNA
40–55	45	0.72	Precursor rRNA
	25–30	1.13	Precursor mRNA and rRNA
55–63	45	1.23	Precursor mRNA

[a] From Georgiev.[24]

with isolated nucleoli showed that both the 45 S and the 35–40 S RNAs labeled rapidly and that they subsequently degraded to 28 and 18 S rRNA. From these data, and with the results of the hybridization experiments, Georgiev concluded that the 45 S RNA was the precursor of 18 S rRNA and that the 35 S was the precursor of the 28 S rRNA. However, Perry,[27] using pulse labeling and actinomycin "chase" in HeLa cells, observed that the 45 S RNA was labeled first and, subsequently, the label appeared in the 35 S RNA. He concluded that the 45 S RNA was degraded to a 35 S RNA plus the 18 S rRNA, which was followed by the breakdown of the 35 S RNA to 28 S rRNA. This has been confirmed by Muramatsu, Hodnett, and Busch[28] in isolated nucleoli from rat liver. Most investigators concur with the observation that ribosomal RNA originates in the nucleolus and that the earlier observations that nucleoli did not contain 18 S rRNA is due to the rapid transfer of 18 S rRNA to the nucleoplasm. The nucleolar synthesis of ribosomal material has been reviewed by Perry.[29]

The DNA-like RNAs (high AU) with high sedimentation coefficients (25–50 S) have been observed[30] and are thought to be precursors of messenger-type RNA.[24] Georgiev, Knichevskaya, and Georgiev[31] isolated a 30 S disk-shaped particle of dimensions $180 \times 180 \times 80$ Å that contained mRNA associated with protein. Deproteinization of this particle yielded an 18 S mRNA. It appears that mRNA retained in the nucleoplasm is in this disklike form.

One of the components in both the nucleoli and nucleoplasm is the 4 S tRNA that apparently originates independently of ribosomal RNA.[27] However, at present the intranuclear sites of tRNA synthesis remain unsettled. Sirlin[32] has presented evidence that tRNA synthesis and methylation occur in the nucleoli of the *Smittia* salivary gland. Busch[25] reported that methylation of a ribosomal-type RNA, but not tRNA, occurs in the nucleoli. On the other hand, the addition of actinomycin D to HeLa cell nuclei, in concentrations that completely block nucleolar RNA synthesis, does not affect the synthesis of tRNA.[27] Woods and Zubay have shown that tRNA is not synthesized in pea seedling nucleoli.[34] To further add to this confusion, a current report by Simon, Glasky, and Rejal[33] documents the methylation of tRNA by a cytoplasmic fraction from rat and cat brain. It is evident that the present data on methylation of tRNA does not relate to the site of tRNA synthesis, and the two events occur at different locales; the synthesis of tRNA is strictly nuclear, but methylation of tRNA is not restricted to nuclei.

IV. NUCLEAR ENZYMES

Great difficulties have been experienced in distinctly localizing enzymes in the nuclei primarily because of cross contamination of enzymes between fractions.[35] Histochemical methods have been of limited value in facilitating enzyme localization mainly because of the limited number of specific reactions available for enzyme identification.

Abood *et al.*[36] reported the presence of the glycolytic pathway in nuclei isolated from homogenates of rat cortical gray matter and spinal cord white matter. Siebert *et al.*[37] also found glycolytic enzymes in bovine brain nuclei isolated in nonaqueous medium. Rappoport, Fritz, and Moraczewski[5] demonstrated the presence of the glycolytic enzymes in neonatal rat brain nuclei isolated from aqueous homogenates. None of these investigators found evidence for oxidative phosphorylation in brain nuclei, hence these nuclei generate ATP by anaerobic glycolysis, similar to the findings in liver nuclei.[38] However, thymus nuclei carry out oxidative phosphorylation,[39] and this may be due to the unique function of the thymocytes as the precursor for subsequent formation of plasma cells.

Barondes isolated a crude particle "aggregate" of RNA polymerase from rat brain nuclei and showed that its requirements for RNA synthesis were similar to those from *E. coli* and other animal tissues.[40] He also reported that the polymerase activity was high in the young rats and decreased in the older animals. However, Bondy and Waelsch[41] claimed that the rat-brain RNA polymerase increased during the growth of the rat. These contradictory claims have been resolved by Fritz, Furusawa, and Rappoport,[42] who confirmed Barondes' findings that RNA polymerase activity is highest in the newborn rat brain nuclei and decreased rapidly until the thirtieth day and, subsequently, remained at this low level of activity in the

TABLE V

Brain Cell Nuclear Enzymes

Enzyme	Tissue	Reference
Adenosine triphosphatase	Guinea pig cerebral cortex	8
	Neonatal rat brain	5
	Rat brain	36
	Bovine brain	3
Adenylate kinase	Bovine brain	3
	Neonatal rat brain	5
Glycolytic enzymes	Bovine brain	3
	Neonatal rat brain	5
ATP:NMN adenyltransferase (NAD pyrophosphorylase)	Guinea pig cerebral cortex	8, 53
5'-Nucleotidase	Dog brain cerebellum	52
Polynucleotide phosphorylase	Catfish brain	74
RNA polymerase	Rat brain (12 day and 9 month)	40
	Bovine brain cerebral cortex and white matter	40
	Rat brain	41
	Rat brain	42
	Guinea pig	8

adult. A report by Murthy and Bharucha[43] describes the presence of DNA polymerase in the rat brain nuclei. Mandel and co-workers have reported the presence of poly C synthetase in rat brain nuclei, and the entire activity was in a particulate fraction.[44] The incorporation of radioactive amino acids into the protein of isolated calf thymus and rat liver nuclei[45] and of isolated nuclei and nucleoli of guinea pig liver[46] has been reported. This activity has not been reported in isolated brain nuclei.

Siebert has listed enzyme activities that have been reported for nuclear fractions and clean nuclei prepared from malignant tissues.[47] Table V lists some enzymes that have been reported in brain nuclear preparations. The basis of these localizations are either through direct enzyme assay in purified nuclei or by the incubation of isolated purified nuclei or nucleoli with radioactive precursors or substrates and demonstration of the resultant radioactive products.

Many enzymes have been reported to be present in brain nuclear fractions, such as β-glucuronidase,[48,49] tyrosine hydroxylase,[50] monaminoxidase,[50] 3',5'-cyclic nucleotide phosphodiesterase,[51] acid and alkaline phosphatases,[50,52] and cathepsin.[48] The presence of these enzymes in the brain nuclei must await confirmatory reports since it is uncertain whether the nuclear preparations are free of contaminating endoplasmic reticulum, microsomes, mitochondria, or lysosomes.

It is now certain that nuclei synthesize proteins as well as nucleic acids. Whether lipids and polysaccharides can be synthesized in the nucleus remains to be demonstrated.

The existing data indicates that NAD is synthesized in the nucleus. Siebert speculated whether the nucleus also can act as a central control over all the NAD-dependent enzymes in the cell.[35] This is a particularly intriguing question since the genetic control of cell growth and metabolism expressed by variation in the kind and amount of protein synthesized is also centered in the nucleus. Kato and Kurokawa[8] have shown that nuclei from guinea pig brain have ATP:NMN adenyltransferase (NAD pyrophosphorylase) activity, and they noted that preparations containing predominantly neuronal nuclei have ten times as much of this adenyltransferase activity as isolated glial nuclei.[53]

V. NUCLEI ISOLATION

Isolated nuclei must conform to the following criteria before they can be used for analytical or enzymological investigations. First, the nuclei must be free of adherent cytoplasm, red cells, and cell debris; second, they must retain their morphological and biochemical integrity (as they appear *in situ*); and, finally, the isolation procedure must yield a representative sample of nuclei from a heterogeneous cell population from such organs as the brain. Purity of the isolated nuclear preparation can be assessed by microscopic examination, either after staining or under phase contrast. Examination of fixed nuclei samples under electron microscopy permits detection of any cytoplasmic or myelin contamination and also allows examination of the nuclear fine structure.

Mitochondrial contamination can be determined by assay for cytochrome oxidase, succinic dehydrogenase, or glutamic dehydrogenase. Acetyl cholinesterase and "transport" ATPase activity in a nuclear preparation are indicative of cell membrane contaminations. Siebert has listed marker enzymes for most cytoplasmic components.[1] Determination of RNA/DNA has been used to follow purification of crude nuclear preparations,[35] since a progressive decrease in this ratio is indicative of reduced cytoplasmic RNA contamination, until the values for this ratio become constant. Nuclear components can be lost through extraction with aqueous buffers and by autolytic degradation of nucleic acids and proteins. Nonaqueous solvents have been used for isolation of nuclei to prevent leaching of nuclear components;[3] however, such organic solvents under some conditions, can denature proteins and thus inactivate nuclear enzymes. Most investigators have used aqueous tissue homogenates for nuclei isolation and under carefully controlled conditions obtained clean, intact nuclei in good yields.

The concentration of sucrose in the medium influences the behavior of brain cell nuclei in subsequent centrifugations.[6,8] Brain nuclei apparently equilibrate with the surrounding environment, exhibiting increased density with increased sucrose concentrations. Generally, homogenization in 0.25–0.40 M sucrose medium yields nuclei with less cytoplasmic contamination in the crude nuclear pellet, but sucrose concentrations greater than 0.40 M can result in significant losses of nuclei.[7]

The pH of homogenizing media influences the fragility and morphology of nuclei and also affects retention of nuclear components. Media at a pH of 5.2–6.2 will cause agglutination of nuclei, and if the pH is above 6.8, nuclei tend to rupture.[7] In general, the pH of 6.3–6.5 is optimal for nuclei isolation.

The presence of Ca^{2+} or Mg^{2+} is essential for the preservation of the nuclear substructure.[4,7,54] Sporn, Wanko, and Dingman noted that K^+ was necessary to prevent swelling of the nuclei.[4] However, Ca^{2+} concentrations of 3 mM cause precipitation of the nucleoplasm.[54] Phosphate and Mg^{2+} concentrations greater than 1.5 mM result in decreased purity of nuclear preparations and cause precipitation of the nucleoplasm.[7,54] Løvtrup-Rein and McEwen report 1.0 mM $MgCl_2$ to be optimal in their system.[7]

Nonionic detergents such as Triton X-100 (iso-octylphenoxypolyethoxyethanol) facilitate solubilization of cytoplasmic membranes. Mitochondria, microsomes, and erythrocytes are disrupted and solubilized by these detergents; however, myelin is resistant to detergent treatment. Rappoport, Fritz, and Moraczewski[5] used Triton X-100 to obtain contaminant-free nuclei from neonatal rat brain. This procedure is simple and, if combined with isopycnic or density gradient centrifugation, should permit isolation of clean nuclei from adult rat brains. Blobel and Potter[55] reported that Triton X-100 removed the outer nuclear membrane during isolation of nuclei from rat liver and, under electron microscopy, the nuclei appear with a single limiting membrane. A similar observation was reported by Hadjiolov, Tencheva, and Bojadjieva-Mikhailova[6] when detergent was used for the isolation of nuclei from cat cerebral cortex. Rappoport, Fritz, and Moraczewski[5] noted that Triton treated neonatal rat brain nuclei undergo swelling but do not rupture when placed in distilled water. Rupture of these nuclei required a concentrated buffer (1 M Tris, pH 7.4) and homogenization. The stability of nuclei prepared in this manner suggests that Triton altered the nuclear membrane. Use of other detergents such as deoxycholate or Tween 40 ruptures nuclei.[55]* Exposure of homogenates or nuclei to Triton must be limited to 15–20 min to prevent nuclear disintegration, particularly neuronal nuclei.[7] Nuclei from neural tissue appear to be more easily ruptured during homogenization than liver nuclei.[4,7] Dounce or Potter-Elvehjem homogenizers with a clearance of 200–250 μ have been reported to be suitable for the preparation of brain nuclei.

Centrifugation of sucrose homogenates at 700–1200 × g for 5–15 min yields a crude nuclear pellet with myelin fragments, capillaries, intact cells, cell debris, erythrocytes, and some mitochondria. Even after homogenization in detergent media, the pellet still contains myelin and cell debris. To free crude nuclear preparations from myelin, it is necessary to use isopycnic or density gradient centrifugation in sucrose solutions of 1.5–2.2 M that removes myelin and other debris.[4,6,56,57] Borkowski, Berbec, and Brzuszkiewicz[58] homogenized rat brain in 1.5 M sucrose containing 5 mM $CaCl_2$ and then centrifuged at 20,000 × g for 60 min to obtain the nuclear pellet. Mandel, Dravid, and Pete[44] homogenized rat brain in 2.2 M sucrose containing 1 mM $MgCl_2$ and 10 mM potassium succinate and, after centrifugation at 78,000 × g for 50 min, obtained clean nuclei. The latter procedure is essentially the one developed by Chauveau, Moule, and Rouiller[59] for isolation of liver nuclei.

Some investigators have used density gradient centrifugation, not only for the purification of nuclei[6] but also to separate neuronal and glial nuclei.[7,8] Attempts to use continuous sucrose gradients for separation of mixed nuclei were not successful because nuclei formed diffuse bands that were difficult to separate.[7] Discontinuous gradients with sharp boundaries separate nuclei into discrete bands. Løvtrup-Rein resuspended the crude nuclear pellet in 2.0 M sucrose and layered it over a discontinuous gradient of 2.2, 2.4, 2.6, and 2.8 M sucrose.[7] Penetration of the nuclei at the interfaces was facilitated by allowing the prepared gradient to stand for 3 hr before use so that the boundaries became slightly diffused. Hadjiolov, Tencheva, and Bojadjieva-Mikhailova[6] layered a cat brain cortex homogenate directly over a gradient of 0.68 and 1.0 M sucrose in wide centrifuge bottles where the area of the interface was sufficiently large to avoid a thick accumulation of cell debris. The resulting nuclear pellet was free of debris.

One of the most unusual techniques for the isolation and analysis of

---

* Casola and Agranoff[74] recently reported that Triton-treated goldfish brain nuclei show degradation of RNA, while use of mixtures of Tween 40 and deoxycholate avoided this difficulty. This indicates that only Triton activated latent RNase.

nuclear components in individual neuronal and glial cells has been developed by Hydén[60,61] and Edström.[62,63] By surgical extraction of individual neurons and glia and subsequent excision of intact nuclei, a number of nuclei (30–50) are collected. The nucleic acids, proteins, or other components are extracted and analyzed by special ultramicroprocedures. These techniques have enabled Hydén and Egyházi[61,64] to investigate changes in ribonucleic acids in the Deiter's cells of the vestibular areas of rat brain before and after training. They were able to find differences in nuclear RNA of trained as compared to untrained rats in both neuronal and glial nuclei. These procedures illustrate that study of individual cells from the brain can yield a great deal of information since masking or dilution of specific molecular changes by components of neighboring heterogeneous cells is completely avoided. The complexity and costs of the equipment required for Hydén's ultramicrotechniques limit its wide use.

Recently, Rose[65] has described a method of separating brain cells and separating enriched neuronal and glial fractions by means of Ficol-sucrose discontinuous density gradient. The glial fraction was free of neurons, but the neuronal fractions contained approximately 11 % glial cells based on the assay of carbonic anhydrase that is concentrated in glia. Similar separation of glia and neurons has been reported by Satake and Abe,[66] and they have currently made improvements in these procedures.[67]

With the capability of isolating clean and intact brain nuclei and the possibility of separating neuronal and glial nuclei, future studies will clarify the glial–neuronal relationships and also may delineate the role of the nucleus in brain function.

VI. REFERENCES

1. G. Siebert, in *Methods in Cancer Research* (H. Busch, ed.), Vol. 11, pp. 287–301, Academic Press, New York (1967).
2. P. Mandel, T. Borkowki, S. Harth, and R. Mardell, Incorporation of ^{32}P in ribonucleic acid of subcellular fractions of various regions of the rat central nervous system, *J. Neurochem.* **8**:126–138 (1961).
3. G. Siebert, Enzyme und substrate der glykolse in isolierten zellkernen, *Biochem. Z.* **334**:369–387 (1961).
4. M. Sporn, T. Wanko, and W. Dingman, The isolation of cell nuclei from rat brain, *J. Cell Biol.* **15**:109–120 (1962).
5. D. A. Rappoport, R. R. Fritz, and A. Moraczewski, Biochemistry of the developing rat brain, I. Soluble enzymes in isolated neonatal brain nuclei, *Biochim. Biophys. Acta* **74**:42–50 (1963).
6. A. A. Hadjiolov, Z. S. Tencheva, and A. G. Bojadjieva-Mikhailova, Isolation and some characteristics of cell nuclei from brain cortex of adult cat, *J. Cell Biol.* **26**:383–393 (1965).
7. H. Løvtrup-Rein and B. S. McEwen, Isolation and fractionation of rat brain nuclei, *J. Cell Biol.* **30**:405–416 (1966).
8. T. Kato and M. Kurokawa, Isolation of cell nuclei from the mammalian cerebral cortex and their assortment on a morphological basis, *J. Cell Biol.* **32**:649–662 (1967).
9. J. D. Schadé and D. H. Ford, *Basic Neurology*, American Elsevier, New York (1965).

10. C. Estable, W. Bernhard, J. Gall, R. Perry, J. Sirlin, and H. Swift, *in International Symposium on the Nucleolus, Its Structure and Function* (W. S. Vincent and O. L. Miller, eds.), Monograph 23, pp. 573–574, National Cancer Institute, Bethesda, Maryland (1966).

11. U. Karlsson, Three-dimensional studies of neurons in the lateral geniculate nucleus of the rat, I. Organelle organization in the perikaryon and its proximal branches, *J. Ultrastruct. Res.* **16**:429–481 (1966).

12. D. W. Fawcett, *An Atlas of Fine Structure, The Cell, Its Organelles and Inclusions,* W. B. Saunders, Philadelphia (1966).

13. W. Bernhard, *in International Symposium on the Nucleolus, Its Structure and Function* (W. S. Vincent and O. L. Miller, eds.), Monograph 23, pp. 13–38, National Cancer Institute, Bethesda, Maryland (1966).

14. H. Ris and B. L. Chandler, *in Cold Spring Harbor Symposium on Quantitative Biology* (L. Frisch, ed.), Vol. 28, pp 1–9, Cold Spring Harbor Laboratory of Quantitative Biology, Cold Spring Harbor (1963).

15. W. Niklowitz, On the submicroscopic structure of the cell nucleus of the Ehrlich ascites carcinoma of the white mouse, *Z. Naturforsch.* **13b**:454–456 (1958).

16. E. Heitz, Heterochromatin, chromocentren, chromomercen, *Ber. Bent. Bot. Ges.* **47**:274–284 (1929).

17. V. C. Littau, V. G. Allfrey, J. H. Frenster, and A. E. Mirsky, Active and inactive regions of nuclear chromatin as revealed by electron microscope autoradiography, *Proc. Natl. Acad. Sci. U.S.* **52**:93–100 (1964).

18. K. W. Cooper, Cytogenetic analysis of major heterchromatic elements (especially XL and Y) *in Drosophilia melanogaster,* and the theory of "Heterochromatin," *Chromosoma (Berlin)* **10**:535–588 (1959).

19. J. H. Frenster, *in The Chromosome: Structural and Functional Aspects* (C. J. Dawc, ed.), Williams and Wilkins, Baltimore (1966).

20. J. H. Frenster, *in The Cell Nucleus—Metabolism and Radiosensitivity,* pp. 27–46, Taylor and Francis, London (1966).

21. E. Harbers and M. Vogt, *in The Cell Nucleus—Metabolism and Radiosensitivity,* pp. 165–172, Taylor and Francis, London (1966).

22. J. H. Frenster, V. G. Allfrey, and A. E. Mirsky, Repressed and active chromatin isolated from interphase lymphocytes, *Proc. Natl. Acad. Sci. U.S.* **50**:1026–1032 (1963).

23. I. B. Zbarsky and G. P. Georgiev, New data on the fractionation of cell nuclei of rat livers and the chemical composition of nuclear structures, *Biokhimiya* **24**:192–200 (1959).

24. G. P. Georgiev, *in Progress in Nucleic Acid Research* (J. N. Davidson and W. E. Cohn, eds.), Vol. 6, pp. 259–351, Academic Press, New York (1967).

25. M. Muramatsu and H. Busch, *in Methods of Cancer Research* (H. Busch, ed.), Vol. 11, pp. 303–359, Academic Press, New York (1967).

26. H. Busch, R. Desjardins, D. Grogan, K. Higashi, S. T. Jacob, M. Muramatsu, T. S. Ro, and W. J. Steele, *in International Symposium on the Nucleolus, Its Structure and Function* (W. S. Vincent and O. L. Miller, eds.), Monograph 23, pp. 193–212, National Cancer Institute, Bethesda, Maryland (1966).

27. R. P. Perry, The cellular sites of synthesis of ribosomal and 4 S RNA, *Proc. Natl. Acad. Sci. U.S.* **48**:2179–2186 (1962).

28. M. Muramatsu, J. L. Hodnett, and H. Busch, *in The Cell Nucleus, Metabolism and Radiosensitivity,* pp. 221–229, Taylor and Francis, London (1966).

29. R. P. Perry, *in Progress in Nucleic Acid Research* (J. W. Davidson and W. E. Cohn, eds.), Vol. 6, pp. 219–257, Academic Press, New York (1967).

30. M. Yoshikawa-Fukada, T. Fukada, and Y. Kawade, Characterization of rapidly labelled ribonucleic acid of animal cells in culture, *Biochim. Biophys. Acta* **103**:383–398 (1965).

31. O. D. Samarino, A. A. Krichevskaya, and G. P. Georgiev, Nuclear ribonucleoprotein particles containing messenger ribonucleic acid, *Nature* 210:1319–1322 (1966).

32. J. L. Sirlin, *in The Cell Nucleus, Metabolism and Radiosensitivity*, pp. 87–95, Taylor and Francis, London (1966).

33. L. N. Simon, A. J. Glasky, and T. H. Rejal, Enzymes in the central nervous system, I. RNA methylase, *Biochim. Biophys. Acta* 142:99–104 (1967).

34. P. S. Woods and G. Zubay, Biochemical and autoradiographic studies of different RNAs. Evidence that transfer RNA is chromosomal in origin, *Proc. Natl. Acad. Sci. U.S.* 54:1705–1712 (1965).

35. G. Siebert and G. B. Humphrey, *in Advances in Enzymology* (F. F. Nord, ed.), Vol. 27, pp. 239–288, Interscience, New York (1965).

36. L. G. Abood, R. W. Gerard, J. Banks, and R. D. Tschirgi, Substrate and enzyme distribution in cells and cell fractions of the nervous system, *Am. J. Physiol.* 168:728–738 (1952).

37. G. Siebert, K. H. Bässler, R. Hannover, E. Adloff, and R. Beyer, Enzymaktiväten in isolierten zellkernen in abhängigkeit von der mitotischen activität, *Biochem. Z.* 334:388–400 (1961).

38. T. E. Conover and G. Siebert, On the occurrence of respiratory components in rat liver nuclei, *Biochim. Biophys. Acta* 99:1–12 (1965).

39. I. Betal and H. M. Klouwen, *in The Cell Nucleus, Metabolism and Radiosensitivity*, pp. 281–291, Taylor and Francis, London (1966).

40. S. H. Barondes, Studies with RNA polymerase from brain, *J. Neurochem.* 11:663–669 (1964).

41. S. C. Bondy and H. Waelsch, Nuclear RNA polymerase in brain and liver, *J. Neurochem.* 12:751–756 (1965).

42. R. R. Fritz, S. Furusawa, and D. A. Rappoport, in preparation.

43. M. R..V. Murthy and A. Bharucha, DNA polymerase of rat brain, *Seventh International Congress of Biochemistry (Tokyo) Abstracts, B-183*, p. 662 (1967).

44. P. Mandel, A. R. Dravid, and N. Pete, Poly C synthetase activity in the particulate fraction of rat brain nuclei, *J. Neurochem.* 14:301–306 (1967).

45. T. Y. Wang, *in The Cell Nucleus, Metabolism and Radiosensitivity*, pp. 243–257, Taylor and Francis, London (1966).

46. R. Maggio, *in International Symposium on the Nucleolus, Its Structure and Function* (W. S. Vincent and O. L. Miller, eds.), Monograph 23, pp. 213–219, National Cancer Institute, Bethesda, Maryland (1966).

47. G. Siebert, Enzymes of cancer nuclei, *Exptl. Cell Res. Suppl.* 9:389–417 (1963).

48. H. Koenig, D. Gaines, T. McDonald, R. Gray, and J. Scott, Studies of brain lysosomes, I. Subcellular distribution of five acid hydrolases, succinic dehydrogenase and gangliosides in rat brain, *J. Neurochem.* 11:729–743 (1964).

49. J. Mordoh, Subcellular distribution of acid phosphatase and β-glucuronidase activities in rat brain, *J. Neurochem.* 12.505–514 (1966).

50. P. L. McGeer, S. P. Bajehi, and E. G. McGeer, Subcellular localization of tyrosine hydroxylase in beef caudate nucleus, *Life Sci.* 4:1859–1867 (1965).

51. E. DeRobertis, G. R. DeLores Arnaiz, M. Alberici, R. W. Butcher, and E. W. Sutherland, Subcellular distribution of adenyl cyclase and cyclic phosphodiesterase in rat brain cortex, *J. Biol. Chem.* 242:3487–3493 (1967).

52. N. Waked and S. Kerr, The distribution of phosphomonesterases and pyrophosphatase in the particulate fractions of dog cerebrum, *J. Histochem. Cytochem.* 3:75–84 (1955).

53. M. Kurokawa, T. Kato, and H. Inamura, Unequal distribution of ATP:NMN adenyltransferase activity among neuronal and glial cell nuclei, *Proc. Japan Acad.* 42:1217–1222 (1966).

54. R. Maggio, P. Siekevitz, and G. E. Palade, Studies on isolated nuclei, I. Isolation and chemical characterization of a nuclear fraction from guinea pig liver, *J. Cell Biol.* **18**:267–291 (1963).

55. G. Blobel and V. R. Potter, Nuclei from rat liver: Isolation method that combines purity with high yield, *Science* **154**:1662–1665 (1966).

56. S. Harth, R. Mardell, T. Borkowski, and P. Mandel, Etude comparee de l'incorporation du ^{32}P et de l'adenine-8-^{14}C dans les fractions subcellulaires des diverse zones du systeme nerveux central, *J. de Physiologie* **53**:362–363 (1961).

57. R. S. Piha, M. Cuenod, and H. Waelsch, *in Protides of the Biological Fluids* (H. Peeters, ed.), Vol. 13, pp. 145–150, Elsevier, Amsterdam (1966).

58. T. Borkowski, H. Barbec, and H. Brzuszkiewicz, Changes in chemical composition of isolated rat brain and liver nuclei, *Acta Biochim. Polon.* **12**:143–149 (1965).

59. J. Chauveau, Y. Moule, and C. H. Rouiller, Isolation of pure and unaltered liver nuclei, morphology and biochemical composition, *Exptl. Cell Res.* **11**:317–321 (1956).

60. H. Hydén, *in Recent Advances in Biological Psychiatry* (J. Wortis, ed.), Vol. 6, pp. 31–53, Plenum Press, New York (1964).

61. H. Hydén, *in Recent Progress in Nucleic Acid Research* (J. N. Davidson and W. E. Cohn, eds.), Vol. 6, pp. 187–218, Academic Press, New York (1967).

62. J.-E. Edström and H. Hydén, Ribonucleotide analysis of individual nerve cells, *Nature* **174**:128–129 (1954).

63. J.-E. Edström, *in Methods in Cell Physiology* (D. M. Prescott, ed.), Vol. 1, pp. 417–447, Academic Press, New York (1964).

64. H. Hydén and E. Egyházi, Changes in RNA content and base composition in cortical neurons of rats in a learning experiment involving transfer of handedness, *Proc. Natl. Acad. Sci. U.S.* **52**:1030–1035 (1964).

65. S. P. R. Rose, Preparation of enriched fractions from cerebral cortex containing isolated, metabolically active neuronal and glial cells, *Biochem. J.* **102**:33–43 (1967).

66. M. Satake and S. Abe, Preparation and characterization of nerve cell perikaryon from rat cerebral cortex, *J. Biochem. (Tokyo)* **59**:72–75 (1966).

67. M. Satake, personal communication.

68. G. P. Georgiev, L. D. Ermolaeva, and I. B. Zbarski, The quantitative relations between protein and nucleoprotein fractions of various tissues, *Biokhimiya* **25**:318–322 (1960).

69. K. Marushige and J. Bonner, Template properties of liver chromatin, *J. Mol. Biol.* **15**:160–174 (1966).

70. D. I. Kurtz and F. M. Sinex, Age related differences in the association of brain DNA and nuclear protein, *Biochim. Biophys. Acta* **145**:840–842 (1967).

71. C. W. Dingman and M. B. Sporn, Studies on chromatin, I. Isolation and characterization of nuclear complexes of deoxyribonucleic acid, ribonucleic acid, and protein from embryonic and adult tissues of the chicken, *J. Biol. Chem.* **239**:3483–3492 (1964).

72. M. Muramatsu and H. Busch, Studies on the nuclear and nucleolar ribonucleic acid of regenerating rat liver, *J. Biol. Chem.* **240**:3960–3966 (1965).

73. H. Daginawala, "Changes in brain nuclear RNA with olfactory stimulation," Ph.D. thesis, The University of Texas Medical Branch, Galveston, 1967.

74. L. Casola and B. W. Agranoff, Studies on RNA in goldfish brain, I. Isolation and labeling, *Brain Res.* **10**:227–238 (1968).

Chapter 12

LYSOSOMES*

Harold Koenig

Neurology Service
V.A. Research Hospital
and Department of Neurology and Psychiatry
Northwestern University Medical School
Chicago, Illinois

In the nervous system, as elsewhere, hydrolytic enzymes with acid pH optima are sequestered in a latent state within a special cytoplasmic granule, the lysosome. This organelle has been implicated in a number of physiological and pathological processes and has attracted considerable attention because of the destructive potential of its enzymes. In this chapter we shall briefly survey the current status of lysosomes with particular reference to the nervous system. For a general review of this subject, the reader is referred to several comprehensive articles.[1-4]

I. BIOCHEMICAL ASPECTS

Although the acid hydrolases are the best known of the lysosomal constituents, nonenzymatic components, notably lipoproteins, also occur in lysosomes. These lipoproteins possess distinctive physicochemical properties and are believed to play an important role in the compartmentation and structural latency of the acid hydrolases within lysosomal particles.[5-9]

A. Lysosomal Enzymes

The lysosomal enzymes feature three important characteristics: (1) They usually have acid pH optima; (2) they are particulate and tend to sediment together when tissue homogenates are fractionated by differential or density gradient centrifugation; and (3) they exhibit structural latency in the particulate state, with disruptive treatments releasing the soluble enzymes in a fully

* The studies in which the author participated were supported in part by grants from the following: National Institutes of Health, U.S. Public Health Service (Nos. NB05509 and NB06838); National Multiple Sclerosis Society (Nos. MS-304 and MS-372); and Atomic Energy Commission [Contract No. AT(11-1)1180].

active form. Several enzymes with pH optima close to neutrality, namely, arylamidase,[10] phosphatidic acid phosphatase,[11] and lysozyme,[12] are lysosomal constituents, but these enzymes also are quite active in acid pH.

1. Enzymatic Components of Neural Lysosomes

From studies involving many tissues it is evident that lysosomal enzymes are capable of hydrolytically splitting virtually all macromolecular tissue components, including nucleic acids, proteins, polysaccharides, and lipids, as well as diverse organic molecules of small molecular size. In mammalian brain the following acid hydrolases have been thus far localized to lysosomes by their sedimentation characteristics, and in some instances by their structural latency: acid phosphatase, β-glucuronidase, β-galactosidase, N-acetyl-β-D-glucosaminidase, acid RNase, acid DNase, aryl sulfatase, cathepsin, sialidase, and cerebroside galactosidase (see Section I,A,2).

A number of other sedimentable acid hydrolases are present in brain whose pH optima range from 3.8 to 5.5. Although the sedimentation and latency properties of these enzymes have not been characterized in brain to date, it is likely that they also are contained within lysosomal particles. Indeed some of these have been shown to be lysosomal constituents in liver and/or kidney. Among these we may mention the glycosidases, β-glucosidase[13] and N-acetyl-β-D-galactosaminidase.[14] The brain glycosidases and sialidase can attack the oligosaccharide sidechains of glycoproteins. Together with ceramidase,[15] an enzyme that hydrolyzes ceramide to sphingosine and fatty acid, and cerebroside galactosidase[16] these enzymes can totally degrade glycosphingolipids, probably in a stepwise fashion.[17]

Brain tissue contains particulate acid hydrolases that split phospholipids. These include a sphingomyelinase,[18] which cleaves sphingomyelin into phosphorylcholine and ceramide, ceramidase,[15] a phospholipase A,[19] and probably a lysolecithinase B.[19] Phospholipid- and sphingolipid-splitting hydrolases occur in lysosomal particles of liver and kidney.[20–22] Phosphatidic acid phosphatase, a neutral hydrolase present in liver lysosomes,[11] occurs in brain.[23] Arylamidases with neutral pH optima, lysosomal constituents in liver and kidney,[10] are present in brain.[24] Acid mucopolysaccharidases such as hyaluronidase also are lysosomal constituents.[25]

2. Sedimentation Properties of Cerebral Lysosomes

Neural lysosomes comprise a heterogeneous population of granules that vary widely in size, structure, physical properties, and chemical composition. This is due partly to the diversity of cell types represented in the nervous system. Thus far efforts to isolate cerebral lysosomes in a morphologically homogeneous state by classical ultracentrifugation procedures have been thwarted by the heterogeneity of lysosomes and the structural complexities of neural tissue. However, fractions can be readily prepared from brain homogenates that are substantially enriched with respect to lysosomal enzymes and lysosomal particles.

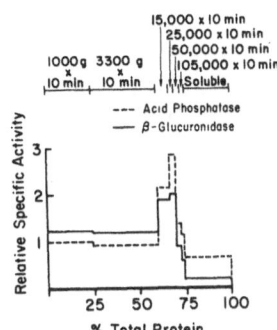

Fig. 1. Subcellular distribution of acid phosphatase and β-glucuronidase activities in rat brain homogenate. Differential centrifugation was by a modification of the method of de Duve et al.[26] Small but significant differences are seen between the two acid hydrolases. From Koenig and Gaines.[27]

The subcellular distribution pattern of lysosomal enzymes in brain homogenates depends upon the technique of fractionation. A "light" mitochondrial fraction can be prepared from 0.25 M sucrose homogenates of rat brain by differential centrifugation[26] in which a number of acid hydrolases exhibit specific activities 2–3.5-fold greater than the whole homogenate. These are acid phosphatase,[27–29] β-glucuronidase,[27] N-acetyl-β-D-glucosaminidase,[29] sialidase,[28] aryl sulfatase,[30] and cerebroside galactosidase.[30] A refinement of this technique gives somewhat improved resolution of lysosomal acid phosphatase and β-glucuronidase,[27] while another scheme for differential centrifugation fails to give a lysosome-enriched fraction[31] (Fig. 1).

Isopycnic centrifugation of particulate fractions over a density gradient affords a more potent means of separating lysosomes from other subcellular particles. Earlier Whittaker,[32] using a simple sucrose gradient system to purify nerve ending particles, noted that β-glucuronidase and acid phosphatase sediment differently and questioned the validity of the lysosome hypothesis as applied to brain. Marks and Lajtha[33] found that acid protease activity is associated with particles of high density and suggested that these particles may be lysosomes.

Koenig et al.[31] first showed that the various acid hydrolases in rat brain sediment together with lysosomal granules when particulate fractions are spun to an equilibrium over a density gradient. Isopycnic centrifugation was performed over a discontinuous density gradient consisting of 0.32, 0.8, 1.0, 1.2, and 1.4 M sucrose (Fig. 2). When a crude mitochondrial fraction of rat

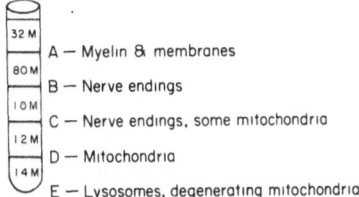

Fig. 2. Sucrose density gradient system for subfractionation of a crude rat brain mitochondrial fraction by isopycnic centrifugation[31] (spun 63,580 × g for 2 hr).

A — Myelin & membranes

B — Nerve endings

C — Nerve endings, some mitochondria

D — Mitochondria

E — Lysosomes, degenerating mitochondria

brain is spun over this gradient system, about 20–40% of the activity of five acid hydrolases is recovered together with about 5% of the protein in a pellet, fraction E (Fig. 3). When the microsomal fraction is fractionated over this gradient, about 20–45% of the acid hydrolases sediment with about 2% of the protein in a pellet, fraction D_1. Fractions E and D_1 also contain the greatest concentration of lysosomal particles, as demonstrated microscopically by staining for acid phosphatase and acid esterase activities, by

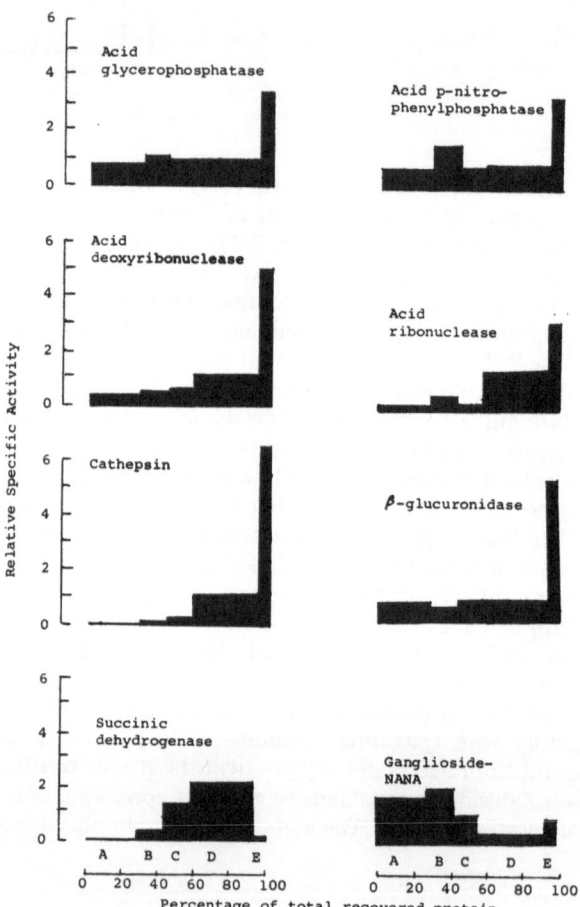

Fig. 3. Histogram of the relative specific activities of acid hydrolases in the mitochondrial subfractions of rat brain prepared by isopycnic centrifugation over the density gradient system shown in Fig. 2. Note the high relative specific activities of all the acid hydrolases and the low relative specific activity of the mitochondrial enzyme, succinate dehydrogenase, in fraction E. From Koenig et al.[31]

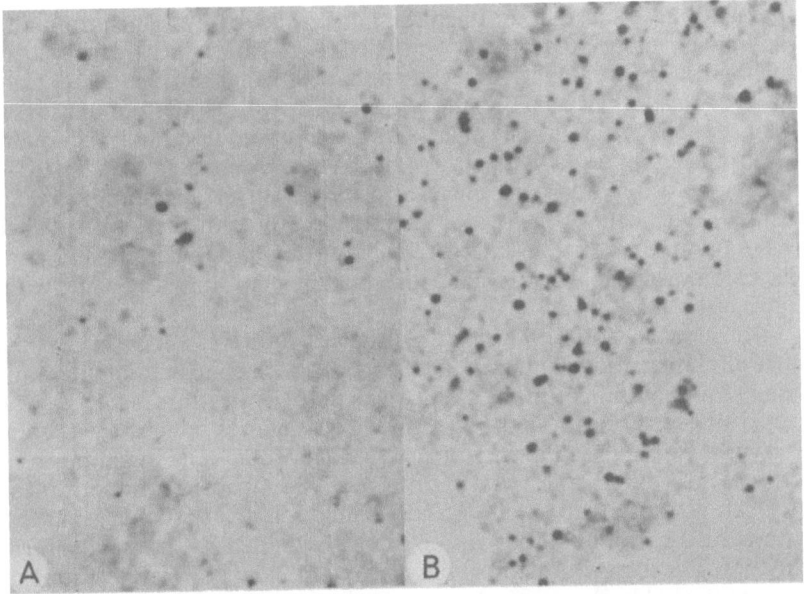

Fig. 4. Photomicrographs of fractions stained for acid phosphatase activity by Gomori's method.[71] (A) Crude mitochondrial fraction. (B) Fraction E. Note the increased concentration of reactive lysosomes in fraction E. × 2400. From Koenig et al.[31]

metachromatic staining with the basic fluorochrome, acridine orange, and by autofluorescence (Fig. 4). Osmiophilic dense bodies identified as lysosomes by electron microscopy are most abundant in fractions E and D_1. Lysosomal particles in fraction E measure about 0.2–0.8 μ, and those in fraction D_1, approximately 0.05–0.25 μ in diameter. Fraction E contains, in addition, numerous abnormal mitochondria and other constituents, while fraction D_1 contains rough endoplasmic reticulum and free ribosomes.

Lysosomes in brain, as in most tissues, tend to be denser than mitochondria. Nevertheless, a substantial portion of the lysosomes in a crude mitochondrial fraction of rat brain, about 35% of the total as measured by acid phosphatase activity, is recovered in the light subfractions A and B. This figure is probably too high, as some dense lysosomes are trapped within axons in fraction A and nerve endings in fraction B.[34]

These results have been largely confirmed by other workers. Hunter and Millson[35,36] and Mordoh,[37] using a similar gradient system to subfractionate mitochondrial pellets from mouse and rat brain respectively, found enhanced β-glucuronidase activity and numerous lysosomal particles by electron microscopy in fraction E. Fraction E also contains a comparable augmentation in the activities of the acid hydrolases, β-galactosidase,[36] N-acetyl-β-D-glucosaminidase,[36] and acid protease.[33] Sellinger and Hiatt,[38] employing a continuous sucrose density gradient for isopycnic centrifugation, found that aryl sulfatase and N-acetyl-β-D-glucosaminidase in rat brain are associated with particles that for the most part are denser than mitochondria. From small differences in the sedimentation characteristics of these hydrolases they deduced the presence of two distinct populations of lysosomes. Disparities in the subcellular distribution of various acid hydrolases have been reported by other workers. These discrepancies indicate that some degree of physical and chemical heterogeneity exists among cerebral lysosomes. This heterogeneity is attributable partly to the multiplicity of cell types in brain, each with its characteristic lysosomal population, and partly to the pleomorphism of lysosomal bodies and their associated enzymes within a single cell type.

The incorporation of foreign substances into lysosomes *in vivo* may induce ultrastructural changes in these particles with concomitant alterations in their sedimentation properties. The weak detergent Triton WR 1339, when injected intraperitoneally in the rat, causes a vacuolation of hepatic lysosomes and a marked reduction in their density on isopycnic centrifugation over a sucrose density gradient.[39] Hypertonic sucrose, injected intraperitoneally in the rat, also produces swelling and vacuolation of hepatic lysosomes,[40] accompanied by a diminution in their density on density gradient centrifugation.[41] The intravital staining of lysosomes by the basic dye, neutral red, causes remarkable changes in lysosomal structure in brain[42,43] and other tissues of the rat, including an hydropic swelling and vacuolation of the lysosomal matrix. In rat brain[43] and liver particulates[44] these structural changes also are reflected in a remarkable diminution in the equilibrium density of lysosomal acid phosphatase activity (Fig. 5).

3. *Structural Latency of Lysosomal Enzymes*

The activities of the classical lysosomal enzymes are largely latent when assayed in isotonic sucrose suspensions of cerebral particulates. A number of disruptive physical and chemical treatments activate and solubilize several lysosomal enzymes *in vitro*. These include freeze-thawing,[29,31,36,45] soni-

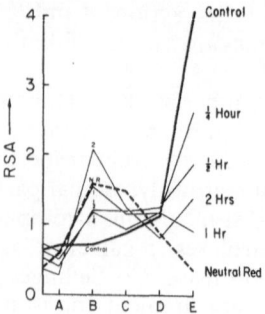

Fig. 5. The effect of neutral red on the distribution of acid phosphatase activity in subfractions prepared from a crude mitochondrial fraction of rat brain by isopycnic centrifugation. Note the shift of lysosomal acid phosphatase from fraction E into lighter fractions peaking at B. The reduction in buoyant density of neutral red-stained lysosomes is completely reversed 18 hr after dye injection. From Koenig.[43]

cation, blenderizing,[29,31,45] hypo-osmotic shock,[31,37,45] thermal activation (37°C, pH 5),[29,31,35,36,45] detergents,[29,31,37,45] and organic solvents.[44] In addition, lysosomal hydrolases can be released from cerebral particulates by low and high pH,[44,46] certain cations,[46,47] and phospholipase C[48] under conditions that do not disrupt the host particles. The mechanism responsible for the structural latency of the lysosomal enzymes is discussed in Section III.

4. Physical and Chemical Properties of Lysosomal Enzymes

Several soluble lysosomal enzymes in brain have been partially purified and their enzymatic properties studied in detail.[13–19,23,24,30,33] However, we must turn to other tissues for information on the physical and chemical characteristics of acid hydrolases. Acid DNase, isolated from hog spleen in a homogeneous state, is a basic protein with an isoelectric point of 10.2.[49] It contains 3% hexosamine and has a molecular weight of about 38,000. β-Glucuronidase from bovine liver contains 3–6% carbohydrate and has a molecular weight of about 280,000.[50] In both acid DNase and β-glucuronidase the dicarboxylic amino acids predominate over the basic amino acids, but they probably occur mainly as the amide derivatives, asparagine and glutamine. The various lysosomal hydrolases of neutrophilic leucocytes, namely, acid RNase, acid DNase, lysozyme, β-glucuronidase, and acid phosphatase, as well as several antibacterial substances, exhibit the electrophoretic property of basic proteins.[51]

A current investigation in the author's laboratory indicates that the lysosomal enzymes of rat kidney and liver are for the most part cationic glycoproteins.[5–7,52] The soluble enzyme protein fraction prepared from purified renal lysosomes contains about 4.7% hexose, 2.0% hexosamine, and 1.6% sialic acid. It is strongly PAS-positive and stains by the alkaline fast green method for histones at pH 10. Approximately 17 protein bands have been resolved from this fraction by polyacrylamide disk gel electrophoresis. The bulk of this protein migrates toward the cathode at pH 4. Only three faint anodic bands are seen at pH 9; a fourth band is acidic lipoprotein.

The amino acid composition of the enzyme fraction is similar to that of hog spleen acid DNase[49] and bovine liver β-glucuronidase.[50] There is a slight predominance of the dicarboxylic amino acids, aspartic and glutamic acid, over the basic amino acids, arginine, lysine, and histidine. However, they probably occur mainly as glutamine and asparagine, since considerable NH_3 is liberated during acid hydrolysis.[52] This would account for the basic character of these proteins. The sulfur-containing amino acids, particularly cysteine and possibly cystine, comprise 2.4% of the total amino acids.

Diverse polyanions reversibly inhibit the activities of some of these cationic enzymes in vitro through electrostatic interaction forces, suggesting that their protonized amino groups play a role in enzymic function.[53,54] The heavy metal cations, Hg^{2+} and Cu^{2+}, inactivate some but not all acid hydrolases, e.g., N-acetyl-β-D-hexosaminidases,[14] phospholipase,[20] arylamidase,[10,24] cerebroside galactosidase,[30] β-glucuronidase,[46] and acid phosphatase.[46] Inactivation is blocked by added glutathione or other sulfhydryl donors in some instances, suggesting that sulfhydryl groups are necessary for enzyme activity. The acid pH optima of these hydrolytic enzymes and their resistance to autolysis may be related to the basic character of the enzyme proteins and the presence of carbohydrate moieties in their molecular structure.

B. Lysosomal Lipids and Lipoproteins

1. Lysosomal Lipids

The lipid composition of neural lysosomes has not been elucidated as yet but probably is similar to that of renal and hepatic lysosomes. These lipids occur largely, if not exclusively, in the form of dense lipoproteins. A morphologically homogeneous fraction of lysosomes isolated from rat kidney by density gradient centrifugation contains the following lipid components (milligrams/milligram protein)[6,7]: phospholipids, 0.113; cholesterol, 0.032; and glycolipid, 0.010. Unsaturated fatty acids (18:1, 18:2, 22:1, 26:2) comprise 38–60% of the total fatty acids in these phospholipids. Liver lysosomes also contain considerable phospholipid and cholesterol.[7,55,56]

2. Lysosomal Lipoproteins

Specific lipoproteins recently have been isolated from lysosomes of rat brain, liver, and kidney in the author's laboratory.

a. Brain Lysosomal Lipoproteins. In an investigation of rat brain lysosomes,[8] mitochondrial subfractions prepared by density gradient centrifugation as described in Section I, A, 2 were sonicated in the presence of the detergent Triton X-100. The soluble lipoproteins were fractionated by flotation ultracentrifugation over KBr into a lighter lipoprotein fraction with a density of 1.08–1.21 and a denser lipoprotein fraction with a density of 1.21–1.35. They were further resolved by polyacylamide gel electrophoresis at pH 9 into a number of acidic lipoprotein bands, each staining for protein by amido Schwarz and for phospholipid by Sudan black B. Only one band closest to the anode from each of the lipoprotein fractions stained metachromatically with acridine orange. These acidic lipoproteins were judged to be lysosomal constituents, as their subcellular distribution in the mitochondrial subfractions paralleled that of the acid hydrolases, the highest concentration of both occurring in fraction E.[8]

b. Renal Lysosomal Lipoproteins. Similar lysosomal lipoproteins have been obtained from morphologically homogeneous fractions of lysosomes isolated from rat kidney by differential and density gradient centrifugation.[5–7,56] These lipoproteins constitute 30%, and the enzyme proteins 70%, of the lysosomal protein. About two-thirds of the lipoprotein is solubilized by sonication with the detergent Triton X-100 (0.1%), the remainder being insoluble. The soluble lipoproteins are separated by flotation ultracentrifugation over KBr into two classes, one with a density of 1.08–1.21 and the other with a density of 1.21–1.35. The lighter lipoprotein fraction is resolved into two acidic bands and the denser fraction into two acidic protein bands by electrophoresis on polyacrylamide gels. These lipoprotein bands are sudanophilic and stain metachromatically with the basic fluorochromes, acridine orange, and rhodamine G IV.

These lipoprotein fractions differ significantly in their lipid composition. The lighter lipoprotein contains 0.38 mg phospholipid and 0.07 mg cholesterol per milligram protein; the molar ratio of phosphatide:cholesterol is 2.9. The major phospholipids are lecithin (and lysolecithin; 35%), sphingomyelin (15%), phosphatidyl ethanolamine (30%), and cardiolipin (10%). The denser lipoprotein contains 0.20 mg phospholipid and 0.043 mg cholesterol per milligram protein, their molar ratio being 2.4. Its phospholipid composition is otherwise similar to that of the lighter lipoprotein.

The protein moieties of the lighter and denser lipoprotein fractions are identical in their amino acid composition. They are characterized by a high ratio of acidic to basic amino acids. The sulfur-containing amino acids, particularly methionine, are quantitatively important components, constituting 3.5% of the total amino acids. They also contain about 3% hexose and 0.7% hexosamine but negligible sialic acid.[7,56]

Barrett and Dingle[9] recently reported in a brief abstract the isolation of a lipoprotein component from rat kidney lysosomes. This component binds ionic nickel, the cationic detergent cetyltrimethylammonium bromide, basic dyes, and the carcinogen dimethylbenzanthracene.

c. Hepatic Lysosomal Lipoproteins. Acidic lipoproteins with similar physicochemical and tinctorial characteristics have been found in lysosomes isolated from rat liver[7,56] by a modification of the method of Wattiaux, Wibo, and Baudhuin.[39] The lighter lipoprotein has a density of 1.08–1.21 and contains 0.3 mg phospholipid per milligram protein. The heavier lipoprotein has a density of 1.21–1.35 and contains 0.34 mg phospholipid per milligram protein. Both possess considerable protein-bound hexose ($\cong 3\%$). These lipoproteins migrate rapidly towards the anode on disk gel electrophoresis at pH 9. They are sudanophilic and stain metachromatically with acridine orange.

d. The Polyanionic Character of Lysosomal Lipoproteins. It should be emphasized that lipoproteins of mitochondria and other structural elements, like the morphologic units from which they derive, *do not* stain metachromatically with acridine orange or other basic dyes. The metachromasia of lysosomal lipoproteins reflects a high concentration of basic dye electrostatically bound to their free phosphoryl and/or carboxyl groups. Since the lysosomal lipoproteins do not seem to differ radically from other lipoproteins, e.g., mitochondrial[57] and erythrocyte membrane[58] lipoproteins, in their lipid or amino acid composition, the polyanionic character of these lipoproteins probably is an attribute of their molecular morphology.

It is likely that the acid hydrolases, which are cationic proteins, form electrostatic complexes with the lipoproteins in the intact lysosomal granules. The significance of this electrostatic binding in the structural latency of the lysosomal enzymes is discussed in Section III, B.

C. Other Lysosomal Components

In addition to the characteristic hydrolases and lipoproteins, lysosomes may contain various minor constituents. Among these are substances that are responsible for the autofluorescence that lysosomes of diverse tissues exhibit.[60] In renal lysosomes a fluorescent substance with the spectral characteristics of pyridine nucleotide (excitation maximum, 360 mμ; emission maximum, 460 mμ) is associated mainly with the soluble enzyme fraction.[5,6] Other fluorescent material, as yet unidentified, occurs in this fraction, and to a lesser extent in the lipoprotein fraction.[7,52] In liver lysosomes flavine fluorescence reportedly predominates over pyridine nucleotide fluorescence,[61] both being associated with the enzyme fraction. Iron-containing compounds, including ferritin, occur in lysosomes of some tissues.[4,62] The heavy metals copper, zinc, and iron in normal human liver have been localized by a histochemical staining method to granules[63] that have the general morphology of lysosomes and lipofuscin pigment. Bilirubin also may occur in lysosomes.[4] The lysosome-like inclusions in certain storage diseases consist predominantly of a single substance, e.g., glycogen in Type II glycogenosis,[64] gangliosides in Tay–Sach's disease,[65] and cerebroside sulfate in metachromatic leukodystrophy.[66] Various foreign substances also may be segregated within lysosomes. These include cationic dyes[67] and drugs,[68] heavy metals,[67] sucrose,[41] the detergent Triton WR 1339,[39] and foreign proteins.[3] This subject will be discussed further in Sections II, C, 1 and IV, B.

II. MORPHOLOGICAL AND CYTOCHEMICAL ASPECTS

Neural lysosomes resemble lysosomes in other tissues in their general morphology and cytochemical properties. These organelles are identifiable in electron micrographs as cytoplasmic "dense bodies" and may be demonstrated in the optical microscope by cytochemical staining reactions for a number of acid hydrolases, as well as for certain nonenzymatic constituents. Several of these components have been localized to the ultrastructural level by electron microscopy. The cytology of neural lysosomes has been reviewed by Novikoff.[69]

A. Ultrastructure of Neural Lysosomes

Lysosomes may be defined cytologically as membrane-limited cytoplasmic bodies with acid hydrolase activity. In normal neural cells dense bodies are the most abundant structures to satisfy this definition. Dense bodies typically consist of a finely granular, compact, electron dense matrix that is surrounded by a single membrane. An electronlucent zone about 70 Å in width generally intervenes between the limiting membrane and the matrix. Various structures may be imbedded in the matrix, including osmiophilic particles, vesicles, "crystalline" or membranous material, and homogeneous masses of variable size and shape. Some dense bodies contain a large electronlucent zone or vacuole that may occupy a third or more of the area and often causes a bulge in their normally circular contour (Fig. 6).

Other structures in neural cells occasionally fulfill the cytological definition of a lysosome. These include some Golgi saccules and vesicles, GERL, multivesicular bodies, and autophagic vacuoles. Autophagic vacuoles are identifiable by the presence of ribosomes, fragments of endoplasmic reticulum, and sometimes mitochondria within the often thickened and irregular limiting membrane. These are encountered only occasionally in normal neural cells. This subject is further discussed in (Section II, B, 1).

B. Enzyme Cytochemistry of Neural Lysosomes

Neural lysosomes are readily demonstrated in frozen section of aldehyde-fixed nervous tissue incubated in appropriate media for the cytochemical staining of various acid hydrolases. These include acid phosphatase, acid DNase, acid RNase, acid esterase, aryl sulfatase, β-glucuronidase, and

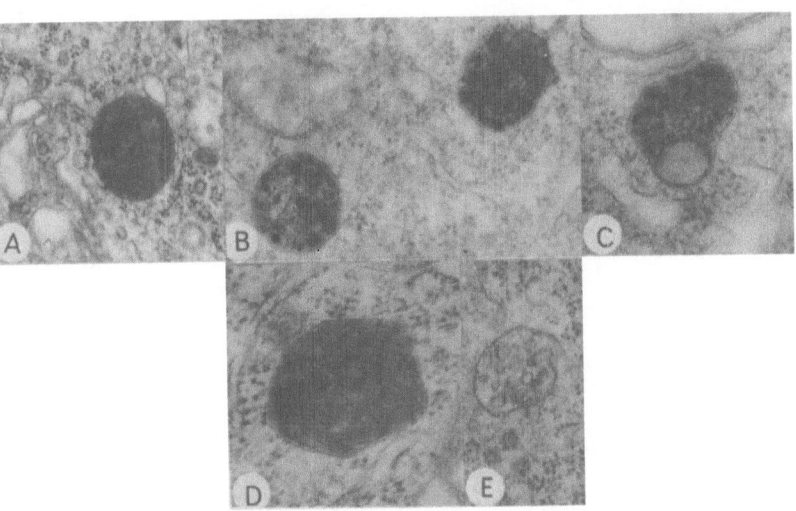

Fig. 6. Electron micrographs of lysosomes in nerve cells of rat cerebral cortex. The membrane-enclosed electron-dense matrix contains osmiophilic droplets (A, B, and C), electronlucent vacuoles (B and C), myelin figures (D), and vesicles (A, B, and C). A multivesicular body is seen in E.

N-acetyl-β-D-glucosaminidase. The histochemistry of the various staining reactions has been comprehensively reviewed.[70]

1. Acid Phosphatase

The principal methods available for the histochemical detection of acid phosphatase activity are: (1) the lead sulfide method of Gomori[71] and (2) coupling azo-dye methods with either phosphate esters of naphthol AS derivatives or α-naphthol phosphate. The simultaneous coupling azo-dye methods do not equal the Gomori method cytologically in the nervous system as they tend to give diffuse, as well as particulate, enzyme product. However, some azo-dye methods, e.g., the Barka–Anderson method,[72] which utilizes hexazonium pararosaniline as a coupling agent, are very sensitive, and the diffuse staining facilitates visual estimation of enzyme activity. Friede and Knoller[73,74] adapted a coupling azo-dye method to quantitative measurement of enzyme activities in brain sections by colorimetry of extracted red azo-dye. Hirsch[75] has assayed acid phosphatase activity in single nerve cells by fluorometric measurement of α-naphthol split from naphthyl phosphate.

The Gomori method utilizes lead ions to trap orthophosphate cleaved from glycerophosphate at sites of enzyme activity. This method has been

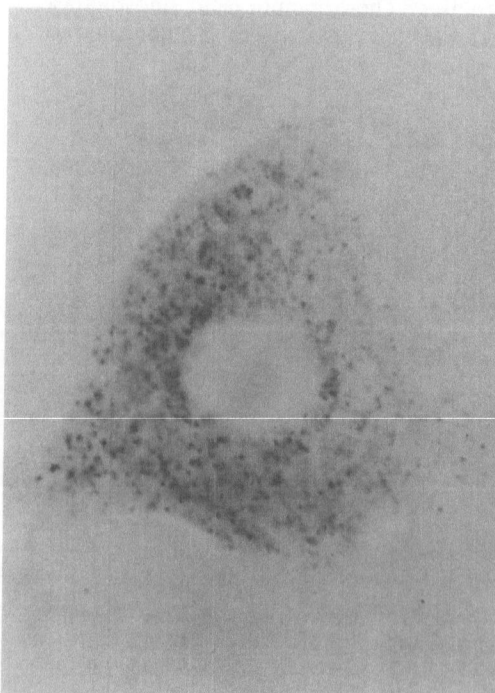

Fig. 7. Spinal motoneuron of cat stained for acid phosphatase activity by Gomori's method.[71] × 750.

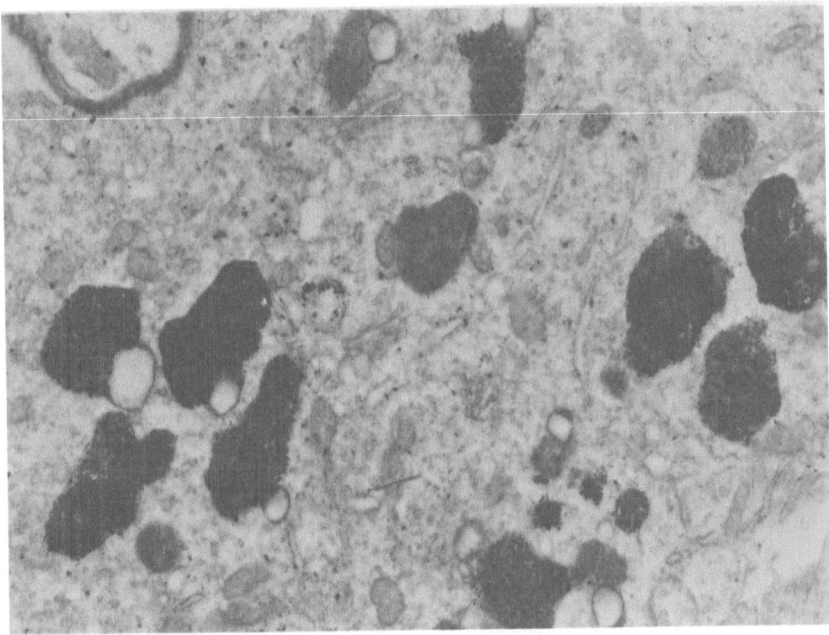

Fig. 8. Electron micrograph of cat spinal motoneuron stained for acid phosphatase activity by Gomori's method.[71] Lead phosphate reaction product occurs in the matrix but is absent from electronlucent vacuoles. × 55,000.

used most widely for the cytochemical demonstration of acid phosphatase activity in tissue sections. It is also useful in monitoring subcellular fractions for lysosomal particles.[31] First applied to aldehyde-fixed frozen sections of nervous tissue by Becker et al.[76] and by Novikoff,[77] it has since been employed in numerous light and electron microscopic studies of normal and pathological nervous tissue (reviewed by Novikoff[69]). Nucleoside monophosphates such as cytidylic acid can be substituted for glycerophosphate and may give crisper localizations.

At the light microscope level acid phosphatase enzyme product, i.e., lead sulfide, occurs mainly in lysosomal particles of neurons, glia, and other neural cells (Fig. 7). In neural cells of older animals, lipofuscin pigment granules also stain for acid phosphatase activity; however, longer periods of incubation are frequently necessary for their demonstration. In occasional cells, lamellar structures belonging to the Golgi apparatus and the GERL of Novikoff may be visibly stained.

In electron micrographs acid phosphatase enzyme product is localized mainly in membrane-delimited dense bodies and in the first one or two saccules on the concave surface of the Golgi complex (Fig. 8). In some neurons acid phosphatase activity occurs in smooth and coated Golgi vesicles near reactive Golgi saccules and occasionally in GERL, agranular membrane systems at the concave surface of the Golgi apparatus.[69,78] Acid phosphatase staining also may be present in vesicles of multivesicular bodies and in autophagic vacuoles. These structures are more frequently reactive under conditions of augmented acid phosphatase synthesis, e.g., sensory ganglion cells during retrograde degeneration, and Schwann cells during anterograde degeneration following axotomy.[78,79]

In dense bodies, enzyme product is invariably associated with the compact, finely granular, osmiophilic matrix. The electronlucent vacuoles present in some dense bodies, which evidently contain lipase-sensitive neutral triglycerides,[80] are devoid of enzyme product.[43,81] In hydropic swelling of neuronal dense bodies due to uptake of the cationic dye, neutral red, enzyme deposits are restricted to the matrix material, none occurring in the often voluminous vacuoles (Fig. 24).[42,43] Similar observations have been made in vacuolated lysosomes of rat liver following intraperitoneal injection of hypertonic sucrose[40] and in swollen lysosomes of mouse pancreas after administration of neutral red.[82] These findings bespeak an intimate association between the enzyme and the lysosomal matrix, i.e., the osmiophilic material that contains the acidic lipoprotein.

Acid phosphatase-reactive lysosomes are more abundant in gray matter than in white matter and are most numerous in the perikarya and dendrites of large multipolar neurons. In the neuropil some lysosomes are found in axons near the axon hillock[86] and in occasional nerve endings.[34,86] However, intraaxonal lysosomes are rare in white matter and peripheral nerve. Both astroglia and oligodendroglia in white and gray matter contain reactive lysosomes. Ependymal cells lining the ventricular system and the central canal of the spinal cord have numerous, relatively fine lysosomes. Epithelial cells of the choroidal plexus contain multitudinous rather coarse lysosomes. Adventitial cells surrounding small blood vessels possess large acid phosphatase-positive lysosomes, as do the cells of the leptomeninges.

2. Acid DNase and Acid RNase

Acid nuclease activity can be demonstrated in neural lysosomes by Vorbrodt's method.[83] This is a modification of Gomori's method for acid phosphatase in which highly purified DNA or RNA is substituted for glycerophosphate as the substrate. The breakdown products of nuclease activity are in turn split by added acid phosphatase, with resultant deposition of lead phosphate at sites of nuclease activity. Omission of acid phosphatase, as well as DNA (or RNA), from the incubation mixture prevents lysosomal staining (Fig. 9).

3. Esterases

The most widely used methods for the cytochemical demonstration of nonspecific esterase activity are (1) the thiolacetic acid method, (2) the naphthol AS acetate coupling azo-dye method, and (3) the indoxyl method employing a substituted indoxyl acetate as substrate. When fixed-frozen

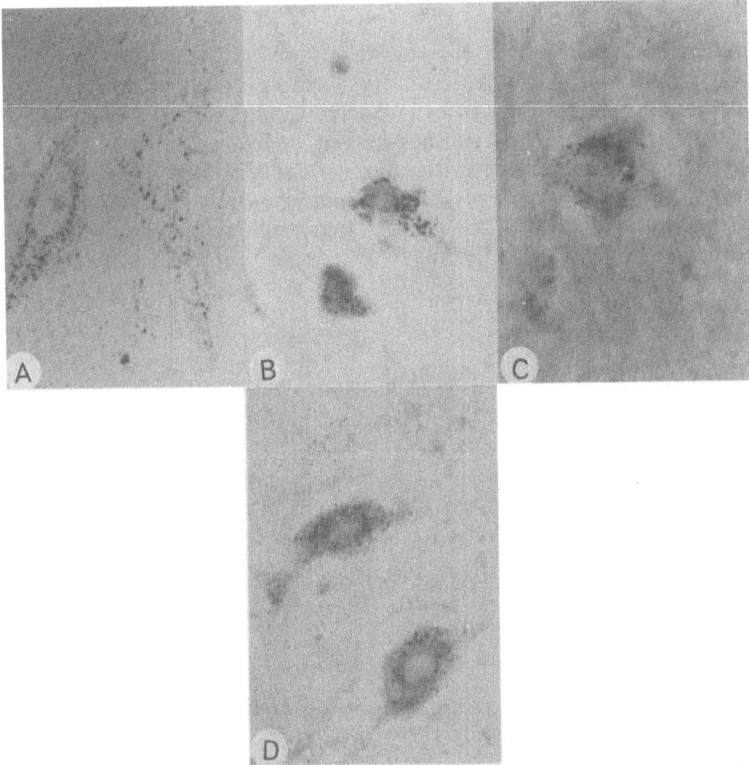

Fig. 9. Rat spinal motoneurons stained for acid hydrolase activities. (A) Acid phosphatase activity by Gomori's method.[71] (B) Acid DNase by Vorbrodt's method.[83] (C) Acid RNase by Vorbrodt's method.[83] (D) Thiolacetic acid esterase activity. × 750.

sections are treated with an organophosphorus inhibitor such as E600 or diisopropyl-phosphofluoridate (10^{-5}–10^{-4} M) prior to incubation with substrate, enzyme product is generally restricted to lysosomal particles (Fig. 9). With the coupling azo-dye procedures considerable diffuse staining may occur in the cytoplasm of some cells. In the thiolacetic acid method, the hydrogen sulfide liberated by esterase activity is precipitated *in situ* as lead sulfide, which can be visualized in the electron microscope.

In general, the lysosomes of oligodendroglia tend to stain more intensely for esterase activity than those of nerve cells.[84–86] However, there is considerable variation in staining intensity among the latter. Lysosomes of large motor neurons usually give strong reactions for esterase activity, while those of small neurons, e.g., cerebral cortex, often stain feebly. Lysosomal esterase activity has been ascribed to cathepsin C[87] and more recently to cathepsin D.[88]

4. Aryl Sulfatase

Goldfischer's method[89] for sulfatase activity is suitable for demonstrating lysosomes in nervous tissue and yields an electron-dense reaction product that can be visualized in the electron microscope. p-Nitrocathecol sulfate is used as a substrate in the presence of lead ions. The lead sulfate deposited by sulfatase activity is subsequently converted to black lead sulfide. Lysosomes of Purkinje neurons and sensory ganglion cells are reactive at pH 5.5, but not at pH 4.2, whereas lysosomes of glia, pericytes, and microneurons are reactive at pH 5.5 and pH 4.2. According to Goldfischer[89] the pH 4.2 reaction may be due to arylsulfatase A and the pH 5.5 reaction may be due to aryl sulfatase B.

5. β-Glucuronidase and N-Acetyl-β-D-Glucosaminidase

A method applicable to nervous tissues employs the substrates naphthol AS-BI-β-glucuronide and N-acetyl-β-D-glucosaminide in a simultaneous coupling azo-dye procedure.[90] Lysosomes of glia and of large motor neurons generally stain more intensely than those of small nerve cells.[86] In a recent electron microscopic study[91] of the β-glucuronidase reaction in rat liver, enzyme product was restricted to the matrix of lysosomes, none occurring in electronlucent vacuoles.

6. Cytochemical Correlation of Enzyme Staining

The presence of various acid hydrolases within the same lysosomal particles has been demonstrated in two ways. In the first approach[92] advantage was taken of the finding that lysosomes exhibit an autofluorescence that is quenched by opaque deposits.[60] Thus the lead sulfide precipitate formed through the activity of any one enzyme based on the Gomori procedure, namely, acid phosphatase, acid DNase, acid RNase, and DFP-resistant thiolacetate esterase, blackens most neural lysosomes and blocks their autofluorescence, indicating that all four of these hydrolases coexist in the same, i.e., autofluorescent, particles.[92] This observation has been corroborated for acid phosphatase and the acid nucleases in sections incubated for these enzyme activities without the ammonium sulfide rinse. In these preparations the opaque lead sulfide is replaced by lead phosphate, which is colorless, translucent, and phase dense, as the enzyme product. Upon examining these sections alternately with dark field and fluorescence optics, it is readily ascertained that granules with reaction product are autofluorescent.[92] In another approach, Novikoff[69] combined the Gomori reaction for acid phosphatase activity, which gives a brown–black color, with a coupling azo-dye method for another enzyme that yields a red color. In the large lysosomes of adventitial cells surrounding small blood vessels of brain and spinal ganglia, he found that acid phosphatase activity was present in the very same granules that contained β-glucuronidase, N-acetyl-β-D-glucosaminidase, and esterase activities.

C. Nonenzymatic Cytochemical Properties of Neural Lysosomes

Lysosomes in nervous tissue exhibit several distinctive cytochemical properties that are attributable to their contained lipoproteins or acid hydrolases and occasionally to other stored components. In the discussion that follows lysosomes are defined as the acid phosphatase-reactive dense bodies, or cytoplasmic granules that correspond to these structures in their general morphology.

1. Vital Staining with Cationic Dyes

Lysosomes of neural and other tissues are rapidly and selectively stained by cationic dyes *in vivo*[67] and *supra vitam*[67] when in the fresh, unfixed state (reviewed by Koenig[43]; Fig. 10). With metachromatic dyes such as acridine orange and toluidine blue, lysosomes exhibit a strong metachromasia indicating the presence of a polyanionic substance, i.e., a chromatrope. This "vital" metachromasia is retained by lysosomes for many hours in cell particulates suspended in cold sucrose media but is promptly lost following fixation. Metachromatic staining does not require energetic or other "vital" cell processes, inasmuch as it occurs at 4°C in cell particulates.

In vitro staining experiments[43,93,97] with acridine orange, a basic fluorochrome, have been carried out to investigate the deduction that the metachromasia of lysosomes is based on an electrostatic binding of basic dyes to the lysosomal particles (Fig. 10). When incubated with dye at neutral pH, lysosomes in particulate suspensions of rat brain emit a yellow to orange fluorescence, denoting metachromatic staining. The intensity of this staining increases with rising pH, lysosomes fluorescing a brilliant red at pH 10–11.

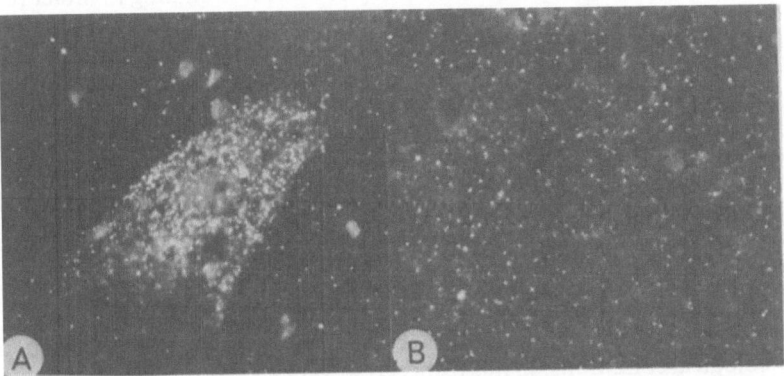

Fig. 10. (A) Rat spinal motoneuron stained *in vivo* by acridine orange (25 mg/100 g) 1 hr before sacrifice. Squash preparation. (B) Lysosome-enriched fraction *E* from rat brain stained *in vitro* with acridine orange. Fluorescence photomicrographs. × 640. From Koenig[43] and Koenig et al.[31]

Acridine orange binding diminishes on the acid side. Metachromatic staining disappears at around pH 5.5, and orthochromatic staining, manifested by a green fluorescence, at pH 4. Metachromatic staining of cerebral lysosomes by acridine orange is depressed or blocked by diverse inorganic and organic cations, including the heavy metals, Ag^+, Cu^{2+}, and Pb^{2+}; the basic polypeptides, protamine, histone, and polylysine; chlorpromazine; and polymixin. The physiological cations Na^+, K^+, and Ca^{2+} exert but little inhibitory effect on lysosomal staining by acridine orange, indicating that they bind feebly to lysosomes. Lysosomes in particulate suspensions of rat liver and kidney behave in a similar manner. These effects of pH and competing cations unequivocally demonstrate the role of electrostatic attraction in the binding of basic dyes to lysosomal particles.

It had been inferred from the effects of lipolytic agents such as ethanol, Triton X-100, and phospholipase C that the chromotrope of the lysosome resides in a special lipoprotein.[43,93] This deduction has now been verified for lysosomes of rat brain,[8] liver,[7,56] and kidney[6,7,56] (see Section I, B, 2). Soluble lipoproteins recently isolated from lysosomes in these organs exhibit the metachromatic staining of the intact particles in the pH range of 5–11. However, aldehyde fixation does not affect the metachromasia of the purified lipoproteins. Hence, the fixation-induced loss of metachromasia of intact lysosomes is probably secondary to the effect of the fixative on the contained hydrolytic enzymes. It is known that aldehyde fixation activates these enzymes[94,95] and simultaneously insolubilizes them *in situ* within the lysosomal particles. Hence, aldehyde-activated hydrolases, themselves cationic molecules, may be free to interact with available acidic sites of the lipoproteins, thereby interfering with the binding of basic dyes to these sites.

Potential acidic, dye-binding sites in lysosomal lipoproteins are the phosphoryl groups of the phospholipids and the carboxylic groups of the polypeptide chain. In a quantitative study of basic dye binding to intact renal lysosomes *in vitro*[95] about 3 moles of neutral red or toluidine blue were bound per mole of phospholipid phosphorus at pH 7. When staining was at pH 3, the molar ratio of bound dye declined to about 0.5; at pH 11, it rose to approximately 15. These results indicate that phospholipid phosphate cannot account for all the dye bound by lysosomes, at least at high pH. Moreover, enzymatic cleavage of phospholipid phosphoesters by phospholipase C, while inhibiting the metachromatic staining of lysosomes to some extent did not abolish it and did not diminish basic dye-binding.[95] It is duly concluded that the carboxylic groups of lysosomal lipoproteins afford anionic sites for basic dye-binding.

Lysosomes of the nervous system and other tissues may be regarded as polyanionic particles as they possess free or available acidic sites under physiological conditions. In this respect lysosomes contrast sharply with other intracellular structures that contain lipoproteins, e.g., mitochondria, or other acidic components, e.g., RNA-rich Nissl bodies, in that the latter are not perceptibly stained by basic dyes *in vivo*. Evidently the anionic groups in these constituents are masked by basic groups of associated proteins. Only when cells are severely injured or dead does the cytoplasm stain diffusely with basic dyes.

Other types of cationic molecules also concentrate within lysosomes. Metallic cations, e.g., Cu^{2+}, Hg^{2+}, Fe^{2+}, Pb^{2+}, and Ag^+, accumulate within lysosomes *in vivo* when given to experimental animals.[67] In Wilson's disease, large quantities of copper are localized to lysosomal particles of liver.[96] Allison and Young have shown that various cationic drugs become concentrated within lysosomes when supplied to living cells grown *in vitro*.[68] These include the antimalarial drugs, quinine and chloroquine; the trypanosomacidal agents, antrycide and stilbamidine; and the anti-inflammatory agents, indomethacin and dexamethasone. The tendency for these and other agents to accumulate within lysosomes may have a bearing on their therapeutic and toxic effects. Moreover, the sequestration of noxious cationic molecules within lysosomes may serve to protect cells from their deleterious effects until they are eliminated.

2. *Osmiophilia*

Lysosomes are conspicuously blackened in tissues fixed primarily or secondarily in cold 1 or 2% osmium tetroxide. Osmium-stained lysosomes are best seen in 0.5–1.0 μ sections of plastic-imbedded tissues when viewed in phase contrast optics. The osmiophilia of lysosomes is due largely to the polyunsaturated fatty acids of lysosomal phosphatides, as it is sharply diminished by prior delipidation[97,98] or bromination[97] of tissue sections. The hydrolytic enzymes also contribute to osmium staining, perhaps through reduction by their sulfhydryl and/or aldehydic groups. Osmium-fixed lysosomes stain metachromatically with toluidine blue and other basic dyes, possibly by oxidation of SH groups to sulfate, and are a prominent feature of "thick" sections of plastic-imbedded tissues. It is clearly evident in the electron microscope that osmiophilic phospholipid, i.e., lipoprotein, is distributed throughout the interior of the lysosomal particle and is not restricted to the membranous envelope.

3. *Phospholipid Stains*

On occasion cytoplasmic particles resembling lysosomes in their general morphology are demonstrated in neurons by Baker's acid hematein stain or by the Sudan black B stain for phospholipid.[97] The coarser lipofuscin pigment granules are more consistently displayed. However, these phospholipid stains are relatively insensitive and capricious. Moreover, they are not specific for lysosomes, as they also stain mitochondria and other structures, e.g., nerve endings and myelin sheaths.

4. *Periodic Acid–Schiff (PAS) Reaction*

The lysosomes of neural and other cells give a positive PAS reaction (Fig. 11).[4,59] This reaction is specific for nonglycogenic carbohydrate, as it

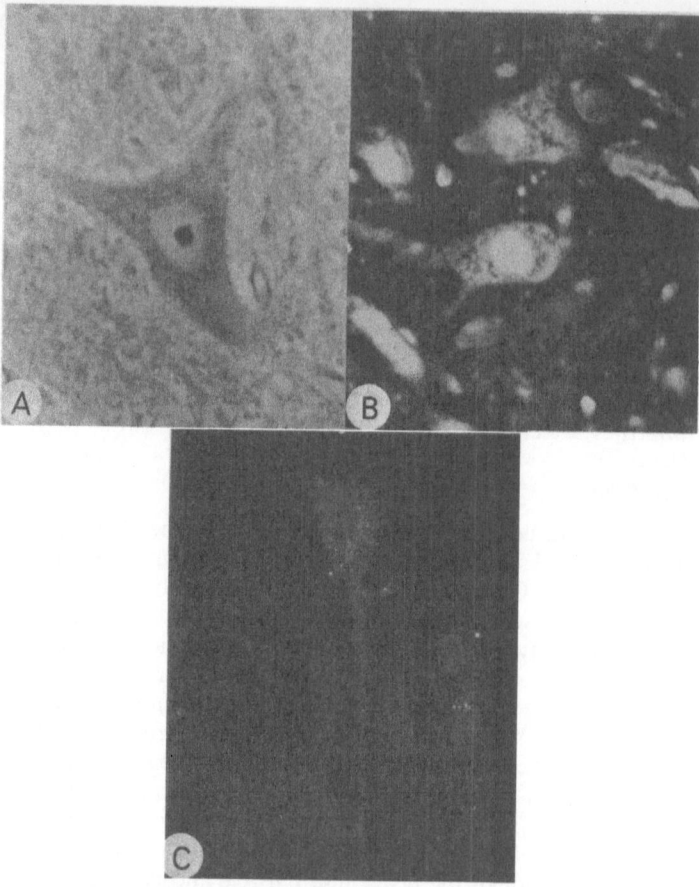

Fig. 11. Rat spinal motoneurons. (A) PAS reaction for carbohydrate. (B) Sulfation-induced metachromasia for carbohydrate. (C) Fluorescence photomicrograph showing autofluorescent lysosomes. × 480.

is blocked by prior acetylation and by omitting periodate oxidation, and is unaffected by diastase digestion. The use of methanamine silver in place of the Schiff reagent confers a dark brown or black stain upon lysosomes and can be used for electron microscope cytochemistry.[99] Lysosomes also exhibit sulfation-induced metachromasia,[59] an independent method for the demonstration of carbohydrates (Fig. 11). This carbohydrate was originally thought to be glycolipid because it is not demonstrable after extracting fresh, but not fixed, tissues in chloroform–methanol (2:1, vol./vol.).[59] Recent findings have rendered this inference untenable. It is now known that the lipoprotein of renal and hepatic lysosomes contains little glycolipid and is not PAS-positive.[5–7,56] However, the enzyme protein fraction from these lysosomes is rich in hexose, hexosamine, and sialic acid and reacts positively by the PAS method.[5-7,52] Thus the acid hydrolases are responsible for the carbohydrate staining reactions of lysosomes. What then accounts for the disappearance of the PAS-positive material from unfixed lysosomes following lipid extraction? Evidently the lipid solvent disrupts the lysosomal particles *in situ*, causing a dispersion of the glycoprotein enzymes within the cytoplasm.

5. *Argentaffin Reaction*

Aldehyde-fixed or unfixed lysosomes of neural tissue and other organs can directly reduce ionic silver to the metallic state, thereby becoming selectively blackened.[100,101] The soluble lysosomal lipoprotein and enzyme fractions obtained from rat kidney and liver lysosomes both exhibit the argentaffin reaction. It has been ascertained with the aid of specific blocking reagents that the polyunsaturated fatty acids and the aldehyde and sulfhydryl groups of the lipoprotein and the sulfhydryl groups of the acid hydrolases are largely responsible for the argentaffin reaction of the intact lysosomes.[102]

6. *Autofluorescence*

Lysosomes in neural and other tissues emit a natural yellow fluorescence when irradiated by ultraviolet light in the neighborhood of 360 mμ (Fig. 15).[60] This autofluorescent material is partly associated with the lipid and the protein components of lysosomal lipoproteins. However, the bulk of fluorescent material is recovered in the soluble enzyme fraction of renal and hepatic lysosomes.[5,6,52] The autofluorescence is partly due to pyridine[5,6] and flavine[61] nucleotides and to other, as yet unidentified, components.

III. STRUCTURAL LATENCY OF THE LYSOSOMAL ENZYMES

The lysosome is an efficient device for the sequestration of the various soluble acid hydrolases in an inert state. The destructive potentialities of these enzymes makes it imperative that their compartmentation and latency *in vivo* be ensured. At the same time lysosomal particles must be sufficiently labile to allow a controlled release of the active enzymes for fulfillment of their biological functions. Two mechanisms have been proposed to account for the structural latency of the lysosomal enzymes, the membrane theory and the matrix binding theory.

A. The Membrane Theory

According to the membrane theory propounded by de Duve and associates,[1,2,26] the lysosome is a simple osmotic sac that is delimited by a membranous lipoprotein envelope (Fig. 12). The impervious membrane serves as a physical barrier to restrict the physical freedom and substrate

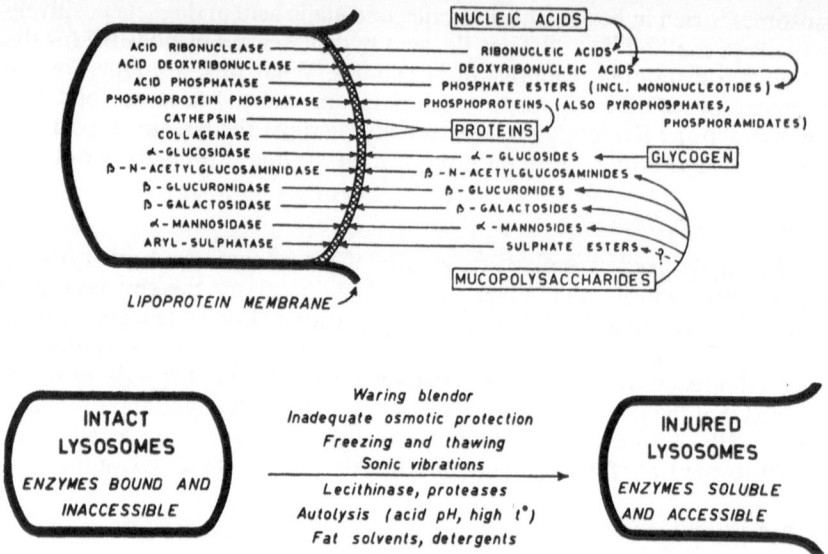

Fig. 12. Membrane theory of the lysosome. From de Duve.[2]

accessibility of the enclosed enzymes that occur in a diffusible, fully active form. The representation of the lysosome as a simple osmotic sac was originally deduced from biochemical experiments purporting to show that acid phosphatase-bearing particles behave as osmotic systems in sucrose media. When isotonic sucrose media are diluted with water the suspended particles rupture, whereupon the contained acid phosphatase is released into solution in a fully active form.[103] The osmotic sac hypothesis seemed to be corroborated by the subsequent finding that osmotic shock, and also other disruptive treatments, liberate the various lysosomal enzymes in the soluble, active form in approximately parallel fashion.[1,26,104] The seemingly contradictory observation that the monovalent salts KCl and NaCl afford only transient osmotic protection to lysosomes was explained by the assumption that these electrolytes, unlike sucrose, rapidly penetrate the lysosomal membrane.[104]

The effectiveness of "hypoosmotic" dilution in releasing lysosomal enzymes from the latent, particulate state has been widely documented. However, the osmotic basis of this phenomenon never has been conclusively demonstrated. The following recent findings are inconsistent with this interpretation and compel a revision of the concept that the lysosome is a simple osmotic bag.

1. The lysosomal membrane does not seem to exhibit the degree of impermeability to sucrose expected of osmotically active particles. Renal lysosomes, when purified by isopycnic centrifugation over a sucrose density gradient, take up considerable quantities of sucrose from the medium.[105]

Hepatic lysosomes are freely permeable to the small organic molecules, sorbose, p-nitrophenol, and glycerol.[106]

2. Lysosomes can display "osmotic" properties within the hypertonic range as well as in the hypotonic range. Renal lysosomes exhibit an increasing release of their bound, latent acid phosphatase and β-glucuronidase activities during stepwise dilution of the suspending medium from 1.75 to 0.3 M sucrose.[107] Liver lysosomes demonstrate an analogous activation of several acid hydrolases when the suspending medium is diluted from 0.7 to 0.3 M sucrose.[108] Electron microscopic study has shown that the first ultrastructural change in renal lysosomes during dilution of the medium is a swelling and increased electronlucency of the matrix. Subsequently lysosomes rupture in increasing numbers concomitant with release of acid hydrolases.[107,109]

These effects of dilution clearly indicate that lysosomes do not behave as simple osmometers in sucrose media. However, they are compatible with the behavior of saclike structures that are permeable to sucrose and are filled with a hydrophilic colloid. During dilution the net transfer of water from the medium into lysosomes occurs probably because the concentration of water in the external medium is greater than that in the particles. Upon imbibing water the sucrose-impregnated lysosomes swell in the manner of a colloidal gel. The distended limiting membrane ultimately ruptures, releasing the lysosomal contents into the medium. A similar sequence of events can be envisaged to explain the dilution-induced disruption of lysosomes prepared in isotonic media.

3. It also has been found that a number of substrates readily penetrate into the interior of renal and hepatic lysosomes under conditions in which their respective enzymes are largely latent.[109] These include β-glycerophosphate, AMP, and CMP, substrates for acid phosphatase; phenolphthalein glucuronide, a substrate for β-glucuronidase; and p-nitrocatechol sulfate, an aryl sulfatase substrate. It must be concluded, therefore, that the membranous envelope of the lysosome does not function as a physical barrier to restrict the access of small molecular weight substrates to their hydrolytic enzymes, as posited by the membrane theory, and thus it cannot account for the latency of these enzymes.

B. The Matrix Binding Theory

In the matrix binding theory advanced by Koenig,[59,97] the lysosome is viewed as a membrane-limited polyanionic lipoprotein granule. Within this granule the hydrolytic enzymes are bound in an inert state by electrostatic attraction forces to the acidic groups of the lipoprotein matrix. This model accounts for the structural latency of the lysosomal enzymes by postulating that their active sites, which probably include basic amino groups, are occluded through ionic linkage to the lysosomal matrix and thus are not available to interact with their substrates. In effect, the acidic lipoprotein functions as an inhibitor of the lysosomal enzymes, the membranous envelope serving mainly to confine the lysosomal contents to discrete granules. This mechanism is comparable to the nonspecific, reversible inhibition of cationic enzymes *in vitro* by a wide variety of macromolecular polyanions or "macroanions" described by Spensley and Rogers[53] and others.[54] Deduced originally from the preferential staining of lysosomes by cation dyes,[59,67] the matrix binding theory has received substantial experimental support.

Cytochemical studies of hydrolase activity at the ultrastructural level are consistent with the view that the soluble enzymes are bound to the lysosomal matrix. Acid phosphatase reaction product, as visualized in the electron microscope, seems to be associated only with the osmiophilic matrix of lysosomes. No enzyme product is found in the normally occurring electronlucent vacuoles of neuronal lysosomes[43,81,86] or in the hydropic vacuoles induced in lysosomes of nervous tissue[42,43] and pancreas[82] by neutral red uptake and in liver lysosomes by sucrose injection.[40] β-Glucuronidase reaction product seems to be absent from electronlucent vacuoles of

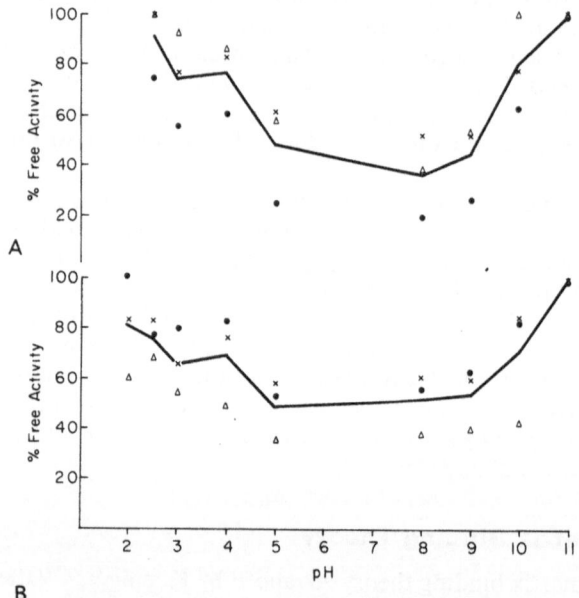

Fig. 13. Effect of pH on the accessibility of lysosomal hydrolases in rat brain mitochondrial fractions. (A) Acid phosphatase activity. (B) β-Glucuronidase activity. Mitochondrial suspensions in 0.25 M sucrose were exposed to buffers of various pH (0.04 M) for 30 min at 4°C and neutralized, and free and total activities were measured according to de Duve *et al.*[26] Each set of symbols represents one experiment. Activation of acid hydrolases occurred at pH extremes without microscopically evident destruction of lysosomal particles. From Koenig.[46,110]

liver lysosome.[91] If the membranous envelope were solely responsible for the containment of the diffusible lysosomal enzymes, as postulated in the membrane theory, enzyme product also should be present in vacuoles as soluble enzyme molecules would tend to diffuse into vacuolar fluid and be retained therein after fixation.

Two lines of evidence implicate electrostatic attraction forces in the binding and the structural latency of the lysosomal enzymes: (1) Under proper conditions the lysosomal enzymes can be released by low pH and by high pH without discernible damage to lysosomal particles.[107,109] Lysosomal acid phosphatase and β-glucuronidase in mitochondrial fractions of rat brain, liver, and kidney in isotonic sucrose are activated by buffering to pH 2–3 or pH 10–11 at 4°C (Fig. 13). Lysosomal particles so treated remain structurally intact when viewed unstained or after acridine orange fluorochroming in the fluorescence microscope. Release of enzymes at low pH is ascribed to suppression of ionization of anionic groups in the lipoprotein matrix. At high pH the protonization of the cationic enzymes would be prevented with resultant dissociation of the electrostatic complex. (2) Certain cationic molecules, e.g., heavy metals and chlorpromazine, may activate lysosomal enzymes in tissue particulates[47] and coincidentally bind to lysosomal particles, as evidenced by impaired staining of the latter by acridine orange.[43,93,97] Enzyme release by heavy metals, e.g., Cu^{2+}, Ag^+, Pb^{2+}, and Hg^{2+} is readily demonstrated in mitochondrial suspensions from rat liver and brain at low concentrations (0.2–1.0 mM),[46,107,109,110] although at higher concentrations these enzymes are inactivated. Metal-treated lysosomes do not appear to be structurally damaged at the light or electron microscopic level[86] (Table I). Interaction of heavy metals with sulfhydryl groups in the lysosomal lipoprotein may be involved in the activation phenomenon, because thiol reagents such as glutathione and L-cysteine in excess prevent the activating effect of Hg^{2+} on acid phosphatase in liver lysosomes.[110]

The capacity of pH extremes and cations to release lysosomal enzymes without concomitant damage to the lysosomal particles indicates that electrostatic bonds are largely responsible for the structural latency of the lysosomal enzymes. Indeed these release phenomena are reminiscent of the elution process in column chromatography, the lysosomal matrix corresponding to a cation exchange resin. In addition, thiol groups may play a role in the structural latency of some hydrolases, possibly through formation of disulfide bonds with thiol groups in lysosomal lipoprotein.

The acidic lipoproteins of the lysosomal matrix possess the proper attributes for forming an electrostatic complex with the cationic lysosomal enzymes (see Section I, B, 2). Correlative experiments with phospholipase C indicate that the phosphoryl group of the lipoprotein phospholipids play an important role in the structural latency of the lysosomal enzymes. Phospholipase C hydrolytically cleaves phosphoesters from phospholipids and sphingomyelin, leaving diglycerides and ceramide respectively. Preincubation with phospholipase C releases lysosomal enzymes in mitochondrial fractions of rat liver,[48,111] brain,[48] and kidney[48] coincident with cleavage of

TABLE I

Activation of Acid Hydrolases in Mitochondrial Fractions of Rat Brain by Heavy Metal Cations[a]

		Activity	
Cation	Concentration molar	Percent free acid phosphatase	β-Glucuronidase
None		34	33
Ag^+	0.05	40	40
	0.50	88	70
	1.0	100	83
Hg^{2+}	0.05	35	40
	0.50	97	53
	1.0	100	65
Cu^{2+}	0.05	36	41
	0.50	43	47
	1.0	73	66

[a] Mitochondrial fractions in 0.25 M sucrose were preincubated for 1 hr at 4°C in the presence of the metallic cation. Free and total enzyme activities were assayed according to de Duve *et al.*[26] Data are mean values of 2–5 measurements. Inhibition of total enzyme activities ranged up to 70%, depending upon the concentration of heavy metal cations. However, this did not obscure the activation phenomenon. From Koenig and Agoro.[46]

phospholipids.[48] When purified renal lysosomes were preincubated with phospholipase C, several acid hydrolases were liberated at a much higher rate than phospholipid phosphate (Fig. 14).[112] The bulk of the latent acid phosphatase and β-glucuronidase, 60–80% of the total, was rendered active when only 7–15% of the lysosomal phospholipid phosphate was split; these enzymes were fully activated when 20–35% of the total phospholipid was split. However, this treatment did not affect the fine structure of

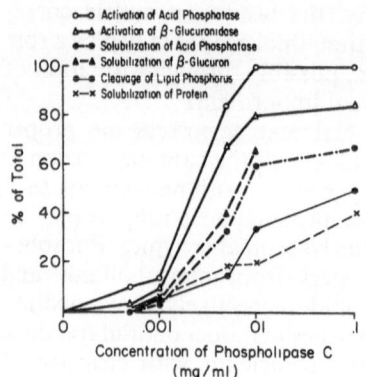

Fig. 14. The effects of phospholipase C on purified lysosomes from rat kidney. The results from a typical experiment are represented in this graph. From Koenig *et al.*[112]

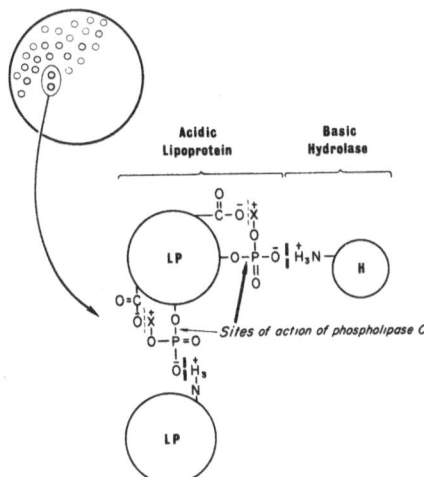

Fig. 15. Matrix binding theory of the lysosome. The basic groups of the phospholipid are represented as being internally neutralized through electrostatic interaction with adjacent carboxylic groups of the lipoprotein, leaving the negatively charged phosphoryl group free for binding the cationic hydrolases (or other lipoprotein macromolecules).

lysosomes as visualized in the electron microscope. Disruption of lysosomal particles became noticeable only when approximately 50% of the lysosomal phospholipid was enzymatically cleaved.[112] From these observations it appears that a portion of the phospholipid phosphate, approximately one-third of the total in renal lysosomes, is involved in the latency of the hydrolases. These phosphoryl groups probably serve as binding sites for the acid hydrolases. The remaining phospholipid phosphate, which is less accessible to the action of phospholipase C, seems to be essential to the structural integrity of the lysosomal particle (Fig. 15).

IV. PHYSIOLOGICAL ASPECTS

It is now quite clear that lysosomes and certain related membrane-bound structures comprise an intracellular digestive system that has been likened to the alimentary tract of higher organisms. The multitude of lysosomes present

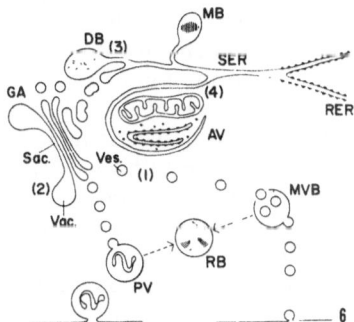

Fig. 16. Suggested origins, forms, and functions of lysosomes. From Novikoff.[69] Abbreviations: AV, autophagic vacuole; DB, dense body; GA, Golgi apparatus; MB, microbody; MVB, multivesicular body; PV, phagocytic vacuole; RB, residual body; RER, rough endoplasmic reticulum; Sac., Golgi saccule; SER, smooth endoplasmic reticulum; Vac., Golgi vacuole; and Ves., Golgi vesicle.

in neurons and other neural cells implies important physiological roles for these organelles, but our knowledge of their roles is still fragmentary. These functions now will be briefly surveyed. For a comprehensive account of this subject the reader is referred to several recent reviews.[3,69]

Two kinds of lysosomes are distinguished: (1) *primary lysosomes*, whose contained enzymes have not been previously involved in a digestive event, and (2) *secondary lysosomes*, which are sites of past or present digestive activity. *Heterophagy* (heterolysis) refers to intracellular digestion of material originating outside the cell and taken up by the process of endocytosis, i.e., through phagocytosis or micropinocytosis. *Autophagy* (autolysis) involves the intracellular digestion of endogenous material (Fig. 16).

A. Formation of Primary Lysosomes

There is considerable indirect evidence to show that primary lysosomes in various cell types are formed in the cisternae of the Golgi complex and probably also in the smooth endoplasmic reticulum. A number of workers have observed acid phosphatase activity in Golgi saccules and in neighboring Golgi vesicles of nerve cells (see Novikoff[69]). Novikoff and associates have called attention to the GERL, which consists of acid phosphatase-reactive, smooth tubular structures in close proximity to acid phosphatase-containing Golgi saccules and vesicles, both smooth and coated.[69,78] GERL is more prevalent in sensory neurons and Schwann cells after axonal section.[78] These workers have suggested that the acid hydrolases synthesized in the ribosomal sites may be conveyed through these channels to dense bodies that form within this membrane-delimited system. This view is supported by the presence of typical dense bodies whose limiting membranes are continuous with smooth endoplasmic reticulum containing a similar dense, acid phosphatase-positive material (see Figs. 42, 43, and 45 in Holtzman, Novikoff, and Villaverde[78]). Novikoff also has suggested that intracellular digestions may occur inside the smooth endoplasmic reticulum of some cells.[69] The monophosphatase activity of the GERL is unusual in two respects: (1) In addition to hydrolyzing nucleoside monophosphates and glycerophosphate at pH 5, it also splits these substrates at a moderate rate at pH 7.2. (2) It splits thymidylic acid more rapidly at pH 5 and pH 7.2 than the other nucleoside monophosphates and glycerophosphate.[69]

This scheme for the origin of primary lysosomes closely resembles the mode of formation of zymogen granules in the pancreas first elucidated by Palade and colleagues[113] and subsequently demonstrated in other types of cells engaged in the manufacture of protein for "export." Neurons evidently utilize the same pathway for the intracellular transport of some newly synthesized protein. Utilizing electron microscope radioautography Droz[114,115] has shown in several types of neurons that ^{3}H-leucine and ^{3}H-lysine are initially incorporated into the protein of the Nissl bodies, maximum radioactivity occurring at 5 min. Peak labeling of protein in the Golgi complex is attained by 20–30 min. Occasional lysosomes become labeled subsequently.

Cohn and co-workers[116] have shown that acid hydrolases are transported via this route to lysosomes in developing mononuclear phagocytes. Bainton and Farquhar[117] have demonstrated this pathway in eosinophil cells. It seems likely that this pathway is also operative for the intracellular transport and compartmentation of lysosomal enzymes in cells of the nervous system.

It is known that the Golgi apparatus is the intracellular site for the synthesis of certain complex carbohydrates from simple sugar precursors, e.g., incorporation of ^{3}H-glucosamine into hyaluronic acid.[118] Inasmuch as the acid hydrolases seem to be glycoproteins (see Section I, A, 4), it is reasonable to suppose that the Golgi complex participates in the formation of these enzymes. Moreover, the smooth endoplasmic reticulum and the Golgi apparatus are known to be important loci for the synthesis of lipids and for the combination of the latter with the acceptor protein to form lipoproteins (reviewed in Jones, Ruderman, and Herera[119]).

The following hypothesis therefore is advanced to account for the cytogenesis of primary lysosomes. The polypeptide portion of the acid hydrolases is elaborated at the ribosomal sites and transferred to the cisternae of the rough endoplasmic reticulum. Subsequently these polypeptides, which presumably are enzymatically inert, are conveyed to the GERL and the Golgi apparatus where they undergo glycosylation to form the completed, active enzymes. Coincidentally the acidic lipoprotein of the lysosomal matrix is assembled in the Golgi complex and/or GERL by the attachment of lipid synthesized *in situ* to the apolipoprotein or receptor polypeptide transported from its site of synthesis in the rough endoplasmic reticulum. The acidic lipoprotein interacts with the newly completed cationic enzyme molecules within the cisternae of the Golgi or GERL to form an electrostatic complex in which the enzymes are inactive. Focal accumulations of lysosomal enzymes and matrix and the membranous encasement subsequently separate from the Golgi and GERL to form primary lysosomes.

B. Heterophagy

The ingestion of exogenous macromolecules by endocytosis seems to be a general property of cells. From studies of various cell types it seems that pinocytotic vesicles containing absorbed protein but no hydrolases, designated *heterophagosomes*, fuse with preexisting lysosomes to become *heterolysosomes*. There is evidence that these secondary lysosomes may acquire additional enzyme activity through fusion with small acid phosphatase-reactive coated vesicles deriving from the Golgi and/or the GERL elements.[69,78,120] Thus these coated vesicles are primary lysosomes involved in the transport of hydrolytic enzymes.

The various nonneuronal cells of the nervous system can ingest macromolecular material through micropinocytosis. Using fluorescence optics, Klatzo and associates[121] observed the incorporation *in vitro* of fluorescein-labeled albumin and globulin into lysosome-like inclusions of microglia, neuroglia, and choroid epithelium. These labeled proteins, when given intravenously in cat,[122] do not enter the brain parenchyma except in areas where the blood–brain barrier is poorly developed, e.g., the area postrema, hypothalamus, and epiphysis. After localized cold injury, labeled protein readily enters edematous brain tissue and becomes concentrated in lysosomal granules of macrophages, astrocytes, oligodendrocytes, and adventitial cells, but not neurons.

Choroid epithelial cells engulf horseradish peroxidase when given by the intravenous and the intraventricular routes. Becker, Novikoff, and Zimmerman[123] in an electron microscopic study of peroxidase uptake by rat choroid plexus found that this foreign protein rapidly passes the capillary wall and enters the choroid cells within coated pinocytotic vesicles. A few minutes later peroxidase activity appears in numerous membrane-bound vesicles, multivesicular bodies, and dense bodies. After a day or two peroxidase activity gradually disappears from cells, presumably by cantheptic degradation within the lysosomal system.

Peroxidase does not seem to be ingested by neurons at the light microscope level. When given intravenously, peroxidase is taken up into lysosomes of pericytes, but not neurons, of rat spinal ganglia.[69] Peroxidase, administered subarachnoidally in cat to bypass the blood–brain barrier, is sequestered within lysosomes of leptomeningeal, adventitial, and glial cells near the pial surface but does not appear in neurons.[43] According to recent electron microscopic studies, however, neurons in toad spinal ganglia[124] and in mammalian brain[125] can ingest the marker proteins, ferritin and peroxidase respectively, by micropinocytosis to a limited extent. Ingested proteins subsequently appear in small vesicles, multivesicular bodies, and dense bodies.

C. Autophagy

The process by which cells sequester and digest portions of their own cytoplasm is termed cellular autophagy or focal cytoplasmic degradation. Such foci, known as *autophagic vacuoles*, can be recognized in the electron microscope by the presence of various cell structures apparently undergoing degeneration in a membrane-enclosed bag. Autophagic vacuoles are encountered only infrequently in normal nerve cells, evidently because focal autolysis proceeds at a rather slow tempo under physiological conditions. Autophagic vacuoles occur in increased numbers in neurons injured by axonal section,[78] x-irradiation,[126] bilirubin,[127] tetrazolium salts,[128] 5-fluoroorotic acid (Fig. 17),[129,130] and retrograde atrophy.[131] Cytolysosomes, large acid phosphatase-reactive inclusions corresponding to autophagic vacuoles, have been found in neurons injured by ischemia,[132] anoxia-ischemia,[133] diphtheria toxin,[134] axonal amputation,[135] x-irradiation,[136] triethyl tin,[69] and actinomycin D (Figs. 18 and 19).[101] During Wallerian degeneration complex autophagic vacuoles form in glia,[137] Schwann cells,[79] and macrophages,[137,138] which evidently contain myelin and other tissue breakdown products.

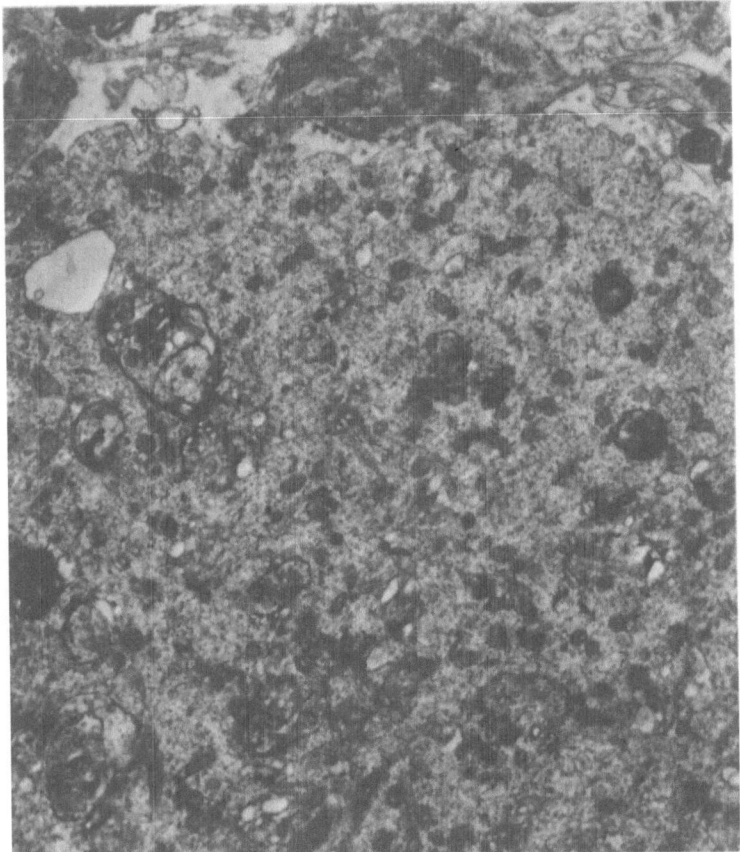

Fig. 17. Electron micrograph showing lumbar motoneuron of cat 7 days after injection of 5-fluoro-orotic acid (15 mg) into the lumbar cistern. The damaged neuron shows a loss of Nissl bodies and a number of membrane-enclosed autophagic vacuoles. × 35,000. From Koenig.[130]

The mode of formation of the autophagic vacuole is not entirely clear. Portions of cytoplasm about to undergo autophagy are delimited by a single- or double-layered membrane that may arise *de novo* or from preformed cytomembranes such as the smooth endoplasmic reticulum.[3,69] Subsequently these structures, designated *autophagosomes*, presumably receive hydrolytic enzymes through merger with primary or secondary lysosomes to become *autolysosomes* or *autophagic vacuoles*. It is likely that autophagy is involved in the turnover of cell constituents, e.g., mitochondria, which seem to be destroyed and renewed in bulk.

Residual bodies are lysosomal bodies containing membranes, grains, myelin figures, amorphous masses, and other structures presumed to be

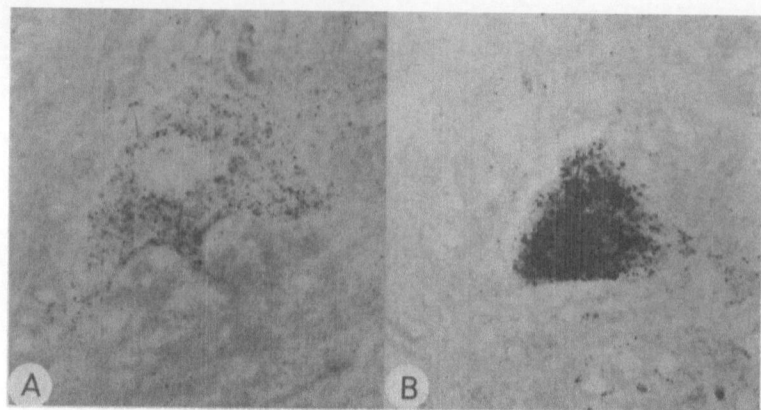

Fig. 18. Acid phosphatase reaction in lumbar motoneurons of cat spinal cord. (A) Control neuron on unoperated side. (B) Experimental neuron 11 days after sciatic nerve section. The latter exhibits a marked increase in reaction product and the presence of cytolysomes.

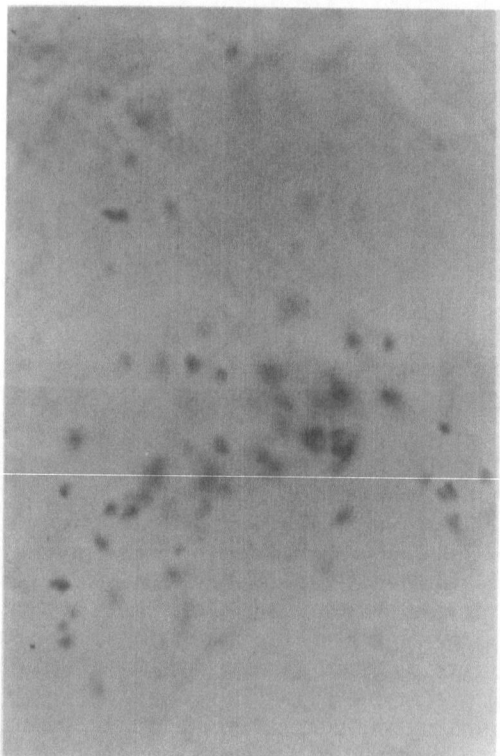

Fig. 19. Acid phosphatase reaction in a lumbar motoneuron of cat spinal cord 10 days after a subarachnoid injection of actinomycin D (2 μg). Numerous cytolysomes are visible. × 1875.

undigested residues of a prior digestive event. Novikoff has suggested that the majority of dense bodies in neurons and other neural cells in fact may be residual bodies.[69]

Lipofuscin pigment granules also constitute a form of residual body. These granules accumulate in neural cells with advancing age and in numerous disease states, including vitamin D deficiency and various neuronal atrophies and degenerations. Although they are readily distinguished from typical lysosomal dense bodies by their coarse size, irregular outline and more complex fine structure, lipofuscin granules share many structural and cytochemical features with lysosomes.[139,140] Lipofuscin granules contain some of the same acid hydrolases, such as acid phosphatase, acid esterase, and acid DNase. In addition, they are osmiophilic, argentaffin, basophilic, PAS-positive, and autofluorescent. However, they are less basophilic than lysosomes on intravital staining, and their PAS-positive and autofluorescent components in unfixed tissues are more resistant to the extractive, i.e., disruptive, action of chloroform-methanol[139] (see Sections II, C, 1, 4, and 6).

The abnormal inclusion bodies in many storage diseases also seem to be a type of residual body. These inclusions generally feature a marked accumulation of a specific chemical component that varies with the disease, e.g., gangliosides in Tay–Sachs disease, cerebroside sulfate in metachromatic leukodystrophy, glucocerebroside in Gaucher's disease, sphingomyelin in Nieman–Pick disease, glycogen in Pompe's disease, and sulfated mucopolysaccharides in Hurler's disease. A genetic deficiency of a specific lysosomal enzyme, recently demonstrated in several of these diseases, is now believed to be responsible for the pathogenesis of this type of disorder. During physiological autophagy the substrates of the deficient enzyme are not digested in the autophagic vacuoles. Consequently, these substances accumulate as membrane-bound residual bodies. A deficiency of aryl sulfatase (cerebroside sulfatase) occurs in metachromatic leukodystrophy,[141] of glucocerebrosidase in Gaucher's disease,[142] of sphingomyelinase in Nieman–Pick disease,[143] of α-glucosidase in Pompe's disease,[144] and of β-galactosidase in generalized gangliosidosis.[145] Ultrastructurally these pathological inclusions are usually filled with parallel arrays of membranous lamellae arranged concentrically, in flat stacks, or chaotically; frequently they exhibit acid phosphatase reaction product attesting to their lysosomal nature.[69,146,147]

Lysosomes with a multilamellated matrix, relatively uncommon in normal neurons, may become numerous during certain intervals following axonal section[78] and x-irradiation.[148] They also occur in the axoplasm of dystrophic axons,[149] in Schwann cells during Wallerian degeneration,[79] and in phagocytes.[137]

D. Structural Lability of Lysosomes

Dense bodies that contain myelin figures or membranous lamellae in their matrix are generally considered to be residual bodies with incompletely digested material, predominantly phospholipids. However, three instances recently have come to our attention in which membranes seem to form directly from the granular matrix of otherwise normal lysosomes: (1) acute neuro-axonal dystrophy caused by fluorocitrate poisoning,[150,151] (2) vital staining of lysosomes by the cationic dye, neutral red,[42,43] and (3) dilution of renal lysosomes isolated in hypertonic sucrose.[109]

A potent inhibitor of the Krebs cycle, fluorocitrate, when injected intrathecally in cat, causes a disgorgement of multitudinous dense bodies and mitochondria from spinal neurons into axons leading to the formation of axonal balloons. Many of the extruded lysosomes are rapidly converted to membrane-filled bodies soon after entering axons and nerve endings (Fig. 20).[151]

Neutral red, given intraperitoneally in rat (0.335 g/kg), enters neural cells by diffusion rather than by endocytosis and causes a hydropic swelling of lysosomes in neural and other cells. Vitally stained lysosomes expand two

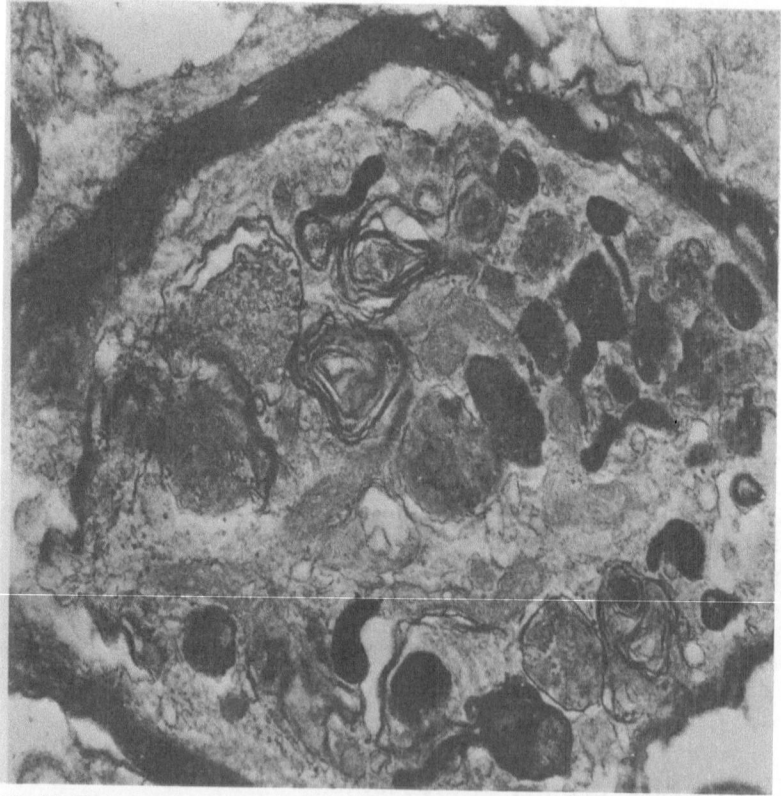

Fig. 20. Electron micrograph of myelinated axon in neuropil of cat lumbar spinal cord 4 hr after a subarachnoid injection of sodium fluorocitrate (10 μg). The axon is filled with multilamellated lysosomes, dense bodies, and degenerating mitochondria. From Koenig. [151]

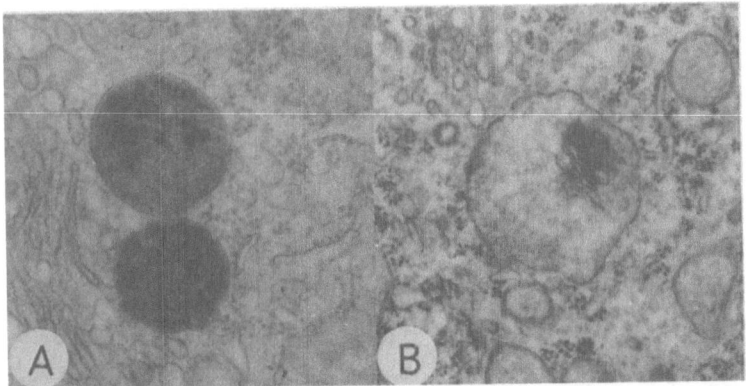

Fig. 21. Electron micrographs of neurons of rat cerebral cortex stained *in vivo* with neutral red. From McDonald and Koenig[42] and Koenig.[43] (A) Control dense bodies. (B) One-half hour after neutral red injection. Lysosome is swollen and matrix is rarefied and contains a stack of flat lamellae. × 36,000.

to three times their normal diameter, a 16–27-fold increase in volume, within 1–2 hr. These lysosomes exhibit radical changes in matrix ultrastructure and also in buoyant density (see Section I, A, 2), which are completely reversible (Figs. 21–23).[42,43] The matrix material assumes various appearances. Within the same lysosomes may be found two or more of the following: seemingly empty vacuoles, attenuated granular material, membranous structures such as vesicles or multiple parallel lamellae arranged concentrically or in flat plates, and masses of electron-dense osmiophilic material. Only the electron-lucent, triglyceride-containing vacuoles of dense bodies are unaffected by dye. The fine structure of these lysosomes reverts toward normal *pari passu* with the disappearance of dye from these particles, complete restoration occurring by 8–10 hr. Despite these marked ultrastructural changes, acid phosphatase activity in altered lysosomes remains associated with the matrix and retains its normal latency toward glycerophosphate (Fig. 24).[43] The limiting membrane of lysosomes seems to be highly elastic, as it remains intact throughout the swelling–contraction cycle.

Remarkable myelin figures may develop in renal lysosomes when stained *in vitro* with neutral red or when the suspending medium is diluted from 1.75 to 0.5 M sucrose.[109] These changes in the fine structure of lysosomes must reflect an alteration in the molecular morphology of the lipoprotein constituents. The binding of cationic dye and the concomitant inhibition of water probably alters the shape of the lipoprotein molecules and initiates a rearrangement of these macromolecules with respect to each other. Multi-lamellated lysosomes are well hydrated, highly anisotropic structures whose lipoprotein macromolecules, which may be linear or flat in shape, are arranged in an orderly manner to form membranes. Dense bodies with a granular matrix are poorly hydrated, isotropic structures in which the lipoprotein macromolecules, possibly globular in configuration, are tightly compressed.

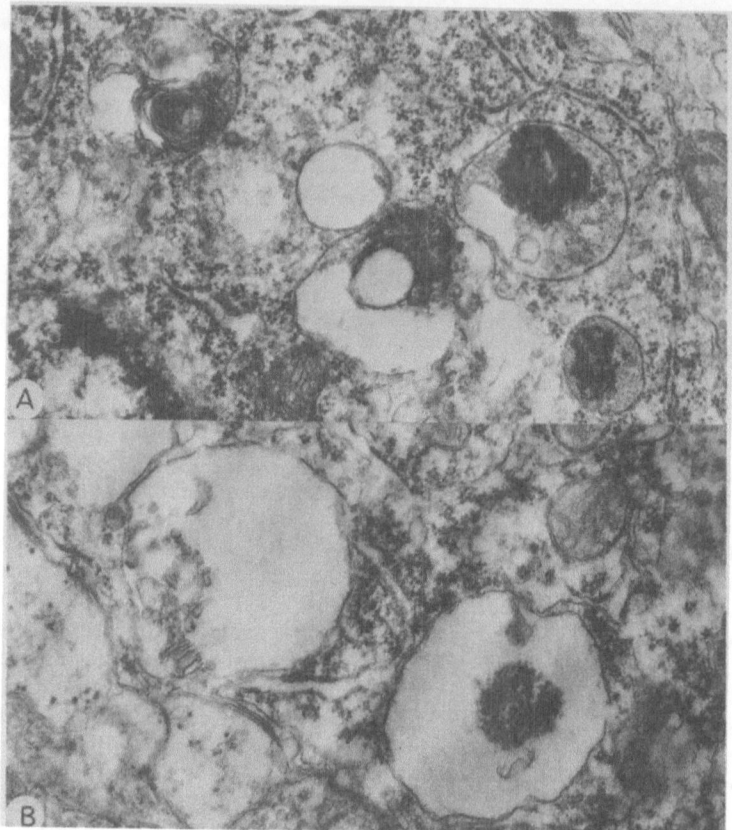

Fig. 22. Electron micrographs of neurons of rat cerebral cortex stained *in vivo* with neutral red. From McDonald and Koenig[42] and Koenig.[43] (A) One hour after neutral red. (B) Two hours after neutral red. Markedly swollen lysosomes show voluminous electronlucent areas. The matrix is granular, lamellated, or vesiculated. × 54,000.

Neutral red granules or vacuoles have been mistakenly identified as residual bodies or autophagic vacuoles in pancreas[82] and liver[152] because of the presence of myelin figures and other membranous structures. The "autophagic vacuoles" produced in macrophages and other cells by chloroquine,[153] an organic cation that is concentrated within lysosomes,[68] closely resemble the neutral red granules in their ultrastructure and in their mode of evolution. Moreover, removal of the drug results in restoration of the normal fine structure of lysosomes without apparent cell injury. It is evident from these observations, therefore, that dense bodies possess a hitherto unrecognized structural plasticity that does not seem to affect the latency of the contained hydrolases.

V. ROLE OF LYSOSOMES IN DISEASE

In a broad sense the pathological consequences of lysosomal dysfunction involve disturbances in intracellular digestion. De Duve[154] has grouped the disorders of lysosomal function into the following four categories: indigestion, dyspepsia, constipation, and perforation. In addition, specific pathophysiologic effects may result from lysosomal dysfunction arising out of special physiological roles that these particles may subserve in neurons and other neural cells.

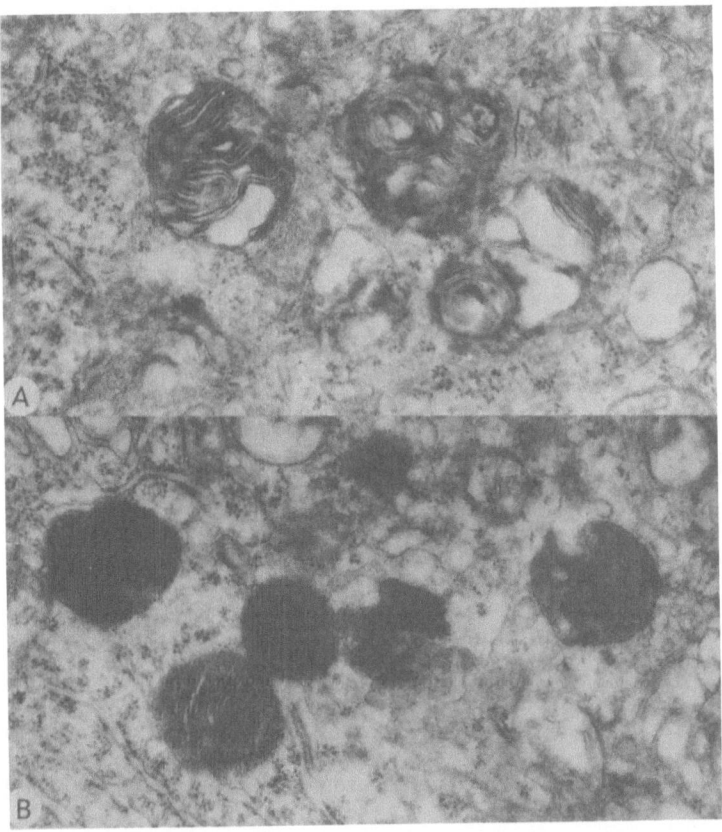

Fig. 23. Electron micrographs of neurons of rat cerebral cortex stained *in vivo* with neutral red. From McDonald and Koenig[42] and Koenig[43] (A) Six hours after neutral red. Lysosomes are multilamellated and still have vacuoles. (B) Seven hours after neutral red. One lysosome is multilamellated, but the others are nearly normal in fine structure. × 54,000.

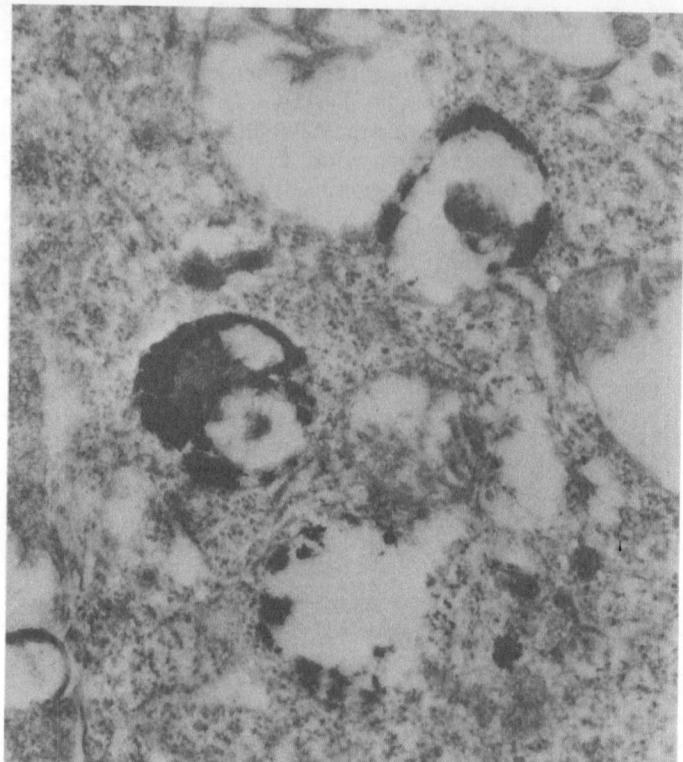

Fig. 24. Electron micrographs of neurons of rat cerebral cortex stained *in vivo* with neutral red. From McDonald and Koenig[42] and Koenig.[43] Acid phosphatase reaction 3 hr after neutral red. Enzyme product is restricted to the matrix, none occurring in the vacuolated areas. × 54,000.

A. Cell Indigestion, Dyspepsia, and Constipation

Under the category indigestion are included those disorders in which indigestible material is taken up into cells by endocytosis. In some instances, e.g., silicosis, asbestosis, and gout, the ingested substances induce a release of lysosomal hydrolases into the cytoplasm with resultant injury or death of the host cells.[155] Cell indigestion is probably an uncommon occurrence in the central nervous system as the latter is protected from foreign macromolecules by the blood–brain barrier.

Cell dyspepsia refers to an accumulation of indigestible material resulting from an enzyme deficiency, as in the storage diseases described earlier (Section III, C). Cell constipation, i.e., the lack of an efficient mechanism for defecation, enhances the effects of cell dyspepsia and cell indigestion. Neurons are particularly vulnerable to these disorders as they seem to be incapable of excreting the remnants of intracellular digestion. Moreover, dyspeptic nerve cells cannot be eliminated and replaced with new neurons through mitotic division. The progressive accumulation of indigestible material causes a ballooning of nerve cells and a crowding of various intracellular structures, e.g., mitochondria and Nissl bodies, with resultant attenuation of these components. Neurons afflicted by an unremitting dyspepsia undergo a progressive compression atrophy, which eventuates in cell death. Early in the evolution of a neuronal storage disease, neurophysiological function may be normal. Later, functional disturbances such as seizure activity appear, followed by a loss of neuronal function. The aging and premature death of nerve cells in senility may be due partly to a progressive accumulation of digestive residues in the form of lipofuscin pigment granules.

B. Leakage of Lysosomal Enzymes

A massive intracellular release of acid hydrolases, as envisaged in the suicide bag hypothesis of de Duve,[1] doubtless would result in autolytic destruction of the host cells, while a gradual leakage of these hydrolases conceivably would cause sublethal cell damage. Agents that promote the release of lysosomal enzymes *in vitro* and/or *in vivo* are known as lysosomal *labilizers*. These include vitamin A; etiocholanolone and other pyrogenic steroids; carbon tetrachloride and similar organic solvents; detergents, lysolecithin, and various surfactants; phospholipases, streptolysins, and bacterial endotoxins; antibodies; cations; and free radical generators, such as ultraviolet light, x-rays, peroxides, and oxidizing agents. Agents that antagonize the effects of lysosomal labilizers and retard the release of lysosomal enzymes, designated lysosomal *stabilizers*, include cholesterol, cortisone and related antiinflammatory steroids, and chloroquine. There is suggestive evidence that the action of these agents on lysosomes may be partly responsible for their biological effects (reviewed by de Duve,[2,154] Allison,[155] and Weissman[156]).

An abnormal release of lysosomal hydrolases *in vivo* has been implicated in the pathogenesis of various disease entities. Among these are auto-immune diseases such as rheumatoid arthritis and generalized lupus erythematosis; progressive muscular dystrophy; poisoning by exogenous agents such as vitamin A, heavy metals, carbon tetrachloride, and bacterial toxins; the pneumoconioses, silicosis and asbestosis; gout; ultraviolet and x-irradiation injury; and chromosomal damage and cancer induced by certain physical and chemical agents (reviewed by de Duve,[2,154] Novikoff,[69] Allison,[155] and Weissman[156]).

Assessment of the pathogenetic role of lysosomal rupture is beset by ambiguities. It is usually difficult to ascertain whether a given release of lysosomal enzymes is causally related to a pathological entity or is a secondary event. Moreover, the techniques used for demonstrating hydrolase activity may not discriminate between enzyme release during life and release that occurs during processing of the tissue sample. An increase in free and soluble hydrolase activities, as demonstrated biochemically in tissue homogenates, may reflect an increased fragility of altered lysosomes to homogenization rather than an actual discharge of these enzymes into the cell cytoplasm *in vivo*. The demonstration of increased hydrolytic activities in the blood serum, in the venous effluent of a diseased organ, or in a special fluid compartment such as the cerebrospinal fluid, is likely to be a more reliable index of an intravital release of lysosomal enzymes if cell death can be excluded as a source of the liberated enzymes. The histochemical method of Bitensky,[94] which employs frozen sections of unfixed tissues, is purported to be a sensitive technique for qualitative assessment of the latency of acid phosphatase. A diffusion of hydrolase reaction product in cytochemical preparations of diseased tissue may denote an *in vivo* release of enzymes, provided that fixation is carefully performed, preferably by vascular perfusion of tissues *in situ*.

Biochemical studies of acid hydrolase activities in neurological diseases have been limited thus far to brain tumors[158] and cerebrospinal fluid.[159,160] β-Glucuronidase activity in the cerebrospinal fluid is moderately increased in acute and extensive demyelinating diseases, Guillain–Barre syndrome, diabetic neuropathy, and tumors[159] and is markedly increased together with acid phosphatase activity in meningeal carcinomatosis.[160] Brain tumors exhibit high β-glucuronidase activities.[158] Increased activities of several acid hydrolases have been reported in brain homogenates of chicks with encephalomalacia due to vitamin E deficiency.[61] Methionine sulfoximine, a potent convulsant, caused a moderate activation of latent acid phosphatase and N-acetyl-β-D-glucosaminidase in rat brain homogenates concomitant with epileptic seizures.[161] However, convulsions produced in rat by the intracerebral injection of fluorocitrate, a Krebs cycle inhibitor, is not accompanied by an activation of latent acid phosphatase and β-glucuronidase in brain homogenates.[162] A release of lysosomal enzymes may be involved in cerebral edema produced by triethyl tin[69] and possibly other agents. The beneficial effect of cortisone and related corticosteroids on cerebral edema is possibly related to the stabilizing action of these agents on cerebral lysosomes. A diffusion of acid phosphatase reaction product occurs in the distal portion of sectioned axons, which may reflect a release of lysosomal hydrolases catalyzing autolysis of axonal components.[79] In the fluorocitrate-poisoned spinal cord of the cat a diffusion of acid phosphatase reaction product has been observed in axonal balloons stuffed with lysosomes and mitochondria.[150]

VI. REFERENCES

1. C. de Duve, *in Subcellular Particles* (T. Hayashi, ed.), pp. 128–159, Roland Press, New York (1959).
2. C. de Duve, *in Ciba Foundation Symposium on Lysosomes* (A. V. S. de Reuck and M. P. Cameron, eds.), pp. 1–31, Little, Brown, Boston (1963).
3. C. de Duve and R. Wattiaux, *Ann. Rev. Physiol.* **28**:435–491 (1966).

4. A. B. Novikoff, *in The Cell* (J. Brachet and A. E. Mirsky, eds.), Vol. 2, pp. 423–488, Academic Press, New York (1961).

5. A. Goldstone and H. Koenig, Acid hydrolases and structural lipoproteins of kidney lysosomes, *J. Histochem. Cytochem.* **16**:511–512 (1968).

6. A. Goldstone, E. Szabo, and H. Koenig, *Biochemical and Histochemical Characterization of the Enzyme and Matrix Components of Renal Lysosomes*, pp. 80–81, Summary Report of the Third International Congress of Histochemistry and Cytochemistry, New York (August 18–22, 1968).

7. A. Goldstone, E. Szabo, and H. Koenig, *Federation Proc.* **28**:266 (1969).

8. R. Mylroie and H. Koenig, in preparation.

9. J. Barrett and J. T. Dingle, A lysosomal component capable of binding cations and a carcinogen, *Biochem. J.* **105**:20P (1967).

10. S. Mahadevan and A. L. Tappel, Arylamidases of rat liver and kidney, *J. Biol. Chem.* **242**:2369–2374 (1967).

11. G. F. Wilgram and E. P. Kennedy, Intracellular distribution of some enzymes catalyzing reactions in the biosynthesis of complex lipids, *J. Biol. Chem.* **238**:2615–2619 (1963).

12. S. Shibko and A. L. Tappel, Rat-kidney lysosomes: isolation and properties, *Biochem. J.* **95**:731–741 (1965).

13. S. Gatt, Enzymic hydrolysis of sphingolipids. Hydrolysis of ceramide glucoside by an enzyme from ox brain, *Biochem. J.* **101**:687–691 (1966).

14. Y. Z. Frohwein and S. Gatt, Isolation of β-N-acetylhexosaminidase, β-N-acetylglucosaminidase and β-N-acetylgalactosaminidase from calf brain, *Biochemistry* **6**:2775–2782 (1967).

15. S. Gatt, Enzymatic hydrolysis of sphingolipids. 1. Hydrolysis and synthesis of ceramides by an enzyme from rat brain, *J. Biol. Chem.* **241**:3724–3730 (1966).

16. A. K. Hajra, D. M. Bowen, Y. Kishimoto, and N. S. Radin, Cerebroside galactosidase of brain, *J. Lipid Res.* **7**:379–386 (1966).

17. S. Gatt, Enzymatic hydrolysis of sphingolipids. V. Hydrolysis of monosialoganglioside and hexosylceramides by rat brain β-galactosidase, *Biochim. Biophys. Acta* **137**:192–195 (1967).

18. Y. Barenholz, A. Roitman, and S. Gatt, Enzymatic hydrolysis of sphingolipids. II. Hydrolysis of sphingomyelin by an enzyme from brain, *J. Biol. Chem.* **241**:3731–3737 (1966).

19. S. Gatt, Y. Barenholz, and A. Roitman, Isolation of rat brain lecithinase-A, specific for the α'-position of lecithin, *Biochem. Biophys. Res. Commun.* **24**:169–172 (1966).

20. A. Mellors and A. L. Tappel, Hydrolysis of phospholipids by a lysosomal enzyme, *J. Lipid Res.* **8**:479–485 (1967).

21. S. D. Fowler, Phospholipid digestion by hepatic lysosomes, *J. Cell Biol.* **35**:41A–42A (1967).

22. N. J. Weinreb, R. O. Brady, and A. L. Tappel, The lysosomal localization of sphingolipid hydrolases, *Biochim. Biophys. Acta* **159**:141–146 (1968).

23. R. E. McCaman, M. Smith, and K. Cook, Intermediary metabolism of phospholipids in brain tissue. II. Phosphatidic acid phosphatase, *J. Biol. Chem.* **240**:3513–3517 (1965).

24. N. Marks, R. K. Datta, and A. Lajtha, Partial resolution of brain arylamidases and aminopeptidases, *J. Biol. Chem.* **243**:2882–2899 (1968).

25. N. N. Aronson, Jr., and E. A. Davidson, Lysosomal hyaluronidase, *J. Biol. Chem.* **240**:PC3222–PC3223 (1965).

26. C. de Duve, B. C. Pressman, R. Gianetti, R. Wattiaux, and F. Appelmans, Tissue fractionation studies. 6. Intracellular distribution patterns of enzymes in rat-liver tissue, *Biochem. J.* **60**:604–617 (1955).

27. H. Koenig and D. Gaines, unpublished observations.

28. A. O. Jibril and P. B. McKay, Lysosomal enzymes in experimental encephalomalacia, *Nature* **205**:1214–1215 (1965).

29. O. Z. Sellinger, D. L. Rucker, and F. de Balbian Verster, Cerebral lysosomes. I. A comparative study of lysosomal *N*-acetyl-β-D-glucosaminidase and mitochondrial aspartic transaminase of rat cerebral cortex, *J. Neurochem.* **11**:271–280 (1964).

30. D. M. Bowen and N. S. Radin, Properties of cerebroside galactosidase, *Biochim. Biophys. Acta* **152**:599–610 (1968).

31. H. Koenig, D. Gaines, T. McDonald, R. Gray, and J. Scott, Studies of brain lysosomes. I. Subcellular distribution of five acid hydrolases, succinate dehydrogenase and gangliosides in rat brain, *J. Neurochem.* **11**:729–743 (1964).

32. V. P. Whittaker, The isolation and characterization of acetylcholine—containing particles from brain, *Biochem. J.* **72**:694–706 (1959).

33. N. Marks and A. Lajtha, Protein breakdown in brain. Subcellular distribution and properties of neutral and acid proteinases, *Biochem. J.* **89**:438–447 (1963).

34. M. K. Gordon, K. G. Bensch, G. G. Deanin, and M. W. Gordon, Histochemical and biochemical study of synaptic lysosomes, *Nature* **217**:523–527 (1968).

35. G. C. Millson, Lysosomal enzymes in normal and scrapie mouse brain, *J. Neurochem.* **12**:461–468 (1965).

36. G. D. Hunter and G. C. Millson, Distribution and activation of lysosomal enzyme activities in subcellular components of normal and scrapie-affected mouse brain, *J. Neurochem.* **13**:375–383 (1966).

37. J. Mordoh, Subcellular distribution of acid phosphatase and β-glucuronidase activities in rat brain, *J. Neurochem.* **12**:505–514 (1965).

38. O. Z. Sellinger and R. A. Hiatt, Cerebral lysosomes. IV. The regional and intracellular distribution of arylsulfatase and evidence for two populations of lysosomes in rat brain, *Brain Res.* **7**:191–200 (1968).

39. R. Wattiaux, M. Wibo, and P. Baudhuin, *in Ciba Foundation Symposium on Lysosomes* (A. V. S. de Reuck and M. P. Cameron, eds.), pp. 176–196, Little, Brown, Boston (1963).

40. D. B. Brewer and D. Heath, Lysosomes and vacuolation of the liver cell, *Nature* **198**:1015–1016 (1963).

41. R. Wattiaux, S. Wattiaux-de Coninck, M-J. Rutgeerts, and P. Tulkens, Influence of the injection of a sucrose solution on the properties of rat-liver lysosomes, *Nature* **203**:757–758 (1964).

42. T. F. McDonald and H. Koenig, Ultrastructural changes in neural lysosomes induced by *in vivo* staining with neutral red, *Anat. Rec.* **151**:385 (1965).

43. H. Koenig, *in Barrier Systems in the Brain* (A. Lajtha and D. H. Ford, eds.), pp. 87–121, Elsevier, Amsterdam (1968).

44. H. Koenig and M. Cavender, unpublished data.

45. H. Beaufay, A. M. Berleur, and A. Dayen, The occurrence of lysosome-like particles in rat brain tissue, *Biochem. J.* **66**:32P (1957).

46. H. Koenig and J. Agoro, unpublished data.

47. H. Koenig and A. Jibril, Acidic glycolipids and the role of ionic bonds in the structure-linked latency of lysosomal hydrolases, *Biochim. Biophys. Acta* **65**:543–545 (1962).

48. H. Koenig and R. Gray, Action of phospholipase C on lysosomes, *J. Cell Biol.* **23**:50A (1964); and unpublished data.

49. G. Bernardi, E. Appella, and R. Zito, Studies on acid deoxyribonucleases. III. Physical and chemical properties of hog spleen acid deoxyribonuclease, *Biochemistry* **4**:1725–1729 (1965).

50. B. V. Plapp and R. D. Cole, Purification and characterization of bovine liver β-glucuronidase, *Arch. Biochem. Biophys.* **116**:193–206 (1966).

51. H. I. Zeya and J. K. Spitznagel, Antibacterial and enzymic basic proteins from leukocyte lysosomes: Separation and identification, *Science* **142**:1085–1087 (1966).

52. A. Goldstone and H. Koenig, Lysosomal enzymes as cationic glycoproteins (1969), submitted for publication.

53. P. S. Spensley and H. J. Rogers, Enzyme inhibition, *Nature* **173**:1190 (1954).

54. P. Bernfeld, in *Metabolic Inhibitors* (R. M. Hochster and J. H. Quastel, eds.), Vol. 2, pp. 437–472, Academic Press, New York (1963).

55. D. Thines-Sempoux, Chemical similarities between the lysosome and plasma membranes, *Biochem. J.* **105**:20P (1967).

56. A. Goldstone, E. Szabo, and H. Koenig, Acidic lipoproteins in lysosomes (1969), submitted for publication.

57. E. A. Napier, Jr., and R. E. Olson, Cellular lipoproteins. I. The isolation of lipoprotein fractions from cellular mitochondria and microsomes, *J. Biol. Chem.* **240**:4244–4252 (1965).

58. S. Bakerman and G. Wasemiller, Studies on structural units of human erythrocyte membrane. I. Separation, isolation, and partial characterization, *Biochemistry* **6**:1100–1113 (1967).

59. H. Koenig, Histological distribution of brain gangliosides: lysosomes as glycolipoprotein granules, *Nature* **195**:782–784 (1962).

60. H. Koenig, The autofluorescence of lysosomes. Its value for the identification of lysosomal constituents, *J. Histochem. Cytochem.* **11**:556–557 (1963).

61. A. L. Tappel, P. L. Sawant, and S. Shibko, in *Ciba Foundation Symposium on Lysosomes* (A. V. S. de Reuck and M. P. Cameron, eds.), pp. 78–108, Little, Brown, Boston (1963).

62. H. Beaufay, D. S. Bendall, P. Baudhuin, and C. de Duve, Tissue fractionation studies. 12. Intracellular distribution of some dehydrogenases, alkaline deoxyribonuclease and iron in rat-liver tissue, *Biochem. J.* **73**:623–629 (1959).

63. T. Kaltenbach and W. Eger, Beiträge zum histochemischen Nachweis von Eisen, Kupfer und Zink in der manschlichen Leber unter besonderer Berucksichtigung des Silbersulfidverfahrens nach Timm, *Acta Histochem.* **25**:329–354 (1966).

64. P. Baudhuin, H. G. Hers, and H. Loeb, An electron microscopic and biochemical study of type II glycogenosis, *Lab. Invest.* **13**:1139–1152 (1964).

65. S. Samuels, S. R. Korey, J. Gonatas, R. D. Terry, and M. Weiss, Studies in Tay–Sachs disease. IV. Membranous cytoplasmic bodies, *J. Neuropathol. Exptl. Neurol.* **22**:81–97 (1963).

66. K. Suzuki, K. Suzuki, and G. Chen, Metachromatic leucodystrophy: isolation and chemical characterization of metachromatic granules, *J. Neuropathol. Exptl. Neurol.* **26**:154–156 (1967).

67. H. Koenig, Intravital staining of lysosomes by basic dyes and metallic ions, *J. Histochem. Cytochem.* **11**:120–121 (1963).

68. A. C. Allison and M. R. Young, Uptake of dyes and drugs by living cells in culture, *Life Sci.* **3**:1407–1414 (1964).

69. A. B. Novikoff, in *The Neuron* (H. Hydén, ed.), pp. 319–377, Elsevier, Amsterdam (1967).

70. P. B. Gahan, in *International Review of Cytology* (G. H. Bourne and J. F. Danielli, eds.), Vol. 21, pp. 1–63, Academic Press, New York (1967).

71. G. Gomori, *Microscopic Histochemistry: Principles and Practice,* University of Chicago Press, Chicago (1952).

72. T. Barka and P. J. Anderson, *Histochemistry: Theory, Practice and Bibliography,* Harper & Row, New York (1963).

73. R. L. Friede and M. Knoller, A quantitative mapping of acid phosphatase in the brain of the rhesus monkey, *J. Neurochem.* **12**:441–450 (1965).

74. R. L. Friede and M. Knoller, Quantitative tests of histochemical methods for phosphomonoesterases in brain tissue, *J. Histochem. Cytochem.* **13**:125–132 (1965).

75. H. E. Hirsch, Acid phosphatase localization in individual neurons by a quantitative histochemical method, *J. Neurochem.* **15**:123–130 (1968).

76. N. H. Becker, S. Goldfischer, W-Y Shin, and A. B. Novikoff, The localization of enzyme activities in the rat brain, *J. Biophys. Biochem. Cytol.* **8**:649–663 (1960).

77. A. B. Novikoff, *in Developing Cell Systems and Their Control* (D. Rudnick, ed.), pp. 167–203, Ronald Press, New York (1960).

78. E. Holtzman, A. B. Novikoff, and H. Villaverde, Lysosomes and GERL in normal and chromatolytic neurons of the rat ganglion nodosum, *J. Cell Biol.* **33**:419–436 (1967).

79. E. Holtzman and A. B. Novikoff, Lysosomes in the rat sciatic nerve following crush, *J. Cell Biol.* **27**:651–669 (1965).

80. R. M. Torack, The nature and significance of compound lipid cytoplasmic bodies in progressive dementia, *J. Neuropathol. Exptl. Neurol.* **26**:151–152 (1967).

81. G. W. Kreutzberg and H. Hager, Electron microscopical demonstration of acid phosphatase activity in the central nervous system, *Histochem.* **6**:254–259 (1966).

82. J. M. Byrne, Acid phosphatase activity in neutral red granules of mouse exocrine pancreas, *Quart. J. Microscop. Sci.* **105**:343–348 (1964).

83. A. Vorbrodt, Histochemical studies on the intracellular localization of acid deoxyribonuclease, *J. Histochem. Cytochem.* **9**:647–655 (1961).

84. K. D. Barron, J. B. Oldershaw, and J. Bernsohn, Hydrolase cytochemistry of retrograde neuronal degeneration in feline lateral geniculate body. With observations on the identification of multiple forms of neural hydrolases having overlapping substrate affinities, *J. Neuropathol. Exptl. Neurol.* **25**:443–478 (1966).

85. D. Schiffer, A. Fabiani, and G. F. Monticone, Acid phosphatase and non-specific esterase in normal and reactive glia of human nervous tissue. A histochemical study, *Acta Neuropathol.* **9**:316–327 (1967).

86. H. Koenig, unpublished observations.

87. R. Hess and A. G. E. Pearse, The histochemistry of indoxylesterase of rat kidney with special reference to its cathepsin-like activity, *Brit. J. Exptl. Pathol.* **39**:292–299 (1958).

88. S. J. Holt, *in Ciba Foundation Symposium on Lysosomes* (A. V. S. de Reuck and M. P. Cameron, eds.), pp. 114–120, Little, Brown (1963).

89. S. Goldfischer, The cytochemical demonstration of a lysosomal aryl sulfatase activity by light and electron microscopy, *J. Histochem. Cytochem.* **13**:520–523 (1965).

90. M. Hayashi, Comparative histochemical localization of lysosomal enzymes in rat tissues, *J. Histochem. Cytochem.* **15**:83–92 (1967).

91. M. Hayashi, T. Shirahama, and A. S. Cohen, Combined cytochemical and electron microscopic demonstration of β-glucuronidase activity in rat liver with the use of a simultaneous coupling azo dye technique, *J. Cell Biol.* **36**:289–298 (1968).

92. H. Koenig, Cytochemical validation of the lysosome hypothesis in brain, *Trans. Am. Neurol. Assoc.* **88**:227–228 (1963).

93. H. Koenig, The staining of lysosomes by basic dyes, *J. Histochem. Cytochem.* **13**:20–21 (1965).

94. L. Bitensky, *in Ciba Foundation Symposium on Lysosomes* (A. V. S. de Reuck and M. P. Cameron, eds.), pp. 362–375, Little, Brown (1963).

95. H. Koenig, J. Agoro, and M. Cavender, unpublished observations.

96. S. Goldfischer and J. Moskal, Electron probe microanalysis of liver in Wilson's disease, *Am. J. Pathol.* **48**:305–315 (1966).

97. H. Koenig, in *Response of the Nervous System to Ionizing Radiation* (T. Haley and R. Snider, eds.), pp. 403–417, Little, Brown (1964).

98. C. T. Ashworth, J. S. Leonard, E. H. Eigenbrodt, and F. J. Wrightsman, Hepatic intracellular osmiophilic droplets. Effect of lipid solvents during tissue preparation, *J. Cell Biol.* **31**:301–318 (1966).

99. A. Rambourg, An improved silver methenamine technique for the detection of periodic acid-reactive complex carbohydrates with the electron microscope, *J. Histochem. Cytochem.* **15**:409–412 (1967).

100. U. Sandbank and N. H. Becker, Ammoniated silver carbonate for the demonstration of lysosomes, *Stain Technol.* **39**:27–31 (1964).

101. H. Koenig, unpublished data.

102. H. Koenig, R. Nayyar, and A. Goldstone, unpublished observations.

103. J. Berthet, L. Berthet, F. Appelmans, and C. de Duve, Tissue fractionation studies. 2. The nature of the linkage between acid phosphatase and mitochondria in rat-liver tissue, *Biochem. J.* **50**:182–189 (1951).

104. F. Appelmans and C. de Duve, Tissue fractionation studies. 3. Further observations on the binding of acid phosphatase by rat-liver particles, *Biochem. J.* **59**:426–433 (1955).

105. H. Koenig, M. Cavender, and E. Szabo, unpublished observations.

106. B. Hainsworth and C. H. Wynn, The permeability of the lysosomal membrane to glycerol, sorbose, and *p*-nitrophenol, *Biochem. J.* **101**:9P (1966).

107. H. Koenig, On the structure-linked latency of the lysosomal enzymes, *J. Histochem. Cytochem.* **15**:767–768 (1967).

108. P. L. Sawant, I. D. Desai, and A. L. Tappel, Factors affecting the lysosomal membrane and availability of enzymes, *Arch. Biochem. Biophys.* **105**:247–253 (1964).

109. H. Koenig, C. Hughes, M. Cavender, and J. Agoro, unpublished results.

110. M. A. Verity and A. Reith, Effect of mercurial compounds on structure-linked latency of lysosomal hydrolases, *Biochem. J.* **105**:685–690 (1967).

111. H. Beaufay and C. de Duve, Tissue fractionation studies, 8. Enzymic release of bound hydrolases, *Biochem. J.* **73**:604–609 (1959).

112. H. Koenig, R. Nelson, D. Gaines, and R. Gray, unpublished observations.

113. L. G. Caro and G. E. Palade, Protein synthesis, storage, and discharge in the pancreatic exocrine cell; an autoradiographic study, *J. Cell. Biol.* **20**:473–495 (1964).

114. B. Droz, Accumulation de protéines nouvellement synthétisées dans l'appareil de Golgi du neurone; étude radioautographique en microscopie électronique, *Compt. Rend. Acad. Sci. Paris* **260**:320–322 (1965).

115. B. Droz, Synthèse et transfert des protéines cellulaires dans les neurones ganglionnaires étude radioautographique quantitative en microscopie électronique, *J. Microscop.* **6**:201–228 (1967).

116. Z. A. Cohn and B. Benson, The *in vitro* differentiation of mononuclear phagocytes. I. The influence of inhibitors and the results of autoradiography, *J. Exptl. Med.* **121**:279–288 (1965).

117. D. G. Bainton and M. S. Farquhar, Segregation and packaging of granule enzymes in eosinophils, *J. Cell Biol.* **35**:6A (1967).

118. P. Barland, C. Smith, and D. Hamerman, Localization of hyaluronic acid in synovial cells by radioautography, *J. Cell Biol.* **37**:13–26 (1968).

119. A. L. Jones, N. B. Ruderman, and M. G. Herera, Electron microscopic and biochemical study of lipoprotein synthesis in the isolated perfused rat liver, *J. Lipid Res.* **8**:429–446 (1967).

120. D. S. Friend and M. G. Farquhar, Functions of coated vesicles during protein absorption in the rat vas deferens, *J. Cell Biol.* **35**:357–376 (1967).

121. I. Klatzo and J. Miquel, Observations on pinocytosis in nervous tissue, *J. Neuropathol. Exptl. Neurol.* **19**:475–487 (1960).

122. I. Klatzo, J. Miquel, and R. Otenasek, The application of fluorescein labeled serum proteins (FLSP) to the study of vascular permeability in the brain, *Acta Neuropathol.* **2**:144–160 (1962).

123. N. H. Becker, A. B. Novikoff, and H. M. Zimmerman, Fine structure observations of the uptake of intravenously injected peroxidase by the rat choroid plexus, *J. Histochem. Cytochem.* **15**:160–165 (1967).

124. J. Rosenbluth and S. L. Wissig, The distribution of exogenous ferritin in toad spinal ganglia and the mechanism of its uptake by neurons, *J. Cell Biol.* **23**:307–326 (1964).

125. N. H. Becker, A. Hirano, and H. M. Zimmerman, Observations of the distribution of exogenous peroxidase in the rat cerebrum, *J. Neuropathol. Exptl. Neurol.* **27**:439–452 (1968).

126. E. B. Masurovsky, M. B. Bunge, and R. P. Bunge, Cytological studies of organotypic cultures of rat dorsal root ganglia following x-irradiation *in vitro*. I. Changes in neurons and satellite cells, *J. Cell Biol.* **32**:467–496 (1967).

127. H. S. Schutta and L. Johnson, Bilirubin encephalopathy in the Gunn rat: A fine structure study of the cerebellar cortex, *J. Neuropathol. Exptl. Neurol.* **26**:377–396 (1967).

128. H. Koenig and S. Jacobson, The mitochondrial lesion produced by tetrazolium salts *in vivo*, *J. Histochem. Cytochem.* **15**:786 (1967).

129. T. McDonald and H. Koenig, unpublished observations.

130. H. Koenig, in *International Review of Neurobiology* (C. C. Pfeiffer and J. R. Smythies, Jr., eds.), Vol. 10, pp. 199–229, Academic Press, New York (1967).

131. K. D. Barron, P. F. Doolin, and J. B. Oldershaw, Ultrastructural observations on retrograde atrophy of lateral geniculate body. I. Neuronal alterations, *J. Neuropathol. Exptl. Neurol.* **26**:300–326 (1967).

132. F. I. Khattab, Alterations in acid phosphatase bodies (lysosomes) in cat motoneurons after asphyxiation of the spinal cord, *Exptl. Neurol.* **18**:133–140 (1967).

133. N. H. Becker and K. E. Barron, The cytochemistry of anoxic and anoxic-ischemic encephalopathy in rats. I. Alterations in neuronal lysosomes identified by acid phosphatase activity, *Am. J. Pathol.* **38**:161–175 (1961).

134. K. D. Barron, Histochemical study of neuronal necrosis produced by intracerebral injection of diphtheria toxin, *Neurology* **11**:714–723 (1961).

135. K. D. Barron and S. Sklar, Response of lysosomes of bulbospinal neurons to axon section, *Neurology* **11**:866–875 (1961).

136. E. H. Kagan, R. H. Brownson, and D. B. Sutter, Radiation-caused cytochemical changes in neurons, *Arch. Pathol.* **74**:195–203 (1962).

137. P. W. Lampert and M. R. Cressman, Fine-structural changes of myelin sheaths after axonal degeneration in the spinal cord of rats, *Am. J. Pathol.* **49**:1139–1155 (1966).

138. J. Escola und E. Thomas, Elektronenmikroscopische Untersuchungen uber die Lokalisation der sauren Phosphatase im Reaktionsbereich experimentell erzeugter Hirngewebsnekrosen, *Acta Neuropathol.* **4**:380–391 (1965).

139. H. Koenig, Neuronal lipofuscin in disease. Its relation to lysosomes, *Trans. Am. Neurol. Assoc.,* 212–213 (1964).

140. T. Samorajski, J. R. Keefe, and J. M. Ordy, Intracellular localization of lipofuscin age pigments in the nervous system, *J. Gerontol.* **19**:262–276 (1964).

141. J. Austin, D. Armstrong, and L. Shearer, Metachromatic form of diffuse cerebral sclerosis: V. The nature and significance of low sulfatase activity, *Arch. Neurol.* **13**:593–614 (1965).

142. R. O. Brady, J. N. Kanfer, R. M. Bradley, and D. Shapiro, Demonstration of a deficiency of a glucocerebroside-cleaving enzyme in Gaucher's disease, *J. Clin. Invest.* **45**:1112–1115 (1966).

143. R. O. Brady, J. N. Kanfer, M. B. Mock, and D. S. Fredrickson, Metabolism of sphingo-myelin. II. Evidence of an enzymatic deficiency in Niemann-Pick disease, *Proc. Natl. Acad. Sci. U.S.* **55**:366–369 (1966).

144. H. G. Hers, α-Glucosidase deficiency in generalized glycogen-storage disease (Pompe's disease), *Biochem. J.* **86**:11–16 (1963).

145. S. Okada and J. S. O'Brien, Generalized gangliosidosis: beta-galactosidase deficiency, *Science* **160**:1002–1004 (1968).

146. B. J. Wallace, L. Schneck, H. Kaplan, and B. W. Volk, Fine structure of the cerebellum of children with lipidoses, *Arch. Pathol.* **80**:466–486 (1965).

147. B. J. Wallace, B. W. Volk, L. Schneck, and H. Kaplan, Fine structural localization of two hydrolytic enzymes in the cerebellum of children with lipidoses, *J. Neuropathol. Exptl. Neurol.* **25**:76–96 (1966).

148. J. Pick, The fine structure of sympathetic neurons in x-irradiated frogs, *J. Cell Biol.* **26**:335–351 (1965).

149. P. W. Lampert, A comparative electron microscopic study of reactive, degenerating, regenerating, and dystrophic axons, *J. Neuropath. Exptl. Neurol.* **26**:345–368 (1967).

150. H. Koenig, in *Brain Damage in the Fetus and Newborn from Hypoxia or Asphyxia* (L. S. James, R. E. Meyers, and G. E. Gaull, eds.), pp. 36–40, Report of the Fifty-seventh Ross Conference on Pediatric Research, Ross Laboratories, Columbus, Ohio (1967).

151. H. Koenig, Acute axonal dystrophy caused by fluorocitrate: the role of mitochondria, *Science* **164**:310–312 (1969).

152. W. S. Morgan, J. Fernando, and M. A. Alousi, Studies of the biological activity of neutral red, *Exptl. Molec. Pathol.* **5**:491–503 (1966).

153. M. E. Fedorko, J. G. Hirsch, and Z. A. Cohn, Autophagic vacuoles produced *in vitro*. I. Studies in cultured macrophages exposed to chloroquine, *J. Cell Biol.* **38**:377–391 (1968).

154. C. de Duve, From cytases to lysosomes, *Federation Proc.* **23**:1045–1049 (1964).

155. A. Allison, Lysosomes and disease, *Sci. Am.* **217**:62–72 (1967).

156. G. Weissman, Labilization and stabilization of lysosomes, *Federation Proc.* **23**:1038–1044 (1964).

157. C. de Duve, R. Wattiaux, and M. Wibo, Effects of fat-soluble compounds on lysosomes *in vitro*, *Biochem. Pharmacol.* **9**:97–116 (1962).

158. N. Allen, Beta-glucuronidase activities in tumors of the nervous system, *Neurology* **11**:578–596 (1961).

159. N. Allen and E. Reagan, β-Glucuronidase activities in cerebrospinal fluid, *Arch. Neurol.* **11**:144–154 (1964).

160. E. C. Shuttleworth and N. Allen, Early differentiation of chronic meningitis by enzyme assay, *Neurology* **18**:534–542 (1968).

161. O. Z. Sellinger and G. D. Rucker, Cerebral lysosomes. III. Evidence for *in vivo* "labilization" by the convulsant DL-methionine-DL-sulfoximine, *Life Sci.* **5**:163–167 (1966).

162. H. Koenig, M. Cavender, and J. Agoro, unpublished observations.

Chapter 13

BRAIN MITOCHONDRIA

Leo G. Abood

*Center for Brain Research
and Department of Biochemistry
University of Rochester
Rochester, New York*

I. INTRODUCTION

The primary objective of this chapter is to present a discussion of the biochemical properties and function of mitochondria in the nervous system, not only in the light of specific investigations on brain mitochondria alone but against the background of extensive knowledge of mitochondria from other tissues. Although mitochondria from neural tissue have certain qualitative differences in their enzymic and chemical constitution from liver and heart muscle mitochondria, the major differences are quantitative. As more investigations are conducted on pure preparations of brain mitochondria, the similarities between mitochondria from brain and those from other tissues have become more apparent. The term "mitochondria" will be used as if it referred to organelles derived from a homogeneous cell population; however, because of the many different cell types in neural tissue with a variety of functions, it is, perhaps, naive to assume that the neural mitochondria are homogeneous in their biochemical and functional characteristics.

For an excellent, detailed account of the general properties of mitochondria from tissues other than brain the reader is referred elsewhere.[1,2]

II. LOCALIZATION OF MITOCHONDRIA

Mitochondria are the cytoplasmic granules containing the enzymes involved in oxidative metabolism coupled to the production of ATP. Inasmuch as the brain utilizes more than 25% of the total body oxygen and glucose, the concentration of mitochondria in brain tissue would be expected to be unusually great. About 15% of the total protein of whole rat brain is mitochondrial, although this figure may be as high as 25% in cerebral gray matter. Although no systematic study utilizing the electron microscope has been made of the regional distribution of mitochondria in the brain, it has been known for some time that the greatest concentration is in the neuron. Mitochondria are present throughout the perikaryon cytoplasm, dendrites, axons, and nerve endings (see Abood and Gerard[3] for review). Such regions as the nerve endings, the axon hillock, and nodes of Ranvier are particularly abundant. The perikaryon of a large neuron, such as a pyramidal or Purkinje cell, may contain many hundred mitochondria; however, there may be many thousands of mitochondria contained within the innumerable nerve endings surrounding the perikaryon membrane. Mitochondria, thus, would seem to be especially concentrated in those regions of the neuron where excitatory events are most frequent, as at the synapses and the axon hillock.

Although mitochondria are found in oligodendroglia and Schwann cells,[4] their presence in normal astrocytes and microglia is sparse. As will be discussed later, during development, regeneration, and under certain pathological conditions the population of mitochondria in the glial cells and less densely populated regions of the neuron can increase markedly.

Schwann cell cytoplasm contains numerous mitochondria more closely packed than subjacent neuron.[5] The mitochondria of the Schwann cell tend to be round rather than elongated and contain dense granules in cristae that resemble those observed in intestinal and renal epithelial mitochondria. It has been suggested that intramitochondrial granules are prominent in cells engaged in ion and water transport.[6] Mitochondria are found most abundant near the nucleus of the Schwann cell but are also present at some distance away. They also are found throughout the length of the axoplasm.

III. MITOCHONDRIAL ULTRASTRUCTURE

Although mitochondria from different tissues are basically similar in their ultrastructure, mitochondria of neural tissue have certain peculiarities the significance of which is unknown. One of the earliest electron micrographic studies of mitochondria was done with neural tissue.[7] Brain, as well as most other, mitochondria are elongated cytoplasmic organelles, $0.3-1.0 \, \mu$ in width and up to $5 \, \mu$ in length, comprised of a surrounding and inner membranous structure. The inner membrane, which is contiguous to the outer limiting membrane, consists of regular fingerlike infoldings called cristae. From serial sections electron microscopy reveals extensive branching or anastomosing of the infoldings arranged longitudinally and parallel to the long axis.[8] The lumen of this complex inner membrane system is continuous throughout and comprised of a soluble or gelatinous matrix of proteins, lipids, and possibly polysaccharide.

With high resolution electron microscopy, utilizing techniques of negative staining, cryofixation at low temperature, or osmium tetraoxide fixation, one observes that the cristae and the inner membranes of the external envelope are comprised of arrays of elementary particles (EP). The EP, which has a diameter of about 100 Å and may number as much as 100,000/ mitochondrion, is a knoblike polyhedral structure attached to a membranous strand via a stem.[9,10] The membranous strand, which is comprised of a bimolecular array of lipids, not only may provide the structural framework for the mitochondria but may be essential to enzymic organization and the process of electron transport coupled to phosphorylation. There does not appear to be complete agreement on the authenticity of the EP.[11] It has been argued that the EP is an artifact arising from the preparative and fixation procedures, since the frequency in the appearance of EP-like structures was found to increase with exposure time of the isolated mitochondria to the suspending media.[11] The mitochondrial membrane may be comprised of a globular substructure of lipoprotein complexes with no distinct separation of phospholipids and proteins.

IV. COMPOSITION OF MITOCHONDRIA

Rat brain mitochondria contain about 35% lipid (dry weight), 57% protein, and 5% ash (magnesium, calcium, sodium, and potassium salts). Of the lipid about 75% is phospholipid, 20% is cholesterol, and the remainder is diglycerides plus triglycerides (Table I). The major phospholipids are lecithin and serine phosphatides (phosphatidal plus phosphatidoyl).[12] Cardiolipin, which has not been detected in other cytoplasmic particulates, is present in brain mitochondria in about the same concentration as in liver or heart mitochondria. Analysis of brain mitochondria reveals 40% to be saturated and 60% unsaturated fatty acids (Table II). The fatty acid composition of brain and liver mitochondria is quite similar, the outstanding difference being

TABLE I

Lipids of Brain Mitochondria[a]

Lipid	Percent total lipid P
Choline phosphatides	38
Ethanolamine phosphatides	26
Serine phosphatides	9
Cardiolipin	8
Inositol phosphatides	4
Sphingomyelin	7
Unidentified	8[b]
Cholesterol	20[b]
Glycerides	5[b]

[a] Data taken from Eichberg, Whittaker, and Dawson,[12] Abood, Kurahasi, and del Cerro,[18] and Abdel-Latif and Abood.[52]
[b] As percent of total lipid by weight.

TABLE II
Fatty Acid Composition of Rat Brain Mitochondria[a]

	Percent total liver	Percent total brain[b]
C14:0	0.6	0.7
C15:0	0.5	0.0
C16:0	16.5	12.1
C16:1	2.2	1.0
C16:2	1.4	0.0
C18:0	16.3	24.5
C18:1	20.9	21.9
C18:2	19.0	3.1
C18:3	0.4	0.0
C21:0	1.5	0.0
C20:4	13.3	11.4
C24:0	0.5	2.5
C22:6	2.7	14.4
C22:?	3.5	0.0
C20:?	0.0	4.2

[a] Data from J. Bernsohn (unpublished).
[b] Data for liver mitochondria given for comparison.

the much higher concentration of linoleic in liver and C22:6 in brain mitochondria. The fatty acid composition of the diet to a limited degree can influence the composition of brain mitochondria, particularly in the case of the polyunsaturated acids.[13] The high concentration of polyunsaturated fatty acids and low cholesterol help maintain mitochondrial membranes in the liquid-expanded state essential for rapid transport. Brain mitochondria have been reported to contain nucleic acid, both in the form of RNA and DNA[14,15]; however, the purity of the mitochondrial preparation in such studies was not sufficient to be certain of their actual presence within mitochondria. With highly purified preparations, only about 0.1 % (dry weight) nucleic acid is found. It is possible that the RNA is in the soluble form and lost during the fractionation. Precise information should come from ultrastructural radioautographic studies of intact cells.

The major components of the mitochondrial enzyme systems consist of (1) the dehydrogenases of the tricarboxylic acid cycle, (2) the electron transport scheme, (3) energy coupling enzymes involved in the synthesis of ATP, (4) a fatty acid oxidase and biosynthetic system, (5) a protein synthesizing system, (6) transaminases, (7) a few glycolytic enzymes, such as hexokinase, and (8) a variety of miscellaneous enzymes such as monoamine oxidase (Table III).

TABLE III

Enzymic Activity of Rat Brain Mitochondria[a]

	Substrate	Activity, μmoles/mg N
ΔP	Pi → ATP	214 (20′)
ΔO[b]	Pyruvate	40
ΔO	α-Ketoglutarate	24
ΔO	Succinate	20
ΔO	Glutamate	24
ΔO	β-Hydroxybutyrate	2.7
Transaminase	Aspartate-2-oxoglutarate	48
Cytochrome oxidase	Fe^{2+} cytochrome c	220
Hexokinase	Deoxyglucose	128 (20′)
Mg^{2+}, activated ATPase	ATP	53
Lactic dehydrogenase	Lactate	30
Monoamine oxidase	Tyramine	90

[a] Data taken from Abood et al.,[19] Brody and Bain,[20] Tanaka and Abood,[23,38] and Abdel-Latif and Abood.[41]
[b] Activity expressed as micromoles instead of usual microatoms.

V. SUBCELLULAR FRACTIONATION OF BRAIN MITOCHONDRIA

Most of our knowledge of the biochemical and physical properties of mitochondria is based on studies with mitochondria isolated from cellular homogenates. Histochemical studies on mitochondria have been few, largely because of the limited resolution of light microscopy and only recent application of the electron microscope to histochemistry. In dealing with isolation procedures it is essential to try to preserve the morphology and biochemical constitution of the mitochondria. Fortunately, the mitochondrion is comprised of an external membrane, or envelope, which is not readily ruptured by the osmotic and other physical forces involved in isolation. Nevertheless, it is not yet possible to determine the extent to which low molecular weight components, particularly soluble ones, may be lost from or absorbed to mitochondria during the fractionation procedure.

In attempting to deal with the problem of the subcellular fractionation of brain tissue one is faced with many problems not encountered in working with most other tissues. To begin with, the brain is a heterogeneous complex of many different types of cells with a wide variety of sizes and shapes. The

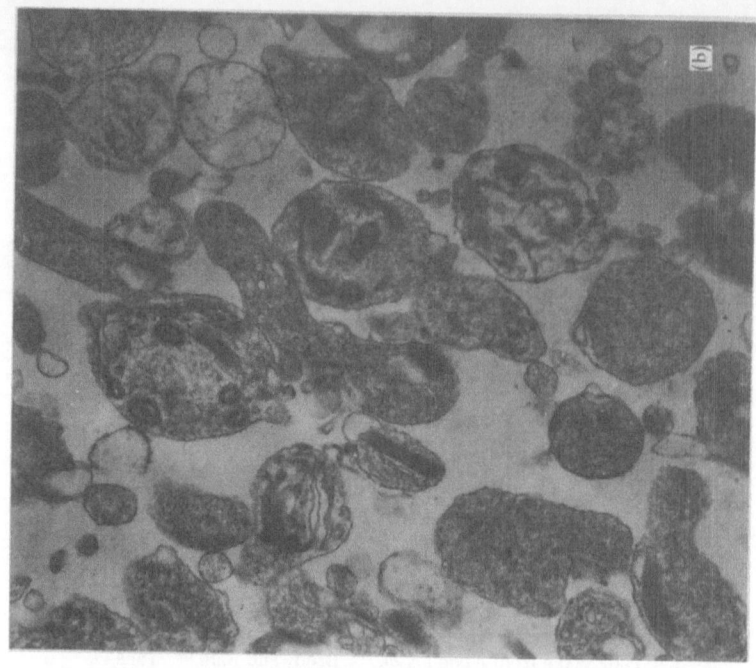

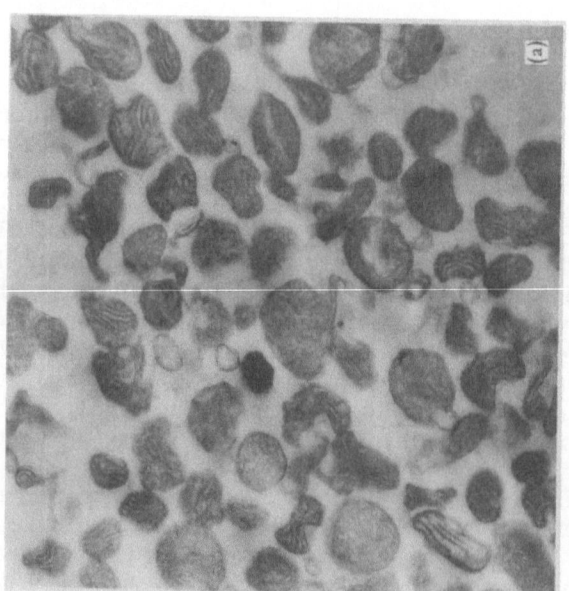

Fig. 1. Electron micrograph of (a) mitochondria and (b) nerve endings isolated from rat brain by Ficoll density gradient centrifugation.[23] Note the presence of mitochondria in nerve endings. × 40,000. Electron micrographs by Manuel P. del Cerro.

principal functional unit of neural tissue, the neuron, is entwined within an intricate network of glial cells and cellular processes, making it virtually impossible in many instances to dissect out an intact neuron from the surrounding neuropil. Although in lower animals there exist nervous systems that are anatomically separable into neuronal and glial elements,[16] the relative size of the material is too small to permit subcellular fractionation by existing techniques. Even in the case of peripheral nerve, the Schwann cell (another type of glia) is too intimately attached to the axon to permit any clear separation of the two.

Most of the published studies employing the technique of differential centrifugation for isolating subcellular constituents have dealt largely with mammalian brain (usually rat). In most instances the whole brain has been used; and although occasionally attempts have been made to work only with cerebral gray matter, the percentage of glial elements as compared to neuronal has not been effectively reduced. Approximately 10% of the total proto- plasmic mass of the whole brain is comprised of roughly equal amounts of neuronal and glial cell bodies, while the remainder is largely dendritic-axonal and glial processes. It is reasonable to assume, therefore, that in dealing with the subcellular components of whole mammalian brain one is concerned largely with the components of cellular processes.

Relatively pure mitochondria have been successfully isolated from laboratory rodents including rat, mouse, rabbit, and guinea pig, and more recently from beef[17] and bullfrog brain,[18] and they all appear to be quite similar in their morphology, lipid composition, and enzymic constitution.

The earliest attempts[19,20] to fractionate brain mitochondria employed the conventional techniques of differential centrifugation in isotonic sucrose (0.25–0.3 M), originally developed for use with liver. Centrifugal speeds of about $10,000 \times g$ were required to separate brain mitochondria in good yields. Although mitochondria could be readily obtained by this procedure, they were contaminated with nerve endings and fragments of myelin, axonal-dendritic processes, and glial processes. Purer mitochondrial prepa- rations could be obtained using hypertonic sucrose (0.8–1.0 M), but exposure to high concentrations of sucrose was deleterious to both mitochondrial morphology and enzymatic activity.

The application to brain tissue of density gradient techniques, employing varying concentrations of sucrose ranging from 2 to 30%, permitted a better resolution of brain mitochondria from nerve endings, myelin, and other cellular processes.[21,22] In order to obtain better preparations for biochemical studies of mitochondria as well as nerve endings, Ficoll, a polysaccharide with only slight osmotic effects, was substituted for sucrose in density gradient centrifugation.[23] Mitochondria prepared in Ficoll show little alteration in their size and ultrastructure, while remaining biochemically unaltered from their original state in isotonic sucrose. Similar resolution of the subcellular fractions could be obtained employing discontinuous instead of continuous Ficoll gradients.[24] Electron micrographs of mitochondria isolated by this procedure are presented in Fig. 1.

VI. SUBMITOCHONDRIAL ENZYME COMPLEXES

When intact rat brain mitochondria are exposed to a variety of surface active agents, they disintegrate into membranous fragments of varying sizes, resolvable by differential centrifugation.[25,26] All the components of the tricarboxylic acid cycle are retained in these particles along with the electron transport scheme and oxidative phosphorylation. Brain mitochondria can be disrupted by 0.05 % lauryl sulfate to such an extent that no cristae or double membrane structure remains. Similar morphological disintegration could be produced by toluene-saturated sucrose solution. Despite the fact that the mitochondria were so badly disrupted and the double membrane structure eliminated, oxidative phosphorylation was not greatly impaired. Mitochondria treated in this drastic manner, however, lose glutamic dehydrogenase and β-hydroxybutyric dehydrogenase activity, indicating that these two enzymes were loosely bound. Mitochondrial ATPase activity, on the other hand, was not affected by treatment with either lauryl sulfate or toluene-saturated sucrose nor was there any activation of mitochondrial ATPase following such treatment. Cytochrome oxidase activity was extremely sensitive to lauryl sulfate but not to toluene treatment. Evidently the lauryl sulfate, but not toluene, removed the phospholipid that is known to be a requirement for cytochrome oxidase activity.

Mitochondria can be disrupted further into various fragments by a combination of procedures involving ultrasonic disintegration, exposure to hypotonic solutions and surfactants, and by freeze-thawing. By differential or density gradient centrifugation the fragments can be resolved into components with relatively distinct ultrastructural and biochemical characteristics. At least three distinct organizational components have been obtained: (1) an insoluble larger membranous fragment, (2) a smaller spherical particulate containing the electron transport and phosphorylating complex, and (3) a membranous component containing a complex of respiratory dehydrogenases.

The electron transport system (ETS) refers to a chain of at least nine oxidation–reduction enzymes responsible for the transfer of electrons from the substrate hydrogens to molecular oxygen (see Fig. 2). On the basis of experiments aimed at reconstruction of ETS from submitochondrial fragments,[27] the system can be visualized as made up of four distinct complexes (Fig. 2). After being removed during the dehydrogenase reaction, the electrons and protons are transferred to NAD (to a second flavoprotein if succinate is the substrate) to form NADH (the reduced form of NAD). A succession of oxidation and reduction steps follow until the electrons are finally transferred by cytochrome oxidase to oxygen. The protons (H^+) move separately in consonance with the electron transport and ultimately react with oxygen to form H_2O. A "difference spectrum" showing the various components of the ETS in a submitochondrial particle of rat brain is presented in Fig. 3.[25]

With the transfer of electrons from NADH and from reduced cytochrome c, ATP is formed from ADP and Pi. In the absence of the phosphorylation of

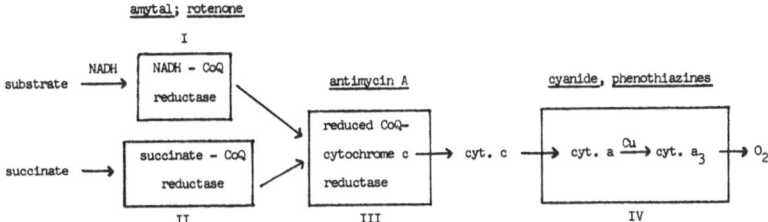

Fig. 2. Components of the electron transport scheme. From reconstitution studies[26] the scheme can be divided into four oxidation–reduction complexes involved in electron transfer from substrate H to molecular oxygen. Specific inhibitors are indicated by underlines.

ADP, electron transfer and oxygen consumption cannot occur. Certain "uncoupling" agents such as DNP can dissociate the link between electron transport and phosphorylation by permitting the flow of electrons and oxygen consumption without the attendant phosphorylation. A scheme (Scheme I) for mitochondrial phosphorylation can be described, where $E \sim I$ represents the nonphosphorylated energy intermediate generated from electron transport and $E \sim P$ is the phosphorylated high-energy intermediate resulting from the reaction of $E \sim I$ with Pi. Subsequent reaction of $E \sim P$ with ADP yields ATP.

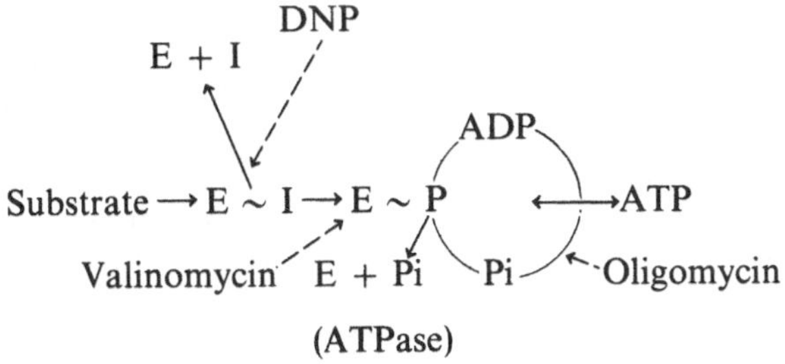

$$E + I$$
$$\text{DNP}$$

$$\text{ADP}$$

$$\text{Substrate} \rightarrow E \sim I \rightarrow E \sim P \longrightarrow \text{ATP}$$
$$\text{Valinomycin} \quad E + Pi \quad Pi \quad \text{Oligomycin}$$

(ATPase)

Scheme I

A variety of specific metabolic inhibitors have been valuable in elucidating this scheme and, particularly, in establishing the existence of the intermediate $E \sim I$. DNP promotes the cleavage of $E \sim I$, thereby preventing the formation of ATP while allowing recycle of the acceptor E and uninterrupted electron transport. Oligomycin interferes with the reaction $E \sim P +$ ADP $\rightleftarrows$ ATP + E by interfering with the hydrolysis of $E \sim P$. DNP, by promoting $E \sim P$ hydrolysis, will restore the inhibition of respiration brought about by oligomycin. Oligomycin can promote phosphorylation in

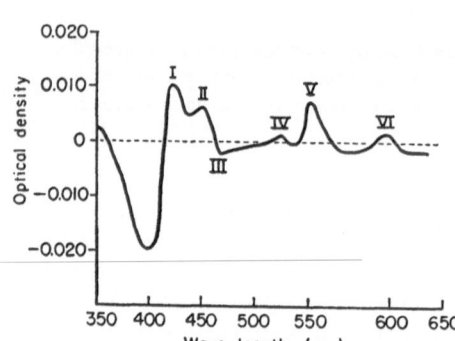

Fig. 3. Difference spectrum of submitochondrial fraction.[25] A cuvette containing 0.02 M potassium pyruvate, 0.001 M potassium malate, 0.005 M ADP, 0.01 M potassium phosphate (pH 0.5), 0.008 M MgCl$_2$, 0.005 M KCN, and submitochondrial fraction (0.5 mg of N) in a total volume of 2.8 ml, representing the reduced enzyme complex, was read against a "blank" cuvette containing everything except the substrates and cyanide. The reduced forms of the carriers are designated by roman numerals: I, γ-peak of cytochrome c; II, γ-peak of cytochrome a and cyanide complex of cytochrome a_3; III, reduced flavoprotein; IV, β-peak of reduced cytochromes b and c; V, α-peak of reduced cytochrome c; and VI, α-peak of reduced cytochrome a and the cyanide complex of cytochrome a_3. (Reprinted by permission of the *Journal of Biological Chemistry*.[25])

a soluble submitochondrial component having high ATPase activity[27] by conserving the phosphorylated intermediate E $\sim$ P.

Amytal and rotenone inhibit the oxidation of NADH (NADH − CoQ reductase). Antimycin A inhibits the oxidation of the reduced NAD-flavoprotein by cytochrome c reductase, but whether the inhibition occurs in the oxidation of reduced CoQ is controversial. The possibility remains that electron transfer from the NAD-flavoprotein (dehydrogenase) may proceed to cytochrome b through a "high potential" flavoprotein without participation of coenzyme Q.[28]

The ETP which is presumably derived from the inner membranes and cristae is comprised primarily of oxidation–reduction proteins. It is also possible to obtain another particle (ETP$_H$) that carries on oxidative phosphorylation in addition to electron transport. The ETP and ETP$_H$, which are cylindrical particles originally believed to be about 450 Å long and 100 Å in diameter,[29] are resolvable into elementary particles and structural lipoproteins.[30] A molecular weight of 10^6 was assigned to EP, which is about one-fourth that originally assigned to ETP. Contained within EP are the four complexes comprising the electron transport chain. When EP is exposed to detergents, followed by ammonium sulfate fractionation, it can be fragmented into the four enzyme complexes of the electron transport chain. By combination of the four complexes the EP is reconstituted with one molecule of each of the four complexes,[26,29] and the sum of the molecular weights of the individual complexes approaches that of EP.[30]

VII. ENERGETIC COUPLING TO ELECTRON TRANSPORT

Since the main function of mitochondria is to generate energy—in the form of ATP—through an elaborate scheme involving electron transport from hydrogen to oxygen, an understanding of the mechanisms linking electron transport and phosphorylation are of fundamental importance. Mitochondria may not only provide energy in the form of ATP, but a number of the intermediate reactions leading to ATP production may be directly linked to ion transport and other exergonic processes in the neuron. Although the precise nature of the mechanisms is not understood and is highly controversial, a number of the more tenable hypotheses will be briefly discussed.

A chain of mitochondrial reactions leads to the formation of an energy-rich component $E \sim I$ (see Scheme I) representing a common energy pool derivable from many energetic reactions (e.g., oxidative phosphorylation, ATPase activity). As indicated earlier the primary function of $E \sim I$ is to phosphorylate ADP through an oligomycin-sensitive system. Another function of $E \sim I$ may be to activate a translocation of protons (H^+) accompanied by a secondary countermovement of cations (M^+), or the primary translocation may be that of cations followed by a countermovement of protons. This may be schematized[28] as follows:

$$\text{Electron transport chain} \longrightarrow E \sim I \longleftarrow \text{ATP}$$

$$\downarrow H^+ \qquad \uparrow M^+$$

$$\uparrow M^+ \qquad \downarrow H^+$$

Another mechanism for ion transport in mitochondria is embodied in the chemiosmotic or proton-gradient scheme.[31] According to this scheme, energy coupling promotes the flow of protons across the membrane which, by creating a potential gradient, promotes the flow of cations in the opposite direction. Again the proton-gradient is dependent upon $E \sim I$. An interesting feature of this hypothesis is that the oxido-reduction components be arranged in series (analogous to a hydrogen-burning fuel cell) so that the current is multiplied and the potential divided. At each step in the electron transport scheme two protons are translocated to yield a total of six for the complete oxidation of one NADH, or two protons are driven outward by each molecule of ATP hydrolyzed. Another feature is that uncoupling agents such as dinitrophenol convert the lipophilic membranous phase (in which the oxido-reductases and ATPase systems are embedded and spatially separate) into a proton conductor, thereby "short-circuiting" the proton current. It has been argued,[28] however, that the pH changes accompanying oxidation in submitochondrial particles are only a small fraction of that predicted by the proton-gradient scheme on the basis of electron transport. Furthermore, it has not been possible to demonstrate with uncoupling agents the necessary pH change predictable from the theory.

Another scheme (Scheme II) for cation transport in mitochondria has been proposed[32] based on the notion that a transmembrane carrier is capable of combining with the alkali metal at the exterior surface while conversion to its conformational state results in cationic release.

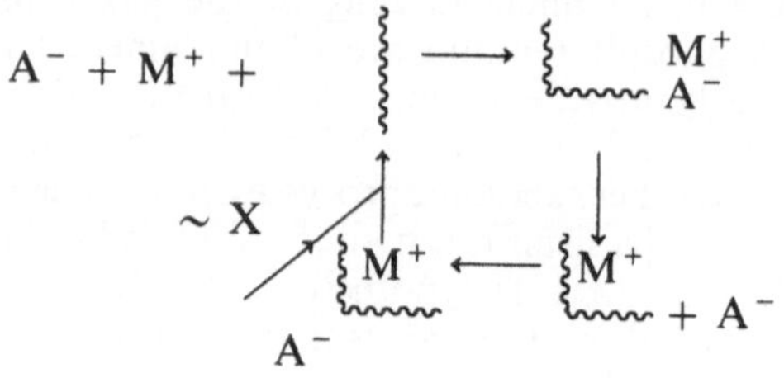

Scheme II

The conversion of the carrier C to its conformer C* is facilitated by a high-energy intermediate $\sim X$ derived from oxidative phosphorylation, whereas the conversion of C* to the C is promoted by agents inducing cation transport such as valinomycin, gramicidin, and parathyroid hormone. Presumably such agents act by forming a metal complex with the cation and transferring it to the transmembrane carrier. Certain inhibitory antibiotics may act by combining with the metal carrier complex and thus prevent $\sim X$ from reacting with the metal carrier complex to release M^+ into the intra-mitochondrial matrix. Since Ca^{2+} is transported into the mitochondria without transport-inducing antibiotics, it may conceivably activate C*. Valinomycin increases K^+ permeability into mitochondria both by increasing mitochondrial permeability and by activation of the energy-requiring transport process.[33]

Respiratory control phenomena may be distributed to a number of sites along the ETS. A theory for energy control and conservation based on secondary and tertiary structural changes recently has been advanced.[28] High-spin transition studies of heme proteins, such as the conversion of ferrimyoglobin to ferrimyoglobin-cyanide, involve conformational changes, not only near the ligand-binding site but also in regions of the protein molecule remote from the heme group. Variations in the Fe-ligand binding create a cooperative interaction along the peptide chain that may then control the reactivity (e.g., with CO) of the heme protein. Conceivably the four cyto-chromes could be linked together into a macromolecular complex where ligand changes associated with oxidation–reduction reactions may involve similar conformational changes.

A preponderance of evidence in favor of any of the proposed mechanisms for energy control coupled to electron transfer is lacking. Nevertheless, each of the hypotheses alludes to chemical mechanisms having either experimental or theoretical validity and may help shed light on this extremely important problem.

A. Other Inhibitors of Energy Coupling to Electron Transfer

A number of psychotropic agents and other heterocyclic aromatic compounds are inhibitors of electron transport and oxidative phosphorylation. Among the more interesting in this group are the phenothiazine derivatives used as psychotherapeutic agents. At relatively low concentrations they inhibit phosphorylation, the DNP-stimulated component of ATPase, and cytochrome oxidase. The mechanism of the action of such heterocyclic aromatic amines on mitochondrial enzyme systems is not clear. One possibility is that they may interact with mitochondrial phospholipids that may be essential to the activity of the enzyme systems involved. Conceivably the lipids may be maintaining the enzymic conformation conducive to optimal activity. The phenothiazines also may directly interact with enzymes to interfere with metal cofactors or again alter molecular configuration. The intimate relationships between various mitochondrial enzymes such as, e.g., the macromolecular complex of the cytochromes, may be readily disturbed by drug–lipid interactions. Complexes also can form between the phenothiazines and ATP,[34] a process that could affect intramitochondrial phosphorylation as well as mitochondrial transfer of ATP.

VIII. THE PHOSPHORYLATIVE ENZYMES

The enzymes leading to the synthesis of ATP coupled to electron transport are believed to involve an enzymic complex of which three components have been described: a DNP-activated ATPase (inhibited by oligomycin and azide), an ATP–Pi exchange reaction (inhibited by DNP, azide, cyanide, and oligomycin), and an ADP–ATP exchange reaction (inhibited by oligomycin, gramicidin, atractylate, and azide).[35] (The exchange reactions are studied by measuring the exchange of ^{32}P between the components.) Evidence for relating the three reactions to oxidative phosphorylation is based largely on the fact that, like phosphorylation, they are sensitive to aging as well as DNP and have an absolute dependence on adenine nucleotides. Certain discrepancies exist, however, in the behavior of oxidative phosphorylation and the three enzymes toward certain inhibitors.[35] Atractylate inhibits the ATP–Pi exchange reaction without appreciably affecting oxidative phosphorylation,[36] whereas cyanide, at concentrations completely inhibitory to electron transport, has no effect on the DNP-activated ATPase. On the other hand, there are arguments for the relationship of these enzymes to oxidative phosphorylation, the most cogent of which is the fact that factors leading to an unmasking of ATPase activity cause a loss of the oligomycin-sensitive component of the ADP–ATP exchange reaction.[37] Conformational changes resulting from physical and chemical treatment can modify the affinity of the ATP-synthetase for ADP and thus convert it into an ATPase. The "coupling" or phosphorylative reactions can be viewed as a group of ATP synthetases linked to the three coupling components of the ETS.

A. ATPase of Brain Mitochondria

When the crude mitochondria-containing fraction of rat brain is fractionated in a Ficoll gradient into myelin, nerve endings, and mitochondria, Mg^{2+}-activated ATPase activity is found in all three fractions. The activity of myelin (also containing axonal–dendritic processes) was enhanced eightfold by Mg^{2+}, while the activity of the nerve ending and mitochondrial fractions was increased fourfold and threefold respectively. Of particular interest is the fact that the ATPase activity of all fractions was increased almost 300% by 0.1 mM DNP, and this DNP-activated component of the ATPase activity is extremely labile to aging at 34°C. Whereas the ATPase activity of the mitochondrial and the nerve ending fractions was inhibited by increasing concentrations of Na^+, the activity of the myelin–axonal fraction markedly increased under the same conditions. On the basis of these and other differences[38] it was possible to infer that there were at least two distinct ATPases in the particulate fractions of rat brain, in addition to the Na^+,K^+-activated one.

A study of the ATPase activity of cruder preparations of rat brain mitochondria had led to the conclusion that brain mitochondria were different from other mitochondria in important respects.[38] Since almost identical findings were obtained with purer mitochondria from rat brain, the differences appear to be real. They are as follows: (1) ATPase of brain mitochondria diminishes with aging, irrespective of the presence of Mg^{2+}, whereas the activity of liver mitochondria increases; and (2) the dinitrophenol-activated component of ATPase activity is extremely unstable, and the pattern of change in activity with aging differs from that of oxidative phosphorylation. If the DNP-activated component of ATPase is related to oxidative phosphorylation, it evidently is distinct from its counterpart in the myelin and other fractions where phosphorylation cannot occur.

By means of electron microscopic techniques applied to histochemical specimens of spinal cord, it was possible to demonstrate the presence of Mg^{2+}-ATPase (presumably the Na^+,K^+-activated) in the mitochondria of the neuropil but not in the perikaryon of the neurons.[39] Since studies on subcellular fractions reveal the Na^+,K^+-specific enzyme to be in nerve endings and not mitochondria, it is likely that the histochemical study was detecting a general ATPase activity.

B. Hexokinase

It has been known for some time that the hexokinase activity of rat brain is associated with a large particulate differing in this respect from other tissues where the enzyme appears in the soluble cytoplasmic fraction.[40] Although it was suspected that the enzyme was within the mitochondria, it was not possible to be certain of its localization because of the heterogeneity of the fractions examined. It now appears that rat brain mitochondria contain at least 60% of the total brain hexokinase activity, while a large portion of the remainder may have been lost from mitochondria during the fractionation procedure. An appreciable amount (20 %) of the hexokinase activity appears, however, in the nerve ending fraction; but whether the enzyme is actually present within the nerve endings is uncertain. Although mitochondria are contained within the nerve endings, it is unlikely their presence alone accounts for the high degree of activity. Although the phosphorylative activity (due entirely to mitochondria) may be tenfold greater in the mitochondrial fraction than in the nerve ending fraction, the difference in hexokinase activity is less than threefold. One possibility is that the nerve ending fraction contains mitochondrial fragments that are minus the full complement of enzymes necessary for oxidative phosphorylation but may still retain hexokinase activity.[41]

IX. FATTY ACID OXIDATION AND SYNTHESIS

In view of the fact that mitochondrial enzymic activity undergoes a change during development, morphological change also may be expected. Fatty acid metabolism undergoes a marked decrease after myelination (15–20 days in the rat brain). As an example, the oxidation of β-hydroxybutyric acid by rat brain mitochondria is three to four times greater at 15 days than at 180 days, being highest at the period of maximal growth and maturation.[42]

Although there are reports that cytochrome oxidase and other respiratory enzymes increase with age, the mitochondrial concentration of respiratory enzymes does not. More recent findings favor the view that brain mitochondria formed during embryonic development have their full complement of respiratory enzymes.[41] If certain enzyme systems, such as the fatty acid oxidases, do decrease or entirely vanish with development, they most likely would be associated with the soluble pool of intramitochondrial enzymes.

Fatty acid oxidation involves the stepwise degradation of long-chain fatty acids to acetyl CoA, which is then fed into the tricarboxylic acid cycle. The initial step involves the formation of an acyl-Co by an acyl thiokinase in the presence of ATP and CoA. After dehydrogenation at the α,β-position, catalyzed by acyl-CoA dehydrogenase (flavoproteins), hydration occurs at the α,β-unsaturated acyl-CoA to form the corresponding β-hydroxyacyl CoA (catalyzed by an enol-CoA hydrase). A second dehydrogenation then occurs (catalyzed by a β-hydroxyacyl CoA dehydrogenase and requiring NAD) and, finally, a thiolysis whereby acetyl CoA is formed.

Fatty acid synthesis occurs by successive additions of acetyl CoA molecules by a reversal of above-mentioned oxidative reactions; however, the terminal dehydrogenation is catalyzed instead by an $NADPH_2$-dependent enoyl-CoA reductase. Evidently fatty acid synthesis within mitochondria is restricted to a few long-chain fatty acids. The majority of fatty acids may be synthesized via the malonyl CoA route, which in certain tissues such as heart and, possibly, embryonic brain may be a mitochondrial enzyme system[43]:

$$CH_3CO \sim CoA + CO_2 + ATP \overset{Mn^{2+}}{\rightleftharpoons}$$

$$ADP + Pi + \underset{\underset{COOH}{|}}{CH_2CO} \sim CoA$$

$$+ 7 \text{ Malonyl CoA}$$

$$+ 14 \text{ NADPH}$$

$$RCH_2CH_2COOH$$

(Long-chain fatty acid)

Since this reaction proceeds much more rapidly with $NADPH_2$ than $NADH_2$ (the former being provided by the pentose shunt pathway of glycolysis), lipogenesis is dependent upon carbohydrate metabolism.

A study of the rat brain thiolase (catalyzing the condensation of acetyl CoA with acyl CoA) with maturation revealed a rapid increase during the first week after birth with essentially little change thereafter.[44] Furthermore, the content of thiolase and β-hydroxyacyldehydrogenase (both mitochondrial enzymes) was surprisingly similar between different parts of the brain. Evidently, therefore, the actual concentration of lipid-synthesizing enzymes does not diminish with brain maturation despite the fact that overall fatty-acid oxidation becomes barely discernible as adjudged by isotopic experiments. A plausible explanation[44] may be that fatty acid synthesis is dependent upon such factors as the availability of acetyl CoA, the electron transport scheme, and numerous coenzymes—factors that are demanded by carbohydrate metabolism. Despite the fact that many of the fatty acid synthesizing enzymes are present in small amounts in brain, the activity of certain CoA deacylases is extremely high, a fact that suggests that fatty acids may be used for carbohydrate metabolism.

Enzymes activating fatty acids, acyl transferases, and the enzymes involved in β-oxidation are localized exclusively in the outer membrane of beef heart mitochondria.[45] Atractyloside is a potent inhibitor of fatty acid oxidation, the inhibition being exerted on the membrane-forming sector of the outer membrane and on the enzymes associated with the detachable sector of outer membrane.

X. PROTEIN SYNTHESIS

The presence of RNA and DNA in brain and other mitochondria has led to a widespread attempt to demonstrate protein synthesis in mitochondria. Since most of these studies have measured the incorporation of radioactive amino acids into mitochondria, one is not fully justified in using the term "protein synthesis" as it is ordinarily used. The first step in amino acid incorporation is the activation of the amino acid by amino acid acyl synthetase prior to its transfer to soluble RNA, each amino acid requiring a specific synthetase and soluble RNA. The arrangement of the activated amino acids onto the ribosomes (attached to the messenger RNA) determines the amino acid sequence leading to a definite peptide. If the equivalent of a ribosomal RNA resides in mitochondria, it does not appear to be susceptible to the same antibiotic inhibitors. Although the amino acid activating enzymes apparently reside in mitochondria from liver, their presence in brain mito-chondria has yet to be demonstrated.

Liver mitochondria, however, do possess the capacity to incorporate amino acids into their structural proteins by a mechanism that in part resembles that of ribosomal protein synthesis. The mitochondrial, like the ribosomal system, requires ATP, while GTP can replace ATP and AMP is inhibitory.[46] Since oligomycin does not inhibit mitochondrial protein synthesis, the high energy intermediate generated by the electron transport system is not obligatory. The basis for distinguishing the mitochondrial from the ribosomal system derives largely from the differential effect of inhibitors. Chloramphenicol only inhibits the mitochondrial system, while cycloheximide only affects the ribosomal system. One explanation for this distinction is that the tRNA of mitochondria differs from that of extra mitochondrial ribo-somes, possibly in its conformation or molecular weight. Chloramphenicol is no longer inhibitory to the RNA once it has been extracted from heart mitochondria.[47] Protein synthesis in brain and other mitochondria is not inhibited by RNAase as is the ribosomal system. It is not yet known whether the RNAase actually penetrates or attacks the mitochondrial RNA.

Although brain mitochondria do appear to carry out amino acid incorporation, it should be emphasized that the level of activity is somewhat less than that present in liver mitochondria. If mitochondrial protein synthesis is associated with mitochondrial replication, the level of activity of neuronal mitochondria may not be as high as that of glial mitochondria where the capacity for regeneration is high. From turnover studies of mitochondrial lipids, the half-life of brain mitochondria was estimated to be as long as 2 years.[48]

In addition to synthesizing mitochondrial components, the mitochondrial nucleic acids may be involved in mitochondrial replication and also serve as regulatory genes in cellular differentiation.[49] The accumulation of large numbers of mitochondria in synaptic endings has led to the speculation that mitochondria may be involved in the production of macromolecular assem-blies corresponding to coding and information storage in molecular switching componentry at the junctional regions, or they might pass through the endoplasmic reticulum membrane and the cisternal cavity into the extra-cellular space.[50] Ultrastructural studies of the retinulae of butterfly photo-receptors reveal a radial arrangement of mitochondria connected by fine tubular channels to the rhabdomeris, which contain the photopigment responsible for the polarization-analyzing property of the insect eye.[51]

XI. GLYCOLYTIC ENZYMES

During the past decade a number of publications have reported that glycolytic (Embden–Meyerhof) enzymes were associated with brain mitochondria. The so-called mitochondrial preparations employed in all these studies were comparatively crude and were subsequently shown to be contaminated with nerve endings and dendritic–axonal as well as glial processes.

With the development of the technique of continuous gradient centrifugation in hypertonic media, it became possible to reexamine more carefully the problem of particulate-bound glycolysis. By employing a Ficoll medium, relatively pure mitochondria could be separated from rat brain that were virtually devoid of glycolytic activity.[23] Insofar as such isolated mitochondria still retained their fine structure, closely resembling mitochondria seen in intact neurons, there was little reason for believing that fractionation procedures could have resulted in a release of glycolytic enzymes normally bound to mitochondria. Furthermore, the oxidative and phosphorylative activity of the mitochondrial preparations was significantly greater than that obtained from cruder mitochondrial preparations.

It would appear, therefore, that particulates other than mitochondria were contributing to the glycolytic activity normally found in the mitochondria-containing fraction obtained from rat brain by conventional fractionation procedures. After subfractionation in a Ficoll gradient, the glycolytic activity appears largely in the fractions containing nerve endings or myelin plus axonal–dendritic fragments. Since treatment of these fractions with water or mild detergents will release most of the glycolytic activity, the enzymes must be contained within the cytoplasmic fluid of the nerve endings. The membranous components remaining after such treatment are devoid of glycolytic activity. On the basis of such studies it is reasonable to conclude that the glycolytic activity found in brain cytoplasmic particulates is associated with the cytoplasmic cell sap retained in the cellular processes and nerve endings, which constitute a major part of the particulates isolated from mammalian brain.

XII. MITOCHONDRIA DURING NEURAL DEVELOPMENT

The turnover of mitochondrial phospholipid and protein increases sharply at the time of neuronal maturation and myelination in the rat brain (age 10–15 days).[52] Although the number of mitochondria also increases rapidly during this period, as evidenced by the increase in mitochondrial enzymes, there does not appear to be any marked change in the enzymic concentration or structural composition of the mitochondria.[41] As mitochondria first appear in the embryonic brain, they seem to have their full complement of enzymes; furthermore, their lipid composition does not vary significantly during the period of neuronal maturation.[52] At that stage in neuronal maturation, where anaerobiosis is replaced by aerobiosis and oxidative enzymes appear, mitochondria become visible in the neuronal cytoplasm. Prior to this stage, oxidative enzymes and mitochondria are sparse.

A problem that remains unresolved is that concerning the extent to which neuronal mitochondria contribute to the overall energy requirements of the neuron. Both the neuron and neuroglia are of ectodermal origin, and only after the differentiation has progressed to the neuroblast stage does neuronal development proceed independently. Although neuronal mitochondria are present in late embryonic development, their number greatly increases just after birth prior to the onset of myelination and neuronal maturation. Lipid metabolism in the oligodendroglia (Schwann cell in the peripheral nervous system) is extremely active during myelination, so that the fatty acid oxidation and lipid synthesis occurring in the neonatal brain may be largely attributable to neuroglial mitochondria. Although lipid metabolism is relatively slow in the mature nervous system, its "reactivation" within the Schwann cell and oligodendroglia can occur after nerve section and neuronal injury.

Transient, focal accumulation of axonal mitochondria occurs during the early stages of Wallerian degeneration.[53] After nerve crush, a striking accumulation of mitochondria was observed in the paranodal region of the axon, accompanying swelling and fragmentation. In the regions of fragmentation of the mitochondria new mitochondria are found in abundance, a finding that suggests that new mitochondria are derived from the old. During the process of demyelination there is a considerable increase in the number of mitochondria in both the astrocytes and oligodendroglia.[54] Normal astrocytes generally contain very few mitochondria, as revealed by both ultrastructural and histochemical studies. It has been reported that following allergic encephalitis or nerve crush, mitochondrial multiplication occurs in peripheral nerve and spinal cord of rabbits.[55] Although the increase in mitochondria has been attributed to the axon growth "cones," they may be due merely to cellular alterations since the accumulation of mitochondria is observed *before* regeneration occurs.[53]

Since lipid and protein metabolism greatly diminish in the brain after myelination and neuronal maturation, it is to be expected that the enzymic profile of brain mitochondria would reflect this change. The rate of amino acid incorporation by brain microsomal systems is greater in the presence of mitochondria from immature rat brain as compared to mitochondria of mature brain or even liver mitochondria.[56] Furthermore, while thyroxine inhibits protein synthesis in the presence of mitochondria from mature brain, it actually stimulates with mitochondria from immature brain and liver.[57] Evidently an unidentified mitochondrial factor, present only in immature brain, interacts with thyroxine to stimulate the transfer of soluble RNA-amino acid into ribosomes.

XIII. PHYSICAL BEHAVIOR OF BRAIN MITOCHONDRIA

Mitochondria from brain as well as from other tissues exhibit swelling–contractile phenomena in response to changes in its external environment.[58] Such agents as KCl, phosphate, and glutathione cause swelling in freshly prepared mitochondria, while ATP induces contraction in swollen mitochondria. In view of the fact that the shrinkage effect of ATP is demonstrable on nerve endings, myelin, and "microsomes" as well as mitochondria, it is unlikely that the effect has any direct relationship with phosphorylation or electron transport. Some investigators have sought to relate the volumetric changes to mitochondrial ATPases which, like muscle actomyosin, may be functioning as a "mechano-enzyme."[59] Since volumetric changes occur in the absence of exogenous ATP, it is not likely that an ATPase is directly responsible. Another explanation may be that ATP and Mg^{2+} are forming complexes with mitochondrial phospholipids and proteins, and such complexes could lead to the extrusion of water and solvated ions in much the same way that greater cross-linkage in ion exchange polymers decreases hydrophilicity and increases ion selectivity. The formation of ATP-Ca^{2+} (or Mg^{2+})-phospholipid complexes recently has been demonstrated.

Isolated rat brain mitochondria, when impaled with the Ling–Gerard ultramicroelectrodes (0.1–$0.2\ \mu$ diameter), exhibit reproducible trans-surface potentials of about -30 mV.[60] Since the electrical measurements required a critically small electrode tip diameter ($< 0.5\ \mu$) and since the potential appeared abruptly on impaling and was sustained, it is likely that such potentials were of individual mitochondria. Cyanide virtually abolished the potential, while replacement of NaCl by KCl significantly decreased the potential.

Attempts have been made to determine the metabolic response of isolated brain mitochondria to electrical stimulation using various types of electrical stimuli.[61] Under certain conditions it is possible to demonstrate an increase in oxidative metabolism associated with an uncoupling of phosphorylation. Apparently, most of these effects can be traced to artifacts largely due to the heavy metal ions of the electrodes used; nevertheless, a portion of the metabolic effects may be attributed directly to the electrical current.

XIV. OTHER FUNCTIONS OF MITOCHONDRIA

Although the main function of neural mitochondria is to synthesize ATP needed for bioelectric phenomena, there are indications that they may play a more direct role in neuronal activity. To begin with, ATP per se may not be the only immediately available energy source, and such intramitochondrial components as the high-energy phosphorylated intermediate E $\sim$ P, as well as the nonphosphorylated E $\sim$ I, may be directly linked to exergonic functional systems. Both intermediates are known to participate in mitochondrial ion transport, and a concentration of mitochondria in such regions as the nerve endings, axon hillock, and nodes of Ranvier could be involved in the maintenance of electrochemical and osmotic gradients essential to excitability. Furthermore, by virtue of their ability to maintain a proton gradient, mitochondria may generate local potential gradients that in turn may create ionic movements and pH differences. Conformational and other changes then may result in various macromolecules within the neuronal membranes to bring about the release of chemical transmitters, inorganic ions, and ATP or to alter enzymic activity.

Mitochondria also are involved in a variety of other metabolic activities related to neural function. The production of transmitter and neurohumoral agents is to a considerable degree a mitochondrial activity. Included in this group are a variety of aliphatic, aromatic, and other types of amines, such as acetylcholine, noradrenaline, histamine, and serotonin. The production and metabolism of a number of amino acids having a regulatory role in excitatory phenomena occur within mitochondria. Among these are glutamic acid, glutamine, and γ-aminobutyric acid, amino acids whose concentrations are controlled by mitochondrial transamination and oxidation.

Since the nervous system is sensitive to virtually any chemical substance, it is conceivable that the complex chemical environment of the cell—a state determined largely by mitochondria—profoundly influences neural activity. A number of the acids of the tricarboxylic acid cycle are capable of complexing with Ca^{2+} and could influence the role of this essential cation in bioelectric phenomena. By regulating the NAD-NADH ratio and the utilization of pyruvic acid, mitochondria indirectly can control glycolytic metabolism. In addition to the concentrations of "functional" intermediates the levels of various inorganic ions, pH, electrochemical gradients, and oxidation–reduction potentials would be influenced by mitochondrial metabolism—either directly or indirectly. The rate of blood flow as well as blood–brain permeability also would be influenced by the metabolic rate, both these processes regulating the entry and removal of substances involved in neural activity.

XV. ORIGIN OF MITOCHONDRIA

Mitochondria, like other cytoplasmic constituents, can originate either from preformed cellular components (membranes or a submicroscopic precursor) or by fission of preexisting mitochondria. Among the cellular components postulated as mitochondrial precursors are the endoplasmic reticulum, the nuclear membrane, the nucleolar membrane, and the cellular membrane (see Lehninger[1] for review). Convincing evidence that mitochondria can arise from any of these membranous components is still lacking. Experiments with neurospora, which incorporate radioactively labeled choline into mitochondrial phosphatidyl choline, provide persuasive evidence that mitochondria arise from division and development of preexisting mitochondria. Electron micrographic studies of the developing brain often reveal the presence of mitochondria with clublike protuberances or constrictive regions, structures that suggest that the mitochondrion may be undergoing cleavage. Another view[62] is that they form as a result of a cytoplasmic process leading to an evagination of the endoplasmic reticulum to form loops that rapidly attenuate and then break off. The dissevered loops thus would consist of a double layer, the inner comprising the cristae. Evidence for this view, however, is lacking.

XVI. REFERENCES

1. A. Lehninger, *The Mitochondrion*, W. A. Benjamin, New York (1965).
2. E. C. Slater, Z. Kaniuga, and L. Wojtczak (eds.), *Biochemistry of Mitochondria*, Academic Press, New York (1967).
3. L. G. Abood and R. W. Gerard, Enzyme distribution in isolated particulates of rat peripheral nerve, *J. Cell. Comp. Physiol.* **43**:379–392 (1954).
4. E. V. Cowdry, The development of the cytoplasmic constituents of the nerve cells of the chick, *Am. J. Anat.* **15**:389–430 (1914).
5. P. L. Williams and D. N. Landon, Parinodal apparatus of peripheral myelinated fibers of mammals, *Nature* **198**:670–673 (1963).
6. J. Rosenbluth, The fine structure of acoustic ganglia in the rat, *J. Cell Biol.* **12**:329–359 (1962).
7. F. S. Sjöstrand, Ultrastructure of cells as revealed by the electron microscope, *Intern. Rev. Cytol.* **5**:455–533 (1956).
8. E. Anderssen and E. A. Cedergren, Ultrastructure of motor-end plate and sarcoplasmic components of mouse skeletal muscle fiber, *J. Ultrastruct. Res. Suppl.* **1** (1959).
9. H. Fernandez-Moran, *in Ultrastructure and Metabolism of the Nervous System* (S. R. Korey, ed.), Vol. 40, ARNMD Series, pp. 235–267, Williams & Wilkins, Baltimore (1962).
10. D. S. Smith, The structure of flight muscle sarcosomes in the blowfly Calliphora erythrocephala (Diptera), *J. Cell Biol.* **19**:115–138 (1963).
11. F. S. Sjöstrand, E. A. Cedergren, and U. Karlsson, Myelin-like figures formed from mitochondrial material, *Nature* **202**:1075–1078 (1964).
12. J. Eichberg, V. P. Whittaker, and R. M. C. Dawson, The distribution of lipids in subcellular particles of guinea pig brain, *Biochem. J.* **92**:91–100 (1964).
13. L. A. Witting, C. C. Harvey, B. Century, and M. K. Horwitt, Dietary alterations of fatty acids of erythrocytes and mitochondria of brain and liver, *J. Lipid Res.* **2**:412–418 (1961).
14. H. G. DuBuy, C. F. T. Mattern, and F. L. Riley, Comparison of the DNA's obtained from brain nuclei and mitochondria of mice and from the nuclei and kinetoplasts of Leishmania enrietti, *Biochim. Biophys. Acta* **123**:298–305 (1966).
15. J. H. Sinclair, B. J. Stevens, N. Gross, and M. Rabinowitz, The constant size of circular mitochondrial DNA in several organisms from different organs, *Biochim. Biophys. Acta* **145**:528–531 (1967).
16. S. W. Kuffler and J. G. Nicholls, Physiology of neuroglia cells, *Ergeb. Physiol. Biol. Chem. Exptl. Pharmakol.* **57**:1–72 (1966).
17. P. Parsons and R. E. Basford, Brain mitochondria. VI. The composition of bovine brain mitochondria, *J. Neurochem.* **14**:823–840 (1967).
18. L. G. Abood, K. Kurahasi, and M. P. del Cerro, Biochemical studies on isolated nerve endings and other particulates of bullfrog brain, *Biochim. Biophys. Acta* **136**:521–532 (1967).

19. L. G. Abood, R. W. Gerard, J. Banks, and R. D. Tschirgi, Substrate and enzyme distribution in cells and cell fractions of the nervous system, *Am. J. Physiol.* **168**:728–738 (1952).

20. T. M. Brody and J. A. Bain, A mitochondrial preparation from mammalian brain, *J. Biol. Chem.* **195**:685–696 (1952).

21. E. de Robertis, A. P. de Iraldi, G. R. de Lores Arnaiz, and C. Gomez, On the isolation of nerve endings and synaptic vesicles, *J. Biophys. Biochem. Cytol.* **9**:229–235 (1961).

22. F. G. Gray and V. P. Whittaker, The isolation of synaptic vesicles from the central nervous system, *J. Physiol.* **153**:35–37 (1960).

23. R. Tanaka and L. G. Abood, Isolation from rat brain of mitochondria devoid of glycolytic activity, *J. Neurochem.* **10**:571–576 (1963).

24. A. A. Abdel-Latif, A simple method for isolation of nerve ending particles from rat brain, *Biochim. Biophys. Acta* **121**:403–406 (1966).

25. L. G. Abood and L. Alexander, Oxidative phosphorylation by an enzyme complex from disrupted brain mitochondria, *J. Biol. Chem.* **227**:717–725 (1957).

26. Y. Hatefi, A. G. Haavik, L. R. Fowler, and D. E. Griffiths, Studies on the electron transfer system. XLII. Reconstitution of the electron transfer system, *J. Biol. Chem.* **237**:2661–2669 (1961).

27. E. Racker, Resolution and reconstitution of the inner mitochondrial membrane, *Federation Proc.* **26**:1335–1340 (1967).

28. B. Chance, C. Lee, and L. Mela, Control and conservation of energy in the cytochrome chain, *Federation Proc.* **26**:1341–1354 (1967).

29. L. Fowler and S. J. Richardson, Studies on the electron transfer system, *J. Biol. Chem.* **238**:456–463 (1963).

30. H. Fernandez-Moran, T. Oda, P. V. Blair, and D. E. Green, A macromolecular repeating unit of mitochondrial structure and function, *J. Cell Biol.* **22**:63–100 (1964).

31. P. Mitchell, Proton-translocation phosphorylation in mitochondria, chloroplasts and bacteria, *Federation Proc.* **26**:1370–1379 (1967).

32. H. A. Lardy, S. N. Graven, and S. Estrada-O, Specific induction and inhibition of cation and anion transport in mitochondria, *Federation Proc.* **26**:1355–1360 (1967).

33. R. S. Cockrell, E. J. Harris, and B. C. Pressman, Energetics of potassium transport in mitochondria induced by valinomycin, *Biochemistry* **5**:2326–2331 (1966).

34. I. Blei, Complex formation between chlorpromazine and ATP, *Arch. Biochem. Biophys.* **109**:321–324 (1965).

35. M. Tuena, A. Gomez-Puyon, A. Pena-Draz, and G. H. Massieu, Studies on ATPase and ATP exchange reactions in brain mitochondria and effect of DNP, *J. Neurochem.* **11**:527–536 (1964).

36. P. V. Vignais, P. M. Vignais, and E. Stanislas, Action of potassium atractylate on oxidative phosphorylation in mitochondria and submitochondrial particles, *Biochim. Biophys. Acta* **60**:284–300 (1962).

37. F. L. Bygrave and A. L. Lehninger, Properties of an oligomycin-sensitive ADP–ATP exchange reaction in intact beef heart mitochondria, *J. Biol. Chem.* **241**:3894–3903 (1966).

38. R. Tanaka and L. G. Abood, Studies on adenosine triphosphatase of relatively pure mitochondria and other cytoplasmic constituents of rat brain, *Arch. Biochem. Biophys.* **105**:554–562 (1964).

39. L. Rechardt and A. Kokko, Electron microscopic observations on the mitochondrial ATPase in the rat spinal cord, *Histochemie* **10**:278–286 (1967).

40. R. K. Crane and A. Sols, The association of hexokinase with particulate fractions of brain and other tissue homogenates, *J. Biol. Chem.* **203**:272–292 (1953).

41. A. A. Abdel-Latif and L. G. Abood, Biochemical studies on mitochondria and other cytoplasmic fractions of rat brain, *J. Neurochem.* **11**:9–15 (1964).

42. C. B. Klee and L. Sokoloff, Changes in D(−) β-hydroxybutyrate dehydrogenase during brain maturation in the rat, *J. Biol. Chem.* **242**:3880–3883 (1967).
43. W. C. Hülsmann, Fatty acid synthesis in heart sarcosomes, *Biochim. Biophys. Acta* **58**:417–429 (1962).
44. F. Lynen, in *Metabolism of the Nervous System* (D. Richter, ed.), pp. 381–395, Pergamon Press, New York (1957).
45. D. W. Allmann, L. Galzigna, R. E. McCaman, and D. E. Green, The membrane systems of the mitochondrion. IV. Localization of the fatty acid-oxidizing system, *Arch. Biochem. Biophys.* **117**:413–422 (1966).
46. L. W. Wheedlon and A. L. Lehninger, Energy-linked synthesis and decay of membrane proteins in isolated rat liver mitochondria, *Biochemistry* **5**:3533–3545 (1966).
47. M. Rabinowitz, J. Sinclair, L. DeSalle, R. Haselkorn, and H. H. Swift, Isolation of DNA from mitochondria of chick embryo heart and liver, *Proc. Natl. Acad. Sci.* **53**:1126–1133 (1965).
48. A. W. Davison and J. Dobbing, Metabolic activity of body constituents, *Nature* **191**:844–848 (1961).
49. J. André and V. Marinozzi, Presence, dans les mitochondries, de particules ressemblant aux ribosomes, *J. Microscopie* **4**:615–626 (1965).
50. H. Swift, N. Kislev, and L. Bogard, Evidence for RNA and DNA in mitochondria and chloroplasts, *J. Cell Biol.* **23**:91A (1964).
51. H. Fernandez-Moran, in *The Neurosciences* (G. C. Quarton, T. Melnechuk, and F. O. Schmitt, eds.), pp. 281–304, Rockefeller University Press, New York (1967).
52. A. A. Abdel-Latif and L. G. Abood, Incorporation of ortho (P³²) phosphate into the subcellular fractions of the developing rat brain, *J. Neurochem.* **12**:157–166 (1965).
53. H. de F. Webster, Transient, focal accumulation of axonal mitochondria during early stages of Wallerian degeneration, *J. Cell Biol.* **12**:361–383 (1962).
54. M. Z. M. Ibrahim and C. W. M. Adams, The relationship between enzyme activity and neuroglia in plaques of multiple sclerosis, *J. Neurol. Neurosurg. Psychiat.* **26**:101–110 (1963).
55. S. A. Luse and D. B. McDougal, Electron microscopic observations on allergic encephalitis in the rabbit, *J. Exptl. Med.* **112**:735–742 (1960).
56. C. B. Klee and L. Sokoloff, Mitochondrial differences in mature and immature rat brain, *J. Neurochem.* **11**:709–716 (1964).
57. L. Sokoloff and C. B. Klee, The effects of thyroid on protein synthesis in brain and other organs, *Assoc. Res. Nerv. Ment. Dis.* **43**:371–386 (1965).
58. F. Chordikian, L. G. Abood, and Noel Howard, Light absorbance changes in pure mitochondria and other particulates of rat brain, *J. Neurochem.* **13**:945–954 (1966).
59. A. L. Lehninger, Water uptake and extrusion by mitochondria in relation to oxidative phosphorylation, *Physiol. Rev.* **42**:467–517 (1962).
60. Gertrude Falk and L. G. Abood, unpublished observations.
61. L. G. Abood and L. Romanchek, Inhibition of oxidative phosphorylation in brain mitochondria by electrical currents and the effect of chelating agents and other substances, *Biochem. J.* **60**:233–238 (1955).
62. J. D. Robertson, in *Regional Neurochemistry* (S. S. Kety and J. Elkes, eds.), pp. 497–534, Pergamon Press, New York (1961).

Chapter 14

THE SYNAPTOSOME

V. P. Whittaker

Department of Biochemistry
University of Cambridge
Cambridge, England
and
New York State Institute for Basic Research in Mental Retardation
New York, New York

I. INTRODUCTION

When brain tissue is homogenized in iso-osmotic aqueous sucrose under conditions of moderate shear force, the clublike central presynaptic nerve terminals are torn away from their axons and seal up to form detached particles[1-3] to which the name *synaptosomes* has been given.[4] These particles retain the morphological features and, as far as we know, the chemical composition (including the transmitter content) of the intact presynaptic terminal. They can be separated from other subcellular particles by conventional fractionation techniques and provide a new type of *in vitro* preparation for the study of many aspects of neuronal function. Synaptosome preparations have been used for investigations of the synthesis, storage, and release of transmitter substances and the effect of drugs and toxins on such processes; as starting material for the attempted identification of new transmitters[5]; for sampling terminal axoplasm in studies of axonal flow[6]; and for studying the permeability properties of the unmyelinated neuronal plasma membrane[7] and the energy metabolism of neuronal cytoplasm.[8]

Synaptosomes can apparently be formed from any type of central nerve terminal. Thus even the mossy fiber endings of the cerebellar cortex are pinched off during homogenization, although on account of their large size the resultant synaptosomes behave anomalously on subcellular fractionation.[9] Synaptosome formation has been observed to occur during the homogenization of the following discrete regions: cerebral neocortex, hippocampus, caudate nucleus,[10] spinal cord,[11] hypothalamus,[12] and hypophysis[13] and in the following species: mouse,[14] rat, guinea pig, rabbit, coypu, cat, dog, sheep,[15] monkey,[16] man,[17] pigeon, and *Octopus*.[18] By contrast, the yield of synaptosomes is very low from

327

peripheral tissues, even richly innervated ones (e.g., vas deferens[19] and electric organ[20]), presumably because the mechanical conditions for the formation of synaptosomes are not met.

Disrupted synaptosomes may be further fractionated and samples of terminal soluble cytoplasm, synaptic vesicles, external synaptosome membranes (ESMs), and intraterminal mitochondria separated for morphological and chemical analysis. In this way information may be obtained concerning the molecular organization of the presynaptic terminal.

II. METHODS OF PREPARATION OF SYNAPTOSOMES AND THEIR CONSTITUENT ORGANELLES

A. Homogenization of Tissue

1. Conditions of Shear

Presynaptic nerve terminals must be regarded as the most stable regions of nerve cells. Any kind of chopping, mincing, or dispersion of the tissue in iso-osmotic media probably permits the survival of a considerable proportion of them as organized structures.

However, there has been no detailed, systematic study of the optimum conditions for the formation of synaptosomes. Since bound acetylcholine is a synaptosome marker, the proportion of total brain acetylcholine remaining in the bound state and recovered in the synaptosome fraction may be used as an index of the yield of synaptosomes under varying conditions of shear force. In a brief study[21] the homogenization conditions originally selected were found to be superior to those using higher rates of shear. These were the use of a smooth-walled Perspex and glass homogenizer pestle and mortar, 0.32 M sucrose as a suspension medium, a clearance of 0.25 mm, and a speed of rotation (at 30-mm pestle diameter) of 840 rpm. Hand homogenization gave the same yield of bound acetylcholine, but a higher proportion sedimented in the nuclear plus debris fraction, suggesting less complete disintegration of tissue. The yield of synaptosomes determined by a morphological method (see Section III) was quite close, using our standard conditions, to the estimated number of nerve terminals in intact tissue,[22] but no comparisons have been made by this method of synaptosome yields under a variety of conditions.

2. Suspension Medium

Synaptosomes are osmotically sensitive structures; thus in work with lower organisms (elasmobranchs[20] and *Octopus*[18]) note must be taken of the osmolarity of the cell environment. In a recent study[18] it was noted that the yield of synaptosomes from *Octopus* brain was very low when the tissue was homogenized in 0.32 M sucrose but very high in 0.8–1.0 M sucrose, 1.1 M glucose, or 0.7 M sucrose–0.33 M urea.

Like other subcellular particles, synaptosomes are caused to coacervate by low concentrations of divalent ions and somewhat higher concentrations of univalent ions.[3] The threshold concentration for coacervation of synaptosomes is 0.5 mg-atoms/liter for calcium and 20 mg-atoms/liter for sodium. Since both sucrose and water may contain calcium it is necessary to use pure reagents and to avoid any appreciable concentration of buffers. Sucrose solutions consisting of analytical grade sucrose and doubly distilled water contain about 12 μg-atoms calcium/liter, which is well below the threshold concentration; the concentration can be reduced to less than 2 μg-atoms/liter by passage through an ion-exchange resin, but no improvement results from this or from the addition of EDTA. Some authors advocate the addition of anticoagulants to the sucrose.[13]

3. Tissue Concentration

Homogenates containing more than 10% wt./vol. of brain tissue do not fractionate satisfactorily, presumably because of the relatively high ionic strength of such homogenates. However, for purposes of analysis, homogenates have sometimes been prepared at a tissue concentration of 20% wt./vol., sampled, and then diluted to 10% wt./vol. with more sucrose before fractionation with satisfactory results.[23]

B. Separation of Synaptosomes from Other Subcellular Structures

1. Forebrain Synaptosomes

Synaptosomes have broadly similar sedimentation properties to mitochondria and to the smaller fragments of myelinated axons, dendrites, and glial processes formed during homogenization. Thus the so-called mitochondrial fraction obtained from sucrose homogenates of brain tissue by differential centrifuging is a mixture of all these components; this fact largely explains the many anomalous properties that were formerly attributed to brain mitochondria.

Synaptosomes may be separated from myelin and free mitochondria by making use of differences in equilibrium density in a sucrose density gradient. Myelin floats on 0.8 M sucrose; mitochondria are denser than 1.2 M sucrose, whereas synaptosomes have an intermediate density. Thus a fraction enriched in synaptosomes and essentially free from myelin and somatic mitochondria may be obtained by centrifuging a crude mitochondrial fraction into a two-step discontinuous gradient consisting of layers of 0.8 and 1.2 M sucrose that have been allowed to diffuse into each other for 1 hr before use. The standard fractionation procedure used in the author's laboratory is shown in Fig. 1.

Minor variations have been made in terminology and procedure according to the requirements of the work; for full details, the original papers should be consulted.[3,23–28] Thus, the crude nuclear pellet is washed twice when it is desired to reduce mitochondrial and synaptosomal contamination of this fraction to a minimum; at other times, preparation of fraction P_2 may be speeded at the expense of a small reduction of yield by washing the crude nuclear pellet only once. Washing more than twice does not reduce the contamination further, perhaps because the residual contamination is due to nerve terminals and mitochondria attached to, or entangled within, lumps of tissue debris.

Fraction P_1 may be further separated by density gradient centrifuging into three subfractions corresponding to those designated in Fig. 1 as A, B, and C. These are referred to as P_1A, P_1B, and P_1C and consist mainly of large myelin fragments, nuclei, and tissue debris, respectively.[26]

The relatively high gravitational field in which fraction P_2 is prepared ensures the sedimentation of small mitochondria and synaptosomes that would otherwise contaminate fraction P_3. This results, however, in considerable microsomal contamination of fraction P_2. When fraction P_2 is used as a source of synaptosomes for further fractionation, microsomal contamination may be reduced by preparing it at a lower speed (e.g., $10,000\,g$ for 20 min) and washing it with 0.32 M sucrose.

When fractions A and B are prepared according to Fig. 1, the supernatants from the final centrifuging are usually discarded. If a complete balance sheet is required, these supernatants designated A_s, B_s are retained for analysis or, alternatively, the final sedimentation of particulate material may be omitted. Sedimentation without dilution lowers the yield of particulate material. Dilution of the synaptosome fraction below about 0.45 M results in morphological damage. The sedimentation procedure was originally devised because of the difficulty of assaying acetylcholine in concentrated sucrose solutions.

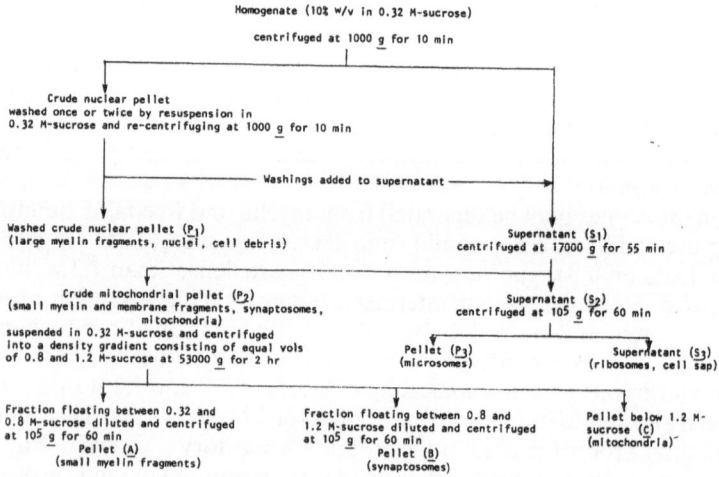

Fig. 1. Scheme for separating synaptosomes.[3] The main morphological components identified in each fraction are given in parentheses; g values are averages.

The presence of large amounts of myelin reduces the efficiency of the separation process, and synaptosomes and mitochondria tend to be held up in the myelin layer. This is particularly evident with spinal cord. It is advisable to remove as much white matter as possible before homogenization: for this reason the smooth cortices of rodents (rat, guinea pig, and coypu) make excellent starting material; the underlying layer of white matter is readily scraped off with a blunt scalpel before homogenizing.

Some workers[29,29a] have omitted differential centrifuging entirely and have applied brain homogenates directly to density gradients. However, the synaptosome preparations obtained in this way have not been critically evaluated. Since nuclei have a somewhat similar equilibrium density and tend to break up in hyperosmotic sucrose with the release of sticky nucleoplasm, it is doubtful whether these synaptosome preparations can be other than heavily contaminated.

The efficiency of the density gradient separation procedure as applied to the crude mitochondrial fraction is thought to depend in part on the differential osmotic dehydration of synaptosomes and somatic mitochondria. As the particles are driven into the density gradient, they are exposed to layers of sucrose of increasing osmotic pressure that cause a progressive loss of water from them. It seems likely that mitochondria are more rapidly or more completely dehydrated than synaptosomes, so that any initial differences in density (due to differences in water content and lipid and protein composition) are thereby enhanced. If synaptosomes exposed to hyperosmotic sucrose are fixed under conditions that do not permit rehydration, they appear to have a bizarre, shrunken appearance, with crenellated outlines and tightly packed cytoplasmic contents. True equilibrium may not be obtained under the conditions of our "standard" procedure, but a 2–5 hr separation period at 53,500 g appears to be optimum. Precise studies of osmotic compartments and water content on the lines of work with liver lysosomes and mitochondria[30] have not been done.

A disadvantage of sucrose density gradients is the damage that may be done to subcellular particles by hyperosmotic sucrose. Thus brain mitochondria isolated in this way are said to lack respiratory control. An alternative is to utilize gradients of carbohydrate or other polymers (e.g., Ficoll) made iso-osmotic with sucrose; these have not given such homogeneous fractions in the author's laboratory as sucrose density gradients, although separations as good as those with sucrose have been reported by others.[31] One difficulty with polymer gradients is that their viscosity increases rapidly with density, so that the approach to density equilibrium is probably much slower than with sucrose. Another disadvantage is that the polymers are not completely standard products and tend to vary considerably in molecular weight, viscosity, and calcium content from one batch to another.

Synaptosomes may be separated into a number of subpopulations by making use of discontinuous gradients with more steps[14,32,33] or continuous gradients.[25,34–36] One group of workers,[33] using discontinuous gradients

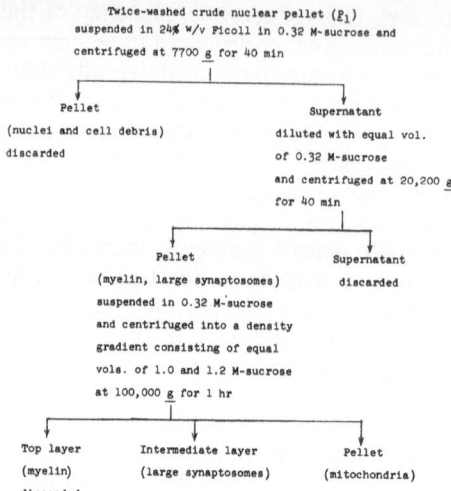

Fig. 2. Scheme for separating synaptosomes derived from cerebellar mossy-fiber endings.[9] The twice-washed crude nuclear pellet (P_1) was prepared from cerebellar cortex as described in Fig. 1.

with sucrose layers of 0.9, 1.0, and 1.1 M sucrose, has suggested that synaptosomes derived from cholinergic neurons may be separated from those derived from noncholinergic endings in this way. Work in the author's laboratory[25,35] has not confirmed this finding. While it is true that the lighter synaptosome fractions tend to be somewhat richer in choline acetylase than in other synaptosome markers, they are also more heavily contaminated with synaptosome "ghosts" and unidentified membrane fragments,[37] structures that are known from other studies to bind choline acetylase nonspecifically in media of low ionic strength[38] (section IV,B,4). It may well be, therefore, that the separation of "cholinergic" from "noncholinergic" endings is more apparent than real.

2. Cerebellar Synaptosomes

The mossy-fiber endings in the granular layer of the cerebellar cortex are 2–5 μ in diameter, considerably larger than endings elsewhere in this or other regions of the central nervous system. They sediment in the nuclear fraction and can be separated[9] by the procedure shown in Fig. 2. The advantage of the preliminary separation in Ficoll is that it eliminates nuclei that would interfere with the separation of the synaptosomes for reasons already mentioned.

C. Subfractionation of Synaptosomes into Their Constituent Organelles

1. Disruption of Synaptosomes

Synaptosomes, once formed, are relatively labile structures and are readily disrupted by a variety of treatments, e.g., warming, vigorous mechanical agitation, supersonic vibrations, detergents, freezing and thawing, incubation with cobra venom, and hypo-osmotic treatment. Most of these treatments result in the prompt liberation of soluble cytoplasmic components, but some of them (shaking, supersonic vibration, incubation with cobra venom, hypo-osmotic treatment, and freezing and thawing) liberate only 50% of the transmitter acetylcholine.[39] This led to the concept of labile-bound and stable-bound forms of transmitter. Morphological[4,40,41] and radiochemical[42] studies have permitted the identification of the labile-bound fraction of acetylcholine released during hypo-osmotic treatment as that fraction of synaptosomal acetylcholine that is present in soluble form in the cytoplasm of the synaptosome, and the stable-bound acetylcholine as the fraction that is associated with synaptic vesicles, probably in local concentrations as high as 0.2 M.[41] The labile-bound fraction is bound only to the extent that the external membrane provides a barrier to the outward diffusion of free cytoplasmic acetylcholine as it does to that of any soluble cytoplasmic constituent.[7]

Hypo-osmotic disruption provides the most suitable procedure so far found for the isolation of the constituent organelles of the synaptosome. Large numbers of synaptic vesicles survive intact; external membranes and intraterminal mitochondria are not unduly fragmented and can be more readily separated from vesicles by centrifuging.

2. Separation of Synaptic Vesicles

a. By Differential Centrifuging.

In one procedure[43] a crude synaptosome preparation (approximately equivalent to fraction P_2 in Fig. 3) is suspended in 10 μM-$CaCl_2$ (10 ml/g of original tissue) and centrifuged at 11,500 g for 20 min to sediment mitochondria and myelin fragments (fraction

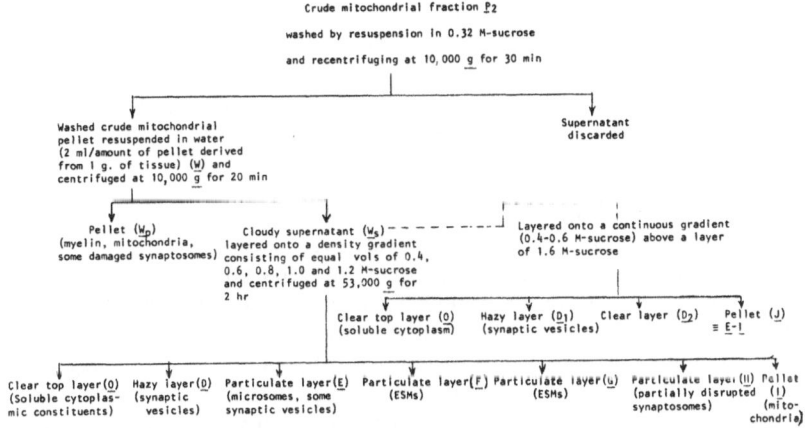

Fig. 3. Scheme for the subfractionation of synaptosomes.[4] The dotted line shows an alternative method when only synaptic vesicles are required.[40]

M_1). The supernatant is centrifuged at 100,000 g for 30 min to give a pellet (M_2), said to consist of free synaptic vesicles mixed with some "curved membranes" and a supernatant (M_3) consisting of soluble cytoplasmic constituents diluted with suspension medium.

While fraction M_2 undoubtedly contains free synaptic vesicles, it is contaminated with considerable amounts of other membranous material and even partly disrupted synaptosomes. This is shown by morphological examination using methods that give adequate preservation and sampling of the material[4] and also by the presence of cholinesterase,[43] Na^+,K^+-activated adenosine triphosphatase,[44] and ganglioside[45] in the fraction, markers for external membranes but absent from more homogeneous synaptic vesicle preparations (see Section III,C and D).[41,44,46–48]

Perhaps a more serious criticism of this preparation is that at the very low ionic strength and relatively low pH of the hypo-osmotic suspension, soluble proteins may be precipitated nonspecifically onto the membranes present in the fraction, leading to incorrect conclusions regarding the composition of synaptic vesicles. This is particularly evident with the soluble enzyme choline acetylase in certain species,[38] notably rat, rabbit, and sheep (see Section IV,B,3,b).[15]

b. By Density Gradient Centrifuging. In the method developed in the author's laboratory[4] (Fig. 3), crude or purified synaptosomes are suspended in water (2 ml/g of original tissue) by sucking up and down in a pipette and layered on a discontinuous or continuous sucrose density gradient. After centrifuging for 2 hr at 53,000 g, a hazy layer is formed in 0.4 M sucrose that consists of a very pure preparation of synaptic vesicles; larger membrane fragments, incompletely disrupted synaptosomes, and mitochondria are carried further down the gradient. The higher ionic strength and pH of this preparation results in less nonspecific adsorption onto membranes.[38]

Synaptic vesicle preparations obtained in this way contain soluble cytoplasmic constituents that have diffused into the hazy layer from the clear band (O) immediately above. Low molecular weight substances may be removed from the fraction by gel filtration through Sephadex G-25 or G-50, after either hypo-osmotic treatment or separation in the gradient. Macromolecular constituents, e.g., soluble proteins, presumably could also be separated from vesicles in this way using low cross-linked gels, but no separations of this kind have been reported: sedimentation of the vesicles by prolonged high-speed centrifuging has, however, been used.[4]

It will be noted in the scheme shown in Fig. 3 that the aqueous suspension of the crude synaptosome fraction (W) is centrifuged at 10,000 g for 20 min to remove the larger mitochondria and myelin fragments before being placed on the gradient. The presence of myelin tends to reduce the efficiency of the density gradient separation, and for the best preparations it is essential to eliminate it as completely as possible. Other ways in which this can be done are by removal of white matter by dissection before homogenization and by using a purified synaptosome preparation (fraction B) rather than the crude fraction (P_2).[46]

The yield of synaptic vesicles using the method described in Fig. 3 has been estimated to be about 12–15% of those present in the original tissue.[41] One source of loss is due to the tendency of vesicles to cohere together as though embedded in a sticky cytoplasm even after the external synaptosome membrane has been disrupted. Such clumps of vesicles sediment to denser regions of the gradient and give rise to a bimodal distribution of the stable fraction of bound acetylcholine. Attempts to increase the yield of vesicles by sonication or warming leads to a heavy contamination of the vesicle fraction by fragments of the external membranes. Recently, suspension in 5 mM-ethylene diamine tetra-acetate-5mM-tris-HCl buffer, pH 7.4, followed by passage through a small column of Sephadex G-50 equilibrated with this suspension medium has been found to improve yields.[8]

Relatively little study has been made of disruptive procedures that do not involve hypo-osmotic treatment as one step. However, it is known that disruption by freezing and thawing does not yield significant amounts of vesicles[4] and that disruption by supersound gives a vesicle fraction heavily contaminated with small fragments of ESM.[46]

3. Separation of Other Synaptosome Components

The discontinuous density gradient method shown in Fig. 3 yields six other bands of material besides that containing vesicles.

Fractions F and G are believed to consist mainly of ESMs (synaptosome ghosts); fraction H contains partially disrupted synaptosomes; fraction I, mitochondria; and fraction O, soluble cytoplasmic constituents. Fraction H contains all the insoluble constituents of the synaptosome: external membranes, synaptic vesicles, and intraterminal mitochondria. Fraction I contains intraterminal mitochondria together with any large somatic mitochondria not sedimented in fraction W_p. Fraction E contains membrane fragments of microsomal dimensions and some synaptic vesicles. These membrane fragments include any microsomes that contaminated the parent fraction and probably also fragments of the ESM of microsomal dimensions.

4. Relationship of Fractions Isolated by Differential and Density-Gradient Centrifuging

It is clear from the method of preparation that fraction W_s is approximately equivalent to the supernatant from fraction M_1; thus fraction O is equivalent to fraction M_3, and fractions $D–I$ (or D_1, D_2, and J) are equivalent to fraction M_2.

III. MORPHOLOGICAL EVALUATION OF FRACTIONS

A. Methods of Preparation for the Electron Microscope

1. Positive Staining

Synaptosome fractions can be prepared for electron microscopy by any of the standard methods of fixing, staining, and embedding nervous tissue that give good preservation of nerve terminals in whole tissue blocks. The following methods have been used: osmium–phosphotungstic acid,[3] permanganate–lead,[22,41] and glutaraldehyde–osmium–lead–uranium.[11] Owing to the particulate nature of the material, penetration of fixative is rapid, and a fixative like permanganate that does not give good results with whole brain tissue is satisfactory with subcellular fractions. Fixation can be carried out on pellets[3] or suspensions. The latter may be infiltrated after being sealed into agar cups[49] or mixed with fixative and subsequently centrifuged to form a pellet that is dehydrated and embedded without further disturbance.[11,22,41] The permanganate method when used with suspensions is particularly convenient because the manganese dioxide formed during fixation seems to act as a cushion during centrifuging and a preparation is obtained in which the structures are well separated without the packing distortions often visible in pellets.[22,40]

Synaptosomes fixed in osmium are fragile—much more so than mito-chondria—and will readily break up if the pellet is handled carelessly. This and the use of methacrylate rather than the epoxy resins as an embedding medium probably explains why synaptosomes remained undetected for so long; inadequately preserved synaptosomes indeed can be seen in electron micrographs of brain mitochondria published before their existence became generally known.

2. Negative Staining

Synaptosomes also can be examined in negative staining.[50] The problem here is to remove the solute of the suspension medium (which would other-wise interfere with the even deposition of negative stain) and to secure sufficient penetration of the stain to show up the characteristic internal organelles (synaptic vesicles and small mitochondria) without bringing about damage so extensive as to make the structures unrecognizable. Fixation with aqueous formaldehyde or osmium tetroxide followed ·by 1–2% sodium phosphotungstate (pH 7.4) has been the most frequently used method; this permits some penetration of stain without osmotic disruption. If iso-osmotic negative stain (e.g., 2% ammonium molybdate) is used without fixation, little or no penetration occurs; if the strongly hypo-osmotic phosphotungstate is used without fixation, extensive disruption and release of cytoplasmic organelles ensues. Some of the clearest pictures of individual synaptosomes that have been obtained were selected images in preparations exposed to mild hypo-osmotic disruption.

Synaptic vesicles may be stained without prior fixation even with hypo-osmotic stains such as 2% sodium phosphotungstate, since they are con-siderably more resistant to hypo-osmotic disruption than synaptosomes.[4] Even after fixation there is little penetration of negative stain, probably because of the presence of a core substance.[19] Prior treatment with sodium phosphate buffer permits the ingress of stain, and the synaptic vesicle membrane then becomes clearly visible.[41] Because of the small size of the vesicles, staining conditions must be optimal in terms of solute removal, particle concentration, and evenness and thickness of stain deposition.

Fractions containing ESMs (F and G) are difficult to stain because of the presence of a macromolecular substance (mucopolysaccharide) that interferes with the even deposition of stain.[46] The staining patterns obtained when this material is present resemble those obtained with Ficoll.

B. Morphological Characteristics of Fractions

1. Synaptosomes

a. In Section. Synaptosomes in section appear as membrane-bounded oval profiles about $0.5\,\mu$ in diameter containing numerous vesicles about 420 Å (SD $\pm$ 50 Å) in diameter and, sometimes, small mitochondria (figures relate to permanganate-fixed material). Not infrequently, lengths of thickened membrane can be seen attached to the ESM with a cleft between about 200 Å wide that is filled with lightly stained striated material. In addition most profiles contain a few larger vesicles (mean diameter 698 Å, SD $\pm$ 54 Å); these are particularly clearly seen in permanganate-fixed material. "Coated" and dense-cored vesicles also are seen within some profiles (Figs. 4–6).

These morphological features occur together, as far as is known, in only one structure in the central nervous system. Taken in conjunction with the high concentration of putative transmitter substances in the synaptosome fraction (Section IV), they provide the basis for the identification of these structures as detached nerve terminals. It is assumed that on homogenization the terminal region of the neuron is "pinched off" and seals up to form a discrete subcellular particle, which carries with it a length of the characteristically thickened postsynaptic membrane. Although the contents of the synaptic cleft are only lightly stained it is evident that the two membranes are securely bonded together; possibly the bonding material is coded for the particular contact made.

In assessing sectioned material, it must be borne in mind that even if all synaptosomes possess post-synaptic attachments and mitochondria as well as synaptic vesicles only certain planes of section will contain all three structural components. Moreover, serial sections show that some synaptosomes do not contain mitochondria; whether these represent mitochondrion-free lobes of larger nerve endings or whether they are derived from endings that are themselves devoid of mitochondria is not known. Although no quantitative studies have been carried out, the number of postsynaptic attachments observed seems less than one would expect if all synaptosomes possessed them. Attachments are more frequently seen in preparations made from regions of the brain (e.g., cerebral cortex and cerebellar cortex) where synaptic contacts are made with presumably easily detached dendritic "spines"; occasionally a complete spine is seen. Fewer attachments are seen in preparations that have been exposed to hyperosmotic sucrose, suggesting that high concentrations of sucrose have a solvent action on the cement in the synaptic cleft.

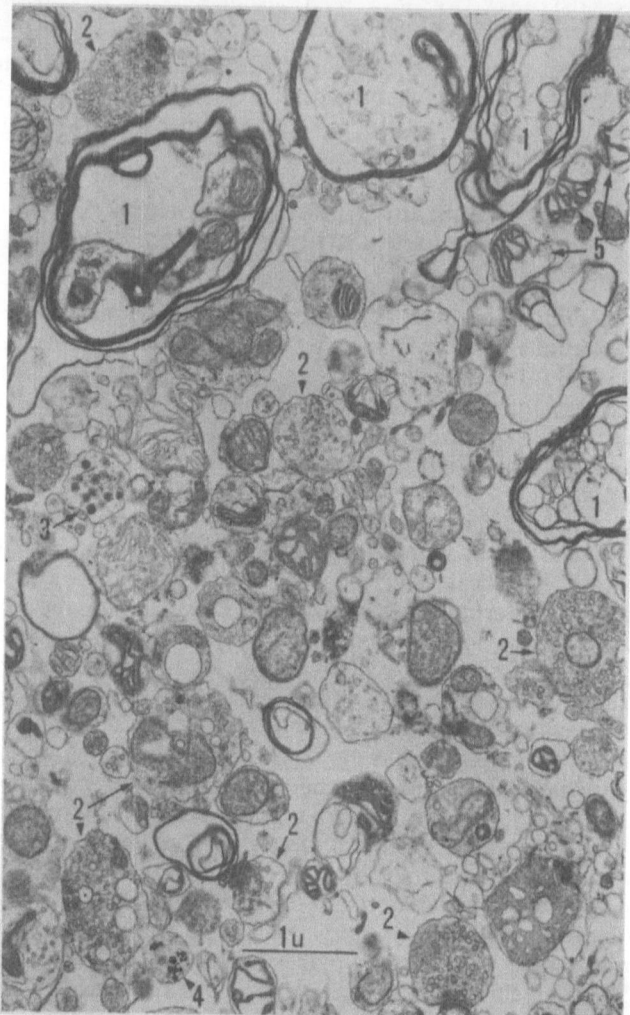

Fig. 4. Fraction P_2 from guinea pig hypothalamus, fixed in suspension in 0.32 M sucrose by addition of five volumes of 2.5% glutaraldehyde in 0.1 M Sørensen's phosphate buffer, pH 7.4. After 15 min the particulate material was collected by centrifuging and the pellet was washed with buffer and fragmented in 1% OsO_4 in phosphate buffer. After 1 hr the fragments of pellet were rapidly dehydrated in a series of ethanol–water mixtures and anhydrous ethanol and embedded in Epon. All steps except embedding were carried out at 0–4°C. Silver-gray sections were stained in uranyl acetate-lead citrate and examined in an AEI EM6B electron microscope using an anticontamination attachment. Note: 1, Myelin fragments with cytoplasmic inclusions; 2, synaptosomes, sometimes with adherent postsynaptic attachments (bottom right) containing synaptic vesicles and small mitochondria; 3, synaptosomes with granulated (dense-cored) vesicles; 4, a glial (?) fragment containing glycogenlike granules; and 5, free mitochondria. This electron micrograph and those in Figs. 5A, 7, and 8 were kindly provided by Dr. L. L. Ross. Magnification ×17,000.

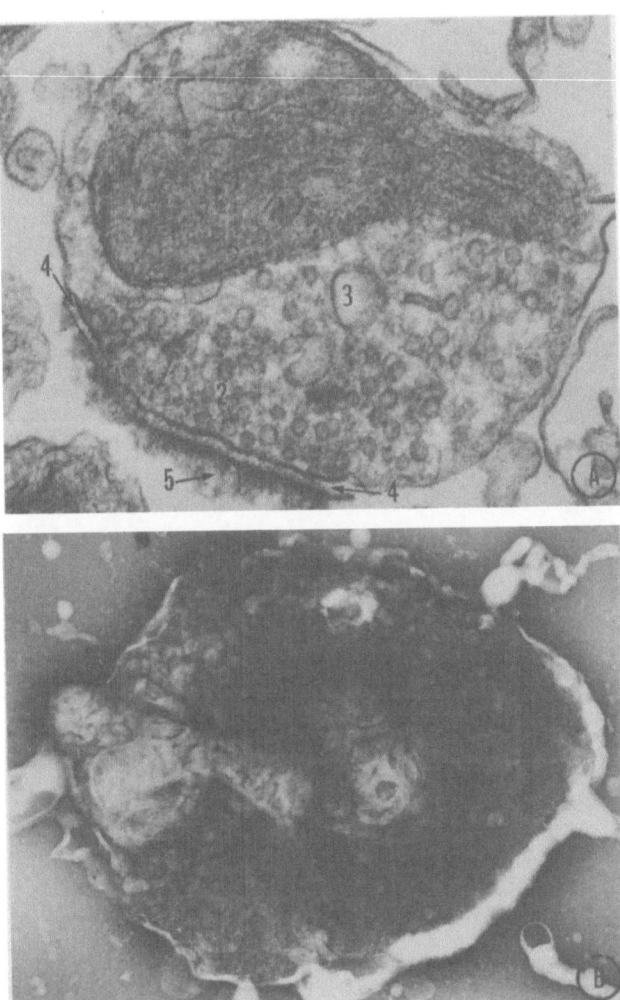

Fig. 5. (A) A synaptosome prepared as in Fig. 1 showing: 1, a mitochondrion; 2, synaptic vesicles; 3, larger vesicles; 4, synaptic cleft containing electron-dense material; 5, fibrillar postsynaptic thickening
("web"). Note the higher electron-density of the apposed pre- and postsynaptic membranes. Magnification ×60,000. (B) A synaptosome in
negative staining (formaldehyde fixed) showing coiled elongated mitochondrion and synaptic vesicles through the thin external membrane.
Magnification ×38,000.

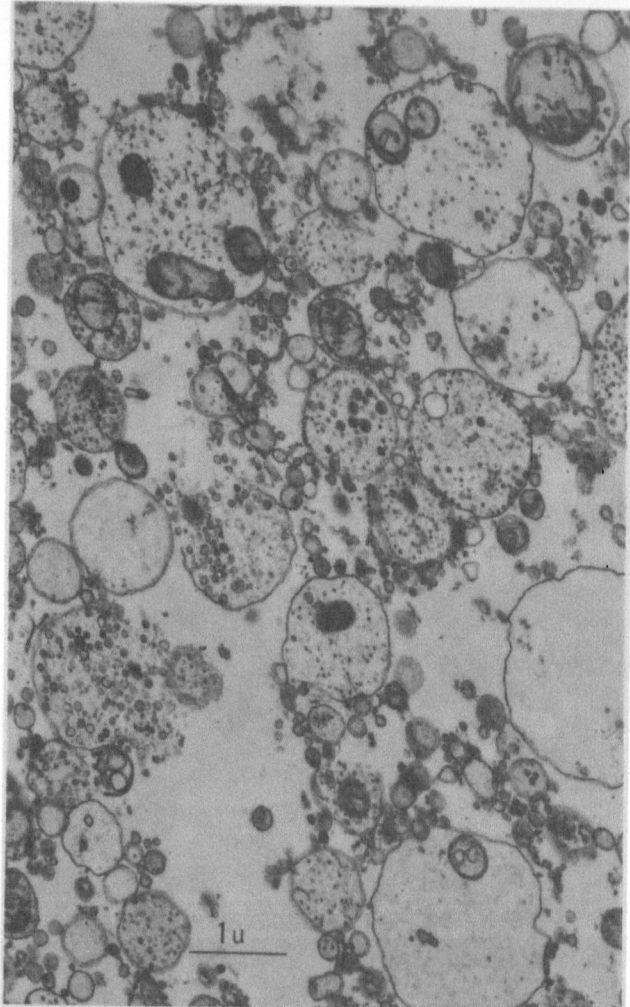

Fig. 6. Fraction P_2 prepared from the head ganglion of *Octopus* in 0.5 M sucrose + 0.33 M urea. Note numerous synaptosomes. Permanganate–araldite. Electron micrograph kindly supplied by Dr. J. G. Jones. Magnification × 14,000.

Sometimes profiles are seen with obvious postsynaptic attachments but few or no synaptic vesicles. These are probably synaptosome "ghosts."

b. In Negative Staining. On fixation the external synaptosome membrane becomes sufficiently permeable to admit some stain, and synaptic vesicles and mitochondria can be seen through the external membrane (Fig.

5B). In stereomicrographs the region containing mitochondria is seen to stand higher on the grid than the parts containing only synaptic vesicles. Many synaptosomes are seen to contain only one elongated, coiled mitochondrion, showing that the adjacent oval mitochondrial profiles often seen in thin section may well represent cross sections of a single mitochondrion winding in and out of the plane of section.[50] Sometimes short lengths of axon are seen attached to synaptosomes; these would be hard to identify in thin section owing to their small diameter and random orientation in the block. On the other hand, postsynaptic adhesions are difficult to identify with certainty in negative staining, and the synaptic cleft is rarely seen, presumably because the requisite orientation is seldom attained.

2. Subfractions of Synaptosomes

a. Synaptic Vesicles. Fraction D (Fig. 7) consists of an almost pure preparation of synaptic vesicles. For the purest preparations, centrifuging is continued for 2–5 hr at 53,000 g and care is taken to exclude membrane fragments that may have collected at the 0.32–0.4 M interface by sectioning the tubes just below this.[41] With these precautions, fewer than 1 vesicle in 300 is in the size range of small microsomes. In a routine preparation, in which interface material is included, contamination may be two to three times this amount. Such a preparation, fixed and stained using the glutaraldehyde–osmium technique, is shown in Fig. 7. The mean synaptic vesicle diameter is 471 Å, SD $\pm$ 77 Å (39 observations). This is a little larger than that formed with permanganate fixed material but is close to the value obtained in negative staining (469 Å, SD $\pm$ 110 Å, 351 observations).[4] It should be mentioned that particles similar in size to the contaminating particles are also seen in synaptosome cytoplasm.

In several places in Fig. 7 structures appear that suggest connections between vesicles. Thin tubular or fibrous connections also are seen in negative stained preparations, together with detached tubular or fibrous material.[47,51]

b. Soluble Cytoplasm. As normally prepared, the clear layer (O), above the band of synaptic vesicles, containing 75% of the soluble cytoplasmic constituents of the synaptosome, is devoid of organized lipoprotein membrane structures, though patches of granular material, probably dried-down soluble protein, are seen in negative staining.[4] However, slight disturbance of the tubes during handling may result in significant numbers of vesicles being thrown up into fraction O; bound acetylcholine is then also found in this fraction.

c. External Membranes. Fractions E and D_2 (Fig. 8A) are heterogeneous: They contain appreciable numbers of synaptic vesicles, a few large membrane fragments similar to those formed in fractions F and G (or J), and numerous fragments of microsomal dimensions. Fraction E probably consists of the smaller fragments of the ESM together with any microsomal contamination present in the starting material.

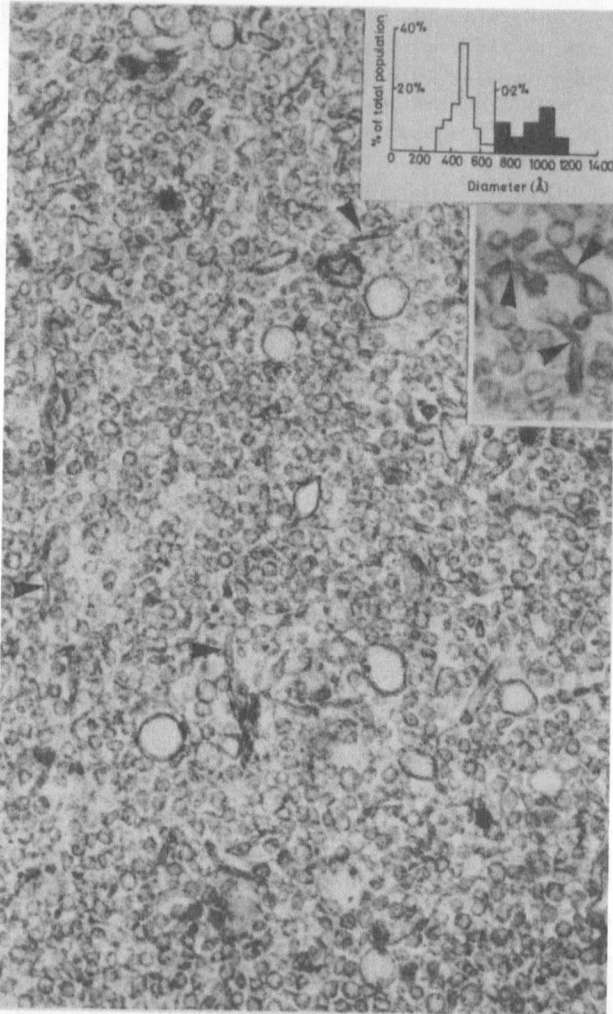

Fig. 7. Fraction *D*, prepared as described in Fig. 4, except that the hypo-osmotic suspension of synaptosomes was placed directly on the gradient. Note numerous synaptic vesicles and a few (about 1:150) larger vesicles similar to structure 3 in Fig. 5A. The top insert is a histogram showing the distribution of vesicle diameters. Note that there are two populations of mean diameter 471 Å, SD ± 77 Å (39) (white blocks), and 950 Å, SD ± 133 Å (9) (black blocks). The second population is plotted on 100 times the scale of the first to render it visible. Note the numerous fibrillar or tubular structures (arrows); these are also seen in negative staining and may be connexions between vesicles. Their structure is shown in more detail in the insert. Magnifications × 50,000 and × 83,000.

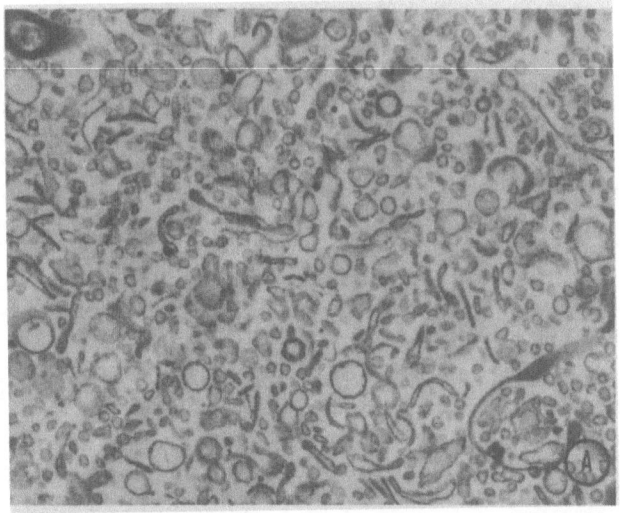

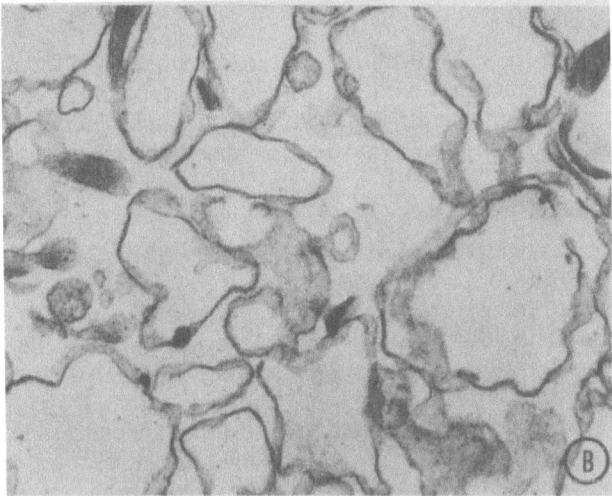

Fig. 8. (A) Fraction *E*, prepared as in Fig. 4. This fraction consists of small- to medium-sized vesicles (400–1500 Å) and some membrane fragments. Studies of marker enzymes show that this fraction contains synaptic vesicles, microsomes, and fragments of ESMs. Magnification × 30,000. (B) Fraction *G*, prepared as in Fig. 4. Note large membrane fragments similar in size to synaptosomes and identified as synaptosome "ghosts" (ESMs). Fraction *F* is similar. Magnification × 42,000.

Fractions *F* and *G* (Fig. 8B) are similar: They are fairly homogeneous and consist mainly of relatively large membranous bags similar in size to synaptosomes. The morphological and chemical characteristics of the fractions are consistent with the assumption that these bags are the "ghosts" of disrupted synaptosomes that have lost their cytoplasm. Negative staining[46] shows more clearly than thin section both the unfragmented ESMs and the fragments of microsomal dimensions that also are present; It further reveals the presence of thickened nonvesicular membranes whose identification is uncertain. These were originally tentatively identified[4] as postsynaptic thickenings, but they could be rolled-up ESMs or even strands of the macro-molecular material already mentioned that interferes with the even deposition of negative stain. The black dots seen in thin section may be cross sections of these nonvesicular structures.

d. Incompletely Disrupted Synaptosomes. The disruptive procedure found to be optimal leaves numerous synaptosomes in a relatively intact state; these migrate to the same position in the gradient as do intact synaptosomes and form a distinct band (fraction *H*). In both thin section and negative staining, however, the vast majority of synaptosome profiles look damaged, and chemical studies show that, in fact, almost all diffusible cytoplasmic material has been lost.

e. Mitochondria. Most mitochondria are recovered in fraction *I*; however, there are numerous small mitochondria in fraction *H* and some incompletely disrupted synaptosomes in fraction *I*. The mitochondria show the morphological alterations typical of exposure to hyperosmotic conditions.

3. *Species Variations*

The morphological descriptions given in the preceding subsections (1 and 2) relate primarily to fractions derived from guinea pig, coypu, and to a lesser extent rat brain. However, the morphology of the primary subcellular fractions and especially of intact synaptosomes from a wide variety of vertebrate species appears to be very similar, regional differences being greater than species differences.

As far as fractions of disrupted synaptosomes are concerned, the range of species studied is smaller. However, dog,[53] rat,[35,38,54] cat,[11] and sheep[15] give similar results to guinea pig and coypu. In the rat, the synaptic vesicles appear to be slightly denser than in the guinea pig[38]; this may be due to the adsorption of more soluble protein into them at the low ionic strength of the preparation. Vesicle preparations from cat[11] and sheep[15] are considerably less homogeneous than those from the rodents, probably because of the greater difficulty in removing myelin.

In spite of these similarities, it is inadvisable to apply methods worked out with one species to another without adequate morphological controls. Indeed, there is sufficient inherent lack of reproducibility in all subcellular fractionation techniques to make morphological controls essential when a new method is being applied, even when the species used is the same as in the original publication.

C. Quantitative Evaluation of Fractions

1. *Yield of Synaptosomes*

An important test of the basic assumption that the vesiculated synaptosome profiles are sections of pinched-off presynaptic nerve terminals would be to compare the number of synaptosomes in a brain homogenate or the fractions derived from it with the number of nerve terminals in the intact tissue from which the homogenate is made. Clearly, if the number of synaptosomes were found to be greatly in excess of the number of nerve endings, doubt would be cast on the whole synaptosome concept and, if considerably less, the synaptosome fraction could not be regarded as representative of nerve endings as a whole. Either way, the validity of using synaptosome for *in vitro* preparations for investigating synaptic function would be questionable.

The application of the test is fraught with difficulty, partly because of the problem of deducing the number of three-dimensional objects from two-dimensional sections and partly because of the sampling problems involved. However, it has been estimated that guinea pig cerebral cortex homogenates contain about 4×10^{11} synaptosomes/g and that the number of nerve terminals in guinea pig cortex is about equal to this.[22]

2. *The Compartments of Synaptosomes*

Statistical analysis of synaptosome profiles provides quantitative information regarding the size of the various compartments in the "average" synaptosome. A synaptosome of mean volume $(0.1 \ \mu^3)$ contains about 73 synaptic vesicles occupying about 4% of the volume; another 24% is taken up by mitochondria and 8% by the external membrane, leaving a cytoplasmic volume of 64% of the total volume.[52]

3. *Homogeneity of Synaptosome Fractions*

As prepared according to the standard method shown in Fig. 1, the synaptosome fraction (*B*) is relatively free from myelin whorls or mitochondria but is contaminated with considerable numbers of membrane fragments. Some of these are imperfectly preserved synaptosomes, but others are of unknown provenance. They range in size from fragments of microsomal dimensions $(0.1–0.2 \ \mu$ in diameter) to large $(0.5–1 \ \mu$ diameter), empty, oval profiles that might be fragments of dendrites or glial processes. Some may be small fragments of myelin, but the low cerebroside content of the preparation (Section IV,D) suggests that there is relatively little myelin contamination.

The presence of nonsynaptosomal material is indicated by the considerable amount of nondiffusible dry solids that cannot be accounted for by

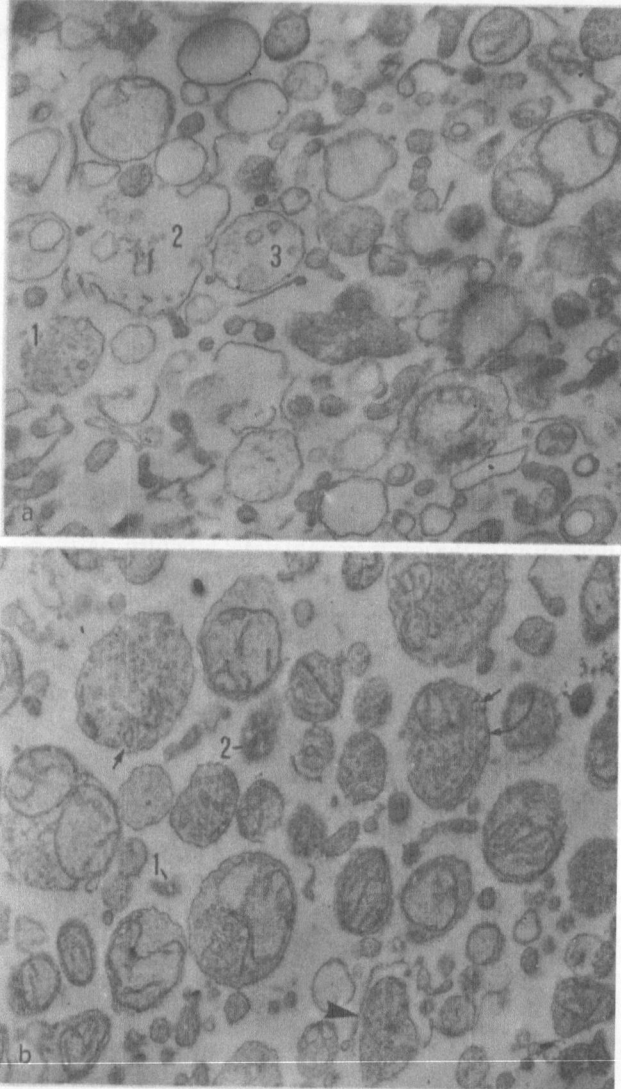

Fig. 9. Two fields from a light synaptosome fraction from the middle of a continuous sucrose density gradient (1.11–1.17 M sucrose). Field a is a region of the block where empty membranous profiles resembling synaptosome "ghosts" are concentrated. Profiles 1, 2, and 3 show intermediate stages between synaptosomes and "ghosts." Field b is a region of the same block showing well-preserved synaptosomes. Note postsynaptic adhesions (large arrow) and large vesicles among smaller ones (small arrows). Black patches (1) and small, ill-defined vesiculated profiles (2) are seen in serial sections to be grazing sections of synaptosomes whose centers lie below or above the plane of section. Magnification (a and b) ×23,000.

the morphological estimate of the synaptosomal volume. Thus the total morphological synaptosomal volume in fraction B from guinea pig cortex has been estimated to be about 31.5 μl/g of original tissue.[7] At an equilibrium density of about 1.15 this is 36 mg/g of original tissue. The non-diffusible dry solids in this fraction are about 22 mg/g of original tissue.[26] This implies a water content of only about 29% if all the dry solids are assumed to originate from synaptosomes. While synaptosomes in equilibrium with 0.8–1.2 M sucrose may well have a relatively low water content (62% assuming a dry density of 1.4), it seems clear that about half of the material in fraction B could be nonsynaptosomal, though not necessarily membranous.

When the crude mitochondrial (P_2) fraction is centrifuged for 2.5 hr into a continuous (0.8–1.6 M) sucrose density gradient, a very much purer synaptosome fraction may be obtained.[35,37] Myelin contamination is insignificant below about 1.0 M; the larger membrane fragments remain in the upper part of the gradient (down to about 1.17 M sucrose), and mitochondrial contamination does not commence until a sucrose concentration of about 1.24 M is reached (Figs. 9 and 10). Thus the zone between 1.17 and 1.24 M containing about 10% of the protein of the parent material constitutes a relatively pure synaptosome preparation, free from myelin and mitochondria and containing only a minor amount of contamination from small membrane fragments.

4. Yield of Synaptic Vesicles

The yield of synaptic vesicles has been evaluated using a "tagging" procedure with polystyrene beads and negative staining.[41,51] A volume of fraction D equivalent to 1 g of original tissue (guinea pig cortex) was estimated to contain about 3.8×10^{12} vesicles, and the yield of vesicles was estimated to be 15–16% of those present in the initial homogenate. This suggests a total number of vesicles of about 2.4×10^{13}/g cortex. This is consistent with the number of synaptosomes produced per gram of cortex and the number of vesicles per synaptosome.[22]

IV. CHEMICAL AND ENZYMIC COMPOSITION OF SYNAPTOSOMES

A. General

The chemical and enzymic analysis of synaptosomes provides a means of studying the molecular composition of the presynaptic region of the nerve cell in isolation from the rest of the CNS. The results so far obtained have been extensively reviewed elsewhere[1,19,47,51,52,55,55a]; the emphasis in this section therefore will be primarily on recent work and problems of interpretation.

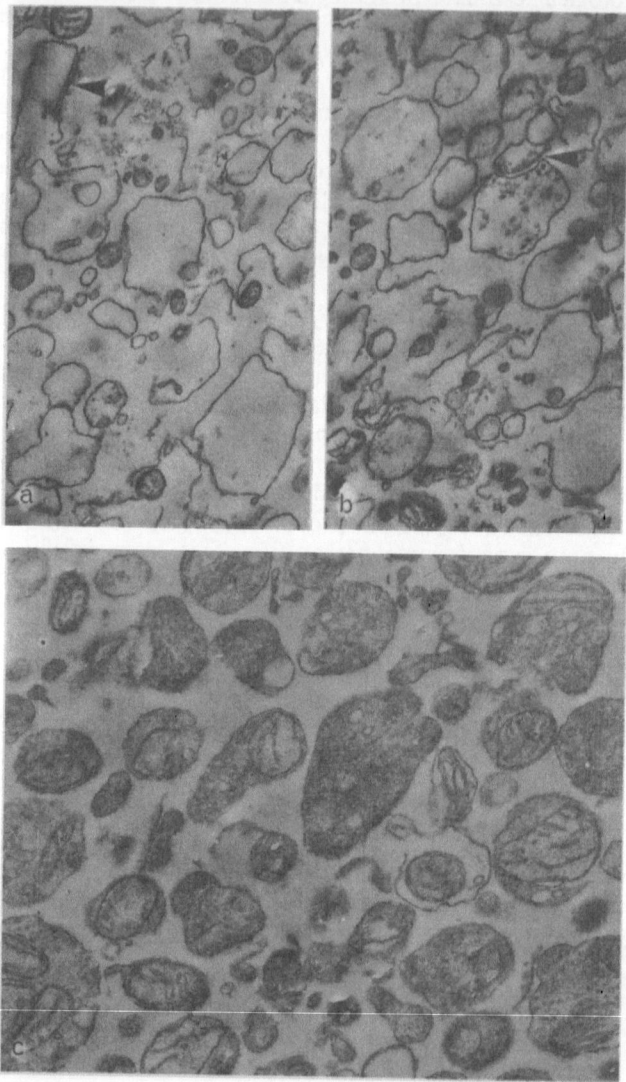

Fig. 10. a and b : Two more fields from the same fraction as that seen in Fig. 9 showing transitional profiles with postsynaptic adhesions (arrows). c: Portion of denser synaptosome fraction (1.24–1.30 M sucrose). Fractions below about 1.17 M sucrose do not contain the membranous contamination seen in Figs. 9a and 10a and b, but mitochondrial contamination begins at about 1.24 M sucrose. Magnification (a, b, and c) × 23,000.

In assessing analytical studies of synaptosomes and other brain fractions the following considerations must be borne in mind:

1. Synaptosome preparations, even when substantially free from myelin and mitochondria, are contaminated to varying extents, as we have seen (Section III,C,3), with membrane fragments of unknown origin for which markers are not yet available. Newer methods, utilizing continuous density-gradient centrifugation, provide purer preparations than those hitherto available but the yield is low[35,37]; however, the use of zonal rotors may overcome this difficulty.

2. All synaptosome preparations are heterogeneous with respect to chemical composition: any given region of the CNS contains neurons utilizing a variety of chemical transmitters; thus the synaptosomes derived from them must likewise show individual variations in transmitter and enzyme content. There is no evidence as yet for any close correlation between chemical type and those morphological characteristics (size and proportion of cytoplasmic volume occupied by mitochondria) that might provide a basis for separation by physical means such as density–gradient centrifuging. Indeed, there well may be as much variation between regions in the size and density of a particular chemical type of synaptosome as there is between synaptosomes of different chemical types within a region.

3. Soluble substances may diffuse out of synaptosomes during preparation; that this happens, e.g., with hydroxytryptamine is shown by the smaller proportion of the amine found in a high-speed supernatant prepared directly from a freshly made homogenate than in the final supernatant fraction (S_3).[25] The transient changes in permeability accompanying osmotic stress may account for the lowered cytoplasmic K^+ and the free K^+ found in synaptosome preparations (see Section V,B,1) and may influence the distribution of transmitters in density gradients.

4. Owing to the coacervating effect of ions on subcellular particles, subcellular fractionation must necessarily be carried out at low ionic strengths. This is particularly true of subsynaptosomal organelles prepared by hypo-osmotic disruption. Some soluble proteins are nonspecifically adsorbed onto membranes under these conditions.[38] The resultant subcellular distribution patterns may be misleading. On the other hand, attempts to fractionate concentrated homogenates (containing more than 10% wt./vol. of tissue) or preparations to which appreciable concentrations of ions (especially divalent ions) have been added may result in heavy cross contamination of fractions.

5. Synthesis or destruction of cell constituents may take place during fractionation. This will be shown by recoveries markedly greater or less than 100%. The recovery of labile substances such as triphosphoinositide[26] or ATP[24] may be as low as 50% even under conditions of rapid fractionation and good refrigeration. In such instances it is difficult to draw firm conclusions regarding the subcellular distribution of the substance being studied.

6. Because a particular substance is found in a particular subcellular fraction, it does not follow that the substance is present in the particles that

mainly constitute the fraction. It may be present in only a proportion of the particles or in a minor morphological constituent in high specific activity. This reservation always must be attached to the statement that a particular substance has a synaptosomal location.

B. Transmitter Content

1. Identity of Central Transmitters

No central chemical transmitter can be identified with complete certainty, but the following are strong contenders: acetylcholine, noradrenaline, dopamine, hydroxytryptamine, γ-aminobutyrate, and glycine. All these compounds are found in appreciable concentrations in the synaptosome fraction. According to current concepts (which, however, have not gone unchallenged at any rate in the peripheral nervous system), each neuron possesses only one transmitter, together with the enzymes required for its synthesis; inactivation of the transmitter, however, may be a function shared by the neurons synthesizing it and upon which it acts. Neurons thus may be classified according to the transmitter they elaborate and, when this is done, a pattern of chemical coding, or chemical specificity, begins to emerge that may have a vital role in suppressing unwanted "cross talk" between different systems.

2. Topographical Considerations

Much evidence (mainly microchemical and histochemical) suggests that transmitters and the enzymes synthesizing them, besides being concentrated in specific nerve terminals, are distributed in lower amounts throughout the length of the neuron. In homogenate in which cell bodies and axons are fragmented but in which nerve endings largely survive as sealed synaptosomes, the distribution of a transmitter between the synaptosomes and the high-speed supernatant will therefore depend (1) on the relative extent to which the cell bodies, axons, and nerve terminals of the neurons containing the transmitter were represented in the original tissue sample and (2) upon the relative concentrations of the transmitter in each portion of the neuron.[52]

It has been fortunate for the development of the synaptosome concept that acetylcholine was the first transmitter whose subcellular distribution in brain tissue was studied. It is now known that most cholinergic neurons in the mammalian brain form part of a great ascending system—the reticular activation system—which terminates on cortical and striate neurons.[56] Thus, in homogenates of cortical tissue, cholinergic neurons are represented mainly by the synaptosomes derived from their numerous endings, to a lesser extent by the unsealed fragments of their axons and the axoplasm which exudes from them, and hardly at all by their cell bodies. Thus 70% of the total acetylcholine of the forebrain remains in a bound form on homogenization; much of this is recovered in the synaptosome fraction (B), and that which is not can be accounted for by synaptosomal contamination of other fractions; the degree of localization is quite comparable to that obtained with mitochondrial markers.

In striking contrast to forebrain, there are few cholinergic terminals in spinal cord; these are restricted, so far as is known, to the recurrent collaterals from motoneurons of the ventral horn terminating on Renshaw cells. In this region, the volume of cholinergic nerve cell bodies is clearly very much greater than that of the endings or the synaptosomes derived from them, and this may well explain one of our earliest observations, namely, that 80–90% of the spinal cord acetylcholine is in the free state in eserinized spinal cord homogenates: The small amount of bound acetylcholine shows, however, a subcellular distribution similar to that in forebrain.[10]

The other extreme situation is found in caudate nucleus, where virtually all the acetylcholine is synaptosomal. This is a part of the brain receiving a massive cholinergic innervation from outside; the axonal cholinergic component is short and the cell-body component is nil.[10]

Interneurons represent a type of neuron in which the endings are all within a very short distance of the cell body; thus soma, axon, and endings are all likely to be found within a piece of tissue containing any one of these components. Interneuron transmitters will thus be expected to show a subcellular distribution pattern resembling that of acetylcholine in eserinized spinal cord preparations, and so differing little from that of any soluble cytoplasmic constituent.[52] Glycine, within the spinal cord, and γ-aminobutyrate, in the cortex, are strong contenders for the role of inhibitory transmitters. In forebrain they both show the type of distribution predicted for the transmitters of an inhibitory interneuron; furthermore, the amount of γ-aminobutyrate in cortical synaptosome preparations is adequate to sustain a transmitter role[5,52] and the bound fraction of the enzyme synthesizing γ-aminobutyrate from glutamate (glutamate decarboxylase, GAD) also has a synaptosomal location.[14,34–36]

3. The Organization of the Cholinergic Neuron

a. Cholinesterase. There is now convincing histochemical evidence[56,57] that in the CNS the enzyme that terminates the transmitter action of acetylcholine, acetylcholinesterase, is synthesized mainly within the cell bodies of the cholinergic cells themselves. In the soma the enzyme is confined within the lumen of the rough-surfaced endoplasmic reticulum; in the axon it is localized on the neuronal plasma membrane; and at the ending it is attached to the external membrane; synaptic vesicles are devoid of the enzyme. This suggests that the enzyme is excreted onto the surface of the plasma membrane and flows down the outside of the neuron, accumulating at the terminal. If the enzyme is functionally always on the outside of the cell, the paradox of the coexistence of free cytoplasmic acetylcholine and cholinesterase in the same cell would be resolved.

The subcellular distribution of acetylcholinesterase is consistent with the histochemical description. Among the primary subcellular fractions of brain, the microsome fraction (derived mainly from the endoplasmic reticulum) has the highest specific activity. However, over half of the total activity is recovered in the crude mitochondrial fraction, and on subfractionation it is seen to be largely associated with synaptosomes. In histochemical studies with synaptosomes, staining has been observed around the ESM[58]; in subfractionation, the fractions containing the highest specific activity are fractions E–G, those containing ESMs, and any microsomal contamination. Fraction D (the isolated vesicle fraction) has a negligible activity.

 b. Choline Acetylase. The subcellular distribution of choline acetylase resembles that of acetylcholine. About 30% (rat, rabbit, and guinea pig) is in the high-speed supernatant in a nonoccluded form; the remainder is occluded and particle-bound and has a synaptosomal location.

On subfractionation of hypo-osmotically disrupted synaptosomes the enzyme behaves as a soluble protein in preparations derived from pigeon and guinea pig, but in rat, rabbit,[38] and sheep[15] much of the enzyme is carried down the gradient into fractions rich in ESMs; very little is bound to vesicles. This phenomenon has been traced to an effect of the low ionic strength of the preparation.[38,59] If the membrane fractions to which the enzyme is bound are resuspended in media of the estimated ionic strength and pH of the cytoplasm of intact synaptosomes, the enzyme is solubilized. The binding and release of the enzyme is reversible,[59] and when bound the enzyme is nonoccluded. At sufficiently low ionic strengths the choline acetylases of all species examined are membrane-bound; however, the rat enzyme is much less soluble at low ionic strengths than the pigeon or guinea pig enzyme, and mixtures can be readily separated by ammonium sulfate precipitation.[59] Similar phenomena are observed with the most basic of the five lactate dehydrogenase isoenzymes also present in synaptosomes and probably account for the nonoccluded microsomal choline acetylase found in goat ventral root preparations.[38] The rebinding of the enzyme at high ionic strengths reported for rabbit preparations[38] was due to the failure adequately to control pH.[59]

 c. Acetylcholine. It is now known that there are at least two forms of acetylcholine in cholinergic neurons, the cytoplasmic and the vesicular. The free supernatant (S_3) form of brain acetylcholine—only seen in preparations to which an anticholinesterase has been added—probably represents the somatic and axonal portions of the cytoplasmic compartment; the "labile-bound" acetylcholine referred to in Section II,C,1 is the sealed-off terminal portion of this fraction. Radiochemical and gel-filtration studies show that the "labile-bound" acetylcholine has all the properties expected of a soluble cytoplasmic small molecule. It has a permeability constant about one-seventh that of K^+. It can be readily labeled by radioactive exchange with added labeled acetylcholine or from precursors such as choline, acetate, pyruvate, or citrate.[8,42]

By contrast, the vesicular fraction, though readily lost, does not show demonstrable labeling *in vitro*, under a wide variety of experimental conditions using synaptosomes or isolated vesicles and both metabolizing and nonmetabolizing systems.[8,42] In intact tissue, preliminary work[60] shows that the vesicular acetylcholine is only very sluggishly labeled from choline compared with the cytoplasmic fraction. It is clear that some rate-limiting step, which could be the synthesis of the vesicle wall or of a lipid-soluble precursor, intervenes in the synthesis of vesicular acetylcholine from choline.

The acetylcholine content of cortical vesicles has been estimated to be about 300 molecules per vesicle. This is an average figure assuming that acetylcholine is distributed evenly throughout the whole vesicle preparation.[41] Since only a small proportion (say 15%) of vesicles are likely to be derived from cholinergic endings, the number of molecules per cholinergic vesicle is probably much higher—about 2000. This would correspond to a solution of acetylcholine of about 0.2 M, and is within the range proposed for the number of acetylcholine molecules per quantum.[41,51] Thus the acetylcholine content of central vesicles seems to satisfy the requirements of the quantal hypothesis.

4. *The Separation of Cholinergic Synaptosomes from Others*

The study of the organization of the cholinergic nerve terminal would be greatly facilitated if cholinergic synaptosomes could be isolated from others. There is some evidence that cholinergic synaptosomes form a somewhat more homogeneous population than the particles containing hydroxytryptamine, dopamine, or noradrenaline. However, on a density gradient, overlap of acetylcholine and hydroxytryptamine was considerable.[25]

In a recent study[35] fraction P_2 was separated on a continuous gradient and lactate dehydrogenase, GAD, choline acetylase, and acetylcholine were followed down the gradient. After 2.5–13 hr, the peak activities of lactate dehydrogenase, GAD, and acetylcholine invariably coincided; however, the densest fractions contained relatively less acetylcholine than lactate dehydrogenase or GAD. This could have been due to loss of cytoplasmic acetylcholine resulting from a transient increase of permeability induced by exposure of the synaptosomes to hyperosmotic conditions (Section V,B,2). Significantly, the peak concentration of γ-aminobutyric acid also lies above that of GAD in sucrose density gradients.[14] Choline acetylase tended to show a peak actually slightly above GAD and acetylcholine. Possibly some soluble choline acetylase had become adsorbed onto the membrane fragments that contaminate the upper part of the gradient at the low ionic strength of the sucrose medium. Similarly, any cholinesterase-rich microsomal contamination will tend to collect in this region of the gradient. Thus there are three separate factors—loss of soluble transmitter in hyperosmotic regions of the gradient, adsorption of enzyme onto membrane fragments, and microsomal contamination—which, operating together, could cause each of the three components of the cholinergic neuron to be partially retained in the upper part of the gradient and so create the impression that cholinergic synaptosomes are being separated from denser noncholinergic synaptosomes. Clearly more evidence will be required to establish that separation is actually occurring.

C. Enzyme Content

Synaptosomes contain a large number of enzymes that may be classified as follows:

1. Enzymes of the intraterminal mitochondria. Relatively little study has been made of intraterminal mitochondria owing to the difficulty of isolating them without damage due to variations in osmotic pressure. However, they seem to resemble somatic mitochondria in their enzymic properties as far as is known.

2. Enzymes of the soluble cytoplasm. These include (a) Enzymes of glycolysis. The synaptosome cytoplasm is believed to contain a full complement of glycolytic enzymes. As discussed in Section V,A, synaptosomes utilize glucose effectively. (b) Enzymes concerned with transmitter synthesis. Some of these have already been discussed. There is reason to believe that most of the enzymes concerned with the last step in the synthesis of putative transmitters, e.g., choline acetylase (acetylcholine) GAD (γ-aminobutyrate), aromatic aminoacid decarboxylase (noradrenaline and hydroxytryptamine), are soluble cytoplasmic enzymes and are not attached to synaptic vesicles. In fairly high Ca^{2+} concentrations, GAD binds to both synaptic vesicles and ESMs, and this binding is not reversible by washing with water—yet GAD appears mainly in the soluble cytoplasmic fraction on hypo-osmotic rupture. It seems unlikely, therefore, that GAD can be bound in this way in the intact synaptosome.[35] (c) Enzymes of protein synthesis. Although presynaptic nerve terminals have not been observed to contain ribosomes, the *in vitro* incorporation of ^{14}C-L-leucine into synaptosomal protein has recently been reported[60a]; two different systems, one attributable to the intraterminal mitochondria, were characterized.

3. Enzymes of the ESM. These include cholinesterase and Na^+,K^+-activated ATPase. Preliminary histochemical studies[58] show that cholinesterase is present on the ESMs of only a limited number of synaptosomes— presumably those derived from cholinergic neurons.

So far as is known, synaptic vesicles contain only one enzyme in low activity, a Mg^{2+}-activated ATPase.[44,46] Purified vesicles do not contain choline acetylase, GAD, Na^+,K^+-activated ATPase or cholinesterase; the presence of the last two enzymes, as we have seen, is a sensitive indicator of contamination of the vesicle fraction by fragments of ESM. The function of the Mg^{2+}-activated ATPase is unknown: Attempts to involve it in transmitter uptake into vesicles have been unsuccessful.

D. Lipid Composition

The lipid composition of synaptosomes has been studied by two groups of workers.[26,61] According to our findings[26] synaptosomes have a higher proportion of their total lipid in the form of phospholipid than any other subcellular organelle except mitochondria. All the usual phospholipids are present: phosphatidyl-choline, -ethanolamine, -serine, and -inositol, ethanolamine plasmalogen, and sphingomyelin (Table I). There is no choline and serine plasmalogen, almost no phosphatidic acid, and very little cerebroside; the distribution of this lipid in the various subcellular fractions shows it to be a marker for myelin (Table II, column 2). Cholesterol is present; the molar ratio cholesterol:phospholipids is 0.43, similar to that of microsomes and considerably lower than that of myelin, where for the purest preparations it is close to 1.0 (Table II, column 3). The very low cerebroside content shows that cholesterol is a true synaptosome constituent and is not present due to myelin contamination.

There is a small amount of the characteristic mitochondrial lipid, cardiolipin, as would be expected from the presence of mitochondria in the synaptosome cytoplasm. The ganglioside content (Table II, column 4) of the fraction is similar to that of microsomes; the ganglioside includes all three types (mono-, di-, and trisialogangliosides), whereas myelin has a preponderance of the monosialoganglioside.

When the analysis is extended to isolated synaptic vesicles and ESMs,[26,47] some interesting differences in composition emerge. As far as phospholipids are concerned (Table I, columns 3 and 4), the ESM fraction is richer in phosphatidic acid and ethanolamine plasmalogen and poorer in phosphatidyl-choline, -ethanolamine, and -serine and in monophosphoinositide and sphingomyelin. More striking differences occur with cholesterol and ganglioside (Table II). The molar ratio cholesterol:phospholipids of ESMs is almost double that of vesicles. High cholesterol:phospholipid ratios have been found to be characteristic of a number of other plasma membrane preparations, in addition to myelin, which is not a typical plasma membrane. The difference in ganglioside content is even more striking; the ganglioside content of fraction G is four times higher than that of D. The low ganglioside content of synaptic vesicles, recently confirmed,[48] makes it unlikely that ganglioside is concerned with transmitter binding as once suggested.[45]

These differences in lipid composition, taken in conjunction with the differences in enzyme composition mentioned in the previous subsection, raise serious difficulties for the reversed micropinocytosis theory of transmitter release.[47,51] They suggest that the vesicles form a membrane system distinct from that of the external membrane, though possibly communicating through it by means of the fine tubules or connections that were mentioned in Section III.

Ganglioside, owing to its strong negative charge, is known to combine avidly with many positively charged organic substances, including pharmacologically active compounds; because of their ganglioside content, preparations of ESMs would be expected to have this property also. There is as yet no evidence that would convincingly relate such binding to the action of receptors; thus the physiological significance of any binding of drugs by preparations containing ESMs is far from clear.

TABLE I

Phospholipid Content of Synaptosomes and Their Constituent Membranes Compared with Other Brain Membranes[a]

Lipid	Synaptosomes (B)	Synaptic vesicles (D)	ESMs (G)	Microsomes (P_3)	Mitochondria (C)	Myelin Large (P_1A)	Small (A)
Phosphatidyl-							
choline	39	41	24	41	40	26	32
ethanolamine	18	15	10	15	23	9	14
serine	13	12	3	12	6	13	14
Monophosphoinositide	4	5	1	4	5	3	3
Phosphatidic acid	1	0	6	0	1	2	1
Cardiolipin	2	1	0	1	11	1	0
Ethanolamine plasmalogen	16	15	20	15	9	26	24
Sphingomyelin	5	11	6	8	4	12	7
Alkylether	2	3	4	2	2	4	2
Recovery	99	94	74	98	101	91	97
Total lipids (percent of dry weight)	46	72		60	32	80	74

[a] The results are percentages of total lipid phosphorus and relate to guinea pig brain. The large myelin fraction was the only fraction containing detectable amounts of choline and serine plasmalogens; these each accounted for 1% of the total lipid phosphorus of that fraction.

TABLE II

Cholesterol, Cerebroside, and Ganglioside Contents of Synaptosomes and Their Constituent Membranes Compared with Other Brain Membranes

Fraction	Cerebrosides: phospholipids, molar ratio	Cholesterol: phospholipids, molar ratio	Ganglioside, μmoles/mg N
Synaptosomes (B)	0.01	0.43	0.14
Synaptic vesicles (D)	0	0.41	0.07
ESMs (G)	0.05	0.72	0.27
Microsomes (P_3)	0.09	0.48	0.14
Mitochondria (C)	0.03	0.15	0.03
Myelin			
Large (P_1A)	0.5	1.03	0.07
Small (A)	0.25	0.73	0.18

Tetanus toxin is also bound by gangliosides; the binding is enhanced by other lipids. In a recent study of the binding capacity of fractions $O–I$, the binding capacity of isolated vesicles was found to be very low and that of the fractions containing ESMs was appreciable.[62]

V. THE SYNAPTOSOME AS A MINIATURE CELL

A. Metabolic Properties

The synaptosome, as a pinched-off cell process, has, within its external membrane, a cytoplasm containing a normal complement of glycolytic enzymes and, frequently, one or more small mitochondria. It thus may be regarded as a small, nonnucleated cell. Recent studies[8] have shown that synaptosomes do, in fact, actively respire when warmed and provided with substrates; they also synthesize high-energy substrates, take up choline by a Na^+-dependent, hemicholinium-sensitive mechanism, and possess a membrane with passive permeability properties similar to those of other nonmyelinated neuronal membranes. These properties permit the integrity of synaptosome preparations to be assessed in a variety of ways.

Unfortunately, synaptosomes autolyze rapidly at 37°C; however, by carrying out experiments at 30°C, reasonable preservation is achieved and the metabolic properties of synaptosomes may be explored and compared with those of mitochondria, brain slices, and other types of preparation.

As might be expected from their cytoplasmic content, synaptosomes respire much more effectively than mitochondria with glucose as a substrate (Table III); the optimum substrate is, however, an equimolar mixture of glucose and succinate. Glucose (Table IV), succinate, and malate alone give rates only about half that obtained with glucose and succinate together; pyruvate, α-oxoglutarate, and glutamate are relatively poor substrates.

TABLE III

Utilization of Glucose and Succinate by Subfractions of a Crude Mitochondrial Fraction from Guinea Pig Cortex

Fraction	Main component	Respiration as percent of total recovered		
		Glucose	Succinate	Glucose + succinate
A	Myelin	13 ± 3	13 ± 5	8 ± 2
B	Synaptosomes	76 ± 2	43 ± 16	56 ± 13
C	Mitochondria	11 ± 5	44 ± 14	35 ± 13

Under optimum conditions the oxygen uptake is from one-third to one-half that of cortex slices and about half of that of whole brain on a protein basis (Table V). The respiratory rate is relatively insensitive to Ca^{2+} or Mg^{2+}. Inclusion of 100 mM-Na^+ or K^+ caused a small increase in respiration (about 20% in the case of K^+); this is smaller than with slices (40–60%); however, as with slices the stimulatory effect of K^+ was abolished by Ca^{2+} (Table IV). The concentration of high-energy phosphase (ATP and creatine phosphate) rises during incubation in the presence of substrates.

B. Permeability of the External Membrane

1. To Small Molecules

Intact synaptosomes retain numerous soluble substances (glycolytic enzymes, K^+, and amino acids) that are typical constituents of cell cytoplasm; rupture of the external membrane by hypo-osmotic shock, addition of detergents, warming, and other disruptive treatments causes the immediate release of these substances and deocclusion of synaptosomal enzymes that are normally inaccessible to their substrates. There is thus biochemical as

TABLE IV

Potassium Stimulation of Synaptosome Respiration[a]

Condition	Respiratory rate as percent of control
No succinate	53
No succinate or glucose	18
Plus K^+ (100 mM)	121
Plus K^+ (100 mM) and Ca^{2+} (6 mM)	77

[a] Control conditions were as in the experiments shown in Table V, with glucose + succinate as substrate.

well as morphological evidence for the presence, around synaptosomes, of a sealed external membrane.

The amount of synaptosomal K^+ and other small ions and molecules can be readily determined by gel-filtration through columns of Sephadex G-25 (bead form) that have been previously equilibrated with sucrose iso-osmotic to the synaptosome preparation.[7] The synaptosomal small molecules emerge in the void volume of the gel, while free small molecules that may have diffused out of the synaptosomes during preparation penetrate into the gel. A paralled run with another sample of the synaptosome preparation through a column of gel equilibrated with 5 mM-tris buffer, pH 7.4, and therefore strongly hypo-osmotic, permits a correction to be made for the small amount of K^+ or other small ions and molecules that are bound to macromolecules and not sequestered within osmotically sensitive structures. Any disruptive procedure that causes rupture of the ESM results in the loss of the osmotically sensitive portion of the void-volume K^+.

Using this gel-filtration technique, it has been established that K^+ and other small ions and molecules diffuse slowly into synaptosomes on standing at 0–5°C. Equilibrium (for K^+) is complete after about 16 hr, i.e., $[K^+]_i$, the concentration of K^+ inside the synaptosome, equals $[K^+]_e$, the concentration in the medium. If at this stage the total osmotically sensitive void-volume K^+, K_T^+, is determined by gel-filtration; then the volume occupied by the osmotically sensitive K^+, viz., $K_T^+/[K^+]_i$, equal to $K_T^+/[K^+]_e$, may be calculated. This is about 28 μl for that amount of synaptosomes isolated from 1 g of tissue, in fairly good agreement with the morphological synaptosome volume determined from electron micrographs (31.5 μl). There is thus reason to believe that the main osmotically sensitive space in a synaptosome preparation is the synaptosome space itself and that the "ghosts" and other membrane fragments present as contamination are either not penetrated by small molecules or, more likely, equilibrate too rapidly to behave as osmotically sensitive, sealed structures. The Na^+ and $[^{14}C]$-D-galactose spaces also have been determined and are close to the K^+ space; by contrast, the space measured with a nonpenetrating macromolecule, $[^{131}I]$-human serum albumin, was only 5% of the K^+ space. The magnitude of these "spaces" is clearly a sensitive indicator of the integrity of the ESM.

The composition of the osmotically sensitive void-volume has been studied for K^+ and other substances as a function of time, osmotic pressure, temperature, and external solute concentration. These measurements have shown that freshly prepared synaptosomes (fraction B, Fig. 1) contain 88 mM-K^+ and 54 mM-Na^+, values close to those reported for slices,[63] though the potassium figures are lower than those reported for peripheral nerve fibers (145 mM for rabbit C fibers[64] and 268 mM for squid axon).[65]

The movement of K^+ across the ESM at 0–5° appears to be a passive process, with a permeability constant of 10^{-8} cm/sec, a value fairly close to that of another nonmyelinated neuronal membrane, the rabbit C fiber.[63] Na^+ has a permeability about 60% of that of K^+, a value consistent with its greater hydrated radius. Presumably, active transport takes place under

TABLE V

Respiration of Synaptosomes[a]

Preparation	Oxygen uptake, microliters O_2/min/100 mg protein
Synaptosomes[b]	22.5, SD $\pm$ 5.2 (20)
Slices[b,c]	60
Whole brain[d]	40–50

[a] The incubation medium for synaptosomes contained 100 mM NaCl, 1.0 mM P, 10 mM $MgCl_2$, 10 mM glucose, 10 mM succinate, and tris buffer, pH 7.4, together with catalytic amounts of ADP, CoA, NAD, NADP, glutathione, lipoic acid, pyridoxal, and thiamine. Measurements were made at 30°C.
[b] Guinea pig cortex.
[c] In Krebs–Ringer solution containing glucose, pyruvate, and fumarate.
[d] Cat, assuming 100 mg protein per gram of wet tissue.

conditions of active metabolism, but so far the sharp rise in permeability and fall in total synaptosome volume that accompany a rise in temperature have prevented a clear demonstration of active transport in synaptosomes.

2. To Substrates

Another way in which the integrity of the ESM may be assessed is by measuring the activity of enzymes that are sequestered within the cytoplasm of the synaptosome. The enzyme most studied so far is lactate dehydrogenase.[8] The distribution of this enzyme down the density gradient after separation of crude synaptosome fractions on a continuous gradient shows that very little is entrapped in myelin and most of the activity can be assigned to synaptosomes.[35] If the synaptosomal lactate dehydrogenase activity is measured under iso-osmotic conditions, the activity recorded is a small fraction of that found immediately after the addition of Triton X-100 to the preparation. Presumably the substrates (pyruvate and $NADH_2$ or lactate and NAD) diffuse only slowly through the ESM, and this greatly slows the rate of hydrogen transfer. Part or all of the activity in iso-osmotic media indeed may be due to free lactate dehydrogenase or to incompletely sealed synaptosomes. Disruptive treatments, as expected, cause deocclusion of the enzyme. Severe disruption is accompanied by solubilization of the enzyme, but small changes in osmotic pressure are accompanied by transient increases in activity that are not associated with release of the enzyme into the surrounding medium: Under these conditions the enzyme can be sedimented at the relatively low speeds required to bring down synaptosomes. Reocclusion takes place spontaneously on standing but can be hastened by the addition of lipids. It is thought that the ESM becomes temporarily more permeable to substrates as the result of osmotically induced movements of water across the membrane.

VI. CONCLUSIONS

This chapter has been concerned with what may be termed the synaptosome hypothesis. To recapitulate, this hypothesis states that when brain tissue is homogenized, presynaptic nerve terminals survive as detached, sealed structures (synaptosomes) that can be separated by subcellular fractionation and that the phenomenon of bound central transmitters is explained by the sequestration, within the detached ending, of the soluble and vesicular components of the terminal cytoplasm. The hypothesis has so far survived all the tests to which it has been submitted since it was first put forward in 1960[1] (for historical review see Whittaker[55]), and in spite of some lingering scepticism[66] it may be regarded as having been generally accepted. Now that synaptosome preparations are so widely used there is, indeed, a danger that their limitations will not always be sufficiently realized, and for this reason an attempt has been made in this chapter to provide a critical assessment of the various methods available for preparing and characterizing synaptosomes and their constituent organelles.

In the course of the work on synaptosomes a number of techniques have been applied in novel ways. These include the use of negative staining for studying mammalian subcellular fractions,[50] the quantitative electron microscopic study of subcellular fractions in thin section[22] or after negative staining,[41] and the application of gel-filtration to the study of subcellular particles.[7] It is believed that the wider application of these methods in conjunction with subcellular fractionation will extend still further the usefulness of this technique.

VII. REFERENCES

1. V. P. Whittaker, The binding of neurochormones by subcellular particles of brain tissue (Fourth International Neurochemical Symposium, Varenna, June 12–17, 1960), in *Regional Neurochemistry: The Regional Chemistry, Physiology and Pharmacology of the Nervous System* (S. Kety and J. Elkes, eds.), pp. 259–263, Pergamon Press, Oxford (1963).
2. E. G. Gray and V. P. Whittaker, The isolation of synaptic vesicles from the central nervous system, *J. Physiol.* **153**:35–37P (1960).
3. E. G. Gray and V. P. Whittaker, The isolation of nerve endings from brain: An electron-microscopic study of cell fragments derived by homogenization and centrifugation, *J. Anat. (London)* **96**:79–88 (1962).
4. V. P. Whittaker, I. A. Michaelson, and R. J. A. Kirkland, The separation of synaptic vesicles from nerve-ending particles ("synaptosomes"), *Biochem. J.* **90**:293–305 (1964).
5. K. Krnjević and V. P. Whittaker, Excitation and depression of cortical neurones by brain fractions released from micropipettes, *J. Physiol.* **197**:288–322 (1965).
6. S. H. Barondes, On the site of synthesis of the mitochondrial protein of nerve endings, *J. Neurochem.* **13**:721–727 (1966).
7. R. M. Marchbanks, The osmotically sensitive potassium and sodium compartments of synaptosomes, *Biochem. J.* **104**:148–157 (1967).
8. R. M. Marchbanks and V. P. Whittaker, Some properties of the limiting membranes of synaptosomes and synaptic vesicles, *Abstr. 1st Intern. Meet. Intern. Soc. Neurochem.*, Strasbourg, p. 147 (1967).

9. M. Israël and V. P. Whittaker, The isolation of mossy fibre endings from the granular layer of the cerebellar cortex, *Experientia* **21**:325–326 (1965).

10. R. Laverty, I. A. Michaelson, D. F. Sharman, and V. P. Whittaker, The subcellular localization of dopamine and acetylcholine in the dog caudate nucleus, *Brit. J. Pharmacol.* **21**:482–490 (1963).

11. L. L. Ross, personal communication.

12. I. A. Michaelson, V. P. Whittaker, R. Laverty, and D. F. Sharman, Localization of acetylcholine, 5-hydroxytryptamine and noradrenaline within subcellular particles derived from guinea pig subcortical brain tissue, *Biochem. Pharmacol.* **12**:1450–1453 (1963).

13. F. S. LaBella and M. Sanwal, Isolation of nerve endings from the posterior pituitary lobe, *J. Cell Biol.* **25**:179–194 (1965).

14. H. Weinstein, E. Roberts, and T. Kakefuda, Studies of subcellular distribution of γ-aminobutyric acid and glutamic decarboxylase in mouse brain, *Biochem. Pharmacol.* **12**:503–509 (1963).

15. S. Tuček, Subcellular distribution of acetyl-CoA synthetase, ATP citrate lyase, citrate synthase, choline acetyltransferase, fumarate hydratase and lactate dehydrogenase in mammalian brain tissue. *J. Neurochem.* **14**:531–545 (1967).

16. H. P. Metzger, M. Cuénod, A. Grynbaum, and H. Waelsch, The use of tritium oxide as a biosynthetic precursor of macromolecules in brain and liver, *J. Neurochem.* **14**:99–104 (1967).

17. A. N. Siakotos, personal communication.

18. D. G. Jones, An electron microscope study of subcellular fractions from *Octopus* brain, *J. Cell. Sci.* **2**:573–586 (1967).

19. V. P. Whittaker, Catecholamine storage particles in the central nervous system (Second Catecholamine Symposium, Milan, July 4–9, 1965), *Pharmacol. Rev.* **18**:401–412 (1966).

20. M. N. Sheridan, V. P. Whittaker, and M. Israël, The subcellular fractionation of the electric organ of *Torpedo*, *Z. Zellforsch.* **74**:291–307 (1966).

21. V. P. Whittaker and G. H. C. Dowe, The effect of homogenization conditions on subcellular distribution in brain, *Biochem. Pharmacol.* **14**:194–196 (1965).

22. F. Clementi, V. P. Whittaker, and M. N. Sheridan, The yield of synaptosomes from the cerebral cortex of guinea pigs estimated by a polystyrene bead "tagging" procedure, *Z. Zellforsch.* **72**:126–138 (1966).

23. I. A. Michaelson and G. H. C. Dowe, The subcellular distribution of histamine in brain tissue, *Biochem. Pharmacol.* **12**:949–956 (1963).

24. M. Nyman and V. P. Whittaker, The distribution of adenosine triphosphate in subcellular fractions of brain tissue, *Biochem. J.* **87**:248–255 (1963).

25. I. A. Michaelson and V. P. Whittaker, The subcellular localization of 5-hydroxytryptamine in guinea-pig brain, *Biochem. Pharmacol.* **12**:203–211 (1963).

26. J. Eichberg, V. P. Whittaker, and R. M. C. Dawson, The distribution of lipids in subcellular particles of guinea-pig brain, *Biochem. J.* **92**:91–100 (1964).

27. J. Mellanby, W. E. van Heyningen, and V. P. Whittaker, Fixation of tetanus toxin by subcellular fractions of brain, *J. Neurochem.* **12**:71–79 (1965).

28. J. L. Mangan and V. P. Whittaker, The subcellular distribution of amino acids in guinea-pig brain, *Biochem. J.* **98**:128–137 (1966).

29. L. T. Potter and J. Axelrod, Intracellular localization of catecholamines in tissues of the rat, *J. Pharmacol.* **142**:291–298 (1963).

29a. A. A. Abdel-Latif, A simple method for isolation of nerve ending particles from rat brain, *Biochim. Biophys. Acta* **121**:403–406 (1966).

30. H. Beaufay and J. Berthet, Medium composition and equilibrium density of subcellular particles from rat liver, *Biochem. Soc. Symp.* **23**:66–85 (1963).

31. M. Kurokawa, T. Sakamoto, and M. Kato, Distribution of sodium-plus-potassium-stimulated adenosine-triphosphatase activity in isolated nerve-ending particles, *Biochem. J.* **97**:833–844 (1965).

32. C. O. Hebb and V. P. Whittaker, Intracellular distributions of acetylcholine and choline acetylase, *J. Physiol.* **142**:187–196 (1958).

33. E. De Robertis, A. P. de Iraldi, G. R. deL. Arnaiz, and L. Salganicoff, Cholinergic and non-cholinergic nerve endings in rat brain. 1. Isolation and subcellular distribution of acetylcholine and acetylcholine esterase, *J. Neurochem.* **9**:23–35 (1962).

34. G. M. J. van Kempen, C. J. van den Berg, H. J. van der Helm, and H. Veldstra, Intracellular localization of glutamate decarboxylase γ-aminobutyrate transaminase and some other enzymes in brain tissue, *J. Neurochem.* **12**:581–588 (1965).

35. F. Fonnum, The distribution of glutamate decarboxylase and aspartate transaminase in subcellular fractions of rat and guinea-pig brain, *Biochem. J.* **106**:401–412 (1968).

36. R. Balázs, D. Dahl, and J. R. Harwood, Subcellular distribution of enzymes of glutamate metabolism in rat brain, *J. Neurochem.* **13**:897–905 (1966).

37. V. P. Whittaker, The morphology of rat forebrain synaptosomes separated on continuous sucrose density gradients, *Biochem. J.* **106**:412–417 (1968).

38. F. Fonnum, The compartmentation of choline acetyltransferase within the synaptosome, *Biochem. J.* **103**:262–270 (1967).

39. V. P. Whittaker, The isolation and characterization of acetylcholine-containing particles from brain, *Biochem. J.* **72**:694–706 (1959).

40. M. K. Johnson and V. P. Whittaker, Lactate dehydrogenase as a cytoplasmic marker in brain, *Biochem. J.* **88**:404–409 (1963).

41. V. P. Whittaker and M. N. Sheridan, The morphology and acetylcholine content of cerebral cortical synaptic vesicles, *J. Neurochem.* **12**:363–372 (1965).

42. R. M. Marchbanks, Compartmentation of acetylcholine in synaptosomes, *Biochem. Pharmacol.* **16**:921–923 (1967).

43. E. De Robertis, G. R. deL. Arnaiz, L. Salganicoff, A. P. de Iraldi, and L. M. Zieher, Isolation of synaptic vesicles and structural organization of the acetylcholine system within brain nerve endings, *J. Neurochem.* **10**:225–235 (1963).

44. M. Germain and P. Proulx, Adenosinetriphosphatase activity in synaptic vesicles of rat brain, *Biochem. Pharmacol.* **14**:1815–1819 (1965).

45. R. M. Burton, R. E. Howard, and J. M. Gibbons, Ganglioside and acetylcholine-containing synaptic vesicles of rat brain, *Abstr. 6th Intern. Congr. Biochem.* V.E.-97 (1964).

46. R. J. A. Hosie, The localization of adenosine triphosphatases in morphologically characterised fractions of guinea-pig brain, *Biochem. J.* **96**:404–412 (1965).

47. V. P. Whittaker, Some properties of synaptic membranes isolated from the central nervous system (New York Academy of Science Conference on "Biological membranes: recent progress" New York, Oct. 4–7, 1965), *N.Y. Acad. Sci.* **137**:982–998 (1966).

48. H. Wiegandt, The subcellular localization of gangliosides in the brain, *J. Neurochem.* **14**:671–674 (1967).

49. V. T. Marchesi, unpublished; for electron micrograph see V. P. Whittaker, The synapse, *Discovery* **23**:7–13 (1962).

50. R. W. Horne and V. P. Whittaker, The use of the negative staining method for the electron-microscopic study of subcellular particles from animal tissues, *Z. Zellforsch.* **58**:1–16 (1962).

51. V. P. Whittaker, The binding of acetylcholine by brain particles in vitro, in *Mechanisms of Release of Biogenic Amines* (U. S. von Euler, S. Rosell, and B. Uvnäs, eds.), Wenner-Gren International Symposium Series, Vol. 5, pp. 147–164, Pergamon Press, Oxford (1966).

52. V. P. Whittaker, The subcellular distribution of amino acids in brain and its relation to a possible transmitter function for these compounds, *in Structure and Function of Inhibitory Neuronal Mechanisms* (C. von Euler, S. Skoglund, and U. Söderberg, eds.), pp. 487–504, Pergamon Press, Oxford (1968).

53. V. P. Whittaker, unpublished observations.

54. L. Austin and I. G. Morgan, Incorporation of ^{14}C-labelled leucine into synaptosomes from rat cerebral cortex *in vitro*, *J. Neurochem.* **14**:377–387 (1967).

55. V. P. Whittaker, The application of subcellular fractionation techniques to the study of brain function, *Progr. Biophys. Molec. Biol.* **15**:39–91 (1965).

55a. I. A. Michaelson, The subcellular distribution of acetylcholine, choline acetyltransferase and acetylcholinesterase in nerve tissue, *Ann. N.Y. Acad. Sci.* **144**:387–407 (1967).

56. P. R. Lewis and C. C. D. Shute, The distribution of cholinesterase in cholinergic neurones demonstrated with the electron microscope, *J. Cell Sci.* **1**:381–390 (1966).

57. M. Brzin, V. M. Tennyson, and P. E. Duffy, Acetylcholinesterase in frog sympathetic and dorsal root ganglia, *J. Cell Biol.* **31**:215–242 (1966).

58. D. G. Jones, reported by V. P. Whittaker, The subcellular fractionation of nervous tissue, *in Structure and Function of Nervous Tissue* (G. H. Bourne, ed.), Vol. 3, pp. 1–24, Academic Press, New York (1969).

59. F. Fonnum, Choline acetyltransferase binding to and release from membranes, *Biochem. J.* **109**:389–398 (1968).

60. L. Chakrin and V. P. Whittaker, The subcellular distribution of [*N-Me-*^{3}H]acetylcholine synthesized by brain *in vivo*, *Biochem. J.* **112** (in press).

60a. I. G. Morgan and L. Austin, Synaptosomal protein synthesis in a cell-free system, *J. Neurochem.* **15**:41–51 (1968).

61. L. M. Seminario, N. Hren, and C. J. Gomez, Lipid distribution in subcellular fractions of the rat brain, *J. Neurochem.* **11**:197–207 (1964).

62. J. Mellanby and V. P. Whittaker, The binding of tetanus toxin by isolated synaptic membranes of the guinea pig, *J. Neurochem.* **15**:205–208 (1968).

63. H. McIlwain, *Biochemistry and the Central Nervous System*, p. 72, Churchill, London (1966).

64. R. D. Keynes and J. M. Ritchie, The movements of labelled ions in mammalian non-myelinated nerve fibres, *J. Physiol.* **179**:333–367 (1965).

65. R. D. Keynes and P. R. Lewis, The sodium and potassium content of cephalopod nerve fibres, *J. Physiol.* **114**:151–182 (1951).

66. R. Vrba, Assimilation of glucose carbon in subcellular rat brain particles *in vivo* and the problem of axoplasmatic flow, *Biochem. J.* **105**:927–936 (1967).

Chapter 15

STRUCTURAL COMPONENTS OF THE SYNAPTIC REGION*

Eduardo De Robertis and Georgina Rodríguez de Lores Arnaiz

Instituto de Anatomía General y Embriología
Facultad de Medicina-Universidad de Buenos Aires
Buenos Aires, Argentina

I. INTRODUCTION

The synaptic region may be defined as the site of contact between two excitable cells having a specific structural, biochemical, and functional differentiation for the transmission of nerve impulses. As suggested by Du Bois Raymond in 1877, synaptic transmission may be either chemical or electrical, and both mechanisms have been found to occur in the peripheral and central nervous systems. However, chemical synapses are by far the more common and are the only ones that will be considered here from the point of view of their structural and biochemical organization.

Hypotheses concerning chemical transmission are based on the assumption that a specific transmitter is synthesized and stored in nerve endings and that it is liberated when the nerve impulse arrives at the terminal. It is also postulated that the transmitter reacts with a chemical receptor situated in the postsynaptic element and that, from this reaction, a change of ionic permeability takes place, with this change inducing a bio-electrical phenomenon. Such a mechanism is found both in excitatory and inhibitory synapses, and the end result depends on the chemical nature of the transmitter, the molecular structure of the receptor, and the ionic species that migrate through the postsynaptic membrane (see Eccles[1]).

A morphological correlate for the complex functions of neurons could be revealed only with the introduction of the high resolution provided by the electron microscope. The first knowledge of the subcellular structures that underlie nerve transmission came in 1954 when De Robertis and Bennett[2]

* The original research contained in this chapter was supported by grants from the Consejo Nacional de Investigaciones Científicas y Técnicas, Argentina, and from the National Institutes of Health, No. NB-06953-02.

found that nerve endings, of chemical synapses, contained a specific component composed of vesicles, about 500 Å wide and remarkably uniform in size. These were designated the synaptic vesicles and interpreted as the sites of storage of the transmitters.[2] Later these morphological entities were considered to represent the quantal release in synaptic transmission.[3]

Further electron microscope studies demonstrated other complexities in the organization of the synaptic region. Particularly relevant were the thickenings of the synaptic membranes and the presence of a synaptic cleft of about 300 Å, crossed by the intersynaptic filaments, and of a system of filaments, the subsynaptic web, projecting into the postsynaptic region (Fig. 1A). Such components constitute what may be called the junctional complex of the synapse. Reviews have been published on these and other morphological aspects of the synapse.[4,5]

This chapter is concerned mainly with our present knowledge of the biochemistry of synaptic vesicles and the membranes of nerve endings. The study of the isolated nerve ending, as a unit of structure and function, is considered elsewhere in this book.

II. ISOLATION OF THE STRUCTURAL COMPONENTS OF THE SYNAPTIC REGION

With the introduction of the cell fractionation methods in the study of brain[6] a more direct approach to the fine localization of transmitters and other active substances, of the enzymes involved in their synthesis or inactivation, and of the chemical receptors was made available. Independently

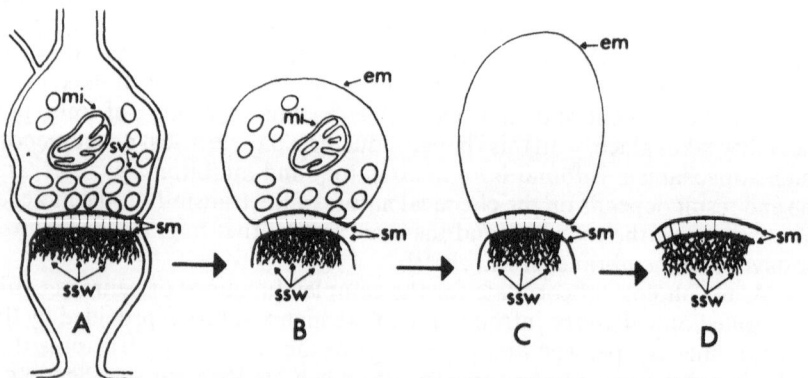

Fig. 1. Diagram of the systematic dissection of a synaptic ending from the cerebral cortex. Synaptic ending as observed (A) *in situ*; (B) after isolation; and (C) after the osmotic shock. (D) The junctional complex left after treatment with a detergent. Ending membrane, em; mitochondria, mi; synaptic vesicles, sv; synaptic membranes, joined by intersynaptic filaments, sm; subsynaptic web, ssw. From De Robertis.[5]

Gray and Whittaker[7] and De Robertis et al.[8] were able to demonstrate that the mitochondrial fraction of brain contained isolated nerve endings (Fig. 1B). These investigations provided the foundation for the neurochemical and neuropharmacological studies that since have been actively carried out both in these laboratories and in others. Whittaker[6] used a sucrose gradient with two steps to separate myelin, nerve endings, and mitochondria. De Robertis et al.[9] used a more complex gradient that permitted the separation of three layers of nerve endings from brain, in addition to myelin and mitochondria. Several studies on the distribution of the biogenic amines and related enzymes demonstrated that two of the nerve ending fractions were essentially aminergic, i.e., rich in biogenic amines, while the other large subfraction was nonaminergic. Recently a fraction of very small nerve endings was isolated from the postmitochondrial fraction of the cerebral cortex.[10] The nonaminergic nerve endings are rich in glutamic acid decarboxylase, the enzyme that synthesizes GABA, and their possible inhibitory nature was postulated.[11]

In 1962 it was demonstrated[12,13] that, if the mitochondrial fraction is homogenized in a hypotonic solution, the nerve endings become swollen and there is rupture of the limiting membrane with release of the enclosed components. After differential centrifugation the synaptic vesicles are preferentially sedimented in subfraction M_2 while the axoplasm is contained in M_3 (Fig. 2). The much bulkier fraction M_1, sedimented at 20,000 g × 30 min, contains myelin, mitochondria, and the nerve ending membranes with the junctional complex attached (Fig. 1C). The different components of this fraction may be isolated on a sucrose gradient[14,15] (Fig. 2). The various subfractions obtained correspond to M_1 0.8, myelin; M_1 0.9, M_1 1.0, and M_1 1.2, nerve ending membranes*; and M_1p, mainly mitochondria. More recently the $M_2 + M_3$ fractions were submitted to a simple gradient centrifugation for obtaining highly purified synaptic vesicles[16] (Fig. 3).

Figure 2 compares the methods used in our laboratory with those described by Whittaker et al.,[17] in which they modified our technique of osmotic shock and isolated the various structural components on a sucrose gradient. The two techniques differ, not only with respect to the treatment after the osmotic shock but also in the proportion of tissue to hypotonic solution and in the previous steps of the purification of the nuclear and mitochondrial fractions. All these factors are of importance in explaining some differences in the distribution and recoveries obtained in the two laboratories. (For a critical discussion on these techniques see McCaman, Rodríguez de Lores Arnaiz, and De Robertis.[18])

---

* The denomination nerve ending membrane replaces those of "synaptic ghosts" or "synaptic membranes" used in previous papers from this laboratory.[12,13] Regarding the problem of terminology, the denomination isolated nerve ending or nerve ending particles used here has been replaced by Whittaker et al.[17] by that of *synaptosomes*. In relation to the use of this word is the following warning: Etymologically synaptosome means a *synaptic body* and may imply a special structure present within the synapse. New names are justified only when they refer to structures or components not described before. When applied to components already known (i.e., nerve ending or synaptic ending) they may introduce confusion.

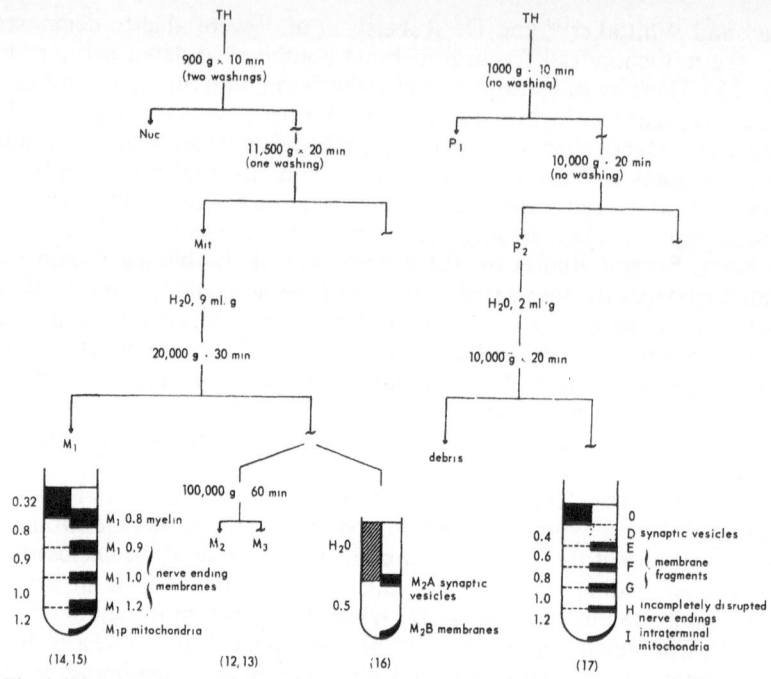

Fig. 2. Diagram of the main techniques used for the separation of synaptic vesicles and nerve ending membranes. Special emphasis is made on the differences between the techniques used by Whittaker, Michaelson, and Kirkland[17] and De Robertis et al.[12-16] These are not only in the final sucrose gradients used but also in the handling of the primary fractions and of the osmotic shock. TH, total homogenate; Nuc and P_1, nuclear fractions; mit and P_2, crude mitochondrial fractions.

A further step in the dissection of the isolated nerve ending has been achieved after a systematic study of the action of the nonionic detergent Triton X-100 on various subcellular fractions of cerebral cortex.[19] The action on different enzymes and on the ultrastructure of the fractions revealed that in low concentrations the detergent had a preferential effect on the limiting membrane of the nerve ending (Fig. 4). The junctional complexes, composed of both synaptic membranes and associated structures, were more resistant to the treatment. If a pure fraction of nerve ending membranes, such as M_1 1.0, was treated with 0.1 % Triton X-100, the sediment contained a mass of isolated junctional complexes (Figs. 1D and 5).

III. SYNAPTIC VESICLES

A. Functional Significance

In 1952, Fatt and Katz demonstrated that in the myoneural junction the transmitter is released in multimolecular packets or quanta, which are

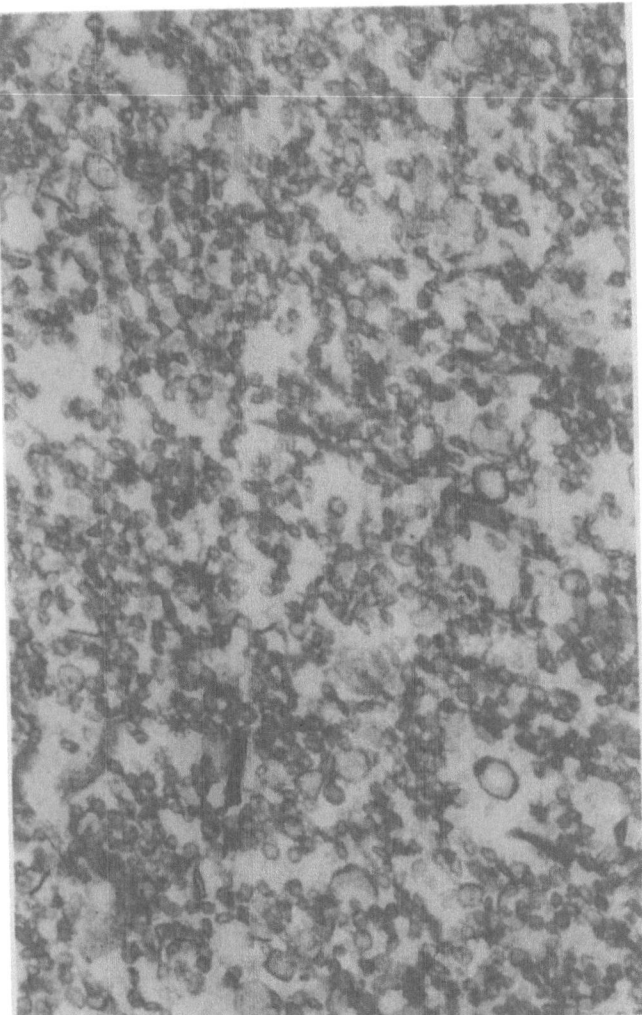

Fig. 3. Electromicrograph of purified synaptic vesicles isolated from the rat cerebral cortex. Notice that while most vesicles are round a few are of elliptical shape. ×80,000.

remarkably uniform in size.[1] A quantum corresponds to a number of ACh molecules simultaneously released at a definite locus of the synapse. Del Castillo and Katz[3] correlated the synaptic vesicles with the quantal release.

After the discovery of the synaptic vesicles[2] different experiments were made to demonstrate their relationship to synaptic transmission. In Wallerian degeneration, the lysis of vesicles was correlated with the alteration in nerve transmission. Depletion of synaptic vesicles was observed in peripheral

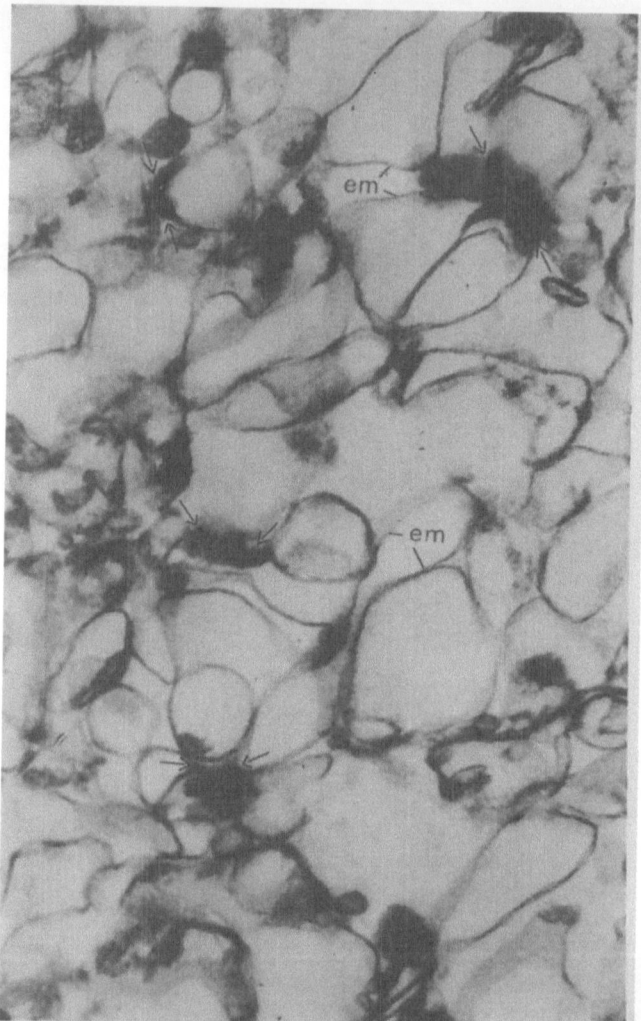

Fig. 4. Electron micrograph of isolated nerve ending membranes, i.e., fraction M_1 1.0, from the rat cerebral cortex. (See description in the text.) em, limiting membrane of the nerve ending; with arrows are indicated some junctional complexes (compare with Fig. 1C). × 75,000.

synapses stimulated at very high frequencies, while an increase in their number was obtained at lower frequencies. This finding suggested that the number of vesicles could result in a balance between the production and release of vesicles.[4] The rate of transmission suggests that the vesicles are used more than once in successive firings.

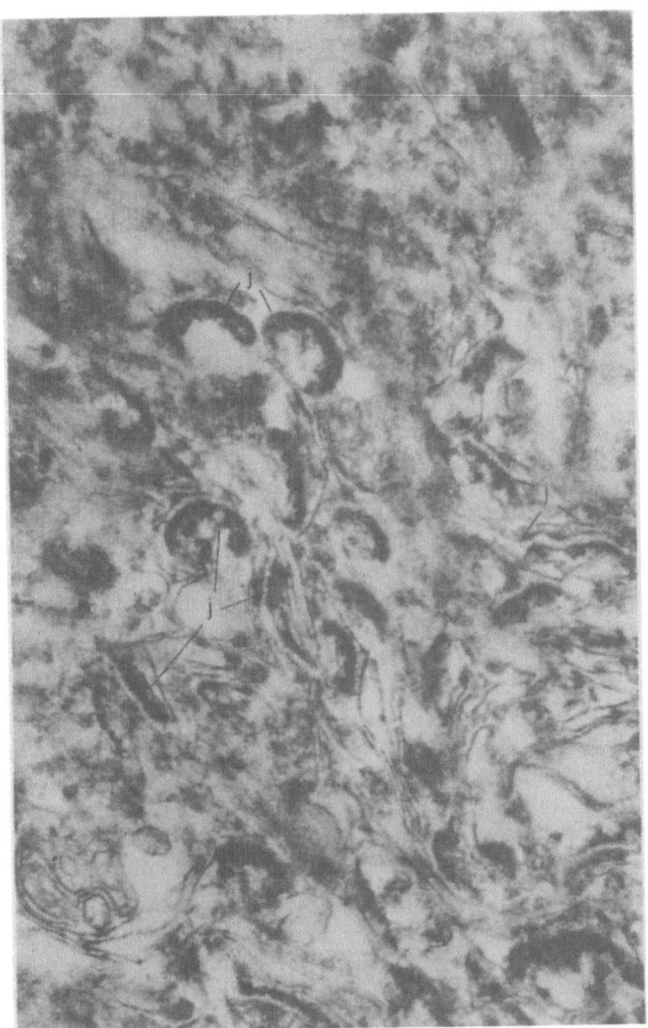

Fig. 5. Electron micrograph of junctional complexes (j) isolated from the rat cerebral cortex after treatment of fraction M_1 1.0 with Triton X-100. (Compare with Fig. 4.) × 70,000.

It is not possible to review here all the contributions related to the ultrastructure and function of synaptic vesicles. Special granulated vesicles were found in adrenergic peripheral endings and, by pharmacological experiments, it was shown that they contain the adrenergic transmitter.[20] More recently this has been corroborated, in our laboratory, by histochemical techniques at the electron microscope level. Granulated vesicles of the pineal gland contain norepinephrine and 5-hydroxytriptamine, while those of the vas deferens have only norepinephrine. In the anterior hypothalamus, large granulated vesicles were observed, and in addition complex and elliptical shapes were found.[20] A basic similarity between the production of synaptic vesicles and the neurosecretory ones (granules) has been postulated.[4] Uchizono[21] has suggested a relationship between elliptical vesicles and inhibitory synapses. The convulsant drug methionine sulphoximine produced a striking decrease in the number of synaptic vesicles, particularly in the fraction of nonaminergic nerve endings.[22]

B. Chemical Composition

1. Proteins

Using acrylamide gel electrophoresis Cotman and Mahler[23] have resolved at least 16 bands from the "insoluble" proteins of the synaptic vesicle fraction. Each band may represent a different protein or protein subunit. This complexity in protein composition may reflect a functional adaptation to other functions than those involved only in the storage of transmitters. At least one of the bands seems to be a proteolipid; another, choline acetylase. With incubation of chopped brain tissue with labeled leucine, it was found that nerve endings synthesize protein without a lag phase. It was postulated that nerve endings do not depend entirely on a somatoaxonal flow for their protein requirements.[24] This problem is related to that of the origin of the synaptic vesicles. De Robertis[4] postulated that they originate by fragmentation of the neurotubules present in the axon, and recently such a transformation has been clearly observed in the regeneration stump of compressed nerve fiber.[5]

Using ^{3}H-leucine intraventricularly the half-life of proteins in synaptic vesicles of rat cerebral cortex has been estimated to be 21 days. A similar value has been found in other subcellular fractions. This long life supports the view that during firing vesicles are used again instead of being destroyed.[25] This is agreement with the results of electrical stimulation mentioned above.[4]

2. Lipids

It has been reported that synaptic vesicles are similar in phospholipid composition to the nerve ending membranes but contain less cholesterol and cerebroside.[26] In the purified synaptic vesicles of fraction M_2A (Fig. 2), a high lipid/protein ratio, 0.82, was found. Furthermore, this fraction is rich in total phospholipids while low in galactolipids. These findings indicate that the lipids, and particularly the phospholipids, are structural components of the vesicles.[27] Another remarkable characteristic of synaptic vesicles is the low content of gangliosides.[16,26,28]

C. Biogenic Amines

1. Acetylcholine

In the CNS, ACh was found in a fraction different from mitochondria,[6] later on demonstrated to be composed of nerve endings.[7,8] ACh and acetylcholinesterase (AChE) were encountered associated with a particular population of nerve endings that was denominated "cholinergic," in distinction to the ACh- and AChE-poor population of nerve endings, which was called "noncholinergic".[9]

The introduction of the osmotic disruption of nerve endings[12] permitted the demonstration that synaptic vesicles had the highest concentration of ACh (Table I). This confirmed the early postulate that they were the site of storage of transmitters. Whittaker, Michaelson, and Kirkland[17] confirmed the localization of ACh in the synaptic vesicles. However, probably because of the above-mentioned differences in technique, their recoveries were low.

The number of ACh molecules contained within a single synaptic vesicle could correspond to a quantum of this transmitter. Based on several assumptions, Birks has calculated this amount to be about 900 ACh per vesicle with a concentration in the vesicular fluid of 0.11 M.[4] Using polystyrene latex beads to determine the number of vesicles isolated from the guinea pig cerebral cortex, the average number of ACh molecules per vesicle was calculated to be about 300. Making allowance for the presence of noncholinergic vesicles, this number was increased to 2000, and the existence of a liquid core filled with a slightly hypertonic ACh solution was suggested.[29]

Early work on particle-bound ACh had shown that about 50% is in a labile form, readily released by relatively mild treatments such as osmotic dilution, freezing and thawing, or dialysis.[6] The rest is more stable and for its release requires damage of the vesicular membrane. The release of ACh

TABLE I

Content of Biogenic Amines in Synaptic Vesicles[a]

Biogenic amines	Fraction		
	M_1	M_2	M_3
Acetylcholine[13]	0.55	2.85	1.20
Noradrenaline[35]	0.40	2.56	1.93
Dopamine[35]	0.46	2.46	1.72
5-Hydroxytryptamine[37]	0.47	1.84	2.31
Histamine[10]	0.39	2.24	2.27

[a] The crude mitochondrial fraction of the brain was shocked osmotically and then centrifuged as indicated in Fig. 2. Fraction M_2 contains the synaptic vesicles. The results are expressed in relative specific concentrations (RSC), which is the percentage of amine recovered divided by the percentage of protein recovered.

from isolated synaptic vesicles is temperature dependent and has a rate proportional to the initial amount bound.[30]

Studies have been made on the accumulation of ACh *in vitro*.[31] It was found that slices of rat striatum in 10 min may increase the intrinsic concentration of ACh six times. Of this, about 20% is osmotically stable or bound, the rest being either free or labile.

These studies are in some way related to the actual mechanisms involved in the physiological release of the transmitter at the terminal.[1,4] It is well known that Ca^{2+} is needed for the release of ACh by nerve impulses. Birks and MacIntosh[32] found that CO_2 and a plasma factor also may be involved, both of them acting by way of Ca^{2+}. It was postulated that the releasing effect of Ca^{2+} would be first exerted on synaptic vesicles attached to the presynaptic membrane, as originally described by De Robertis and Bennett.[2]

2. Catecholamines and 5-Hydroxytryptamine

The subcellular distribution of catecholamines in the primary fractions and in nerve endings was previously reviewed.[20] The first report that norepinephrine was present in synaptic vesicles from brain was presented at the Galesburg meeting on biogenic amines on January 1963.[33] The finding of axons and nerve endings containing granulated vesicles in the anterior hypothalamus led to the isolation of such structures and then to the separation of the vesicular fraction.[34] This appeared, under the electron microscope, to be rather heterogeneous and to contain between 10 and 20% of granulated vesicles. In such a fraction the content of norepinephrine is about nine times higher than in a similar one from total brain (Table II).

Dopamine, in spite of being more soluble, is also concentrated in the vesicular fraction[35] (Table I). These results indicate that the synaptic vesicles are the main stores for catecholamines in brain. While in our laboratory it was not possible to demonstrate the localization of 5HT in

TABLE II

Norepinephrine Content of Synaptic Vesicles in Total Brain and Hypothalamus[a]

Fraction	Total brain		Hypothalamus	
	ng/g	RSC	ng/g	RSC
M_1	32.4	0.40	730	0.67
M_2	34.2	2.56	300	1.94
M_3	57.5	1.93	500	1.58

[a] Results are expressed in nanograms per gram wet tissue and relative specific concentration (RSC), i.e., percentage recovered amine divided by percentage recovered protein. For description of fractions, see Fig. 2. Data from De Robertis et al.[34]

synaptic vesicles,[36] Maynert and Kurijama[38] have reported on the accumulation and release of catechol- and indolamines by the vesicular fraction in different conditions such as differences in temperature, amine concentration, and action of reserpine. A transport system was suggested to be operating both at the nerve-ending particle and the synaptic vesicle. The amines may be concentrated against a concentration gradient, and the formation of a nondiffusible complex within the vesicle was postulated.[38] That the storage or binding of catechol- and indolamines to the subcellular structures may differ is suggested by experiments showing a different effect of x-ray irradiation and of certain drugs on these two pools of monoamines.[39]

3. Histamine

The subcellular distribution of histamine in brain has been studied using fluorescence and bioassay methods, and its localization in nerve endings was demonstrated.[10] Kataoka and De Robertis[10] confirmed the above findings and demonstrated that in the rat cerebral cortex histamine is concentrated in very small nerve endings, which may be found in the mitochondrial fraction and especially in the postmitochondrial fraction. A heavy microsomal pellet, mic-20 (separated at $20,000\,g$ for 30 min), showed the highest content of this amine. After the osmotic shock, of both the mitochondrial fraction and of mic-20, a significant concentration in the synaptic vesicles was found (Table I). This finding suggests that histamine may play a synaptic role in central synapses. An active incorporation of histamine in the mitochondrial fraction at 37°C was observed, which may be suppressed at 0°C.

D. Enzymes

So far only a few enzymes have been associated with the synaptic vesicles. Several enzymes related to the synthesis of biogenic amines have been found to be present in the axoplasm of the nerve ending. De Lores Arnaiz and De Robertis[40] demonstrated that 5-hydroxytryptophan-decarboxylase, an enzyme related to the decarboxylation of dihydroxyphenylalanine and 5-hydroxytryptophan, was localized in nerve endings, but after the osmotic shock it became entirely soluble. Glutamic acid decarboxylase, present in the nonaminergic fraction of nerve endings, also became soluble after their osmotic disruption. Salganicoff and De Robertis[11] made the interesting observation that with a small concentration of Ca^{2+} in the medium the synaptic vesicles bound considerable amounts of the soluble enzyme. This suggested that *in vivo* the enzyme could be attached to the outer surface of the vesicular wall. It was suggested that the synaptic vesicles of the glutamic acid decarboxylase-rich nerve endings could store GABA and that this amino acid could function as an inhibitory transmitter.[11] Glutamic acid decarboxylase, GABA, and glutamic acid were reported to be bound to synaptic vesicles in mouse brain.[41]

De Robertis and Rodríguez de Lores Arnaiz

TABLE III

Cholineacetyltransferase in Synaptic Vesicles in Brain of Four Species[a]

Fraction	Rat			Rabbit			Guinea pig			Pigeon		
	P	ChAc		P	ChAc		P	ChAc		P	ChAc	
	Percent	Percent	RSA	Percent	Percent	RSA	Percent	Percent	RSA	Percent	Percent	RSA
M_1	67	49	0.73	68	52	0.77	68	53	0.78	64	19	0.30
M_2	15	45	3.00	15	34	2.26	14	22	1.57	16	15	0.94
M_3	19	7	0.37	16	15	0.94	18	26	1.45	20	66	3.30
Percent recorded	97	93		101	92		98	84		97	92	

[a] Results are expressed as RSA (see Table I). P denotes protein. For structure of fraction see Fig. 2. Data from McCaman, Rodríguez de Lores Arnaiz, and De Robertis.[18]

1. Choline Acetyltransferase (ChAc)

Of considerable importance is the localization of ChAc, the enzyme that synthesizes ACh. In the submitochondrial fractions of rat brain it was found that ChAc had a very similar distribution to ACh and AChE.[9] After the osmotic shock ChAc was found to be highly concentrated in the vesicular fraction M_2. Here a relative specific activity of 5.6 (as compared to 3.6 for ACh) was found, while in M_1 it was only 0.7 and in M_3 0.9.[13] In the guinea pig brain, Whittaker, Michaelson, and Kirkland[17] found that after the osmotic shock ChAc is mainly soluble, not associated with the synaptic vesicles. These conflicting results were clarified by McCaman, de Lores Arnaiz, and De Robertis,[18] who demonstrated that they were due to differences in the species and also in the techniques used in both laboratories. It was confirmed that ChAc is bound to the vesicular fraction in the rat and to a lesser extent in the rabbit and guinea pig, while in the pigeon it was mainly soluble (Table III). The low yields of ChAc in synaptic vesicles, found by Whittaker, Michaelson, and Kirkland,[17] may be explained on the basis of incomplete osmotic shock and of severe cross contamination due to the lack of washings of the nuclear and mitochondrial fractions (Fig. 2). It is interesting that no species differences for the enzymes AChE, 5-hydroxy-tryptophandecarboxylase, and MAO were found.[18] The association of ChAc with synaptic vesicles in rat brain was confirmed in other laboratories.[23,42] Species differences also were found between the sheep, dog, and cat, in which ChAc remained mainly bound, and the guinea pig and pigeon, in which the enzyme was more soluble.[43] To interpret these species differences, it may be postulated that ChAc is present in the axoplasm and bound in varying degree to the outer surface of the synaptic vesicle.

Species differences also have been found in the case of the enzymes related to the synthesis of the precursor of the ACh, acetyl-CoA. Of these, the particulate portion of ATP-citrate lyase was found preferentially associated to the vesicular fraction of the sheep brain.[44]

2. Cation-Stimulated Phosphohydrolases

Several cations, i.e., Na^+, K^+, Ca^{2+}, and Mg^{2+}, may influence the levels of free and bound ACh, and it is conceivable that they exert their action through activation of phosphohydrolases. Albers, de Lores Arnaiz, and De Robertis,[45] reported that the Na^+-K^+-activated and Mg^{2+}-dependent ATPases, as well as the K^+-activated and Mg^{2+}-dependent phosphatases, were found in negligible amounts in the synaptic vesicles, while they were concentrated in the nerve ending membranes. Similar results were obtained by others.[46–48] However, a special type of phosphohydrolase, which is dependent on Mg^{2+} and Ca^{2+}, was found in the fraction of synaptic vesicles.[47,48]

Since ATP and Mg^{2+} stimulate the binding of $^{45}Ca^{2+}$ to a preparation of nerve endings,[49] and in view of the importance of Ca^{2+} in transmitter release, it is expected that the presence of a Mg^{2+} and Ca^{2+} ATPase in synaptic vesicles may be related to the storage and release of ACh. However, more information is needed to ascertain the physiological significance of phosphohydrolases in synaptic vesicles.

IV. NERVE ENDING AND SYNAPTIC MEMBRANES

The study of the nerve ending membrane with the attached junctional complex is of particular interest in our understanding of the primary functions of neurons. Being a continuation of the axolemma, the limiting membrane of the ending regulates the passage of ions, metabolites, and other small molecules, maintaining a special milieu for the metabolic activities of this compartment. Furthermore, across this membrane ionic fluxes take place that originate the resting and action potentials. Since the junctional complex includes the subsynaptic membrane belonging to the second neuron, its isolation presumably may lead to a better knowledge of the chemical receptor involved in nerve transmission.

It has been pointed out that osmotic shock of the nerve endings, used to separate the synaptic vesicles,[12,13] also may be employed to isolate the nerve ending membranes.[14,15] In the method used in this laboratory the nerve ending membranes are separated in fractions M_1 0.9, M_1 1.0, and M_1 1.2, while in that of Whittaker, Michaelson, and Kirkland[17] membrane fragments are found in fractions E, F, and G (Fig. 2).

Under the electron microscope the nerve ending membranes are characterized by their size, which corresponds to that of the nerve terminals, and the observation, in many profiles, of the attached junctional complexes (Figs. 1C and 4). The various membrane populations, in the gradient of M_1, are separated by their specific gravities, which depend on the lipid/protein ratio.[27] These membranes are also characterized by other differences in chemical composition, particularly in the content of gangliosides, by a group of membrane-bound enzymes and by their different binding capacity for radioactive biogenic amines and blocking agents.

A. Chemical Composition

1. Proteins

As mentioned above, Cotman and Mahler[23] were able to dissolve the structural proteins of brain membranes and to separate them by acrylamide gel electrophoresis. As in the case of the synaptic vesicles, 16 bands were resolved in nerve ending membranes and also in mitochondria. However, it was difficult for the authors to evaluate the purity of the nerve ending membrane fraction. Austin and Morgan[24] have observed that nerve ending membranes are relatively rich in RNA, this component being higher than in other constituents of nerve endings. This fact may be of importance in any further study of the biosynthesis of structural proteins in membranes.

2. Lipid Molar Ratios

In the nerve ending membranes separated in fraction G (Fig. 2), the cerebroside content is very low and the ganglioside content is relatively high.[26] In an extensive study of the lipid composition of membranes isolated from the rat cerebral cortex, Lapetina, Soto, and De Robertis[27] have found that each one has a specific pattern that is especially well reflected in the molar ratios of the different lipid species and the amino acids of proteins and proteolipids (Table IV). It is interesting to observe that, while myelin (i.e., M_1 0.8) has the typical phospholipid/cholesterol/galactolipid ratio of 2.4:2.4:1, the three layers of nerve ending membranes have a ratio of approximately 5.4:4.2:1. Table IV also shows the high content of phospholipid and cholesterol in the purified vesicular fraction M_2A.

The ^{32}P incorporation into phospholipids of the different synaptic structures has been studied.[50] After subarachnoidal administration of ^{32}Pi the rate of incorporation in total phospholipids is practically the same in all the membranous subcellular fractions. Of the various phospholipids that are present in synaptic vesicles and nerve ending membranes, only phosphatidyl-serine + phosphoinositides may have a higher turnover.

3. Gangliosides

Gangliosides are found concentrated in the CNS, particularly in gray matter. They have been associated with the maintenance of electrical activity and the transport of cations and with receptors to tetanus toxin and certain biogenic amines. While most of the gangliosides were previously found in the microsomal fraction, presumably associated with fragments of dendrites, the use of milder homogenization procedures[9] led to the demonstration that gangliosides are concentrated in the aminergic group of nerve endings.[51]

When the nerve ending membranes are isolated from the M_1 fraction (Fig. 2), gangliosides appear to be highly concentrated in M_1 0.9 and M_1 1.0, i.e., in the AChE-rich membranes, while they are much less concentrated in M_1 1.2.[16] The high concentration of gangliosides in nerve ending membranes also has been reported in other laboratories.[26,28]

B. Enzymes

1. Acetylcholinesterase

The study of the subcellular distribution of AChE is of considerable interest in light of the existing evidence concerning the cytochemistry of this enzyme. Koelle[52] has shown that AChE is present along the entire length of cholinergic neurons, while in the noncholinergic neurons the histochemical reaction is slight or lacking. These and other studies have suggested the existence of a functional portion of the enzyme, externally oriented with respect to the neuronal membrane, which hydrolyzes the ACh liberated during synaptic transmission, and of an internal AChE, which is related to the inner membranes of the endoplasmic reticulum and should sediment

TABLE IV

Molar Ratios of Lipids and Protein Amino Acids in Membranes of the Cerebral Cortex

Fraction[a]	Total phospholipids	Cholesterol	Galactolipids	Plasmalogens	Protein amino acids	Proteolipid protein amino acids
M_2 A	10.8	6.7	1	2.3	142	11
M_1 0.8	2.4	2.4	1	0.7	29	7
M_1 0.9	5.4	4.2	1	0.5	63	7
M_1 1.0	5.4	4.2	1	1.1	83	7
M_1 1.2	5.3	3.8	1	1.6	103	7
$M_1 p$	14.4	4.7	1	2.7	415	27

[a] For description of fractions see Fig. 2. Data from Lapetina, Soto, and De Robertis.[27]

with the microsomes. (For literature on the distribution of AChE and pseudocholinesterase in brain fractions, see De Robertis.[4])

More than half of the AChE of the rat brain was sedimented in the crude mitochondrial fraction. This, after gradient centrifugation, was distributed mainly in the fractions of aminergic nerve endings. The nonaminergic nerve endings were AChE poor.[9] After osmotic disruption of the nerve endings[12,13] most of the mitochondrial AChE remained in the bulk fraction M_1 (Fig. 2). In the early isolation of synaptic vesicles[12,13] the relative specific activity of AChE in the M_2 fraction of synaptic vesicles was rather high, but with further purification of the vesicles the AChE of the M_2A fraction became very low or practically absent.[16] In fact most of the AChE of M_2 was found as a contaminant of small membrane fragments in M_2B (Fig. 2). Similar findings on the localization of AChE have been reported.[26]

After the gradient centrifugation of the M_1 fraction, the AChE was distributed mainly in the M_1 0.9 and M_1 1.0 subfractions, which contain nerve ending membranes (Table V).[15] Myelin (M_1 0.8) and mitochondria (M_1p) are low in AChE, and also subfraction M_1 1.2. The latter probably belongs to the nonaminergic endings of the cerebral cortex.[5] It is shown later that both types of membranes also may be differentiated by their binding capacity with cholinergic blocking agents.

The use of low concentrations of Triton X-100 to dissolve most of the nerve ending membrane leaving the junctional complex intact (Fig. 1D) has permrtted a further knowledge on the localization of AChE. This treatment produces solubilization of about 70% of the enzyme present in the M_1 0.9 and M_1 1.0 fractions, indicating that the main portion of the AChE is located presynaptically.[19] It is interesting that in autonomic ganglia AChE also is presynaptic. According to Koelle[52] this localization is related to protection of the terminal from the depolarizing action of ACh coming from adjacent nerve endings.

2. Cation-Stimulated Phosphohydrolases

Nucleotide phosphohydrolases act as energy transducers, liberating energy from high phosphate bonds that is then used in the active transport of ions and other energy-consuming transformations. Increasing emphasis has been given to this group of enzymes since Skou[53] demonstrated that they may be activated by Na^+ and K^+.

Albers, de Lores Arnaiz, and De Robertis[45] studied the Mg^{2+}-dependent and the Na^+-K^+-activated ATPases, as well as the Mg^{2+}-dependent and the K^+-activated p-nitrophenylphosphatases. The last enzyme, according to previous investigations, could be involved in the final step of the action of ATPase. These enzymes, which were also distinguished by their different sensitivity to ouabaine, were found in the microsomal fraction and were even more concentrated in the nerve ending fractions obtained in the gradient of mitochondria. It was interesting that, while the Na^+-K^+-activated ATPase was localized preferentially in the aminergic fractions of

TABLE V
Distribution of Enzymes in Nerve Ending Membranes[a]

Subfraction	Structure	AChE	Na^+-K^+ ATPase	K^+ p-Nitrophenyl phosphatase	Glutamine synthetase	Adenyl cyclase	Phosphodiesterase	MAO
M_1 0.8	Myelin	1.68	1.37	0.56	0.66	1.06	1.32	0.44
M_1 0.9	Nerve ending membranes	3.22	2.28	2.41	0.96	2.04	1.77	0.33
M_1 1.0	Nerve ending membranes	2.13	3.16	1.39	2.04	2.46	2.68	0.23
M_1 1.2	Nerve ending membranes	0.98	1.40	2.53	1.74	1.95	1.03	0.65
$M_1 p$	Mitochondria	0.15	0.17	0.30	0.73	0.31	0.43	1.56

[a] The results are expressed as in Table I. Data from Rodríguez de Lores Arnaiz et al.[15] and De Robertis et al.[57]

nerve endings, the K^+-activated phosphatase was more conspicuous in the nonaminergic fraction. It also was shown that, after the osmotic shock of the ending, these phosphohydrolases were associated mainly with the nerve ending membranes, present in fraction M_1.[45] Independently Hosie[46] reached similar conclusions and demonstrated that the Na^+-K^+-activated ATPase was related to the membranes sedimenting in the E–H fractions (Fig. 2). The specific activity was highest in fraction G, and AChE was found not to follow exactly the distribution of Na^+-K^+ ATPase, being mainly in the smaller membrane fragments.

The localization of Na^+-K^+ ATPase and K^+ phosphatase has again been investigated in the gradient on M_1.[15] It was found that while the former followed AChE, although with a slightly wider distribution, the K^+ phosphatase was more widely spread into the AChE-poor nerve ending membranes of M_1 1.2 (Table V). This is further support for the concept that Na^+-K^+ ATPase and K^+ phosphatase are two distinct enzymes with different structural associations.[45] The Mg^{2+}-dependent p-nitrophenylphosphatase has been found to be activated by noradrenaline and dopamine.[54]

3. Glutamine Synthetase

Another membrane-bound enzyme that has an important role in brain is glutamine synthetase. This is mainly concentrated in microsomes,[11,55] but as shown by de Lores Arnaiz, Alberici, and De Robertis,[15] there is a definite portion bound to nerve ending membranes. In the gradient of M_1, glutamine synthetase has a localization reminiscent of AChE and Na^+-K^+ ATPase, but with a higher concentration in the AChE-poor nerve ending membranes (Table V). Synthesizing glutamine, an amino acid that easily crosses the nerve ending membranes,[11] glutamine synthetase may regulate the concentration of this amino acid within nerve endings, controlling its transformation into glutamate and GABA. Methionine sulphoximine, a drug that produces seizures and inhibits glutamine synthetase, also alters considerably the nonaminergic nerve ending.[22] Having such a strategic position in nerve ending membranes this enzyme could play a synaptic role in those terminals utilizing glutamate and GABA mechanisms.

4. Adenyl-Cyclase and Cyclic Phosphodiesterase

Of great interest is the distribution of adenyl cyclase, which by catalyzing the production of cyclic AMP is involved in the regulation of various cell activities.[56] This enzyme reaches a concentration highest of any tissue in the cerebral cortex. The intracellular level of cyclic AMP depends on the activity of adenyl cyclase and cyclic phosphodiesterase, the enzyme that catalyzes the hydrolysis of the $3'P$ bond.

De Robertis et al.[57] have shown that the adenyl cyclase of the rat cerebral cortex is practically all particulate and is distributed equally between the mitochondrial and microsomal fractions. This enzyme is concentrated in the aminergic fractions of nerve endings, including those small endings found in the postmitochondrial fraction (mic-20), which are rich in histamine.[10] This last point should be correlated with the fact that in slices of the brain cortex histamine increases considerably the production of cyclic AMP.[58] After the osmotic disruption of the nerve endings, little adenyl cyclase was found in synaptic vesicles, most of it being present in nerve ending membranes (Table V).

While partly free, the particulate portion of cyclic phosphodiesterase is also bound to the same nerve ending membranes (Table V). The localization of these two enzymes favors the view that cyclic AMP may be important in regulatory functions within the nerve ending, but their exact nature remains unknown. As mentioned later, the localization of adenyl cyclase in nerve ending membranes also may be of interest in relation to aminergic receptors. In fact, in other tissues this enzyme is considered to be part of the β-adrenergic receptor.[56]

5. *Position of Active Groups of Enzymes in Nerve-Ending Membranes*

In a close subcellular compartment, such as the nerve ending, the activity of the enzymes may be related to the penetrability of substrates across the membrane and to the position of the active groups of the membrane-bound enzymes. It was observed when incubating nerve endings in media of different osmotic concentrations that the activity varied of the enzymes related to the metabolism of glutamate, glutamine, and GABA. While some substrates became readily available to the enzymes, others needed the osmotic disruption of the nerve ending and even a treatment with detergents to obtain maximal activity.[11]

In the case of adenyl cyclase,[57] little activity was observed in nerve endings assayed in hypertonic or isotonic sucrose, and this reached a maximum at 0.03 M sucrose, a hypotonic concentration that disrupts the nerve ending membranes (Fig. 6). This result suggested that adenyl cyclase has the active groups directed toward the inner surface of the nerve ending membrane. Na^+-K^+-activated ATPase, an enzyme that uses the same nucleotide as substrate, showed maximal activity at all osmolarities, and similar findings were obtained with AChE. These results suggested that both enzymes were on the outer surface of the membrane. In avian erythrocytes, it had been previously shown that adenyl cyclase is located in the inner side of the membrane while the catalytic side of ATPase is externally placed.[56] In the case of phosphodiesterase, about one-third of the enzyme of the nerve endings was active in an isotonic solution, and the full activity could be demonstrated only in hypotonic conditions.[57]

C. Chemical Receptor

1. *Receptor Properties of Nerve-Ending Membranes*

A basic assumption of the chemical theory of synaptic transmission is that the subsynaptic membrane should contain the chemical receptors for the

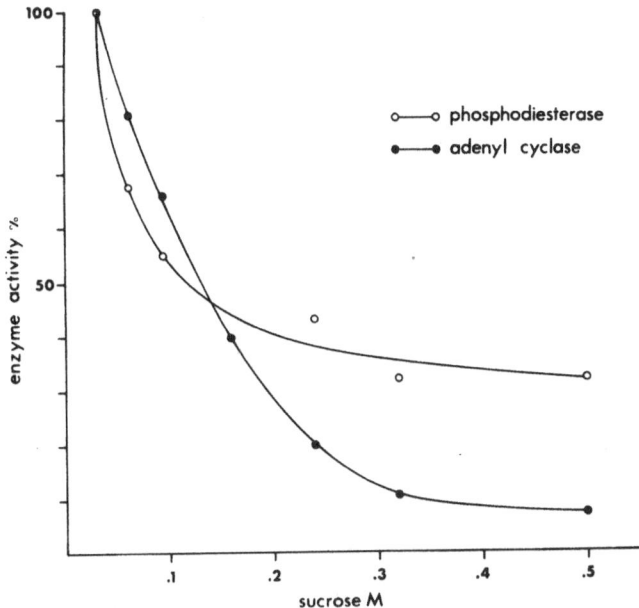

Fig. 6. Effect of different osmotic concentrations on the activity of adenyl cyclase and cyclic phosphodiesterase. From De Robertis et al.[57]

various transmitters. Since this membrane remains attached to the nerve ending membrane, it was expected that this also should have the receptor properties. These have been investigated mainly with the use of radioactive drugs that act as blocking agents for the transmitters and also with some labeled biogenic amines.[14,19,59] Using an *in vitro* technique in which the subcellular fractions of the rat cerebral cortex were exposed to minimal concentrations (i.e., 10^{-7}–10^{-6} M) of dimethyl-^{14}C-d-tubocurarine, this drug became preferentially bound to the AChE-rich nerve ending membranes of fraction M_1 0.9 and M_1 1.0 (Table VI). Similar findings were obtained with methyl-^{14}C-hexamethonium and ^{3}H-alloferine.[59] The AChE-poor membranes of fraction M_1 1.2 showed much less binding capacity. These results indicated that AChE, and the cholinergic receptor, were associated in the same type of nerve ending membrane. In high concentrations atropine, eserine, and the natural transmitter ACh interfered with the binding of these cholinergic blocking agents. More recently, the nerve ending membranes of the basal ganglia and brain stem were found to bind preferentially the α-adrenergic blocking agent [14]C Sy28 [N-α-naphthylmethyl-N-ethyl-β-bromoethylamine][60] and ^{14}C dibenamine, the β-blocking drug ^{14}C propanolol, and also ^{3}H chlorpromazine, ^{14}C LSD, and ^{14}C serotonin. Some of these findings should be correlated with the distribution of adenyl cyclase in similar membranes of the cerebral cortex.[57]

2. Receptor Properties of Junctional Complexes

As mentioned in the section on techniques, the use of a mild detergent permitted the isolation of the junctional complex (Fig. 1D).[19] While Triton X-100 had no effect on ChAc and on the mitochondrial enzyme MAO, the detergent caused considerable solubilization of AChE and other membrane-bound enzymes. In addition there was a considerable loss of proteins but not of the proteolipids. Investigating the receptor properties with dimethyl-^{14}C-d-tubocurarine it was found that the isolated junctional complexes maintained the original binding capacity of the nerve ending membranes (Fig. 7). These findings support the interpretation that the cholinergic receptor properties are localized in the junctional complexes, probably at the subsynaptic membrane, while the membrane-bound enzymes, e.g., AChE, have a wider distribution that includes the limiting membrane of the nerve ending. Furthermore, they demonstrate that the binding sites for cholinergic blocking agents and AChE are in two separate macromolecular entities.[19]

3. Separation of Receptor Proteins

The in vitro uptake of radioactive blocking agents by the nerve ending membranes provided a good technique to attempt the isolation of the molecular species involved in the binding. De Robertis, Fiszer, and Soto[61] found that after an extraction with chloroform:methanol—which produced complete inhibition of AChE—the dimethyl ^{14}C-d-tubocurarine was recovered in the organic phase. The residual protein, which comprised more

TABLE VI

Binding Capacity for Cholinergic Blocking Agents in Nerve Ending Membranes[a]

Subfraction	Structure	AChE	[14]C DMTC	[14]C MHM	[3]H Alloferin
M_1 0.8	Myelin	1.64	2.14	2.92	3.86
M_1 0.9	Nerve ending membranes	3.40	4.16	4.44	4.04
M_1 1.0	Nerve ending membranes	3.45	6.88	4.76	4.37
M_1 1.2	Nerve ending membranes	1.44	3.00	2.52	2.89
$M_1 p$	Mitochondria	0.38	1.60	0.72	1.87

[a] Comparison between distribution of AChE expressed in RSA as in Table I and cholinergic binding capacity expressed as specific binding ratio (counts) per milligram protein in fraction divided by counts per milligram protein in total particulate; dimethyl ^{14}C-d-tubocurarine (^{14}C-D MTC); methyl ^{14}C hexamethonium (^{14}C MHM). Data from De Robertis et al.[14] and Azcurra and De Robertis.[59]

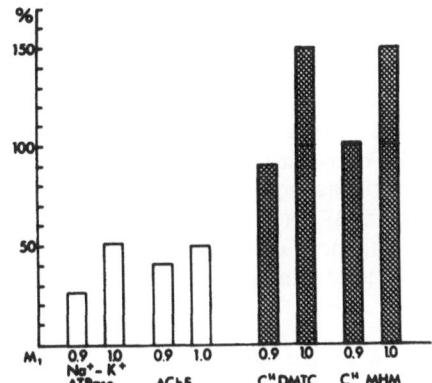

Fig. 7. Diagram showing the percentage of Na^+-K^+ ATPase, AChE, uptake of dimethyl-^{14}C-d-tubocurarine (^{14}C DMTC) and methyl-^{14}C-hexamethonium (^{14}C MHM) after Triton X-100 treatment of the M_1 fraction. In each case the untreated control is 100%. From DeRobertis, Azcurra, and Fiszer.[19]

than 90% of the total, had practically no radioactivity. After water partition of the organic phase—a treatment that separates the gangliosides and other water-soluble molecules—few counts went to the aqueous phase. With the use of thin-layer chromatography it could be demonstrated that the radioactivity remained at the point of origin, together with the proteolipids, while very little was recovered in the various lipids. Furthermore, with column chromatography with Sephadex-L20, the dimethyl-^{14}C-d-tubocurarine was eluted, together with a protein peak. All these findings suggest that the receptor for this cholinergic blocking agent is a proteolipid that in itself is only a small proportion of the total protein of the nerve ending membrane[27] and of the junctional complex. Experiments with purified myelin demonstrated that, although in these membranes there is the highest concentration of proteolipids, the binding capacity was very small. This is in agreement with the concept that the chemical receptor is a special type of proteolipid present in the nerve ending membrane and probably localized in the subsynaptic membrane.[61]

Essentially similar findings have been obtained with the α-adrenergic blocking agent ^{14}C Sy28, in the nerve ending membranes isolated from basal ganglia and brain stem of the rat and cat.[60]

D. Antigenic Properties

Immunochemical studies may give indirect information on the cytochemical organization of the synaptic region. Antiserums against nerve endings isolated from the cerebral cortex of rabbits and cats were prepared by injection into rabbits.[62] They were then studied by complement fixation, and their lytic effect on isolated nerve endings was studied, *in vitro*, under the electron microscope. These antiserums, when topically applied to the visual cortex of the cat, produced high-voltage epileptiform discharges. More recently a more potent antiserum against the nerve ending membranes of fraction M_1 1.0 was prepared and, after absorption with brain mitochondria and myelin, was tried *in vitro*. In the presence of complement this antiserum produced disruption of the nerve endings with liberation of vesicles and had no visible effect on myelin and mitochondria. Acting on a living preparation of molluscan ganglia, this antiserum, in the presence of complement, produced a sequence of bioelectrical changes, i.e., blocking of potentials induced by ACh, reduction and disappearance of spontaneous spike potentials, disappearance of synaptically induced potentials, reduction of resting potential, and increase in membrane conductance, which ended in complete impairment of neuronal functions. At this last stage neurons showed considerable damage at the ultrastructural level.[63]

V. NERVE ENDING MITOCHONDRIA

Since MAO is localized in brain mitochondria,[64] this enzyme has been used as a marker to study the development of this organoid within nerve endings. Barondes[65] has inactivated MAO with β-phenylisopropyl-hydrazine and followed the reappearance of this enzyme in the isolated nerve endings. In addition he has studied the synthesis of the intraterminal soluble proteins. The results obtained were interpreted as being due to axoplasmic flow. However, some of the mitochondrial protein could be synthesized *in situ* at the nerve ending. Furthermore, it has been found that there are no differences in nucleic acid content between mitochondria in nerve endings and in perikarya.[24]

Biochemical differences between mitochondria present in nerve endings and in neuronal or glial perikarya were found. They were related to the concentration of enzymes involved in the metabolism of amino acids, to the regulatory action exerted by substrates on some of these enzymes, and to the differential sensitivity of NAD-dehydrogenases.[11]

VI. CONCLUDING REMARKS

The two main morphological components of chemical synapses—the synaptic vesicles and the junctional complexes—have been isolated and studied in their ultrastructure and chemical organization. Synaptic vesicles with morphological and cytochemical differences have been described. They contain the highest concentration of biogenic amines, amino acids, and other unknown active substances. These findings support the concept that synaptic vesicles are the quantal units in the storage and release of the transmitter at the synaptic junction.

The separation of the nerve ending membrane and of the junctional complex, which includes the subsynaptic membrane, have permitted the study of enzymes related to the transport of ions, the inactivation of transmitters, and the regulation of metabolic cycles. The binding capacity of these membranes for radioactive transmitters, blocking agents, and other drugs has led to the isolation of special proteins that are responsible for the receptor properties of the synapse. Between the membranes forming the synaptic vesicles and those limiting the nerve endings there are differences in molecular population, especially reflected in the molar ratios of lipid and protein amino acids and in the membrane-bound enzymes.

These studies should be pursued with a more dynamic approach, which tries to learn the synthesis and turnover rates of the molecular constituents of the synaptic vesicles, the nerve ending membranes, and the chemical receptor. It also should be considered that such structural components are integrated within a more complex subcellular unit—the nerve or synaptic ending—in which the enzymes for the synthesis of the transmitters, for the metabolism of amino acids, and for those involved in the glycolytic and citric acid cycles as well are localized. In addition to consideration of the morphological compartments and subcompartments comprised within such a unit, account should be taken of the existence of pools of free amino acids and electrolytes, the relative concentration of metabolites, the permeability of the membranes, and the localization of the systems furnishing the substrates within the living cell. It is only with this holistic approach that it will be possible to envision the synapse as a self-regulatory unit in brain endowed with the capacity of transmitting nerve impulses and of other much less-known functions that may be the bases of the processes of conditioning and learning.

VII. REFERENCES

1. J. C. Eccles, *The Physiology of Synapses*, Springer Verlag, Berlin (1964).
2. E. De Robertis and H. S. Bennett, Submicroscopic vesicular component in the synapse, *Federation Proc.* **13**:35 (1954); Some features of the submicroscopic morphology of synapses in frog and earthworm, *J. Biophys. Biochem. Cytol.* **1**:47–58 (1955).
3. J. del Castillo and B. Katz, Biophysical aspects of neuromuscular transmission, in *Progress in Biophysics* (J. A. Butler and B. Katz, eds.), Vol. 6, pp. 121–170, Pergamon Press, London (1956).
4. E. De Robertis, *Histophysiology of Synapses and Neurosecretion*, Pergamon Press, New York (1964).
5. E. De Robertis, Ultrastructure and cytochemistry of the synaptic region, *Science* **156**:907–914 (1967).
6. V. P. Whittaker, The isolation and characterization of acetylcholine-containing particles from brain, *Biochem. J.* **72**:694–706 (1959).
7. E. G. Gray and V. P. Whittaker, The isolation of synaptic vesicles from the central nervous system, *J. Physiol.* (*London*) **153**:35–37P (1960).
8. E. De Robertis, A. Pellegrino de Iraldi, G. Rodríguez de Lores Arnaiz, and C. J. Gómez, On the isolation of nerve endings and synaptic vesicles (Meeting of the Sociedad Argentina de Biología, Mendoza, October, 1960), *J. Biophys. Biochem. Cytol.* **9**:229–235 (1961).
9. E. De Robertis, A. Pellegrino de Iraldi, G. Rodríguez de Lores Arnaiz, and L. Salganicoff, Cholinergic and non-cholinergic nerve endings in rat brain. I. Isolation and subcellular distribution of acetylcholine and acetylcholinesterase, *J. Neurochem.* **9**:23–35 (1962).

10. K. Kataoka and E. De Robertis, Histamine in isolated small nerve endings and synaptic vesicles of rat brain cortex, *J. Pharmacol. Exptl. Therap.* **156**:114–125 (1967).
11. L. Salganicoff and E. De Robertis, Subcellular distribution of glutamic decarboxylase and gamma-aminobutyric alpha-ketoglutaric transaminase, *Life Sci.* **2**:85–91 (1963); Subcellular distribution of the enzymes of the glutamic acid, glutamine and γ-aminobutyric acid cycles in rat brain, *J. Neurochem.* **12**:287–309 (1965).
12. E. De Robertis, G. Rodríguez de Lores Arnaiz, and A. Pellegrino de Iraldi, Isolation of synaptic vesicles from nerve endings of the rat brain, *Nature* **194**:794–795 (1962).
13. E. De Robertis, L. Salganicoff, L. M. Zieher, and G. Rodríguez de Lores Arnaiz, Acetylcholine and cholinacetylase content of synaptic vesicles, *Science* **140**:300–301 (1963); E. De Robertis, G. Rodríguez de Lores Arnaiz, L. Salganicoff, A. Pellegrino de Iraldi, and L. M. Zieher, Isolation of synaptic vesicles and structural organization of the acetylcholine system within brain nerve endings, *J. Neurochem.* **10**:225–235 (1963).
14. E. De Robertis, M. Alberici, G. Rodríguez de Lores Arnaiz, and J. M. Azcurra, Isolation of different types of synaptic membranes from the brain cortex, *Life Sci.* **5**:577–582 (1966).
15. G. Rodríguez de Lores Arnaiz, M. Alberici, and E. De Robertis, Ultrastructural and enzymatic studies of cholinergic and noncholinergic synaptic membranes isolated from brain cortex, *J. Neurochem.* **14**:215–225 (1967).
16. E. G. Lapetina, E. F. Soto, and E. De Robertis, Gangliosides and acetylcholinesterase in isolated membranes of the rat brain cortex, *Biochim. Biophys. Acta* **135**:33–43 (1967).
17. V. P. Whittaker, I. A. Michaelson, and R. J. A. Kirkland, The separation of synaptic vesicles from nerve ending particles ("synaptosomes"), *Biochem. J.* **90**:293–303 (1964).
18. R. E. McCaman, G. Rodríguez de Lores Arnaiz, and E. De Robertis, Species differences in subcellular distribution of choline acetylase in the CNS. A study of choline acetylase, acetylcholinesterase, 5-hydroxytryptophan decarboxylase, and monoamine oxidase in four species, *J. Neurochem.* **12**:927–935 (1965).
19. E. De Robertis, J. M. Azcurra, and S. Fiszer, Ultrastructure and cholinergic binding capacity of junctional complexes isolated from rat brain, *Brain Res.* **5**:45–56 (1967).
20. E. De Robertis, Adrenergic endings and vesicles isolated from brain, *Pharmacol. Rev.* **18**:413–424 (1966).
21. K. Uchizono, Characteristics of excitatory and inhibitory synapses in the central nervous system of the cat, *Nature* **207**:642–643 (1965).
22. E. De Robertis, O. Z. Sellinger, G. Rodríguez de Lores Arnaiz, M. Alberici, and L. M. Zieher, Nerve endings in methionine sulphoximine convulsant rats, a neurochemical and ultrastructural study, *J. Neurochem.* **14**:81–89 (1967).
23. C. W. Cotman and H. R. Mahler, Resolution of insoluble proteins in rat brain subcellular fractions, *Arch. Biochem. Biophys.* **120**:384–396 (1967).
24. L. Austin and I. G. Morgan, Incorporation of ^{14}C-labelled leucine into synaptosomes from rat cerebral cortex "in vivo," *J. Neurochem.* **14**:377–387 (1967).
25. K. von Hungen, H. R. Mahler, and W. J. Moore, Turnover of protein and RNA in synaptic subcellular fractions of rat brain, *J. Biol. Chem.* **243**:1415–1423 (1968).
26. V. P. Whittaker, Some properties of synaptic membranes isolated from the central nervous system, *Ann. N.Y. Acad. Sci.* **137**:982–998 (1966).
27. E. G. Lapetina, E. F. Soto, and E. De Robertis, Lipids and Proteolipids in isolated subcellular membranes of rat brain cortex, *J. Neurochem.* **15**:437–445 (1968).
28. H. Weigant, The subcellular localization of gangliosides in the brain, *J. Neurochem.* **14**:671–674 (1967).
29. V. P. Whittaker and M. N. Sheridan, The morphology and acetylcholine content of isolated cerebral cortical synaptic vesicles, *J. Neurochem.* **12**:363–372 (1965).

30. L. A. Barker, J. Amaro, and P. S. Guth, Release of acetylcholine from isolated synaptic vesicles. I. Methods for determining the amount released, *Biochem. Pharmacol.* **16**:2181–2187 (1967).

31. A. Sattin, The synthesis and storage of acetylcholine in the striatum, *J. Neurochem.* **13**:515–524 (1966).

32. R. I. Birks and F. C. MacIntosh, Acetylcholine metabolism of a sympathetic ganglion, *Can. J. Biochem. Physiol.* **39**:787–827 (1961).

33. E. De Robertis, Electron microscope and chemical study of binding sites of brain biogenic amines, *in Progress in Brain Research* (H. E. and W. A. Himwich, eds.), Vol. 8, pp. 118–136, Elsevier, Amsterdam (1964).

34. E. De Robertis, A. Pellegrino de Iraldi, G. Rodríguez de Lores Arnaiz, and L. M. Zieher, Synaptic vesicles from rat hypothalamus. Isolation and norepinephrine content, *Life Sci.* **4**:193–201 (1965).

35. L. M. Zieher and E. De Robertis, Distribución subcelular de noradrenalina y dopamina en cerebro de rata, *VI Congreso de la Asociación Latinoamericana de Ciencias Fisiológicas, Viña del Mar, Chile*, November, 1964, p. 150 (1964).

36. L. M. Zieher and E. De Robertis, Subcellular localization of 5-hydroxytryptamine in rat brain, *Biochem. Pharmacol.* **12**:596–598 (1963).

37. E. W. Maynert, R. Levi, and A. J. De Lorenzo, The presence of norepinephrine and 5-hydroxytryptamine in vesicles from disrupted nerve-ending particles, *J. Pharmacol. Exptl. Therap.* **144**:385–392 (1964).

38. E. W. Maynert and K. Kuriyama, Some observations on nerve ending particles and synaptic vesicles, *Life Sci.* **3**:1067–1087 (1964).

39. Dj. Palaić and Z. Supek, Liberation of brain 5-hydroxytryptamine and noradrenaline by X-ray treatment in the new born and adult rat, *J. Neurochem.* **13**:705–709 (1966).

40. G. Rodriguez de Lores Arnaiz and E. De Robertis, 5-Hydroxytryptophan decarboxylase activity in nerve endings of the rat brain, *J. Neurochem.* **11**:213–219 (1964).

41. E. Roberts and K. Kuriyama, Biochemical–physiological correlations in studies of the γ-aminobutyric acid system, *Brain Res.* **8**:1–35 (1968).

42. J. K. Saelens and L. T. Potter, Subcellular localization of choline acetyltransferase in rat brain cortex, *Federation Proc.* **25**:451 (1966).

43. S. Tuček, On the subcellular localization and binding of choline acetyltransferase in the cholinergic nerve endings of the brain, *J. Neurochem.* **13**:1317–1327 (1966).

44. S. Tuček, Subcellular distribution of acetyl-CoA synthetase, ATPcitrate lyase, citrate synthase, choline acetyltransferase, fumarate hydratase and lactate dehydrogenase in mammalian brain tissue. *J. Neurochem.* **14**:531–545 (1967).

45. R. W. Albers, G. Rodríguez de Lores Arnaiz, and E. De Robertis, Sodium-potassium-activated ATPase and potassium-activated p-nitrophenylphosphatase: a comparison of their subcellular localizations in rat brain, *Proc. Natl. Acad. Sci. U.S.* **53**:557–564 (1965).

46. R. J. A. Hosie, The localization of adenosine triphosphatases in morphologically charac terized fractions of guinea-pig brain, *Biochem. J.* **96**:404–412 (1965).

47. M. Germain and P. Proulx, Adenosintriphosphatase activity in synaptic vesicles of rat brain, *Biochem. Pharmacol.* **14**:1815–1819 (1965).

48. K. Kadota, S. Mori, and R. Imaizumi, The properties of ATPase of synaptic vesicles fraction. *J. Biochem. (Tokyo)* **61**:424–432 (1967).

49. H. Yoshida, K. Kadota, and H. Fujisawa, Adenosine triphosphate dependent calcium binding of microsomes and nerve endings, *Nature* **212**:291–292 (1966).

50. E. G. Lapetina, G. Rodríguez de Lores Arnaiz, and E. De Robertis, ^{32}P incorporation into different membranous structures separated from rat cerebral cortex, *J. Neurochem.* **16**:101–106 (1969).

51. L. M. Seminario, N. Hren, and C. J. Gómez, Lipid distribution in subcellular fractions of the rat brain, *J. Neurochem.* **11**:197–207 (1964).

52. G. B. Koelle, Evidence for differences in primary fractions of acetylcholinesterase at different synapses and neuroeffector junctions, *in Regional Neurochemistry* (S. S. Kety and J. Elkes, eds.), pp. 312–323, Pergamon Press, Oxford (1961).

53. J. C. Skou, Enzymatic basis for active transport of Na^+ and K^+ across cell membrane, *Physiol. Rev.* **45**:596–617 (1965).

54. J. Clausen and B. Formby, Effect of noradrenaline on phosphatase activity in synaptic membranes of the rat brain, *Nature* **213**:389–390 (1967).

55. O. Z. Sellinger and I. De Balbian Verster, Glutamine synthetase of rat cerebral cortex: intracellular distribution and structural latency, *J. Biol. Chem.* **237**:2836–2844 (1963).

56. E. W. Sutherland, I. Øye, and R. W. Butcher, The action of epinephrine and the role of the adenyl cyclase system in hormone action, *Recent Progr. Hormone Res.* **21**:623–642 (1965).

57. E. De Robertis, G. Rodríguez de Lores Arnaiz, M. Alberici, R. W. Butcher, and E. W. Sutherland, Subcellular distribution of adenyl cyclase and cyclic phosphodiesterase in rat brain cortex, *J. Biol. Chem.* **242**:3487–3493 (1967).

58. S. Kakiuchi and T. W. Rall, Effects of norepinephrine and histamine on levels of adenosine $3',5'$ phosphate (3,5-AMP) in brain slices, *Federation Proc.* **24**:150 (1965).

59. J. M. Azcurra and E. De Robertis, Binding of dimethyl-^{14}C-d-tubocurarine, methyl ^{14}C-hexamethonium and ^{3}H alloferine by isolated synaptic membranes of brain cortex, *Intern. J. Neuropharmacol.* **6**:15–26 (1967).

60. S. Fiszer and E. De Robertis, Subcellular distribution and nature of the α-adrenergic receptor in the CNS. *Life Sci.* **7**:1093–1103 (1968).

61. E. De Robertis, S. Fiszer, and E. F. Soto, Cholinergic binding capacity of proteolipids from isolated nerve-endings membranes, *Science* **158**:928–929 (1967).

62. E. De Robertis, E. G. Lapetina, J. Pecci Saavedra, and E. F. Soto, *In vivo* and *in vitro* action of antisera against isolated nerve ending of brain cortex, *Life Sci.* **5**:1979–1989 (1966).

63. F. Wald, A. Mazzuchelli, E. Lapetina, and E. De Robertis, The effect of antiserum against nerve ending membranes from cat cerebral cortex on bioelectrical activity of mollusc neurons, *Exptl. Neurol.* **21**:336–345 (1968).

64. G. Rodríguez de Lores Arnaiz and E. De Robertis, Cholinergic and non-cholinergic nerve endings in the rat brain. II. Subcellular localization of monoamine oxidase and succinate dehydrogenase, *J. Neurochem.* **9**:503–508 (1962).

65. S. H. Barondes, On the site of synthesis of the mitochondrial protein of nerve endings, *J. Neurochem.* **13**:721–727 (1966).

Chapter 16

PERIPHERAL NERVE

Giuseppe Porcellati

Department of Biological Chemistry
University of Pavia
Pavia, Italy

I. INTRODUCTION

The neurochemistry of the peripheral nerve (PN), its chemical composition and enzyme content, and its basic metabolic processes and their regulation need to be known to establish whether these aspects are specific for the PNS and whether they may give us more insight into the relationship of metabolism to function.

Owing to its small size and rather feeble metabolism, PN is often a difficult material to handle for chemical and biochemical exploration. Most of the published analyses have been necessarily made on whole, epineurium-free PN, which consists of the axon, investing myelin, Schwann cells, and other membrane structures. It would be desirable, obviously, to eliminate inhomogeneities of the experimental system. The problem has been approached by carrying out studies with unmyelinated PN, which, however, is not entirely free from myelin[1,2] and does contain satellite cells. Another approach has been recently made by preparing PN myelin.[3-5] It is difficult, however, to obtain a myelin completely free from axonal fragments and axolemmatic membranes.

Some workers have found it useful to adopt for their studies the giant fibers of the squid and cuttlefish, which vary from 0.3 to 1 mm in diameter; fibers of this type are present also in lobsters. The size of the giant axons gives great experimental scope; a clear separation of axoplasm and sheath can be achieved simply by extruding the axoplasm from a cut end of the nerve. The Mauthner axon of the carpfish also provides uncontaminated axon and sheath.

Biochemical investigations of PN have seldom considered the role of the satellite cells and usually have been referred to axons or to fibers as a whole. Knowledge of the detailed biochemistry of the satellite cell in a PN is scanty, and probably the studies carried out on PNS undergoing Wallerian degeneration have offered the most abundant and reliable results by providing important information about localization and activity of enzymes and chemical composition of normal PNS.

Various important surveys on the biochemistry of PN have appeared in recent years, partially covering the field of this subject matter.[6–8] No review or recent monograph has yet appeared that treats completely the chemistry and biochemistry of PNS. This chapter tries to collect the most relevant and recent literature in this field. Technical problems of handling and processing PN structures for chemical and enzymic studies are not discussed in this survey.

II. THE CHEMICAL CONSTITUTION OF THE PERIPHERAL NERVE

The quantitative microchemical techniques developed in recent years for the coordinate microassay of various chemical components and enzyme activities of brain tissue have been very seldom applied, because of analytical difficulties, to studies on PN. Perhaps only ultramicrochemical analyses have been more frequently used in this connection.[2,9] Analyses often have been carried out on the whole length of a PN, although the existence in the nerve of various biochemical concentration gradients is clearly established. The values that are reported here often refer to the total length of the nerve and apply to the whole PN, myelin sheath, or whole axons, since with few exceptions subfraction separations from PN have not been carried out.

Gross composition (total N, total P, water, and ash) is not reported; the reader may consult other contributions.[7,10,11] Levels for total protein and lipid, however, are included. All the values (ranges or average data) are expressed as micromoles, micrograms, or milligrams per gram of wet weight and have often been recalculated from previously published figures.

A. Protein

1. *Characterization and Content in PN*

a. Total Protein. Large amounts of protein have been demonstrated histochemically by color reactions for amino acid in axoplasm and adjacent structures. The problem of distinguishing, by biochemical means, the PN myelin protein from the protein of other morphological components of the nerve trunk is largely unsolved, and values of total protein content (Table I) usually refer to the whole PN. The protein of the squid axon amounts to about 4%, which is 40% of the dry substance. Nearly 55% of the protein of the squid giant fiber is in the sheath:[14] the value (Table I) is high because it includes the collagen, which is about 45% of the dry wt.[13] Moreover, 75% of the total protein of the sheath is connective protein and 25% is sheath protein located in the cellular component.[14] Additional data on nerve protein content have been published.[15]

TABLE I

Protein Content of Peripheral Nerve[a]

Material	Protein content	Reference
Mammalian peripheral nerves	110–150	7
Cat peripheral nerves	109–123	10
Rabbit tibial	139 ± 2.4	9[b]
Ox intradural spinal roots	111	3
Hen sciatic	157	12
Frog peripheral nerve	67–75	10
Squid axoplasm	32.4 ± 1.2	13
Squid axon sheath	30.7–32.7	13

[a] Values have been converted to milligrams per gram wet weight of epineurium-free tissue by adopting the approximate factors reported in the footnote for Section II,B,1,a.
[b] Value divided by 2.5, according to the dry weight value.[9]

b. Phosphoprotein. Phosphoprotein content of cat sciatic nerve ranges from 0.02 to 0.03 mg P/g wet wt. of tissue,[16] while it is ten times higher in guinea pig sciatic.[17] A proximodistal gradient of phosphoprotein in PN has been found[17] and confirmed.[6]

c. Collagen. The collagen constitutes 40–45% of the dry weight of a PN; its nitrogen content represents about 35% of the total nitrogen.[18] Collagen is essentially located in the myelin sheath, and amounts in rabbit sciatic nerve to 32 mg/g wet wt.[18] and to rather similar levels in cat PN.[16] It is thought to be inactive in terms of enzymic activity. By omitting collagen, the protein content of the noncollagenous component of the sheath of various nerves normally ranges to about 3%.[13]

d. Soluble Protein. Earlier work on the characterization of the soluble proteins of PN has been reviewed,[7] and the conditions necessary for their maximal extraction have been discussed.[22] Some proteins have been isolated from the squid giant fiber axoplasm.[19,20] In *Loligo pealii* a fibrous-like protein forms about 10% of the protein of the nerve axoplasm and amounts to about 2.6% of the dry weight. A soluble protein has been isolated and analyzed from lobster nerve also.[21] Electrophoretic separation and composition of soluble protein from vertebrate PN has been studied in various animal species,[22,23] and the albumin fraction has been further isolated from cow PN and purified. Saline-soluble protein from cat sciatic nerve accounts for nearly 20% of the total protein.[22] A specific soluble protein (S-100), with a molecular weight of 30,000, has been found in rabbit and rat sciatic and vagus nerves,[24] forming 0.5–3% of the total soluble protein of the nerve.

e. Proteolipid and Phosphatidopeptides. Proteolipids are lipoproteins whose properties, nature, possible structure, and distribution have been recently reviewed.[25,26] The content in PNS is reported in Table II.

The proteolipid content of various PN appears similar. Human spinal roots contain about 4 times the amount of proteolipids present in human sciatic herves and about 15 times that found in brachial plexus. The trypsin-digestible "acid-soluble proteolipid" is considered a true proteolipid and not a chemical artifact.[26] Its content in bovine spinal roots is at least 10 times higher than that of the classical proteolipid.

No clear evidence as yet exists about the occurrence of phosphatido-peptides in PNS, nor does data exist that might clarify possible relationships of phosphatidopeptides with other known protein or protein-lipid complexes of PN, except perhaps with neurokeratin.

f. Neurokeratin and Other Residual Insoluble Proteins. Residual insoluble proteins are the proteins not extractable by lipid solvents and aqueous solution.* Neurokeratin, the more important component of this fraction, can be prepared essentially free from collagen and mucopolysaccharide.[31] An important protein fraction of PN is also the so-called trypsin-resistant protein residue (TRPR).[15,29] The relationships between neurokeratin and TRPR in PNS have been investigated,[15,31−33] and there is discrete evidence that now indicates that in PN neurokeratin and TRPR are the same entity.[25,26,33]

_____

* The term residual protein is often intended to mean protein not extractable by lipid solvents and therefore will include also the water-soluble proteins. Probably residual protein is a more appropriate reference for expression of enzyme activity in PN than is the total protein.

TABLE II

Proteolipid and Neurokeratin Content in PNS[a]

Component	Tissue	Amounts	Reference
Proteolipid protein	Horse S	0.63	27
	Rat S.	0.72	15
	Bovine S	0.45	28
	Bovine S	0.6–1.1	29
	Human S	1.15	30
	Human BP	0.30	30
	Human SR	3.3–4.7	30
	Human SR	1.77	31
	Bovine SR	1.03	31
"Acid-soluble proteolipid protein"	Bovine SR	10	26
Neurokeratin (TRPR)	Rat S	5.5	15
	Bovine SR	4–6	33

[a] Data expressed as milligrams per gram wet weight of tissue. Proteolipid protein is 70–80% of the total proteolipid. For abbreviations, see Table III.

Proteolipid protein and neurokeratin (TRPR) in both CNS and PNS probably have the same amino acid composition[33]: both resist hydrolysis with NaOH and digestion with trypsin; both have high sulfur content and the same relationship to the myelin sheath; and both contain discrete amounts of tryptophan.[33] The only variation resides in the different solubilities of their lipid complexes, since proteolipid is extracted from fresh nervous tissue by chloroform–methanol, whereas TRPR is not. Probably lipid is bound to these proteins in a very different way, which affects the solubility properties of the compounds as they exist in PN *in vivo*. Unlike the proteolipid, TRPR is much more concentrated in PN (Table II).

A chloroform–methanol insoluble, trypsin-digestible protein occurs in PN,[15,26] which is different, by definition, from both proteolipid protein and neurokeratin. Phospholipid and cerebroside are probably attached to this protein, which presumably occurs in the PN myelin sheath as a polysaccharide–lipid–protein complex. This protein is different from the trypsin-digestible, "acid-soluble proteolipid protein,"[26] since it is insoluble in chloroform–methanol.

2. Content in Different Nerves and in Nerves of Various Animal Species

The electrophoretic pattern of PN-soluble proteins of man, cat, cow, ox, and other species is substantially similar,[23,24] and no variation in proteolipid protein content exists among horse, cow, and rabbit sciatic nerves[27] and between anterior and posterior human spinal roots.[30]

3. Differences Between PN and Other Nervous Tissues

Total protein content of PN is lower than that of whole brain. Squid axoplasm protein content is about 3%[13] as compared to whole brain (10%), and that of rabbit tibial nerve is lower (347 ± 6 mg/g dry wt.) than that of optic nerve (437 ± 11).[9] CNS possesses very little collagen (3 mg/g wet wt.),[7] of the order of 0.2% of the total protein[7,16] and probably only of arterial origin.

The electrophoretic pattern of the extractable proteins of PN is different from that of brain white matter, brain gray matter, cerebellum, brain stem, and spinal cord.[22,23] The proportion of the albumin fraction is much greater in PN than in CNS.[22,34] The same fraction is more abundant in ox spinal roots than in CNS but is less concentrated than in sciatic nerve.[22]

Proteolipid protein is much more concentrated in brain white matter, spinal cord, and other regions of the brain stem than in the PNS,[6,15,25–31,33] which contains at most one fifteenth to one twentieth of the concentration of cerebral white matter and optic nerve.[25,27,28] The anterior and posterior spinal roots contain proteolipids at concentrations intermediate between spinal white matter and PN.[30] There are probably qualitative differences between myelin from different areas or differences in the extractability of the proteolipids. It also can be supposed that proteolipids are not derived wholly

TABLE III

Total Lipid and Phospholipid in PN[a]

Nerve	Total lipid	Total phospholipid	Nerve	Total lipid	Total phospholipid
Mammalian PN			Rabbit S	199 ± 13[37]	98 ± 13[37]
Human S	167–215[37]	22–139[7]	Rabbit T	219 ± 9[36]	137 ± 2[9]
Monkey S	23–185[10]	30–43[38]	Rabbit V		15–20[2]c
Bovine S		1.67[1]b	Hen S	161[41]	52–63[38]
Bovine SR	175[37]	44[37]	Lobster C	20–23[43]d	10.6–17.4[43]d
Bovine SP	222[3]	14[37]	Lobster NC	24.2[43]d	11.7[43]d
Sheep S	26[37]	1.67[1]b	Crab C		0.37[1]b
Dog S		51.4–72.7[39]	Crab L		0.35[1]b
Cat S	197[40]	59–73[39]	Squid giant fiber		3.34[44]
Guinea pig S		81 ± 11[17]	Squid giant (axoplasm)		1.80[44]
Hen S	198[42]	76.2[41]	Squid giant (axon sheath)		10.30[44]

a Values expressed as milligrams per gram wet weight (see footnote for Section II,B,1,a). Abbreviations: S, sciatic; SR, spinal roots; SP, splenic; T, tibial; V, vagus; C, claw nerve; L, leg nerve; NC, nerve cord; F, femoral; and BP, brachial plexus.
b Milligrams P over grams wet weight.
c Approximate value (see footnote for Section II,B,1,a).
d Original value divided by 6.9, according to the dry weight value.[37]

from myelin but also from some other components of the nervous system. Differences in CNS and PN proteolipid content are not paralleled by corresponding variation in the neurokeratin levels (Table II). The proteolipid protein content is about ten times that of neurokeratin in CNS, whereas it is about one sixth that of neurokeratin in spinal roots and sciatic nerves.[15,26,31] Probably the "myelin-staining" protein material of PN, seen by histochemical methods, is neurokeratin, while that in CNS tracts is almost totally proteolipid protein.[15]

B. Lipids

1. Characterization and Content in Peripheral Nerve

a. Total Lipid and Phospholipid. Values for total lipid and phospholipid of PN in adult animal have been tabulated in previous surveys.[7,10,11] Additional data, converted to terms of milligrams per gram wet weight,* are reported in Table III.

Values among the various nerves are not exceedingly different. Unmyelinated or lightly myelinated nerves, as those of crab, lobster, and the splenic and vagus nerves, contain about one sixth of the total lipid or phospholipid of myelinated PN. The lipid content of the whole giant fiber of the squid stellar nerve and of its sheath considerably exceeds that of the axoplasm.[44]

The sum of cholesterol (CHOL), total phospholipid, and cerebrosides (CER) normally amounts to about 95% of the total lipid. The remainder may be triglyceride; this lipid has been found, however, only in trace amounts in lobster[43] and chicken[41] nerves. In bovine spinal roots[3] and bird sciatic nerves[35] it is present only in amounts of 9.8 mg/g wet wt. of tissue (4.4% of the total lipid) and 23.8 mg/g, respectively. Much higher values of triglyceride content (127 and 68.6 mg/g wet wt. in chicken and cat sciatic nerves, respectively), together with small amounts of diglyceride and monoglyceride, have been reported,[40,42] probably because of a larger contribution of the epineurial and other adjacent structures.

b. Cholesterol. CHOL is found in large amounts in PN (Table IV). By carefully removing epineurium, connective, and adipose tissues, higher values may be obtained[9]; this difference cannot be attributed to technique alone. Myelinated PN contains a far greater concentration of CHOL than unmyelinated nerve. This is obviously related to the abundance of CHOL in

---

* Values of dry weights of myelinated PN normally range between 300 and 370 mg/g wet wt., while those for frog PN is lower (about 250 mg/g wet wt.). Corresponding values for unmyelinated PN (splenic, vagus, or crab and lobster PN) are of the order of 140–160 mg/g wet wt. If a fat-free dry weight is considered, the values range between 300 and 350 mg/g of total dry weight or between 100 and 130 mg/g of wet weight. References may be found in various works.[9–11,35–37] Consequently, in order to have values of dry weight converted to wet weight of tissue, we have divided published values by 6.7–7.0, 4 and 3, for unmyelinated nerves, frog PN, and myelinated nerves or roots, respectively. To have values of fat-free dry weight converted to wet weight, published values can be divided by 7.5–8.5. Published dry weight or fat-free dry weight values have been adopted, when available.

TABLE IV

Cholesterol Content of PN[a]

Nerve	Content	Nerve	Content
Mammalian PN	11–48[7]	Hen S	44.3[41]
Human S	12–15[39]	Hen S	34.2[35]
Human F	14–15[39]	Hen S	29.5–30.5[45]
Human T	13 ± 2.4[11]	Hen S	20.1[1]
Monkey S	30.9[1]	Rabbit V	8.5[2]b
Bovine SR	48.3–49.3[3]	Cow SP	3.51[1]
Cow S	23[37]	Cow SP	3.5[37]
Dog S	24.3–30.7[39]	Lobster L	2.70[1]
Cat S	28.7–31.9[39]	Lobster NC	2.46[43]c
Cat S	32.6[40]	Lobster C	2.08–2.75[43]c
Rabbit S	26.7–36.5[39]	Lobster C	2.47[1]
Rabbit S	41 ± 4[37]	Crab C	2.77[1]
Rabbit S	18.91[1]	Crab L	2.74[1]
Rabbit T	58 ± 1[9]b	Squid giant fiber	0.84[44]
Sheep S	19.7[1]	Squid giant (axoplasm)	0.34[44]
		Squid giant (axon sheath)	3.1[44]

[a] Values expressed in milligrams per gram wet weight of tissue (see footnote for Section II,B,1,a). For abbreviations, see Table III.
[b] Original value divided by 7.88, according to a fat-free dry weight per gram weight of 127.
[c] Original value divided by 6.9, according to the dry weight value.[37]

the peripheral myelin sheath,[4,37,39] a finding that has been confirmed by the isolation of almost pure myelin from brain and PN.[3–5] A molar ratio of CHOL:phospholipid:CER close to 1:1:0.5, either in CNS or in PN myelin, has been confirmed by several authors. These proportions are of great importance in determining the properties of living membranes.

CHOL content of squid axoplasm is very low[44] as compared to that of the axon sheath. Moreover, it is not equally distributed throughout the whole length of hen PN.[45]

No sterols other than CHOL have been detected in PN. Small quantities of CHOL esters exist in adult PN (1.2 and 2.8 mg/g wet wt. in hen sciatic nerve,[35,42] 0.04 mg/g in cat sciatic,[40] and trace amounts in spinal roots).[3] Other workers have failed to detect sterol esters in PN.[41,43,45]

c. *Phospholipids and Glycolipids.* Histochemical localization of phospholipid and glycolipid in PN has been discussed.[6] Early interesting biochemical findings have been reviewed.[7,9,10,11,40,43] More recent analyses are partially described in Tables V and VI. Other data may be found elsewhere.[1,40,44,46]

The values of sphingomyelin (SP) do not represent only choline-containing phospholipids, since ethanolamine-containing SP is known to occur in nervous tissues. SP often contains small amounts of plasmalogen

TABLE V
Lipids of PN of Various Species[a]

Nerve	SP	LEC	PhE	EPL	PhS	PA	CA	MPI	DPI	TPI	AEP	CER
Human S[38,39]	16–23	3.5–4.4		11–13								11–25
Bovine SR[3]	28.5	30.3	34.3		18.3							27.3
Ox S[37]	15	8										34
Ox SP[1,37]b	4[a]	4.2[a]	0.02	0.05	0.045	0.002	0.002	0.015	0	Traces	0.006	2.7[a]
Monkey S[1]b	0.50	0.22	0.07	0.45	0.27	0.003	0	0.009	0.006	0.053	0.06	20.4[a]
Dog S[39]	29.1–32.4	7.2–10.3										19.7–41.5
Cat S[40]	17.5	17.5	8.2	16.7	17			3				14.2
Rabbit T[9]c	46	20										38
Rabbit S[1,37]b	0.38	0.26	0.10	0.48	0.23	0.026	0	0.022	0.002	0.05	0.05	39[a]
Rabbit V[2]c	4.9	7										10.5
Sheep S[1]b	0.43	0.17	0.06	0.37	0.25	0.04	0	0.05	0.009	0.043	0.032	17.8[a]
Crab L[1]b	0.04	0.11	0.025	0.05	0.02	0.01	0.001	0.01	0	0.001	0.054	0
Lobster L[1]b	0.07	0.13	0.03	0.08	0.03	0.02	0.002	0.003	Traces	0.003	0.061	0

[a] Values expressed as milligrams per gram wet weight (see footnote for Section II,B,1,a). For abbreviations, see Table III and the text.
[b] Values in milligrams P per gram wet weight.
[c] Original values divided by 7.88, according to fat-free dry weight per gram wet weight of 127.
[d] Micromoles per gram wet weight.

TABLE VI

Phospholipid and Glycolipid in Hen PN[a]

Lipid	(1)[b]	(35, 38, 46)[b]	(42)	(41, 47)[b]
SP	0.37	0.27	6.3	27.2[a]
LEC	0.24	0.60	5.8	
CPL	0.021	0.04		
PhE	0.07	0.23	9.5[d]	
EPL	0.50	20–27[a]		0.47
PhS	0.19	0.41	3.5	
SPL	0			0.10
PA	0.021	0.05		
CA	0.004	0.014		
PG	0			
MPI	0.014 ⎫			
DPI	0.001 ⎬	0.15		
TPI	0.024 ⎭			
AEP	0.07			
CER	18.2[c]	23.5[a]	13.3	41.4[a]

[a] Values expressed in milligrams per gram wet weight (see footnote for Section II,B,1,a), unless otherwise stated. For abbreviations, see Table III and the text.
[b] Values in milligrams P per gram wet weight.
[c] Micromoles per gram wet weight.
[d] PhE + lysophosphatidylethanolamine.

and alkyl ether phospholipids (AEP). In most PN, SP occurs in higher proportions than cerebrosides (CER),[3,9,37,39] although not consistently.[4] Values for CER also comprise gangliosides (GAN), if present; true GAN content has never been studied in PN. Owing to the fact, however, that the glycolipids of myelinated PN are chiefly CER, the contribution of GAN to the total glycolipid content of PN is negligible. The plasmalogen is chiefly an ethanolamine plasmalogen (EPL), though discrete amounts of serine plasmalogen (SPL) and traces of choline plasmalogen (CPL) also have been determined in PN.[1,46,47] The amounts of SPL range between 0.003 and 0.04 mg P/g wet wt. in various nerves,[1] and CPL is present in crab, lobster, and chicken PN.[1,46] High levels of lecithin (LEC) occur in PN[3,40,42,46]; this finding would confirm that, contrary to earlier assumptions, LEC is also a myelin component. The sum of the choline-containing phospholipids (LEC, SP, and CPL) in myelinated and unmyelinated nerves is constant.[1]

Cardiolipin (CA) has been found in small amounts in PN.[1,3,46] Traces of other lipids, such as lysolecithin, galactosyl diglyceride, phosphatidyl glycerol (PG), lysophosphatidylethanolamine, ceramide, and hydrocarbons also have been found in PN.[1,3,38,40,42,46] Diphosphoinositide (DPI) and triphosphoinositide (TPI) have been predominantly localized in PN myelin,[1] although small amounts of these lipids also exist in apparently unmyelinated nerves.[1] Monophosphoinositide (MPI), phosphatidic acid (PA), and phosphatidylethanolamine (PhE) do not seem to be particularly located in the myelin, unlike the phosphatidylserine (PhS). The importance of phospholipid and CHOL for the structure of the myelin membrane has been discussed.[48]

Lipids form 33% of the dry substance of mammal nerve axoplasm, with LEC forming 1.1% wet wt.; CER, 0.35%; and GAN, 0.5%. The axoplasm of squid nerve has been found to contain 0.2% wet wt. of phospholipid. Squid giant axons contain 0.74 mg/g wet wt. SP:[44] the sheath possesses a much larger amount (2.9 mg) than the axoplasm (0.27 mg). Axon phospholipid probably derives from the neuronal cell, by axoplasmic flow. A proximodistal decrease in lipid content is present in rat PN.

d. Lipid Fatty Acids and Free Fatty Acids. Fatty acid and fatty aldehyde of total lipid and individual lipids of PN have been determined.[3,40,42] Small amounts of free fatty acids (1.64 mg/g wet wt. in cat PN) are present in nerve structures.[40,42]

2. Differences Among Nerves of Various Animal Species

The phospholipid concentration of human sciatic nerve is lower than that of other corresponding myelinated nerve (Tables IV–VI). Beef PN contains lower amounts of neutral lipids and higher amounts of phospholipids than cat and chicken PN,[40] and cat PN has a higher amount of the "cephalin" fraction than beef and chicken nerve.[40] Unmyelinated nerve contains much lower amounts of total lipid, individual phospholipids, and CHOL.[1,2,44] SP, CER, PhS, EPL, SPL, and TPI are present, in fact, in very low amounts in unmyelinated nerves, being more concentrated in PN myelin.[1,3–5,27,37,39] Within each group of myelinated or unmyelinated nerve, the concentrations of total phospholipid and of CHOL are similar; on the other hand, the molar ratio of total phospholipid to CHOL is higher in unmyelinated than in myelinated nerves.[1] CER occurs in very low amounts in the crustacean PN,[1,44] which, on the contrary, contains higher quantities of AEP (12–14% of the total phospholipid) than mammalian or avian nerves (1.4–4%).[1] Vagus nerve contains rather large amounts of phospholipids; its sphingolipid fraction is made up almost presumably of GAN.[2] Differences in the fatty acid composition of lipid and phospholipid from rat, cat, and chicken PN[40] and from the garfish olfactory (unmyelinated) and trigeminal (myelinated) PN[49] have been reported.

3. Differences Between PN and Other Nervous Tissues

Total lipid and phospholipid content is generally higher in spinal cord than in PN but is lower in whole brain than in nerves. The values of the individual lipids of myelinated PN resemble those of spinal cord and brain white matter, while those of unmyelinated PN or axoplasm resemble gray matter. The amounts of CER, SP, PhS, EPL, SPL, TPI, and CHOL in PN are higher and those of LEC and PhE are lower than in gray matter.[41,46,47] CPL, which is present in low amounts in PN, is nearly absent in cerebral tissue. Rabbit tibial nerve contains approximately 20% more total lipid than optic nerve.[9] Both nerves do not differ for the "cephalin," CER and CHOL content, whereas SP levels are 159% greater and LEC is 22% less abundant in the tibial than in the optic nerve.[9] Rather similar findings have been described for splenic, sciatic, and optic PN of a variety of animal species.[3,37,39] Vagus nerve is the only PN that bears some similarity to central white matter.[9,37] The fatty acid pattern of lipid from myelinated nerves resembles that of brain white matter, while that from unmyelinated PN resembles gray matter.[49]

C. Carbohydrates

A high concentration of glycogen (0.3–2.2 g/100 g wet wt.) is present in crab nerves.[6] The content of rabbit tibial PN is 1.10 μmole $\pm$ 0.04/g wet wt. and that of Glu and Fru is 3.67 $\pm$ 0.25 and 1.15 $\pm$ 0.25 μmole/g wet wt., respectively.[36]

Mucoproteins and uronic acid-containing mucopolysaccharides are probably present in PN. The occurrence of unsulfonated hyaluronic acid has been histochemically demonstrated in the axoplasm and between the myelin and neurilemma sheath of amphibian and mammalian species, particularly at the nodes of Ranvier.[50] Chondroitinsulfuric acid esters are probably present in PN, but histochemical evidence for any polysaccharides in myelin appears very weak, apart from Wolman's histochemical identification of a weak acid mucopolysaccharide in lipid-extracted myelin of PN, bound to lipid and protein structures.[6]

D. Minerals

Ion content and concentration in PN have been reviewed in recent years[6,7,11] and do not need to be extensively treated here. Values for the whole PN are, in addition, of little interest, unless extracellular space is determined. Another complication is the Schwann cell layer, which may have an internal composition different from that of the axon itself. Better results have been obtained with giant axons or extruded axoplasm.[51,52] In the medial giant axon of the ventral nerve cord of the crayfish, the following ion concentrations have been found (mM $\pm$ SE): Na, 17.4 $\pm$ 0.4; K, 265 $\pm$ 1.8; and Cl, 12.7 $\pm$ 0.4.[51] Ratios are similar to those found in other invertebrate nerves,[53] but Na and Cl concentrations are much lower than those found in earlier investigations in several animal species. The major inorganic electrolytes in PN also have been examined by Lewis.[54]

E. Nucleic Acids and Nucleotides

Values for RNA and DNA content in whole PN have little interest, owing to the contribution made to axonal material by the Schwann cell and the myelin sheath. Avian[35] and cat[16] sciatic nerves contain 180 and 51 μg

P-RNA/g wet wt., respectively, while avian,[35] cat,[16] rabbit sciatic,[9] and rabbit vagus nerves[2] possess 104, 58, 55, and 162 μg P-DNA/g wet wt., respectively. Guinea pig sciatic nerve contains 194 μg P-RNA + P-DNA/g wet wt.[17] DNA and RNA purine and pyrimidine bases, and nucleotide components of RNA from whole bullfrog sciatic nerves have been determined.[55]

RNA has been demonstrated in the Mauthner axon and myelin sheath from growing goldfishes,[56,57] in the motor axons of the cat XIth cranial nerve root,[58] and in other PN structures. The total amounts of RNA decrease proximodistally throughout the axon,[56,57] whereas its concentrations decrease proximodistally only in the first two-thirds of the axon and then increase again.[57] RNA concentration in Mauthner giant axon axoplasm (0.03–0.07% of wet wt.) is about one twentieth to one fortieth that of perikarya,[56,57] but the total amount of RNA per one axon is three times that in the perikaryon (6000 pg axonal RNA against 2000 pg of nuclear RNA)[56]; myelin sheath RNA content averages to about 8000 pg.[57] Concentration in the XIth cranial nerve axon is from 0.0015 to 0.0042% (80–225 pg/10 cm of axon),[58] and this value is one five-hundredth to one one-thousandth that of cell bodies. In rabbit nerve axons, a higher value (35–40 pg/cm axon) has been found,[59] which corresponds to a concentration of about 0.006% (one two-hundredth to one four-hundredth of the nerve cell content).

Axonal and myelin RNA is of ribosomal type.[57–59] Local protein synthesis by denuded myelin-free axons,[59] Mauthner axons,[60] nerve endings, and synaptosomes has been reported.

F. Amino Acid and Related Compounds

1. Content in Peripheral Nerve

Various reports have appeared on the content of free amino compounds in PN,[7,8,10,11,61–67] and technical problems of their extraction have been discussed. Table VII gives some data.

GSH, Arg, His, Ile, Lys, Tyr, Orn, and Phe occur in very low amounts in hen PN.[65] N-acetyl-aspartic acid (NAA) and N-acetyl-α-aspartyl-glutamic acid (NAAGA) are present in PN (< 0.1 μmole and 1–1.5 μmole/g wet wt., respectively),[61,64,68] and N-acetyl-histidine occurs in fish, amphibian, and reptile nerves (2.05–17.5 μg/g wet wt.) but not in those of selachian species, cyclostoms, mammals, and birds.[69] Cys, Dopa, met, pro, trp, and ethanolamine are probably all present in PN, but in undetectable amounts, and GABA is absent or present only in traces.[61,64,65] A phosphopeptide has been found in hen sciatic nerve (0.63 μmole asp equivalents/g wet wt.),[47,70] and peptides also have been detected in squid giant nerve axoplasm.[8] Glycine betaine (73.65–119.11 μmole/g fresh axoplasm) and homarine (20.36–21.40) occur in squid giant nerve,[8,52] as well as cysteic acid amide (0.42–4.86 μmole/g), met sulfoxide, and citrulline (0.32 and 0.43 μmole/g, respectively).[8,67] Choline also has been detected in hen, frog, and rabbit PN.

TABLE VII

Amino Compounds (AC) in PN[a]

(AC)	Rat S[64]	Rat V[64]	Hen S[65]	Frog S[66]c	Crab PN[54]	Lobster PN[54]	Cuttlefish[54,61]	Squid[52,54,67]d	
Asp	0.68 ± 0.06	1.97 ± 0.21	1.1	0.20	138	112	28.5, 82	62.5	111.3
								79.1	
Glu	0.72 ± 0.09	1.66 ± 0.14	1.7	0.45	35	25	17.5, 39	13.5	27.6
								21.2	
Gln			0.50						
GABA	<0.1	<0.1	1.5[b]				<0.1		
Gly			0.56		<5	35	<5	11.6	11.0
Ala			0.44		33	33	21	14.0	
								8.6	9.8
Ser			0.34					9.0	
								4.0	1.0
Thr			0.20					5.7	
Tau			0.52		65	12	103	2.0	0.32
								76.0	33.1
Leu			0.14					106.7	
Val			0.15					2.9	0.21
Arg								2.4	0.53
Lys								3.5	4.3
Met								2.6	0.22
Orn								0.54	0.32
Phe								1.9	0.32
Tyr								0.65	0.16
								0.76	0.32

[a] Values in micromoles per gram wet weight. For abbreviations, see Table III.
[b] Inconsistently found.
[c] Desheathed peripheral nerve.
[d] Giant nerve axoplasm. First column, *Loligo pealii*; second column, *Dosidicus gigas*.

2. Differences Among Nerves

GABA is nearly absent in vertebrate PN,[61,64,65] but it is present in crustacean PN.[71] Inhibitory axon fibers of lobsters possess much higher concentrations (0.1 M) than excitatory axons (less than 1% of that amount).[71]

Much larger concentrations of free amino acid exist in invertebrate[8,52,54,61,67] than in vertebrate PN.[8,62–66] In squid axoplasm, of the 520 μmole/g of total base, 72% are balanced by organic acids: 15% by Asp and Glu, and 42% by isethionic acid, the hydroxy analog of Tau, which is also present in large amounts.[52] Differences in amino acid content exist between *Loligo pealii* and *Dosidicus gigas* axoplasm.[8,52,67]

Amino acid content of rat spinal roots is similar to that of rat sciatic nerve, while it is higher in rat vagus nerve.[64] In the rabbit unmyelinated vagus, however, total acid-soluble amino-N content (190 μmole/g wet wt.) is lower than in tibial PN (232 μmole/g).[2]

3. Differences Between PN and Other Nervous Tissues

Brain tissue of vertebrates contains higher concentrations of Glu, Gln. GABA, and NAA than PN.[8,61–66] The content of other amino compounds does not vary noticeably among vertebrate brain, spinal cord, and PN[8] or is even higher in PN.[63,65] A much lower amount of N-acetyl-histidine has been found in PN than in brain and spinal cord,[69] whereas almost similar concentration of NAAGA has been measured.[68] A phosphopeptide compound found in nervous tissues is more concentrated in PN than in brain, though less than in spinal cord.[47,70] Higher levels of acid-soluble amino-N are present in optic nerve than in tibial PN.[9]

G. Phosphorus-Containing Compounds

Levels of phosphorylated compounds in PN have been reported.[7,8,10,11,62,70,72] Methods of extraction have been discussed.[36,73]

Levels of total acid-soluble P and Pi range between 280–520 and 50–250 μg P/g wet wt. of nerve, respectively.[7,10,11,17,72,74] There is no difference in the content of rabbit vagus and tibial nerves,[2] whereas strong variation exists for Pi content between rat (21.9 $\pm$ 1.6 mg P/100 g wet wt.) and cat (5.28 $\pm$ 0.24) sciatic nerves.[74] Abood and Gerard have found 2.5 times more phosphorylated compounds in spinal than in sciatic PN of frog.[75] ATP, ADP, and AMP have been determined in PN, with conflicting quantitative results (μmole/Kg wet wt.): ATP (1260 and 1030 in rabbit tibial nerve,[36,73] 760 and 860 in rat sciatic nerve,[74,75] 680 in rat spinal nerve,[75] 162–183 in cat sciatic,[74] and 1000–2000 in frog sciatic nerve[76]); ADP (146 and 150 in rabbit tibial,[36,73] 200 and 1110 in rat sciatic,[74,75] 370 in rat spinal nerve,[75] and 166–300 in cat sciatic[74]); and AMP (19 and 30 in rabbit tibial,[36,73] 300 and 1190 in rat sciatic,[74,75] 500 in rat spinal nerve,[75] and 255–274 in cat sciatic[74]). ATP is metabolically related to PN creatine phosphate, whose content is 1–1.5 μmole/g wet wt. of nerve.[36,73] The concentrations of ATP, ADP, AMP, and creatine phosphate of rabbit tibial nerve are about one third of the levels of mouse brain and half the level of rabbit vagus.[36] NAD content of frog sciatic nerve has been reported[76]: 85% of the coenzyme is in the oxidized state.

Thiamine pyrophosphate levels are two times higher in rat brain and spinal cord than in sciatic nerve.[77] Values of Glu-6-P (45 μmole/Kg wet wt.), Fru-1,6-dP (10 μmole), and α-glycerophosphate (58–86 μmole) content in PN have been reported.[36,46,73] The concentration of the first two phosphoric esters is about one third that of the brain,[36] while that of glycerophosphate is about one sixth. Levels of hen PN phosphorylserine (0.04 μmole/g wet wt.), phosphorylethanolamine (0.13–0.19), phosphorylcholine (0.09–0.11), glycerylphosphorylethanolamine (0.15), and glycerylphosphorylcholine (0.16–0.26) have been reported[46,62,70,72] and found to be generally higher than in brain and spinal cord.[70] Extracts of hen PN contain the L-serine ethanolamine phosphate,[47] and various unidentified P-containing compounds.[72]

H. Other Compounds

Lactate concentration in rabbit tibial nerve is 1.04–1.11 μmole/g wet wt.,[36,73] while that of malate in bullfrog PN is very low (1×10^{-10} mole/mg nerve protein).[66] Total creatine of rabbit tibial nerve amounts to 3.17 μmole/g[73] and total free and phosphorylated thiamine, to 1.5–1.6 μg/g.[77] The thiamine content of PN is less than one third that of CNS,[77] and ascorbic acid is one fourth to one fifth that of central white matter.[6]

ACh is particularly abundant in the presynaptic axoplasm. It is sequestered within special granules of the PN, bound to a storage protein and stored in the synaptic vesicles. The amount of ACh in PN[11,78,79] varies enormously with the state of the animal and with the part and type of the nerve under investigation. There are some sympathetic nerves that contain up to 200 nmole ACh/g. Normally, cat preganglionic autonomic trunks, sympathetic ganglia, ventral roots, and predominantly motor nerves all contain high amounts of ACh (10–40 μg/g), mixed nerves less, and sensory nerves very little (< 1 μg/g).[79] ACh levels in cat and rabbit ventral roots (78.1 μmole $\pm$ 22.7 SD and 48.0 $\pm$ 19.0/kg wet wt., respectively) and in sciatic nerves (16.6 $\pm$ 7.3) have been recently reported, together with the presence along the trunks of a concentration gradient.[80]

Low amounts of catecholamines (CAT) are present in PN; probably they occur mostly in autonomic or vegetative nerves. Noradrenaline (NA) is the main CAT in mammalian nerves,[81,82] whereas in frogs, toads, and salamanders the CAT is represented mainly if not absolutely by adrenaline (A).[81] 3-Hydroxytyramine (dopamine) also is present.[82]

Quantitative estimations of CAT in PN and distribution studies of NA and A have been carried out.[79,82,83] In most nerves A constitutes only about 2% of the total CAT content, with few exceptions. The NA levels vary greatly among nerves; from 10 ng/g wet wt. in dog ventral roots to 62 ng/g in dorsal roots; 100 ng/g in vagus nerve; and 200–1000 ng/g, 4000 ng/g, and up to 15,000 ng/g in saphenous, splanchnic, and splenic nerve, respectively.[82] NA is barely detectable in cat sciatic nerve.[84] Adrenergic axons have the CAT stored in special granular structures[6,82] of 0.05–0.2 μ diameter. 5-HT is present in very low amounts (about 10 ng/g wet wt.) in cat sciatic nerve[84] and in barely detectable amounts in PN of other vertebrates.[81,85] It is certainly present, on the contrary, in invertebrate nerves.[81,86]

Histamine content of cutaneous sensory nerves has been reported: preganglionic and postganglionic autonomic nerves contain high concentrations (2–30 μg/g wet wt.), but the content is certainly related to the presence of various mast cells in the PN perineurium.[87] Homogenates of ox splenic nerve may contain up to 100 μg of histamine per gram wet weight[82] which does not seem to be bound to a particular sedimentable nerve fraction. Dog dorsal or ventral roots do not contain detectable amounts of histamine.

Substance P, which is a polypeptide of neurobiological interest probably concentrated in the synaptosomal fraction, has been found in the PN of various animals, ranging in dog PN from 5 up to 40 units/g.[88]

III. ENZYMES OF PERIPHERAL NERVES

A. Occurrence and Localization in Nerve

1. Introduction

Biochemical studies of enzymic activities of PN are few compared to those made on CNS. Probably most of the nerve enzymes originate in the nerve cell body, although recent studies have demonstrated that axons synthesize enzyme and nonenzyme protein locally, this mechanism being at least concerned with the renewal of the enzyme content of the axolemmal membrane.[60,89] A large part of the enzyme content of the myelin sheath is synthesized by the Schwann cells, which in addition build up their own enzyme material. PN myelin seems devoid of most enzymes,[32] though various histochemical reports indicate that several of them are present.[90,91] Various enzymic activities of PN have been observed first during investigations on Wallerian changes.[92]

2. Hydrolases and Transferases

Both true (EC 3.1.1.7) and pseudocholinesterase (EC 3.1.1.8) activity, which we will indicate as AchE and chE, respectively, have been found in PN[92,93]: the ratio of their activities is not always clear, although it has been noted recently that in most nerve 87–96% of the total cholinesterase activity is due to AchE.[93] No clear-cut difference in AchE activity exists between sensory and mixed nerves, whereas the activity of dog and cat autonomic nerves is much higher than that of somatic nerves.[93] Also, adrenergic fibers and nerve trunks, containing many adrenergic axons, show variable AchE and chE activity.[93] A progressive decrease of AchE along PN in a proximodistal direction has been detected.[93]

The observation that chE is largely confined to the Schwann cell of myelinated nerve and not to the axon proper[94] has been confirmed.[6,92,95] AchE is wholly neuronal in origin,[92,95] occurring within PN axons and neurilemma sheaths and not in the Schwann cells.[90,95] Contrary to histochemical findings, no cholinesterase activity has been found in the PN myelin.[32] AchE has been observed in the axonal membrane structure and not in the axoplasm of squid giant axons. More recently, the occurrence of the enzyme at the surface membrane of the axons, in the elements of the endoplasmic reticulum and in the pre- and postsynaptic membranes, has been established by various technical means,[6,96−98] while chE has been located in the dense layers of the myelin lamellae.[98]

Choline acetylase (EC 2.3.1.6) activity (ChAc) has been observed in several PN,[99] showing a proximo-distal activity gradient.[99] The enzyme is more concentrated in ventral than in dorsal roots[6] and is localized in the axonal structure,[92,99,100] probably attached to the microsomal membranes.[101] Acetyl-CoA synthetase (EC 6.2.1.1), on the other hand, is probably located in the Schwann cell[100]; it seems therefore that ACh is formed only by the concerted action of enzymes from Schwann cells and axoplasm.

Aspecific carboxyesterases (EC 3.1.1.1) exist in PN,[92,93,102] probably located in the Schwann cell[102] but not in myelin.[32] Other workers report that the enzyme is axonal in location and shows a proximodistal concentration gradient.[92]

Various phosphoesterases have been found in PN, such as alkaline phosphatase, acid phosphatase, glucose-6-phosphatase, AMPase, inosine diphosphatase, ATPases, CTP phosphatase, PPi phosphatase, and others.[6,92] Various phosphoesterases have been localized in PN myelin,[90] but subsequent work has not confirmed the finding.[32]

Contrary to earlier work,[90,92] alkaline phosphatase has been located in PN myelin,[32] at the nodes of Ranvier and incisures of Schmidt-Lanterman,[103] and in the vascular endothelium of the PN[90]; and acid phosphatase and AMPase, in axons[44] and probably in the lysosome-like particles of Schwann cells[104] and axons. Phosphatases hydrolyzing lipid phosphate esters and nucleoside diphosphate esters have been reported to occur in PN.[105,106]

A Na^+,K^+-activated, Mg^{2+}-dependent ATPase occurs in the membranes of the endoplasmic reticulum and in the neuronal membranes of PN.[107] An ATPase of undetermined type, probably a Ca^{2+}-activated ATPase,[6] is 100 times more concentrated in the sheath than in the axoplasm of the squid.[108] A Mg^{2+}-activated ATPase has been found in rabbit, dog, and cat sciatic nerves, being mostly localized in the microsomal fraction, in the Schwann cell, and along the axon but not in the myelin sheath.[109]

Peptidyl peptide hydrolase (EC 3.4.4.), acting at neutral pH, has been found in PN[12] and is equally distributed along the whole length of the nerve. The enzyme is particularly associated with PN myelin[15,32,110,111]; accordingly, the unmyelinated lobster leg nerve contains no neutral proteinase activity.[111] A peptidyl peptide hydrolase, acting at acidic pH, also has been observed in myelinated[110] and unmyelinated PN.[47] Acid proteinase appears to be associated with lysosome particles.[110,112] Aminopeptidase (EC 3.4.1.2; EC 3.4.1.3) activity has been observed in PN, being in part located in myelin fractions and cellular components.[2,32,113]

Glycoside hydrolases (EC 3.2.1.) and pentosyltransferases (EC 2.4.2.) have been detected in rabbit and cat PN, as have β-glucuronidase (EC 3.2.1.31), β-galactosidase (EC 3.2.1.23), and β-glucosidase (EC 3.2.1.21).[2,92,114] Also adenosine deaminase (EC 3.5.4.4) and purine nucleoside phosphorylase (EC 2.4.2.1) have been found.[2,114]

Hexokinase (EC 2.7.1.1), glycogen phosphorylase (EC 2.4.1.1), and aspartate aminotransferase (EC 2.6.1.1) have been found and measured in the squid axoplasm, together with glucosephosphate isomerase (EC 5.3.1.9) and oxido-reductases.[13] The enzymes are also present in the squid axon sheath, which seems to depend on the citric acid cycle much more than the axoplasm does.[13] Aspartate aminotransferase also occurs in rabbit myelinated and unmyelinated fibers.[115]

GABA-aminotransferase (EC 2.6.1.19) has not been detected in most mammalian PN, with some exceptions.[116] This enzyme, together with succinic semialdehyde dehydrogenase (EC 1.2.1.17), is on the contrary very active in the lobster axons.[117] The GABA-cycle enzymes, together with glutamate decarboxylase (EC 4.1.1.15), are probably neuronal and are not associated with axis cylinders or myelin.

Catechol-O-methyl transferase (EC 2.1.1.6), which is the major enzyme in the CAT inactivation, and histamine methyl transferase (EC 2.1.1.8) are widely and equally distributed in sensory, motor, parasympathetic, and sympathetic PN.[118]

3. Oxido-Reductases and Lyases

A detectable proximodistal gradient of O_2 uptake exists in rat sciatic nerve, which may partially explain the greater vulnerability of the distal segment of a PN. High oxido-reductase activity in a PN normally exists at the nodes of Ranvier, in the bodies of neurilemma cells, and in the axonal mitochondria, while lower activity is present inside the axon and in the perineural epithelium.[6] Schwann cells, together with the axon, contain various oxidative enzymes.[32] The occurrence of oxidative enzymes in PN myelin[91] has not been confirmed.[32]

Biochemical investigations have indicated the presence in PN of several oxido-reductases.[2,6,13,14,32,115] Glycerophosphate dehydrogenase has been reported to be axonal in distribution[2] and less concentrated in the vagus than in the tibial nerve.[2] Enzymes of Krebs cycle are present both in myelinated and unmyelinated nerves of the rabbit,[115] with a close association of aspartate aminotransferase and malate oxidoreductase activity.

Isocitrate oxidoreductase activity is higher in unmyelinated fibers[115] and is low in squid giant axoplasm.[13] Cytochrome oxidase and succinic oxidoreductase are both present in lobster leg and claw nerves and in squid giant fibers,[14] which possess, however, very low activity of the second enzyme. The sheath cells of the squid axon lack glutamate oxidoreductase[13] but depend much more than the axoplasm on the citric acid cycle; correspondingly, the sheath accounts for more than 50% of the total respiration of a squid giant fiber.[14]

Many oxidoreductases of PN act in the mitochondrial structures.[7,25] Subcellular particles from rat sciatic and spinal nerves contain the system of oxidative phosphorylation.[75] Bullfrog, lobster, and rabbit PN possess all the enzymic machinery for the conversion of CO_2 to Krebs cycle intermediates and related amino acids,[66] the conversion being more evident in the myelinated sciatic nerve of the frog than in the unmyelinated nerve of the walking legs of the lobster. Lobster nerve also converts acetate, pyruvate, and citrate into acetyl-CoA and ACh.[119]

Tyroxine hydroxylase occurs in bovine splenic nerve,[120] together with the dopamine β-hydroxylase (EC 1.14.2.1), which converts dopamine into NA; the latter enzyme is located in the specific NA-storing granules.[120] Monoamine oxidase (EC 1.4.3.4) of PN[121] probably acts either on tyramine or 5-HT.[121] The enzyme, which is indicative of adrenergic function, also is found in predominantly cholinergic autonomic neurons and in sensory and motor cutaneous nerves.[122] Succinic semialdehyde dehydrogenase has been discussed in Section III,A.2.

Lactate and malate dehydrogenase isoenzymes occur in PN.[23,34,123,124] Variation of the isoenzymic pattern of lactate dehydrogenase has been observed among different cutaneous and muscular nerves[123] and between PN and CNS.[23,124]

Fumarate hydratase (EC 4.2.1.2), mostly concentrated at the nodes of Ranvier, is present in myelinated and unmyelinated nerves.[115] Hydroxytryptophan carboxylyase (EC 4.1.1.28), probably similar to DOPA carboxylyase (EC 4.1.1.26), has been studied in the rabbit trigeminal nerve and spinal root.[121] DOPA-carboxylyase[79,82] is particularly concentrated in the postganglionic nerve trunks and sympathetic ganglia[82] and is not localized in particulate nerve fractions.[120] Histidine decarboxylase (EC 4.1.1.22) is present in postganglionic autonomic fibers.[87]

Glutamate carboxylyase (EC 4.1.1.15), which generates GABA, is probably a neuronal enzyme.[116] It is present in low amounts in rabbit sciatic nerve[125] and other vertebrate PN,[6,64] being, on the contrary, fairly active in lobster PNS.[126]

4. Protein- and Lipid-Synthesizing Enzymic Systems

Invertebrate and vertebrate peripheral axons probably have the enzymic system involved in local protein synthesis and also a DNA-dependent RNA synthesis.[59,60,89] The system probably is located within the axonal membranes and axoplasmic particulate fractions[59,89] and also is present in the surrounding myelin sheath.[60] Schwann cells are also active in this connection.

PN possesses the enzymic system that synthesizes neutral lipid, CHOL, and phospholipid from various precursors.[46,74,127,128] Choline kinase (EC 2.7.1.32), phosphatidic acid phosphoesterase (EC 3.1.3.4), cholinephosphotransferase (EC 2.7.8.2), glycerolphosphate acyltransferase (EC 2.3.1.15), and probably cholinephosphate cytidylyltransferase (EC 2.7.7.15) all have been found in PN.[106,128-131] The enzymic system that forms phosphoinositides also is present in PN,[127] as well as the mechanisms for regulating fatty acid synthesis.[132]

B. Levels of Enzymic Activities in Peripheral Nerve

1. Enzyme Content in Nerves

Absolute values for enzymic activities of PN and myelin-free axons have been reported by various workers. To integrate most of the data in biochemical terms, however, it would be necessary to take into account the Michaelis constant and local concentrations of metabolites and coenzymes. Most of the findings that are listed here will suffer therefore from these limitations.

Quantitative levels of AchE,[89,93,96] chE,[95] aliesterase,[93,102] ChAc,[100] and acetyl-CoA synthetase[100] have been published. Acid phosphatase,[2,104,114] alkaline phosphatase,[114] some phosphomonoesterases,[105] phosphodiesterase,[106] and ATPase[107] activities also have been determined. Peptide hydrolases also have been estimated,[2,12,15,32,110] as well as glycoside hydrolases.[2,114] Additional enzymic activities, such as those of hexokinase,[13] glycogen phosphorylase,[13] aspartate aminotransferase,[13,115] glucosephosphate isomerase,[2,13] GABA-aminotransferase,[116] histidine-N-methyl transferase,[118] purine nucleoside phosphorylase,[2] aldolase,[2] choline kinase,[2,128,129] phosphatidic acid phosphatase,[130] phosphocholine glyceride transferase,[128,131] cholinephosphate cytidylyltransferase,[106,128] glycerolphosphate acyltransferase,[128] glutamate carboxylyase,[125,126] 5-hydroxytryptophan carboxylyase,[121] and fumarase,[2,115] have been determined in PN. Of the dehydrogenases, quantitative values may be found for PN isocitric oxidoreductase, glycerophosphate dehydrogenase, lactic dehydrogenase, malic dehydrogenase, "malic enzyme," glucose-6-phosphate dehydrogenase, NAD-dependent isocitric dehydrogenase, 6-phosphogluconic oxidoreductase, monoamine oxidase, cytochrome oxidase, and succinic dehydrogenase.[2,13,14,34,115,121,124]

2. Variations Among Nerves and Between PN and Other Nervous Tissues

α-Glycerophosphate dehydrogenase activity of vagus nerve is only 20% that of rabbit tibial PN.[2] Adrenaline-methyltransferase and histamine-methyltransferase activities are quantitatively similar in sensory, motor, parasympathetic, and sympathetic PN of the monkey.[118] Large variations exist for isocitric dehydrogenase, lactic dehydrogenase, and phosphogluco-isomerase between vertebrate PN and squid giant axoplasm.[2,13]

Glutamate carboxylyase activity of whole brain is much higher than in PN,[125] as are monoamine oxidase, 5-hydroxytryptophan carboxylyase,[121] and GABA-oxoglutarate aminotransferase.[116] The activities of various enzymes are higher in optic nerve and central white matter than in rabbit tibial PN.[2] The properties of some enzymes of giant axoplasm are similar to those of rabbit brain.[13]

IV. PERIPHERAL NERVE MYELIN CONSTITUTION

A. Protein and Lipid Content

Work on CNS myelin isolation has appeared, but isolation of PN myelin has been accomplished by only a few workers.[3–5,133] The problem of inter-preting results obtained with whole PN and PN myelin is still formidable.

It is very difficult, at the moment, to state what relationship the PN proteins bear to the PN myelin protein. Probably TRPR, proteolipid protein, and trypsin-digestible protein are present in myelin, but experimental evidence indicates that at least the first two proteins are not solely myelin proteins.[25,26,29] Proteins of PN myelin recently have been studied and found to be almost insoluble in chloroform–methanol and partly digestible with trypsin.[133] No comparison has been made between the saline-extract-able protein of PN and those of the corresponding myelin. Recently, basic proteins of PN myelin have been separated in at least four different electro-phoretic bands of almost equal intensity.[133]

Although various works have indirectly indicated the abundance in myelin of CER, EPL, SP, PhS, TPI, and CHOL, there is no direct proof that these lipids are peculiarly myelin lipids. The lipid composition of CNS myelin of various animal species is similar, whereas the lipid:protein ratio is variable.[134] This assumption also may be valid for PN myelin. Although whole intradural spinal roots contain some triglyceride, the corresponding myelin has only traces of that material.[3] PN myelin contains only few enzymes.[32]

B. Comparison Between PNS and CNS Myelins

Proteolipid protein appears to be a myelin constituent of CNS, whereas the corresponding trypsin-resistant protein of PN is probably TRPR.[15,31] Protein-digestible protein is present in both CNS and PNS myelin, though in greater concentrations in the latter. Owing to the fact that proteolipid protein and TRPR are probably similar,[31,33] variations in the protein composition of CNS and PNS myelin are therefore more of a quantitative than a qualitative kind. Probably the qualitative difference, as regards protein constitution,[27,28] resides in the different way in which protein is linked to lipid. It has been recently observed, however, that also the amino acid pattern and electrophoretic properties of PNS and CNS myelin proteins are different.[133]

It has been found that, although the phospholipid composition of PNS and CNS myelin is constant,[4,27] the phospholipid to CHOL ratio in PN myelin is higher than in CNS myelin.[4] Recently, it has been observed, however, that also the lipid composition of PNS myelin differs from that of the CNS[3]: PNS myelin contains twofold higher proportions of SP and correspondingly lower proportions of CER than CNS myelin, as also observed in works on whole PN.[10,37,39] Also, fatty acid and fatty aldehyde composition of PNS myelin lipid is rather different from that of the corresponding lipid of CNS myelin.[4]

PNS myelin contains higher proteinase and alkaline phosphatase activity than CNS myelin, which, on the other hand, possesses higher leucine aminopeptidase activity.[32]

V. DEVELOPMENTAL CHANGES AND NUTRITIONAL ASPECTS

Developmental studies on the variation of the concentration of substances in PN with the age of the animal are scanty, and so are the enzymic changes. Certainly, the rate of growth of a PN is regulated by the rates of synthesis and of catabolism of PN proteins: in this connection, a general increase of both acid and neutral proteinase activities has been found in rat PN according to the development from birth to young adult levels.[47] However, it is not fully known what the time relationship is of enzymic activity changes between nerve cells and the long peripheral axonal process during development. No histochemical investigations have been made on proteins, proteolipids, and TRPR of PN during development and myelination. The acidic protein S-100, found in adult nerve tissue,[24] is absent at birth.[135] In chicken and rat PN a greater rate of amino acid incorporation into protein has been described in immature animals,[136] and this result parallels the relative decrease of proteinase activity of PN during development.[47] Concurrently, there is a change of the RNA content of PN but not of DNA content.[136]

Lipid compositional changes, described in brain tissue during development,[6] have been studied very little in PN and its myelin.

Almost nothing is known about developmental changes in enzymes in PN, except for changes of cholinesterase, alkaline phosphatase, AMPase, diaphorase, and some oxidative enzymes. Also, glycogen phosphorylase activity seems to develop just at the time of the appearance of the myelin sheath.[6] Quantitative changes of isoenzymic patterns of lactate dehydrogenase have been studied in PN during development and maturation,[124] as well as those of enzymes related to lipid biosynthesis.[130,131]

The amino acid pattern of adult chicken PN is not appreciably altered by starvation.[47] The effect of starvation on chicken and rat PN lipid content and fatty acid pattern has been studied.[41,46,137] Fatty acid composition of PN lipids has also been examined in regard to different natural diets.[40,49] Finally, seasonal variations have been observed for the rate of CO_2 fixation into Krebs cycle intermediates and related amino acids in the lobster nerve.[138]

VI. REFERENCES

1. A. Sheltawy and R. M. C. Dawson, The polyphosphoinositides and other lipids of peripheral nerves, *Biochem. J.* **100**:12–18 (1966).
2. R. E. McCaman, in *Ultrastructure and Metabolism of the Nervous System* (S. R. Korey, A. Pope, and E. Robins, eds.), pp. 169–181, Williams and Wilkins, Baltimore (1962).
3. J. S. O'Brien, E. L. Sampson, and M. B. Stern, Lipid composition of myelin from the peripheral nervous system. Intradural spinal roots, *J. Neurochem.* **14**:357–366 (1967).
4. M. J. Evans and J. B. Finean, The lipid composition of myelin from brain and peripheral nerve, *J. Neurochem.* **12**:729–734 (1965).
5. L. A. Horrocks, Composition of myelin from peripheral and central nervous systems of the squirrel monkey, *J. Lipid Res.* **8**:569–577 (1967).
6. R. L. Friede, *Topographic Brain Chemistry*, Academic Press, New York (1966).
7. R. J. Rossiter, in *Neurochemistry* (K. A. C. Elliott, I. H. Page, and J. H. Quastel, eds.), pp. 10–54, Charles C. Thomas, Springfield, Illinois (1962).
8. H. H. Tallan, in *Amino Acid Pools* (J. T. Holden, ed.), pp. 471–485, Elsevier, Amsterdam (1962).
9. R. E. McCaman and E. Robins, Quantitative biochemical studies of Wallerian degeneration in the peripheral and central nervous systems. I. Chemical constituents, *J. Neurochem.* **5**:18–31 (1959).
10. W. S. Spector (ed.), *Handbook of Biological Data*, W. B. Saunders, Philadelphia (1956).
11. G. B. Ansell, in *Biochemists' Handbook* (C. Long, E. J. King, and W. M. Sperry, eds.), pp. 661–665, E. and F. N. Spon, London (1961).
12. G. Porcellati and B. Curti, Proteinase activity of peripheral nerves during Wallerian degeneration, *J. Neurochem.* **5**:277–282 (1960).
13. N. R. Roberts, R. R. Coelho, O. H. Lowry, and E. J. Crawford, Enzyme activities of giant squid axoplasm and axon sheath, *J. Neurochem.* **3**:109–115 (1958).
14. J. M. Foster, Enzymatic studies of mitochondria and other constituents of lobster and squid nerve fibres, *J. Neurochem.* **1**:84–90 (1956).
15. C. W. M. Adams and N. A. Tuqan, Histochemistry of myelin. II. Proteins, lipid-protein dissociation and proteinase activity in Wallerian degeneration, *J. Neurochem.* **6**:334–341 (1961).
16. J. E. Logan, W. A. Mannell, and R. J. Rossiter, Estimation of nucleic acids in tissue from the nervous system, *Biochem. J.* **51**:470–482 (1952).
17. A. J. Samuels, L. L. Boyarsky, R. W. Gerard, B. Libet, and M. Brust, Distribution, exchange and migration of phosphate compounds in the nervous system, *Am. J. Physiol.* **164**:1–15 (1951).
18. M. Abercrombie and M. L. Johnson, Collagen content of peripheral nerve, *J. Neurol. Neurosurg. Psychiat.* **9**:113–125 (1946).
19. M. Maxfield and R. W. Hartley, Dissociation of the fibrous protein of nerve, *Biochim. Biophys. Acta* **24**:83–87 (1957).

20. F. Huneeus-Cox, Electrophoretic and immunological studies of squid axoplasm proteins, *Science* **143**:1036–1037 (1964).

21. B. A. Koechlin and H. D. Parish, The amino acid composition of a protein isolated from lobster nerve, *J. Biol. Chem.* **205**:597–604 (1953).

22. A. V. Palladin, Ya. V. Belik, N. M. Polyakova, and T. P. Silich, *in Problems of the Biochemistry of the Nervous System* (A. V. Palladin, ed.), pp. 3–17, Pergamon Press, London (1964).

23. A. Lowenthal, *in Protides of the Biological Fluids* (H. Peeters, ed.), Vol. 13, pp. 197–199, Elsevier, Amsterdam (1966).

24. B. W. Moore, A specific nervous tissue protein, *Intern. Neurochem. Conf., Oxford* (July 25–30, 1965), p. 82.

25. J. Folch-Pi, *in Protides of the Biological Fluids* (H. Peeters, ed.), Vol. 13, pp. 21–34, Elsevier Amsterdam (1966).

26. F. Wolfgram, Macromolecular constituents of myelin, *Ann. N.Y. Acad. Sci.* **122**:104–115 (1965).

27. J. B. Finean, J. N. Hawthorne, and J. D. E. Patterson, Structural and chemical differences between optic and sciatic nerve myelins, *J. Neurochem.* **1**:256–259 (1957).

28. J. Folch-Pi, M. B. Lees, and S. Carr, Studies of the chemical composition of the nervous system, *Exptl. Cell Res. Suppl.* **5**:58–71 (1958).

29. L. Amaducci, A. Pazzagli, and G. Pessina, The relation of proteolipids and phosphatido-peptides to tissue elements in the bovine nervous system, *J. Neurochem.* **9**:509–519 (1962).

30. L. Amaducci, The distribution of proteolipids in the human nervous system, *J. Neurochem.* **9**:153–160 (1962).

31. F. Wolfgram and A. S. Rose, A study of some component proteins of central and peripheral nerve myelin, *J. Neurochem.* **8**:161–168 (1961).

32. C. W. M. Adams, A. N. Davison, and N. A. Gregson, Enzyme inactivity of myelin: Histochemical and biochemical evidence, *J. Neurochem.* **10**:383–395 (1963).

33. F. Wolfgram and A. S. Rose, The amino acid composition of central and peripheral nerve neurokeratin, *J. Neurochem.* **9**:623–627 (1962).

34. A. Lowenthal, M. van Sande, D. Karcher, and J. Richard, *in Comparative Neurochemistry* (D. Richter, ed.), pp. 139–148, Pergamon Press, London (1964).

35. P. J. Heald, H. G. Badman, B. F. Frunival, and P. A. L. Wight, Chemical changes in nerves from birds affected by fowl paralysis, *Poultry Sci.* **43**:701–710 (1964).

36. M. A. Stewart, J. V. Passonneau, and O. H. Lowry, Substrate changes in peripheral nerve during ischaemia and Wallerian degeneration, *J. Neurochem.* **12**:719–727 (1965).

37. G. Branté, Studies of the lipids in the nervous system, *Acta Physiol. Scand.* suppl. 63, **18**:1–186 (1949).

38. G. R. Webster, Studies on the plasmalogens of nervous tissue, *Biochim. Biophys. Acta* **44**:109–116 (1960).

39. A. C. Johnson, A. R. McNabb, and R. J. Rossiter, Lipids of peripheral nerve, *Biochem. J.* **43**:578–584 (1948).

40. J. F. Berry, W. H. Cevallos, and R. R. Wade, Jr., Lipid class and fatty acid composition of intact peripheral nerve and during Wallerian degeneration, *J. Am. Oil Chem. Soc.* **42**:492–500 (1965).

41. C. D. Joel, H. W. Moser, G. Majno, and M. L. Karnovsky, Effects of bis-(monoisopropyl-amino)-fluorophosphine oxide (Mipafox) and of starvation on the lipids in the nervous system of the hen, *J. Neurochem.* **14**:479–488 (1967).

42. J. F. Berry and W. H. Cevallos, Lipid class and fatty acid composition of peripheral nerve from normal and organophosphorus-poisoned chickens, *J. Neurochem.* **13**:117–124 (1966).

43. A. G. Richards, *in Neurochemistry* (K. A. C. Elliott, I. H. Page, and J. H. Quastel, eds.), pp. 818–843, Charles C. Thomas, Springfield, Illinois (1955).

44. W. C. McMurray, J. D. McColl, and R. J. Rossiter, *in Comparative Neurochemistry* (D. Richter, ed.), pp. 101–107, Pergamon Press, London (1964).

45. C. H. Williams, H. J. Johnson, and J. L. Casterline, Cholesterol content of spinal cord and sciatic nerve of hens after organophosphate and carbamate poisoning, *J. Neurochem.* **13**:471–474 (1966).

46. G. Porcellati and M. A. Mastrantonio, Phospholipid metabolism of peripheral nerves during demyelination by organophosphorus compounds. *Ital. J. Biochem.* **13**:332–352 (1964).

47. G. Porcellati, unpublished results.

48. J. S. O'Brien, Stability of the myelin membrane, *Science* **147**:1099–1107 (1965).

49. R. J. Light and D. M. Easton, Saponifiable fatty acids of myelinated and unmyelinated nerve fibers of the garfish, *J. Neurochem.* **14**:141–142 (1967).

50. L. G. Abood and S. K. Abul-Haj, Histochemistry and characterization of hyaluronic acid in axons of peripheral nerve, *J. Neurochem.* **1**:119–125 (1956).

51. B. G. Wallin, Intracellular ion concentrations in single crayfish axons, *Acta Physiol. Scand.* **70**:419–430 (1967).

52. B. A. Koechlin, On the chemical composition of the axoplasm of squid giant fibers with particular reference to its ion pattern, *J. Biophys. Biochem. Cytol.* **1**:511–529 (1955).

53. F. J. Brinley, Jr., Sodium, potassium, and chloride concentrations and fluxes in the isolated giant axon of *Homarus, J. Neurophysiol.* **28**:742–772 (1965).

54. P. R. Lewis, The free amino acids in invertebrate nerve, *Biochem. J.* **52**:330–338 (1952).

55. L. G. Abood, E. Goldman, and V. Lipman, Metabolic studies on phospholipids and nucleic acid of frog nerve during excitation, *J. Neurochem.* **2**:318–325 (1958).

56. J. E. Edström, D. Eichner, and A. Edström, The ribonucleic acid of axons and myelin sheaths from Mauthner neurons, *Biochim. Biophys. Acta* **61**:178–184 (1962).

57. A. Edström, The ribonucleic acid in the Mauthner neuron of the goldfish, *J. Neurochem.* **11**:309–314 (1964).

58. E. Koenig, Synthetic mechanisms in the axon. II. RNA in myelin-free axons of the cat, *J. Neurochem.* **12**:357–361 (1965).

59. E. Koenig, Synthetic mechanisms in the axon. IV. *In vitro* incorporation of ^{3}H precursors into axonal protein and RNA, *J. Neurochem.* **14**:437–446 (1967).

60. A. Edström, Amino acid incorporation in isolated Mauthner nerve fibre components, *J. Neurochem.* **13**:315–321 (1966).

61. Y. Tsukada, K. Uemura, S. Hirano, and Y. Nagata, *in Comparative Neurochemistry* (D. Richter, ed.), pp. 179–183, Pergamon Press, London (1964).

62. I. Montanini and G. Porcellati, Protein metabolism of peripheral nerves during demyelination by organophosphorus compounds, *Ital. J. Biochem.* **13**:230–239 (1964).

63. G. Porcellati and S. Luciani, Distribuzione di amino-composti nel sistema nervoso centrale e periferico, *Boll. Soc. Ital. Biol. Sper.* **36**:213–215 (1960).

64. Y. Nagata, Y. Yokoi, and Y. Tsukada, Studies on free amino acid metabolism in excised cervical sympathetic ganglia from the rat, *J. Neurochem.* **13**:1421–1431 (1966).

65. G. Porcellati and R. H. S. Thompson, The effect of nerve section on the free amino acids of nervous tissue, *J. Neurochem.* **1**:340–347 (1957).

66. L. J. Côté, S. C. Cheng, and H. Waelsch, CO_2 fixation in the nervous system. I. CO_2 fixation in the sciatic nerve of the bullfrog, *J. Neurochem.* **13**:271–279 (1966).

67. G. G. J. Deffner, The dialyzable free organic constituents of squid blood; a comparison with nerve axoplasm, *Biochim. Biophys. Acta* **47**:378–388 (1961).

68. E. Miyamoto and T. Tsujio, Determination of *N*-acetyl-α-aspartylglutamic acid in the nervous tissue of mammals, *J. Neurochem.* **14**:899–903 (1967).

69. V. Erspamer, M. Roseghini, and A. Anastasi, Occurrence and distribution of *N*-acetyl-histidine in brain and extracerebral tissues of poikilothermal vertebrates, *J. Neurochem.* **12**:123–130 (1965).

70. G. Porcellati and S. Luciani, Gli esteri fosforici fosfolipidici del sistema nervoso periferico, *Boll. Soc. Ital. Biol. Sper.* **36**:216–219 (1960).

71. E. A. Kravitz and D. D. Potter, A further study of the distribution of Gaba between excitatory and inhibitory axons of the lobster, *J. Neurochem.* **12**:323–328 (1965).

72. G. Porcellati, The levels of some free nitrogen-containing phosphate esters in nervous tissues, *J. Neurochem.* **2**:128–137 (1958).

73. M. A. Stewart and G. I. Moonsammy, Substrate changes in peripheral nerve recovering from anoxia, *J. Neurochem.* **13**:1433–1439 (1966).

74. W. L. Magee, J. F. Berry, M. Magee, and R. J. Rossiter, Chemical studies of peripheral nerve during Wallerian degeneration. X. *In vitro* incorporation of radioactive inorganic phosphate into phosphatides and acid-soluble phosphorus compounds, *J. Neurochem.* **3**:333–340 (1959).

75. L. G. Abood and R. W. Gerard, Enzyme distribution in isolated particulates of rat peripheral nerve, *J. Cell Comp. Physiol.* **43**:379–393 (1954).

76. P. Greengard, F. Brink, Jr., and S. P. Colowick, Some relationships between action potential, oxygen consumption and coenzyme content in degenerating peripheral axons, *J. Cell. Comp. Physiol.* **44**:395–420 (1954).

77. P. M. Dreyfus, The quantitative histochemical distribution of thiamine in normal rat brain, *J. Neurochem.* **4**:183–190 (1959).

78. C. O. Hebb and K. Krnjević, in *Neurochemistry* (K. A. C. Elliott, I. H. Page, and J. H. Quastel, eds.), pp. 452–521, Charles C. Thomas, Springfield, Illinois (1962).

79. U. S. von Euler, in *Neurochemistry* (K. A. C. Elliott, I. H. Page, and J. H. Quastel, eds.), pp. 426–439, Charles C. Thomas, Springfield, Illinois (1955).

80. C. A. N. Evans and N. R. Saunders, The distribution of acetylcholine in normal and in regenerating nerves, *J. Physiol. (London)* **192**:79–92 (1967).

81. B. B. Brodie, D. F. Bogdanski, and L. Bonomi, in *Comparative Neurochemistry* (D. Richter, ed.), pp. 367–377, Pergamon Press, London (1964).

82. U. S. von Euler, in *Metabolism of the Nervous System* (D. Richter, ed.), pp. 543–552, Pergamon Press, London (1957).

83. E. Florey, in *Neurochemistry* (K. A. C. Elliott, I. H. Page, and J. H. Quastel, eds.), pp. 673–693, Charles C. Thomas, Springfield, Illinois (1962).

84. E. G. Andersen and L. O. Holgerson, The distribution of 5-hydroxytryptamine and norepinephrine in cat spinal cord, *J. Neurochem.* **13**:479–485 (1966).

85. J. H. Welsh, in *Comparative Neurochemistry* (D. Richter, ed.), pp. 355–366, Pergamon Press, London (1964).

86. P. Correale, The occurrence and distribution of 5-hydroxytryptamine in the central nervous system of vertebrates, *J. Neurochem.* **1**:22–31 (1956).

87. G. B. West, in *Metabolism of the Nervous System* (D. Richter, ed.), pp. 578–581, Pergamon Press, London (1957).

88. B. Pernow, Studies on substance P. Purification, occurrence and biological actions, *Acta Physiol. Scand.*, suppl. 105, **29** (1953).

89. E. Koenig, Synthetic mechanisms in the axon. I. Local axonal synthesis of acetyl-cholinesterase, *J. Neurochem.* **12**:343–355 (1965).

90. F. Wolfgram and A. S. Rose, The histochemistry of neurokeratin in normal and degenerating sciatic nerve, *Neurology* **10**:365–371 (1960).

91. H. B. Tewari and G. H. Bourne, Histochemical studies on the distribution of oxidative enzymes in the cerebellum of the rat, *J. Histochem. Cytochem.* **10**:619–627 (1962).

92. R. J. Rossiter, *in Chemical Pathology of the Nervous System* (J. Folch-Pi, ed.), pp. 207–230, Pergamon Press, London (1961).

93. L. Lubińska, S. Niemierko, B. Oderfeld, and L. Szwarc, The distribution of acetylcholinesterase in peripheral nerves, *J. Neurochem.* **10**:25–41 (1963).

94. G. B. Koelle, The elimination of enzymatic diffusion artifacts in the histochemical localization of cholinesterases and a survey of their cellular distributions, *J. Pharmacol. Exptl. Therap.* **103**:153–171 (1951).

95. J. B. Cavanagh, R. H. S. Thompson, and G. R. Webster, The localization of psuedo-cholinesterase activity in nervous tissue, *Quart. J. Exptl. Physiol.* **39**:185–197 (1954).

96. S. Niemierko and L. Lubińska, Two fractions of axonal acetylcholinesterase exhibiting different behaviour in severed nerves, *J. Neurochem.* **14**:761–769 (1967).

97. H. Kewitz and F. Welsch, Separation of a cholinesterase containing membrane fraction from unmyelinated nerves, *Naunyn-Schmiedebergs Arch. Exptl. Pathol. Pharmakol.* **258**:1–10 (1967).

98. P. Kása and B. Csillik, Electron microscopic localization of cholinesterase by a copperlead-thiocholine technique, *J. Neurochem.* **13**:1345–1349 (1966).

99. C. O. Hebb and A. Silver, Gradient of choline acetylase activity, *Nature* **189**:123–125 (1961).

100. J. F. Berry and R. J. Rossiter, Chemical studies of peripheral nerve during Wallerian degeneration. VIII. Acetic thiokinase and choline acetylase, *J. Neurochem.* **3**:59–64 (1958).

101. C. O. Hebb and A. Silver, *in Protides of the Biological Fluids* (H. Peeters, ed.), Vol. 13, pp. 179–180, Elsevier, Amsterdam (1966).

102. J. B. Cavanagh and G. R. Webster, On the changes in ali-esterase and pseudo-cholinesterase activity of chicken sciatic nerve during Wallerian degeneration and their correlation with cellular proliferation, *Quart. J. Exptl. Physiol.* **40**:12–23 (1955).

103. B. Pinner, J. F. Davison, and J. B. Campbell, Alkaline phosphatase in peripheral nerves, *Science* **145**:936–938 (1964).

104. R. O. Weller and R. S. Mellick, Acid phosphatase and lysosome activity in diphtheritic neuropathy and Wallerian degeneration, *Brit. J. Exptl. Pathol.* **47**:425–434 (1966).

105. G. Porcellati and B. Curti, L'effetto della demielinizzazione sperimentale sulla attività di alcune fosfatasi dei nervi periferici, *Biochim. Appl.* **7**:42–56 (1960).

106. J. F. Berry and J. D. Coonrad, Hydrolysis of nucleoside diphosphate esters in peripheral nerve during Wallerian degeneration, *J. Neurochem.* **14**:245–255 (1967).

107. H. S. Bachelard and G. D. Silva, The Na^+,K^+-activated adenosine triphosphatase in degenerating peripheral nerve, *Arch. Biochem. Biophys.* **117**:98–105 (1966).

108. B. Libet, ATPase in nerve, *Federation Proc.* **7**:72 (1948).

109. A. V. Palladin and O. M. Rojmanova, *1st Intern. Meeting Intern. Soc. Neurochem.*, Strasbourg, p. 165 (July 23–28, 1967).

110. G. Porcellati, *in Protides of the Biological Fluids* (H. Peeters, ed.), Vol. 13, pp. 115–126, Elsevier, Amsterdam (1966).

111. C. W. M. Adams and N. A. Tuqan, The histochemical demonstration of protease by a gelatin-silver film substrate, *J. Histochem. Cytochem.* **9**:469–472 (1961).

112. A. Lajtha and N. Marks, *in Protides of the Biological Fluids* (H. Peeters, ed.), Vol. 13, pp. 103–114, Elsevier, Amsterdam (1966).

113. C. W. M. Adams and G. G. Glenner, Histochemistry of myelin. IV. Aminopeptidase activity in CNS and PNS, *J. Neurochem.* **9**:233–239 (1962).

114. M. Arnaki and S. Weissbarth, Studies in demyelination. II. Enzymatic patterns in central nervous system and sciatic nerve in experimental allergic encephalomyelitis, *Proc. Soc. Exptl. Biol. Med.* **116**:210–212 (1964).

115. D. B. McDougal, E. M. Jones, and U. I. Sila, *in Ultrastructure and Metabolism of the Nervous System* (S. R. Korey, A. Pope, and E. Robins, eds.), pp. 182–188, Williams and Wilkins, Baltimore (1962).

116. F. N. Pitts, C. Quick, and E. Robins, The enzymic measurement of Gaba-α-oxoglutaric transaminase, *J. Neurochem.* **12**:93–101 (1965).

117. Z. W. Hall and E. A. Kravitz, The metabolism of Gaba in the lobster nervous system. II. Succinic semialdehyde dehydrogenase, *J. Neurochem.* **14**:55–62 (1967).

118. J. Axelrod, P. D. MacLean, R. W. Albers, and H. Weissbach, *in Regional Neurochemistry* (S. S. Kety and J. Elkes, eds.), pp. 307–311, Pergamon Press, London (1961).

119. R. Nakamura and S.-C. Cheng, *1st Intern. Meeting Intern. Soc. Neurochem.*, *Strasbourg*, p. 160 (July 23–28, 1967).

120. L. Stjärne and F. Lishajko, Localization of different steps in noradrenaline synthesis to different fractions of a bovine splenic nerve homogenate, *Biochem. Pharmacol.* **16**:1719–1728 (1967).

121. R. E. McCaman, M. W. McCaman, J. M. Hunt, and M. S. Smith, Microdetermination of monoamine oxidase and 5-hydroxytryptophan decarboxylase activities in nervous tissues, *J. Neurochem.* **12**:15–23 (1965).

122. K. Yasuda and W. Montagna, Histology and cytochemistry of human skin. XX. The distribution of monoamine oxidase, *J. Histochem. Cytochem.* **8**:356–366 (1960).

123. I. A. Brody, Lactate dehydrogenase isoenzymes: a difference between cutaneous and muscular nerves, *J. Neurochem.* **13**:975–978 (1966).

124. V. Bonavita, *in Protides of the Biological Fluids* (H. Peeters, ed.), Vol. 13, pp. 163–172, Elsevier, Amsterdam (1966).

125. I. P. Lowe, E. Robins, and G. S. Eyerman, The fluorimetric measurement of glutamic decarboxylase and its distribution in brain, *J. Neurochem.* **3**:8–18 (1958).

126. E. A. Kravitz, Enzymic formation of Gaba in the peripheral and central nervous system of lobsters, *J. Neurochem.* **9**:363–370 (1962).

127. G. Porcellati, Organofosforici demielinizzanti e lipogenesi del nervo periferico, *Acta Neurol.* **22**:168–172 (1967).

128. G. Porcellati, The effect of organo-phosphorus compounds on nerve phospholipid metabolism, *Progr. Biochem. Pharmacol.* **3**:49–58 (1967).

129. R. E. McCaman, Intermediary metabolism of phospholipids in brain tissue. I. Microdetermination of choline phosphokinase, *J. Biol. Chem.* **237**:672–676 (1962).

130. R. E. McCaman, M. Smith, and K. Cook, Intermediary metabolism of phospholipids in brain tissue. II. Phosphatidic acid phosphatase, *J. Biol. Chem.* **240**:3513–3517 (1965).

131. R. E. McCaman and K. Cook, Intermediary metabolism of phospholipids in brain tissue. III. Phosphocholine-glyceride transferase, *J. Biol. Chem.* **241**:3390–3394 (1966).

132. Ann H. Hughes and S. G. Eliasson, Synthesis of cholesterol and fatty acids in fractions of peripheral nerve, *J. Clin. Invest.* **39**:111–115 (1960).

133. F. Wolfgram, *1st Intern. Meeting Intern. Soc. Neurochem. Strasbourg*, p 220 (July 23–28, 1967).

134. M. L. Cuzner, A. N. Davison, and N. A. Gregson, The chemical composition of vertebrate myelin and microsomes, *J. Neurochem.* **12**:469–481 (1965).

135. L. E. Rasmussen, Soluble proteins of postnatal rat brain, *Intern. Neurochem. Conf.*, *Oxford*, p. 95 (July 25–30, 1965).

136. D. F. Matheson and J. B. Cavanagh, Protein synthesis in peripheral nerve and susceptibility to diphtheritic neuropathy, *Nature* **214**:721–722 (1967).

137. M. E. Smith, The effect of fasting on lipid metabolism of the central nervous system of the rat, *J. Neurochem.* **10**:531–537 (1963).

138. S.-C. Cheng and P. Mela, CO_2 fixation in the nervous system. II. Environmental effects on CO_2 fixation in lobster nerve, *J. Neurochem.* **13**:281–287 (1966).

139. C. Gueuning and G. L. A. Graff, Sciatic nerve phospholipids in the common rabbit, *Compt. Rend. Soc. Biol.* **161**:965–970 (1967) (II,B,1,c).*

140. A Sheltawy and R. M. C. Dawson, The metabolism of polyphosphoinositides in hen brain and sciatic nerve, *Biochem. J.* **111**:157–166 (1969) (II,B,1,c; II,B,3).

141. M. G. Doane, Fluorometric measurement of pyridine nucleotide reduction in the giant axon of the squid, *J. Gen. Physiol.* **50**:2603–2622 (1967) (II,G).

142. S. Hori, Zetler's satellite polypeptides of substance P in subcellular particles of bovine peripheral nerves, *Japan. J. Physiol.* **18**:746–771 (1968) (II,H).

143. M. Canessa-Fischer, F. Zambrano, and V. Riveros-Moreno, Properties of the ATPase system from the sheath of squid giant axons, *Arch. Biochem. Biophys.* **122**:658–663 (1967) (III,A,2).

144. M. T. Sabatini, R. Dipolo, and R. Villegas, Adenosine triphosphatase activity in the membranes of the squid nerve fiber, *J. Cell. Biol.* **38**:176–183 (1968) (III,A,2).

145. P. Laduron and F. Belpaire, Transport of noradrenaline and dopamine-β-hydroxylase in sympathetic nerves, *Life Sci.* **7**:1–9 (1968) (III,A,3).

146. M. G. Larrabee and F. J. Brinley, Incorporation of labelled phosphate into phospholipids in squid giant axons, *J. Neurochem.* **15**:533–546 (1968) (III,A,4).

147. G. Porcellati, Neurotoxic organophosphorus compounds and lipid metabolism, *Advan. Exptl. Med. Biol.* **4** (1969), (III,A,4).

148. G. Porcellati, La reazione fosforilcolina : citidililtrasferasica nella degenerazione mielinica sperimentale, *Acta Neurol.* **24**:179–187 (1969) (III,A,4).

149. W. J. Hendelman and R. P. Bunge, Radioautographic studies of choline incorporation into peripheral nerve myelin, *J. Cell. Biol.* **40**:190–208 (1969) (III,A,4).

150. A. Dahlström and J. Jonason, Dopa-decarboxylase activity in sciatic nerves of the rat after constriction, *European J. Pharmacol.* **4**:377–383 (1969) (III,B,1).

151. C. W. M. Adams, Y. H. Abdulla, D. R. Turner, and O. B. Bayliss, Subcellular preparation of peripheral nerve myelin, *Nature (London)* **220**:171–173 (1968) (IV,A).

152. F. Wolfgram and K. Kotorii, The composition of the myelin proteins of the peripheral nervous system, *J. Neurochem.* **15**:1291–1296 (1968) (IV,A ; IV,B).

153. G. Porcellati, in *Protein Metabolism of the Nervous System* (A. Lajtha, ed.), Plenum Press, New York (1970), (IV,A).

154. M. C. MacBrinn and J. S. O'Brien, Lipid composition of optic nerve myelin, *J. Neurochem.* **16**:7–12 (1969) (IV,A).

155. N. Miani, C. Cavallotti, and A. Caniglia, Synthesis of ATP by myelin of spinal nerves of rabbit, *J. Neurochem.* **16**:249–260 (1969) (IV,A).

---

*References 139–155 have been added in proof. The numbers between parentheses at the end of these references indicate the relative section of this chapter to which the reference applies.

Chapter 17

NUCLEIC ACID AND PROTEIN METABOLISM OF THE AXON

Edward Koenig

Department of Physiology
State University of New York at Buffalo
Buffalo, New York

I. THE AXON AS A SPECIALIZED APPENDAGE

A. Prevailing Set Regarding Axonal Protein Synthesis

Several factors have contributed to the generally held notion that the axon depends wholly upon its cell body for all of its soluble and structural proteins. Historically, the nineteenth century experiments of Waller and Türck, demonstrating the importance of continuity between axon and cell body for preserving structural integrity of the axon, were perhaps most compelling in structuring subsequent thinking about the axon's dependency. A second factor was the concept of axoplasmic flow, a proposed mechanism whereby the perikaryon supplies the axon with its metabolic needs. This was a logical hypothesis that followed directly from the studies involving axonal degeneration. Indeed, Barker states in his textbook published in 1899,[3]: "To explain the influence of the cell body upon the fibre, Goldscheider has advanced a very ingenious hypothesis. He suggests that it is most probable that there is an actual transport of a material, *perhaps a fermentlike substance*, from the cell along the whole course of the axone to its extremity, and first through the influence of this chemical body the axone is enabled to make use for its nutrition of the material placed at its disposal in its anatomical course." The phenomenon of axoplasmic flow is now well established, owing to the classic experiments of Weiss and co-workers that stimulated the large number of studies that followed subsequently (see Chapter 18, Vol. II of this *Handbook*). However, in view of the paucity of information at that time and the lapse of more than 40 years before the phenomenon was finally demonstrated, Goldscheider may be credited with having had profound intuitive insight.

A third factor, which perhaps more than any other has been responsible for the belief that the axon is dependent on the soma for all proteins, is the apparent lack of ultrastructural evidence for protein synthesizing machinery. Electron microscopists, beginning with Palay and Palade,[35] have repeatedly failed to observe ribosomes in the axons of differentiated neurons. These studies have simply verified at a more refined level the long-known absence of Nissl substance in the axon, a discovery originally made by Schaffer in 1893.[37] Microspectroanalysis of axons[34] also have indicated a lack of ultraviolet absorption. However, whereas RNA has not been demonstrated in differentiated axons by electron microscopy or by ultraviolet microscopy, there is evidence suggesting that ribosomes are present, nevertheless, in the undifferentiated neuroblast axon. Ultraviolet absorption was observed throughout neuroblasts, including their axons, of developing nervous tissues of chick[23] and guinea pig.[24] Moreover, ribosomes have been observed in axons of cultured sympathetic ganglion cells from explanted human neuroblastoma, which also exhibit RNAase-sensitive fluorescence with acridine orange.[18]

A point arising from Hughes' study of developing chick spinal cord[23] was a loss of ultraviolet absorption from axons after their outgrowth. The loss of ultraviolet absorption, presumably due to a degradation of RNA, would seem to signal the end of the axon as a simple extension of the cell body, as is the case with dendrites, and the beginning of the axon as a specialized, highly differentiated appendage. The biological significance of this differentiation is unknown and can be only a matter of conjecture. An obvious possibility, however, concerns logistical problems inherent in nuclear–cytoplasmic relationships and the need to preserve close, physical proximity for optimum interaction.

II. EVIDENCE FOR AUTOCHTHONOUS AXONAL PROTEIN SYNTHESIS

A. Indirect Evidence

The first suggestive evidence that the axon can synthesize protein appeared from studies that used an indirect approach. The objective was to study the synthesis of the enzyme, acetylcholinesterase (AChE), which is associated with the axon but not with its satellite cell, the Schwann cell. The experimental design entailed an initial, irreversible inactivation of a major proportion of AChE by an organophosphorus anticholinesterase compound. The rate of enzyme restoration then was studied along segments of peripheral nerve. This type of experiment was undertaken independently in the laboratories of Koelle[32,33] and Waelsch[5,6] but employing different organophosphorus inhibitors and different species. The basic proposition tested was whether the restoration of AChE along peripheral nerve would occur at a uniform rate, which would indicate a local synthesis, or whether the restoration would occur such that a proximo-distal gradient would appear during early stages of restoration; the latter result would suggest that the enzyme was being supplied to the axon from the perikaryon. The findings from both laboratories were consistent with that of a local synthesis.

The question of reactivation of AChE after inhibition by organo-phosphorus anticholinesterases was considered by Koenig and Koelle[33] and by Clouet and Waelsch[6]; the possibility of reactivation was regarded as unlikely. Subsequent studies with inhibitors of protein synthesis[26,28] have supported this conclusion.

In these latter studies, local axonal synthesis of AChE was demonstrated in two ways. Sawyer[36] had shown earlier that transection of a peripheral nerve would result in an enduring increase of AChE in the central stump. When examined in the hypoglossal nerve of the cat, the AChE activity was observed to increase by an average of fivefold in the distal 3 mm of the central stump 24 hr later, irrespective of whether most of the preexisting enzyme was inactivated before sectioning by diisopropylfluorophosphate (DFP) or not.[26] Local treatment of the central stump with puromycin on one side reduced the 24-hr increase to half of that of the untreated, contra-lateral central stump.

In a second experiment, a significant restoration of AChE after DFP was demonstrated in a segment of nerve that had been decentralized.[28] In another group of animals this restoration was inhibited by 5-fluororotic acid, a pyrimidine antimetabolite. The restoration of enzyme in the intact, contra-lateral nerve was also severely impaired. The mechanism by which fluororotic acid achieved inhibition of AChE synthesis is unknown.

B. Direct Evidence

1. Histoautoradiographic Evidence

The giant axon of the Mauthner neuron, traversing the medulla and spinal cord of fish, has been under study for several years by J.-E. and A. Edström and their co-workers. The combination of this very favorable axonal material, which can be isolated and stripped of its surrounding myelin, and autoradiographic techniques provided the first direct evidence for incorporation of labeled amino acids into axonal protein, both *in vivo* and *in vitro*.[12] A. Edström observed that in 24 hr following an intracranial in-jection of ^{14}C-lysine into goldfish, there was a proximo-distal radioactivity gradient that was displaced distally with time. These results were consistent with a proximo-distal movement of labeled protein from the cell body; however, this movement was superimposed upon local synthetic activity that could be demonstrated autoradiographically in distal portions of the axon, quite removed from the labeled, moving front of axoplasmic proteins. In addition, *in vitro* experiments, in which spinal cord segments were incubated with ^{14}C-lysine and constituent axonal segments were isolated subsequently for autoradiography, showed linear uptake over the 4-hr period investigated; the incorporation was reduced 90–100% with puromycin.

The use of an intact isolated axon for purposes of autoradiography would seem to have distinct advantage over that of sectioned nerves in demonstrating local incorporation. This holds particularly if the incorpora-tion is restricted to the membrane. In autoradiographic studies, involving light and electron microscopy, where negative findings with respect to local incorporation in the axon have been reported,[7,8,9] the membrane profile of either transverse or longitudinal sections represents such a small area relative to that of the total membrane that the probability of observing a reduced silver grain overlying the proportion of membrane area being sampled would seem to be remote indeed. On the other hand, autoradiographic studies of Singer and Salpeter[38,39] have indicated rapid labeling *in vivo* by tritiated amino acids, mainly ^{3}H-histidine, of axons and myelin sheaths of intact and decentralized sciatic nerves of newts. The disparity in findings between the two groups of workers has been attributed to differences in fixation procedure.[4] However, the questions that have been raised are still unsettled, and further work will be necessary to resolve the matter. Irrespectively, the observations by Singer and Salpeter indicate that amino acids can be transferred from the endoneural space into the axon via the myelin sheath, or at least at selective regions of the sheath. These findings are consistent with studies on *in vitro* incorporation of ^{3}H-leucine into axonal protein, using direct chemical analysis and radiometry (see below).

2. *Evidence from Direct Chemical Analysis*

Direct measurement of amino acid incorporation into axonal protein became possible when axons could be isolated free of myelin[27] and methods had been developed to analyze tritium-labeled protein in the nanogram range.[31] These studies were all carried out on the spinal accessory nerve (XI cranial nerve) root of the rabbit and entailed the incubation of excised nerve root or root segments in a synthetic medium containing ^{3}H-leucine; protein determination and subsequent radiometry were carried out on solubilized, myelin-free axons after incubation. The *in vitro* system was very stable, and uptake still could be demonstrated after an initial *in vitro* pre-incubation of 18 hr; however, an undiminished incorporation could be demonstrated only for the first 10 hr *in vitro*. Similar evidence for the stable nature of the axonal protein synthesizing system can be seen from A. Edström's results with the Mauthner axon,[12] where a linear uptake was observed for at least 4 hr. These observations are consistent also with the capacity of the decentralized nerve to synthesize AChE in the several-day period anteceding Wallerian degeneration.[28]

Inhibitors had variable effects on ^{3}H-leucine incorporation. Puromycin inhibited uptake by 69% after 5 hr of incubation.[30] Chloramphenicol had no apparent effect.[29] Actinomycin D also had no measurable effect, either when the nerve was preincubated for 18 hr with it and then tested for ^{3}H-leucine uptake or when it was included in the incubation mixture.[29,30]

In an interesting investigation, Austin and Morgan[2] studied the *in vitro* incorporation of ^{14}C-leucine into suspensions of chopped cerebral cortex of rat. Various purified subcellular fractions were isolated subsequently for analysis. Of significance to the problem of membrane protein synthesis was the demonstration that purified subfractions of synaptosomes, including synaptosome ghosts, incorporated leucine readily. Puromycin produced moderate inhibition, whereas chloramphenicol caused a rather substantial inhibition. The latter finding was not in agreement with experiments on spinal accessary axons.[29] In any case, however, these results are important in adding further support to the concept of a local membrane machinery.

Inhibition of uptake by puromycin, observed in the several studies relating to axonal protein synthesis thus far, indicate the involvement of ribosomes or their functional axonal equivalent. Although ribosomes as discrete morphological entities are conspicuously absent from axons, their functional equivalent may be sequestered within the plasma membrane and consequently be indiscernible. Unfortunately, for the present this theoretically important question must remain speculative.

Actinomycin D did not have an apparent effect on ^{3}H-leucine incorporation in axons from rabbit spinal accessory but was shown to have a very significant action on the *in vitro* incorporation of labeled amino acids in the Mauthner axon of the carp.[13] The level of inhibition after only 1 hr was 58%, which was quite substantial and suggested to Edström that local protein synthesis in the Mauthner axon is dependent upon a "local" DNA. The basis for the differences in the findings of the two axonal systems is unknown. Certain observations do lead to the impression that the Mauthner neuron is peculiar in some respects. For example, the administration of thyroxine to immature tadpoles causes antithetical effects on neurons of the medulla, i.e., Mauthner cells involute, a normal occurrence during metamorphosis with the accompanying loss of dominance of the tail, while neighboring cells are stimulated to hypertrophy.[40] It is possible that the effect may not be inducible in fish. Still another distinction is the considerably greater RNA content and RNA concentration in the Mauthner compared to either cat or rabbit spinal accessory axons (see below). However, other experimental evidence from mammalian nerves does support the hypothesis that a "local" or "peripheral" DNA-dependent mechanism is operative in the axon, but it may not be concerned in this case with specifying proteins (see below).

III. EVIDENCE FOR AXONAL RNA

A. Results of RNA Analysis

1. RNA Content and Concentration

The first unequivocal demonstration of RNA in an axon was achieved by Edström, Eichner, and Edström,[17] who used the sensitive ultramicroanalytical procedures of J.-E. Edström[15,16] and the Mauthner axon of goldfish. Early estimates of concentration was high, owing to shrinkage[17]; subsequent analysis[10] yielded an average concentration of 0.05 % (wt./vol.). A similar value of 0.06 % was reported also for the axon of the lobster stretch receptor neuron.[19] In his study on the Mauthner fiber,[10] A. Edström observed a proximo-distal gradient in the absolute amount of RNA along the axon and along the subjacent myelin sheath. Contents in both structures paralleled each other, with the axon showing only 75 % of that of the myelin sheath. Concentration of axonal RNA varied along the length, however, showing a minimum in the intermediate region and higher values in proximal and distal portions.

The axon of the cat spinal accessory nerve root was the first mammalian axon to be shown to contain RNA.[27] An estimate of average concentration was 0.003 %, and later estimates[29] in the rabbit yielded twice this value. However, such estimates were crude at best, owing to errors due to contamination by an acid-soluble material believed to be polypeptides (Koenig, unpublished). In the earlier study there was also a loss of absorption by pyrimidines because of their reaction with hydroxylamine. Thus, the Mauthner axon has about 50–100 times more RNA per unit length and 10–20 times greater concentration than that of the two mammalian motor axons analyzed. The significance of such differences is unknown but may relate to the apparent DNA dependence of protein synthesis in the Mauthner that could not be demonstrated in the rabbit spinal accessory (see above). There is the possibility that most of the axonal RNA is not associated with the axoplasm in mammalian axons but rather with the axolemma. Should this prove to be the case, RNA concentration would be a meaningless parameter.

A special case where axonal RNA was found in high concentration is the rat treated with β-β-iminodipropionitrile (IDPN).[21] Animals, effectively treated, develop axonal balloons in the initial segment of spinal motoneurons, which are characterized by an entangled meshwork of neurofibrils and mitochondria. Axonal balloons are 4 times the volume of cell bodies on the average, and RNA concentration is 0.23 % or about one twenty-fifth of that of the cell body.[21] The basis for the axonopathy produced by IDPN is not understood, but the proximity of axonal balloon to perikaryon might underlie the high concentration of RNA in this instance.

In their study on synaptosomes from rat cortex, Austin and Morgan[2] analyzed RNA in purified subfractions. They found in unfractionated synaptosomes, 12 μg/mg protein; in synaptosome ghosts, 20 μg/mg protein; in a membrane subfraction "E," 14 μg/mg protein; and in a membrane subfraction "F," 21 μg/mg protein. In comparison, the microsomal fraction contained 40 μg/mg protein.

2. RNA Base Composition

Analysis of axonal RNA base composition has not shown any unusual base proportions, although contamination with protein often obscures accurate determination of pyrimidines. For this reason the adenine:guanine (A/G) ratio has been used both by Edström for the Mauthner axon[10] and by Koenig for the cat spinal accessory.[27] On the basis of the A/G ratio, there appears to be very little difference between the two axonal systems, notwithstanding the marked differences in RNA concentration, i.e., A/G ratio for Mauthner, 0.56[10] and that for spinal accessory, 0.57.[27] The A/G ratio is characteristic of ribosomal RNA, but such a functional designation should not be made on the basis of A/G ratio alone. Other evidence (see below) indicates that not all axonal RNA can be extracted quantitatively by ribonuclease digestion, at least in the spinal accessory; this resistant fraction could influence the overall base composition if it were relatively large with unusual base ratios. However, its relative size is yet to be determined.

3. Induced Changes in RNA

Two experimental approaches have been used to induce quantitative and qualitative changes in axonal RNA: (1) increased functional activity[25] and (2) axonal regeneration following axotomy.[11] In the former case goldfish were rotated in the plane of their longitudinal axes for 30 min, thereby increasing the vestibular input to the two Mauthner neurons bilaterally. The animals were killed at intervals of 30, 60, and 180 min after stimulation. A rapid, significant *decrease* in RNA occurred after 30 min, persisted without change until 60 min, and was restored to control level by 180 min. The myelin sheath, to the contrary, exhibited no significant loss, but an apparent increase that was not significant.

Transection of the Mauthner axon produced significant changes in the A/G ratios in the central portion of axon and myelin sheath RNA.[11] Already 12 hr after axotomy, changes in the ratio of purines were very significant; maximum differences from that of control appeared between 2 and 3 days following sectioning and returned to control by 23 days. No significant difference in RNA content was noted at 2 or 30 days after axotomy, despite the notable alteration of the A/G ratio at 2 days.

B. Evidence for Local DNA-Dependent RNA Synthesis

1. Incorporation of RNA Precursors

On *a priori* grounds it would seem unlikely that there should be RNA synthesis in the axon since in eukaryotes this function is generally believed to be restricted to the cell nucleus. Nonetheless, ^{3}H-orotic acid and ^{3}H-adenine were shown to be incorporated into axonal RNA of rabbit spinal accessory *in vitro*; the incorporation was actinomycin-D sensitive.[29] The labeled RNA could be operationally divided into 3 fractions: (1) rapidly RNAase digestible, (2) slowly RNAase digestible, and (3) RNAase resistant (Koenig, unpublished). In a typical experiment of 5 hr duration, the first fraction constituted about 20 % of the total radioactivity incorporated and was released during the first 2 hr of RNAase digestion (0.4 mg/ml) at 37°C. The second fraction contributed 20–30 % of the radioactivity and was released at a decreasing rate over the subsequent 22 hr. Another 15–20 % of radioactivity was released with exhaustive deoxyribonuclease (DNAase) digestion, suggesting DNA associated RNA, perhaps incompletely synthesized. The remaining 30–35 % of radioactivity was associated with the insoluble axonal residue. Acid hydrolysis (e.g., 1 N HCl) at 37°C showed kinetics of release similar to RNAase digestion and left 45–55 % of the radioactivity with the residue after 24 hr.[30] The results suggested a highly stabilized RNA by virtue of its being conjugated, perhaps with another RNA (i.e., double stranded).

Experiments to ascertain the subcellular localization of the newly synthesized RNA indicated it to be extramitochondrial (Koenig, unpublished). Specific radioactivity, based on protein but owing to incorporation of RNA precursors (viz., ^{3}H-orotic acid and ^{3}H-adenine), were compared in purified whole nerve mitochondria and axons from the same incubation mixture. The axons showed twice the specific radioactivity. Exhaustive digestion with RNAase and DNAase removed 82% of the radioactivity from mitochondria but only 62% from the axonal material. Since mitochondria constitute a small fraction of total axonal mass, the results indicated that most of the radioactive product was extramitochondrial in the axon.

Some experiments by Austin, Bray, and Young[1] also suggested an incorporation of RNA precursors locally. ^{3}H-orotic acid, injected intrathecally in the chicken, was observed to be displaced at successive intervals in the sciatic nerve as part of the acid-soluble pool. A progressive increase in labeling of RNA along the nerve could be interpreted as local synthesis. ^{14}C-leucine, on the other hand, did not appear in the acid-soluble pool along the nerve.[1]

Experiments in the immediate future probably will center around characterizing the newly synthesized axonal RNA in terms of its molecular weight and the localization of the DNA template. Some preliminary experiments by Edström and Edström[31] on the Mauthner axon already indicate the synthesis of a low molecular weight product, sedimenting in the 4S region. The localization of the DNA template, no doubt, will offer some difficulties; the two obvious possibilities are extra- and/or intra-axonal, neither being mutually exclusive. In the former case, this would imply the satellite cell (e.g., Schwann cell or oligodendrocyte) along the preterminal axon or a cell of epithelial origin at sensory axonal terminals. A likely intra-axonal localization is the mitochondrion; other sites (e.g., membrane) certainly cannot be excluded.

IV. SPECIFICATION AND SYNTHESIS OF MEMBRANE PROTEINS

A. An Argument for a Local Membrane Machinery

Certain theoretical considerations favor an *a priori* conclusion that the neuronal plasma membrane, of which the axolemma is an integral part, contains an intrinsic, local protein-synthesizing machinery. The fundamental premise states that the membrane is functionally differentiated on a *regional* basis. This follows from the consideration that there are two general types of bioelectrical potentials that can be elicited from excitable tissues, viz., (1) so-called slow potentials, which are not propagated, whose initiation is graded and whose decay is exponential in time and space owing to the passive electrical properties of the membrane and (2) action potentials, which are propagated, whose initiation is all or none and whose termination is actively brought about. The active processes, involving selective changes in Na^+ permeability of the membrane during the action potential, have been characterized by Hodgkin and Huxley in their classical studies as Na^+ activation and Na^+ inactivation (i.e., increase and decrease in Na^+ conductance, respectively); these mechanisms are *not* present apparently at sites exhibiting nonpropagated, graded potentials.[20] From the Q_{10} value for the Na^+ activation mechanism[22] it can be inferred that the active process involves an "enzyme-like behavior," e.g., protein conformational change. It follows, then, that here is an example of at least one protein (or proteins) that is present in certain regions of the membrane and not in others. Examples of sites exhibiting graded potentials are subsynaptic regions and sensory axonal endings (i.e., receptors), whereas an obvious example of a site giving rise to a propagated response is the preterminal axon.

It is clear from such considerations that the membrane is composed of functionally contiguous regions, physical continuity notwithstanding. The proposed basis for the functional distinction is the presence or absence of specific membrane-associated proteins, in particular delimited membrane areas. The question then poses itself as to how a centralized, cytoplasmic protein-synthesizing machinery can direct the incorporation of specific functional proteins into discretely circumscribed areas of the excitable membrane. There is no simple way of testing the validity of the assumption that a centralized machinery specifies (i.e., immediately) and synthesizes proteins for membrane protein turnover. However, the question of whether a centralized machinery is obligatory can be tested. In this regard the axon has provided a means for testing this proposition insofar as the axon constitutes a specialized appendage of the cell that contains no ribosomes. The hypothesis that protein turnover in the neuronal plasma membrane is achieved by a local machinery, distinct from the major cytoplasmic center, was advanced earlier.[26] However, it is still a moot point as to whether the machinery for protein synthesis in the axon is axoplasmic or axolemmal in its localization, but there is some evidence in the mammalian axon suggestive of the latter. A large proportion of the labeled axonal RNA is firmly associated with the insoluble axonal residue and can be released only after virtual solubilization, i.e., gel-like state (Koenig, unpublished). Also, Austin and Morgan,[2] in their analysis of synaptosomes, showed very significant amounts of RNA associated with membrane subfractions, including ghosts.

B. Regulation of Local Machinery

In the mammalian axon, the evidence at present indicates that protein synthesis is supported by metabolically stable messenger RNA (*m*RNA), i.e., actinomycin D has no effect on amino acid incorporation and decentralized nerves continue to synthesize AChE for several days (see above). In the Mauthner axon, on the other hand, the evidence clearly points to a very significant dependence upon labile, DNA-dependent *m*RNA. In view of the uptake of RNA precursors by mammalian axons that is actinomycin-D sensitive, the question arises as to its functional significance. Some insight is offered by experiments that were designed to test the effects of actinomycin D on AChE synthesis along the cat hypoglossal nerve.[28]

A local, topical application of actinomycin D to a segment of nerve *in situ* produced a 40% increase in AChE activity 2 days later, limited essentially to the region treated with actinomycin D. The increase was blocked by either puromycin or 5-fluororotic acid, both of which had been shown to inhibit AChE synthesis.[26,28] These results were interpreted to indicate that the synthesis of specific functional proteins of the membrane (e.g., AChE) is not controlled at transcription (i.e., through regulation of *m*RNA synthesis) but at translation; the control being through a local DNA-dependent, labile product interacting with stable *m*RNA.[30]

The theoretical implications of the foregoing hypothesis is that it predicts that the membrane is everywhere omnipotential or omnivalent in terms of its functional properties because the information specifying membrane proteins is redundant throughout the membrane. The translation of encoded information, however, is postulated to be controlled by local, extramembrane DNA-dependent mechanisms, which could be extracellular as well as intracellular. Thus, by this mechanism circumscribed regions of the membrane can become functionally differentiated; the corollary to this is that it should be possible to bring about a transduction of a given membrane region, under appropriate conditions, from one functional type to that of another.

V. REFERENCES

1. L. Austin, J. J. Bray, and R. J. Young, Transport of proteins and ribonucleic acid along nerve axons, *J. Neurochem.* **13**:1267–1269 (1966).
2. L. Austin and I. G. Morgan, Incorporation of ^{14}C-labelled leucine into synaptosomes from rat cerebral cortex *in vitro*, *J. Neurochem.* **14**:377–387 (1967).
3. L. F. Barker, *The Nervous System and Its Constituent Neurones*, p. 307, Appleton-Century-Crofts, New York (1899).
4. S. H. Barondes, Synaptic plasticity and axoplasmic transport, *Neurosci. Bull.* **5**:365–370 (1967).
5. D. H. Clouet and H. Waelsch, Recovery of cholinesterases in the frog nervous system after inhibition, *in The Regional Chemistry, Physiology and Pharmacology of the Nervous System* (S. S. Kety and J. Elkes, eds.), pp. 243–247, Pergamon Press, New York (1961).
6. D. H. Clouet and H. Waelsch, Amino acid and protein metabolism of the brain.—VIII. The recovery of cholinesterase in the nervous system of the frog after inhibition, *J. Neurochem.* **8**:201–215 (1961).
7. B. Droz, Rate of newly synthesized proteins in neurons, *in The Use of Radioautography in Investigating Protein Synthesis* (C. P. Leblond, ed.), pp. 159–175, Academic Press, New York (1965).
8. B. Droz and C. P. Leblond, Migration of proteins along the axons of the sciatic nerve, *Science* **137**:1047–1048 (1962).
9. B. Droz and C. P. Leblond, Axonal migration of proteins along the axons of the sciatic nerve, *J. Comp. Neurol.* **121**:325–345 (1963).
10. A. Edström, The ribonucleic acid in the Mauthner neuron of the goldfish, *J. Neurochem.* **11**:309–314 (1964).

11. A. Edström, Effect of spinal cord transection on the base composition and content of RNA in the Mauthner nerve fibre of the goldfish, *J. Neurochem.* **11**:557–559 (1964).

12. A. Edström, Amino acid incorporation in isolated Mauthner nerve fibre components, *J. Neurochem.* **13**:315–321 (1966).

13. A. Edström, Inhibition of protein synthesis in Mauthner nerve fibre components by actinomycin-D, *J. Neurochem.* **14**:239–243 (1967).

14. A. Edström and J.-E. Edström, Identification and properties of RNA from Mauthner nerve fibre components of the goldfish, in *Macromolecules and the Function of the Neuron* (Z. Lodin and S. P. R. Rose, eds.), pp. 103–110, Excerpta Medica, Amsterdam (1968).

15. J.-E. Edström, Quantitative determination of ribonucleic acid in the micro-microgram range, *J. Neurochem.* **3**:100–106 (1958).

16. J.-E. Edström, Extraction, hydrolysis and electrophoretic analysis of ribonucleic acid from microscopic tissue units (microphoresis), *J. Biophys. Biochem. Cytol.* **8**:39–43 (1960).

17. J.-E. Edström, D. Eichner, and A. Edström, The ribonucleic acid of axons and myelin sheaths from Mauthner neurons, *Biochim. Biophys. Acta* **61**:178–184 (1962).

18. M. Goldstein, Long term tissue culture of neuroblastoma.—IV. Growth differentiation and function during the development of an established line of neuroblastoma, *J. Natl. Cancer Inst.*, in press.

19. W. Grampp and J.-E. Edström, The effect of nervous activity on ribonucleic acid of the crustacean receptor neuron, *J. Neurochem.* **10**:725–731 (1963).

20. H. Grundfest, Electrophysiology and pharmacology of different components of bioelectric transducers, in *Sensory Receptors, Cold Spring Harbor Symp. Quant. Biol.* **30**:1–14 (1965).

21. H. A. Hartmann, J. Lin, and M. C. Shively, RNA of nerve cell bodies and axons after beta, beta-iminodiproprionitrile, *Acta Neuropathol.* **11**:275–281 (1968).

22. A. L. Hodgkin, A. F. Huxley, and B. Katz, Measurement of current-voltage relations in the membrane of the giant axon of *Loligo*, *J. Physiol.* (*London*) **116**:424–448 (1952).

23. A. Hughes, The development of the neural tube of the chick embryo. A study with the ultraviolet microscope, *J. Embryol. Exptl. Morphol.* **3**:305–325 (1955).

24. A. Hughes and L. B. Flexner, A study of the development of cerebral cortex of the foetal guinea pig by means of the ultraviolet microscope, *J. Anat.* (*London*) **90**:386–394 (1956).

25. B. Jakoubek and J.-E. Edström, RNA changes in the Mauthner axon and myelin sheath after increased functional activity, *J. Neurochem.* **12**:845–849 (1965).

26. E. Koenig, Synthetic mechanisms in the axon.—I. Local axonal synthesis of acetylcholinesterase, *J. Neurochem.* **12**:343–355 (1965).

27. E. Koenig, Synthetic mechanisms in the axon.—II. RNA in myelin-free axons of the cat, *J. Neurochem.* **12**:357–361 (1965).

28. E. Koenig, Synthetic mechanisms in the axon.—III. Stimulation of acetylcholinesterase by actinomycin-D in the hypoglossal nerve, *J. Neurochem.* **14**:429–435 (1967).

29. E. Koenig, Synthetic mechanisms in the axon.—IV. *In vitro* incorporation of ^{3}H-precursors into axonal protein and RNA, *J. Neurochem.* **14**:437–446 (1967).

30. E. Koenig, Intrinsic protein synthesizing mechanisms in the axon as bases for renewal and local functional differentiation of membrane, in *Macromolecules and the Function of the Neuron* (Z. Lodin and S. P. R. Rose, eds.), pp. 121–128, Excerpta Medica, Amsterdam (1968).

31. E. Koenig, A method for determining total protein of isolated cellular elements and corresponding tritium radioactivity, *J. Cell. Biol.* **38**:562–573 (1968).

32. E. Koenig and G. B. Koelle, Acetylcholinesterase regeneration in peripheral nerve after irreversible inactivation, *Science* **132**:1249–1250 (1960).

33. E. Koenig and G. B. Koelle, Mode of regeneration of acetylcholinesterase in cholinergic neurons following irreversible inactivation, *J. Neurochem.* **8**:169–188 (1961).

34. J. A. Nurnberger, B. Edström, and B. Lindström, A study of the ventral horn cells of the adult cat by two independent cytochemical microabsorption techniques, *J. Cell. Comp. Physiol.* **39**:215–251 (1952).

35. S. L. Palay and G. E. Palade, The fine structure of neurons, *J. Biophys. Biochem. Cytol.* **1**:69–88 (1955).

36. C. H. Sawyer, Cholinesterase in degenerating and regenerating peripheral nerve, *Am. J. Physiol.* **146**:246–253 (1946).

37. K. Schaffer, Kurze Anmerkung über die morphologische Differenz des Axencylinders im Verhältnisse zu den protoplasmatischen Fortsatzen bei Nissl's Färbung, *Neurol. Centralbl. (Leipzig)* **12**:849–851 (1893).

38. M. Singer and M. M. Salpeter, Transport of tritium-labelled *l*-histidine through the Schwann and myelin sheaths into the axon of peripheral nerves, *Nature* **210**:1225–1227 (1966).

39. M. Singer and M. M. Salpeter, The transport of ^{3}H-*l*-histidine through the Schwann and myelin sheath into the axon, including a reevaluation of myelin function, *J. Morphol.* **120**:281–315 (1966).

40. P. Weiss and F. Rossetti, Growth responses of opposite sign among different neuron types exposed to thyroid hormone, *Proc. Natl. Acad. Sci. (U.S.)* **37**:540–556 (1951).

Chapter 18

AXOPLASMIC TRANSPORT

Samuel H. Barondes

Departments of Psychiatry and Molecular Biology
Albert Einstein College of Medicine
Bronx, New York

When an axon is severed, the portion separated from the perikaryon degenerates and there is progressive regeneration from the perikaryal stump. This observation is the basis for the concept that the viability of the axon is dependent upon transport of materials from the nerve cell body. It illustrates not only that the axon cannot survive without perikaryal materials but also that the nerve cell body, even in adulthood, has the capacity to synthesize and transport all these critical constituents. In the past few years there has been increasing study of the details of axoplasmic transport. Although it would appear that a major function of this process is the replacement of axonal materials that have been degraded or secreted, increasing attention is being directed to the possibility that materials are transported selectively for regulation of aspects of axonal or nerve ending function. The possibility that important biosynthetic processes can occur in the axon and nerve ending independent of the perikaryon also is being investigated.

The purpose of this chapter is to consider some of the methods for the study of axoplasmic transport and to call attention to some aspects of this phenomenon that recently have been discovered. An exhaustive review and bibliography have not been attempted. More complete presentations of some of the material considered here and a more detailed bibliography will be found in the report of a recent conference on this subject.[1] A review of the extensive contributions of Paul Weiss[2] is included in that report. Several other relatively recent reviews[3,4] also may be consulted.

I. METHODS FOR STUDYING AXOPLASMIC TRANSPORT

A. Nerve Transection

The earliest method for study of the transport of axonal materials from the perikaryon was by observation of axonal regeneration after transection of a nerve. Regeneration was found to proceed at rates in the range of 1 or a few millimeters per day[3] in a variety of preparations. Little information could be obtained with this method since the perikaryon was rendered abnormal by the transection and because the observed rate of regeneration was limited by the rate of transport of the most slowly migrating axonal constituents that tended to obscure the existence of transport of some materials at high rates.

B. Nerve Constriction Methods

It was shown by Weiss and Hiscoe[5] that after application of a ligature around a nerve, the diameter of the nerve on the perikaryal side of the ligature increased, whereas that on the peripheral side decreased. After release of the ligature the peripheral segment resumed its previous size at a rate of approximately 1 mm/day. Considerable research on axoplasmic transport was stimulated by these observations. In many such studies the axons are only constricted, but in others there is actually some transection due to crushing. The nature of materials accumulating at the perikaryal side of the constriction and their rate of accumulation have been studied with a variety of methods. Evidence has been presented that mitochondria[6] and catecholamine-containing vesicles[7,8] accumulate proximal to the constriction. Other studies have shown an accumulation of enzymes,[9,10] phospholipids,[11] and radioactive proteins[12] proximal to the ligation that indicates that all these materials migrate into the axon from a proximal site. Attention also has been directed toward accumulation of materials on the peripheral side of the constriction. In many studies evidence for some accumulation of enzymes,[13] catecholamines,[7] or other markers distal to the ligature have been found. This has been taken as evidence for the existence of some retrograde transport of axoplasm—i.e., transport of axoplasmic constituents toward the perikaryon. The degree of accumulation of materials on the distal side of the ligature is always much smaller than that on the proximal side, and there has been some controversy about the possibility that the distal accumulation may be artifactual and may not actually indicate retrograde transport.

C. Microscopic Examination of Living Axons

Direct inspection of the movement of materials in living axons has been achieved by a number of investigators using a variety of optical systems.[2,14] This approach is limited in two ways: First, only gross displacements, either of particles or of columns of fluid, can be observed. Second, the materials that lend themselves best to microscopic examination are axons grown in tissue culture that are not obscured by myelin, but the ambiance in tissue culture is markedly different from that in the mature nervous system. Therefore, observations made on tissue culture may be extrapolated to the mature animal only with considerable qualification. Nevertheless, direct visual studies of axoplasmic movement have proved helpful in that they illustrate the very active movement that occurs within axoplasm, the frequent saltatory transport of particles (which remain stationary for long periods and then suddenly jump long distances within the axon), and the presence of bidirectional streaming, i.e. movement of axonal materials both toward and away from the cell body, often simultaneously, within the same axon.

D. Segmental Analysis of a Nerve After Administration of a Radioactive Substance in the Perikaryal Region

A number of investigators have applied ^{32}P and radioactive amino acids in the region of an accumulation of nerve cells (e.g., hypoglossal nucleus[11] and ventral horn of spinal cord[15]). They examined a series of segments of a nerve at increasing distances from the site of injection for the presence of radioactive material (phospholipids or proteins) at a number of intervals after administration of the radioactive precursor. In this manner the progressive migration of these materials has been studied. Care must be taken not to mistake intraaxonal radioactive materials from those in the periaxonal tissues, but this problem is largely overcome by the localized administration of the radioactive precursors near a collection of nerve cell bodies.

E. Autoradiographic Studies

There have been a large number of autoradiographic studies of the distribution of radioactive protein along the axon at intervals after administration of radioactive amino acid (e.g., Droz and LeBlond[16] and Lasek[17]). In these studies, unlike the segmental studies, the precursor may be given systemically rather than in the region of nerve cell bodies since the presence of material in the axoplasm can be distinguished from that in periaxonal elements. In this way the rate of migration of radioactive proteins or other substances has been directly studied not only in peripheral nerve but also in axons within the central nervous system. More recently, autoradiographic studies with the electron microscope have been used to refine these observations.[18]

F. Studies of the Appearance of Materials at Nerve Endings

Studies[19–22] of the transport of materials to the nerve endings in the central nervous system have been made by observing the rate of appearance of radioactive protein in a nerve ending fraction isolated by a slight modification of the procedure of Whittaker et al.[23] at intervals after intracerebral injection of radioactive amino acid. C^{14}-leucine incorporation into the protein of all subcellular fractions of brain except isolated nerve endings was completed within 1 hr of its administration, and the specific activity of the protein in these fractions declined thereafter. In contrast, the specific activity of the protein of isolated nerve endings, particularly that "soluble" protein released by lysis with water, increased progressively in the ensuing hours and days. The progressive increase in the specific activity of the protein of nerve endings is due to transport of protein to this site after its synthesis elsewhere —presumably in the perikaryon. The soluble protein obtained by lysis of isolated nerve endings is usually used for these studies since there is very little contaminating soluble protein from other parts of the brain, whereas there may be more extensive contamination with particulate materials. To minimize variability due to the intracerebral injection technique and to attempt to correct for turnover of protein during the period studied, the ratio of the specific activity of the soluble protein of nerve endings to that of the soluble protein of whole brain may be calculated, and the progressive increase in this ratio may be taken as a measure of the rate of appearance of soluble protein at nerve endings.

The rate of appearance of radioactive protein at nerve endings is a measure only of the average rapidity with which the perikaryon can transport "information," in the form of proteins, to nerve endings. It is not a measure of the rate of axoplasmic transport of proteins since the axonal lengths in the mouse cerebrum are heterogeneous and since the average axonal length is unknown. However, the method can be employed for studying the effects of various experimental manipulations on the rate of axoplasmic flow by comparing the amount of labeled soluble protein in isolated nerve endings at some given time (e.g., 1 day) after administration of ^{14}C-leucine in untreated mice and in mice treated in some manner during this period. For such studies it is best to effect the manipulation being studied one or more hours after administration of the radioactive amino acid (when its incorporation into protein has been completed) so that the manipulation would have no effect on radioactive amino acid incorporation into protein but only on transport to nerve endings. The rapidity with which such studies can be completed and the brief duration during which the manipulation must be maintained (1 day or less) make this method particularly useful for such studies.

Studies with this method have indicated that there is relatively rapid "communication" between perikaryon and some nerve endings in mouse cerebrum since some "soluble" protein appears at some nerve endings in the mouse cerebrum within a few hours of its synthesis,[22] presumably in the perikaryon. Transport of protein associated with mitochondria[20,21] and synaptic vesicles[21] to nerve endings also has been shown by studying the kinetics of appearance of labeled protein in these subfractions of isolated nerve endings after administration of ^{14}C-leucine. Local synthesis of macromolecules at nerve endings also has been suggested by the finding that there is rapid incorporation of ^{14}C-leucine into some mitochondrial protein of nerve endings[20,21] and also of ^{14}C-glucosamine into macromolecules at nerve endings.[24] The method also has been used to study the transport to the nerve endings of a specific enzyme, monoamine oxidase,[20] by observing its rate of regeneration in nerve endings and in mitochondria from whole brain at a number of times after the administration of an irreversible inhibitor of this enzyme.

A method for the study of the transport of glutamate to the neuromuscular junction of the snail after its application in or on nerve cells also has recently been described.[25,26] Since the length of the axons in this system are known, rates of transport may be calculated. This method has been applied to studies of the effects of nerve stimulation and of pharmacological agents on axoplasmic transport. Since the cell bodies, axons, and nerve endings are all accessible in this system it appears to be particularly well suited for such studies. Transport of labeled proteins to the nerve endings of the optic nerve after application of radioactive amino acids in the eye has recently been studied with autoradiographic techniques.[27]

II. RATES OF AXOPLASMIC TRANSPORT

A. Low Rates of Transport

Since transected nerves regenerate at rates in the range of one or a few millimeters per day,[3] and since the nerve segment peripheral to a constriction resumes its normal diameter at a similar rate,[5] it has been concluded that a significant component of axoplasm is transported at this relatively low rate. Further evidence for transport of some materials at this rate is provided by studies of the transport of radioactively labeled protein in the central nervous system,[16] in peripheral nerves,[12,17] and in the optic nerve.[27] Weiss has proposed that this slowly moving material is not transported within the axon but rather that the entire axon continues to grow out from the nerve cell body throughout life.[2]

B. Evidence for Higher Rates of Axoplasmic Transport

Considerable evidence recently has been accumulated that indicates some axoplasmic constituents are transported in the axoplasm at rates substantially greater than 1 mm/day. Miani found that ^{32}P-labeled phospholipids are transported in axons at rates of up to 7 cm/day.[10] Lasek, in studies of the transport of radioactive-labeled protein in axons, showed not only slow transport of axoplasmic protein but also considerable transport of protein at rates in the range of 20 cm/day.[17] Grafstein also has found evidence for transport of proteins at both low and considerably higher rates in the optic nerve of the goldfish.[27] Dahlstrom, studying the accumulation of catecholamine-containing-granules at a nerve constriction, concluded that these particles were transported at rates of many centimeters per day.[7,8] Extremely high rates of transport of radioactive glutamate from nerve cell bodies to muscle (centimeters per hour) have been found in the snail by Kerkut.[25,26] High rates of movement of axoplasmic constituents have been found by microscopic examination of axons in cultures of tissue from the nervous system.[14] These studies all indicate that some materials are rapidly transported peripherally. Furthermore, it does not appear that there is a single low rate and a single high rate of transport. Rather, the studies suggest that there are many rates at which materials may be transported. Transport of protein into the central axon of the dorsal root ganglion is significantly slower than into the peripheral axon.[28] Therefore, the processes that regulate transport operate differently in two axons from the same neuron. Furthermore, there appears to be transport of materials at different rates within the same axon. This is most clearly seen in tissue culture where some particles are transported in a saltatory manner, whereas other particles in the same axon do not migrate at all.[14] Concurrent bidirectional streaming of material within the same axon also has been observed, and the existence of retrograde migration is also suggested by the finding of the accumulation of a number of materials on the peripheral side of a nerve ligation.[3,7,13] Therefore, it appears that materials may be transported at a variety of rates not only in different axons but also within the same axon. The mechanisms for regulating the rates of transport of different materials and the physiological consequences of such regulation are presently unknown.

III. THE NATURE OF TRANSPORTED MATERIALS

A. Materials Transported from the Perikaryon

Many different types of molecules have been shown to be transported in the axoplasm, and the specific nature of those studied has been dictated by available methodology. The transport of proteins has been extensively demonstrated, and there is evidence that this includes protein associated with mitochondria[6,20] and synaptic vesicles[8,21] as well as soluble protein. Of the proteins, a number of specific enzymes have been shown to be transported in the axoplasm as have low molecular weight materials like phospholipids and catecholamines. However, not all perikaryal constituents are transported into the axoplasm. This is most readily apparent in the case of ribosomes, which are abundant in the perikaryal cytoplasm but which have not been detected by electron microscopy in the axon or the nerve ending. Kerkut has presented evidence that there is actually great selectivity in what is transported, even in the case of small molecules.[26] He finds that application of radioactive glutamate either into snail brain or directly into snail neurons results in rapid transport of this radioactive amino acid to the neuromuscular junctions of the snail. In contrast, radioactive xylose, even if administered within the cells, is not transported to the nerve endings. The fact that there is selection of materials that are transported suggests that regulation of the composition and function of the axon and the nerve ending may be achieved by a mechanism that controls transport of specific perikaryal materials.

B. Materials Transported from Other Sites

All constituents of the axoplasm are not transported from the perikaryon. The presence of mitochondria and multiple enzymes in nerve endings suggests that there is considerable metabolism at this site, and it seems almost certain that the glucose required for energy is transported by diffusion from the capillaries into nerve endings and axoplasm. Likewise, there is evidence for substantial synthesis of norepinephrine in nerve endings,[29] presumably from precursors that entered the nerve ending through the circulation. The question of transport of macromolecules into the axoplasm from other sites recently has been investigated by Singer and Salpeter.[30] They find that in the newt there is transport of radioactive histidine into axons that have been severed from their cell bodies. The evidence suggests that there is transport of amino acids into the axon from periaxonal elements, and it seems possible that some of the histidine-containing material transported into the axons may be bound in protein, although this is certainly not conclusively shown by these studies. Therefore, transfer of small molecules into the axon from non-perikaryal sources seems certain whereas transfer of large molecules has not been established.

C. Evidence for Macromolecular Biosynthesis Within Axons and Nerve Endings

Although it is generally accepted that small molecules may be synthesized within the axon, the possibility of synthesis of macromolecules in the axoplasm or in the cytoplasm of the nerve ending only recently has received serious attention. The obvious high concentration of microsomes in the perikaryon and their absence in the axoplasm of nerve endings has suggested that the latter are completely dependent on the perikaryon for new protein. However, recent studies have indicated that the axoplasm and nerve ending may be capable of synthesizing macromolecules. These studies to some extent have been encouraged by work on *acetabularia*,[31] a unicellular organism, which like the neuron has a long cytoplasmic process emanating from the cell body in which the nucleus resides. There is clear evidence for protein and RNA synthesis in this cytoplasmic process, largely if not exclusively in chloroplasts. The evidence for macromolecular synthesis in axoplasm is not as good, but recent work provides some support for this. Studies with the Mauthner axon of the goldfish have suggested the presence of considerable RNA in this axon,[32] and there is evidence that there may be synthesis of RNA and protein[33] within the axoplasm in this unusual neuron. This axon is atypical since it is very large and also makes electrical synapses. Therefore, it may not be an appropriate model for mammalian axons. However, Koenig has presented evidence that suggests there may be very small amounts of RNA in mammalian axons[34] and that protein synthesis may be directed by this RNA.[35] He also has reported experiments that suggest that there is synthesis of acetylcholinesterase within the axon.[36] Evidence suggesting that there may be synthesis of some mitochondrial protein at nerve endings has been presented,[20] although it was also shown that some mitochondrial protein is transported to the nerve ending in the axoplasm. *In vitro* incorporation of amino acids into protein by a nerve ending fraction has been reported recently.[45,46] However, only limited conclusions may be drawn from all these studies since it is difficult to be completely certain that there has not been significant contamination by other brain constituents.

Recent studies have shown that incorporation of some carbohydrate into macromolecules of nerve endings may occur independently of the perikaryon.[24] Upon intracerebral administration of ^{14}C-glucosamine to mice there is substantial rapid incorporation of glucosamine and possibly sialic acid into a macromolecular constituent of both the soluble and particulate fractions obtained by lysis of isolated nerve endings. These studies show that, although protein synthesis at nerve endings at best is limited, structural changes in proteins may be effected at nerve endings by biosynthetic modification of these proteins either by incorporation of carbohydrates into the peptide chain itself or into an attached polysaccharide. The possibility of a local nerve ending mechanism for regulation of protein structure and function through biosynthetic modification is suggested by these studies.

IV. FATE OF TRANSPORTED MATERIALS

A substantial amount of the material transported in the axon probably replaces material that has been degraded. Proteins, e.g., are actively degraded and resynthesized in the nervous system, and the presence of proteolytic enzymes in the axoplasm and at nerve endings[37] suggests that degradation occurs at these sites as well. Some materials are transported for secretion at the nerve endings. Such materials may be secreted into the general circulation. This is the case with posterior pituitary peptide hormones that are synthesized in the perikarya of some hypothalamic nerve cells and then migrate to the nerve endings in the posterior pituitary and are secreted into the bloodstream.[38] There is also some evidence that amino acid-containing materials, possibly proteins, are secreted directly into muscle where they may in turn subserve some regulatory function.[39]

In addition to these other functions, axoplasm is transported for the purpose of axonal and nerve ending proliferation and hypertrophy. This is clearly the case in the maturing animal in which axons grow out from the perikaryon. A similar potentiality exists in adulthood and is manifested after nerve section. It also is possible that alterations in interneuronal relationships in the adult central nervous system may occur by the outgrowth of axons or nerve endings.[40] There is evidence that this occurs in neuromuscular innervation when new sproutings from an intact nerve may form additional synapses with a muscle whose other innervation has been interrupted.[41] It is also possible that similar changes occur in the brain and that they might mediate the storage of changes due to learning.[40]

V. MECHANISM OF AXOPLASMIC TRANSPORT

One possible means by which axoplasm may be propelled distally is by pressure due to continuing synthesis of materials in the nerve cell body. This is unlikely from hydrodynamic considerations[2]; and the fact that when tandem ligations are made axoplasm moves distally between the two ligations[2,7] is strong evidence against this possibility. The finding that marked inhibition of cerebral protein synthesis with acetoxycycloheximide has no apparent effect on the transport of protein to nerve endings[21] also suggests that continued protein synthesis is not required to propel existent protein peripherally. Another mechanism that has been considered is that peristaltic activity possibly driven by energy from periaxonal elements might propel materials distally.[2] Ochs[42] has shown that after freezing stretched nerves the axons have a wavy shape similar to that seen in peristalsis in living systems. Weiss[2] has found evidence for what appears to be peristaltic activity in microscopic examinations of adult nerves that were stripped of surrounding tissue. An alternative proposal, which is more consistent with some of the observations in tissue culture, is that the mechanism for transport is intraaxonal. In tissue culture it is observed that many particles move in a saltatory manner without evidence of a peristaltic wave. In addition, simultaneous bidirectional streaming within the same axon has been observed that would not be possible if contraction of the entire axon were responsible for movement. These studies suggest that there may be a contractile protein within the axon that, by reversible changes in configuration, may generate the mechanical energy needed for the transport of axoplasm. Some materials might be propelled suddenly for long distances by a configurational change in a very long chain of this protein, and bidirectional movements could be due to localized structural changes in portions of the axoplasm. Recently Borisy and Taylor[43] showed that squid axoplasm and mammalian brain are extremely rich in a protein that binds colchicine and that, from other evidence, is believed to be associated with microtubules or filaments. Such structures are abundant in the axon. Since a similar protein appears to be associated with microtubules of sperm tail, cilia, and the mitotic apparatus, and since this protein may be involved in movement in these structures, the possibility that this protein plays some role in the transport of axoplasm is being considered.[44]

There presently is little information on the effect of changes in neuronal function on axoplasmic transport. Kerkut[26] has presented evidence that electrical stimulation of snail neurons accelerates the transport of glutamate (which may be a neurotransmitter in this organism) from the perikaryon to the myoneural junction. However, Dahlstrom[8] thus far has been unable to demonstrate any significant acceleration of transport of catecholamines after nerve stimulation. Therefore, the effect of neuronal activation requires further study. Aging or maturation of the nervous system has been shown to decrease the rate of transport to protein. Protein is transported significantly more slowly in adult rats than in immature rats[16] and appears less rapidly in nerve endings from 1-year-old mice than in those from young adult (3-month-old) mice.[21] The mechanism for these changes is unknown.

VI. CONCLUSION

The axon is protoplasm in an unusual gross conformation. It is the link through which electrical and chemical information are transported from the perikaryon to the nerve ending and synapse. Studies of axoplasmic transport suggest that the axon should not be viewed as an inert wire connecting two important but distant structures. Rather it appears that there may be rapid transport of materials from some nerve cell bodies to some nerve endings and that many active processes, both mechanical and chemical, may occur within the axon and nerve ending. Regulation of these processes may be an important means for regulating the function of the nervous system.

VII. REFERENCES

1. S. H. Barondes and F. E. Samson, *Axoplasmic Transport, Report of a Work Session,* Neurosciences Research Program Bulletin (1967).*

2. P. Weiss, *in Axoplasmic Transport, Report of a Work Session,* Neurosciences Research Program Bulletin (1967).

3. L. Lubinska, Axoplasmic streaming in regenerating and in normal nerve fibers, *in Progress in Brain Research* (M. Singer and J. P. Schade, eds.), Vol. 13, pp. 1–71, Elsevier, Amsterdam (1964).

4. R. L. Friede, Convection of cytoplasm and its constituents within nerve cells, *in Topographic Brain Chemistry*, pp. 388–400, Academic Press, New York (1966).

5. P. Weiss and H. B. Hiscoe, Experiments on the mechanism of nerve growth, *J. Exptl. Zool.* **107**:315–395 (1948).

6. P. Weiss and A. Pillai, Convection and fate of mitochondria in nerve fibers: Axonal flow as vehicle, *Proc. Natl. Acad. Sci.* **54**:48–56 (1965).

7. A. Dahlström, The transport of noradrenaline between two simultaneously performed ligations of the sciatic nerves of rat and cat, *Acta Physiol. Scand.* **69**:158–166 (1967).

8. A. Dahlström, *in Axoplasmic Transport, Report of a Work Session,* Neurosciences Research Program Bulletin (1967).

9. R. L. Friede, Transport of oxidative enzymes in nerve fibers. A histochemical investigation of the regenerative cycle in neurons, *Exptl. Neurol.* **1**:441–466 (1959).

10. G. W. Kreutzberg and W. Wechsler, Histochemische Untersuchungen oxydativer Enzyme am regenerierenden Nervus ischiadicus der Ratte, *Acta Neuropathol.* **2**:349–361 (1963).

11. N. Miani, Mechanism of Neural Regeneration, *in Progress in Brain Research* (M. Singer and J. P. Schade, eds.), Vol. 13, pp. 115–126, Elsevier, Amsterdam (1964).

12. H. Koenig, The synthesis and peripheral flow of axoplasm, *Trans. Am. Neurol. Assoc.* **83**:162–164 (1958).

13. L. Lubinska, S. Niemierko, B. Oderfeld, L. Swarc, and J. Zelena, Bi-directional movements of axoplasm in peripheral nerve fibers, *Acta Biol. Exptl., Vars.* **23**:239–247 (1963).

14. W. O. Burdwood, reported by R. D. Allen, *in Axoplasmic Transport, Report of a Work Session,* Neurosciences Research Program Bulletin (1967).

15. S. Ochs, Axoplasmic flow in neurons, *in Macromolecules and Behavior* (J. Gaito, ed.), Appleton-Century-Crofts, New York (1966).

16. B. Droz and C. P. Leblond, Axonal migration of proteins in the central nervous system and peripheral nerves as shown by radio-autography, *J. Comp. Neurol.* **121**:325–345 (1963).

17. R. J. Lasek, *in Axoplasmic Transport, Report of A Work Session,* Neurosciences Research Program Bulletin (1967).

---

*Qualified persons may receive a copy of this report from George Adelman, Librarian, Neurosciences Research Program, 280 Newton Street, Brookline, Massachusetts 02146.

18. B. Droz, Synthesis et transport des proteines cellulaires dans les neurones ganglionnaires: etude radioautographique quantitative en microscopie electronique, *J. Microscop.* **6**:201–228 (1967).
19. S. H. Barondes, Delayed appearance of labeled protein in isolated nerve endings and axoplasmic flow, *Science* **146**:779–781 (1964).
20. S. H. Barondes, On the site of synthesis of the mitochondrial protein of nerve endings, *J. Neurochem.* **13**:721–727 (1966).
21. S. H. Barondes, Further studies of the transport of proteins to nerve endings, *J. Neurochem.* **15**:343–350 (1968).
22. S. H. Barondes, Rapid appearance of soluble protein at nerve endings, *Comm. Behav. Biol.*, Part A, **1**:179–181 (1968).
23. V. P. Whittaker, A. Michaelson, and R. J. A. Kirkland, The separation of synaptic vesicles from nerve-ending particles (synaptosomes), *Biochem. J.* **90**:293–303 (1964).
24. S. H. Barondes, Incorporation of radioactive glucosamine into macromolecules at nerve endings, *J. Neurochem.* **15**:699–706 (1968).
25. G. A. Kerkut, A. Shapiro, and R. J. Walker, The transport of labeled material from CNS to muscle along a nerve trunk, *Comp. Biochem. Physiol.* **23**:729–748 (1967).
26. G. A. Kerkut, in *Axoplasmic Transport, Report of a Work Session*, Neurosciences Research Program Bulletin (1967).
27. B. Grafstein, Transport of protein by goldfish optic nerve fibers, *Science* **157**:196–198 (1967).
28. R. J. Lasek, Axoplasmic transport in cat dorsal root ganglion cells: As studied with L-leucine-H³, *Brain Res.* **7**:360–377 (1968).
29. J. Glowinski and L. Iversen, Regional studies of catecholamines in the rat brain. III. Subcellular distribution of endogenous and exogenous catecholamines in various brain regions, *Biochem. Pharmacol.* **15**:977–987 (1966).
30. M. Singer and M. M. Salpeter, The transport of ³H-L-histidine through the Schwann and myelin sheath into the axon, including a re-evaluation of myelin function, *J. Morphol.* **120**:281–316 (1966).
31. J. Brachet, Protein synthesis in the absence of the nucleus, *Nature* **213**:650–655 (1967).
32. A. Edström, The ribonucleic acid in the Mauthner neuron of the goldfish, *J. Neurochem.* **11**:309–314 (1964).
33. A. Edström, Amino acid incorporation in isolated Mauthner nerve fiber components, *J. Neurochem.* **13**:315–321 (1966).
34. E. Koenig, Synthetic mechanism in the axon. II. RNA in myelin free axons of the cat, *J. Neurochem.* **12**:357–361 (1965).
35. E. Koenig, Synthetic mechanisms in the axon. IV. In vitro incorporation of ³H-precursors into axonal protein and RNA, *J. Neurochem.* **14**:437–446 (1967).
36. E. Koenig, Synthetic mechanisms of the axon. I. Local axonal synthesis of acetylcholinesterase, *J. Neurochem.* **12**:343–355 (1965).
37. A. Lajtha and N. Marks, Cerebral protein breakdown, in *Protides of the Biological Fluids* (H. Peeters, ed.), pp. 103–114, Elsevier, Amsterdam (1966).
38. H. Heller and R. B. Clark (eds.), *Neurosecretion*, Academic Press, New York (1962).
39. I. M. Korr, P. N. Wilkinson, and F. W. Chornock, Axonal delivery of neuroplasmic components to muscle cells, *Science* **155**:342–345 (1967).
40. S. H. Barondes, in *Axoplasmic Transport, Report of a Work Session*, Neurosciences Research Program Bulletin (1967).
41. M. V. Edds, Jr., Collateral regeneration of residual motor axons in partially denervated muscles, *J. Exptl. Zool.* **113**:517–552 (1950).
42. S. Ochs, Beading phenomona of myelinated nerve fibers, *Science* **139**:599–600 (1963).

43. G. Borisy and E. W. Taylor, The mechanism of action of colchicine. Binding of colchicine-^{3}H to cellular protein, *J. Cell. Biol.* **34**:525–533 (1967).

44. E. W. Taylor, *in Axoplasmic Transport, Report of a Work Session*, Neurosciences Research Program Bulletin (1967).

45. L. Austin and I. G. Morgan, Synaptosomal protein synthesis in a cell-free system, *J. Neurochem.* **15**:41–51 (1968).

46. L. A. Autilio, S. H. Appel, P. Pettis, and P. L. Gambetti; Biochemical studies of synapses in vitro: I. Protein synthesis, *Biochemistry* **7**:2615–2622 (1968).

Chapter 19

COMPARTMENTATION OF AMINO ACID METABOLISM

S. Berl
Department of Neurology
College of Physicians and Surgeons
Columbia University
New York, New York

and D. D. Clarke
Chemistry Department
Fordham University
New York, New York

I. INTRODUCTION

Compartmentation of metabolic events has received increasing attention in recent years. In the area of amino acid metabolism heterogeneous functional pools have been postulated for microorganisms,[1-3] plants,[4,5] and animal tissues.[6-8] The concept of metabolic compartmentation was suggested because of observations that could not be explained by the simple precursor–product relationship as described by Zilversmit, Entenman, and Fishler.[9]

This chapter is limited to a discussion of the biochemical consequences attributable to the compartmentation of glutamate and associated metabolites in brain. The mathematics involved in the kinetic analysis of compartments will not be considered here nor the detailed theoretical assumptions involved in such considerations. A number of articles are available in the literature that give considerable attention to these aspects; some of the more useful ones are those of Solomon,[10] Robertson,[11] Reiner,[12] and Aronoff.[13]

In general two types of compartments are recognized. One is the physiological or anatomical compartment that can be described in anatomical terms as, e.g., the plasma, extracellular and intracellular fluids, a whole organ such as brain, or a particular cell or subcellular structure. The second type of compartment is the metabolic compartment that can be described only in biochemical terms until it can be related to anatomical substructures. As pointed out by Solomon,[10] a compound formed in a biological system itself may be considered as a compartment and subjected to the same analysis as would be applied to physiological compartments. The localization of metabolic compartments within physiologic or anatomic boundaries is still to be solved since, in order to do this, one would need to obtain samples from each compartment during the course of the experiment. Methods to accomplish this have not been developed in the study of brain metabolism. Therefore, this chapter is restricted to the consideration of metabolic or biochemical compartmentation.

A. Definition

By definition metabolic compartmentation refers to the presence in a tissue of more than one distinct pool of a given metabolite. These separate pools are not in rapid equilibrium with each other but maintain, more or less, their own integrity, turnover rates, and flux rates.

B. Criteria

In a single-compartment system, following the administration of tracer quantities of an isotopically labeled compound serving as precursor to a metabolic product, as the specific activity (amount of isotope per unit weight of compound) of the precursor falls that of its product rises. With time the specific activity of the product becomes equal to or slightly greater than that of its precursor.[9] However, if the specific activity of the product rises within minutes to values 4–5 times as great as that of its precursor, it is most unlikely to be a one compartment system. Implicit here is the assumption that the product–precursor relationship between the compounds is supported by the evidence. In addition, when the peak of the specific activity of the product is greater than the peak of the specific activity of its precursor, the system cannot be described as a single compartment. What is most likely is that a small, metabolically active pool of the precursor is being converted to the product. Upon extraction of the tissue, the larger, less active pool of precursor would then dilute the specific activity of the smaller pool to a value less than that of the product. In the glutamate–glutamine system of brain this was confirmed by computer analysis by Garfinkel.[14,15]

It also should be emphasized that while the above criteria strongly suggests compartmentation, the converse is not necessarily true. That is, if the specific activity of the product is less than that of its precursor, compartmentation of metabolism cannot be excluded. The size of the pools, the rates at which they are penetrated, and their relative turnover rates may very well mask their presence as viewed from this vantage point.

II. *IN VIVO* STUDIES

A. Intravenous Administration of ^{14}C-Glutamic Acid

The intravenous administration of large amounts of glutamic acid to rats, raising the blood level 30–50-fold, failed to produce an increase in the concentration of glutamic acid in brain.[16] This suggested the presence of a blood–brain barrier to the passage of this amino acid. This concept received confirmation when tracer amounts of ^{14}C-glutamic acid (uniformly labeled) were administered into the tail vein of rats and mice[17] (Table I). Although a small but significant amount of radioactivity did enter the brain and was distributed in glutamic acid, glutamine, γ-aminobutyric acid, and glutathione, the specific activities (counts per minute per micromole) of the glutamate and glutamine were only a fraction of that present in other organs. In several of these short-time experiments the specific activity of plasma glutamine exceeded that of brain or liver glutamic acid or glutamine.

Elimination of the liver as a major source of plasma glutamine by tieing off the inferior vena cava and abdominal aorta[18,19] (Table II) produced essentially the same results. In addition, the specific activity of the brain glutamine was greater than that of brain glutamic acid. The degree of restriction of the circulation can be seen from the difference between the specific activities in the muscles of the front legs and the hind legs.

The authors explained the high specific activity of plasma glutamine as possibly due to cellular compartmentation of its synthesis. They suggested that several separate pools of glutamate coexist in tissue, each metabolized at its own rate. The glutamine is synthesized from ^{14}C-glutamate that has not mixed with the total organ glutamic acid, and part of the newly formed amide is returned to the blood again before being equilibrated with the organ

TABLE I

Specific Activity of Compounds After Injection of ^{14}C-Glutamic Acid[17]a

Duration of experiment, min	Organ	Specific activity of compounds, counts/min/μmole			
		Glutamic acid	Glutamine	-Amino-butyric acid	Glutathione
2	Plasma	12,000	1,900		
	Brain	15	13	3	8
	Liver	1,100	510		38
	Red blood cells	1,900	220		65
	Kidney	704	390		340
5	Plasma	6,000	1,900		
	Brain	90	36	33	
	Liver	1,200	850		

[a] Four adult mice were pooled for each point. They were injected with ^{14}C-glutamic acid intravenously. 0.5 μC and 7 μg glutamic acid was injected in each animal corresponding to about 30% of the plasma glutamic acid content and 4% of blood volume. All values were corrected for counts introduced by blood remaining in the excised organ.

TABLE II

Specific Activity in Rat Organs with Restricted Blood Flow[18,19]a

		Counts/min/μmole	
	Organ	Glutamic acid	Glutamine
Unrestricted circulation	Plasma	520,000	38,000
	Red blood cells	28,000	8,200
	Lung	68,000	34,000
	Heart	7,800	5,300
	Muscle (front legs)	3,800	2,000
	Brain	300	550
Restricted circulation	Liver	1,800	790
	Kidney	17	190
	Muscle (hind legs)	100	54

a The inferior vena cava and the abdominal aorta were tied off in rats and 0.1 ml ^{14}C-glutamic acid (5 μC and 1 μmole) was injected in the subclavian vein; the animals were sacrificed after 5 min.

glutamine. The endoplasmic reticulum would be the most likely site for such metabolic activity.[18,19] They also concluded that under conditions in which the liver is removed from the circulation the brain compartment for glutamine synthesis from glutamic acid is stimulated. This is consistent with the observations of Flock et al.[20] that, following hepatectomy, brain glutamine rises without a proportional increase in the glutamine content of the blood.

B. Direct Application of ^{14}C-Glutamic Acid and ^{14}C-Aspartic Acid to the Brain

Roberts, Rothstein, and Baxter[21] were the first to show that ^{14}C-glutamic acid administered into the cisterna magna is rapidly converted to glutamine, γ-aminobutyric acid, and aspartic acid. Conversely, the intracisternal administration of 1-^{14}C-γ-aminobutyric acid led to rapid labeling of glutamic acid and glutamine. In the latter case after 27 min the glutamine had the highest label next to that of γ-aminobutyric acid.

Others[18,22] investigated the time sequence of conversion of glutamate to glutamine following intracisternal administration of tracer amounts of ^{14}C-glutamate in the rat (Table III). Within 5 min the ratio of specific activities of glutamine to that of glutamic acid* was approximately 5 while the total counts in the glutamate plus glutamine remained unchanged during this period of time. It should also be noted that the peak specific activity of the glutamine reached at 5 min was almost twice that of the highest glutamic acid value observed at 25 sec. These studies with the rat (Table IV) and the monkey (Table V) showed that such rapid conversion of glutamate to

* Synonymous with relative specific activity (RSA).

TABLE III

Cerebral Glutamic Acid Metabolism After Intracisternal Administration of [^{14}C] Glutamic Acid[22]a

Time, min	Substrates	Specific activity, counts/μmole/min			Ratio Glu-NH$_2$/Glu	Total counts in 1 g fresh tissue	
		Plasma	Brain	Liver	Brain	Brain	Brain total (Glu + Glu-NH$_2$)
1/4	Glu[b]		7,700		0.43	77,000	92,000
	Glu-NH$_2$[b]		3,300			15,000	
1	Glu	180,000	5,500	200	1.3	51,000	84,000
	Glu-NH$_2$	1,200	7,300	66		33,000	
2	Glu	130,000	3,200	800	3.8	32,000	91,000
	Glu-NH$_2$	2,200	12,000	600		59,000	
5	Glu	34,000	2,700	700	5.2	27,000	97,000
	Glu-NH$_2$	6,100	14,000	760		70,000	
15	Glu	8,900	2,200	190	5.0	22,000	77,000
	Glu-NH$_2$	6,500	11,000	280		55,000	
30	Glu	3,300	2,100	180	3.5	21,000	58,000
	Glu-NH$_2$	870	7,300	240		37,000	
60	Glu	1,500	1,900	61	2.6	19,000	44,000
	Glu-NH$_2$	140	5,000	130		25,000	

a In each experiment two rats were injected via the cisterna magna with uniformly labeled [^{14}C] glutamic acid (0.02 ml of a solution containing 40 μC and 4 μmole/ml in 0.9% saline). Each figure is the average of two experiments.
b Glu, glutamic acid; Glu-NH$_2$, glutamine.

glutamine probably occurred in all brain areas as well as the spinal cord. The GABA was also significantly labeled as was the glutamic acid of the GSH.

Potter and Van Harreveld[23] showed that ^{14}C-glutamate applied to the surface of the brain is rapidly converted to glutamine. However, the authors did not investigate their specific activities. Berl[24] also studied the penetration into the cortex of ^{14}C-aspartate and ^{14}C-glutamate applied directly to the surface of the brain of the cat (Table VI). After 10 min the tissue was cut into 3 or 4 horizontal sections and each was analyzed separately. With either

TABLE IV

Cerebral Metabolism of Glutamic Acid After Intracisternal Administration of [^{14}C] Glutamic Acid[22]a

Brain area	Counts/min/μmole				Relative specific activity, glutamic acid = 1.0		
	Glutamic acid	Gluta-mine	Gluta-thione	γ-Amino-butyric acid	Gluta-mine	Gluta-thione	γ-Amino-butyric acid
Cerebellum	7,000	37,000	7,200	2,400	5.3	1.0	0.34
Pons medulla	14,000	70,000	16,000	3,200	5.0	1.1	0.23
Posterior superior cerebrum	1,600	5,100	1,600	460	3.2	1.0	0.29
Anterior superior cerebrum	260	1,400	380	80	5.4	1.5	0.31
Posterior inferior cerebrum	4,600	18,000	4,300	560	3.9	0.93	0.12
Anterior inferior cerebrum	1,800	8,700	1,500	230	4.8	0.83	0.13

a For experimental details see Table III. Rats were killed after 2 min. Average of four experiments.

substrate as the precursor, the glutamic acid, glutamine, and aspartic acid in all sections of the cortex were labeled. The RSA of glutamine was significantly greater than 1 in all sections, but the highest ratio was present in the section of cortex beneath the topmost one, 0.5–1.0 mm beneath the surface of the brain. This was true even though the highest concentration of radioactivity was in the outermost section of the cortex.

There can be little doubt that the conversion of aspartic acid to glutamic acid via the Krebs cycle and then further to glutamine is an intracellular process. It also appears that evidence for compartmentation of glutamic acid metabolism can be obtained at all depths of the cortex and that it is greatest beneath the surface of the brain in an area most dense in neuronal cells.

C. Intravenous Administration of ^{15}N-Ammonium Acetate and ^{14}C-Sodium Bicarbonate

A study of the detoxification of ammonia by brain tissue provided additional means for the study of compartmentation of the glutamic acid–glutamine system in this organ.[25,26]

^{15}N-ammonium acetate was infused at a constant rate into the carotid artery of the cat for periods of 8–25 min. The results of a typical experiment are given in Table VII. The ^{15}N content of the amide group of cerebral

TABLE V

Cerebral Glutamic Acid Metabolism in the Monkey After Intracisternal Administration of [^{14}C] Glutamic Acid[22]a

Brain area	Counts/min/μmole			
	Glutamic acid	Glutamine	Glutathione	γ-Amino-butyric acid
Frontal cortex	430	1,600	370	51
Parietal cortex	82	160	77	24
Temporal cortex	540	5,300	780	310
Occipital cortex	76	170	82	(9)
Lenticular nucleus	440	1,000	490	46
Thalamus	210	370	190	22
Hypothalamus + optic chiasm	4,300	6,300	3,300	340
Caudate nucleus	56	100	48	(1)
Cerebellum	5,600	21,000	3,700	3,000
Mesencephalon	2,400	12,000	2,800	440
Pons	14,000	78,000	11,000	4,200
Medulla	26,000	110,000	15,000	6,400
Frontal lobe white matter	220	170	120	(16)
Corpus callosum	220	480	200	64
Rest of brain	380	510	280	110
Spinal cord, upper part	66,000	310,000	32,000	18,000
Spinal cord, lower part	25,000	52,000	10,000	1,400
Blood	10,000	310	270	

a Java monkey (*Macacca irus*) was injected via the cisterna magna with 0.4 ml of [^{14}C] glutamic acid (50 μC and 5 μmole/ml in physiological saline). Under light ether anesthesia the skull was opened, and 10 min after injection the brain was removed and dissected (total time after injection 20 min). Spinal cord removed 35 min after injection.

glutamine was higher than that of liver or blood glutamine. This indicated that the glutamine was synthesized in the brain. The level of glutamine in brain increased considerably. Therefore, the high specific activity of the amide group was probably due not only to an exchange with free ammonia but to a net synthesis of glutamine; this occurred because doses of ^{15}NH$_3$ much greater than tracer amounts were administered. This is in accord with the evidence obtained from hepatectomized animals.[20] The α-amino group of brain glutamine had 10 times the specific activity found in brain glutamic acid. Since ^{15}NH$_3$ was administered by continuous infusion, this result could not be attributed to redistribution of label during the decay period of an initial dose of isotope. Therefore, the glutamine must have been derived from a compartment of glutamic acid that was not in equilibrium with the total tissue content of this acid unless glutamine was synthesized appreciably by a pathway that did not require glutamic acid as the immediate

TABLE VI

Labeling of Glutamic Acid, Glutamine, and Aspartic Acid Following Application of ^{14}C-Glutamic Acid and ^{14}C-Aspartic Acid to Surface of Brain of Adult Cat[24]a

Substrate applied	Glutamic acid		Glutamine		Aspartic acid		Ratio of specific activities of glutamine to glutamic acid
	Amount, µmoles/g	Specific activity, cpm/µmole	Amount, µmoles/g	Specific activity, cpm/µmole	Amount, µmoles/g	Specific activity, cpm/µmole	
1. ^{14}C-Glutamic acid							
Section a	11.0	23,000	6.2	45,900	3.4	2,000	2.00
Section b	12.2	5,760	6.0	21,500	3.7	970	3.73
Section c	9.9	580	6.3	1,410	2.7	140	2.43
2. ^{14}C-Aspartic acid							
Section a	11.2	37,500	5.1	103,200	2.6	230,000	2.75
Section b	11.5	4,750	5.4	25,800	2.9	25,700	5.43
Section c	10.5	1,330	4.8	2,040	2.7	5,300	1.53
3. ^{14}C-Aspartic acid							
Section a	10.9	40,100	5.4	68,300	3.0	158,200	1.70
Section b	12.4	8,800	5.8	26,600	2.7	57,600	3.02
Section c	11.7	11,200	5.7	28,300	2.6	63,700	2.53
Section d	11.3	4,100	5.5	11,100	2.5	21,200	2.71

[a] The brain cortex was cut into three or four horizontal sections, each approximately 0.5–1.0 mm thick. The amount of substrate applied was 1.8 µC in 0.020 ml (205 µC/µmole) in Experiment 1 and 1.8 µC in 0.020 ml (164 µC/µmole) in Experiments 2 and 3. The experimental time period was 10 min.

TABLE VII

Isotopic Concentration in Glutamic Acid and Related Metabolites After ^{15}N-Labeled Ammonia Infusion into the Cat[25]a

	Blood		Brain cortex		Liver	
	μM/g tissue	^{15}N at. % excess	μM/g tissue	^{15}N at. % excess	μM/g tissue	^{15}N at. % excess
Ammonia	2.8	81.9	1.8	65.7	4.6	42.6
Glutamine amide	0.32	21.4	7.8	38.7	0.83	28.7
Glutamine α-amino	0.32	1.8	7.8	8.6	0.83	5.4
Glutamic acid	0.08	3.1	9.3	0.87	1.6	13.6
GABA			1.2	0.22		

a Nitrogen-15 ammonium acetate (99.6 at. % excess) 1 mM/ml infused into carotid artery of a succinylcholine paralyzed 3.3-kg cat at a uniform rate for 24 min (total dose, 16 mmoles). Experiment terminated at 25 min by bleeding from the carotid artery; cortex and liver frozen *in situ* with solid carbon dioxide.

precursor. If glutamic acid was utilized for the synthesis of the additional glutamine in the brain, the former amino acid was apparently made in this organ and not derived from the blood, since the ^{15}N content of the α-amino group of brain glutamine (8.6%) was higher than that of blood glutamic acid (3.1%).

It is of interest to compare the data from brain with that obtained from liver.[26] In the latter organ the α-amino group of glutamine had a lower ^{15}N concentration than that of the α-amino group of glutamic acid. However, particularly in shorter time experiments, frequently the α-amino group of aspartic acid had a higher ^{15}N specific activity than that of glutamic acid. The results suggest the existence of pools of highly labeled glutamic acid, glutamine, or both, which transaminate with oxaloacetate and do not exchange rapidly with the remainder of the liver glutamic acid or glutamine. These pools may be more directly in the pathway toward urea formation.

The study of ammonia metabolism in brain suggested that a mechanism exists in this organ for the replenishment of 4-carbon units of the Krebs cycle utilized for glutamine formation. Carbon dioxide fixation in brain was a good candidate for such a role. If it occurred it should yield additional information on the compartmentation of glutamate metabolism in brain. Upon intracarotid infusion of ^{14}C-sodium bicarbonate the amino acids showed considerable radioactivity and the specific activity of the glutamine was greater than or equal to the specific activity of the glutamic acid[27–29] (Table VIII). When ammonium acetate was administered simultaneously with

TABLE VIII

^{14}C Incorporation into Components of the Nonprotein Fraction of Brain Cortex During $NaH^{14}CO_3$ Infusion[29]a

Duration, min ·	counts/min/μmole (μmole/g)		
	3.5	8.5	18
Glutamic acid	320(9.5)	840(8.5)	890(10.2)
Glutamine	570(5.3)	930(7.6)	1640(5.7)
Aspartic acid	2320(2.7)	3860(2.6)	7110(2.3)
α-Oxoglutarate	730(0.0025)	1770(0.001)	1070(0.006)
Pyruvate	410(0.006)	1320(0.004)	820(0.020)
Tissue $CO_2 \times 10^{-3}$	40.0(10.8)	33.0(13.2)	29.0(14.2)

a 1.7 ml of the bicarbonate solution (0.06 M, 0.29 mC/ml in saline) was given during the first 10 sec to bring the specific activity of blood CO_2 to a precalculated level. Infusion was then continued at a rate of 0.2 ml/min to maintain blood CO_2 at a constant specific activity. The solution was administered into the inferior vena cava through a cannula inserted via the femoral vein.

radioactive bicarbonate, there was an increase in the specific activity of glutamine relative to that of aspartic acid as well as relative to that of glutamic acid (Table IX). The specific activity of glutamine also was high relative to that of oxoglutarate, particularly in the experiments with ammonium acetate. This suggests that the oxoglutarate in cerebral cortex is also compartmented into at least two pools.[28,29] A similar conclusion was reached by Gaitonde[30] as a result of studies of the metabolism of ^{14}C-glucose in rat brain.

D. Metabolism of ^{14}C-Labeled Acetate, Butyrate, Propionate, Glucose, Lactate, and Glycerol

It had been shown by Busch that within minutes after administration of ^{14}C-acetate[31] or pyruvate[32] almost all the radioactivity in brain can be accounted for by amino acids; the ^{14}C-pyruvate entered the brain after prior conversion to lactate. O'Neal and Koeppe[33,34] in their study of precursors *in vivo* of glutamate, glutamine, and aspartate used ^{14}C-labeled acetate, butyrate, propionate, glucose, lactate, and glycerol. They confirmed the earlier studies on the metabolism of administered ^{14}C-glutamic acid and also confirmed the reports of Cremer[35] and Gaitonde[30] that when labeled glucose is administered to rats the specific activity of glutamate of brain is always greater than that of glutamine. In addition they reported that when glucogenic precursors were used—such as glucose, lactate, or glycerol—the

TABLE IX

^{14}C Incorporation into Components of the Nonprotein Fraction of Brain Cortex During $NaH^{14}CO_3$ and Ammonium Acetate Infusion[29]a

Duration, min	counts/min/μmole (μmoles/g)	
	15.5	16
Glutamic acid	1520(10.4)	1320(8.4)
Glutamine	7860(7.3)	5370(7.6)
Aspartic acid	5900(2.1)	2330(1.7)
α-Oxoglutarate	3190(0.03)	2540(0.02)
Pyruvate	840(0.14)	1070(0.06)
Tissue $CO_2 \times 10^{-3}$	55(11.4)	38(15.3)
NH_3	(6.9)	(4.3)

[a] See Table VIII for method of administration. The ammonium acetate was administered as a 2.5 M solution containing $NaH^{14}CO_3$ (0.06 M, 0.29 mC/ml). It was infused at a rate of 0.2 ml/min immediately following the rapid injection of the 1.7 ml of the $NaHCO_3$ solution.

specific activity of glutamine was always less than that of glutamic acid and, conversely, when glutamate, glutamine, or ketogenic precursors were used— such as acetate, butyrate, or propionate—the specific activity of glutamine was always greater than that of glutamic acid. They suggested that the data could be explained best by assuming two pools of tricarboxylic acid cycle intermediates, both accessible to pyruvate but only one accessible to meta- bolites that show compartmentation. Such a scheme also would account for the two pools of oxoglutarate suggested by Waelsch et al.[29] and Gaitonde.[30] More than one type of tricarboxylic acid cycle was also suggested by Van den Berg et al.[36,37] following their studies in mice carried out with acetate-1-^{14}C and -2-^{14}C in comparison with glucose-1-^{14}C, -2-^{14}C, and -6-^{14}C. They suggested that glutamine derived from acetate is coupled to the tricarboxylic acid cycle in a different manner than glutamine derived from glucose. The rapid labeling of glutamate, glutamine, aspartate, and γ-aminobutyric acid in brain following intravenously administered 1-^{14}C-acetate was used by Berl and Frigyesi[38] to study the turnover time of these amino acids in the brain of the cat. As early as 1 min after administration of the 1-^{14}C-acetate the specific activity of glutamine in the areas investigated (cortex, thalamus, and caudate nucleus) was greater than that of glutamic acid. The decay curves for the four amino acids were distinctly diphasic. The first phase of the curves had a range of half-lives of 11–18 min for the four amino acids in the three brain areas. The second phase of the curves had half-lives of 3–4 hr. The radioactivity in the glutamine decayed somewhat more slowly than did that of the glutamic acid.

E. Leucine Metabolism

Roberts and Morelos[39] reported that ^{14}C-leucine, uniformly labeled, when administered to rats intravenously is converted in the brain in large measure to glutamic acid, aspartic acid, glutamine, and GABA. This is consistent with the established pathway of leucine degradation to acetyl CoA,[40] which then enters the Krebs cycle. Their data indicated that the glutamine may be of a higher specific activity than the glutamic acid. Berl and Frigyesi,[41] in their studies of the cat, confirmed these findings. In addition they found that in cortex, hippocampus, thalamus, pons, and medulla the RSA of glutamine was greater than 1 following ^{14}C-L-leucine-u.l. administration. In contrast, in the cerebellum and caudate nucleus the RSA of glutamine was less than 1. Since 1-^{14}C-acetate administration resulted in glutamine with an RSA greater than 1 in the caudate nucleus and cerebellum as well as in the other brain areas, the authors postulated that in the former two areas the acetyl-CoA formation from leucine must be taking place in a different compartment from that in which acetyl-CoA is derived from acetate.

III. DEVELOPMENTAL ASPECTS

The maturational development of the glutamate–glutamine compartmentation system in cerebral cortex was studied by the application of ^{14}C-glutamic acid-u.l. to the surface of the brains of kittens of different ages.[24] Through the third week of life the specific activity of glutamine remained below that of glutamic acid (Fig. 1). This relationship then reversed and the specific activity of the glutamine became greater than that of glutamic acid. The plot of the ratios of the specific activities of glutamine to glutamic acid (Fig. 2) showed that during the first 2 weeks after birth they remained at about 0.4, rose to about 0.8 at 3 weeks of age, sharply increased to about 2.75 during the fourth week, and approached ratios of 4 in animals 6 weeks of age.

The development of glutamate compartmentation in other areas of the kitten brain was studied by administration of ^{14}C-glutamic acid-u.l. into the cisterna magna.[42] The youngest animals investigated were 2 days of age. In the hippocampus of these animals, the specific activity of glutamine achieved values more than twice that of glutamic acid. In the cerebellum, mesodiencephalon, and brainstem this relationship did not become established until 6 weeks of age.

In ontogenetic studies of the brain of cat it has been shown that in the neocortex, morphological[43,44] and electrophysiological development[45] of neurons and synaptic organization proceeds rapidly during the first postnatal month. During the latter part of this period and continuing afterward elaboration of dendritic and axonal plexuses occurs. Morphophysiological studies of the immature hippocampus indicated that neuronal and synaptic organization in this structure are considerably more mature at birth than

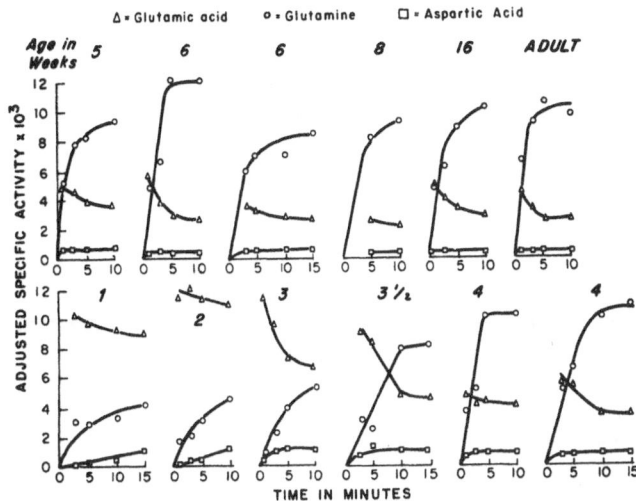

Fig. 1. Specific activities (counts per minute per micromole) of glutamic acid, glutamine, and aspartic acid in cerebral cortex following application of L-glutamic-^{14}C acid, uniformly labeled (20 μl, 1.8 μC: 120–205 μC/μmole), to the surface of the brains of kittens of different ages.[24]

neuronal elements in neocortex.[46] On the other hand, Purkinje cells and other neural elements in the cerebellum do not acquire complete maturational characteristics until 6 weeks postnatally.[45] Thus, in three types of cortex for which ontogenetic data are available there appears to be some correlation between morphophysiological development and development of amino acid compartmentation.

Studies to determine the extent of participation of glutamine synthetase in the development of glutamic acid–glutamine compartmentation also have been undertaken.[47] Only in the neocortex did the increase in enzyme activity coincide with that of the development of compartmentation of glutamic acid metabolism (Fig. 3). This was not true to the same extent for the hippocampus, cerebellum, brainstem, or meso-diencephalon. It was suggested that glutamine synthetase activity is only one factor in the development of the compartmentation of glutamic acid metabolism.

IV. *IN VITRO* STUDIES

A. Metabolism of ^{14}C-Labeled Glutamate, Aspartate, Acetate, Bicarbonate, and GABA

The experiments described up to now in which glutamine of higher specific activity than glutamate is obtained after the administration of various

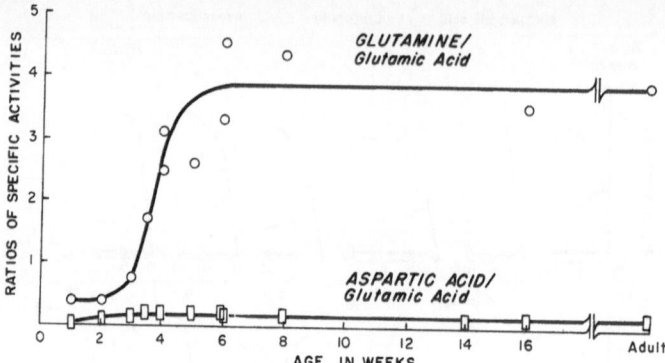

Fig. 2. Ratios of the specific activities of glutamine and aspartic acid to that of glutamic acid at the 10–15-min time points following application of L-glutamic-^{14}C acid to the surface of the cerebral cortex of kittens.[24]

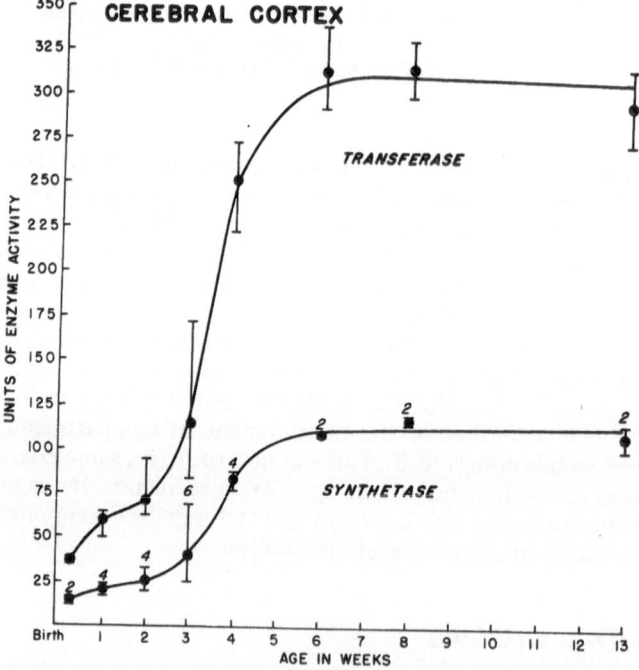

Fig. 3. Glutamine synthetase and transferase activity of kitten cerebral cortex. Units of enzyme activity per gram of tissue wet weight are given. 1 unit = 1 μmole of γ-glutamylhydroxamate formed per hour. The range of values, the average value, and the number of animals for each point are indicated.[47]

labeled precursors were carried out on living animals. Attempts to demonstrate compartmentation of glutamate metabolism by this means in an isolated system were generally unsuccessful until recently. Thus, Cremer[35] reports that she was unable to observe this phenomenon with brain slices using uniformly labeled glutamate-[14]C. Our early attempts to demonstrate RSA's of glutamine greater than 1 with brain cortex slices were unsuccessful when using aspartate-U-^{14}C as tracer.

Insofar as all the *in vivo* experiments had suggested that the compartmentation of the glutamate–glutamine system occurred at the cellular level, it was expected that this phenomenon ought to be observable in tissue slice preparations. A systematic study of various incubation conditions showed that preparation of tissue slices at 0°C led to a large inhibition of the incorporation of radioactivity from aspartate into glutamine while having little effect on the labeling of glutamate. This inhibition could be reversed by preincubation of tissue slices prepared at 0°C in Krebs–Ringer medium at 37° for 10 min or more followed by transfer to fresh medium before addition of the tracer substance.[48] This is shown by the data presented in Table X. Alternatively, brain cortex slices prepared at room temperature could be used directly to demonstrate compartmentation of the glutamate–glutamine system. However, even here preincubation and transfer to fresh medium led to further increases in the RSA of glutamine (Table X).

When glutamic acid-U-^{14}C was used as the tracer substance similar results to those obtained with aspartate were observed although the magnitude of the increases in RSA of glutamine in response to the method of preparation of the slices or the preincubation and transfer effects was much smaller.

If the incubations of the tissue slices were carried out in a modified Krebs–Ringer medium in which the K^+ level of the medium was increased to 27 mM while the Na^+ concentration was reduced to maintain osmolarity, the differences between slices prepared at room temperature and in the cold no longer could be seen whether glutamate or aspartate was the tracer metabolite. The effect of preincubation and transfer still could be seen, although it was considerably reduced (Table X). The use of Krebs–Ringer bicarbonate or phosphate media gave essentially similar results. In a few of the early experiments the levels as well as specific activities of the amino acids in the medium were measured. The levels were quite low particularly of glutamic and aspartic acids, and therefore the specific activities were not very reliable. Sodium bicarbonate-^{14}C also gave glutamine with an RSA greater than one (Table X). Here, even slices prepared in the cold and not transferred to fresh medium gave RSA's of glutamine greater than one. Transfer to fresh medium increased these values fourfold. These results demonstrate appreciable CO_2 fixation in brain tissue slices, an observation previously reported only *in vivo*.[29]

Another ion that had a marked effect on the RSA of glutamine was Ca^{2+}. When Ca^{2+} was omitted from the medium the RSA of glutamine fell to values of 0.2 ± 0.1 with either glutamate or aspartate as the labeled precursor.

TABLE X

Relative Specific Activity of Glutamine from U-^{14}C-Aspartic Acid or U-^{14}C-Glutamic Acid in Guinea Pig Brain Cortex Slices $\pm$ SD[48]a

| | Relative specific activity[b] | | | |
| | Cold | | Room temperature | |
^{14}C-substrate	Untransferred	Transferred	Untransferred	Transferred
	5.2 mM K$^+$			
Aspartic acid	0.40 ± 0.23	2.80 ± 1.19	1.25 ± 0.33	3.41 ± 0.88
	$(N = 8)$	$(N = 9)$	$(N = 9)$	$(N = 10)$
Aspartic acid	0.94 ± 0.20^c		2.40 ± 0.32^c	
	$(N = 2)$		$(N = 2)$	
Glutamic acid	0.65 ± 0.19	1.67 ± 0.51	0.92 ± 0.17	1.70 ± 0.34
	$(N = 4)$	$(N = 6)$	$(N = 4)$	$(N = 7)$
Sodium bicarbonate[d]	1.18 ± 0.15	5.87 ± 1.02	1.36 ± 0.10	4.08 ± 0.01
	$(N = 2)$	$(N = 2)$	$(N = 2)$	$(N = 2)$
	27 mM K$^+$			
Aspartic acid	0.62 ± 0.36	0.81 ± 0.44	0.63 ± 0.31	1.03 ± 0.28
	$(N = 3)$	$(N = 4)$	$(N = 4)$	$(N = 4)$
Glutamic acid	1.08 ± 0.13	1.54 ± 0.17	1.20 ± 0.13	1.57 ± 0.30
	$(N = 4)$	$(N = 4)$	$(N = 4)$	$(N = 4)$

[a] The tissue and slices were kept in an ice bath or at room temperature until incubated at 37°C. The medium used was Krebs–Ringer phosphate or Krebs–Ringer bicarbonate (three experiments) containing 1% glucose. Approximately 100 mg of slices and 2.5 ml of medium containing 0.5 μC of radioactive tracer (> 200 μC/μmole) were used. The transferred slices were preincubated at 37° for 10 min, transferred to fresh medium before addition of isotope, and incubated for 10 min. Glutamic acid = 1.

[b] Relative specific activity based on counts per minute per micromole of amino acid.

[c] Preincubated for 10 min but *not* transferred to fresh medium before addition of isotope.

[d] Phosphate buffer replaced with Tris buffer, pH 8.

In this respect the results obtained with low calcium concentration do not resemble those obtained with media of high K$^+$ concentration. Kini and Quastel[49] had observed that both conditions produced similar increases in oxygen utilization over controls and suggested that Ca^{2+} and K$^+$ played reciprocal roles in regulating respiration in brain slices. More recently Chan and Quastel,[50] in studying the action of tetrodotoxin, showed that the K$^+$ and Ca^{2+} effects on respiration can be differentiated. This latter observation is in agreement with the results reported above.

Acetate-1-^{14}C and glutamate-1-^{14}C also demonstrated the compartmentation of the glutamate–glutamine system. Glutamate-1-^{14}C allowed direct observation of glutamine formation without the added complications of cycling through the tricarboxylic acid cycle, which needs to be considered when glutamate-U-^{14}C is the tracer. In general the results obtained with glutamate-1-^{14}C were quite similar to those with glutamate-U-^{14}C; hence, the effects of cycling appear to be minimal when ^{14}C-glutamate is the tracer substrate. With acetate-1-^{14}C as the tracer relatively high RSA's of glutamine were obtained in the incubations of 10 min duration. These ratios decreased with time (Table XI). The effects of preparation of slices at 0°C were not observed with this tracer, in contrast to the results with aspartate and glutamate. Incubation in medium containing 27 mM K$^+$ lowered the RSA of glutamine in the 10-min experiment to that in the experiments of longer duration so that no change in the ratio with time was observed.

Balazs[51] has obtained indications for the compartmentation of the glutamate–glutamine system using 1-^{14}C-GABA as the tracer metabolite. In guinea pig brain cortex slices incubated in glucose–saline the RSA of the glutamine was greater than one during the entire incubation (maximal values were greater than two). Increasing the concentration of the K$^+$ in the incubation medium to 66 mM caused a release of about a third of the tissue content of the GABA into the medium; only about 3–5% of other amino acids were released.[52] The compartmentation of glutamate and glutamine was affected by the high concentration of K$^+$. The RSA of glutamine decreased immediately after the addition of K$^+$ from a value greater than one to less than one.[53]

Thus the demonstrations of compartmentation of glutamate metabolism in tissue slices can be carried out with results similar to those obtained *in vivo* and without the complications of conversions of the added tracer to another labeled metabolite by the liver or other organs before entering the brain.

B. Studies with Metabolic Inhibitors

It is also possible to study the effect of metabolic inhibitors on the compartmentation of amino acid metabolism in brain slices. Fluoroacetate, a known convulsant agent, has been studied extensively for its effect on energy and amino acid metabolism. It has been shown that fluoroacetate at 1 mM

TABLE XI

Relative Specific Activity of Glutamine from 1-^{14}C-Acetate in Guinea Pig Brain Cortex Slices ± SD[71]a

Incubation time, min	Cold	Room temperature	
	5.2 mM K$^+$	5.2 mM K$^+$	27 mM K$^+$
10	6.09 ± 1.09 (N = 2)	6.28 ± 0.90 (N = 4)	3.26 ± 0.27 (N = 4)
20	4.92 ± 1.52 (N = 2)	3.79 ± 0.54 (N − 3)	3.17 ± 0.44 (N − 4)
30	4.04 ± 0.81 (N = 2)	3.62 ± 0.42 (N = 3)	3.00 ± 0.25 (N = 4)

a See Table X for incubation conditions. All slices were preincubated and transferred to fresh medium before addition of 2.0 μC of 1-^{14}C-acetate, sodium (58 μC/μmole). Glutamic acid = 1.

concentration has no measurable effect on the oxygen uptake or $^{14}CO_2$ formation from 6-^{14}C-glucose in cerebral cortex slices.[54] It markedly inhibits the incorporation of radioactivity from ^{14}C-glucose into glutamine,[54] and it inhibits $^{14}CO_2$ production from 1-^{14}C-acetate.[55] Although fluorocitrate is known to be a potent inhibitor of aconitase and hence of the citric acid cycle, the fact that fluoroacetate, which would be expected to yield fluorocitrate in tissues, does not lead to a significant lowering of the respiratory rate of brain slices has led Lahiri and Quastel[54] to suggest that the effect of fluoroacetate on brain may be explained best by its effect on glutamine synthesis rather than on inhibition of the citric acid cycle.

We have confirmed the observations of Lahiri and Quastel[54] that indicate that fluoroacetate inhibits incorporation of radioactivity from glutamate into glutamine. In addition, with all the tracer substances used in our work, viz., aspartate, glutamate, and acetate, fluoroacetate at 1 mM brings about a considerable decrease in the RSA of glutamine to values of 0.2–0.4. This would support the contention of Lahiri and Quastel that fluoroacetate is exerting its effect on ammonia metabolism by preventing the formation of glutamine. Since fluoroacetate itself does not inhibit glutamine synthetase they suggested that perhaps fluorocitrate is the inhibitor of this enzyme.[54]

It seems more consistent with the overall picture of compartmentation developed here that fluoroacetate is affecting the turnover of the small pool of glutamate in brain that is active in glutamine formation.

The fact that acetate, leucine, and other precursors of acetyl CoA give glutamine with RSA greater than one and glucose and nonketogenic precursors give glutamine with RSA less than one indicates that there must be separate pools of acetyl CoA and, in fact, separate citric acid cycles.[33,34,36,37] Furthermore, if acetate thiokinase is not uniformly distributed in mitochondrial populations of cerebral cortex, a suggestion[56] that has received considerable support from studies on the distribution of enzymes in cell particulates separated in a continuous density gradient,[57] the activation of acetate may be confined to a metabolic compartment that labels that small pool of glutamic acid active in glutamine synthesis and probably forms part of a separate citric acid cycle that may be looked upon as a "synthetic cycle." We suggest that fluoroacetate is converted to fluorocitrate and produces its inhibitory effect largely in this "synthetic cycle," which contains the small pools of Krebs cycle intermediates. Thus, fluoroacetate can have a profound effect on the labeling of glutamine without a corresponding effect on the labeling of glutamate. Furthermore, the inhibition of acetate-1-^{14}C oxidation by fluoroacetate but not that of glucose-6-^{14}C, reported by Gonda and Quastel,[55] agrees very well with this interpretation of at least two citric acid cycles.

In an attempt to gain further support for this idea, viz., that fluoroacetate may selectively inhibit the smaller pool of the citric acid cycle in brain, the effect of fluoropyruvate was studied. Fluoropyruvate (1 mM) did not lead to a change in the RSA of glutamine when 1-^{14}C-acetate was the precursor. These contrasting effects of fluoroacetate and fluoropyruvate seem to support this hypothesis. However, fluoropyruvate is known to react readily with sulfhydryl compounds[58] with the liberation of fluoride ion, and therefore its effect may be more complex. In fact, 1 mM NaF produced results similar to those obtained with fluoropyruvate.

Ouabain at the 10 μM level lowered the RSA of glutamine to values of 0.2–0.3. The levels of the amino acids in the tissue were also lowered considerably. This leakage of amino acids into the medium in the presence of ouabain also has been reported by Cremer.[59] Thus ouabain appears to act on at least two membrane sites, one involved in the maintenance of the tissue levels of the amino acids and another involved in the maintenance of their compartmentation. Rose[60] also has argued for multiple sites of action for ouabain. Fluoroacetate, Ca^{2+}, and K^+ appear to have more limited sites of action. Although the details of compartmentation remain hazy, it may be possible through choice of inhibitors and activators as well as tissue preparations to sharpen the focus of the picture.

V. THEORETICAL CONSIDERATIONS

In order to obtain experimental data on the size of the pools it is essential to sample the pools during an experiment. Since it has not yet proven feasible to sample the various tissue compartments involved, the mathematical analysis has been approached by the use of computer simulated models by Garfinkel.[14,15] Rather broad assumptions are made as to the number of pools, pool sizes, and rate constants and the experimental data used to choose the model or models that show the best fit. Further experiments may then be suggested to aid in narrowing the choice of models.

The experimental conditions so far used have limited the consideration of compartmentation to metabolic definition although, no doubt, at least some of the pools have anatomic boundaries. In consideration of anatomic localization two general categories of compartmentation can be envisioned, cellular and intracellular. At the cellular level it is conceivable, since one deals with a mixed population of cells, that compartmentation is a reflection of preferential penetration of the cells or cellular extensions by one or another of the administered tracer compounds. Compartmentation then would be determined by the type of cell into which any given metabolite could enter in preference to another type of cell. Thus glucose, pyruvate, or lactate may be able to enter all cells while acetate, aspartate, glutamate, or leucine may be able to enter only some cells or parts of cells. While such compartmentation may very well contribute to the total metabolic picture, and the extent to which it does cannot as yet be evaluated, it is unlikely to be the entire explanation. This conclusion is supported by the fact that $^{15}NH_4^+$ and $H^{14}CO_3^-$ also exhibit the effects of metabolic compartmentation; it is unlikely that these ions can enter only into some cells and not into others.

It also may be considered that the compartments consist of two or more cell types, one of which contains only a low level of glutamic acid but rapidly converts the amino acid to glutamine, resulting in a compound of high specific activity. The other cell type (or parts of cells such as dendrites or axons) could contain substantial quantities of glutamate and relatively less glutamine with little capacity to make it and thus would dilute the isotopic glutamic acid to a greater extent. However, there is no evidence to date for such disproportionate distribution of glutamic acid and glutamine within the elements of cerebral tissue.[61,62]

In the mature cortex of the cat the highest ratios of specific activities of glutamine to glutamate were attained in the layers of cortex most dense in neuronal cells beneath the surface of the brain.[24] During ontogenesis, compartmentation continues to increase during the period of active proliferation of glial cells.[24] There also is some data from the monkey that suggests that the corpus callosum, a tissue composed of white matter and devoid of neuronal cells, also may have a compartmented glutamic acid–glutamine system[22] (cf. Table V). Thus, both neuronal and nonneuronal cells may contribute, in varying degrees, to the composite picture of compartmentation in the central nervous system. It would seem, therefore, that intracellular compartmentation, i.e., the relative metabolic contributions of subcellular structures and their associated enzymes, may be of dominant importance.

There is increasing evidence that the mitochondrial population in tissue may not be homogeneous in enzyme content. Salganicoff and De Robertis[63] have reported a heterogeneous distribution of enzymes in mitochondrial preparations from brain tissue including an enrichment of the glutamic acid decarboxylase enzyme in the mitochondria isolated from nerve endings. Similar conclusions as to the heterogeneity of the mitochondrial population of brain tissue is supported by the work of others.[36,37,56,57,64] Heterogeneity of mitochondrial enzyme distribution is not unique to brain. It also has been described for liver mitochondria[65] and may be generally true for all tissues.

Another consideration that may be of basic importance to compartmentation is the anatomic relationship of the subcellular structures. Thus homogenates of guinea pig cerebrum in Ringer's solution fortified with glucose do not show compartmentation of the glutamate–glutamine system. In such preparations the specific activity of the glutamine does not rise above that of the glutamic acid. These observations have been made after incubation of brain homogenates with ^{14}C-acetate, ^{14}C-aspartate, or ^{14}C-glutamate as the tracer metabolite. The anatomic relationship of the mitochondria to the endoplasmic reticulum may be of prime importance for the maintenance of compartmentation. In homogenates of tissue all structures may be intermingled sufficiently to allow pools of glutamate of high and low specific activity to become equally available for conversion to glutamine. The endoplasmic reticulum has been described as the structure with which the highest glutamine synthetase activity is associated.[66]

A number of experiments have pointed to the existence of more than one pool of α-ketoglutarate in brain tissue.[28–30,67] As indicated previously the Acetyl-CoA from glucose, lactate, and glycerol[33,36,37] behaves differently from that obtained from leucine, acetate, and butyrate[33,39,41] as measured by the RSA of glutamine in brain. This has led to the suggestion that there exist in brain at least two separate pools of citric acid cycle intermediates. Since the mitochondria have been shown to be the site of operation of the citric acid cycle, heterogeneity of the mitochondria would fit well with the concept of more than one citric acid cycle that show preferential utilization of certain metabolites. Several schema have been presented to describe the entry of various precursors into these cycles.[30,33,68] Figure 4 represents our present speculations, based on our data and published data of others as well as numerous discussions with colleagues, mainly those who had worked in the laboratory of the late Dr. H. Waelsch. We assume that one citric acid cycle is composed of relatively larger pools of intermediates that interchange rapidly with a large pool of glutamic acid and a small pool of glutamine; the other cycle, composed of relatively smaller pools of intermediates, interchanges rapidly with the small pool of glutamic acid that equilibrates with the large pool of glutamine. The cycle containing the large pool of intermediates may be mainly involved in energy generation and would be well protected from depletion of its intermediates by the large stabilizing pool of glutamic acid. The cycle with smaller pools would better serve in a synthetic

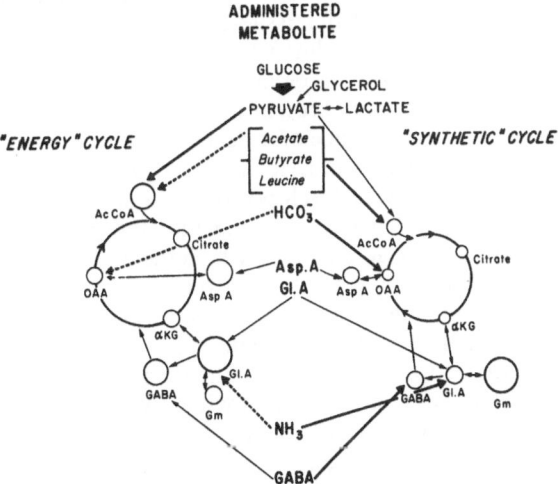

Fig. 4. Schematic description of compartmentation of amino acid metabolism in brain based on two separate tricarboxylic acid cycles. The size of the circles indicate only that some pools are considered to be larger than others. They are not intended to describe the true relative sizes of the various pools; this information is not known. The thick, thin, and dotted lines indicate the decreasing order of preferred metabolic pathways. Gl. A, glutamic acid; Gm, glutamine; Asp. A, aspartic acid; OAA, oxaloacetate; and αKG, α-ketoglutarate.

capacity, e.g., ammonia detoxification, protein synthesis, and possibly maintenance of transmitter substances. For convenience the former may be
designated as the "energy cycle" and the latter as the "synthetic cycle." It
should be emphasized that this convenient designation does not exclude the
probability that each of these metabolic cycles consists of more than one
pool of one or more metabolites or that there may be additional complete
cycles.

The concept of at least two pools of glutamic acid, one consisting of the
major portion of the total tissue content of this amino acid and another of
relatively smaller proportions (of the order of 10%), which is the precursor
for the synthesis of the major portion of glutamine, arose from the early
observations with ^{14}C-glutamate and ^{14}C-aspartate that the RSA of glutamine in brain achieves values as high as five or more within minutes after
intracisternal administration of the labeled metabolites.[22] These findings
were supported by many subsequent studies with intravenously administered
^{15}N-ammonium acetate,[25,26] ^{14}C-sodium bicarbonate,[27-29] ^{14}C and
^{3}H-acetate,[33,34,36-38,69] and ^{14}C-leucine.[39,41] In the ^{14}C-bicarbonate
studies the specific activity of the blood bicarbonate was maintained by
constant infusion until the animal was sacrificed. Here, in addition to the
amino acids, the keto acids were also studied. The specific activity of the
α-ketoglutarate was close to that of the glutamic acid and glutamine but
considerably lower than that of the tissue CO_2. It would be anticipated that
the specific activity of the α-ketoglutarate should approach that of the CO_2
in the tissue, which was many times greater than that of the glutamine, if only
one pool of α-ketoglutarate was present in the tissue. Therefore, it became
necessary to postulate the presence of at least two pools of α-ketoglutarate,
one of which is small and heavily labeled and is the precursor of the small pool
of glutamate. The same idea was developed by other investigations from other
types of experiments.[30,67] It was assumed that the small pool of α-keto-
glutarate was part of a citric acid cycle in which the other intermediates were
also of relatively smaller size, while the converse was assumed for the large
pool of α-ketoglutarate. The concept of at least two citric acid cycles into
which entry of a given metabolite may be different also may be applied to an
understanding of the action of inhibitors. The theory implies that metabolic
inhibitors may interfere to a greater extent with one cycle than with another.
The inhibitory effects of fluoroacetate on brain metabolism have presented
a number of paradoxes to investigators. On the one hand it has no effect on
the rate of respiration or CO_2 formation from glucose,[54] and on the other
hand it does inhibit CO_2 formation from acetate.[55] Although citrate
accumulation occurs in the brains of animals poisoned with fluoroacetate,
the levels of ATP and phosphocreatine remain normal.[70] In addition,
glutamine formation from glucose,[54] acetate,[71] glutamate,[54,71] and
aspartate[71] is greatly inhibited by fluoroacetate, while glutamic acid formation is not inhibited. These apparently conflicting data fall into an
explainable pattern if one assumes that the fluoroacetate is acting mainly on
the citric acid cycle which contains the smaller pools of intermediates. Thus,

if the block in citrate metabolism by fluorocitrate formation occurred essentially in the latter cycle the effective increase in the citrate level in that cycle would be much greater than would be apparent from a measure of the total tissue content of citrate. It also would produce a negligible effect on the total oxygen uptake by the tissue or CO_2 production from glucose since the latter is oxidized mainly via the "energy cycle." If acetate as well as fluoroacetate is activated mainly in the "synthetic cycle," inhibition of $^{14}CO_2$ formation from labeled acetate as well as selective inhibition of glutamine synthesis by fluoroacetate becomes readily explainable.

The major direction for future work no doubt will be the correlation of metabolic compartments with cellular and subcellular structures. The work on mitochondrial heterogeneity is a promising beginning in this direction.

VI. REFERENCES

1. R. J. Britten, R. B. Roberts, and E. F. French, Amino acid adsorption and protein synthesis in Escherichia coli, *Proc. Natl. Acad. Sci. U.S.* **41**:863–873 (1955).
2. D. B. Cowie and B. P. Walton, Kinetics of formation and utilization of metabolic pools in the biosynthesis of protein and nucleic acid, *Biochim. Biophys. Acta* **21**:211–226 (1956).
3. D. B. Cowie and F. T. McClure, Metabolic pools and the synthesis of macromolecules, *Biochim. Biophys. Acta* **31**:236–245 (1959).
4. F. C. Steward, G. S. Bidwell, and E. W. Yemm, Protein metabolism, respiration and growth, *Nature* **178**:734–738 (1956).
5. D. H. MacLennan, H. Beevers, and J. L. Harley, "Compartmentation" of acids in plant tissues, *Biochem. J.* **89**:316–327 (1963).
6. A. Korner and H. Tarver, Studies on protein synthesis *in vitro*—VI. Incorporation and release of amino acids in particulate preparations from livers of rats, *J. Gen. Physiol.* **41**:219–231 (1957).
7. N. M. Green and D. A. Lowther, Formation of collagen hydroxyproline *in vitro*, *Biochem. J.* **71**:55–66 (1959).
8. D. M. Kipnis, E. Reiss, and E. Helmreich, Functional heterogeneity of the intracellular amino acid pool in mammalian cells, *Biochim. Biophys. Acta* **51**:519–524 (1961).
9. D. B. Zilversmit, C. Entenman, and M. C. Fishler, On the calculation of "turnover time" and "turnover rate" from experiments involving the use of labeling agents, *J. Gen. Physiol.* **26**:325–331 (1943).
10. A. K. Solomon, Compartmental methods of kinetic analysis, *in Mineral Metabolism* (C. L. Comar and F. Bronner, eds.), Vol. 1, Part A, pp. 119–167, Academic Press, New York (1960).
11. J. S. Robertson, Theory and use of tracers in determining transfer rates in biological systems, *Physiol. Rev.* **37**:133–154 (1957).
12. J. M. Reiner, The study of turnover rates by means of isotopic tracers, *Arch. Biochem. Biophys.* **46**:53–99 (1953).
13. S. Aronoff, Pools, cycling and turnover time, *in Techniques of Radiobiochemistry*, pp. 75–93, The Iowa State University Press, Ames, Iowa (1956).
14. D. Garfinkel, Computer simulation of steady-state glutamate metabolism in rat brain, *J. Theoret. Biol.* **3**:412–422 (1962).

15. D. Garfinkel, A simulated study of the metabolism and compartmentation in brain of glutamate, aspartate, the Krebs cycle and related metabolites, *J. Biol. Chem.* **241**:3918–3929 (1966).

16. P. Schwerin, S. P. Bessman, and H. Waelsch, The uptake of glutamic acid and glutamine by brain and other tissues of the rat and mouse, *J. Biol. Chem.* **184**:37–44 (1950).

17. A. Lajtha, S. Berl, and H. Waelsch, Amino acid and protein metabolism of the brain—IV. The metabolism of glutamic acid, *J. Neurochem.* **3**:322–332 (1959).

18. A. Lajtha, S. Berl, and H. Waelsch, Compartmentalization of glutamic acid metabolism in the central nervous system, *in Inhibition of the Nervous System and λ-Aminobutyric Acid* (E. Roberts, ed.), pp. 460–467, Pergamon Press, New York (1960).

19. S. Berl, A. Lajtha, and H. Waelsch, Studies of glutamic acid metabolism in the central nervous system, *in Chemical Pathology of the Nervous System* (J. Folch-Pi, ed.), pp. 361–369, Pergamon Press, London (1961).

20. E. V. Flock, M. A. Block, J. H. Grindlay, F. C. Mann, and J. L. Bollman, Changes in free amino acids of brain and muscle after total hepatectomy, *J. Biol. Chem.* **200**:529–536 (1953).

21. E. Roberts, M. Rothstein, and C. F. Baxter, Some metabolic studies of γ-aminobutyric acid, *Proc. Soc. Exptl. Biol. Med.* **97**:796–802 (1958).

22. S. Berl, A. Lajtha, and H. Waelsch, Amino acid and protein metabolism—VI. Cerebral compartments of glutamic acid metabolism, *J. Neurochem.* **7**:186–197 (1961).

23. R. L. Potter and A. Van Harreveld, The effect of metrazol on the glutamate metabolism of cerebral cortex, *J. Neurochem.* **9**:105–112 (1962).

24. S. Berl, Compartmentation of glutamic acid metabolism in developing cerebral cortex, *J. Biol. Chem.* **240**:2047–2054 (1965).

25. G. Takagaki, S. Berl, D. D. Clarke, D. P. Purpura, and H. Waelsch, Glutamic acid metabolism in brain and liver during infusion with ammonia labeled with nitrogen-15, *Nature* **189**:326 (1961).

26. S. Berl, G. Takagaki, D. D. Clarke, and H. Waelsch, Metabolic compartments *in vivo*: Ammonia and glutamic acid metabolism in brain and liver, *J. Biol. Chem.* **237**:2562–2569 (1962).

27. S. Berl, G. Takagaki, D. D. Clarke, and H. Waelsch, Carbon dioxide fixation in the brain, *J. Biol. Chem.* **237**:2570–2573 (1962).

28. C. A. Rossi, S. Berl, D. D. Clarke, D. P. Purpura, and H. Waelsch, Rate of CO_2 fixation in brain and liver, *Life Sci.* **10**:533–539 (1962).

29. H. Waelsch, S. Berl, C. A. Rossi, D. D. Clarke, and D. P. Purpura, Quantitative aspects of CO_2 fixation in mammalian brain *in vivo*, *J. Neurochem.* **11**:717–728 (1964).

30. M. K. Gaitonde, Rate of utilization of glucose and compartmentation of α-oxoglutarate and glutamate in rat brain, *Biochem. J.* **95**:803–810 (1965).

31. H. Busch, Studies on the metabolism of acetate-1-^{14}C in tissues of tumor-bearing rats, *Cancer Res.* **13**:789–794 (1953).

32. H. Busch, Studies on the metabolism of pyruvate-2-^{14}C in tumor-bearing rats, *Cancer Res.* **15**:365–374 (1955).

33. R. M. O'Neal and R. E. Koeppe, Precursors *in vivo* of glutamate, aspartate and their derivatives of rat brain, *J. Neurochem.* **13**:835–847 (1966).

34. R. M. O'Neal, R. E. Koeppe, and E. I. Williams, Utilization *in vivo* of glucose and volatile fatty acids by sheep brain for the synthesis of acidic amino acids, *Biochem. J.* **101**:591–597 (1966).

35. J. E. Cremer, Amino acid metabolism in rat brain studied with ^{14}C-labeled glucose, *J. Neurochem.* **11**:165–185 (1964).

36. C. J. Van den Berg, P. Mela, and H. Waelsch, On the contribution of the tricarboxylic acid cycle to the synthesis of glutamate, glutamine and aspartate in brain, *Biochem. Biophys. Res. Commun.* **23**:479–484 (1966).

37. C. J. Van den Berg, Lj. Kržalic, and P. Mela, *The Existence of Two Pathways from Acetyl-CoA to Glutamate in the Central Nervous System*, p. 21, First International Meeting of the International Society for Neurochemistry, Strasbourg (1967).

38. S. Berl and T. L. Frigyesi, The turnover of glutamate, glutamine, aspartate and GABA labeled with 1-^{14}C-acetate in caudate nucleus, thalamus and motor cortex (cat), *Brain Res.* 12:444–455 (1969).

39. S. Roberts and B. S. Morelos, Regulation of cerebral metabolism of amino acids—IV. Influence of amino acid levels on leucine uptake, utilization, and incorporation into protein *in vivo, J. Neurochem.* 12:373–387 (1965).

40. A. Meister, *in Biochemistry of the Amino Acids*, 2d ed., Vol. II, p. 744, Academic Press, New York (1965).

41. S. Berl and T. L. Frigyesi, Comparison of cerebral regional metabolism of ^{14}C-leucine following third ventricle and intravenous administration in the cat, *J. Neurochem.* 16: 405–416 (1969).

42. S. Berl and D. P. Purpura, Regional development of glutamic acid compartmentation in immature brain, *J. Neurochem.* 13:293–304 (1966).

43. C. R. Noback and D. P. Purpura, Postnatal ontogenesis of neurons in cat neocortex, *J. Comp. Neurol.* 117:291–308 (1961).

44. K. Voeller, G. D. Pappas, and D. P. Purpura, Electron microscope study of development of cat superficial neocortex, *Exptl. Neurol.* 7:107–130 (1963).

45. D. P. Purpura, R. J. Shofer, E. M. Housepian, and C. R. Noback, Comparative ontogenesis of structure–function relations in cerebral and cerebellar cortex, *in Progress in Brain Research, Growth and Maturation of the Brain* (D. P. Purpura and J. P. Schade, eds.), Vol. 4, pp. 187–221, Elsevier, Amsterdam (1964).

46. D. P. Purpura, Relationship of seizure susceptibility to morphologic and physiologic properties of normal and abnormal immature cortex, *in Neurological and Electrophysiological Correlative Studies in Infancy* (P. Kellaway, ed.), p. 117, Grune and Stratton, New York (1964).

47. S. Berl, Glutamine synthetase. Determination of its distribution in brain during development, *Biochem.* 5:916–922 (1966).

48. S. Berl, W. J. Nicklas, and D. D. Clarke, Compartmentation of glutamic acid metabolism in brain slices, *J. Neurochem.* 15:131–140 (1968).

49. M. M. Kini and J. H. Quastel, Effects of veratrine and cocaine on cerebral carbohydrate amino acid interrelations, *Science* 131:412–414 (1960).

50. S. L. Chan and J. H. Quastel, Tetrodotoxin: Effects on brain metabolism *in vitro, Science* 156:1752–1753 (1967).

51. R. Balazs, personal communication.

52. Y. Machiyama, R. Balazs, and D. Richter, Effect of K$^+$ stimulation on GABA metabolism in brain slices *in vitro, J. Neurochem.* 14:591–594 (1967).

53. R. Balazs, Y. Machiyama, and D. Richter, *Metabolism of GABA in Cerebral Cortex*, p. 13, First International Meeting of the International Society for Neurochemistry, Strasbourg (1967).

54. S. Lahiri and J. H. Quastel, Fluoroacetate and the metabolism of ammonia in brain, *Biochem. J.* 89:157–163 (1963).

55. O. Gonda and J. H. Quastel, Transport and metabolism of acetate in rat brain cortex *in vivo, Biochem. J.* 100:83–94 (1966).

56. C. J. Van den Berg and P. Mela, On the heterogeneity of mitochondrial pathways from acetyl-CoA to glutamate in the central nervous system, *Fourth Meeting Federation European Biochem. Soc.*, p. 48 (1967).

57. A. Neidle, C. J. Van den Berg, and A. Grynbaum, Heterogeneity of rat brain mitochondria isolated in continuous sucrose gradients, *J. Neurochem.* 16:225–234 (1969).

58. R. A. Peters and R. J. Hall, On the reaction of fluoropyruvate with some -SH substances, *Biochim. Biophys. Acta* 26:433–434 (1957).

59. J. E. Cremer, Studies on brain-cortex slices. The influence of various inhibitors on the retention of potassium ions and amino acids with glucose or pyruvate as substrate, *Biochem. J.* **104**:223–228 (1967).
60. S. P. R. Rose, Effects of ouabain, glutamate and cations on phosphate incorporation in brain slices, *Biochem. Pharmacol.* **14**:589–601 (1965).
61. S. Berl and H. Waelsch, Determination of glutamic acid, glutamine, glutathione and γ-aminobutyric acid and their distribution in brain, *J. Neurochem.* **3**:161–169 (1958).
62. J. L. Mangan and V. P. Whittaker, The distribution of free amino acids in subcellular fractions of guinea pig brain, *Biochem. J.* **98**:128–137 (1966).
63. L. Salganicoff and E. De Robertis, Subcellular distribution of the enzymes of the glutamic acid, glutamine and γ-aminobutyric acid cycles in rat brain, *J. Neurochem.* **12**:287–309 (1965).
64. R. Balazs, D. Dahl, and J. R. Harwood, Subcellular distribution of enzymes of glutamate metabolism in rat brain, *J. Neurochem.* **13**:897–905 (1966).
65. R. W. Swick, J. L. Stange, S. L. Nance, and J. F. Thomson, The heterogeneous distribution of mitochondrial enzyme in normal rat liver, *Biochemistry* **6**:737–744 (1967).
66. O. Z. Sellinger and F. de Balbian Verster, Glutamine synthetase of rat cerebral cortex: Intracellular distribution and structural latency, *J. Biol. Chem.* **237**:2836–2844 (1962).
67. G. Simon, J. B. Drori, and M. M. Cohen, Mechanism of conversion of aspartate into glutamate in cerebral-cortex slices, *Biochem. J.* **102**:153–162 (1967).
68. H. Waelsch, Compartmentalized biosynthetic reactions in the central nervous system, *in Regional Neurochemistry* (S. S. Kety and J. Elkes, eds.), pp. 57–64, Pergamon Press, New York (1961).
69. W. J. Nicklas, D. D. Clarke, and S. Berl, Labeling of glutamate, glutamine and aspartate in brain from ^{14}C-aspartate and -acetate, *Federation Proc.* **26**:388 (1967).
70. N. D. Goldberg, J. V. Passonneau, and O. H. Lowry, Effects of changes in brain metabolism on the levels of citric acid cycle intermediates, *J. Biol. Chem.* **211**:3997–4003 (1966).
71. W. J. Nicklas, D. D. Clarke, and S. Berl, The effect of fluoroacetate on amino acid metabolism in brain, *Federation Proc.* **27**:463 (1968).

Chapter 20

METABOLIC AND ANATOMICAL SPECIALIZATION WITHIN THE RETINA

Richard N. Lolley

Neurochemistry Laboratories
Veterans Administration Hospital
Sepulveda, California
and
Department of Anatomy
UCLA School of Medicine
Los Angeles, California

I. INTRODUCTION

The retina comprises the photosensitive and neural integrative portion of the eye. It develops embryologically as an ectopic portion of the primitive forebrain, and it never loses its similarity to the central nervous system. The inner layers of the mature retina show a lamanellar organization comparable to that of the gray matter in the cortex. The optic nerve shows the compartmentation by pia and the absence of sheaths of Schwann that characterize white matter in brain[1]; thus, the optic nerve has the characteristics of a central nervous system fiber tract.

The retina is composed of a pigmented epithelium and a neural retina.[1,2] The gross and fine structural anatomy of the retina have been studied extensively, and excellent reviews are available.[2–5] This discussion is directed to the neural retina, and other publications should be consulted for information concerning the structural and functional characteristics of the pigmented epithelium of the retina.[5–9]

The neural retina is composed of three cellular layers (Fig. 1). The outermost layer of cells, lying closest to the pigment epithelium, is composed of the photoreceptor cells, and the innermost layer, lying closest to vitreous, is populated by the retinal ganglion cells. Sandwiched between the two layers is a discrete stratum of nuclei that belong to the bipolar, Müller, amacrine, and horizontal cells. The three layers can be subdivided, and a total of ten sublayers can be identified[2] between the photoreceptor outer segments (choroidal border of the retina) and the nerve fibers of the ganglion cells (vitreal border of the retina). Glial elements separate the neural cell bodies

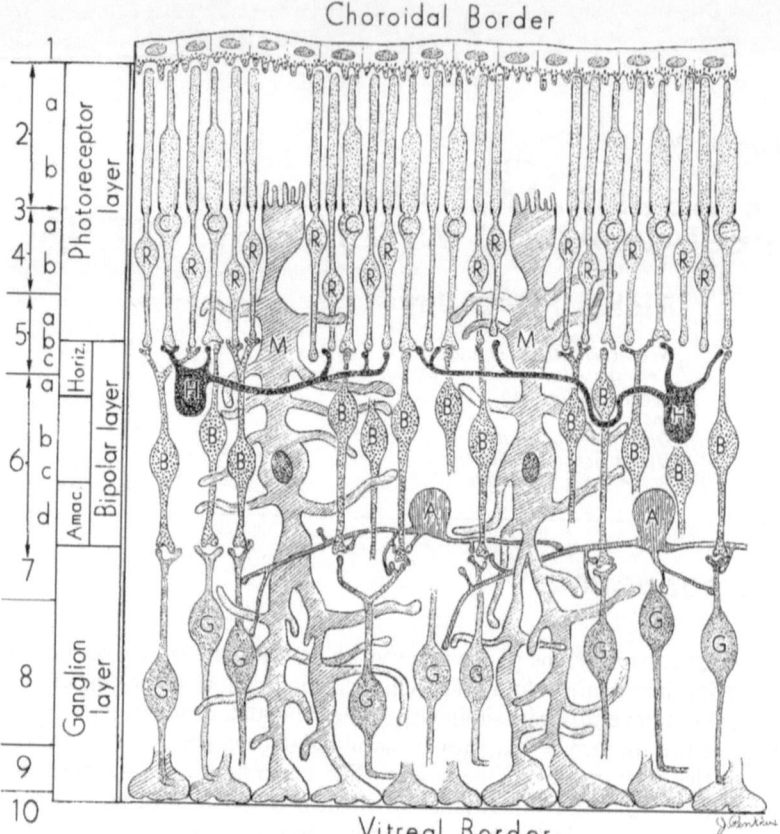

Fig. 1. A scheme of the major constituents of the vertebrate retina and their synaptic relationships. The layers are (1) pigment epithelium; (2a) outer receptors and (2b) inner segments of the rods and cones; (3) outer limiting membrane; (4) outer nuclear layer; (5) outer synaptic layer; (6) inner nuclear layer; (7) inner synaptic layer; (8) ganglion cells; (9) optic nerve fibers; and (10) inner limiting membrane. The cells are designated as (R) rods, (C) cones, (H) horizontal cells, (B) bipolar cells, (A) amacrine cells, (G) ganglion cells, and (M) Müller cells.[2,10]

and their axonal and dendritic extensions. The Müller cells are the largest and most glial-like elements in the retina.[10] The cell body of the Müller cell extends from the vitreal border of the retina to the base of the photoreceptor outer segment. Müller cell processes ramify within the interior of the retinal structure to encompass the neural elements and the capillaries that penetrate the inner retinal layers of some species. Very few astroglia, oligodendroglia, or micro glia are present in mammalian retina.[10,11]

The functional properties of the retina are unique because within the tissue light is received, the visual stimulus is encoded into a complicated electrophysiological message, and the information is transmitted to the higher centers of the brain. Thus, the retina receives, modifies, and transmits visual information. The act of reception occurs within the outer segments of the photoreceptor cells, the ability for information processing resides within the more complex middle region of the retina, and the ganglion cells transmit the neural signals to the lateral geniculate body.[12]

The retina performs its highly important functional task under what might appear to be a serious metabolic handicap; in all species studied, the outer half of the retina is avascular. In rabbit and guinea pig the entire retina is avascular.[5,13] Many general properties of the retina are shared by different species, but metabolic and enzymatic differences have been observed between species that possess different degrees of retinal vascularity. Within a given species the distribution of the blood supply, and therefore the availability of nutritional support of the retina, has been suggested to influence the metabolic properties of the retinal tissue.[14] The avascular retina represents the simplest level of retinal vascularity. For this reason the rabbit retina shall be utilized to establish a metabolic profile for an avascular retina. When data derived from the study of the retina of other animals are cited, the species will be specifically identified.

II. METABOLISM

A. Carbohydrate

Glucose (Table I) is the primary substrate of metabolism in the adult retina, and it is utilized in abundance. The blood supply within the capillaries of the choroid contributes the major proportion of glucose to the retina, with rapid transport of glucose across the blood–retinal barrier.[15,16] A small reservoir of glucose is maintained within the inner layers of the retina in the form of glycogen, and these areas possess the enzymatic mechanisms for mobilization of glycogen to glucose.

Carbohydrate metabolism within the isolated retina has three unusual characteristics: retina converts glucose to lactic acid at a prodigious rate, it consumes oxygen more rapidly than any other tissue, and it produces large quantities of lactic acid in the presence of oxygen (aerobic glycolysis).[17] The rate of formation of lactic acid in the presence of oxygen is only about one-fourth to one-half less than its rate of production in a nitrogen atmosphere.

The retinal rate of utilization of glucose and oxygen, and of production of carbon dioxide and lactic acid, exceeds that reported for cerebral cortex. In the retina, aerobic lactic acid production exceeds oxygen consumption by a factor of 2, while the reverse is true for cerebral cortex.[18] Moreover, the brain cortex has about half the rate of respiration of retina and produces

TABLE I

Substrates and Cofactors of Metabolism in The Retina[a]

		μmole Substrate/g wet wt.		
Glycogen	23.0		Lactate	23.0
Glucose	4.0		Pyruvate	0.12[F]
Glucose-6 P	0.3		Glutamate	6.4, 3.2[F]
U D P-Glucose	0.3		GABA	2.0[F]
Fructose di-P	0.03			
		μmole Cofactor/g wet wt.		
Pi	3.6		NAD	0.21, 0.29,[r] 0.11[F]
ATP	1.6, 2.3[F]		NADH$_2$	0.04, 0.04,[r] 0.03[F]
ADP	0.8[F]		NADP	0.01, 0.01,[r] 0.02[F]
AMP	0.4[F]		NADPH$_2$	0.02, 0.02,[r] 0.03[F]
PCr	1.5, 4.6[F]			

[a] The levels of substrates and cofactors in the whole retina of rabbit (no superscript), rat (r), and frog (F) were converted to a wet weight standard by dividing dry weight measurements by 5.5.[149] Rabbit and rat retinas were obtained after the animals were beheaded,[20,23,24,31] and frog retinas were obtained from eyes frozen *in vivo*.[25]

lactic acid aerobically at 15% and anaerobically at 35% of the rate of adult retina.[18]

1. Anaerobic and Aerobic Glycolysis

The rate at which isolated retina produces lactic acid is surprisingly high. When Warburg first observed this characteristic of retina he believed that it was a secondary response of the tissue to physical damage that was produced during isolation. Subsequently, investigators have shown that this interpretation was wrong and that a high rate of lactic acid production is typical of all retinal tissue.[19,20] The Embden–Meyerhof pathway of glucose metabolism participates in the conversion of glucose to lactic acid in retina, and the overall rate of this reaction is probably controlled by the enzymes P-fructokinase and hexokinase, as it is in other nervous tissue.[21] These control steps are operative under both aerobic and anaerobic conditions.

Pyruvate, which is produced by the stepwise degradation of glucose through the Embden–Meyerhof pathway, does not accumulate in the tissue but is converted to carbon dioxide and water via the Krebs cycle or to lactic acid by the enzyme lactic dehydrogenase. The level of lactate that accumulates in the tissue under steady-state conditions is an index of the distribution of pyruvate metabolism between the Krebs cycle and lactic dehydrogenase. Aerobically most tissues convert the majority of pyruvate to carbon dioxide and water, while anaerobically lactic acid increases within the tissue, but adult retinal tissue accumulates lactic acid both aerobically and anaerobically. The mechanism by which the prodigious rate of lactic acid production is achieved and maintained in the presence of oxygen is not entirely understood. The enzymatic pathway for the production of lactic acid is more than likely the same during both the periods of anaerobic and aerobic glycolysis, but the factors that control the rate of lactic acid production differ. Therefore, the rate of lactic acid production under anaerobic conditions may be considered as the maximum level of activity that the Embden–Meyerhof pathway will support, and the rate of lactic acid production in the presence of oxygen indicates the extent to which oxidative reactions compete with lactic dehydrogenase for pyruvate and $NADH_2$. As a result of this added restraint, the rate of aerobic glycolysis is reduced from that observed under anaerobic conditions.

Under anaerobic conditions, the rate of lactic acid production in the isolated retina is controlled by the availability of substrate and cofactors or the rate of some enzymatically mediated step in the reaction sequence. The primary substrate, glucose, does not appear to be limiting *in vivo* or *in vitro*,[20,22] and significant levels of the cofactors NAD, ADP, and inorganic phosphate have been measured in rabbit, monkey, and frog retina.[23–25] The specific activity of many of the enzymes of glycolysis also have been measured in rabbit and rat retina, e.g., hexokinase, glucose-6-P dehydrogenase, P-fructokinase, aldolase, glyceraldehyde-P dehydrogenase, and lactic dehydrogenase.[26–29] Unfortunately, the enzymes and metabolites of glycolysis have not been measured under the same conditions so that the rate-limiting step cannot be directly assessed. Indirect evidence can be obtained, however, about the rate-controlling step in glycolysis by correlating the change in rate of anaerobic glycolysis during maturation of the retina with the change in the capacity of the enzymes of glycolysis during the same period. As one might expect, there are tremendous changes in the enzyme content of the retina during development and an equally impressive increase in the rate of anaerobic glycolysis. The amount of hexokinase, P-fructokinase, and glyceraldehyde-P dehydrogenase increases in rat retina, the amount of glucose-6-P dehydrogenase decreases, and the amount of total lactic dehydrogenase remains nearly constant.[26,27] The profile of lactic dehydrogenase isozymes changes, however.[26,30] The greatest period of enzyme change in the rat was noted between 4 and 20 days after birth.[26,27] During this period the total NAD content of the tissue changed very little.[31] The rate of anaerobic glycolysis increased by 70–80% between 4 and 20 days of postnatal life in the rat, and the increased rate of glycolysis closely paralleled in time the increased content of P-fructokinase within the retina. The temporal correlation of these two changes implies that the rate of anaerobic lactic acid production in the retina is limited by the overall rate of the Embden-Meyerhof pathway.

The presence of oxygen depresses the rate of lactic acid production by 25–50%. The degree of inhibition by oxygen is much less than that reported for other tissues. Correspondingly, the Krebs cycle appears to compete less effectively for pyruvate in the retina than it does in other tissues. Again the mechanisms are not clearly defined, but studies of aerobic glycolysis reveal that both a Pasteur effect and a Crabtree effect can be observed with isolated retinal tissue.[32] In exhibiting the latter effect, retinal metabolism resembles that of embryonic tissues, tumor cells, and leucocytes.[33] In these tissues the availability of ADP and inorganic phosphate limits the respiratory rate because aerobic glycolysis and oxidative reactions occur in the same cellular compartment and they compete for the ADP and inorganic phosphate within the compartment.[34,35] The Crabtree effect decreases in the retina with increasing age of the animal.[33] This implies that either ADP and inorganic phosphate levels are sufficiently high in mature retina so that they fail to limit the respiratory rate of the tissue or that glycolysis and respiration become separated so that they no longer compete for the same metabolic pool of ADP and inorganic phosphate.

It is difficult to decide accurately between these alternative explanations. However, the available data seem to support the concept that glycolysis and respiration are specialized within different areas of the adult retina. ADP and inorganic phosphate levels have not been determined during the course of maturation of the retina, but Matschinsky has measured their content and distribution in the adult retina.[20,23] ADP and inorganic phosphate levels are greatest within the inner layers of the retina where the contents of Krebs cycle enzymes are greatly diminished and the enzymes of glycolysis are abundant.

The spatial compartmentation of glycolysis and oxidative reactions has been suggested to exist within the photoreceptor cell and between the neural and glial elements of the retina. For example, the mature photoreceptor cell does not show the Crabtree effect[32] and the quantitative histochemistry of this area of the retina indicates that the enzymatic capacity for oxidative reactions is separated from the enzymatic capacity for glycolysis.[28,29] Correspondingly, electron microscope studies show that the location of mitochondria coincides with that found for the enzymes of the Krebs cycle by histochemical studies.[5] In addition to this intracellular compartmentation, there may be a metabolic compartmentation between the neuronal and glial elements.[18] Glial cells actually can be shown in relief by a histochemical stain for lactic dehydrogenase.[36] It is possible, therefore, that a major proportion of aerobic glycolysis in the retina is produced by a tissue compartment that is poorly supplied with the capacity for respiration and therefore is little affected by oxygen, while another compartment may function in the complex manner to which we are accustomed from the study of other tissues.

The ability of anaerobic glycolysis to support light-activated electrical activity in the retina has been studied in vitro.[37] Both the light-evoked potential shift recorded directly from the retina and the light-evoked compound action potential recorded in the optic nerve fail immediately upon withdrawal of oxygen. The retinal response is markedly diminished and the nerve response abolished by 3 min of anoxia.[38] It appears that the ganglion cells are more sensitive to oxygen deprivation than are the specialized photoreceptors and other retinal elements and that glycolysis alone cannot support retinal function.

2. Respiration

The oxidative capacity of the rabbit retina develops progressively after birth, and the increased capacity for glucose oxidation and respiration accompanies cellular growth and differentiation of the tissue, since cellular division terminates between 2 and 4 days of postnatal life.[33,39,40] The low rate of glucose oxidation in the immature retina increases by a factor of 6 between 8 and 20 days after birth, whereupon it stabilizes at the adult level.[33,41] Mature retina incubated *in vitro* consumes oxygen at a high rate and simultaneously oxidizes glucose carbons to carbon dioxide.[33]

The primary avenue of glucose oxidation in the retina is the Krebs cycle, and this system, in conjunction with the mitochondrial electron transport system, produces carbon dioxide and water while synthesizing ATP from ADP and inorganic phosphate. Electron microscope studies of the retina indicate that there are fewer mitochondria per unit of tissue than in the cerebral cortex. Mitochondria that are present are capable of coupling glucose oxidation to ATP synthesis and of exhibiting respiratory control by ADP and inorganic phosphate.[42] However, they are unable to oxidize extra mitochondrial $NADH_2$ directly, or through the β-hydroxybutyrate or α-glycero-P cycles.[42,43]

Both isolated retina and retinal mitochondria can utilize carbon substrates other than glucose.[33,42,44,45] Pyruvate, α-ketoglutarate, succinate, and glutamate can be readily oxidized by isolated mitochondria. Pyruvate carbon rapidly entered the Krebs cycle of most tissue, and this may be the case in retina when pyruvate production is located in the cellular vicinity of mitochondria. In some situations pyruvate destined for lactic acid appears to be separated from that in the Krebs cycle.[33] However, if steady-state levels of pyruvate are considered in the tissue, then the metabolism of pyruvate may be considered as a fulcrum over which the rates of respiration and lactic acid production are balanced.

Under anaerobic conditions $NADH_2$ accumulates as pyruvate is produced by the Embden–Meyerhof sequence. Without oxidized NAD the flow of triose phosphate through the sequence would stop, but activity normally continues since tissue levels of NAD are maintained within functional limits by the enzymatic reduction of pyruvate to lactate.[31] The same mechanism appears to be operative under aerobic conditions, but in the presence of oxygen the oxidation of $NADH_2$ is shared between LDH and the mitochondrial electron transport system. Thus, pyruvate is linked to oxidation–reduction reactions that are capable of supplying oxidized NAD to the tissue under either anaerobic or aerobic conditions.

When conditions are appropriate, pyruvate readily enters the Krebs cycle. Indeed, it may enter the cycle by two different reactions. First, pyruvate can be converted to Acetyl-CoA, which then condenses to form citric acid. This reaction sequence reduces NAD. Second, pyruvate can absorb CO_2 and enter the cycle as malic acid. This reaction oxidizes $NADPH_2$. The latter reaction proceeds very rapidly in retina.[46] The importance of the malic enzyme reaction in retinal metabolism has not been clearly defined, but it may be very great since the reaction can potentially regulate the local concentration of CO_2 and of $NADPH_2$ (an analogous reaction to that with LDH and $NADH_2$) as well as offer an alternative route into the Krebs cycle.

There are reasons to believe that the components of the Krebs cycle in the retina are identical to those found in other tissues because most of the enzymes required for the cycle have been found in retinal homogenates or identified histochemically. Furthermore, the Krebs cycle is operative within the isolated retina because labeled CO_2 is rapidly produced from glucose that is labeled specifically in carbon C_6. However, less CO_2 is produced from carbon C_6 of glucose than from carbon C_3 and C_4,[33] and a sizable proportion of the C_6 radioactivity that is not released as CO_2 is found in glutamate and aspartate. Cohen[33] reported that 20% of the glucose carbon that enters the Krebs cycle ends up in glutamate, with a smaller amount in aspartate. Glutamate and aspartate are linked to the Krebs cycle by transaminase reactions of which glutamate–aspartate transaminase is particularly active in retina.[28] It appears that glutamate can readily be converted into Krebs cycle intermediates. From a physiological point of view it should be remembered also that glutamate can be rapidly produced from glutamine in the retina.[47] The potential importance of glutamine as a carbon reservoir for the Krebs cycle is indicated by *in vitro* experiments that show that the ammonium content of frog and cattle retina increased during illumination when incubated without substrate.[48] The addition of glucose, lactate, and pyruvate to the incubating media prevented the increase in ammonium. Thus, a system of readily reversible reactions dynamically links together respiration, amino acid, and ammonia metabolism.

In vivo, carbohydrate polymers other than glycogen have been identified.[49,50] Hall, Ocumpaugh, and Young[51] have studied the incorporation of ^{35}S-sulfate into mucopolysaccharides of the rat retina by autoradiographic and biochemical techniques. The sulfated mucopolysaccharides occur in the synaptic and optic nerve fiber layers but are most heavily concentrated in the photoreceptor layer.[52]

The incubating conditions that will most nearly reproduce the *in vivo* environment of the rabbit retina are no doubt complex, since the glucose supply enters the retinal tissue from only the choroidal surface of the retina and the location within the retina of metabolic enzymes has developed in conjunction with this polar arrangement.[20] The design of *in vitro* experiments has generally ignored the tissue organization (see Ames[37] for exception), and it seems necessary to recall that the *in vitro* procedures have produced variable and often conflicting reports about the metabolic capacity of retinal tissue.

The *in vivo* capacity of the retina for glycolysis and respiration is still uncertain since isolated retina displays different rates of glycolysis and respiration when it is incubated in different types of media. For example, the rate of glycolysis and respiration in a complete medium buffered with bicarbonate exceeded that of a comparable medium buffered by phosphate.[53-55] The reasons for the different rates of glycolysis and respiration are unknown, but the level of carbon dioxide in retinal tissue may influence the metabolism of retinal pyruvate and, therefore, the balance between glycolysis and respiration.[56]

The metabolic rate of isolated retina does not appear to be influenced by physiological stimulation of the retina even though isolated retinal preparations are readily excitable by light. Ames has shown that a complete bicarbonate mixture is capable of supporting *in vivo* levels of intracellular water and electrolytes as well as the excitability properties of isolated retinal tissue,[37,57] but these parameters have not been measured with phosphate buffer in the incubating media. Furthermore, the intracellular levels of water and electrolytes are unchanged by an exposure to flashing light. These experiments, however, do not indicate to what extent the respiratory rate of the tissue is influenced by physiological activity. Cohen[33] measured the respiration of isolated retina in phosphate buffered medium. He found that the respiratory rate was not affected by high or low levels of potassium in the incubating medium or by steady levels of illumination, but the respiratory rate was increased by decreasing the normal level of calcium in the incubating fluid. The latter experiments suggest that retinal metabolism differs from that of cerebral cortex in its response to elevated concentrations of potassium in the incubating fluid,[58] but the meaning of this lack of metabolic response cannot be assessed with any certainty until the variability that exists in the metabolic capacity of retina in different incubating fluids is clarified. From the many similarities that exist between retina and cerebral cortex, one intuitively suspects that the respiratory rate of the isolated retina can be influenced by the level of neural activity within the tissue. Still, the only evidence that supports this contention is the observation that electrical activity in the isolated retina is dependent upon the presence of oxygen.

3. *Hexose Monophosphate Shunt*

In isolated rabbit retina, most of the carbon dioxide produced from glucose is formed in the Krebs cycle.[59] However, when electron acceptors are added to the incubating medium a high capacity of glucose oxidation via the hexose monophosphate shunt can be shown, and this effect is best displayed under anaerobic conditions.[33] The hexose monophosphate shunt activity is measurable in both the mature and immature retina,[31,33,60] but the relative importance of the shunt for the utilization of glucose declines as the retina matures.[31,33] The decreased activity of the hexose monophosphate shunt in the adult retina, however, is not caused by a reduction in the enzymatic capacity of the retina but rather the capacity of the shunt appears to be restricted by the rate of oxidation of $NADPH_2$ in the tissue. This is indicated by the observation that the activity of the hexose monophosphate shunt in the retina can be greatly accelerated *in vitro* by the addition of pyruvate or phenazine methosulfate to the incubating medium, and it is known that these compounds readily accept electrons from $NADPH_2$. In this process the latter compound is six times more efficient than pyruvate in stimulating the preferential oxidation of C_1 carbons via the hexose monophosphate shunt.[33]

Numerous reactions may participate in the removal of electrons from $NADPH_2$ *in vivo*, and the requirements for $NADPH_2$ will depend upon the total activities of all these reactions. During the growth phase of retinal development, nucleic acid and protein synthesis utilize large quantities of $NADPH_2$ in the manufacture of cellular constituents. Correspondingly, the hexose monophosphate shunt appears to exhibit its highest capacity during the early stages of postnatal development.[32] The ratio of carbon dioxide that is produced from C_1 and C_6 of specifically labeled glucose decreased from 2.9 to 1.7 between 6 and 14 days of postnatal life and approaches a value of nearly 1.0 in the adult retina.[31,32]

The adult retina possesses the enzyme capacity for a relatively high rate of hexose monophosphate shunt activity, but this capacity is not normally realized due to the low rate of removal of electrons from $NADPH_2$. However, the reactions that drain $NADPH_2$ from the hexose monophosphate shunt in the retina in themselves may be highly important biochemical events. In addition to the important role of supporting the steady-state synthesis of cellular protein and nucleic acid, specific reactions involving $NADPH_2$ may have special significance for certain functions peculiar to the retina. For example, Futterman has shown that photoreceptor segments can couple reduction of vitamin A aldehyde to the oxidation of $NADPH_2$.[61] This reaction may be highly important in the maintenance of visual pigment levels in the photoreceptor outer segments. Matschinsky found by quantitative histochemical methods that the steady-state ratio of $NADP/NADPH_2$ of the photoreceptor outer layers was inverted from that found in other retinal layers and in neural tissue in general (Fig. 2D).[24] His finding indicates that mechanisms exist for the rapid oxidation of $NADPH_2$ in the photoreceptor outer segment layer. Numerous investigators have suggested that this oxidation, apparently by the rapid turnover of NADP, is coupled with reduction of NADP in the hexose monophosphate shunt. Horecker expresses the opinion that NADP is phylogenetically a younger system than NAD and that the reactions coupled to NADP oxidation–reduction reactions are separated from those that supply cellular energy in the form of ATP.[62] Such specialized division of labor could exist in the photoreceptor cell, but as yet there are so little data upon the specific role of the hexose monophosphate shunt in the retina that most generalizations remain hypothetical.

4. *Biochemical Energy*

Carbohydrate metabolism serves two general functions in cellular metabolism: It produces building blocks from which nucleic acids, proteins, and lipids are constructed, and it provides energy in a biochemically useful form. The energy so obtained is stored in retina as ATP as well as compounds that are readily convertible to ATP (P-creatine and $NADH_2$), and as $NADPH_2$, which possesses chemical reducing power.[62] ATP is rapidly produced from

P-creatine and $NADH_2$ by transphosphorylization and oxidative phosphorylization, respectively, and as ATP the energy is utilized in cellular reactions. $NADPH_2$, on the other hand, is only sluggishly oxidized by molecular oxygen, and this oxidation does not yield ATP.[62] Under aerobic conditions, therefore, NAD is in equilibrium with molecular oxygen, while NADP is not. Correspondingly, retinal NAD (Table I) exists primarily in its oxidized form, while NADP is predominantly reduced.[25,31] In retina, therefore, as in other tissues, ATP and $NADPH_2$ are complementary energetic compounds that are maintained by the metabolism of carbohydrate.

ATP breakdown accompanies many biochemical reactions within the retinal tissue. Kornberg[63] generalizes that most of the important reactions in which macromolecules are synthesized involve the formation of inorganic pyrophosphate from ATP. The enzyme inorganic pyrophosphatase exists in high concentration in the retina.[46] Other reactions utilize only the terminal phosphate of ATP and produce ADP and inorganic phosphate. The latter class of reactions that hydrolyze only the terminal phosphate group of ATP are of particular interest in neural tissue since they appear to be linked with the active transport of cations across neural membranes.[58] Skou has described the properties of the enzyme system Na^+-K^+ ATPase in brain tissue, where its activity is markedly activated by appropriate concentrations of sodium and potassium ions and where this activation is inhibited by the cardiac glycoside ouabain.[64] A Na^+-K^+ ATPase has been studied in retina that appears to have slightly different concentration optima for sodium and/or potassium from those reported for brain tissue[65–67] but which is inhibited by small concentrations of ouabain.[66,67] As found in brain, the activity of this enzyme increases rapidly during the period of maturation in which the vertebrate retina develops an adult pattern of electrical discharge.[27,68,69] It appears that there is a relationship between the Na^+-K^+ ATPase activity and the electrical discharge capabilities of the retina. This relationship has been demonstrated *in vitro* by Frank and Goldsmith[70] and *in vivo* by Langham, Ryan, and Kostelnik. Langham injected small quantities of ouabain near the vitreal border of the retina and observed that the electrical discharge pattern of the retina decreased shortly after the injection. Lolley observed the same relationship between the energy supply and functional activity of the retina. Using the opposite experimental approach, he showed that ATP and P-creatine levels decreased upon exposing the retina to flash illumination *in vivo*.[25] It appears that the biochemical energy stores of retina are utilized both for maintenance and for functional reactions and that the proportion of energy that is utilized for neural excitation and neural transmission phenomena *in vivo* is related to the form and intensity of illumination to which the retina is exposed.

B. Nucleic Acids and Protein

During embryogenesis and for a short time after birth, the cells of the mouse retina promenade between the inner layers of the retina, where their

DNA replicates, and the outer surface of the retina, where the cells divide.[39] This ritual begins to subside between second and sixth days of postnatal life, and thereafter there is no DNA replication or cellular division within the retina.[39] By this time the retina contains its full DNA complement. The total DNA content of rabbit retina is about 800 mg %,[71,72] and this level of DNA is not markedly different from that reported for neural tissue.[71] The ratio of DNA/RNA, on the other hand, is not that typically found in other tissues. The DNA/RNA ratio of rabbit and cat retina (range 3.6–7.0) is reversed from that found in gray matter (0.6) and liver (0.2).[71,74] An explanation of this unusual relationship has been found from the study of individual ganglion and photoreceptor cells of the rabbit retina.[72] A striking difference was noted between the two cell types in their content of RNA; photoreceptor cells and ganglion cells contain 0.65 $\mu\mu$g and 50 $\mu\mu$g RNA, respectively. If both cells are diploid and contain 7 $\mu\mu$g DNA/cell,[75] the DNA/RNA ratio of the photoreceptor and ganglion cell is 11 and 0.14, respectively. Therefore, the high DNA/RNA ratio of the whole retina results from the low content of RNA within the rod cell, relative to its DNA, and to the vast number of rod cells in the retina.[74]

While the distribution of DNA and RNA in some areas of the retina is unusual, the involvement of retinal DNA and RNA in protein synthesis appears to be normal, with DNA acting as the genetic information code and RNA serving as the messenger between the nucleus and the ribosomes of the cytoplasm.[76] The different types of RNA have not been studied in retina, but information has been obtained about the intracellular mobility of RNA by autoradiographic studies. Bok[77,78] has shown that cytidine-^{3}H is rapidly incorporated into RNA of the rat and frog photoreceptor cell. RNA is synthesized in the nucleus and begins to appear in the cytoplasm within 1 hr after the animal was injected.

From the study of RNA and protein synthesis of other tissues, it is known that some of the RNA, which is synthesized in the nucleus, is ultimately incorporated into ribosomes, and from this location RNA arranges amino acids in some predescribed sequence.[76] The amino acids for protein synthesis are usually present in the cytoplasmic fluid. So far glutamic acid, glutamine, aspartic acid, and glycine as well as taurine and β-aminoisobutyric acid have been identified in rat retina.[79] The full complement of amino acids appears to be available, however, since the incorporation of radioactive amino acid into retinal protein is rapid in retina, both in vitro[60] and in vivo.[60,80,81] Droz[80] and Young[81] have studied the dynamics of protein synthesis within the rat, mouse, and frog photoreceptor cells. They found that radioactive amino acids were incorporated into protein molecules primarily within the inner segments of the photoreceptor cells. A small proportion of the newly synthesized protein appeared to remain within the area of the inner segment, but the majority of the protein migrated through the shaft that connects the inner with the outer segment and became incorporated into the structure of the outer segment. Young[82] has shown with great clarity, by electron microscope radioautography, that the migrating protein is located within the lamellar disk of the outer segment. The radioactive protein then migrates sclerally as a band within the rod outer segment. Upon reaching the outer limits of the rod segment, the protein is apparently engulfed by the pigment epithelium and the radioactivity becomes dispersed.[76]

Within limits, therefore, the rate of protein displacement within the outer segment is an index of the rate of protein synthesis within the rod inner segment, and the time required for the band of radioactivity to traverse the length of the outer segment indicates the interval of time in which the complete outer segment is renewed. The photoreceptor outer segment of rats and mice is renewed every 9 days,[81] while the process requires 5 to 7 weeks in the frog rods.[81] The renewal period becomes shorter for the frog as the ambient temperature increases, and a 10°C rise reduces the migration time within the outer segment by one-half.[81] Furthermore, darkness retards and high-intensity illumination increases protein synthesis,[81] as indicated by the rate with which the radioactive band of protein migrates.

Upon continued illumination, some aspect of protein synthesis appears to fail, and structural damage of the photoreceptor outer segments occurs.[83–85] Noell[86] has shown that albino rats, but not a pigmented strain, suffer retinal damage during a 24-hr exposure to ordinary fluorescent light. The injury does not appear to be heat-activated[86] but rather more specifically related to light and, for this reason, the process which produces this lesion is a fascinating area for further study.

C. Lipid

The study of lipid metabolism in the retina has been sadly neglected even though the retina is rich in lipid.[87] Histochemical staining techniques indicate that the concentration of lipid is greatest in the outer segments of the photoreceptor cell, with decreasing amounts in the inner segment of that cell, the fiber layers, and nuclear layers of the retina.[88] The latter three areas are unexplored, but the composition and metabolism of lipids in the photoreceptor outer segment have been investigated.

The most recent estimate of the lipid content of cattle rod outer segments suggests that 60% of their dry weight is lipid[89] and that 80–90% of the lipid exists as phospholipid.[89] The corresponding values for rabbit and frog rod outer segment are somewhat lower.[5,90] Seven polar fractions of lipid have been separated by chromatographic techniques from cattle outer segment material, and they show that outer segments are richly supplied with phosphatidyl-ethanolamine and lecithin but are very deficient in cardiolipin.[89] Several neutral lipid fractions were also observed, and at least two of these fractions possess the capacity to oxidize methylene blue.[91] None of these fractions, however, appear to be coenzyme Q10 or Q6,[91] even though they possess strikingly similar oxidation–reduction potentials.[92] It is interesting that the outer segment membrane is very deficient in cholesterol and devoid of ganglioside.[93] Radioautographic studies reveal that some phospholipids, sphingomyelin, and lecithin are renewed in the adult retina.[76,82]

Of the lipids that have been identified in the outer segment structure, only vitamin A has been assigned a specific functional role. There exists an extensive literature from studies of vitamin A metabolism,[94–96] and the relationship of vitamin A to the visual pigment.[97–102] Wald states that the only action of light in vision is the conversion of 11-*cis* vitamin A aldehyde (retinal) to the *all-trans* configuration.[98] In that fashion the trigger is released and visual events begin. The trigger must be reset, however, if vision is to continue, and one must ask, "How is the visual pigment resynthesized?" This interesting and pressing problem appears to be focused upon the question of how *all-trans*-retinal is reisomerized to 11-*cis* retinal *in vivo*.

It is reported and generally believed that retinal is lost from the outer segment during intense illumination, *in vitro*, and ultimately finds its way into the pigment epithelium.[103] Enroute[76,104] or in the pigment epithelium itself,[103] retinal is converted to vitamin A alcohol (retinol). Furthermore, isomerization of retinal to the 11-*cis* configuration may take place in the pigment epithelium. During dark adaptation, the retinal (*trans* or 11-*cis*) must return to the outer segment before combining with the protein, opsin, to form a visual pigment. *In vivo*,[103,105] as well as *in vitro*,[106] the rate at which the levels of the visual pigment rhodopsin is restored after a period of light adaptation is slow and restoration may require an hour. Under conditions which light-adapt the retina, rhodopsin is not completely depleted, but rather the content of rhodopsin is reduced in proportion to the intensity of light, and a new steady state is achieved between the rate at which the visual pigment is bleached and the rate at which the pigment is synthesized.[107] The turnover of retinal must be quite rapid under light-adapting conditions, and it has been shown that light can reisomerize at least part of the *all-trans* retinal to the 11-*cis* configuration.[102,108,109] These studies suggest that there may be at least two pathways for the metabolism of retinal: One may involve the pigment epithelium, and the other may occur locally in the outer segment. The mechanisms that control the isomerization of retinal and the rate of visual pigment synthesis are still unknown, but they may be highly important because Donner[105] has found that the log threshold (excitability) is linearly related to the log rate of visual pigment regeneration, when the threshold is expressed in terms of the relative number of quanta absorbed. The transducer mechanism appears to monitor the change in visual pigment synthesis per time, and it is sensitive to the differential of this reaction rate.

III. CYTOARCHITECTURE

The retina is a tissue composed of numerous cellular types. The physiological response of this tissue to light is appreciated by everyone, and in this regard the neural elements are often considered as the sole constituent of the retina. However, in recent years physiological studies have shown that the glial elements may participate in the visual phenomena in a rather dynamic fashion. Thus, the picture that emerges is one of cooperative coexistence between the cellular units. From this point of view it is meaningful to evaluate the relative contribution of the retinal layers to the general metabolic capacity of the retina and to define, where possible, the metabolic capacity of each cellular type.

A. Distribution of Substrates, Cofactors, and Enzymes of Metabolism Within the Layers of the Retina

Lowry *et al.* have shown by quantitative histochemical techniques that the enzymes of metabolism are not uniformly distributed across the thickness of the rabbit retina and that the enzymes associated with mitochondrial respiration are separated from those involved in glycolysis.[20,46] In the quantitative histochemical technique, the layers of the retina are individually dissected from dried slices of retina, weighed, and assayed.[28] Histochemical staining techniques also have been used to assess the qualitative distribution of oxidative and glycolytic reactions. The data generally agree with the conclusions derived from the quantitative histochemical method.[110,111,112] Malic dehydrogenase illustrates the distribution patterns found for the enzymes involved in oxidative metabolism, while P-fructokinase represents the distribution pattern found for the enzymes which participate in glycolysis (Fig. 2A). It is interesting that the distribution of hexokinase more closely resembles the profile reported for the oxidative enzymes than that of glycolysis.[46] Glucose-6-P-dehydrogenase (entrance to hexose monophosphate shunt) is distributed primarily within the photoreceptor and bipolar layers (Fig. 2B). Thus, the predominant metabolic system changes as one proceeds from the choroidal border of the retina to the vitreous so that oxidative reactions become less important than glycolysis. This metabolic profile probably reflects an adaption to the angio-architecture of the rabbit retina.[20,23]

A knowledge of the distribution of the potential activity for these enzymes, however, does not indicate to what extent these enzymes are being used. The distribution of metabolites and changes in the levels with function are also needed. Glucose (Fig. 2C) is the retina's primary substrate of carbohydrate metabolism, and its concentration is highest at the choroidal and vitreal borders of the retina. This pattern of distribution indicates that glucose diffuses into the retina from both surfaces. Of the two surfaces, however, the choroidal appears to be by far the greatest contributor.[20] In contrast, the concentration of lactate is high at only the vitreal border of the retina and the lactate concentration decreases sharply in the outer layers of the retina. This distribution pattern of lactate, the major product of metabolism in the retina, suggests that there is active aerobic glycolysis in the inner retinal layers that are most isolated from the blood supply. It also

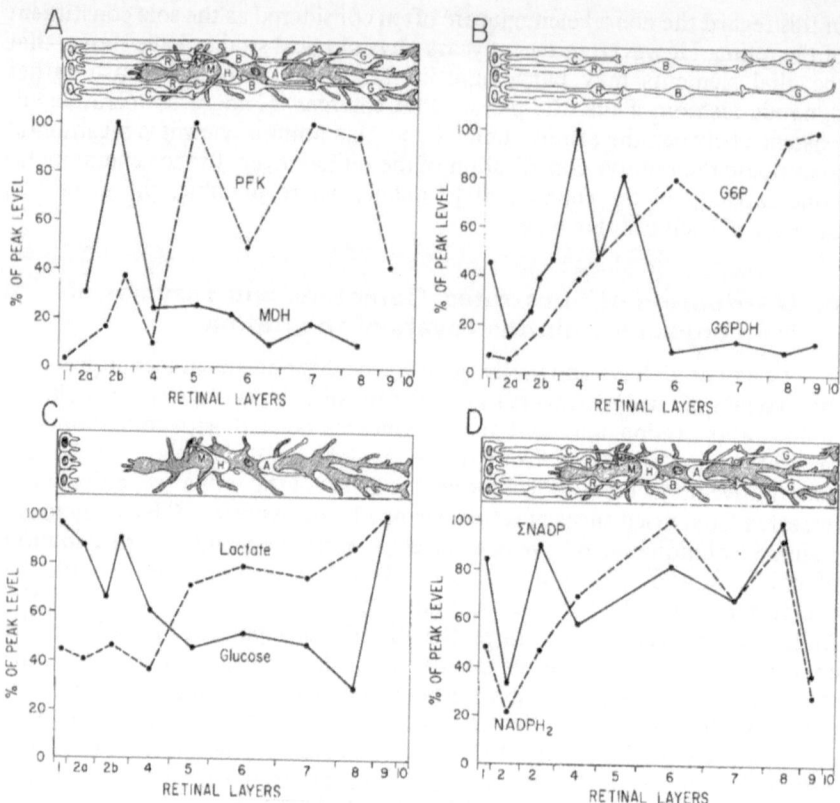

Fig. 2. A: Malic dehydrogenase (MDH) and P-fructokinase (PKF) in rabbit retina. Peak levels are, respectively, 199 and 31 mmoles product produced per gram of fat-free dry weight per hour.[28,29,46] Points have been connected for better visualization, but this does not imply gradual transition of values from one layer to the next. An anatomical scheme is included above A, which is subdivided in B and C into the neuronal and nonneuronal cellular types, without regard to contents of the figure. B: Distribution of glucose-6-P dehydrogenase (G6PDH) and glucose-6-P (G6P) in rabbit retina. Peak levels are, respectively, 20 mmoles of substrate oxidized per gram of fat-free dry weight per hour and 3.9 μmoles/g dry wt.[20,29] C: Distribution of lactate and glucose in layers of rabbit retina. Peak levels are, respectively, 204 and 11.5 μmoles/g dry wt.[20] D: Profiles of total NADP (ΣNADP) and NADPH₂ in rabbit retina. Peak levels are, respectively, 0.271 and 0.171 μmoles/g dry wt.[24]

suggests that lactate diffuses from the area of production to the blood supply of the choroid and is lost from the retina in this manner. In a similar pattern, glucose-6-phosphate levels rise steeply from the choroid to the vitreal border of the retina (Fig. 2B) and, in general, the glucose-6-phosphate profile follows the glycogen distribution in the different layers. Since phosphorylase follows the same rule, it seems probable that the enzyme, acting on its sub-strate, is responsible for the elevated glucose-6-phosphate levels where they occur.[23] Comparative measurements of retina from rabbit and monkey indicate that the enzymes and the levels of intermediates of the glycogen system to a certain extent are inversely related to the degree of capillarity, yet there is a general polarity of chemical organization that is independent of the blood supply.[20,23]

The efficiency with which retinal metabolism has adapted to the unique micro environment of the retina is indicated by the high levels of ATP and P-creatine in retinal tissue.[25] In monkey retina, the greatest shares of these compounds are located in the middle and inner layers of the retina. The photoreceptor cell is relatively deficient in them in both the monkey and frog.[20,23,113] The profile of inorganic phosphate parallels that of the energy-rich compounds in monkey and rabbit retina. As in other tissues, the level of inorganic phosphate is rapidly elevated by brief periods of ischemia so that it appears as if all layers of the retina are closely linked to aerobic meta-bolism.[23]

The profile of distribution of total NAD, total NADP, $NADH_2$, and $NADPH_2$ are similar in both rabbit and monkey retina.[23,24] Total NAD levels are higher in the inner layers, but throughout the retina about 10% of the total NAD system is in the reduced form.[24] In contrast, the total NADP content is quite constant throughout the layers of the retina, but the degree of reduction varies among the layers (Fig. 2D). Approximately 60% of the total NADP is reduced in the inner layers of the retina, while only about 25% is reduced in the inner segment of the photoreceptor cell.

B. Synaptic Transmitter Agents

Two discrete synaptic regions exist within the retina. In one, the photo-receptor cells communicate with the bipolar, and in some species the hor-izontal, cells. This region has been extensively studied,[5,11,114,115] and a three-dimensional montage of the synapse in it has been constructed from serial sections of the guinea pig retina.[116] The retinal rod-bipolar synapse is a highly convoluted and complex synapse, and both pre- and post-synaptic terminals possess vesicles.[115] No chemical transmitter has been identified in this region, and the mechanism for synaptic transmission is still unresolved.

The second, or inner synaptic, layer is anatomically more heterogeneous and contains a conglomerate mixture of amacrine, bipolar and ganglion cell extensions. Synaptic relationships have been described for this area using the light microscope[2,3,117] and the electron microscope,[5,11,117–120] and a three-dimensional construct has been reported by Dowling and Boycott.[121] Much also is known about the neural transmitters of this area. Adrenergic and cholinergic transmitters have been found in this region and, in addition, it has been suggested that certain amino acids may function as transmitter agents within the inner synaptic zone.

The adrenergic retinal neurons have been identified by the fluorescence technique of Falck and Hillarp[119] and quantitated by specific chemical methods.[123,124] Dopamine is the predominant catecholamine in the retina. Dark-adapted rabbit retina contains 0.1–2 μg dopamine/g wet wt.[123,124] The concentration of dopamine is reduced by 60% in the light-adapted retina.[125] Reserpine decreases the catecholamine level of retina since after reserpinization no adrenergic structures exhibit fluorescence by the Falck and Hillarp technique.[126] Unlike the rabbit retina, frog retina appears to contain sizable quantities of 5-hydroxytryptamine (0.2–0.6 μg/g).[27] Three adrenergic fiber layers have been identified in the inner synaptic region of the rabbit retina: an outer adrenergic fiber layer at the border between the inner nuclear and inner synaptic layer, a middle adrenergic fiber layer in the middle of the inner synaptic layer, and an inner adrenergic fiber layer at the border between the inner synaptic layer and the ganglion cell layer.[126] Similarly, three types of adrenergic nerve cells have been found: outer adrenergic cells in the outer adrenergic fiber layer, eremite cells within the middle adrenergic fiber layer, and alloganglionic cells with a position and appearance resembling some of the ordinary nonadrenergic cells of the ganglion cell layer.[126] The latter cells may be adrenergic afferent neurons to the lateral geniculate.[124]

The retina contains large amounts of acetylcholine, while the optic nerve has but a small content of this transmitter.[128] The enzyme that produces acetylcholine, choline acetylase, is correspondingly abundant in retina,[129] and it is practically absent in the optic nerve.[130] The enzyme reaction that destroys acetylcholine specifically, acetylcholine esterase, has been localized exclusively in the inner synaptic region,[112] using histochemical techniques. Furthermore, the activity of retinal acetylcholinesterase is influenced by the level of incident illumination.[131] These findings strongly suggest that acetylcholine is a synaptic transmitter in the retina and, within the limits of reliability for histochemical techniques to differentiate between true and pseudocholine esterase, that cholinergic terminals are restricted to the inner synaptic region.

Data has accumulated that suggests that the inner synaptic region of the retina is particularly sensitive to glutamate.[132] In an *in vivo* situation, Noell injected small quantities of glutamate into the ganglion cell layer of the retina and found that the rate of discharge of these cells increased.[133] Furthermore, *in situ* application of the proper combination of exciting and inhibiting amino acids have been shown to enhance only the light-induced spikes from the frog ganglion cell without giving rise to any spontaneous discharge.[134]

In addition to influencing the electrophysiological output of the retina, glutamate disrupts the structural integrity of the inner layers of the retina. Ames has shown that isolated retina accumulates glutamate from the incubating medium, and electrolytes and water accompany glutamate into the intracellular spaces of the retina.[57] Wald and De Robertis[135] studied the effect of glutamate upon the submicroscopic structure of the turtle retina.

They found that the inner synaptic layer was preferentially altered; the outer synaptic layer was unaffected. *In vivo*, high glutamate levels similarly disrupt the retinal structure since mice injected with large amounts of glutamate develop histological lesions in the inner layers of the retina.[136,137]

GABA appears to oppose the excitatory effect of glutamate upon the ganglion cells of the retina.[133] It is concentrated in the retina[138,139] but is absent from the optic nerve.[140] The middle layers of the retina are particularly rich in GABA,[138] which may be preferentially located in the amacrine cells.[141] Through GABA, the amacrine cells could exert an inhibitory influence upon retinal activity. In this regard, dark-adapted and light-adapted retinas exhibit markedly different levels of sensitivity to light,[12] and the concentration of GABA also changes as a result of light and dark adaption. Frog retina possesses 50 % more GABA in the light-adapted than in the dark-adapted state.[138]

C. Neural–Glial Relationship

Typical neural elements are found in the retina that are interconnected by synaptic junctions; they are the photoreceptor, bipolar and ganglion cells of all vertebrate retina.[2,3] Correspondingly, the Müller cells are universally classified as glial elements.[6] The remaining cellular types of the retina (horizontal and amacrine cells) appear to be of an intermediate class, and their anatomical classification is species-dependent.[11,117–119,142]

The combined metabolic and enzymatic properties of the neural and glial elements have been described in the different layers of the retina by quantitative histochemical techniques, but this method of analysis cannot differentiate the metabolic capacity of the two elements. A more selective method is required for this differentiation and, in this regard, histochemical staining techniques and electrophysiological experiments have proved useful.

Studies of the histochemical staining properties of many enzyme systems have been reported for the retina,[112,143,144] and these studies have often been able to assess the relative concentration of enzymes within the neural and glial cells. One of the most striking contrasts between neural and glial cells is in the histochemical staining properties of carbonic anhydrase.[145] The reaction products of the enzyme are visibly concentrated in glia while neurons are unstained. A similar relationship has been suggested for brain tissue.[146] Furthermore, the enzymes of the glycolytic pathway appear to be concentrated in the Müller cell, and these cells contain the major store of glycogen.[144,147,148] The rate with which the Müller cell mobilizes and then resynthesizes glycogen following an experimental injury to the retina is impressive,[144] particularly since glycogen in the retina is resistant to extensive periods of retinal ischemia.[23]

Sjostrand has suggested from electron microscopic studies that the large Müller cell could represent the extracellular space in the retina.[5,116] Ames disagrees with this suggestion because when he used a cation elution technique to evaluate the extra- and intracellular space of the retina, he found that it was quite unlikely that glial intracellular fluid could account for any appreciable portion of the rapidly elutable extracellular fluid.[149] The glial elements, therefore, must contain appreciable quantities of cation, and in this respect they resemble glia of leech brain.[150]

The horizontal and amacrine cells differ from both the neural elements and the Müller cells in most species studied.[5,118,119,142] Cajal classified the horizontal cells as a "short axon cell" and the amacrine cell as a "no axon neuron."[3] They respond to light and exhibit a dynamic pattern of metabolism.[151] The metabolic properties were assessed by adding chemical compounds to the fluid in which isolated fish retina were incubated and observing the effect of these compounds upon the electrical response of these cells to light. The cells do not generate spike potentials, instead they show a shift in their DC potential when light is presented (S-potential).[152,153] The total functional and metabolic profile of these cells has led Svaetichin[154] to suggest that the horizontal and amacrine cells represent a new class of cells that he has labeled "controller cells." Considerable electrophysiological information has been obtained that indicates that these cells are present in all vertebrate retina. The S-potential of this class of cells exhibits a high temperature dependence[155] and susceptibility to low levels of ammonia, carbon dioxide, alcohols, and volatile anesthetics.[156,157] It is also affected by anoxia and metabolic inhibitors.[158] Although at present there is no conclusive evidence as to the nature of the coupling between aerobic metabolism and the membrane potential of an S-potential cell, it is likely that active ion transport processes are involved.[158]

IV. PHOTORECEPTOR CELL

The photoreceptor cells constitute the neural ectodermal layer of the retina. The rod and cone organelles of these cells are the photosensitive end organs. They extend beyond the outer limiting membrane of the retina into the potential space that exists between the retina proper and the pigment epithelium.[116] This space is analogous to the ventricles of the central nervous system.[1]

The mechanisms by which the visual pigment molecules of the rod and cone organelles, after absorbing incident radiation, initiate a nervous response in the retina and optic nerve provide some of the most fascinating problems in sensory physiology. At the dark-adapted threshold of vision, a rod may be stimulated by the absorption of a single quantum of light.[102] Although not every quantum absorbed need excite, at least one out of two does so.[102] Thus, the transduction process is highly efficient.

The whole photoreceptor cell is probably involved in the transduction process. This process is frequently subdivided for the purpose of discussion into three operational phases, i.e., initiation of excitation by the absorbing molecule, followed by amplification and transfer of this excitation over a considerable distance to the synapse of the photoreceptor cell. Numerous original and review articles have been published about the chemical composition and functional properties of the photoreceptor cell.[83,159–161] Articles by McConnell,[162,163] Hagins,[164] Abrahamson,[165] Enoch,[166,167] Fujishita,[168] and Lolley[169] are recommended to the interested reader. This literature is rapidly expanding, and this problem soon may advance beyond the descriptive phase. Already proposals have been published that suggest numerous physical and biochemical mechanisms for this operation, i.e., wave guide properties,[166] excitation migration,[170,171] photoconduction,[161] and biochemical photomultiplication.[83,103] These different proposals appear to possess one important assumption in common that has not been clearly stated, i.e., the events or reactions that lead eventually to neural excitation are distinct from the changes that occur in the visual pigment, even though they share a common origin. For example, rhodopsin absorbs a quantum of light and events begin. One can follow the photochemically induced changes in the visual pigment that appears to proceed down, rather than up, the energy scale.[172] Or one can follow the absorbed packet of energy that may excite electrons, cause structural changes in protein configuration, and be released as fluorescence or as heat. After the initial event triggered by the absorption of light, the visual pigment changes and energy transfer may be quite spatially separated and independent events. Of course, they may not be totally independent even if they are separated because the outer segment, particularly that of the rod, is a paracrystalline structure.[159] Nonetheless, the events that take place within the different reaction sequences would be related to time, but each would not necessarily proceed at the same rate. Consequently, correlation of the length of time required to initiate excitation after the absorption of a quantum of light, with the changes that have occurred in the visual pigment path, is an accurate but not necessarily a meaningful correlation due to the fact that the visual pigment changes may not initiate the excitation. This assumption bears heavily on our conceptual as well as our operational approach to the problem of the visual-excitation mechanism. Therefore, it is imperative that this concept be evaluated in the photoreceptor outer segment.

V. EFFECT OF ILLUMINATION UPON RETINAL NEUROCHEMISTRY

When placed in absolute darkness, the *in vivo* retina initially maintains a considerable amount of spontaneous activity[173] that decreases until the retina is electrically quiet. If flashing light is now presented to the retina, a volley of electric discharges occur at the onset and termination of each flash of light.[4,174,175] As the flashing rate increases, the ability of the retina to distinguish individual flashes disappears and the illumination appears continuous, a condition called "flicker-fusion."[176] Correspondingly, the on–off pattern of electrical discharge is lost and the light-adapted retina displays a random pattern of spontaneous activity.[105] Many visual situations have been studied that indicate that the pattern of electric discharge in the optic nerve is normally more complex than the simple situation stated above[177,178] and that the complex pattern of discharge in the optic nerve arises from an interplay of excitatory and inhibitory events.[175]

The electric activity of neural tissues, thus far studied, reflects the movement of sodium and potassium across neural membranes.[57,179] A cell loses potassium and gains sodium during the action potential, but these changes are quickly counterbalanced by mechanisms that favor a high concentration of intracellular potassium and low intracellular sodium. The restorative reactions utilize cellular energy, e.g., ATP, and by increasing the intracellular levels of ADP and inorganic phosphate they influence glucose metabolism.[21,180] Therefore, excessive neural activity alters both the cation levels and the rate of glucose metabolism within neural tissue.

The steady-state level of intracellular electrolytes in the retina (Table II) is very similar to that reported for other neural tissues.[57] The change of these levels with physiological stimulation of the retina by light *in vitro*, however, has not produced the dynamic effects that one sees by electrically stimulating cerebral cortex slices.[57,181] Yet, there are some indications that sodium and potassium are associated with electric activity in the retina. Furukawa and Hanawa[182] and Hamasaki[183] have shown that the electroretinogram of isolated retina is dependent upon sodium ions in the incubating media, and Bukser and Diamond[184] indicate that radioactive sodium accumulates in the retina upon exposure to continuous illumination. Potassium levels of the incubating medium also effects the electroretinogram *in vitro*.[185] These sketchy data are strengthened by the observation that glucose and oxygen are required for the maintenance of both electric activity in the retina and its characteristic levels of cations.[38,57] It may be presumed that the neural action potential reflects the movement of potassium and sodium down their respective concentration gradients in the retina.

The effect of *in vivo* illumination of the retina has been studied in the dark and light adapted state as well as conditions of flash illumination. In contrast to *in vitro* preparations of retina,[33,57] these different states of

TABLE II

Intracellular Concentration (mM) of Electrolytes in the Rabbit Retina[a]

Na$^+$	27	Ca^{2+}	1.4
K$^+$	151	Mg^{2+}	12.1
Cl$^-$	21		

[a] Concentrations computed by Ames[57] from data that was obtained by an elution technique (millimoles electrolyte per kilogram dry weight times kilogram dry weight per kilogram H_2O).

illumination that exhibit quite different profiles of physiological activity, influence the metabolism of the retina to measurably different degrees. Considering the dark adapted state as a control state to which conditions for light adaption can be compared, Lolley[25] notes that the levels of ATP and P-creatine are influenced by *in vivo* illumination of the frog retina. The P-creatine level is slightly increased after 30 min exposure to normal laboratory fluorescent light, while the ATP and NAD levels are unaltered. The changes observed in the retinal metabolites are more impressive with flashing light. The P-creatine and ATP content of the tissue is decreased, while NAD levels are unchanged by 30 min of continuous flash illumination. The kinetics of these changes indicate that metabolism is altered dramatically in the first few minutes after exposure to flashing light (Fig. 3). Furthermore, the profile of change of high-energy phosphates is reminiscent of that reported for the electric stimulation of cerebral cortex slices.[186] Thus, it would appear that *in vivo* physiological stimulation of the retina by light increases the rate of ATP utilization and indirectly the rate of glucose metabolism.

A relationship between the functional activity and the metabolism of retinal tissue has been shown by numerous *in vitro* experimental procedures. Microchemical[187] and histochemical staining[188–192] studies have indicated that illumination enhances the succinate oxidase and succinate dehydrogenase activity of retina. It is possible that illumination increases the rate at which new enzyme proteins are produced.[131,193] The RNA content[194,195] and the level of protein synthesis[196] in the ganglion cells of the

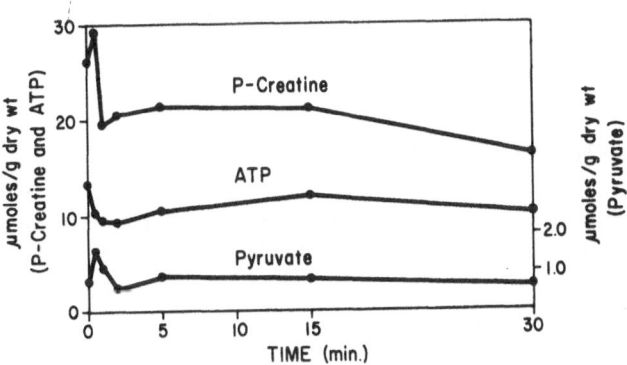

Fig. 3. Change in content of P-creatine, ATP, and pyruvate during continuous flash illumination of the *in vivo* frog retina. Light of 14,000 candlepower peak intensity, 10 μsec duration, was presented at a frequency of 5/sec for the time indicated. The frogs were immediately frozen. Eyes were sectioned at −20°C and lyophilized. Retinal samples were microdissected, weighed, and assayed enzymatically. The curves represent the average of three experiments.

frog retina are increased by illumination. Similar conditions of illumination also induce the cyclic oxidation of pyridine nucleotides in frog retina.[197]

The biochemical mechanisms involved with cation transport in neural tissues have been linked to the general metabolism of nerve cells.[56,198] A similar relationship appears to be operative in retinal tissue. Carbohydrate metabolism and oxygen uptake of pig retina were decreased by the absence of sodium ions from the incubating medium or by the inclusion of ouabain.[199] The decreased level of oxygen consumption was partially relieved by dinitrophenol, and these data have been interpreted to indicate the regulatory role of Na^+-K^+ ATPase in carbohydrate metabolism by the control of ADP levels within the tissue.[199]

The interrelationship between function and metabolism in the retina does not appear to be unlike that found in the other neural tissues, even though the retina is highly specialized in both categories. This specialization may not be uncommon in localized neural centers, even those within the central nervous system, but comparison cannot be made of the retina and any specific area of the central nervous system because so little is known about the structural, functional relationship within the brain. Much is known, however, about the retina, and the studies of retinal anatomy and biochemistry complement the work of electrophysiology and the more specialized researches in vision. The retina and the higher visual centers integrate visual information, and in this capacity the retina contributes handsomely. The problems in neurology and neurochemistry that remained unsolved or undented, such as synaptic transmission, neural–glial relationship, and information processing, are equally troublesome to the understanding of vision. These problems might be profitably studied in the retina, and the retina may prove to be a simplified model of the central nervous system.

VI. ACKNOWLEDGMENT

Bibliographic aid has been received from the UCLA Brain Information Service, which is a part of the Neurological Information Network of the NINDB and supported under Contract Ph-43-66-59.

VII. REFERENCES

1. D. G. Cogan, *Neurology of the Visual System*, Charles C. Thomas, Springfield, Illinois (1966).
2. S. Polyak, *The Vertebrate Visual System*, The University of Chicago Press, Chicago, Illinois (1957).
3. R. Y. Cajal, *Histologie du systeme nerveux de l'homme et des vertebrates*, Madrid, Consejo Superior de Investigaciones Cientificas (1952).

4. J. H. Prince and D. G. McConnell, in The Rabbit in Eye Research (J. H. Prince, ed.), pp. 385–448, Charles C. Thomas, Springfield, Illinois (1964).

5. F. S. Sjostrand and S. E. Nilsson, in The Rabbit in Eye Research (J. H. Prince, ed.), pp. 449–513, Charles C. Thomas, Springfield, Illinois (1964).

6. R. A. Allen, Isolated cilia in inner retinal neurons and in retinal pigment epithelium, J. Ultrastruct. Res. 12:730–747 (1965).

7. T. Samorajski, J. R. Keefe, and J. M. Ordy, Morphogenesis of photoreceptor and retinal ultrastructure in a sub-human primate, Vision Res. 5:639–648 (1965).

8. V. C. Glocklin and A. M. Potts, The metabolism of retinal pigment cell epithelium. II. Respiration and glycolysis, Invest. Ophthalmol. 4:226–234 (1965).

9. M. H. Bernstein, in The Structure of the Eye (George K. Smelser, ed.), pp. 139–150, Academic Press, New York and London (1961).

10. K. Meller and P. Glees, The differentiation of neuroglia-Müller-cells in the retina of chick, Z. Zellforsch. 66:321–332 (1965).

11. G. M. Villegas, Ultrastructure of the human retina, J. Anat. London 98:501–513 (1964).

12. R. Granit, in Handbook of Physiology (John Field, ed.), Section I, Neurophysiology, Vol. I, pp. 693–712, American Physiology Society, Washington, D.C. (1959).

13. J. H. Kinoshita, Selected topics in ophthalamic biochemistry, Arch. Ophthalmol. 72:554–572 (1964).

14. W. K. Noell, Cell physiology of the retina, J. Opt. Soc. Am. 53:36–48 (1963).

15. F. M. Matschinsky, J. V. Passonneau, and O. H. Lowry, Quantitative histochemistry of metabolites of glycolysis in retina, J. Histochem. Cytochem. 13:707 (1965).

16. J. G. Cunha-Vaz, Studies on the permeability of the blood–retinal barrier. III. Breakdown of the blood–retinal barrier by circulatory disturbances, Brit. J. Ophthalmol. 50:505–516 (1966).

17. O. Warburg, Über die Klassifizierung tierischer Gewebe nach ihrem Stoffwechsel, Biochem. Z. 184:484–488 (1927).

18. F. H. Adler, Physiology of the Eye, 3d ed., pp. 558–583, Mosby, St. Louis (1965).

19. J. Brotherton, Studies on the metabolism of the rat retina with special reference to retinitis pigmentosa, I. anaerobic glycolysis, Exptl. Eye Res. 1:234–245 (1962).

20. F. M. Matschinsky and D. B. McDougal, Quantitative histochemistry of enzymes and glycolytic intermediates of retina, pancreas, and organ of Corti, in 6th Intern. Congr. Clin. Chem., Munich, 1966, Vol. 3: Progress in Clinico-Chemical Methods, pp. 71–86, Karger, Basel/New York (1968).

21. O. H. Lowry, J. V. Passonneau, F. X. Hasselberger, and D. W. Schulz, Effect of ischemia on known substrates and co-factors of the glycolytic pathway in brain, J. Biol. Chem. 239:18–30 (1964).

22. H. deE. Webster and A. Ames, III, Reversible and irreversible changes in the fine structure of nervous tissue during oxygen and glucose deprivation, J. Cell Biol. 26:885–909 (1965).

23. F. M. Matschinsky and O. H. Lowry, unpublished observation.

24. F. M. Matschinsky, Quantitative histochemistry of nicotinamide adenine nucleotides in retina of monkey and rabbit, J. Neurochem. 15:643–657 (1968).

25. R. N. Lolley, Light-induced alterations in retinal pyruvate and high-energy phosphates, in vivo, submitted to J. Neurochem. (1969).

26. V. Bonavita, in Biochemistry of the Retina (Clive N. Graymore, ed.), pp. 5–13, Academic Press, London and New York (1965).

27. V. Bonavita, R. Guarneri, and F. Ponte, Neurochemical studies on the inherited retinal degeneration of the rat—III. Hexokinase, phosphofructokinase and ATPase in the developing retina, Vision Res. 7:51–58 (1967).

28. O. H. Lowry, N. R. Roberts, and C. Lewis, The quantitative histochemistry of the retina, J. Biol. Chem. 220:879–892 (1956).

29. O. H. Lowry, N. R. Roberts, D. W. Schulz, J. E. Clow, and J. R. Clark, Quantitative histochemistry of retina, II. Enzymes of glucose metabolism, *J. Biol. Chem.* **236**:2813–2820 (1961).

30. C. Graymore, Possible significance of the isoenzymes of lactic dehydrogenase in the retina of the rat, *Nature* **201**:615–616 (1964).

31. C. N. Graymore, *in Biochemistry of the Retina* (Clive N. Graymore, ed.), pp. 83–90, Academic Press, London and New York (1965).

32. L. H. Cohen and W. K. Noell, *in Biochemistry of the Retina* (Clive N. Graymore, ed.), pp. 36–50, Academic Press, London and New York (1965).

33. L. H. Cohen and W. K. Noell, Glucose metabolism of rabbit retina before and after development of visual function, *J. Neurochem.* **5**:253–276 (1960).

34. R. Wu and I. Racker, Regulatory mechanisms in carbohydrate metabolism. IV. Pasteur effect and Crabtree effect in ascites tumor cells, *J. Biol. Chem.* **234**:1036–1041 (1959).

35. N. Glick, E. Gillespie, and P. G. Scholifield, Oxidation of glucose and other substrates by Ehrlich ascites tumor cells, *Can. J. Biochem.* **45**:1401–1412 (1967).

36. T. Kuwabara, *in Biochemistry of the Retina* (Clive N. Graymore, ed.), pp. 93–98, Academic Press, London and New York (1965).

37. A. Ames, III, *in Biochemistry of the Retina* (Clive N. Graymore, ed.), pp. 22–30, Academic Press, London and New York (1965).

38. A. Ames, III, and B. S. Gurian, Effects of glucose and oxygen deprivation on function of isolated mammalian retina, *J. Neurophysiol.* **26**:617–634 (1963).

39. R. L. Sidman, *in The Structure of the Eye* (George K. Smelser, ed.), pp. 487–506, Academic Press, London and New York (1961).

40. A. A. Volokhov and N. N. Shilyagina, Ontogenic development of function in cortical and subcortical parts of visual system, *Federation Proc. Trans. Suppl.* **25**:T221–226 (1966).

41. C. Graymore, Metabolism of the developing retina, III. Respiration in the developing normal rat retina and the effect of an inherited degeneration of the retinal neuro-epithelium, *Brit. J. Ophthalmol.* **44**:363–369 (1960).

42. S. Papa, N. E. Lofrumento, A. G. Secchi, and E. Quagliariello, Utilization of respiratory substrates in calf-retina mitochondria, *Biochim. Biophys. Acta* **131**:288–294 (1967).

43. C. Graymore and M. Towlson, Levers of soluble NAD-linked α-glycerophosphate dehydrogenase and lactic dehydrogenase in rat retina, *Exptl. Eye Res.* **2**:48–52 (1963).

44. S. Futterman and J. H. Kinoshita, Metabolism of the retina, I. Respiration of cattle retina, *J. Biol. Chem.* **234**:723–726 (1959).

45. P. Joanny, J. Corriol, and A. Turc, Influence de differents substrats sur la repartition ionique alcaline de la retine de boeuf *in vitro*, *J. Physiol.* **57**:635–636 (1965).

46. O. H. Lowry, *in Morphological and Biochemical Correlates of Neural Activity* (M. M. Cohen and R. S. Snider, eds.), pp. 178–191, Harper & Row, New York (1964).

47. H. A. Krebs, Metabolism of amino acids. IV. The synthesis of glutamine from glutamic acid and ammonia, and the enzyme hydrolysis of glutamine in animal tissues, *Biochem. J.* **29**:1951–1969 (1935).

48. A. Pirie and R. van Heyningen, *Biochemistry of the Eye*, p. 213, Blackwell, Oxford (1956).

49. E. R. Berman, Isolation of neutral sugar-containing mucopolysaccharides from cattle retina, *Biochim. Biophys. Acta* **101**:358–360 (1965).

50. A. I. Yartakovskaia, The protein polysaccharide composition of the retina, *Vestn. Oftal'mol.* **78**:62–65 (1965).

51. M. O. Hall, D. E. Ocumpaugh, and R. W. Young, The utilization of ^{35}S-sulfate in the synthesis of mucopolysaccharides by the retina, *Invest. Ophthalmol.* **4**:322–329 (1965).

52. D. E. Ocumpaugh and R. W. Young, Distribution and synthesis of sulfated mucopolysaccharides in the retina of the rat, *Invest. Ophthalmol.* **5**:196–203 (1966).

53. F. N. Craig and H. K. Beecher, The effect of carbon dioxide tension on tissue metabolism (retina), *J. Gen. Physiol.* **26**:473–478 (1943).

54. L. Hopkinson and M. Kerly, The effect of monoiodoacetate on the aerobic metabolism of ox retina *in vitro*, *Biochem. J.* **72**:22–27 (1959).

55. J. Corriol, P. Joanny, and A. Simonpaoli, Consommation d'oxygene, glycolyse aerobie et gradients ioniques alcalins de la retine de boeuf *in vitro*, *J. Physiol.* **56**:327–328 (1964).

56. H. Waelsch, Biochemical interpretations of neurophysiological vectors, *Perspectives Biol. Med.* **9**:165–186 (1965).

57. A. Ames, III, Y. Tsukada, and F. B. Nesbett, Intracellular Cl^-, Na^+, K^+, Ca^{++}, Mg^{++}, and P in nervous tissue; Response to glutamate and to changes in extracellular calcium, *J. Neurochem.* **14**:145–159 (1967).

58. H. McIlwain, *Chemical Exploration of the Brain*, Elsevier, Amsterdam, London, and New York (1963).

59. W. K. Noell, Differentiation metabolic organization and viability of the visual cell, *A.M.A. Arch. Ophthalmol.* **60**:702–733 (1958).

60. H. W. Reading, in *Biochemistry of the Retina* (Clive N. Graymore, ed.), pp. 73–82 (1965).

61. S. Futterman, Metabolism of the retina, III. The role of reduced triphosphopyridine nucleotide in the visual cycle, *J. Biol. Chem.* **238**:1145–1150 (1963).

62. B. L. Horecker, Pathways of carbohydrate metabolism and their physiological significance, *J. Chem. Educ.* **42**:244–253 (1965).

63. A. Kornberg, in *Horizons in Biochemistry* (M. Kasha and B. Pullman, eds.), pp. 251–264, Academic Press, New York (1962).

64. J. C. Skou, Enzymatic basis for active transport of Na^+ and K^+ across cell membranes, *Physiol. Rev.* **45**:596–617 (1965).

65. Y. Sekoguti, On the ATPase activities in the retina and the rod outer segments, *J. Cell Comp. Physiol.* **56**:129–136 (1960).

66. S. L. Bonting, L. L. Caravaggio, and M. R. Canady, Studies on sodium–potassium-activated adenosine triphosphatase, X. Occurrence in retinal rods and relation to rhodopsin, *Exptl. Eye Res.* **3**:47–56 (1964).

67. M. E. Langham, S. J. Ryan, and M. Kostelnik, The Na,K ion dependent adenosinetriphosphatase of the retina and the mechanism of visual loss caused by cardiac glycosides, *Life Sci.* **6**:2037–2047 (1967).

68. S. L. Bonting, L. L. Caravaggio, and P. Gouras, The rhodopsin content, electroretinogram and rod structure in the rat, *Exptl. Eye Res.* **1**:14–24 (1961).

69. L. L. Caravaggio and S. L. Bonting, The rhodopsin cycle in the developing vertebrate retina. II. Correlative study in normal mice and in mice with hereditary retinal degeneration, *Exptl. Eye Res.* **2**:12–19 (1963).

70. R. N. Frank and T. H. Goldsmith, Effects of cardiac glycosides on electrical activity in the isolated retina of the frog, *J. Gen. Physiol.* **50**:1585–1606 (1967).

71. G. Ehrlich and Z. Dische, Content of desoxyribo and ribonucleic acids of retina under various conditions, *Soc. Exptl. Biol. Med. Proc.* **74**:40–42 (1950).

72. Y. Tsukada, K. Uyemura, and T. Matsutani, Metabolism of amino acid and nucleic acid on isolated rabbit retina, *Advan. Neurol. Sci.* (*Shinkei Ken Kyu No Shrimpo*) **10**:210–218 (1967).

73. P. Mandel, H. Rein, S. Harth-Edel, and R. Mardell, in *Comparative Neurochemistry*, pp. 149–166, Macmillan, New York (1964).

74. E. Koenig, Observations on selected isolated retinal elements and an analysis of rod cell RNA of the rabbit, *J. Cell. Biol.* **34**:265–274 (1967).

75. J. E. Edström and J. Kawiak, Microchemical deoxyribonucleic acid determination in individual cells, *J. Biophys. Biochem. Cytol.* **9**:619–626 (1961).

76. R. W. Young, *in The Retina: Morphology, Function and Clinical Characteristics* (R. Allen, M. Hall, and B. Straatsma, eds.), University of California Press, Los Angeles (1967), in press.

77. D. Bok, RNA and DNA metabolism in rat photoreceptors, *Anat. Rec.* **154**:320 (1966).

78. D. Bok, RNA metabolism in the red rods of the frog retina, *Anat. Rec.* **157**:217 (1967).

79. J. Brotherton, Studies on the metabolism of the rat retina with special reference to retinitis pigmentosa. II. Amino acid content as shown by chromatography, *Exptl. Eye Res.* **1**:246–252 (1962).

80. B. Droz, Dynamic condition of proteins in the visual cells of rats and mice as shown by radioautography with labeled amino acids, *Anat. Rec.* **145**:157–167 (1963).

81. R. W. Young, The renewal of photoreceptor cell outer segments, *J. Cell Biol.* **33**:61–72 (1967).

82. R. W. Young, personal communication.

83. W. K. Noell, *in Biochemistry of the Retina* (Clive N. Graymore, ed.), pp. 51–72, Academic Press, London and New York (1965).

84. R. A. Gorn and T. Kuwabara, Retinal damage by visible light, *Arch. Ophthalmol.* **77**:115–118 (1967).

85. B. S. Fine and W. J. Geeraets, Observations on early pathologic effects of photic injury to the rabbit retina, *Acta Ophthalmol.* **43**:684–691 (1965).

86. W. K. Noell, V. S. Walker, B. S. Kang, and S. Berman, Retinal damage by light in rats, *Invest. Ophthalmol.* **5**:450–473 (1966).

87. R. D. Lillie, Histochemical studies on the retina, *Anat. Rec.* **112**:477–495 (1952).

88. E. Raviola and G. Raviola, Ricerche istochimiche sulla retina di coniglio nel corso dello sviluppo postnatale, *Z. Zellforsch.* **56**:552–572 (1962).

89. S. Fleischer and D. G. McConnell, Preliminary observations on the lipids of bovine retinal outer segment disks, *Nature* **212**:1366–1367 (1966).

90. R. N. Lolley and H. H. Hess, The retinal rod outer segment of the frog: Detachment, isolation, phosphorus fractions and enzyme activity, *J. Cell Physiol.* **73**:9–23 (1969).

91. A. G. E. Pearse and D. G. McConnell, Cytochemical localization of redox compounds in isolated bovine retinal outer segment disks, *Nature* **212**:1367–1368 (1966).

92. A. G. Everson Pearse, *in Biochemistry of the Retina* (Clive N. Graymore, ed.), pp. 110–114, Academic Press, London and New York (1965).

93. H. H. Hess, personal communication.

94. H. Linschitz, Studies on the photochemistry of rhodopsin and chlorophyll, *Radiation Res. Suppl.* **2**:182–195 (1960).

95. U. Gloor and O. Wiss, Fat-soluble vitamins, *Ann. Rev. Biochem.* **33**:313–330 (1964).

96. A. L. Koen and C. R. Shaw, Retinol and alchol dehydrogenase in retina and liver, *Biochim. Biophys. Acta* **128**:48–54 (1966).

97. J. E. Dowling and I. R. Gibbons, *in The Structure of the Eye* (George K. Smelser, ed.), pp. 85–100, Academic Press, New York and London (1961).

98. G. Wald, P. K. Brown, and I. R. Gibbons, Visual excitation: A chemoanatomical study, *Symp. Soc. Exptl. Biol.* **16**:32–57 (1962).

99. R. Hubbard, D. Bownds, and T. Yoshizawa, The chemistry of visual photoreception, *Symp. Quant. Biol.* **30**:301–315 (1965).

100. J. J. Wolken, Lipids and the molecular structure of photoreceptors, *J. Am. Oil Chem. Soc.* **43**:271–274 (1966).

101. W. E. J. Phillips, Interrelationship of vitamin A and C on tissue ubiquinones and sterols of rats and guinea pigs, *Can. J. Biochem.* **45**:749–756 (1967).

102. C. D. B. Bridges, *in Comprehensive Biochemistry* (Marcel Florkin and Elmer H. Stotz, eds.), Vol. 27, pp. 31–78, Elsevier, Amsterdam, London, and New York (1967).

103. G. Wald, *in The Structure of the Eye* (George K. Smelser, ed.), pp. 101–115, Academic Press, New York and London (1961).

104. J. S. Andrews and S. Futterman, Metabolism of the retina. V. The role of microsomes in vitamin A esterification in the visual cycle, *J. Biol. Chem.* **239**:4073–4076 (1964).
105. K. O. Donner and T. Reuter, The dark-adaptation of single units in the frog's retina and its relation to the regeneration of rhodopsin, *Vision Res.* **5**:615–632 (1965).
106. H. J. A. Dartnall, *The Visual Pigments*, Wiley, New York (1957).
107. R. A. Weale, High light intensities and photo-chemical reactions of human visual pigments in situ, *Symp. Quant. Biol.* **30**:335–343 (1965).
108. T. Reuter, The synthesis of photosensitive pigments in the rods of the frog's retina, *Vision Res.* **6**:15–38 (1966).
109. T. P. Williams, Photoreversal of rhodopsin bleaching, *J. Gen. Physiol.* **47**:679–689 (1964).
110. A. G. E. Pearse, *in The Structure of the Eye* (George K. Smelser, ed.), pp. 53–72, Academic Press, New York and London (1961).
111. O. Eranko, M. Niemi, and E. Merenmies, *in The Structure of the Eye* (George K. Smelser, ed.), pp. 159–172, Academic Press, New York and London (1961).
112. M. Niemi, *in Neurochemistry* (C. W. M. Adams, ed.), pp. 599–621, Elsevier, Amsterdam, London, and New York (1965).
113. R. N. Lolley, personal observation.
114. W. K. Stell, Correlation of retinal cytoarchitecture and ultrastructure in Golgi preparations, *Anat. Rec.* **153**:389–397 (1965).
115. A. I. Cohen, Some electron microscopic observations on inter-receptor contacts in the human and macaque retinae, *J. Anat.* **99**:595–610 (1965).
116. F. S. Sjostrand, *in The Structure of the Eye* (George K. Smelser, ed.), pp. 1–28, Academic Press, New York and London (1961).
117. G. Raviola and E. Raviola, Light and electron microscopic observations on the inner plexiform layer of the rabbit retina, *Am. J. Anat.* **120**:403–426 (1967).
118. J. E. Dowling and B. B. Boycott, Neural connections of the retina: Fine structure of the inner plexiform layer, *Symp. Quant. Biol.* **30**:393–402 (1965).
119. A. Selvin de Testa, Morphological studies on the horizontal and amacrine cells of the teleost retina, *Vision Res.* **6**:51–59 (1966).
120. A. Pelligrion de Iraldi and G. J. Etcheverry, Granulated vesicles in retinal synapses and neurons, *Z. Zellforsch.* **81**:283–296 (1967).
121. J. E. Dowling and B. B. Boycott, Organization of the primate retina: Electron microscopy, *Proc. Roy. Soc. B* **166**:80–111 (1966).
122. B. Falck, N. A. Hillarp, G. Thieme, and A. Torp, Fluorescence of catecholamines and related compounds condensed with formaldehyde, *J. Histochem. Cytochem.* **10**:348–354 (1962).
123. J. Haggendal and T. Malmfors, Identification and cellular localization of the catecholamines in the retina and the choroid of the rabbit, *Acta Physiol. Scand.* **64**:58–66 (1965).
124. B. D. Drujan, J. M. Diaz Borges, and N. Alvarez, Relationship between the contents of adrenaline, noradrenaline and dopamine in the retina and its adaptational state, *Life Sci.* **4**:473–477 (1965).
125. C. W. Nichols, D. Jacobowitz, and M. Hottenstein, The influence of light and dark on the catecholamine content of the retina and choroid, *Invest. Ophthalmol.* **6**:642–646 (1967).
126. B. Ehinger, Adrenergic retinal neurons, *Z. Zellforsch.* **71**:146–152 (1966).
127. J. H. Welsh, *in Comparative Neurochemistry*, pp. 355–366, Macmillan, New York (1964).
128. F. C. MacIntosh, The distribution of acetylcholine in the peripheral and central nervous system, *J. Physiol.* **99**:436–442 (1941).
129. C. O. Hebb, Choline acetylase in mammalian and avian sensory systems, *Quart. J. Exptl. Physiol.* **40**:176–186 (1955).
130. W. Feldberg and M. Vogt, Acetylcholine synthesis in different regions of the central nervous system. *J. Physiol.* **107**:372–381 (1948).

131. P. H. Glow and S. Rose, Activity of cholinesterase in the retina with different levels of physiological stimulation, *Australian J. Exptl. Biol. Med. Sci.* **44**:65–72 (1966).

132. E. De Robertis, *Histophysiology of Synapses and Neurosecretion*, Macmillan, New York (1964).

133. W. K. Noell, The visual cell: Electric and metabolic manifestations of its life processes, *Am. J. Ophthalmol.* **48**:347–370 (1959).

134. K. Kishida and K. I. Naka, Amino acids and the spikes from the retinal ganglion cells, *Science* **156**:648–650 (1967).

135. F. Wald and E. De Robertis, The action of glutamate and the problem of the "extracellular space" in the retina. An electron miscroscope study, *Z. Zellforsch.* **55**:649–661 (1961).

136. D. R. Lucas and J. P. Newhouse, The toxic effect of sodium L-glutamate on the inner layers of the retina, *A.M.A. Arch. Ophthalmol.* **58**:193–201 (1957).

137. A. M. Potts, *in Biochemistry of the Retina* (Clive N. Graymore, ed.), pp. 155–162, Academic Press, London and New York (1965).

138. L. T. Graham, Jr., R. N. Lolley, and C. F. Baxter, Effect of illumination upon levels of γ-aminobutyric acid and glutamic acid in frog retina, *in vivo*, *Federation Proc.* **27**:463 (1968).

139. K. Yamamoto, H. Aono, and K. Kunimitsu, GABA shunt of TCA cycle in·the retina, *Acta Soc. Ophthalmol. Jap.* **68**:48–52 (1964).

140. E. Florey, *in Inhibition in the Nervous System and Gamma-aminobutyric Acid* (Eugene Roberts, ed.), pp. 202–206, Pergamon Press, New York, Oxford, London, and Paris (1960).

141. E. Roberts and K. Kuriyama, Biochemical-Physiological correlations in studies of the γ-aminobutyric acid system, *Seitai No Kagaku, Jap. Life Sei., Tokyo* **15**:2–27 (1967).

142. E. Yamada and T. Ishikawa, The fine structure of the horizontal cells in some vertebrate retinae, *Symp. Quant. Biol.* **30**:383–392 (1965).

143. A. G. E. Pearse, *in The Structure of the Eye* (George K. Smelser, ed.), pp. 53–72, Academic Press, New York and London (1961).

144. T. Kuwabara, *in Biochemistry of the Retina* (Clive N. Graymore, ed.), pp. 93–98, Academic Press, New York and London (1965).

145. E. Korhonen and L. K. Korhonen, Histochemical demonstration of carbonic anhydrase activity in the eyes of rat and mouse, *Acta Ophthalmol.* **43**:475–481 (1965).

146. T. H. Maren, Carbonic anhydrase: Chemistry, physiology and inhibition, *Physiol. Rev.* **47**:595–781 (1967).

147. G. Raviola, E. Raviola, and M. T. Tenconi, Sulla organizazione dello strato granulare esterno e della membrana limitante esterna nella retina de coniglio, *Z. Zellforsch.* **70**:532–553 (1966).

148. K. Mizuno, Histochemical studies on the glia cell in the retina and optic nerve, *Acta Soc. Ophthalmol. Jap.* **68**:1567–1573 (1964).

149. A. Ames, III, and F. B. Nesbett, Intracellular and extracellular compartments of mammalian central nervous tissue, *J. Physiol.* **184**:215–238 (1966).

150. J. G. Nicholls and S. W. Kuffler, Na and K content of glial cells and neurons determined by flame photometry in the central nervous system of the leech, *J. Neurophysiol.* **28**:519–525 (1965).

151. K. Negishi and G. Svaetichin, Oxygen dependence of retinal S-potential producing cells, · *Science* **152**:1621–1623 (1966).

152. G. Svaetichin, The cone action potential, *Acta Physiol. Scand.*, Suppl. 106 **29**:565–600 (1953).

153. P. Witkovsky, A comparison of ganglion cell and S-potential response properties in carp retina, *J. Neurophysiol.* **30**:546–561 (1967).

154. G. Svaetichin, M. Laufer, G. Mitarai, R. Fatehchand, E. Vallecalle, and J. Villegas, *in The Visual System* (R. Jung and N. Kornhuber, eds.), pp. 445–456, Springer, Berlin (1961).

155. K. Negishi and G. Svaetichin, Effects of temperature on S-potential producing cells and on neurons, *Pfluegers Arch.* **292**:206–217 (1966).
156. K. Negishi and G. Svaetichin, Effects of anoxia, CO_2 and NH_3 on S-potential producing cells and on neurons, *Pfluegers Arch.* **292**:177–205 (1966).
157. K. Negishi and G. Svaetichin, Effects of alcohols and volatile anesthetics on S-potential producing cells and on neurons, *Pfluegers Arch.* **292**:218–228 (1966).
158. R. Fatehchand, G. Svaetichin, K. Negishi, and B. Drujan, Effects of anoxia and metabolic inhibitors on the S-potential of isolated fish retinas, *Vision Res.* **6**:271–283 (1966).
159. M. F. Moody, Photoreceptor organelles in animals, *Biol. Rev.* **39**:43–86 (1964).
160. S. L. Bonting and A. D. Bangham, On the biochemical mechanism of the visual process, *Exptl. Eye Res.* **6**:400–413 (1967).
161. W. Lohmann, The importance of sulfur and iron in the retina as determined by paramagnetic resonance studies, *Experientia* **20**:1–6 (1964).
162. D. G. McConnell, The isolation of retinal outer segment fragments, *J. Cell Biol.* **27**:459–473 (1965).
163. D. G. McConnell, Chemical theories, *Physiol. Bull.* **61**:252–261 (1964).
164. W. A. Hagins, Electrical signs of information flow in photoreceptors, *Symp. Quant. Biol.* **30**:403–418 (1965).
165. E. W. Abrahamson and S. E. Ostroy, The photochemical and macromolecular aspects of vision, *Progr. Biophys. Mol. Biol.* **17**:181–215 (1967).
166. J. M. Enoch, Physical properties of the retinal receptor and response of retinal receptors, *Psychol. Bull.* **61**:242–251 (1964).
167. J. M. Enoch and L. E. Glismann, Physical and optical changes in excised retinal tissue, *Invest. Ophthalmol.* **5**:208–221 (1966).
168. S. Fujishita, Photo-release of oxygen from rhodopsin solution, *Jap. J. Physiol.* **16**:576–583 (1966).
169. R. N. Lolley, ATPase activity of retina and photoreceptor outer segments, submitted to *J. Cell Physiol.* (1969).
170. W. A. Hagins and W. H. Jennings, Energy transfer with special reference to biological systems, *Disc. Faraday Soc.* **27**:180 (1960).
171. P. A. Liebman, *In situ* microspectrophotometric studies on the pigments of single retinal rods, *Biophys. J.* **2**:161–178 (1962).
172. R. Hubbard, The stereoisomerization of 11-cis-retinal, *J. Biol. Chem.* **241**:1814–1818 (1966).
173. R. W. Rodieck and P. S. Smith, Slow dark discharge rhythms of cat retinal ganglion cells, *J. Neurophysiol.* **29**:942–953 (1966).
174. T. Tomita, Electrical activity in the vertebrate retina, *J. Opt. Soc. Am.* **53**:49–57 (1963).
175. S. W. Kuffler, Discharge patterns and functional organization of mammalian retina, *J. Neurophysiol.* **16**:35–68 (1953).
176. E. Dodt and J. B. Walther, Photic sensitivity mediated by visual purple, *Experientia* **14**:142–143 (1958).
177. D. N. Spinelli, Visual receptive fields in the cat's retina: Complications, *Science* **152**:1768–1769 (1966).
178. M. Weingarten and D. N. Spinelli, Retinal receptive field changes produced by auditory and somatic stimulation, *Exptl. Neurol.* **15**:363–376 (1966).
179. A. L. Hodgkin, The ionic basis of electrical activity in nerve and muscle, *Biol. Rev.* **26**:339–409 (1951).
180. R. D. Hill and P. D. Boyer, Inorganic orthophosphate activation and adenosine diphosphate as the primary phosphoryl acceptor in oxidative phosphorylation, *J. Biol. Chem.* **242**:4320–4323 (1967).

181. H. McIlwain, *Biochemistry and the Central Nervous System*, 3d ed., Churchill, London (1966).
182. T. Furukawa and I. Hanawa, Effects of some common cations on the electroretinogram of the toad, *Jap. J. Physiol.* **5**:289–300 (1955).
183. D. I. Hamasaki, Effects of sodium ion concentration on the electroretinogram of isolated retina of the frog, *J. Physiol.* **167**:156–168 (1963).
184. S. Buckser and H. Diamond, Increase in sodium concentration of the isolated frog retina after light stimulation, *Biochem. Biophys. Res. Commun.* **23**:240–242 (1966).
185. D. I. Hamasaki, The electroretinogram after application of various substances to the isolated retina, *J. Physiol.* **173**:449–458 (1964).
186. P. J. Heald, *Phosphorus Metabolism of Brain*, Pergamon Press, New York (1960).
187. M. H. Epstein and J. S. O'Connor, Enzyme changes in isolated retinal layers in light and darkness, *J. Neurochem.* **13**:907–911 (1966).
188. J. M. Enoch, The use of tetrazolium to distinguish between retinal receptors exposed and not exposed to light, *Invest. Ophthalmol.* **2**:16–23 (1963).
189. W. L. Fowlks and D. E. Peterson, Substrate inhibition of tetrazolium salt reduction in dark adapted retinae, *Proc. Soc. Exptl. Biol. Med.* **118**:491–494 (1965).
190. M. A. Ostrovskii, Investigation of certain links in the photoenzymochemical chain of processes in photoreceptors, *Federation Proc. Trans. Suppl.* **25**:T227–230 (1966).
191. Y. A. Vinnikov, Structural and cytochemical organization of receptor cells of the sense organs in the light of their functional evolution, *Federation Proc. Trans. Suppl.* **25**:T34–42 (1966).
192. T. P. Lukashevick, Changes in the activity of enzymes of the succinoxidase system in the photoreceptors of vertebrates kept in different conditions of illumination, *Dokl. Akad. Nauk USSR* **145**:798–801 (1963).
193. G. Maraini, F. Carta, R. Franguelli, and M. Santori, Effect of monocular light-deprivation on leucine uptake in the retina and the optic centers of the newborn rat, *Exptl. Eye Res.* **6**:299–302 (1967).
194. A. L. Byzov, Functional properties of different cells in the retina of cold-blooded vertebrates, *Symp. Quant. Biol.* **30**:547–558 (1965).
195. I. A. Utina, The changes of the amount of RNA in the horizontal and amacrine cells of the frog retina in various conditions of illumination. *Dokl. Akad. Nauk USSR* **157**:1216–1218 (Eng 511–514) (1964).
196. I. U. Chentsov, V. L. Boroviagin, and B. I. Brodskii, Submicroscopic morphology of retinal neurons as a reflection of specific changes in their metabolism, *Biophysics* **6**:61–70 (1961).
197. W. Sickel, Respiratory and electrical responses to light stimulation in the retina of the frog, *Science* **148**:648–651 (1965).
198. R. Whittam and K. P. Wheeler, The sensitivity of a kidney ATPase to ouabain and to sodium and potassium, *Biochem. Biophys. Acta* **51**:622–624 (1961).
199. M. V. Riley, in *Biochemistry of the Retina* (Clive N. Graymor, ed.), pp. 149–154, Academic Press, London and New York (1965).

Chapter 21

METABOLIC INFORMATION DERIVED FROM RADIOAUTOGRAPHY

B. Droz

Département de Biologie
Commissariat à l'Energie Atomique
Saclay, France

I. INTRODUCTION

Radioautography is an application of the nuclear tracer method that allows both the *detection* and the *location* of radioactive atoms in biological structures. Radioautographs performed at various time intervals after the administration of a substance tagged with a radioactive atom make it possible to trace the incorporation and migration of the label through various regions of the nervous system and even through various organelles of a given cell. Thus, as noted by Leblond (1965), "Radioautography stands at the crossroads of biochemistry and morphology."

A. Principle of the Method

Radioactive atoms, especially those emitting beta rays of low energy (^3H, ^{14}C, ^{35}S, etc.), may be visualized by their action on a nuclear photographic emulsion covering a tissue section (Fig. 1). The silver halide crystals of the emulsion act as microdetectors of radiations and, therefore, the radioautographic reaction pinpoints radioactive sources contained in tissue sections (see review in Gross *et al.*[1] and Rogers[2]).

B. Nature of the Labeled Molecules to Be Detected

The radioautographic detection of a given substance depends primarily on its physico-chemical properties. Let us consider four cases.

1. *Newly Synthesized Macromolecules*

The label that tags a soluble molecule (precursor) is incorporated into an insoluble macromolecule (product). For instance, when labeled amino acids are introduced into a system that is synthesizing proteins, labeled amino

505

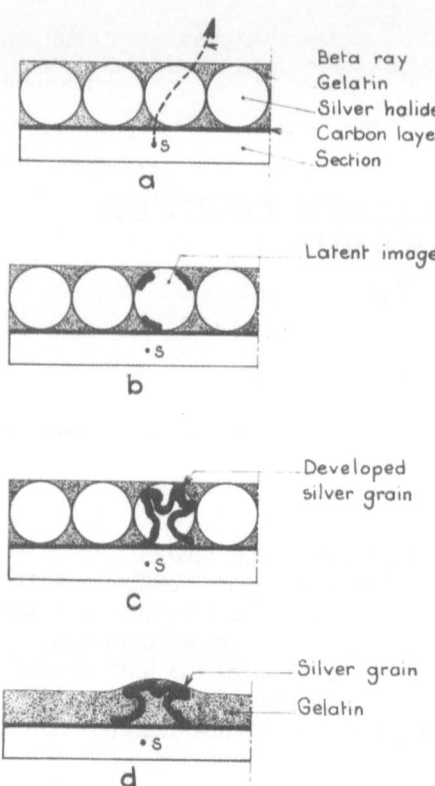

Beta ray
Gelatin
Silver halide crystal
Carbon layer
Section

a

Latent image

b

Developed
silver grain

c

Silver grain

Gelatin

d

Fig. 1. Diagrammatic representation of the radioautographic reaction as revealed with the electron microscope. At the top (a), the section that contains radioactive sources (s) emitting beta rays of low energy is coated with an homogeneous layer of nuclear emulsion (silver halide crystals and gelatin). A thin layer of carbon is interposed between the section and the emulsion. When a beta ray hits a silver halide crystal (a), the ionization results in the deposition of metallic silver in the silver halide crystal, producing the so-called "latent image" (b). When acted upon by the photographic developer (c), a filament of metallic silver appears in silver halide crystal bearing a latent image and not in the virgin crystals. After thiosulfate fixation (d), intact silver halide crystals are dissolved, whereas silver filaments of the developed grains are preserved.

acids are incorporated into the growing polypeptide chains, giving rise to proteins. Hence, these newly synthesized proteins in turn become radioactive and may be detected by the radioautographic reaction (Figs. 2–9). This result, however, is obtained only if several conditions are fulfilled: (1) The newly formed protein must be preserved in the course of tissue preparation; (2) substances representing intermediary steps of synthesis and catabolites of the precursor must constitute only a minimal fraction of the radioactivity contained in tissue sections; (3) free amino acids as well as amino acids adsorbed on preexisting structures must be washed out in the fixative. Therefore, *it is extremely important to analyze biochemically the nature of the product responsible for the radioautographic reaction* and to rule out an artefactual retention by linkages other than peptide bonds.[3-5]

2. Bound Molecules

The injected compound is a labeled molecule that can be specifically bound to a macromolecular structure. For example, norepinephrine-^{3}H, introduced into a lateral ventricle of the brain is picked up and accumulated

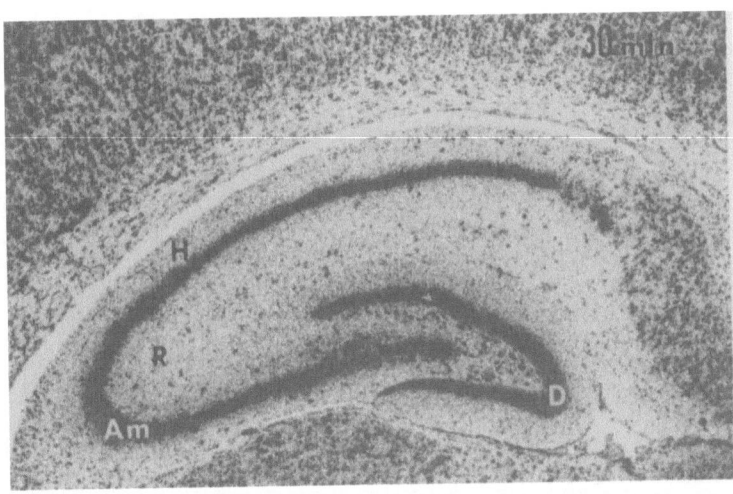

Fig. 2. Figures 2–5 show unstained radioautographs of hippocampus of 50-g rats injected with a single dose of methionine-^{35}S and sacrificed at various time intervals thereafter. At 30 min, the radioautographic reaction is intense over the nerve cell bodies of Ammon's pyramidal cells (Am) and the cells of the gyrus dentatus (D). The reaction is rather weak over the alveus hippocampi (H) and over the stratum radiatum (R), which contain, respectively, the axons and the dendrites of the Ammon's pyramidal cells. The weak reaction seen over these structures and over the fimbria is due mainly to glial cells.

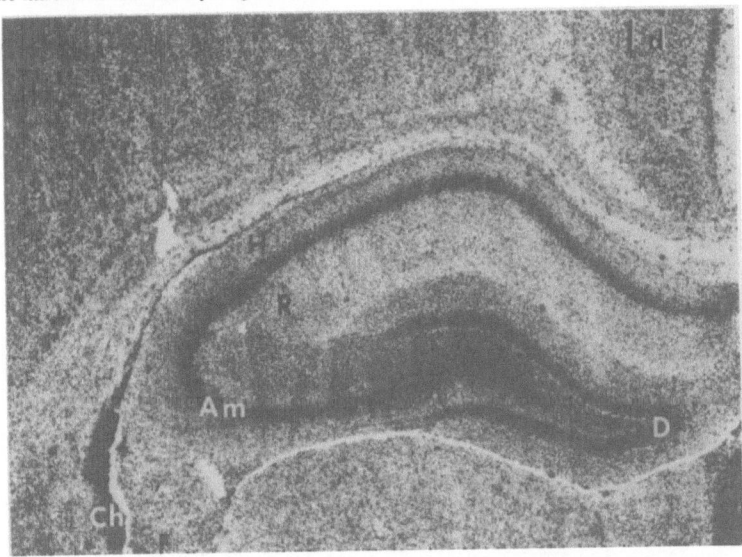

Fig. 3. At 24 hr after injection, the radioautographic reaction is still strong over Ammon's pyramidal cells (Am) and the cells of the gyrus dentatus (D), but now a moderate reaction is seen over the alveus hippocampi (H) and stratum radiatum (R), especially in the proximity of nerve cell bodies. At the lower left, note an intense radioautographic reaction over the choroid plexus (Ch).

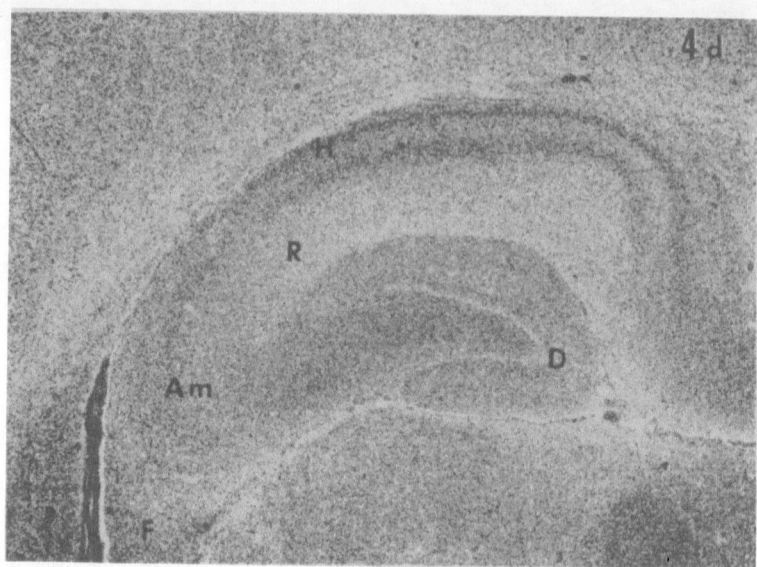

Fig. 4. At 4 days, the reaction has decreased considerably over Ammon's pyramidal cells and the cells of the gyrus dentatus, but not over the alveus hippocampi. The fimbria now shows an increased and diffuse radioautographic reaction.

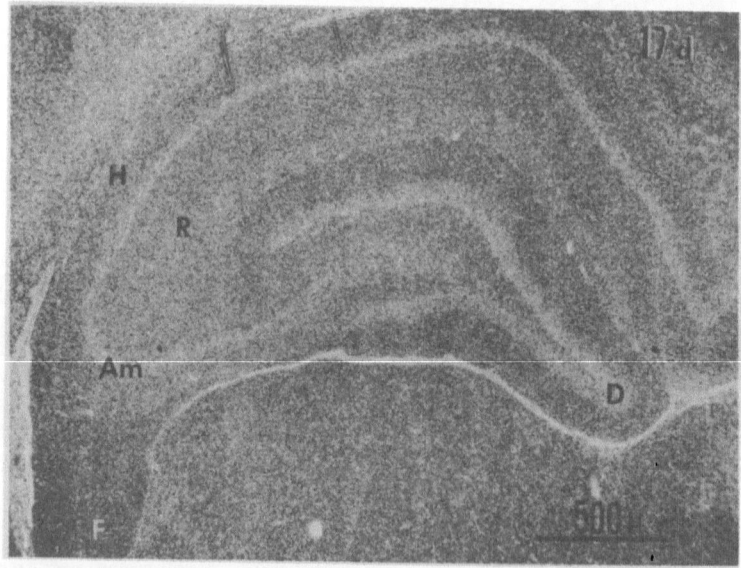

Fig. 5. At 17 days, the group of Ammon's pyramidal cells (Am) and the cells of the gyrus dentatus (D) appear as light bands, whereas the radioautographic reaction is strong over the fimbria (F). (From Droz and Leblond.[17])

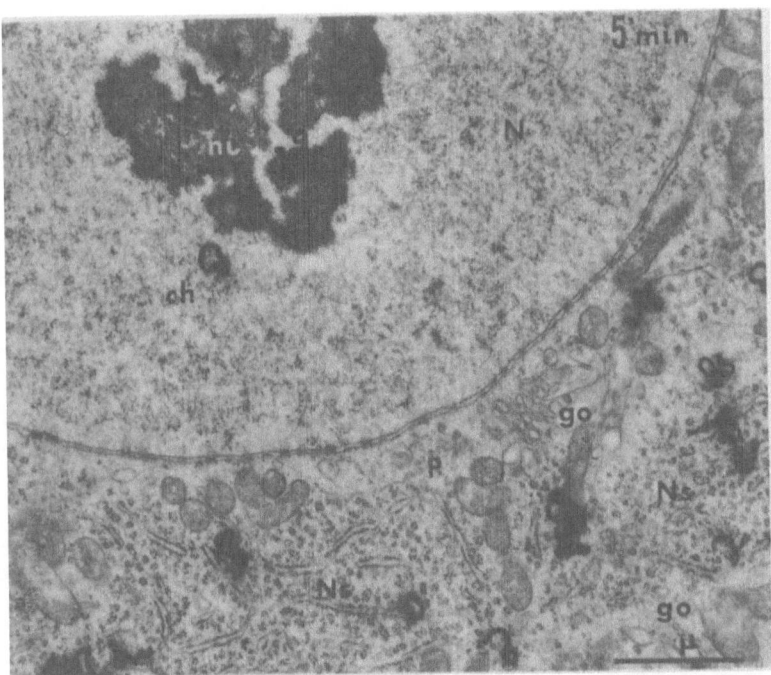

Fig. 6. Electron microscope radioautographs are shown in Figs. 6 and 7 of ciliary ganglion cells of chicken sacrificed 5 min (Fig. 6) and 20 min (Fig. 7) after an intravenous injection of L-leucine-^{3}H (specific activity, 60 C/mM) at the dose of 83 μC/g body weight. In the nucleus (N) of the ganglion cell, the reaction appears at 5 min over both nucleolus (nu) and chromatin (ch). In the perikaryon (P), most of the silver grains are distributed over the Nissl substance (Ns), rich in ribosomes. Golgi complexes (go) are devoid of label.

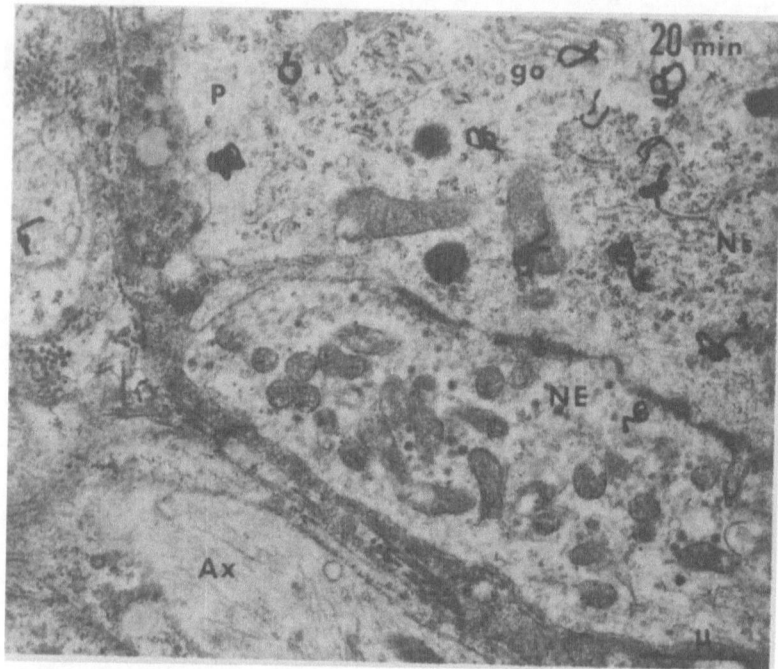

Fig. 7. At the periphery of the perikaryon, the reaction persists over the Nissl substance, but at 20 min silver grains are seen over vacuoles and sacs of the Golgi complex. Axons (Ax) and nerve endings (NE) display only a minimal number of silver grains (about twice above background.) (From Droz and Koenig, unpublished.)

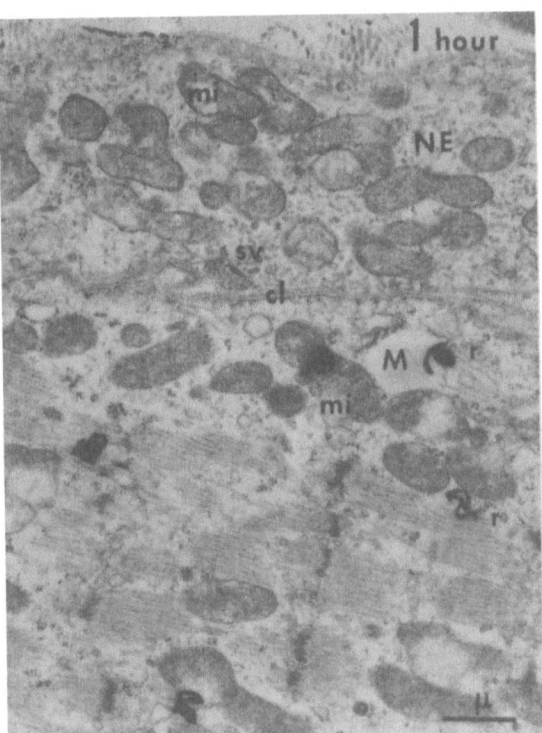

Fig. 8. Electron microscope radioautographs are shown in Figs. 8 and 9 of neuromuscular junctions in the ciliary muscle of a chicken sacrificed 1 hr (Fig. 8) and 3 days (Fig. 9) after the injection of L-leucine-^{3}H. By 1 hr, the radioactivity in the striated muscle (M) seems to be associated with ribosomes (r) and mitochondria (mi). The nerve ending (NE), which contains numerous synaptic vesicles (sv) and mitochondria, is free of label. Note the synaptic cleft (cl).

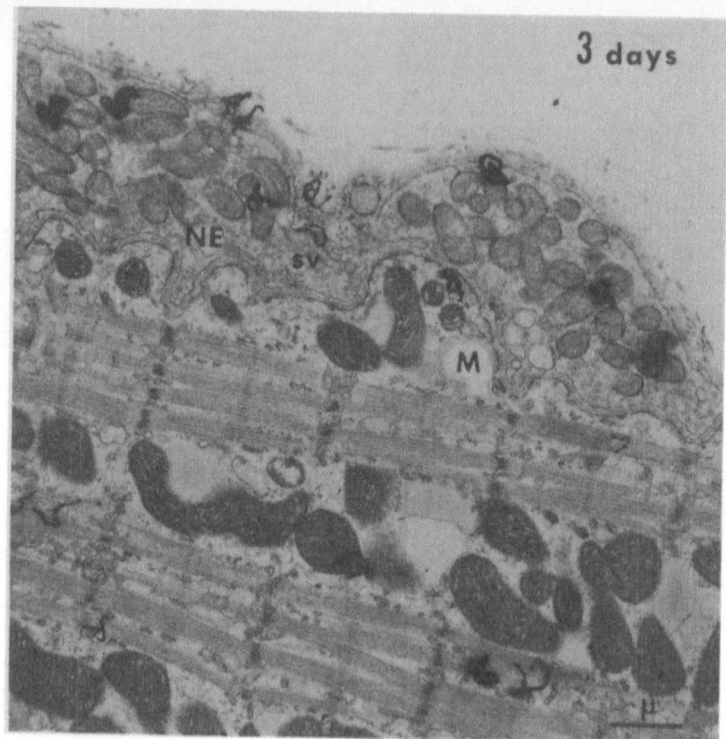

Fig. 9. Three days later the radioactivity invades nerve endings previously unlabeled. Silver grains are distributed over mitochondria and areas rich in synaptic
vesicles. (From Droz, Koenig, and Young.[45])

in a few neurons and nerve endings (Figs. 12 and 13) presumably by linkage
with specific molecules. Here again it is essential to determine biochemically
the specificity of the radioautographic reaction.[38]

3. *Liposoluble Compounds*

The labeled substance, soluble in lipid solvents, may be preserved in
frozen tissue sections (Fig. 10). Under these conditions, labeled fatty acids,
cholesterol, phospho- and sulpholipids may be detected by either "contact"
or "stripping" techniques.[1,2]

4. *Water-Soluble Components*

These components, such as labeled ions, free amino acids, or nucleotides,
are washed out in the course of histological fixation. Several techniques,
however, have been proposed to study their distribution by radioautography,
but losses and displacements of the label cannot be completely eliminated.[1,2]

C. Radioautographic Techniques

In addition to physico-chemical properties of the radioactive substance, the degree of resolution required for locating the label influences the choice of a radioautographic technique.

1. Anatomical Localization

The distribution of a label in whole brain sections is easily obtained by the "contact" technique (Fig. 10) introduced by Lacassagne and Lattes (1924). The contact method, in which a section is closely apposited to a photographic film (Kodirex), has the advantage of simplicity but gives a rather poor resolution because of the residual interspace between specimen and emulsion.[1]

2. Cellular Localization

Owing to the heterogeneity of the cell population in the nervous system, it is often necessary to identify cell types containing the label and to determine its intracellular turnover. For this purpose the "coating" technique, devised by Bélanger and Leblond (1946), may be helpful. It consists in covering radioactive sections with a fluid emulsion and provides an intimate contact between sections and the nuclear emulsion (e.g., Kodak NTB, NTB_2; Ilford G_5, K_5). The technical procedure described by Kopriwa and Leblond[6] is especially recommended to obtain a precise and reproducible localization of the radioactivity in paraffin or epon-embedded sections; the resolution achieved with tritium is about 1 μ. The "stripping" film technique

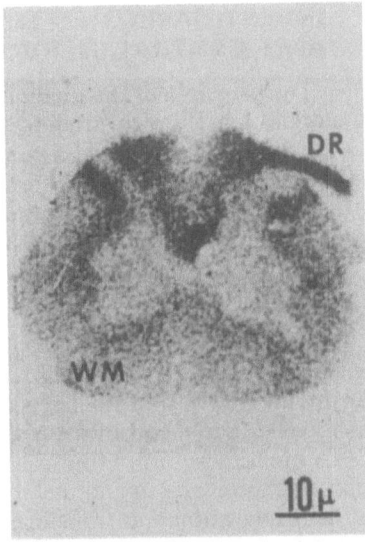

Fig. 10. Figures 10 and 11 show radioautographic detection of transferred cholesterol. Contact radioautography is shown of the spinal cord of a rat fed with cholesterol-^{14}C for 60 days. The label predominates in the white matter (WM). Note the intense reaction in the dorsal root (DR). (From Chevallier and Petit.[26])

introduced by Pelc (1947) is particularly suitable to radioautograph frozen sections in which lipid components must be retained: The sections are covered with a film peeled off from a nuclear plate such as Kodak AR 10 (see details of the technique in Rogers[2]).

3. Subcellular Localization

The intracellular pathways followed by a radioactive substance may be traced by means of electron microscope radioautography in thin sections of either whole cells or cell fractions obtained by differential centrifugation. The electron microscope radioautographic technique used in our laboratory is mainly derived from the principles proposed by Granboulan.[7] Epon-embedded sections are placed on a glass slide previously covered with a celloidin membrane, stained with uranium, and lead and carbon coated.[8] The section-bearing slides are dipped into the melted emulsion Ilford L4 diluted 1:4, which coats the sections with a monolayer of silver halide crystals. After a 1–3-month exposure, radioautographs are processed in Microdol X and the celloidin membrane is peeled off the glass slide. The radioautographs, placed on a grid, are directly examined under the electron beam after dissolution of the celloidin membrane.[9]

To relate a developed silver grain to its radioactive source, it has been calculated that a disintegrated atom of tritium must be located within a circle of about 0.2μ radius from the center of the silver grain and has in fact the greatest probability of being confined within a smaller circle (see discussion in Granboulan,[7] Salpeter and Bachmann,[8] and Droz[9]).

II. APPLICATION OF RADIOAUTOGRAPHY FOR INVESTIGATING METABOLISM IN THE PERIPHERAL AND CENTRAL NERVOUS SYSTEM

The principle of the method consists of introducing a small number of molecules labeled with a radioactive atom, the fate of which is then followed in the nervous system as a function of time. Assuming that the labeled molecules are mixed with the unlabeled homologous molecules and behave like them, the distribution of the label reflects the distribution of the homologous molecular population at any time.

A. DNA

Brain DNA was found to turn over at an extremely low rate by biochemical determination. Radioautography following the administration of thymidine-^{3}H was particularly appropriate to locate and identify proliferating cells that incorporate tritium-labeled thymidine into newly formed DNA molecules prior to mitosis. In the central nervous system, multiplying glial cells were labeled and therefore could be traced eventually in the course of their migration and transformation.[10,11] Nerve cells possessing a long axon were never found to be labeled; however, a few nerve cells having a short axon may divide after birth in the granular layer of the olfactory bulb, gyrus dentatus, and cerebellum.[12]

B. RNA

The synthesis, transport, and fate of RNA were studied by radio-autography in nerve and glial cells. After administration of a labeled precursor of RNA, it was necessary to treat the sections with DNAse, since precursors of RNA also may give rise to DNA. When cytidine-^{3}H was injected, ganglion cells showed an intense incorporation of the label into the RNA of chromatin and nucleolus, at early intervals (20 min). Then, 1–3 days later, the perikaryal RNA became radioactive by transfer from the nucleus. The absence of radioactivity in control sections treated with RNAse confirmed that the label was effectively incorporated into RNA.[13]

The use of adenine-^{14}C or orotic-^{14}C acid exhibited a similar pattern in neurons of the central nervous system. Labeled RNA was synthesized in the nucleus and then transported into the perikaryon and the base of dendrites. None migrated to the axon. Among the glial cells, a strong incorporation was noted in the dark nuclei of oligodendrocytes.[14]

C. Protein

Synthesis of protein was demonstrated to take place in nerve cells by means of radioautography following an injection of methionine-^{35}S or tritium-labeled amino acids (see review in Droz[15,16]). With time, newly formed proteins disappeared from nerve cell bodies (Figs. 2–5 and 16) and invaded regions that were previously unlabeled.[17] In neurons, proteins are synthesized in the nucleus, perikaryon (Figs. 6 and 7), and base of dendrites, but none in the axon (or, if any, an exceedingly small amount[18]). Quantitative electron microscope radioautography[9] showed that one fraction of proteins newly synthesized in the Nissl substance migrates to and accumulates in the Golgi complex (Fig. 15) which gives rise to lysosomes and components of the smooth-surfaced endoplasmic reticulum. Another fraction that bypasses the Golgi complex probably would contribute to the elaboration of neurofilaments and neurotubules present in the neuroplasm. Then newly formed proteins are engulfed into the axon hillock and, later, into the axon proper, along which they move (Fig. 14) toward nerve endings at a rate of 0.6–0.9 mm/day in adult and 2–2.5 mm/day in young growing rats.[15,16] Nerve endings, which do not seem able to synthesize significant amounts of protein (Figs. 7 and 8), are ultimately invaded by labeled proteins mostly related to mitochondria and regions containing synaptic vesicles (Fig. 9). Thus, nerve cell processes and especially the axon are continuously supplied with new proteins and organelles manufactured in the perikaryon.

D. Enzyme Activity

The distribution of acetylcholinesterase (AcChase) at motor end plates was recently investigated by means of light[19] and electron microscope[20]

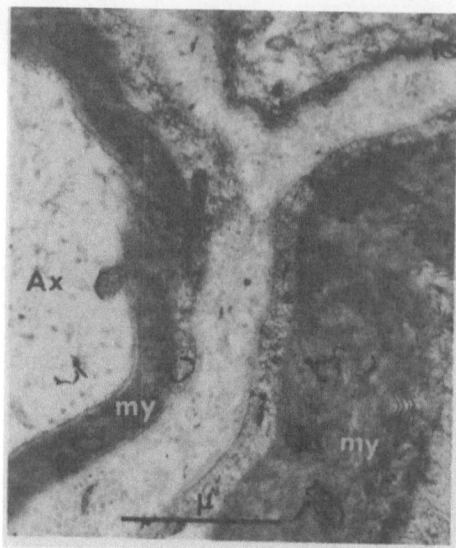

Fig. 11. Electron microscope radio-autography of myelinated axons in the dorsal roots of a rat fed with choles-terol-^{3}H for 10 days. To enhance the preservation of the alcohol-soluble material, glutaraldehyde-osmium fixed tissues were dehydrated at 4°C through the series of ethanol for only 5 min and then directly embedded in epon for polymerization. Most of the silver grains overlay the piled membranes of the myelin sheath (my). A few grains are seen over the axoplasm (Ax) and seem to be related to the smooth-surfaced endoplasmic reticulum. (From Chevallier and Droz, unpublished.)

radioautography. Tissues were first incubated in nonradioactive di-isopropylfluorophosphate (DFP) to phosphorylate all-sensitive sites, and then AcChase was specifically reactivated by pyridine-2 aldoxime methiodite. Subsequently the tissues were treated with tritium-labeled DFP, which reacted then with the only reactivated AcChase and radioautographed. Quantitative analysis of the scattered silver grains showed that most of the AcChase activity was located in the subneural apparatus of motor end plates.

E. Complex Carbohydrate

Synthesis of complex carbohydrates was reported to be associated with membranes fractions in the central nervous system.[21,22] After an intra-venous injection of galactose-^{3}H, radioautographs of spinal ganglion cells showed, at 5 min, that most of the label was concentrated in the piled sacs of the Golgi complex.[23] By 1 hr the label seemed to be associated with "coated" vesicles that would contain carbohydrates.[24] It is postulated that such complex carbohydrates are involved in the renewal of cell membrane and/or coat.

F. Lipid

The incorporation of sulfate-^{35}S into sulfolipids was observed in regions of the brain rich in bundles of myelinated fibers by comparison of radio-autographed frozen sections and lipid-extracted sections.[25]

The turnover of cholesterol in the brain results from a transfer from plasma and also from a local synthesis. Transferred radioactive cholesterol was localized (Figs. 10 and 11) in the myelinated bundles of the nervous system of rats fed for several days with labeled cholesterol.[26] Cholesterol recently synthesized from intraventricularly injected mevalonic-^{14}C acid was biochemically studied, and its distribution in brain was established by "contact" radioautography.[27]

The synthesis of phosphatidylinositol was shown to take place in ganglion cells of the superior cervical ganglion and to be stimulated by incubation with acetylcholine and eserine. Radioautographs provided evidence that the stimulated incorporation of myo-inositol-^{3}H was confined to the perikaryon and not to the plasma membrane alone.[28]

G. Catecholamine and Indoxylamine

In the peripheral nervous system, radioautography combined with electron microscopy was used for locating neurotransmitters in various peripheral neurons after administration of norepinephrine-^{3}H[29–32] or 5-hydroxytryptophan-^{3}H.[30–34] Most of the label was found to be concentrated in axonal processes containing dense core vesicles, except in the intestine.[30,32]

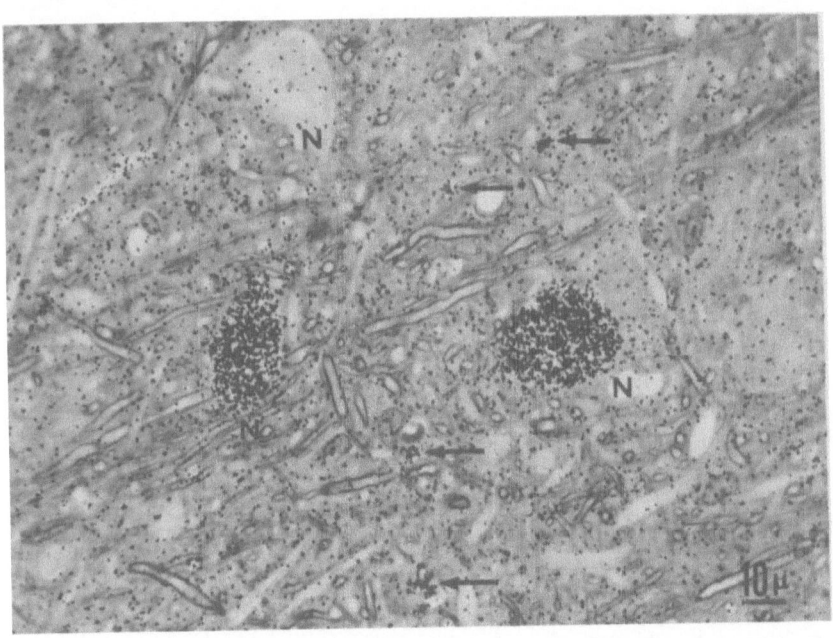

Fig. 12. Radioautographic localization is shown in Figs. 12 and 13 of labeled catecholamines in the central nervous system of a rat perfused for 3 hr in a lateral ventricle with 500 μC of DL-norepinephrine-^{3}H (specific activity, 8.8 C/mM). This 2-μ thick section of the substantia nigra shows two heavily labeled neurons (N) and another unlabeled one at the top. Clumps of silver grains (arrows) are seen in the neuropil and probably correspond to nerve endings. (From Descarries and Droz.[38])

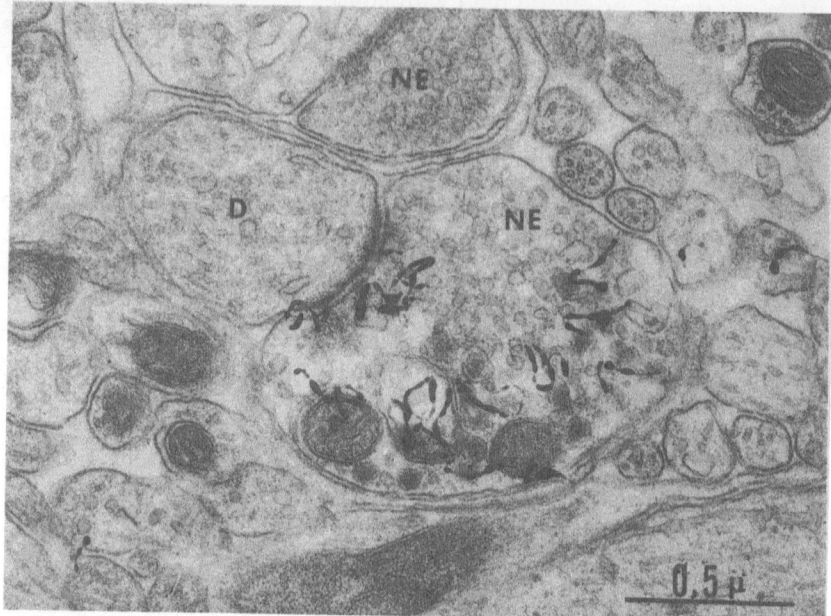

Fig. 13. Electron microscope radioautograph of the locus coeruleus. A presynaptic nerve ending (NE) visible in the middle of the picture has accumulated the label, whereas another one, (at the top) and a dendrite (D) are unlabeled. (Material prepared by L. Descarries.)

In the central nervous system the blood–brain barrier precludes the entry of biogenic amines. The intraventricular route was used by Glowinski[35] to study the gross distribution of labeled catecholamines in brain by "contact" radioautography. The use of 1-μ thick sections or of the electron microscope allowed a better resolution of the radioautographic localization of catecholamines or indoxylamines in a few neurons or nerve endings of the hypothalamus[36,37] caudate nucleus, substantia nigra, and locus coeruleus (Figs. 12 and 13).[38]

III. WHAT RADIOAUTOGRAPHY ADDS TO NEUROCHEMISTRY

The nervous system is composed of a great variety of cells—neurons, astrocytes, oligodendrocytes, and microgliocytes—which differ from each other in morphology, metabolism, and function. Moreover, one part of a cell also may differ from another one, e.g., the ultrastructure and the synthetic capacity of the axon is different from that of the perikaryon. By biochemical means one can isolate, from homogenates of a region of the brain, crude fractions of subcellular particles that can be purified and chemically analyzed. Owing to the structural complexity of the central and peripheral nervous system, it is extremely difficult for neurochemists to relate a chemically defined labeled compound to a structurally defined cell type, except by radioautography.

Kinetic studies involving the use of radioautography as a means of detection of a label indeed may supplement neurochemical data.

A. Sites of Incorporation

Very soon after the administration of a labeled substance, the first structures, which display a radioautographic reaction, may be considered to be the sites of incorporation of the label (Figs. 2 and 6), once an artefactual retention has been ruled out. These early labeled structures, however, may remain radioactive for a relatively long time after the injection, when the rate of disappearance of the label from the precursor pool is low or when the label is recycled from the product to the precursor.

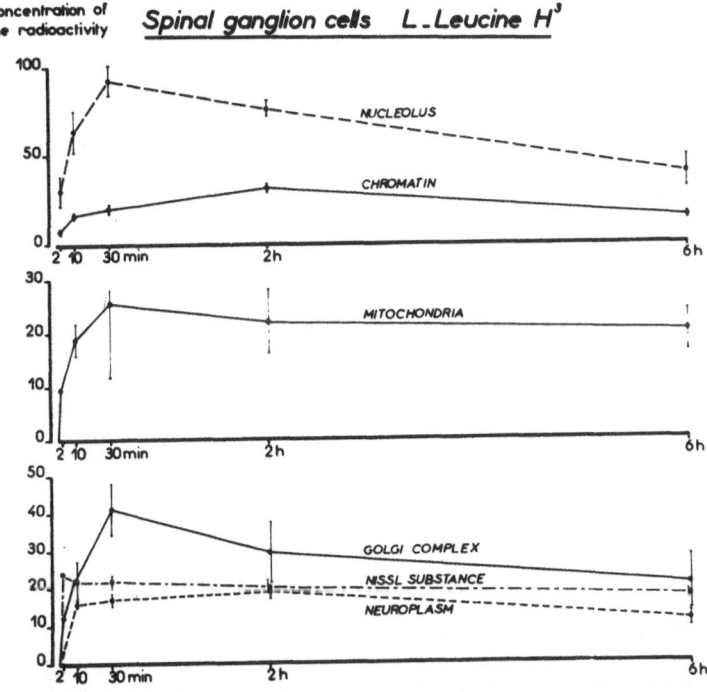

Fig. 14. Time curves of the concentration of the label in various cell compartments of spinal ganglion cells in 250-g female rats. The nucleus shows a first peak of radioactivity in the nucleolus and a second one in the chromatin. In the perikaryon, the mitochondria display an early concentration of the label, which persists at a high level. The radioactivity reaches a first peak in the Nissl substance and, later, other maximal values in the neuroplasm and in the Golgi complex. This fact indicates that proteins elaborated in the Nissl substance are eventually transferred to the neuroplasm and to the Golgi complex. (From Droz.[9])

B. Pathways of Migration

Subsequent appearance of the label in structures that show little or no radioactivity immediately after the injection may be due to a migration of newly synthesized products (Figs. 3, 4, 5, and 9). However, it is often difficult to decide whether the label appearing in a given cell structure originates from a local and slow incorporation or, on the contrary, has migrated by fast transfer from another structure. This problem often may be solved by the analysis of the curves of radioactivity concentration that may indicate successive peaks in a series of cell structures (Fig. 14).

C. Velocity of Migration

The velocity at which the label moves in tiny structures may be directly estimated by radioautographs taken at successive time intervals. For instance, migration of labeled protein could be both visualized and measured in

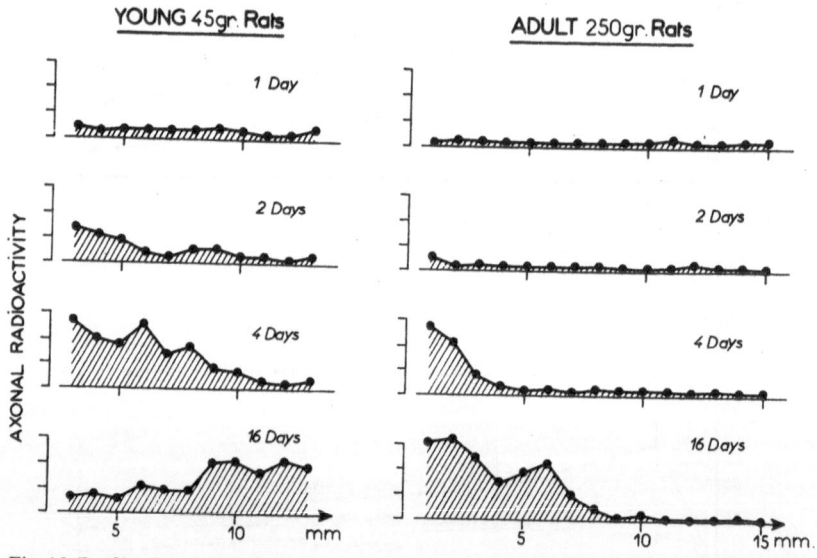

Fig. 15. Profile of the radioactivity recorded along axons of the hypoglossal nerves in young 45-g rats and adult 250-g female rats given multiple injections of leucine-³H (every 3 hr for a 24-hr period) and sacrificed 1, 2, 4, and 16 days after the first injection of the series. Radioautographs of hypoglossal nerve cross sections were prepared at every millimeter, starting from the hypoglossal canal, taken as origin. Grain counts were made over individual axons. The average number of silver grains per axon were plotted against the distance in millimeters from the hypoglossal canal. By 1 day, a uniformly low number of silver grains was counted over the axons. Then, after 2–4 days, the label appeared in proximal regions of the axons. Finally, at 16 days, the radioactivity extended to more distal regions. Note that the velocity of migration is faster in young (2.2 mm/day) than in adult animals (0.6 mm/day).

Fig. 16. Decay curves of radioactive proteins in three types of nerve cell body. Top: Labeled proteins in Purkinje cells of the cerebellum after leucine-^{3}H injection to rats. Middle: Labeled proteins in pyramidal cells of the cornu Ammonis in the hippocampus after leucine-^{3}H injection to rats. Bottom: Labeled proteins in ganglionic cells of the semilunar ganglion of the trigeminal nerve after injection of arginine-^{3}H to mice. The radioactivity concentration is expressed as the number of silver grains per unit area (1). To determine the grain concentration, the contours of the nerve cell body are outlined with a pencil on paper by means of a camera lucida; the silver grains are drawn in and counted. The outlined nerve cell body is then cut out and weighed to measure its surface. After subtraction of the background, the grain concentration (x) is plotted on a semilog paper against time after the injection (t). The right part of the curve has a straight line portion, the slope of which corresponds to the rate $(-k_2)$ at which one class of proteins turns over,

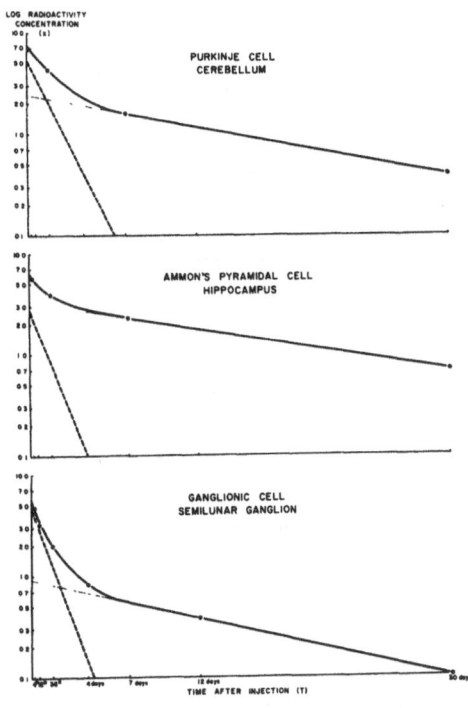

i.e., 5.8–7.2% day. By extrapolating back to time 0, this line in turn may be subtracted from the corresponding portion of the original curve. The difference yields a straight line of steep slope $(-k_1)$, indicating that 71–111% of the second class of protein turn over in 1 day in the perikaryon. Thus, at least two classes of protein may be distinguished: a rapidly turning over one of rate $-k_1$ and a slowly turning over one of rate $-k_2$, such that $x = x_1 e^{-k_1 t} + x_2 e^{-k_2 t}$ in which x_1 and x_2 may be graphically determined by extrapolating to 0 time. Since $-k_1$ and $-k_2$ measure the rate of decay of the two components, their turnover times are, respectively, $1/k_1$ and $1/k_2$. The rapidly turning over proteins, which leave the cell body in about 1 day, are referred to as "migratory" proteins, and the slowly turning over proteins that sojourn about 2 weeks in the perikaryon are called "sedentary" proteins. (From Droz and Leblond.[17])

various axons (Fig. 15) and in outer segments of photoreceptors, as distinct from the radioactivity of the surrounding structures.[15,17,39,44]

D. Intracellular Turnover Rate

The turnover rate of a substance such as protein may be determined in various types of nerve cells[15,17] or photoreceptors[40] by means of decay curves (Fig. 16). Since several assumptions are required to analyze the kinetics of the label, these measurements cannot provide an accurate estimation but rather an order of magnitude of the turnover rate.

E. Functional Activity

It could be hoped that radioautography would help to establish a relationship between nerve cell activity and protein or RNA metabolism. After administration of various drugs such as reserpine, diallylbarbituric acid, insulin,[41] and chloropromazine,[42] the incorporation of methionine-^{35}S was considerably depressed in most neurons of the brain. Recently an increased protein metabolism was observed in most of the nerve cell bodies in the brain after submitting rats to a stress reaction,[43] but this effect was reduced in animals preadapted to the experimental conditions. In fact, a modification of the incorporation of a labeled precursor into nerve cells in various functional states may be due to a defect of penetration of the free precursor or an alteration of the cellular respiration as well as to a regulation of the mechanism of protein synthesis.

In conclusion, radioautography is one of the most productive tools to visualize the transport of labeled product through various structures of the nervous system. Once a substance responsible for a radioautographic reaction has been chemically identified, radioautography offers neurochemists the possibility of tracing, in complex biological structures the life history of molecules, from their sites of synthesis to their sites of degradation.

IV. REFERENCES

1. J. Gross, R. Bogoroch, N. J. Nadler, and C. P. Leblond, The theory and methods of the radioautographic localization of radioelements in tissues, *Am. J. Roentgenol. Radium Therapy* **65**:420–458 (1951).
2. A. W. Rogers, *Techniques of Autoradiography*, Elsevier, Amsterdam (1967).
3. B. Droz and H. Warshawsky, Reliability of the radioautographic technique for the detection of newly-synthesized protein, *J. Histochem. Cytochem.* **11**:426–435 (1963).
4. M. Bergeron and B. Droz, Analyse critique des conditions de fixation et de préparation des tissus pour la détection radioautographique des protéines néoformées, en microscopie électronique, *J. Microscopie* **7**:201–228 (1968).
5. Th. Peters and C. A. Ashley, An artefact in radioautography due to binding of free amino acids to tissue by fixatives, *J. Cell Biol.* **33**:53–60 (1967).
6. B. Kopriwa and C. P. Leblond, Improvements in the coating technique of radioautography, *J. Histochem. Cytochem.* **10**:269–284 (1962).
7. P. Granboulan, Comparison of emulsions and techniques in electron microscope radioautography, *in The Use of Radioautography in Investigating Protein Synthesis* (C. P. Leblond and K. B. Warren, eds.), Vol. 4, pp. 43–63, Academic Press, New York (1965).
8. M. M. Salpeter and L. Bachmann, Assessment of technical steps in electron microscope autoradiography, *in The Use of Radioautography in Investigating Protein Synthesis* (C. P. Leblond and K. B. Warren, eds.), Vol. 4, pp. 23–41, Academic Press, New York (1965).
9. B. Droz, Synthèse et transfert des proteines cellulaires dans les neurones ganglionnaires; étude radioautographique quantitative en microscopie électronique, *J. Microscopie* **6**:201–228 (1967).
10. I. Smart and C. P. Leblond, Evidence for division and transformations of neuroglial cells in the mouse brain as derived from radioautography after injection of thymidine-^{3}H, *J. Comp. Neurol.* **116**:349–368 (1961).

11. O. R. Hommes and C. P. Leblond, Mitotic division of neuroglia in the normal adult rat, *J. Comp. Neurol.* **129**:269–278 (1967).

12. J. Altman, DNA metabolism and cell proliferation, in *Handbook of Neurochemistry* (A. Lajtha, ed.), Vol. II, pp. 137–182, Plenum Press, New York (1969).

13. C. P. Leblond and M. Amano, Synthetic activity in the nucleolus as compared to that in the rest of the cell, *J. Histochem. Cytochem.* **10**:162–174 (1962).

14. H. Koenig, An autoradiographic study of nucleic acid and protein turnover in the mammalian neuraxis, *J. Biophys. Biochem. Cytol.* **4**:785–792 (1958).

15. B. Droz, Fate of newly-synthesized proteins in neurons, in *The Use of Radioautography in Investigating Protein Synthesis* (C. P. Leblond and K. B. Warren, eds.), Vol. 4, pp. 159–175, Academic Press, New York (1965).

16. B. Droz, Protein metabolism in nerve cells, in *International Review of Cytology* (G. H. Bourne and J. F. Danielli, eds.), Vol. 25, pp. 363–390, Academic Press, New York (1969).

17. B. Droz and C. P. Leblond, Axonal migration of proteins in the central nervous system and peripheral nerves as shown by radioautography, *J. Comp. Neurol.* **121**:325–346 (1963).

18. E. Koenig, Nucleic acid and protein metabolism of the axon, in *Handbook of Neurochemistry* (A. Lajtha, ed.), Vol. II, pp. 423–434, Plenum Press, New York (1969).

19. E. A. Barnard and K. Ostrowski, Autoradiographic methods in enzyme cytochemistry. II. Studies on some properties of acetylcholinesterase in its sites at the motor end-plate, *Exptl. Cell Res.* **36**:28–42 (1964).

20. M. M. Salpeter, Electron microscope radioautography as a quantitative tool in enzyme cytochemistry. I. The distribution of acetylcholinesterase at motor end plates of a vertebrate twitch muscle, *J. Cell Biol.* **32**:379–389 (1967).

21. L. S. Wolfe, The distribution of gangliosides in subcellular fractions of guinea-pig cerebral cortex, *Biochem. J.* **79**:348–355 (1961).

22. E. G. Brunngraber and B. D. Brown, Fractionation of brain macromolecules. II. Isolation of protein-linked sialomucopolysaccharides from subcellular, particulate fractions from rat brain, *J. Neurochem.* **11**:449–459 (1964).

23. B. Droz, L'appareil de Golgi comme site d'incorporation du galactose-^{3}H dans les neurones ganglionnaires spinaux chez le rat, *J. Microscopie* **6**:419–424 (1967).

24. A. Rambourg, Détection de glycoproteines en microscopie électronique: coloration de la surface cellulaire et de l'appareil de Golgi par un mélange acide chromique-phosphotungstique, *Compt. Rend. Acad. Sci. (Paris)* **268**:1426–1428 (1967).

25. L. Olivier, *Aspects histologiques de la fixation du soufre dans l'organisme*, Thèse de Doctorat en Médecine, Faculté de Médecine, Paris (1956).

26. F. Chevallier and L. Petit, Incorporation of cholesterol into the central nervous system and its autoradiographic localization, *Exptl. Neurol.* **10**:250–254 (1966).

27. F. Chevallier, C. Gautheron, and L. Petit, unpublished.

28. L. E. Hokin, Autoradiographic localization of the acetylcholine-stimulated synthesis of phosphatidylinositol in the superior cervical ganglion, *Proc. Natl. Acad. Sci.* **53**:1369–1376 (1965).

29. E. D. Wolfe, L. T. Potter, K. C. Richardson, and J. Axelrod, Localizing tritiated norepinephrine in sympathetic axons by electron microscopic autoradiography. *Science* **138**:440–442 (1962).

30. J. Taxi and B. Droz, Etude de l'incorporation de noradrénaline-^{3}H et de 5-hydroxytryptophane-^{3}H dans les fibres nerveuses du canal déférent et de l'intestin, *Compt. Rend. Acad. Sci. (Paris)* **263**:1237–1240 (1966).

31. J. Taxi and B. Droz, Etude de l'incorporation de noradrénaline-^{3}H et de 5-hydroxytryptophane dans l'épiphyse et le ganglion cervical supérieur, *Compt. Rend. Acad. Sci. (Paris)* **263**:1326–1329 (1966).

32. J. Taxi and B. Droz, Localisation d'amines biogènes dans le système neurovégétatif périphérique. Etude radioautographique en microscopie électronique après injection de noradrénaline-³H et de 5-hydroxytryptophane-³H, *in Neurosecrétion* (F. Stutinsky, ed.), pp. 191–202, Springer Verlag, Berlin (1967).

33. M. D. Gershon and L. L. Ross, Radioisotopic studies of the binding, exchange and distribution of 5-hydroxytryptamine synthesized from its radioactive precursor, *J. Physiol.* **186**:451–476 (1966).

34. M. D. Gershon and L. L. Ross, Location of sites of 5-hydroxytryptamine storage and metabolism by radioautography, *J. Physiol.* **186**:477–492 (1966).

35. M. Reivich and J. Glowinski, An autoradiographic study of the distribution of ¹⁴C-norepinephrine in the brain of the rat, *Brain* **90**:633–646 (1967).

36. G. K. Aghajanian and F. E. Bloom, Localization of tritiated serotonin in rat brain by electron microscopic autoradiography, *J. Pharmacol. Exptl. Therap.* **156**:23–30 (1967).

37. G. K. Aghajanian and F. E. Bloom, Electron microscopic localization of tritiated norepinephrine in rat brain: Effect of drugs, *J. Pharmacol. Exptl. Therap.* **156**:407–416 (1967).

38. L. Descarries and B. Droz, Incorporation de la noradrenaline-³H dans le système nerveux central du Rat: étude radioautographique en microscopie électronique, *Compt. Rend. Acad. Sci. (Paris)* **266**:2480–2482 (1968).

39. R. W. Young, The renewal of photoreceptor cell outer segments, *J. Cell Biol.* **33**:61–72 (1967).

40. B. Droz, Dynamic condition of proteins in the visual cells of rats and mice as shown by radioautography with labeled amino acids, *Anat. Rec.* **145**:157–167 (1963).

41. S. Flanigan, E. R. Gabrielli, and P. D. McLean, Cerebral changes revealed by radio-autography with ³⁵S labeled-L-methionine, *Arch. Neurol. Psychiat.* **77**:588–594 (1957).

42. J. Verne, B. Droz, L. Olivier, and A. Rambourg, Remarques sur l'action de la chlorpromazine sur la synthèse des protéines dans le tissu nerveux, *Ann. Histochim.* **7**:227–231 (1962).

43. J. Altman, Behavioral influences on the utilization of ³H-leucine by the brain, *in Protides of the Biological Fluids* (H. Peeters, ed.), Vol. 13, pp. 127–136, Elsevier, Amsterdam (1966).

44. R. W. Young and B. Droz, The renewal of protein in retinal rods and cones, *J. Cell Biol.* **39**:159–184 (1968).

45. B. Droz, H. L. Koenig, and R. W. Young, Radioautographie en microscopie électronique et renouvellement des protéines axonates dans les terminaisons motrices et sensitives, *in Electron Microscopy* (D. S. Bocciarelli, ed.), Vol. II, pp. 523–524, Tipografia Poliglotta, Vaticana, Rome (1968).

Chapter 22

ENZYME HISTOCHEMISTRY: APPLICATIONS AND PITFALLS

C. W. M. Adams

Sir William Dunn Professor of Pathology
Guy's Hospital Medical School
London University
and
Honorary Consultant Morbid Anatomist
Guy's Hospital
London, Great Britain

I. INTRODUCTION

The purpose of this brief review is to interpret technique rather than discuss particular biological or pathological problems. As two recent books have been concerned with the results of enzyme histochemistry in neurobiology and neuropathology,[1,2] it seems pointless and premature again to summarize the considerable body of evidence that has accumulated in these applied fields. It is now more appropriate briefly to consider what is the value of histoenzymic methods and how reliable are the commonly used histochemical enzyme techniques in research studies on the nervous system.

II. VALUE OF METHODS

The criticism is frequently leveled at "slide" histochemists that their techniques are nonquantitative and of little value in a biochemical sense. Nevertheless, the kinetic methods introduced by Glenner[3] and his associates do allow approximate quantitation to be made on microscopic preparations. Furthermore, elution methods allow certain histochemical enzyme reaction products to be extracted from the section and estimated by spectrophotometric methods. Such methods are primarily conducted on histological sections and are clearly unrelated to the quantitative histochemical and cytochemical techniques introduced by Linderstrøm-Lang[4] and applied extensively in neurochemistry by Lowry, Pope, and their associates.

The main use of "slide" histochemistry is undoubtedly for localizing enzyme activity in certain cells or tissues. Over the last decade finer localization has been obtained by adapting certain histochemical methods for electron microscopy (see the review by Scarpelli and Kanczak[5]), but a major problem here is the complex compromise between adequately fixing the tissue to preserve some ultrastructure and at the same time not inhibiting all enzyme activity.

III. RELIABILITY OF METHODS

Having briefly outlined the applicability of enzyme histochemistry, it is now necessary to consider the reliability of the methods. There are essentially five main technical groups of histochemical enzyme methods:

1. The tetrazolium techniques, where the reduced product (formazan) is precipitated in the tissues at the site of reductase activity.
2. The metallic methods, where the enzyme product combines with a metal to form an insoluble salt.
3. The naphthol azo-coupling techniques, where naphthol or a derivative is liberated by enzymic action and is then visualized by coupling with a diazonium salt.
4. The autogram methods, where a film of native substrate is presented to the section and enzymic activity is subsequently demonstrated as a negative image after staining the residual substrate.
5. Miscellaneous techniques.

A. Tetrazolium Techniques

These are used for demonstrating various dehydrogenases, $NADH_2$ and $NADPH_2$ reductases, and monoamine oxidase (MAO). Nitroblue tetrazolium and its tetra derivative are now the most widely used electron acceptors in these reactions:

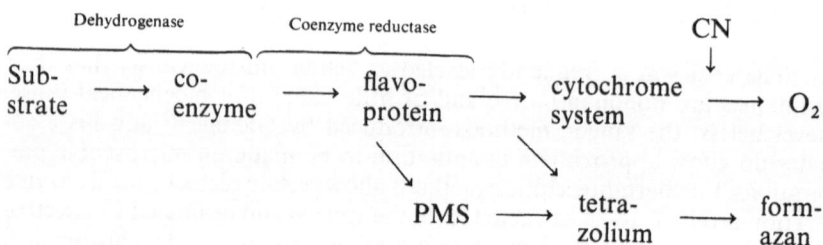

Incubation is conducted in the presence of cyanide in order to deviate electron transfer away from the cytochrome system toward the tetrazolium salt. In the case of MAO, however, tetrazolium is reduced nonenzymically by the formed aldehyde,[3] so cyanide is not required here.

There are three major difficulties with this group of methods: (1) tissue $NADH_2$-tetrazolium reductase must be present, otherwise specific dehydro-

genase activity cannot be demonstrated; (2) many of the specific dehydrogenases are highly soluble; and (3) a slight "blank" reaction ("nothing
dehydrogenase" activity) is frequently observed.

The first of these difficulties can be circumvented by using phenazine
methosulfate (PMS) as an intermediary electron acceptor between the reduced
coenzyme-flavoprotein complex and the tetrazolium salt. It is essential,
however, that cyanide is added under these conditions because PMS also
appears to stimulate the cytochrome system.[6,7]

The problem of dehydrogenase solubility is very serious in histochemistry. Certain dehydrogenases are leached out of the tissue section and
then reduce NAD or NADP in the incubating medium. Thus, substrate
($NADH_2$ or $NADPH_2$) is formed for the coenzymes reductases that remain
in the tissue (Fig. 1), with the result that the specific dehydrogenase is falsely
localized (coenzyme reductase artefact).[8,9] When PMS is added to the
incubating medium, the position is as bad because reduced tetrazolium

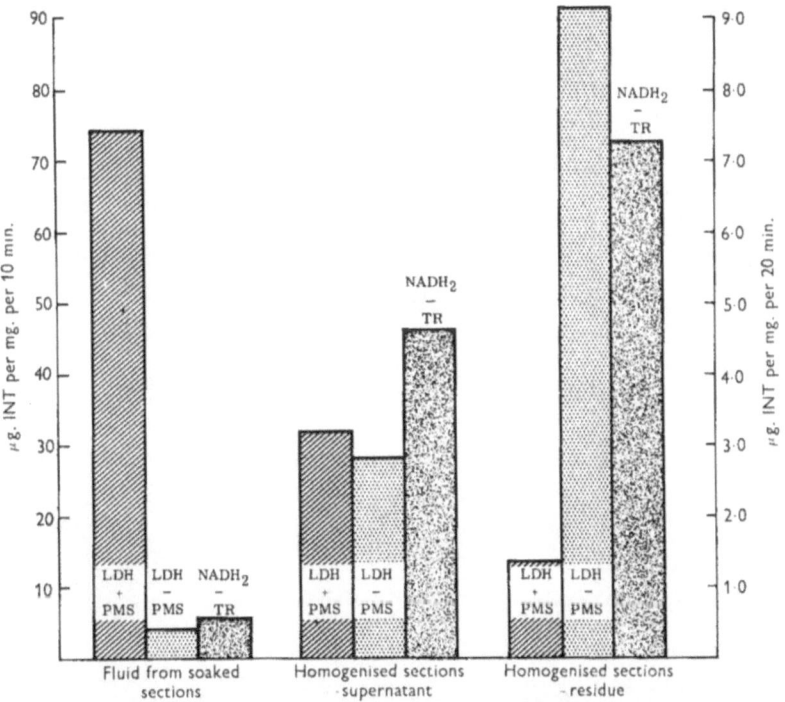

Fig. 1. Leakage of lactic dehydrogenase (LDH + PMS) from dummy-incubated, unfixed,
cryostat sections of human aorta. Conversely, note essential preservation of $NADH_2$-
tetrazolium reductase ($NADH_2$-TR) within the sections. In the absence of phenazine
methosulfate, residual LDH activity (LDH-PMS) is of the same order as that of
$NADH_2$-TR. LDH + PMS calibrated on left, other enzymes on right. Reproduced from
Saudek et al.[17] by courtesy of the editor of J. Pathol. Bacteriol.

TABLE I

Rat Brain Enzyme Activity with Different Histochemical Systems

Enzyme	Incubation method			
	"Standard" procedure[a]	With PMS[b]	With PMS on gel film[c]	Diffusion into PMS medium[b]
NADH$_2$- tetrazolium reductase	+ + +		+[d]	
NADPH$_2$- tetrazolium reductase	+ +		+[d]	
Succinic dehydrogenase	+ +	+ + +	±	0
Lactic dehydrogenase (NAD)	+ +	+ + +	+ + +	+
α-Glycerophosphate dehydrogenase (NAD)	±	+	±	+
Glucose-6-phosphate dehydrogenase (NADP)	+	−	+	+
6-Phosphogluconic dehydrogenase (NADP)	±	−	±	+

[a] Cryostat sections incubated with specific substrate, coenzyme (where relevant), cyanide, and nitroblue tetrazolium; sections post-rinsed with acetone.
[b] The same as for a with added phenazine methosulfate; sections pre- and post-rinsed with acetone.
[c] The same as for b but incubating medium enclosed in a gel film.[11]
[d] Gel film but no PMS.

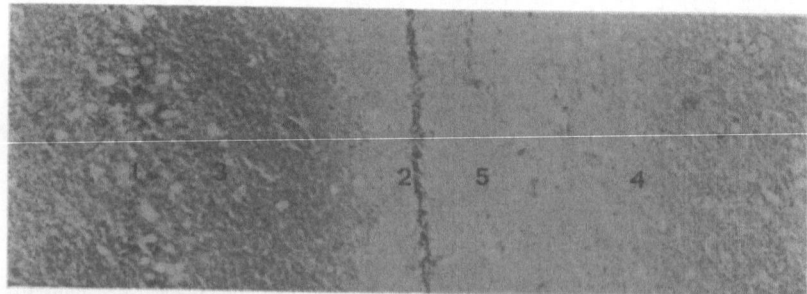

Fig. 2. Standard histochemical reaction for lactic dehydrogenase (LDH). Note strong reaction in pyramidal cells of hippocampus (1) and ependymal cells of lateral ventricle (2) and slight reaction in hippocampal (3) and cortical (4) neuropil. No reaction occurred in myelinated fibers, but moderate reaction did occur in oligodendrocytes of corpus callosum (5). Rat brain, lactate-cyanide-nitroblue tetrazolium, post-rinsed with acetone. × 236.

(formazan) is formed in the incubating medium and is avidly absorbed by lipid droplets within the tissue section.[10] Unfortunately, such lipid droplets are frequently small and have been mistaken for mitochondria. This last problem can be circumvented by extracting the tissue with a preliminary rinse with acetone in order to remove such hydrophobic globular lipid.[10] In order to reduce diffusion of soluble dehydrogenases, many workers have recommended that plasma expanders should be added to the incubating medium, but better results seem to be obtained by incubating the section with substrate embedded in a gelatin film.[11]

Fixation with various aldehydes has been used by many workers[12–15] to prevent enzyme diffusion, but such treatment causes differential inactivation of enzymes and other effects that have been discussed by Glenner.[3]

Succinic dehydrogenase,[15A] NADH$_2$-tetrazolium reductase[16] (Fig. 1), and NADPH$_2$-tetrazolium reductase appear to be almost insoluble, because they are bound to the mitochondrion or endoplasmic reticulum. Conversely, cytoplasmic enzymes—such as lactic (Fig. 1), 6-phosphogluconic and glucose-6-phosphate dehydrogenases—are highly soluble.[11,16,17] On the other hand, it is claimed that various soluble dehydrogenase activities can be detected histochemically in different specific sites in the nervous system[18–19A] and that such enzymes become less soluble during brain development.[20] It can be argued that such differences mean that a substantial amount of "soluble" dehydrogenase activity is retained in CNS cells, otherwise equal activity (NADH$_2$ reductase artefact) would be encountered in all cells endowed with coenzyme reductases. However, taking the converse view, Glenner[3] questioned whether localization of a soluble enzyme is particularly meaningful when a large part of its activity already has been leached out.

Clearly the present somewhat vague attitude to dehydrogenase solubility in histochemical brain preparations is thoroughly unsatisfactory. It is highly desirable that histochemists should either measure the extent of solubility during "dummy" incubation,[15A–17] or should observe the amount of formazan that is formed in the incubating medium in the presence of PMS.[8,20]

In a recent study[20A] lactic dehydrogenase (LDH), glucose-6-phosphate dehydrogenase (G6PD), and 6-phosphogluconic dehydrogenase (6PGD) all were found to diffuse out of sections of unfixed adult rat brain that had been incubated in the presence of PMS, as shown by the formation of formazan in the incubating medium. Little or no succinic dehydrogenase (SDH) and α-glycerophosphate dehydrogenase (αGPD) diffused out of the tissue under these conditions. The histochemical results are summarized in Table I. These results confirm that brain G6PD and 6PGD are almost entirely soluble (i.e., the PMS slides were essentially negative). These enzymes are not even partially retained unless the tissue section is incubated on a gelatin film; their apparent reaction in the "standard" histochemical procedure is an example of the above-mentioned "coenzyme reductase artefact." Conversely, SDH was relatively insoluble, as shown by the absence of formazan from the

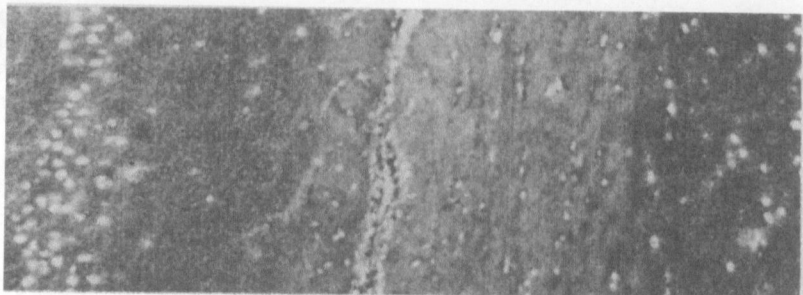

Fig. 3. Phenazine methosulfate (PMS) reaction of LDH. Note less reaction in hippocampal pyramidal (1) and ependymal (2) cells. Stronger reaction in neuropil (3, 4). No change in myelin and oligodendrocytes of corpus callosum (5). Rat brain, lactate-PMS-cyanide-nitroblue tetrazolium, pre- and post-rinsed with acetone. × 236.

incubating medium; its weak reaction when incubated with PMS on gelatin indicates that it is so tightly bound within the cell that it is unable to approach the external film of substrate. α-Glycerophosphate dehydrogenase was intermediate in these respects between SDH and the pentose shunt dehydrogenases. This feature would be consistent with a partly particulate and partly cytoplasmic intracellular distribution of this enzyme.

Rather more LDH activity was apparent in the brain with the PMS method than with the "standard" incubation procedure (Figs. 2 and 3). Moreover, such activity was differently distributed than in slides subjected to the "standard" technique. More activity appeared in the "neuropil" (axonal, dendritic, and glial processes) and less in neuronal perikarya and ependymal cells (cf. Figs. 2 and 3). In the absence of preliminary acetone extraction some formazan appeared to be nonspecifically absorbed by myelin lipids with the PMS–LDH technique (Fig. 4); such artefactual absorption was minimized by preliminary acetone extraction (Fig. 3). As the distribution of "LDH activity" with the "standard" technique (Fig. 2) was virtually

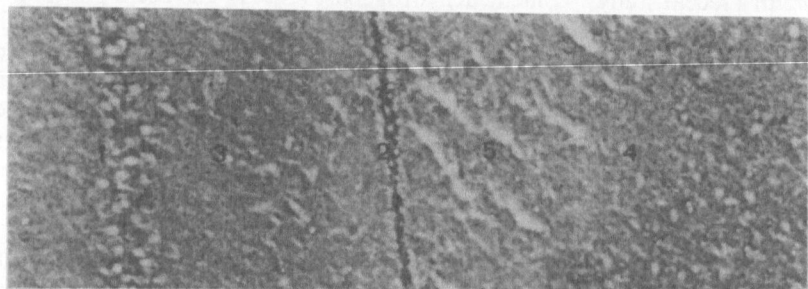

Fig. 4. Same technique and tissue as Fig. 3, but sections were not prerinsed with acetone. Note artefactual staining of myelin (5).

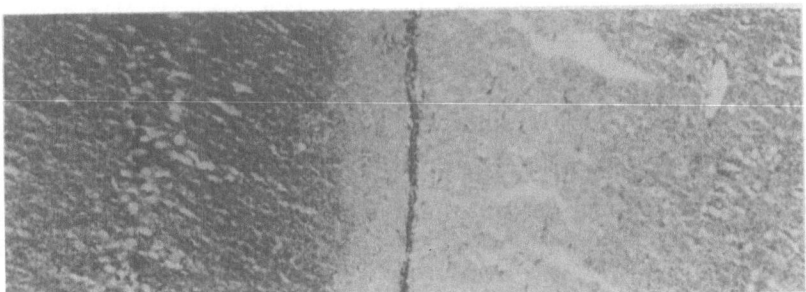

Fig. 5. NADH$_2$-tetrazolium reductase. Note similar distribution to that of LDH in Fig. 2. NADH$_2$-cyanide-nitroblue tetrazolium, postrinsed with acetone. × 236.

identical to that of NADH$_2$-tetrazolium reductase (Fig. 5), the results with the "standard" method can reasonably be attributed to the above-mentioned "coenzyme reductase artefact." On the other hand, activity revealed with PMS in acetone-extracted sections appears to localize LDH that has not been leached out of the section during incubation. However, these results do not show how much and what isozymic fraction of LDH is retained in the section[3]: quantitative histochemical and zymogram studies in the future may clarify this problem.

The blank histochemical reaction with tetrazolium salts ("nothing dehydrogenase" activity[21]) no longer causes much practical confusion, because such activity is slight when the incubating pH is kept around 7.2 and only small amounts of coenzyme are added to the medium (see Adams[1] for references). The cause of this blank reaction remains obscure.

B. Metallic Techniques

These techniques depend on the precipitation of an insoluble metallic salt formed between the enzymic reaction product and a suitable metal ion. Typical examples are Gomori's[22] acid and alkaline phosphatase methods, Wachstein and Meisel's[23] ATPase method, Koelle and Friedenwald's[24] cholinesterase methods, and Gomori's[22] Tween lipase methods. With the first three techniques the precipitated salt is lead phosphate, whereas in the last two examples it is, respectively, copper thiocholine and a lead soap. These metallic salts are suitably electron-dense for ultrastructural localization but, for light microscopy, they require to be converted to brown–black metallic sulfides by treatment with dilute ammonium sulfide.

$$
\begin{array}{ccccc}
\text{AB} & + & \text{M}^{2+} & \longrightarrow & \text{A} + \quad \text{MB} \\
\text{Substrate} & & \text{Metal ion} & & \qquad\quad \text{Insoluble} \\
& & & & \qquad\quad \text{product}
\end{array}
$$

$$
\text{MB} + (\text{NH}_4)_2\text{S} \longrightarrow \quad \text{MS} \quad + (\text{NH}_4)_2\text{B}
$$
$$
\text{Black sulfide}
$$

A number of problems arise with these metallic methods: (1) nonspecific metal binding, (2) wide substrate specificity, and (3) tissue penetration.

Nonspecific nuclear binding of Pb^{2+} has been noted with both Wachstein and Meisel's ATPase[25] and Gomori's alkaline phosphatase[26] methods; myelin also has a moderate affinity for metallic ions. Preexisting tissue Ca^{2+} also may cause difficulty in that it undergoes base exchange with Pb^{2+} and, hence, gives a false positive reaction. This last problem can be circumvented by preliminary treatment of the tissue with EDTA, while nonspecific metallic binding in general can be recognized by examining sections incubated in medium without substrate (i.e., the "blank" control).

Substrate specificity is not easy to establish with some of these metallic methods. Gomori's[22] acid phosphatase technique with β-glycerophosphate seems to demonstrate only a fraction of the molecular species or isoenzymes that can be displayed in brain with naphtholic phosphate substrates[27,28] (see Section III,C). It is thought that three or four different types of phosphatase activity are demonstrated with ATP as substrate,[29, 30] but opinions differ as to whether or not the Pb^{2+} in the incubating medium inactivates mitochondrial ATPase.[29,31,32] Moses and Rosenthal[33] consider that at least part of the apparent location of ATPase in cell membranes is due to lead-catalyzed hydrolysis of ATP, but Novikoff[34] denies this view.

McClurkin[35] claims that "sodium-pump" ATPase can be demonstrated by the different activity shown before and after adding Na^+ to the incubating medium. However, Torack[36] pointed out that "sodium-pump" ATPase is inactivated by Pb^{2+} and cannot be demonstrated with the Wachstein–Meisel substrate. Moreover, McClurkin's enzyme does not exhibit the correct inhibition–activation characteristics for the "pump" enzyme.[37]

The specificity of brain 5′-nucleotidase activity in a histochemical system has been carefully worked out by Scott.[37A] The enzyme hydrolyzes a wide range of mononucleotides but does not split dinucleotides, trinucleotides, or ribose-5-monophosphate. As the hydrolysis of various monophosphate esters is differentially inhibited by certain metal ions, Scott concluded that brain 5′-nucleotidase activity may be due to a range of isoenzymes.

Penetration of substrate presents a considerable problem when teased nerve preparations are stained by Koelle and Friedenwald's[24] copper thiocholine method or Karnovsky and Roots'[38] modification with ferricyanide. Axons are only stained at their cut ends,[39] presumably because Cu^{2+} and ferricyanide ions fail to penetrate the perineurium and myelin.[40,41]

Differential penetration into the tissue section is another hazard that Glenner[3] has fully discussed. He emphasized that the penetration rate of an inhibitor or capture (precipitating) reagent towards the enzyme center must be as fast as that of the substrate. If the sequence of arrival at the enzyme center is not synchronous, then artefactual results theoretically could be obtained. Thus, if the substrate penetrates to the enzyme site before the inhibitor, the latter would appear to fail in its action, while if the substrate penetrates faster than the capture reagent, the enzymic reaction product would diffuse.[3] We have only had experience with this problem in using cholinesterase inhibitors in conjunction with the Karnovsky–Roots ferricyanide–thiocholine method. In practice, just adding the inhibitor to the incubating medium produced the same results as preincubating the tissue section with inhibitor.[39]

C. Azo-Naphthol Techniques

With these methods a naphthol-bound substrate is hydrolyzed and the liberated naphthol is then coupled with a stabilized diazonium or tetrazonium salt:

$$N\phi \cdot R + R'N{=}N \cdot Cl \longrightarrow R + N\phi \cdot N{=}N \cdot R'$$

| Naphthol substrate | Diazonium salt | | Split product | Colored precipitated azo-naphthol dye |

These techniques are commonly used for acid phosphatase[42] (naphthyl α-phosphate, -ASTR phosphate, or -ASBl phosphate), aminopeptidase[43] (L-leucyl-β-naphthylamine), nonspecific esterases (α-naphthyl acetate, -AS acetate or -ASLC acetate), and glucuronidase[44] (naphthol-ASBl-β-D-glucosiduronic acid). Better localization is usually obtained with the various above-mentioned substituted naphthol compounds than with either α-naphthol or β-naphthol derivatives. A wide variety of these substrates have been synthesized and assessed for histochemical use by A. M. Seligman and his associates at Baltimore.

The major disadvantages with these methods are (1) inhibition of enzyme activity by the coupling reagent (2) nonspecific absorption of liberated naphthol or naphthylamine by lipids, (3) the unphysiological nature of the substrate.

With some azo-naphthol methods the coupling compound can be applied after incubation (postcoupling technique). However, with most of these methods it is necessary to add the coupling reagent to the incubating medium, because the naphtholic product diffuses and can be precipitated only by "simultaneous coupling" as soon as it is formed. Under these circumstances certain enzymes are inhibited by the coupling diazonium compound. For example, aminopeptidase in the CNS and peripheral nerve is inhibited by diazonium salts,[45] as is another amino-peptidase in rat kidney.[46]

Another problem with azo-naphthol techniques, particularly with the postcoupling sorts and those involving the use of α-naphthyl derivatives, is that the reaction product may be absorbed by lipids.[45,47,47A] This may cause particular difficulty with myelin, neuronal lipofuscin granules, and other lipid deposits in the nervous system.

The artificiality of naphthol and naphthylamine substrates may lead the reader to wonder just what type of enzyme activity is revealed by the substrates. Nevertheless, the use of inhibitors and activators does allow the characteristics of most of such enzyme activities to be accurately compared with those against more physiological substrates. In fact, the convenience of naphthol substrates, particularly in respect of their chromogenic properties, has led to their wide use in biochemical estimation of esterase (naphthyl esters) and aminopeptidase (aminoacid naphthylamides) activities.

D. Autogram Techniques

In order to overcome the above-mentioned artificiality of certain histo-chemical substrates, a number of techniques have been introduced in which tissue sections are incubated on or under films of native substrate (see review[48]). Such methods include techniques for DNAse, RNAse,[49] proteinase,[50,51] and fibrinolysin-activator.[52] The proteinase method[50] has permitted proteolytic activity at pH 5.6 and 7.4 to be localized in various parts of the normal and diseased nervous system.[53-56]

In general these methods suffer from the limitation that intracellular localization is quite impossible. Enzyme activity can be localized to groups of cells or, with luck, to solitary cells, but not to structures within cells. Enzymes demonstrated by these methods obviously must diffuse out of their cellular sites to react with the external substrate film. This basic requirement imposes the unavoidable limitations of imprecise cellular localization and inaccessibility of enzyme that is tightly bound to intracellular particles. Nevertheless, these methods are histologically useful, even though they completely fail at the cytological level.

E. Miscellaneous Techniques

The histochemical technique for cytochrome oxidase depends on the oxidation of an aromatic amine (e.g., p phenylene-diamine) by oxidized cytochrome c. The formed semiquinone is then allowed to react either with excess amine to form an azine dye or with an added α-naphthol derivative to form an indoaniline dye.[57] These reaction products are lipid soluble, so the possibility of artefactual localization to fine lipid droplets must be excluded.

In Holt's esterase method,[58] indoxyl is released from 5-bromo-indoxyl acetate; on oxidation with potassium ferricyanide two molecules of indoxyl are condensed to form dark-blue 5,5'-dibromoindigo. Unfortunately, this reaction product is also slightly lipid soluble.[59] For reasons that are obscure, we have noticed that esterolytic activity against this substrate proceeds more slowly than against esters of α-naphthol and its derivatives.

Apart from the azo-naphthol method (see Section III,C), β-glucuroni-dase activity also has been demonstrated by hydrolysis of 8-hydroxyquino-line-β-glucuronide. Liberated 8-hydroxyquinoline is chelated with Fe^{3+}, which is then converted to ferric ferrocyanide (Prussian blue) by treatment with potassium ferrocyanide. However, doubt has been expressed about the enzymic nature of the reaction in that Fe^{3+} is nonspecifically bound by tissues[60] (viz., by acid mucopolysaccharides). However, the validity of the method has been strongly defended by Fishman and co-workers.[61]

IV. CONCLUSIONS

In this brief and superficial account of enzyme histochemistry, attention has been directed toward the validity and limitations of a number of commonly used histoenzymic methods. Many techniques have not been discussed, because either they are not relevant to neural tissue or their physiological significance is not fully established. Moreover, this review is intended to illustrate rather than exhaust the topic. More detailed information about the validity of various histoenzymic techniques may be obtained from the reviews cited above. [3,9]

The specificity of enzymes demonstrated by histochemical methods is often hard to establish, but progress is being made by using inhibitors and activators. The biochemist can separate enzymes by their solubility, electrophoretic and other physical properties, but the histochemist is unable to carry out such preliminary purifications. Moreover, as stated by Glenner, [3] histochemical enzyme systems are more complex than most biochemists would care to work with and certainly more complex than most histochemists realize. The solubility of coenzymes and other factors, differential solubility of isoenzymic components, and cellular disruption at the cut surfaces of tissue sections constitute formidable hazards when working with histoenzymic methods.

A problem common to both histochemical and biochemical enzyme studies is the question of the physiological significance of the results. Study of a particular enzyme in a metabolic pathway does not necessarily indicate the overall activity of the pathway, because some other enzyme may act as a rate-limiting metabolic bottleneck. It follows that histochemical study of a particular enzyme or group of enzymes may not reveal the physiological activity of a pathway unless the rate-limiting enzyme is identified and demonstrated. As a counsel of perfection it would be desirable to use histochemical methods that display the end product of a metabolic route. So far this ideal has only been realized with Takeuchi's amylophosphorylase[62] and UDPG-glycogen synthetase[63] methods, where the *synthesized* polysaccharide is displayed in the tissue section by means of iodine.

In spite of the many reservations expressed in this chapter, histoenzymic methods are clearly of great value in localizing enzyme activity. Histochemical methods, when used in conjunction with quantitative biochemical methods and electron microscopy, yield far more information than can be obtained with each discipline in isolation.

V. ACKNOWLEDGMENTS

I am indebted to Mrs. O. B. High and Mr. R. S. Morgan for preparation and photography of the histochemical preparations.

VI. REFERENCES

1. C. W. M. Adams, *in Neurohistochemistry* (C. W. M. Adams, ed.), pp. 253, 332, 403, 437, and 488, Elsevier, Amsterdam (1965).
2. R. L. Friede, *Topographic Brain Chemistry*, Academic Press, New York (1966).
3. G. G. Glenner, *in Neurohistochemistry* (C. W. M. Adams, ed.), pp. 109–160, Elsevier, Amsterdam (1965).
4. K. Linderstrøm-Lang, Distribution of enzymes in tissue and cells, *Harvey Lectures Ser.* **34**:214–245 (1939).
5. D. G. Scarpelli and N. M. Kanczak, Ultrastructural cytochemistry: principles, limitations and applications, *Intern. Rev. Exptl. Pathol.* **4**:55–126 (1965).
6. I. A. Brody and W. K. Engel, Effects of phenazine methosulfate in histochemistry, *J. Histochem. Cytochem.* **12**:928–929 (1964).
7. J. S. Mathiesen and S. I. Mellgren, Some observations concerning the role of phenazine methosulfate in histochemical dehydrogenase methods, *J. Histochem. Cytochem.* **13**:408–409 (1965).
8. E. Farber, Control studies on the histochemical localization of specific DPN-linked dehydrogenases, *J. Histochem. Cytochem.* **10**:657–658 (1962).
9. A. B. Novikoff, *in Histochemistry and Cytochemistry* (R. Wegmann, ed.), pp. 465–481, Pergamon Press, Oxford (1963).
10. J. W. Hitzeman, Observations on the subcellular localization of oxidative enzymes with Nitroblue tetrazolium, *J. Histochem. Cytochem.* **11**:62–70 (1963).
11. Z. Lojda, Remarks on histochemical demonstration of dehydrogenases. II. Intracellular localization, *Folia Morphol.* (*Warszawa*) **13**:84–96 (1965).
12. M. S. Burstone, Histochemical demonstration of acid phosphatases with naphthol AS phosphates, *J. Natl. Cancer Inst.* **21**:523–539 (1958).
13. A. B. Novikoff, The intracellular localization of lactic dehydrogenase studied by biochemical and staining methods, *J. Histochem. Cytochem.* **6**:397 (1958).
14. D. D. Sabatini, K. Bensch, and R. J. Barrnett, Cytochemistry and electronmicroscopy. The preservation of cellular ultrastructure and enzymatic activity by aldehyde fixation, *J. Cell Biol.* **17**:19–58 (1963).
15. J. C. McAlpine, Histochemical survival of diaphorase activity in formalin-fixed tissues stored for 18 months in cold gum-sucrose, *J. Histochem. Cytochem.* **13**:296 (1965).
15a. G. R. N. Jones, Quantitative histochemical studies on the succinate-neotetrazolium reductase system, *Exptl. Cell Res.* **43**:268–280 (1966).
16. C. D. Saudek, C. W. M. Adams, and O. B. Bayliss, The quantitative histochemistry and cytochemistry of lactic dehydrogenase and $NADH_2$-tetrazolium reductase in human aortic wall, *J. Pathol. Bacteriol.* **92**:265–279 (1966).
17. P. B. Gahan and M. Kalina, The validity of using neotetrazolium for studying labile, NADP-linked dehydrogenases in histological sections; a quantitative study, *Biochem. J.* **96**:11P–12P (1965).
18. E. Thomas and A. G. E. Pearse, The fine localization of dehydrogenases in the nervous system, *Z. Zellforsch. Abt. Histochem.* **2**:266–282 (1961).
19. R. L. Friede, L. M. Fleming, and M. Knoller, A comparative mapping of enzymes involved in hexosemonophosphate shunt and citric acid cycle in brain, *J. Neurochem.* **10**:263–277 (1963).
19a. M. J. Blunt, C. P. Wendell-Smith, P. B. Paisley, and F. Baldwin, Oxidative enzymes in macroglia and axons of cat optic nerve, *J. Anat.* (*London*) **101**:13–26 (1967).
20. V. Janata, J. Fischer, L. Jilek, and V. Malik, The diffusion of lactate, succinate and glutamate dehydrogenase from cryostat sections of the brain of rats into the incubating medium during ontogeny, *Acta Histochem.* **26**:28–35 (1967).

20a. C. W. M. Adams and O. B. Bayliss, unpublished observations.

21. H. Zimmermann and A. G. E. Pearse, Limitations in the histochemical demonstration of pyridine nucleotide-linked dehydrogenases ("nothing dehydrogenase"), *J. Histochem. Cytochem.* **7**:271–275 (1959).

22. G. Gomori, *Microscopic Histochemistry*, Chicago University Press, Chicago (1952).

23. M. Wachstein and E. Meisel, Histochemistry of hepatic phosphatases at a physiologic pH, *Am. J. Clin. Pathol.* **27**:13–23 (1957).

24. G. B. Koelle and J. S. Friedenwald, A histochemical method for localizing cholinesterase activity, *Proc. Soc. Exptl. Biol. (N.Y.)* **70**:617–622 (1949).

25. H. W. Deane, Nuclear location of phosphatase activity: fact or artifact?, *J. Histochem. Cytochem.* **11**:443–444 (1963).

26. A. B. Novikoff, The validity of histochemical phosphatase methods on the intracellular level, *Science* **113**:320–325 (1951).

27. P. J. Anderson, S. K. Song, and N. Christoff, *in Proceedings of the Fourth International Congress of Neuropathology* (H. Jacob, ed.), Vol. 1, pp. 75–79, Thieme, Stuttgart (1962).

28. K. D. Barron, J. Bernsohn, and A. R. Hess, Zymograms of neural acid phosphatases. Implications for slide histochemistry, *J. Histochem. Cytochem.* **12**:42–44 (1964).

29. D. G. Freiman and N. Kaplan, Studies on the histochemical differentiation of enzymes hydrolyzing adenosine triphosphate, *J. Histochem. Cytochem.* **8**:159–170 (1960).

30. H. Barden and S. S. Lazarus, Histochemical characteristics of adenosine triphosphate dephosphorylating enzymes in rabbit pancreas, *J. Histochem. Cytochem.* **11**:578–589 (1963).

31. A. B. Novikoff, J. Drucker, W.-Y. Shin, and S. Goldfischer, Further studies of the apparent ATPase activity of cell membranes in formol-calcium-fixed tissues, *J. Histochem. Cytochem.* **9**:434–451 (1961).

32. M. Wachstein, M. Bradshaw, and J. M. Ortiz, Histochemical demonstration of mitochondrial ATPase activity in tissue sections, *J. Histochem. Cytochem.* **10**:65–74 (1962).

33. H. L. Moses and A. S. Rosenthal, On the significance of lead catalysed hydrolysis of nucleoside phosphates in histochemical systems, *J. Histochem. Cytochem.* **15**:354–355 (1967).

34. A. B. Novikoff, Enzyme localizations with Wachstein–Meisel procedures: real or artefact, *J. Histochem. Cytochem.* **15**:353–354 (1967).

35. I. T. McClurkin, A method for the cytochemical demonstration of sodium-activated adenosine triphosphatase, *J. Histochem. Cytochem.* **12**:654–658 (1964).

36. R. M. Torack, *in Neurohistochemistry* (C. W. M. Adams, ed.), pp. 161–188, Elsevier, Amsterdam (1965).

37. J. McD. Tormey, Significance of the demonstration of ATPase noted for active transport, *Nature (London)* **210**:820–822 (1966).

37a. T. G. Scott, The specificity of 5′-nucleotidase in the brain of the mouse, *J. Histochem. Cytochem.* **13**:657–667 (1965).

38. M. J. Karnovsky and L. Roots, A "direct-coloring" thiocholine method for cholinesterases, *J. Histochem. Cytochem.* **12**:219–221 (1964).

39. C. W. M. Adams, R. T. Grant, and O. B. Bayliss, Cholinesterases in peripheral nervous system. I. Mixed, motor and sensory trunks, *Brain Res.* **5**:366–376 (1967).

40. C. Bell, Use of the direct coloring thiocholine technique for demonstration of intracellular neuronal cholinesterases, *J. Histochem. Cytochem.* **14**:567–570 (1966).

41. R. T. Grant, O. B. Bayliss, and C. W. M. Adams, Cholinesterases in peripheral nervous system. II. Motor, sensory and sympathetic nerves in rabbit ear perichondrium and rat cremaster muscle, *Brain Res.* **6**:457–474 (1967).

42. T. Barka and P. J. Anderson, Histochemical methods for acid phosphatase using hexazonium pararosaniline as coupler, *J. Histochem. Cytochem.* **10**:741–753 (1962).

43. G. G. Glenner, L. A. Cohen, and J. E. Folk, The enzymatic hydrolysis of aminoacid-β-naphthylamides. I. Preparation of aminoacid and dipeptide β-naphthylamides, *J. Histochem. Cytochem.* **13**:57–64 (1965).

44. W. H. Fishman and S. S. Goldman, A post-coupling technique for β-glucuronidase employing the substrate, naphthol AS-BI-β-D-glucosiduronic acid, *J. Histochem. Cytochem.* **13**:441–447 (1965).

45. C. W. M. Adams and G. G. Glenner, Histochemistry of myelin. IV. Aminopeptidase activity in CNS and PNS, *J. Neurochem.* **9**:233–239 (1962).

46. M. M. Nachlas, M. M. Friedman, and A. M. Seligman, New observations on discrepancies in the histochemical localization of leucine aminopeptidase, *J. Histochem. Cytochem.* **10**:315–323 (1962).

47. B. Monis, K.-C. Tsou, and A. M. Seligman, Development of a histochemical method for α-D-galactosidase and its distribution in the rat, *J. Histochem. Cytochem.* **11**:653–661 (1963).

47a. F. Wolfgram, A note on the use of the azo dye methods in the enzyme histochemistry of nervous tissue, *J. Histochem. Cytochem.* **9**:171–175 (1961).

48. R. Daoust, Histochemical localization of enzyme activities by substrate film methods: Ribonucleases, deoxyribonucleases, proteases, amylase, and hyaluronidase, *Internat. Rev. Cytol.* **18**:191–221 (1965).

49. R. Daoust, Modified procedure for the histochemical localization of ribonuclease activity by the substrate film method, *J. Histochem. Cytochem.* **14**:254–259 (1966).

50. C. W. M. Adams and N. A. Tuqan, The histochemical demonstration of protease by a gelatin-film substrate, *J. Histochem. Cytochem.* **9**:469–472 (1961).

51. L. Cunningham, Histochemical observations of the enzymatic hydrolysis of gelatin films, *J. Histochem. Cytochem.* **15**:292–298 (1967).

52. A. S. Todd, Histological localization of fibrinolysin activator, *J. Pathol. Bacteriol.* **78**:281–283 (1959).

53. C. W. M. Adams and N. A. Tuqan, Histochemistry of myelin. II. Proteins, lipid-protein dissociation and proteinase activity in Wallerian degeneration, *J. Neurochem.* **6**:334–341 (1961).

54. G. Benetato, E. Gabrielescu, and I. Boros, The histochemistry of cerebral proteases in experimental allergic encephalitis, *Rev. Roumaine Physiol.* **2**:379–384 (1965).

55. G. Benetato, E. Gabrielescu, L. Stoenescu, and A. Bordeianu, The histochemistry of proteases in the nervous system in the course of the stimulation process, *Rev. Roumaine Physiol.* **2**:13–22 (1965).

56. C. W. M. Adams, *in Symposium on Nucleic Acids and Proteins and the Functions of the Nervous System* (Z. Lodin and S. P. R. Rose, eds.), pp. 111–120, Excerpta Medica Foundation, Amsterdam (1968).

57. M. S. Burstone, Modifications of histochemical techniques for the demonstration of cytochrome oxidase, *J. Histochem. Cytochem.* **9**:59–65 (1961).

58. S. J. Holt and R. J. F. Withers, Studies in enzyme histochemistry. V. An appraisal of indigogenic reactions for esterase localization, *Proc. Roy. Soc. Ser. B* **148**:520–532 (1958).

59. C. W. M. Adams, *Vascular Histochemistry*, p. 18, Lloyd-Luke, London (1967).

60. D. T. Janigan and A. G. E. Pearse, The mechanism of tissue staining by ferric hydroxyquinoline methods for β-glucuronidase, *J. Histochem. Cytochem.* **10**:719–730 (1962).

61. W. H. Fishman, S. S. Goldman, and S. Green, Several biochemical criteria for evaluating β-glucuronidase localization, *J. Histochem. Cytochem.* **12**:239–251 (1964).

62. T. Takeuchi, Histochemical differentiation of branching enzyme (amylo-1,4→1,6-transglucosidase) in animal tissues, *J. Histochem. Cytochem.* **6**:208–216 (1958).

63. T. Takeuchi and G. G. Glenner, Histochemical demonstration of uridine diphosphate glucose-glycogen transferase in animal tissues, *J. Histochem. Cytochem.* **9**:304–316 (1961).

Chapter 23

NEUROCHEMISTRY OF INVERTEBRATES

G. A. Kerkut

Department of Physiology and Biochemistry
Southampton University
Southampton, England

I. INTRODUCTION

Although the Invertebrates make up over 97% of all known species of animals, their neurochemistry has been somewhat neglected, in part because of the difficulty of obtaining the required volume of nervous tissue from small invertebrate animals and also the lack of appreciation of the advantages that the invertebrate preparations offer. Perhaps the present account will help to indicate some of the advantages of working on invertebrate preparations.

II. AMINO ACIDS IN NERVE

The large size of the invertebrate axons, which in the giant axons of the squid *Loligo* can reach 900 μ in diameter, has allowed the extrusion of the axoplasm and the chemical analysis of the pure axoplasm. Two of the few analyses made of axoplasm are shown in Table I, that for squid axoplasm and for crab axon. There are interesting differences between these two nerves, particularly the high concentration of isethionate in squid axon and its absence in crab axon. The crab axon contains a high concentration of aspartate.

With the increased sensitivity of methods for amino acid analysis it soon should be possible to obtain an analysis for the amino acids of the single giant neurons of the sea hare *Aplysia*, where the cells can reach 800 μ in size, and possibly also the snail *Helix*, where the neurons are up to 200 μ in diameter.

The crustacean nerves differ from vertebrate nerve trunks in that they are not so tightly bound by connective tissue and it is fairly easy to dissect out single axons. The crustacea share another advantage with other arthropoda in that they have special axons that are inhibitory. This is different from the situation in the vertebrates, where the inhibitory axons lie within the CNS and are not easily obtained.

TABLE I

Analysis of Axoplasm, Values in
Milliequivalents per Kilogram

	Squid[41] Loligo pealii	Crab[47] Carcinus maenas
K	344	260
Na	65	152
Ca	7	13
Mg	20	23
Cl	140	145
Aspartate	65	138
Glutamate	10	35
Taurine	77	65
Glycine	14	5
Alanine	9	33
Isethionate	220	
Total P	24	45

Kravitz, Kuffler, and Potter[44] dissected out the 60-μ diameter fibers from the leg of the lobster *Homarus americanus* and obtained up to 5 m of axon. They separated the inhibitory axons from the excitatory axons and found that the inhibitory axon had up to 1 % of its wet weight as GABA, whereas the excitatory axon had only 0.01 %. This is of interest because GABA is most probably the inhibitory transmitter at the crustacean nerve–muscle junction. GABA was shown to be highest in the inhibitory axon, to be released on stimulation of the inhibitory nerves and not from the excitatory nerves, to be inhibited selectively by muscle,[42,43,44,45,49] and to inhibit the contraction of the muscle.

Otsuka, Kravitz, and Potter[50] have extended the method to the dissection of selected known cells from the lobster abdominal ganglion, and the GABA was analyzed in the single cell bodies. Since the cell body is free from synaptic connections, the analysis would be of cell cytoplasm and not of dendritic cytoplasm and synaptic vesicles. Control cells were examined in histological section to check on the extent to which adhering tissue remained on the dissected isolated cell body.

These workers were able to determine electrophysiologically for a given group of neurons whether they were excitatory or inhibitory. The cell bodies of some of these neurons were constant in position in the ganglion. The GABA content of the single dissected free neurons fell into two groups, those with a content of 5.4–10.3 $\times$ 10^{-11} moles of GABA and those with a content of 0.5 $\times$ 10^{-11}. The high GABA content was found only in the inhibitory neurons, while the low GABA was found in the excitatory neurons. The glutamate content of both sets of neurons was about the same level.

Figure 1, taken from the paper of Otsuka, Kravitz, and Potter,[50] shows a physiological map of the third ganglion from a 0.5-kg lobster, on which the excitor cells are marked in black and the inhibitory cells in white. The figure is also a chemical map on which the cells that contain more than 2×10^{-11} moles of GABA are marked in white while those that contained no detectable GABA are marked in black. There is a good correlation between the GABA-containing cells and the inhibitory cells.

Florkin and Schoffeniels[17] have shown that there can be a considerable alteration in the intracellular amino acid content of muscles and nerves of animals that are subject to changes in osmotic pressure. The intracellular amino acid content of muscles from the crab *Carcinus maenas* was 294 mOsm in animals in sea water, but 182 mOsm in animals in 50% sea water. The major changes were in the glycine and proline content. Somewhat similar changes can take place in the amino acid content of crab nerve, and it was concluded that the intracellular osmotic regulation was not dependent on a hormonal mechanism but that the amino acid content of the cell is in

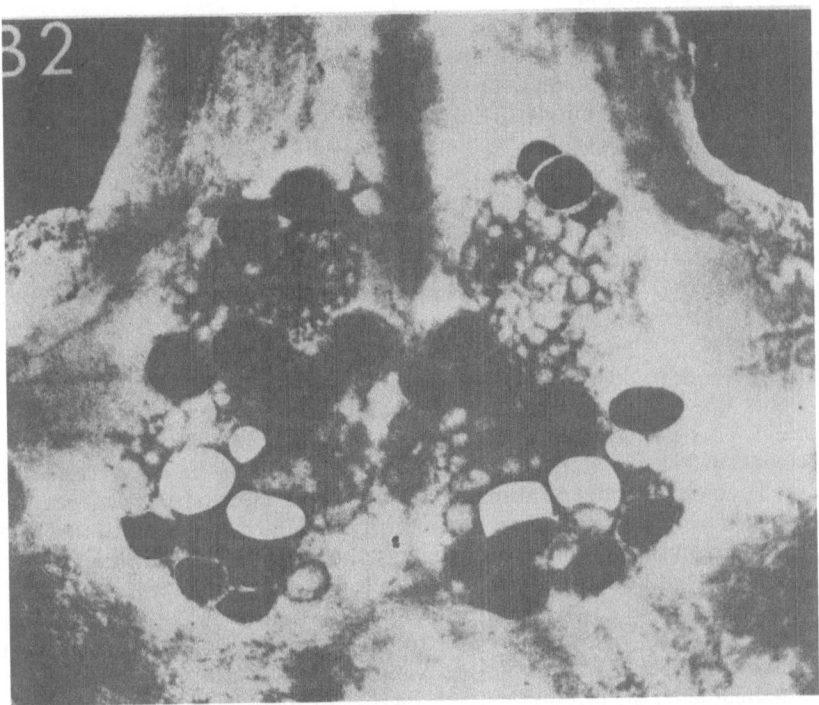

Fig. 1. Map of the cells in the third thoracic ganglion of the lobster showing the presence of cells that had a high (more than 2×10^{-11} moles) GABA content (marked white). There are six such cells shown here. The other black cells had no detectable GABA content. The white cells were shown electrophysiologically to be inhibitory, while the black cells were excitatory.

some way regulated by the cationic content of that cell. This change in the intracellular amino acid composition and content of nerve cells is not unique to invertebrates. Baxter[3] has shown that there is a comparable change in the amino acid content in the nervous system of amphibians subject to osmotic stress.

The organic composition of axoplasm could be expected to vary according to the metabolic state of the neuron, but there may be other differences in amino acid content (such as the GABA content of inhibitory neurons) that reflect the physiological role that the neuron plays.

III. IONIC CONTENT OF NERVES

The ionic content of nerve cells is normally considered to be constant and "straightforward," with a high internal potassium ion concentration and a low internal sodium and chloride ion concentration. The levels of the sodium and potassium ion within the cell are controlled by the linked sodium–potassium pump.

Keynes[40] showed that the chloride concentration of squid axons was higher than that established by a passive movement of the chloride ions along the electrochemical gradient. The inward flux of ^{36}Cl was reduced by 0.2 mM DNP, although the outward flux of chloride was not affected by DNP. Ouabain had no effect on either the inward or the outward fluxes, indicating that the inward flux was not linked to the sodium pump. A similar chloride pump was suggested by Strickholm and Wallin[56] from their measurement of the chloride concentration within the giant axons of the crayfish. The internal chloride concentration was 35 mM, which was five times greater than that expected (7.3 mM) from a passive distribution across the membrane following the resting potential.

Oomura[48,66] (and personal communication) suggested that there could be a difference in the chloride concentration of snail neurons, depending on whether the cells were hyperpolarized (H) or depolarized (D) by acetylcholine. This was checked by Kerkut and Meech,[32] who developed a glass electrode with a tip blocked with silver that proved to be sensitive to the internal chloride concentration of the cell. This electrode and a normal potassium acetate-filled electrode were inserted into the neuron, and the potential difference between the two was shown to be proportional to the internal chloride concentration. The H cells had a chloride concentration of 8.7 ± 0.4 mM (n = 25) while the D cells had a chloride concentration of 27.5 ± 1.5 mM (n = 25). The point to be stressed here is that there is a significant difference in the chloride composition of these nerve cells, which can be related to the physiological properties of these cells.

Strumwasser[57,58] studied the activity of single nerve cells within the ganglia of *Aplysia* for several days and showed that certain cells become active at specific times of the day. One particular cell is called the "parabolic burster" (PB) neuron—because it spontaneously emits a series of action potentials in bursts, the intervals between the impulses approximate to a parabola. The cell has another and longer term rhythm in that its activity recorded over 24 hr is greatest at dawn. This 24-hr rhythm can be affected by the previous daylight treatment of the animals. The timing of each burst is an endogenous process determined by factors within the neurons and not by synaptic activity from outside. The parabolic pattern of activity can be altered by injecting chloride ions into the cell. The particular pattern of activity shown by this neuron and its relation to the internal chloride concentration, and the chloride pump emphasizes the possible differences between neurons of the one ganglion with regard to ionic balance and ionic properties.

IV. CHEMICAL STUDIES ON SINGLE NEURONS

A. Crayfish Stretch Receptor

These sensory cells that lie on some crustacean muscle fibers were first described by Alexandrowicz.[1, 13, 16] The stretch receptors (Fig. 2) are about 600 μ long and 100 μ wide and, when dissected free from surrounding tissue and placed in Van Harrevelds saline, can remain alive for more than 20 hr.

Giacobini, Handelman, and Terzuolo[20] placed the stretch receptors in a Cartesian diver and increased the frequency of action potentials in the neuron by means of an increased external potassium ion concentration. They found an increased rate of respiration when there was a higher frequency of action potentials. There was no change in the ATP level in cells given some

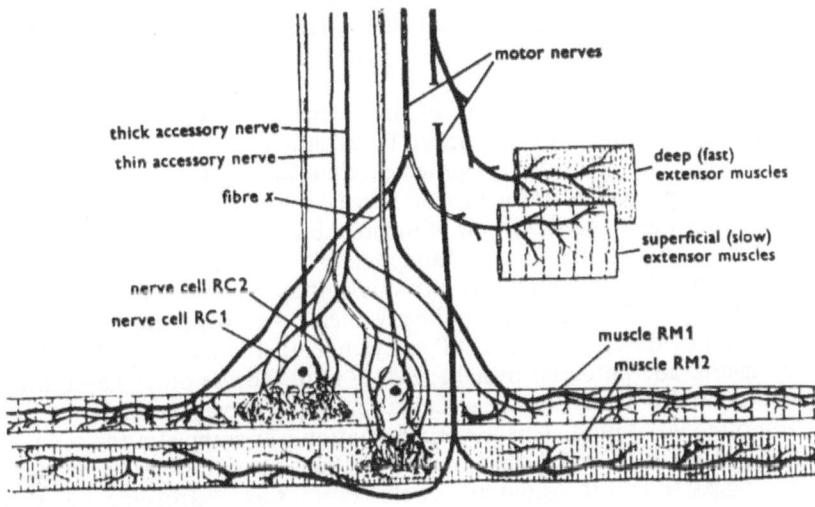

Fig. 2. The crayfish stretch receptor. Two of these are shown ending on muscle fibers. The cells are up to 600 μ long and 100 μ wide.[1] They can be examined fluorometrically.[61]

150,000–250,000 action potentials, and it was thought that this was because the first demand of the neuron would be on arginine phosphate rather than ATP.[60]

The large size of the neuron has enabled microfluorimetric estimation of the concentration of reduced pyridine nucleotides (PNH) of the cell when placed in various substrates.[61] The concentration of PNH was increased by α ketoglutarate, glutamate, dicoumarol, or ouabain and was decreased by gluconate, oxaloacetate or GABA. Metabolic inhibitors affecting ATP formation (monoiodoacetate) and uncoupling agents (dicoumarol 2.4 DNP) did not affect the generation of action potentials, perhaps because crustacean tissue contains cytochrome b_5, which is resistant to cyanide in the microsomal fraction.[55]

The concentration of RNA has been determined in the crustacean stretch receptor by Grampp and Edström.[21] The cells, whose volume was 500,000 μ^3, contained 2800 $\mu\mu g$ RNA; this is a value of 0.5% which is similar to that of Dieter's cell. There was no difference in the total amount of RNA between control cells and those that had given 100,000 action potentials, but there was a small but significant change in the base ratios of adenine:uracil (A:U) and purine:pyrimidine (P:P). In the controls the A:U was 1.18; in the active cells it was 1.28. In the controls the P:P was 1.16; in the active cells it was 1.24.

B. *Aplysia* and *Helix* Neurons

A spectrophotometric analysis of the changes in 20 μ regions of *Aplysia* neurons was carried out by Chalazonitis and Arvanitaki.[9] They followed the changes in succinic dehydrogenase and the cytochrome systems within the neuron and showed that in the *Helix* neuron the most active region was at the axon hillock, which also had the highest cytochrome h activity.[10]

Chalazonitis[8] made a study of the chemopotentials—the changes in electrical activity of nerve cells elicited by variations in the concentration of CO_2, O_2, H^+, or OH^- and suggested mechanisms by which the chemical and electrical properties of the neuron may be interrelated.

The previously accepted view of the membrane potentials of nerve cells was that the resting potential was mainly set up by the ionic imbalance across the membrane. Even though this was due to the long-term activity of the sodium pump and hence was indirectly metabolic, when the pump was switched off by the addition of DNP or cyanide, there was no significant fall in the membrane potential over the first 5 min. On the other hand, after injection of sodium ions into the snail neuron, rapid hyperpolarization of the neuron often was found. This hyperpolarization was stopped if the cell was treated with ouabain or if the external potassium concentration was reduced. It was suggested that the rapid hyperpolarization was electrogenic, i.e., that the movement of the ions or the activity of the pump set up a hyperpolarization of up to 15 mV, and if the pump was stopped the potential fell 15 mV.[38,67,68] This would give the "metabolic systems" a more direct role in the establishment of membrane potentials and thus perhaps also in the finer control of the sensitivity or "setting" of the nerve cell.

V. SYNOPTIC APPROACH

This section summarizes some of the work that is being carried out at Southampton, which may indicate the use of several methodological approaches to a few selected invertebrate preparations.

A. Ionic Variability Within Neurons

It already has been mentioned that it is possible to make chloride-sensitive electrodes and insert them into snail neurons. There is a difference in the chloride concentration of the H and the D cells, and there also appears to be a difference in the internal potassium ion concentration of a selected neuron in the snail brain throughout the year,[32] from an estimation by extrapolation from the Nernst equation. It should be fairly simple to make electrodes from potassium-sensitive glass and make direct determination of the intracellular potassium ion concentration.[67]

B. Chemical Transmission at the Nerve–Muscle Junction

The best-known transmitters at the vertebrate nerve–muscle junctions are acetylcholine and noradrenaline, and there is not much evidence that other chemicals can be transmitters. In the invertebrates, however, there are some nerve–muscle junctions where the transmitter is most probably not acetylcholine or noradrenaline.

The evidence already has been discussed that GABA could be the inhibitory transmitter at the lobster nerve–muscle junction. GABA was shown to be present in the inhibitory nerves and to be released on stimulation and taken up by the muscle system.

The nature of the excitatory transmitter is still under discussion, but a reasonable case can be made that glutamate is the most likely chemical. The end-plate region of the crustacean muscle is sensitive to the addition of glutamate. Takeuchi and Takeuchi[59] showed that 4.5×10^{-16} M would depolarize the end-plate region. Kerkut et al.[31] showed that in three different invertebrate preparations (*Periplaneta*, *Carcinus*, and *Helix*) stimulation of the nerve trunk led to the release of a ninhydrin-positive substance at the muscle perfusate, which was identified as glutamate. Glutamate applied to these muscles made them contract,[33,39,69] and the ninhydrin-positive material when chromatographed and applied to the muscles also made them contract. Glutamate can be taken up by the muscle systems.[25]

C. Isolated CNS–Nerve Trunk–Muscle Preparations

The snail had a low-pressure haemocoelic blood system. It therefore seemed likely that the CNS could survive for many hours when isolated and kept in Ringer solution. The nerve cells in the isolated snail brain often show electrical activity for up to 72 hr after isolation in Ringer.

A preparation of the isolated snail brain–nerve trunk–muscle was set up, and a lanolin barrier was placed between the CNS and the muscle (Fig. 3). The nerve trunk penetrated the lanolin barrier. This preparation was in effect two pools, a CNS pool and a muscle pool, connected by the nerve trunk. When the snail brain was stimulated, glutamate appeared in the muscle perfusate, the amount being proportional to the number of stimuli given. If ^{14}C glutamate was placed in the brain compartment and left to incubate for 3 hr, electrical stimulation of the brain produced a release of ^{14}C glutamate in the muscle compartment. The amount liberated was proportional to the number of stimuli given to the CNS.

If ^{14}C glucose or ^{14}C alanine was incubated in the brain compartment for 3 hr and the brain was stimulated, a small amount of labeled material appeared in the muscle compartment, but this was glutamate and not glucose or alanine. If labeled xylose was placed in the brain compartment, no labeled material appeared in the muscle compartment.

This preparation takes advantage of the special nature of the nerve cell in that the cell body lies in the CNS compartment and the long axon can act as a pipeline connecting the brain compartment to the muscle compartment. Thus, what the nerve cell body will take up and what it will transmit along the nerve trunk could be studied.

Also, labeled material put into the muscle compartment appeared in the CNS.

This type of preparation also has been applied to the frog nerve cord–sciatic nerve gastrocnemius muscle preparation. Labeled material could be

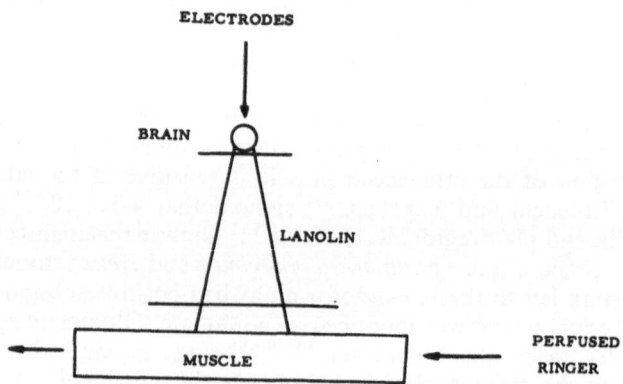

Fig. 3. Isolated CNS nerve trunk–muscle preparation. The CNS is put in one compartment and the muscle in another. The two compartments are linked by the nerve trunk. Radioactive material placed in the CNS compartment after some time can be liberated in the muscle compartment when the CNS is stimulated.

carried from the nerve cord to the muscle and from the muscle to the nerve cord. There was some evidence that the material from the muscle to the nerve cord could travel along the motor pathways since it reached the nerve cord even when the dorsal roots had been cut, and there was some indication that material could be transferred from a muscle cell to a nerve cell.[33]

The isolated preparation of the CNS–nerve trunk–muscle may allow a new experimental approach to the study of the chemical organization of the CNS. Such a system will make possible the labeling of just a few cells and, hence, by the choice of suitable chemicals and localized application, the chemical heterogeneity and the chemical specialization of known neurons and the manner in which the chemical properties of neurons change with time and experience can be studied.

D. Chemical Transmitters in the CNS

The invertebrates have nerve cell bodies lying peripherally in the CNS that are easily accessible to microelectrodes and drug application. There are some cells that are particularly large and/or conspicuous which can be found from one preparation to the next. The position of some identified cells in ganglia are shown in Figs. 4–8, which show the location of cells in the ganglia of *Aplysia*, *Helix*, *Periplaneta*, and *Hirudo* (sea hare, snail, cockroach, and leech). It is important from the experimental point of view to be able to work on known specified cells, since the reaction of a cell to a given chemical may vary according to the cell. In order to get results it is necessary to repeat the experiment on the same neuron. This is an advantage of the invertebrate preparations; there are very few vertebrate preparations (Mauthner cell in fish) on which one can work on a known specified neuron in each experiment.

It is possible to apply different techniques to the same invertebrate preparation and see how the results fit together. The catecholamines and indolalkyamines can be extracted from the ganglia of the leech (*Hirudo*) and analyzed chromatographically and spectrophotofluorimetrically. The results indicate that the main material is 5HT.[34] The Hillarp Falck formaldehyde fluorescent technique shows that in each ganglion there are six cells that fluoresce yellow and contain 5HT, two of which are the Retzius cells, and are about 60 μ in diameter (Fig. 6). The Retzius cells can be seen in the living preparations, and microelectrode study of them showed that they are stimulated by addition of acetylcholine and inhibited by the action of 5HT.[39] Here is a cell that probably contains 5HT and is itself sensitive to 5HT.[35] There is also some evidence that the miniature end-plate potentials at the leech nerve–muscle junction are stimulated by acetylcholine and inhibited by 5HT.

The fluorescence technique also has shown that some neurons in the snail *Helix* ganglia fluoresce green (because of the presence of DOPAmine, as shown by extraction and chromatography), whereas other cells fluoresce yellow because of 5HT. A few neurons are particularly interesting because they contain both green and yellow fluorescent granules. If the snail brain is

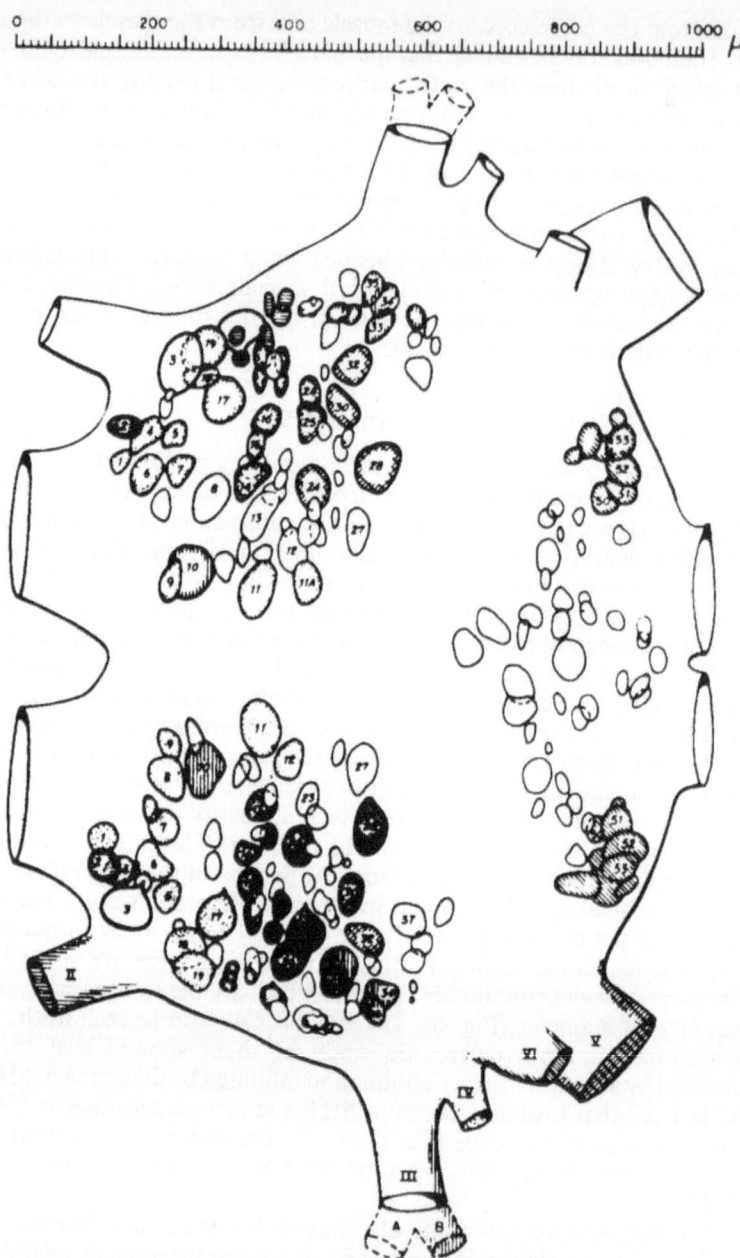

Fig. 4. Third thoracic ganglion of cockroach. Specific cells can be identified, and their peripheral connections are known.[12]

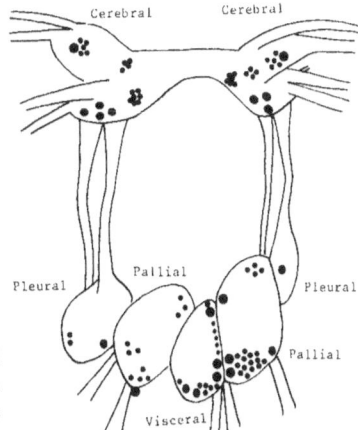

Fig. 5. Diagram of the ganglia in the brain of the snail *Helix aspersa*. The position of some cells whose pharmacological properties are known or which are known to contain DOPAmine or 5HT is shown.

pretreated with DOPA and then examined by the formaldehyde technique, these special cells are found to contain only green granules but no yellow granules. If the brain is pretreated with 5HTP, the cells contain yellow

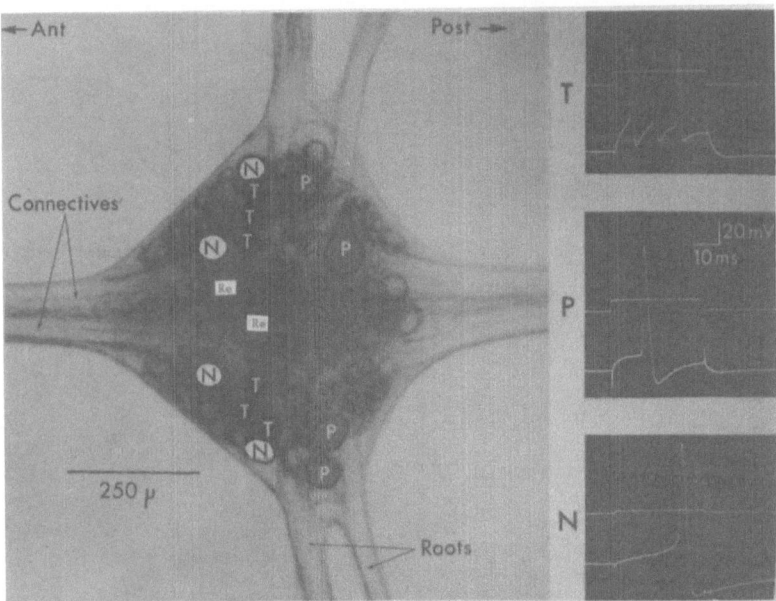

Fig. 6. Photograph of cells in the ganglion of the leech *Hirudo*. Some of the cells are identified by the specific nature of their electrical activity. The position of certain groups of cells (T P N) is shown together with the type of electrical activity shown by them. In addition, the two larger Retzius cells (Re) are marked. Figure kindly provided by Dr. J. G. Nicholls.[70]

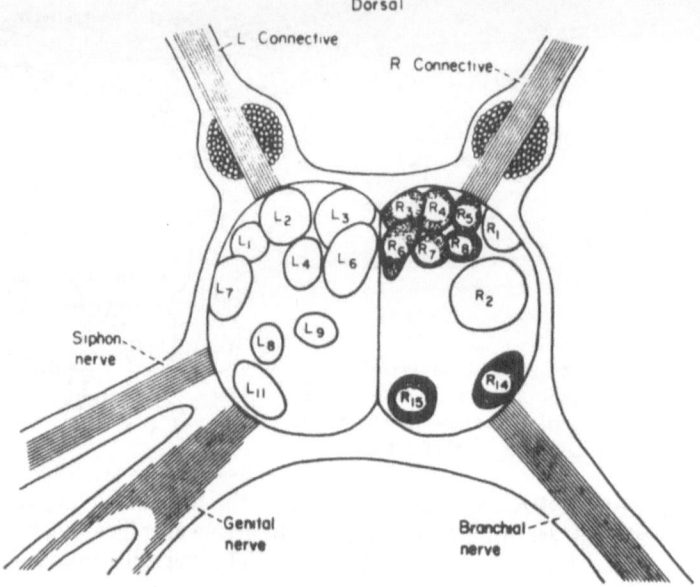

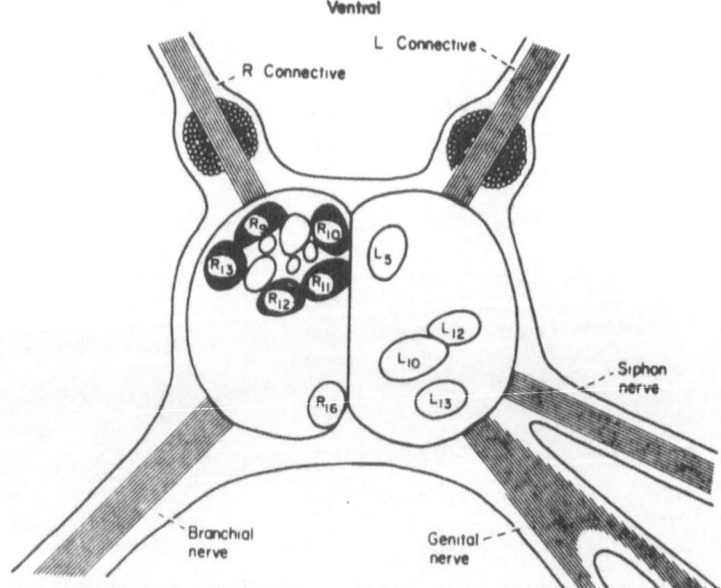

Fig. 7. The location of 29 identified cells in the abdominal ganglia of *Aplysia*. The cells can be identified on coloration and electrophysiological activity.[11,71,72]

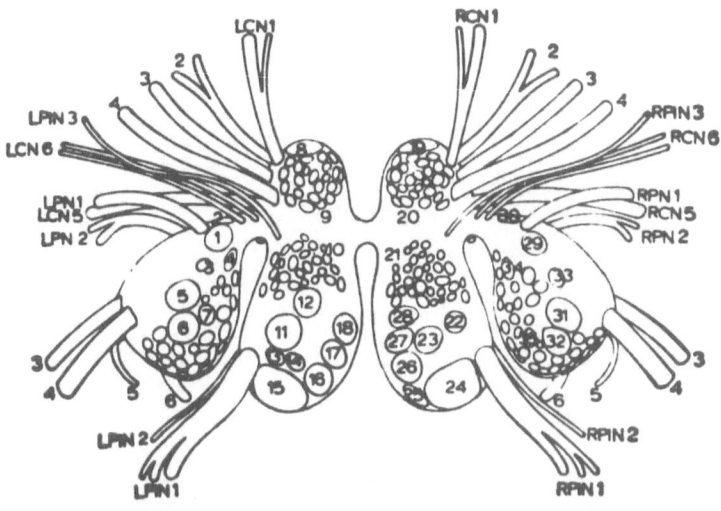

Fig. 8. Location of known cells in the brain of the nudibranch mollusc *Tritonia*. The cells have known peripheral connections, and in some cases their behavioral role is known.[64,73]

granules but no green granules.[36] The nerve cells are able to take up DOPA and convert it to DOPAmine. They can also take up 5HTP and convert it to 5HT. It is not clear whether some cells are normally able to take up both DOPA and 5HTP and form two different transmitters from the one cell. There is evidence that one neuron can have different sites on the membrane sensitive to 5HT or acetylcholine.[19]

E. Comparative Biochemical Studies of the Nervous System

One way that the comparative biochemistry of the nervous system can be studied is to measure the uptake and conversion of various labeled substrates by different invertebrate brains (*Helix, Carcinus, Periplaneta*, and *Mytilus*; Fig. 9) to see how the chemical differences reflect the physiological differences.[25] For example, the fact that a high concentration of GABA is produced from glutamate by the cockroach CNS but no GABA is produced by the snail brain may indicate the role of GABA as an inhibitor in the one brain but not in the other.

It is also possible by genetic selection to obtain strains of one animal species that have definite behavioral characteristics. This has been carried out for rats, mice, and *Drosophila*, and behavioral strains of these animals are in most cases commercially available. We have studied the nervous systems of high-activity and low-activity rats[26] and high-activity and low-activity mice[52] and have shown that in the low-activity animals there is a higher GABA/glutamate ratio than in the active strains. Work is now in

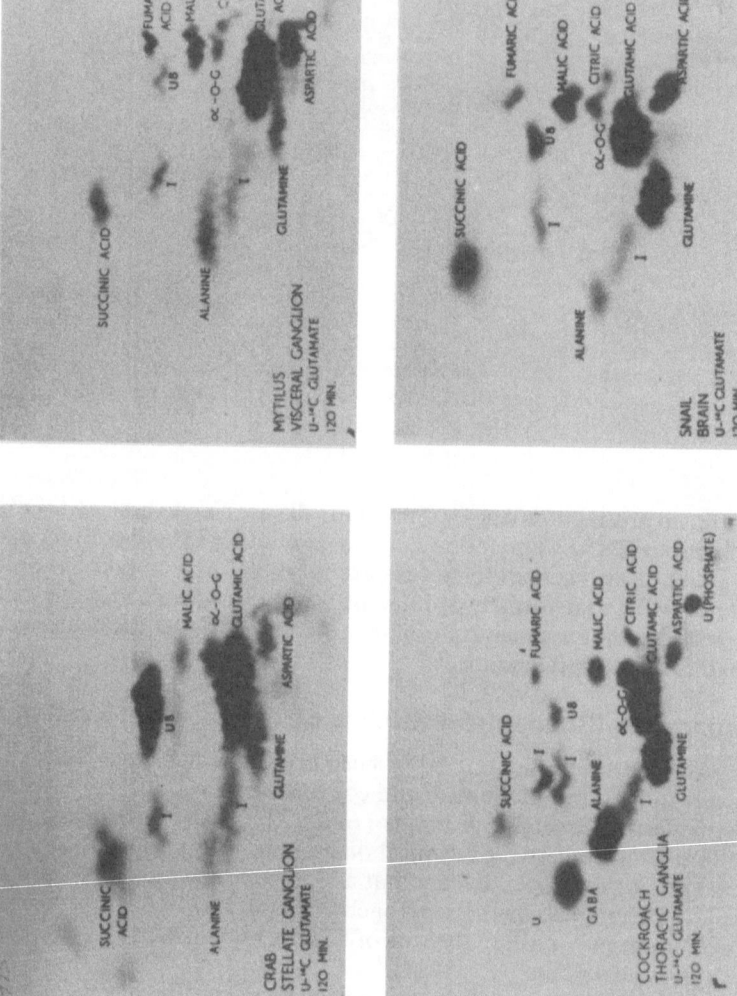

Fig. 9. The conversion of labeled glutamate by the brains of the crab, mussel, cockroach, and snail. Note the well-developed GABA spot in the cockroach and its absence in the snail. A small amount is present in the crab.

progress on the nervous systems of different behavioral strains of *Drosophila*. It should be possible to breed strains of *Drosophila* not only for selected behavioral strains but also for selected strains with high levels of substrates or enzyme activity in the CNS.

Though the approach is still somewhat crude, it should further the appreciation of some of the biochemical differences in the nervous system that underly the behavioral patterns of the whole animals.

VI. NEUROSECRETION

The majority of invertebrates have some nerve cells that give a positive staining reaction with chrome haematoxylin and phloxin or with paraldehyde fuchsin. Within these neurosecretory cells are secretory droplets that are either discharged directly or conveyed along to axon to be discharged or stored.

The subject of neurosecretion has been reviewed by Gabe[18] and for invertebrates by Bern and Hagedorn.[4] There is little precise chemical knowledge about the nature of the secretory materials produced by the neurons. It is a subject of both theoretical and also economic importance, because knowledge about the material produced by the neurosecretory cells of e.g., nematodes might allow us to prepare substitutes that would affect the molting pattern of these worms and help in their biological control.

A. Insect Brain–Hormone System

In the Insects the hard outer skin is periodically shed, and the animal takes up water or air and so increases its volume before the new skin sets hard again. This shedding of the skin is necessary for insect growth and is called *ecdysis*. Ecdysis is brought about by the release of a hormone, *ecdysone*, from the prothoracic gland. This hormone was isolated by Butenandt and Karlson[29] and recently was shown to have the structural formula shown in Fig. 10. The release of ecdysone is controlled by the activity of neurosecretory cells in the insect brain (Fig. 11). The secretions of these cells pass along the axons and are stored in the corpora cardiaca. The substance stored is a trophic hormone and when released will stimulate the prothoracic glands to liberate ecdysone into the bloodstream, which then brings about the shedding of the insect's skin. The situation is analogous to that of the vertebrates, where the hypothalamus sends material along the axons to the neurohypophysis. These substances are released into the bloodstream, where they will then affect the target organs.

The insects have another structure, the corpora allata, which produces a juvenile hormone that keeps the insect in the larval or nonadult stage. A series of substances with a similar physiological action to juvenile hormone has been synthesized by Law, Yuan, and Williams[46] from trans farnesolic acid and ethanolic HCl. These substances will block hatching, metamorphosis,

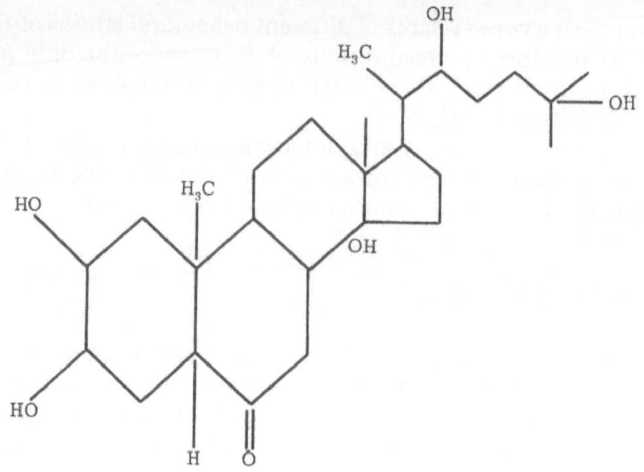

Fig. 10. The chemical structure of the insect hormone ecdysone.

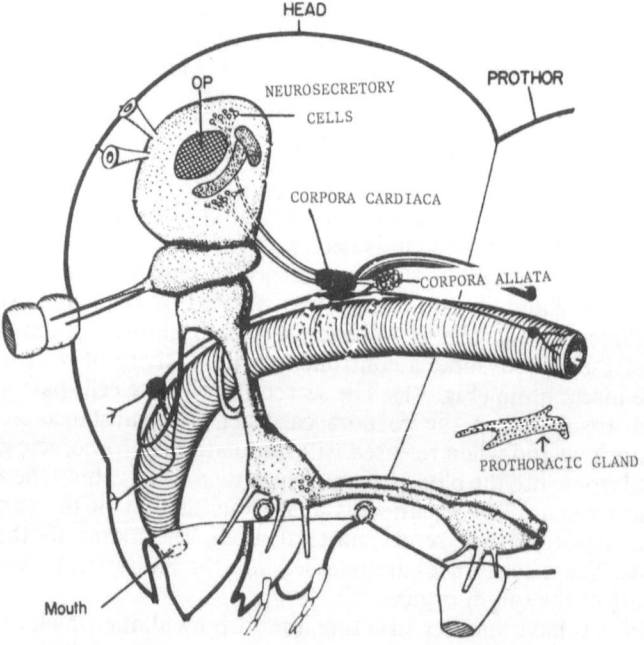

Fig. 11. Diagram of the head of an insect showing the position of the neurosecretory cells in the brain, the corpora cardiaca and the corpora allata.

and development of sexual maturity in the human body louse *Pediculus humanis*[62] and will also block metamorphosis in the mosquito *Aedes aegypti*.

The corpora allata secrete a gonadotropic hormone that affects the activity of the ovaries, and the activity of the corpora allata is affected by the activity of neurosecretory cells in the brain. There is some indication that the suboesophageal ganglia can liberate material into the bloodstream that can affect the rhythmic locomotory activity of insects.[22,53,5,6]

The brain activity controlling these systems is summarized in Fig. 12. In most cases the nature of the chemical systems involved is unknown.

VII. LEARNING IN INVERTEBRATE GANGLIA

Much interest has been paid to the study of learning. Though in the past most attention was paid to a study of the connections between nerve cells and the relationship between the central nervous system and computers, more recently interest has shifted toward the chemical theories of learning. A major difficulty has been the selection of a suitable experimental preparation.

The chemical differences between the brain from an untrained animal and that from a trained animal may be swamped by the normal chemical differences in the cells that have not taken part in the learning. The problem is how to increase the signal/noise ratio. One method is to locate the region of the brain where learning has taken place. Young[65] and Wells[63] have shown in the *Octopus* that one can localize the learning system to one lobe of the *Octopus* brain and even to a few thousand nerve cells. More recently other preparations have been developed where the neurochemist may have a greater chance in locating exactly which five or six neurons have actually been changed in the learning process.

Work has begun on several new preparations that might allow a critical chemical analysis to be carried out on the few neurons that have actually learned.

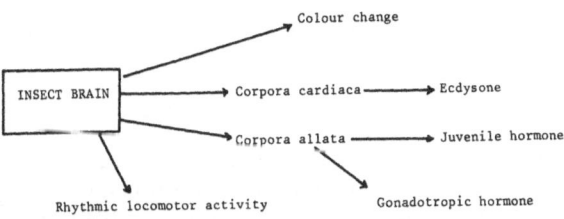

Fig. 12. Diagram of the manner in which the insect brain affects the various activities of the animal.

A. Learning in Isolated Insect Ganglia

Horridge[23] showed that it was possible to train a headless cockroach or locust not to allow a limb to drop below a certain position to avoid an electric shock. He arranged as a control a second animal (R) in series with the first, which would get a shock every time the first animal (P) was shocked. The difference between the P and R animals was that R would get a shock regardless of the position of its limbs whereas P would get a shock when its limb was in a certain position (P = position, R = random). Horridge found that after about 10 min the P unlike the R animal learned to keep its leg away from the contact and in the new position. This learning could last for several hours. It could be made to disappear if the ventral connectives to the ganglia were given a series of shocks in the range of 1–10/sec.

Eisenstein and Cohen[14] adapted the preparation to consist of just one thoracic segment and a pair of legs, which could still learn to adjust the position of the leg. There is some indication that addition of puromycin affects the ability of the preparation to "learn."

The preparation was extended by Hoyle,[24] who analyzed the electrical activity that occurred in the nerves to and from the leg. Hoyle showed that the preparation could be trained to control the frequency of action potentials coming along the motor nerve from the ganglion, by giving an appropriate "punishment" when the frequency deviated from the one experimentally chosen.

Cohen and Jacklet[12] have carried out degeneration studies on the axons of the metathoracic ganglion of the cockroach and have shown that when the motoraxon is cut there appears within 12 hr a change in the RNA pattern around the nucleus of that cell body. In this way it is possible to trace the location of the cell body belonging to specific axons. It also then is possible to map out the position of the cells in the ganglion and show that there is a fairly constant and repeated pattern with a strong bilateral symmetry to the ganglion (Fig. 4). Cohen and Jacklet found that there also were instances of transneuronal degeneration of specific neurons.

The electrophysiological procedure that is necessary for setting up a "learning process" in a given simple system is known[24]; one also can determine which cell bodies are involved in this process, and it should not be too difficult to see if there are chemical and electrophysiological changes taking place in these neurons during and after the learning process. The preparation has the advantage that one would be studying only the 10 or 20 specified neurons that took part in the learning procedure.

B. Long-Term Changes in *Aplysia* Ganglia

Strumwasser[57,58] has shown that the electrical activity in certain neurons in *Aplysia* displays a rhythmic pattern, the cell becoming active at a specific time during the 24 hr, the time depending on the synchronization of the 12 hr light–12 hr dark that the whole animal had previously been treated to.

Another study of the relationship between neuronal activity and the behavior of the whole animal has been carried out by Tauc and his colleagues in Paris. If the head or tentacle of *Aplysia* is stimulated by touch or by letting a number of drops of sea water fall on the head, the tentacle is retracted. The retraction decreases (habituation) with continued stimulation but returns after a short period of rest or if an adjacent region of the head is scratched. The preparation has been electrophysiologically analyzed by Bruner and Tauc.[7] They recorded the electrical activity in the large cell in the left pleural ganglion and showed that stimulation of the head region led to an EPSP in this neuron; the EPSP diminished in size with successive stimulation of the head and recovered in size after a period of rest or if the skin was gently scratched. The behavior of the EPSP resembled the retraction response of the tentacle.

A similar pattern of EPSP could be brought about by stimulating nerve trunks; stimulating the cerebral nerve brought about a diminution of the EPSP, which could be restored by stimulating an adjacent nerve trunk or connective. This allowed an analysis of the behavioral response of the animal in terms of electrical changes in a specified cell following electrical stimulation along known pathways. The change that takes place in the EPSP does not appear to be located at the postsynaptic membrane since the conductance of the membrane does not change during the habituation process. It is possible that the change in the EPSP could be due to a diminution in the amount of transmitter released.

Kandel and Tauc[27,28] applied conditioning techniques with known inputs onto the cells in the abdominal ganglion of *Aplysia* and found that the EDSP could be facilitated for up to 40 min following the conditioning stimulation. The problem of heterosynaptic facilitation in *Aplysia* also has been investigated by Baumgarten and Jahan-Parvar.[2]

These studies on learning in *Aplysia* and the cockroach indicate that it is possible to work on a few neurons and show that some type of learning has taken place. In the case of *Aplysia* one can say at which neuron the output (EPSP) takes place. The problem now would seem to be in determining the chemical changes that have taken place in a few neurons following the conditioning stimulus so that the chemical effect can last for 40 min or be deleted by a few seconds stimulation along another source.

VIII. ADVANTAGES OF INVERTEBRATE PREPARATIONS

In summary, the Invertebrates appear to offer several advantages for neurochemical investigation:

1. Some invertebrates have large, single neurons or axons. This makes it possible to carry out chemical analysis on single cells or spectrophotometric studies on a part of a cell.
2. Some invertebrates have inhibitory axons that run peripherally or

central inhibitory neurons that can be dissected free. The chemical analysis of such systems may help determine the nature of the inhibitory chemical transmitters.

3. Invertebrates may develop certain specialities (GABA or glutamate as transmitters) that are not so clearly shown in the vertebrates.

4. The large neurons make it possible to show in the leech that the Retzius cells probably contain and liberate 5HT, and that they themselves are sensitive to 5HT.

5. One can study the same neuron in each animal of a given species, thus reducing the variability of the system. The neurons that have taken part in some learning system would then allow a greater change of identification of the chemicals involved in memory.

6. In some invertebrates and lower vertebrates it is possible to set up an isolated CNS–nerve trunk–muscle preparation and study the transfer of labeled material from the CNS muscle. This method allows the localized labeling of functional systems in the CNS.

IX. GENERAL CONCLUSIONS

A. Chemical Heterogeneity of the Nervous System

The studies on the neurochemistry of invertebrates indicate that one should not consider all neurons as being chemically identical. There are marked differences in cells concerning the ionic composition, the amino acid composition, the transmitter chemicals, and the response of the cells to drug application. For this reason it is important to know on which cell one is working and, if possible, to carry out the chemical analysis on a specified neuron. The invertebrates offer considerable scope for the study of the basic neurochemistry of single neurons and of the chemical changes in known neurons that take place during behavioral and physiological changes.

B. Chemical Relationship Between Cells

By studying the transfer of labeled material from one cell to another, say, muscle to nerve or nerve to nerve, it should be possible to understand the manner in which the multicellular body is organized and the biochemical features that allow it to remain as an integrated working system.

The invertebrates allow such studies because with their large neurons it is possible to inject labeled materials into single cells and see how the materials are transferred or incorporated into other cells. These studies must be made on a background of general neurochemical knowledge, but they indicate a natural development of the subject toward a link between the neurochemical, neurophysiological, and behavioral studies of the nervous system.

X. REFERENCES

1. J. S. Alexandrowicz, Receptor organs in thoracic and abdominal muscles of Crustacea, *Biol. Rev.* **42**:288–326 (1967).
2. R. V. Baumgarten and B. Jahan-Parvar, Problem of heterosynaptic facilitation in *Aplysia californica, Pfluegers Arch. Physiol.* **295**:328–346 (1967).
3. C. F. Baxter and C. L. Ortiz, Amino acids and the maintenance of osmotic equilibrium in brain tissue, *Life Sci.* **5**:2321–2329 (1966).
4. H. A. Bern and I. H. Hagedorn, *in Structure and Function of the Nervous System of Invertebrates* (T. H. Bullock and G. A. Horridge, eds.), Vol. I, pp. 356–429, W. H. Freeman, San Francisco (1965).
5. J. Brady, Control of circadian rhythm of activity in the cockroach. 1. The role of the corpora cardiaca, brain and stress, *J. Exptl. Biol.* **47**:153–164 (1967).
6. J. Brady, Control of the circadian rhythm of activity in the cockroach. 2. The role of the suboesophageal ganglion and the ventral nerve cord, *J. Exptl. Biol.* **47**:165–178 (1967).
7. I. Bruner and L. Tauc, *in Nervous and Hormonal Mechanisms of Integration* (G. M. Hughes, ed.), *Symp. Soc. Exptl. Biol.* **20**:457–476 (1966).
8. N. Chalazonitis, Chemopotentials in giant nerve cells *(Aplysia fasciata), in Nervous Inhibition* (E. Florey, ed.), pp. 179–194, Pergamon Press, Oxford (1961).
9. N. Chalazonitis and A. Arvanitaki, Chromoproteides et Succinoxydase dans divers grains isolables du cytoplasm neuronique. *Arch. Sci. Physiol.* **10**:291–319 (1956).
10. N. Chalazonitis and M. Gola, Analyses microspectrophotometriques relatives a quelques catalyseurs respiratoires dans le neurone isolé *(Helix pomatia), Compt. Rend. Soc. Biol. (Paris)* **158**:1908–1914 (1964).
11. R. E. Coggeshall, E. R. Kandel, I. Kupperman, and R. Waziri, A morphological and functional study on a cluster of identifiable neurosecretory cells in the abdominal ganglia of *Aplysia californica, J. Cell Biol.* **31**:363–368 (1966).
12. M. J. Cohen and J. W. Jacklet, Neurons of insects; RNA changes during injury and regeneration, *Science* **148**:1237–1239 (1965).
13. C. Edwards, Physiology and pharmacology of the crustacean stretch receptor, *in Inhibition and the Nervous System and Gamma amino butyric acid* (E. Roberts, ed.), pp. 386–408, Pergamon Press, New York (1960).
14. E. M. Eisenstein and M. J. Cohen, Learning in an isolated prothoracic insect ganglion, *Animal Behav.* **13**:104–108 (1965).
15. E. M. Eisenstein and G. H. Krasilovsky, Studies of learning in isolated insect ganglia, *in Invertebrate Nervous Systems* (C. A. G. Wiersma, ed.), pp. 239–332, University of Chicago Press, Chicago (1967).
16. C. Eyzaguirre, Excitation and inhibitory processes in crustacean sensory nerve cells, *in Nervous Inhibition* (E. Florey, ed.), pp. 295–317, Pergamon Press, New York (1961).
17. M. Florkin and E. Schoffeniels, Euryhalinity and the concept of physiological radiation, *in Studies in Comparative Biochemistry* (K. A. Munday, ed.), pp. 6–40, Pergamon Press, London (1965).
18. M. Gabe, *Neurosecretion,* Pergamon Press, Oxford (1966).
19. H. Gerschenfeld and E. Stefani, Acetylcholine and 5 HT receptors in neurones of the land snail, *Cryptomphallus aspersa, J. Physiol. (London)* **191**:14–15P (1967).
20. E. Giacobini, E. Handelman, and C. A. Terzuolo, Isolated neuron preparation for studies of metabolic effects at rest and during impulse activity, *Science* **140**:74–75 (1963).
21. W. Grampp and J. E. Edström, The effect of nervous activity on ribonucleic acid of the crustacean receptor neuron, *J. Neurochem.* **10**:725–731 (1963).

22. J. Harker, *The Physiology of Diurnal Rhythms*, Cambridge University Press, New York (1964).

23. G. A. Horridge, Learning of leg position by the ventral nerve cord of headless insects, *Proc. Roy. Soc. B* **157**:33–52 (1962).

24. G. Hoyle, Neurophysiological studies on "learning in headless insects," *in The Physiology of the Insect Central Nervous System* (J. E. Treherne and J. W. L. Beaumont, eds.), pp. 203–232, Academic Press, New York (1965).

25. A. K. Huggins, J. T. Rick, and G. A. Kerkut, A comparative study of the intermediary metabolism of L-glutamate in muscle and nervous tissue, *Comp. Biochem. Physiol.* **21**:23–30 (1967).

26. A. K. Huggins, J. T. Rick, and G. A. Kerkut, The comparative production of gamma amino butyric acid in the Maudsley reactive and non reactive strains of rat, *Comp. Biochem. Physiol.* **20**:1009–1012 (1967).

27. E. R. Kandel and L. Tauc, Heterosynaptic facilitation in neurones of the abdominal ganglion of *Aplysia depilans*, *J. Physiol.* **181**:1–27 (1965).

28. E. R. Kandel and L. Tauc, Mechanisms of heterosynaptic facilitation in the giant cells of the abdominal ganglion of *Aplysia depilans*, *J. Physiol.* **181**:28–37 (1965).

29. P. Karlson and C. E. Sekeris, Ecdysone, an insect steroid hormone and its mode of action, *Recent Progr. Hormone Res.* **22**:473–502 (1966).

30. G. A. Kerkut, Biochemical aspects of invertebrate nerve cells, *in Invertebrate Nervous Systems* (C. A. G. Wiersma, ed.), pp. 5–37, University of Chicago Press, Chicago (1967).

31. G. A. Kerkut, L. D. Leake, A. Shapira, S. Cowan, and R. J. Walker, The presence of glutamate in nerve-muscle perfusates of *Helix*, *Carcinus* and *Periplaneta*, *Comp. Biochem. Physiol.* **15**:485–502 (1965).

32. G. A. Kerkut and R. W. Meech, The internal chloride concentration of H and D cells in the snail brain, *Comp. Biochem. Physiol.* **19**:819–832 (1966).

32a. G. A. Kerkut, A. Shapira, and R. J. Walker, The effect of acetylcholine, glutamate and gamma amino butyric acid on the contractions of the perfused cockroach leg, *Comp. Biochem. Physiol.* **16**:37–48 (1965).

33. G. A. Kerkut, A. Shapira, and R. J. Walker, The transport of labelled material from CNS ⇌ Muscle along a nerve trunk, *Comp. Biochem. Physiol.* **23**:729–748 (1967).

34. G. A. Kerkut, C. B. Sedden, and R. J. Walker, The effect of dopa, α-methyldopa and reserpine on the dopamine content of the brain of the snail, *Helix aspersa*, *Comp. Biochem. Physiol.* **18**:921–930 (1966).

35. G. A. Kerkut, C. B. Sedden, and R. J. Walker, Cellular localization of monoamines by fluorescence microscopy in *Hirudo medicinalis* and *Lumbricius terrestris*, *Comp. Biochem. Physiol.* **21**:687–690 (1967).

36. G. A. Kerkut, C. B. Sedden, and R. J. Walker, Uptake of DOPA and 5 hydroxytryptophan by monoamine forming neurones in the brain of *Helix aspersa*, *Comp. Biochem. Physiol.* **23**:159–162 (1967).

37. G. A. Kerkut, C. B. Sedden, and R. J. Walker, A fluorescence microscopic and electrophysiological study of the giant neurones of the ventral nerve cord of *Hirudo medicinalis*, *J. Physiol.* **189**:83–85 (1967).

38. G. A. Kerkut and R. C. Thomas, An electrogenic sodium pump in snail nerve cells, *Comp. Biochem. Physiol.* **14**:167–183 (1965).

38a. G. A. Kerkut and R. J. Walker, The effect of L-glutamate, acetylcholine and gamma amino butyric acid on the miniature end plate potentials and contractures of the coxal muscles of the cockroach *Periplaneta americana*, *Comp. Biochem. Physiol.* **17**:435–454 (1966).

39. G. A. Kerkut and R. J. Walker, The action of acetylcholine, DOPAmine and 5 hydroxytryptamine on the spontaneous activity of the cells of Retzius of the leech *Hirudo medicinalis*, *Brit. J. Pharmacol. Chemotherap.* **30**:644–654 (1967).

40. R. D. Keynes, Chloride in the squid giant axon, *J. Physiol.* (*London*) **169**:690–705 (1963).
41. B. A. Koechlin, On the chemical composition of the axoplasm of squid giant nerve fibres with particular reference to its ion pattern, *J. Biophys. Biochem. Cytol.* **1**:511–529 (1955).
42. E. A. Kravitz, Enzymatic formation of GABA in the peripheral and central nervous system of lobsters, *J. Neurochem.* **9**:363–370 (1962).
43. E. A. Kravitz and D. D. Potter, A further study of the distribution of GABA between excitatory and inhibitory axons of the lobster, *J. Neurochem.* **12**:323–328 (1965).
44. E. A. Kravitz, S. W. Kuffler, and D. D. Potter, GABA and other blocking compounds in Crustacea. III. Their relative concentrations in separated motor and inhibitory axons, *J. Neurophysiol.* **26**:739–751 (1963).
45. E. A. Kravitz, S. W. Kuffler, D. D. Potter, and N. M. Van Gelder, GABA and other blocking compounds in Crustacea. II. Peripheral nervous system, *J. Neurophysiol.* **26**:729–738 (1963).
46. J. H. Law, C. Yuan, and C. M. Williams, Synthesis of a material with a high juvenile hormone activity derived from farnesic acid, *Proc. Natl. Acad. Sci. U.S.* **55**:576–577 (1966).
47. P. R. Lewis, The free amino acids of invertebrate nerve, *Biochem. J.* **52**:330–338 (1952).
48. Y. Oomura, in *The Physiology of Synapses* (J. C. Eccles, ed.), p. 200, Springer Verlag, Berlin (1964).
49. M. Otsuka, L. L. Iversen, Z. W. Hall, and E. A. Kravitz, Release of gamma aminobutyric acid from inhibitory nerves of lobster, *Proc. Natl. Acad. Sci. U.S.* **56**:1110–1115 (1966).
50. M. Otsuka, E. A. Kravitz, and D. D. Potter, Physiological and chemical architecture of a lobster ganglion with particular reference to gamma amino butyrate and glutamate, *J. Neurophysiol.* **30**:725–752 (1967).
51. J. T. Rick, A. K. Huggins, and G. A. Kerkut, The comparative production of γ-amino butyric acid in the Maudsley reactive and non-reactive strains of rat, *Comp. Biochem. Physiol.* **20**:1009–1012 (1967).
52. J. T. Rick and G. A. Kerkut, The comparative production of GABA in bidirectionally bred behavioural strains, *Abstract of First International Meeting of International Society for Neurochemistry*, p. 177, Strasbourg (1967).
53. S. K. de F. Roberts, Circadian activity rhythms in cockroaches. III. The role of endocrine and neural factors, *J. Cell Physiol.* **67**:473–486 (1966).
54. H. Roller, K. H. Dahm, C. C. Sweely, and B. M. Trost, The structure of the juvenile hormone, *Angew. Chem.* **6**:179–180 (1967).
55. L. Rossini, H. P. Cohen, E. J. Handelman, S. Lin, and C. A. Terzuolo, Measurements of oxido-reduction processes and ATP levels in an isolated crustacean neurone, in *Biological Membranes, Ann. N.Y. Acad. Sci.* **137**:864–876 (1966).
56. A. Strickholm and B. G. Wallin, Intracellular chloride activity of crayfish giant axons, *Nature* (*London*) **208**:790–791 (1965).
57. F. Strumwasser, The demonstration and manipulation of a circadian rhythm in a single neurone, in *Circadian Clocks* (J. Aschoff, ed.), pp. 442–462, North Holland, Amsterdam (1965).
58. F. Strumwasser, Types of information stored in single neurons, in *Invertebrate Nervous Systems* (C. A. G. Wiersma, ed.), pp. 291–319, University of Chicago Press, Chicago (1967).
59. A. Takeuchi and N. Takeuchi, The effect on crayfish muscle of iontophoretically applied glutamate, *J. Physiol.* (*London*) **170**:296–317 (1964).
60. C. A. Terzuolo, E. J. Handelman, and L. Rossini, An isolated crustacean neuron preparation for metabolic and pharmacological studies, in *Invertebrate Nervous Systems* (C. A. G. Wiersma, ed.), pp. 55–64, University of Chicago Press, Chicago (1967).
61. C. A. Terzuolo, B. Chance, E. J. Handelman, L. Rossini, and P. Schmelzer, Measurements of reduced pyridine nucleotides in single neuron, *Biochim. Biophys. Acta* **126**:361–372 (1966).

62. J. W. Vinson and C. M. Williams, Lethal effects of synthetic juvenile hormone on the human body louse, *Proc. Natl. Acad. Sci. U.S.* **58**:294–297 (1967).

63. M. J. Wells, Learning in the *Octopus*, *in Nervous and Hormonal Mechanisms of Integration* (G. M. Hughes, ed.), *Symp. Soc. Exptl. Biol.* **20**:477–508 (1966).

64. A. O. D. Willows, Behavioural acts elicited by the stimulation of single identifiable brain cells, *Science* **157**:570–574 (1967).

65. J. Z. Young, *A Model of the Brain*, Clarendon Press, Oxford (1964).

The following references were added in proof:

66. Y. Oomura, H. Ooyama, and M. Sawada, Intracellular chloride concentration in the *Onchidium* neurone, *Proc. 24 Intern. Physiol. Congr. Washington*, No. 990 (1968).

67. R. C. Thomas, Membrane current and intracellular sodium changes in snail neurone during extrusion of injected sodium, *J. Physiol. (London)* **201**:495–514 (1969).

68. G. A. Kerkut and B. York, The oxygen sensitivity of the electrogenic sodium pump in snail neurones, *Comp. Biochem. Physiol.* **28**:1125–1134 (1969).

69. R. Beranek and P. L. Miller, The action of iontophoretically applied glutamate on insect muscle fibres, *J. Exptl Biol.* **49**:83–93 (1968).

70. J. G. Nicholls and D. A. Baylor, Specific modalities and receptive fields of sensory neurons in CNS of the Leech, *J. Neurophysiol.* **31**:740–756 (1968).

71. W. T. Frazier, E. R. Kandel, I. Kupferman, R. Waziri, and R. E. Coggeshall, Morphological and functional properties of identified neurones in the abdominal ganglion of *Aplysia californica*, *J. Neurophysiol.* **30**:1287–1351 (1957).

72. E. R. Kandel, W. T. Frazier, R. Waziri, and R. E. Coggeshall, Direct and common connections among identified neurons in *Aplysia*, *J. Neurophysiol.* **30**:1352–1376 (1957).

73. A. O. D. Willows, Behavioural acts elicited by stimulation of single identifiable nerve cell, *in Physiological and Biochemical Aspects of Nervous Integration* (F. D. Carlson, ed.), pp. 217–244, Prentice-Hall, Englewood Cliffs, N. J. (1968).

SUBJECT INDEX

Acetate
  metabolism in brain, 456–457, 462–463
Acetazolamide
  effect on transport in choroid plexus,
    64–65
Acetylcholine
  cytoplasmic and vesicular, 352–353
  distribution in neurons, 350–351
  as neurotransmitter, 112, 369, 373, 374
  in peripheral nerve, levels, 408
  and related enzymes in SC, 88–89
  in retina, 490
  and Retzius cells, 547
  in spinal cord, 88–89
  in stretch receptor neuron of crayfish
    variation with activity, 212–213
  in synaptic vesicles, 373–374
  as synaptosome marker, 328, 333
  vesicular and cytoplasmic, 352–353,
    373–374
Acetylcholinesterase
  distribution
    in spinal cord, 88–89
    subcellular, 351–352, 379–382
  in peripheral nerve, 409–410
  in retina
    localization, 490
  in spinal cord
    developmental changes, 89
    distribution, 88–89
  in stretch receptor neuron of crayfish
    activity
      assay methods, 212
      variation during activity, 212, 213
  synthesis in neuron, site, 351, 424–425
Acid hydrolases
  and disease, 293–294
  in neural lysosomes, 256
Acid phosphatase
  histochemistry of lysosome, 266–268
Actinomycin D
  effect on SRN activity, 205–206
Active transport
  of cations, 214, 223, 313–314
Adenyl cyclase
  localization in nerve endings, 382–384
ADP–ATP
  levels
    luciferase method of determination,
      220
    in peripheral nerves, 407–408

ADP–ATP (cont.)
  in retina, 476
    during functional activity, 495–496
  metabolism
    in mitochondria, 310–311, 313–316
    coupled to electron transport,
      313–314
    phosphorylating enzymes, 315–316
    in stretch receptor neuron, 214–215,
      218, 226
  in retina
    as biochemical energy, 482–483
    distribution, 489
    level, 476
    during functional activity, 495–496
Alanine
  in invertebrate axoplasm, 540
Alloganglionic cells
  function, 490
  location, 490
Amacrine cells
  function, 491–492
  location, 489
  metabolism, 492
Amino acids
  in brain cells
    glia and neurons compared, 187,
      188–189
  compartmentation of metabolism,
    447–469
    criteria, 448
    defined, 448
    developmental aspects, 458–459
    studies
      in vitro, 459–465
      in vivo, 448–458
    theoretical considerations, 465–469
  in invertebrate axoplasm
    levels, 539–540
  in invertebrate nerve
    in axoplasm, 539–540
    and osmotic pressure, 541–542
  and ionic flux, 128
  in spinal cord
    concentration, 82–83
    distribution, 82–83
    enzymes of metabolism, 84
    transport, 83–84
γ–aminobutyrate (see GABA)
Ammonia metabolism
  in brain, 452–455